UINTA FLORA

A Guide to the Vascular Plants of the Uinta Basin and Uinta Mountains

Dryas octopetala
(Mountain avens)

SHEREL GOODRICH
and ALLEN HUBER

Edited by STEVE W. CHADDE

2016

ABBREVIATIONS

q.v.   which see
ssp.   subspecies
var.   variety
n.   north
e.   east
s.   south
w.   west
ne.   northeast
nw.   northwest
se.   southeast
sw.   southwest
c.   central
Co., Cos.   County, Counties
mt., mtn.   mount, mountain
mtns.   mountains
AUF   *A Utah Flora*
FNA   *Flora of North America*
IMF   *Intermountain Flora*
TPD   *The Plants Database*

UINTA FLORA
A Guide to the Vascular Plants of the Uinta Basin and Uinta Mountains

Uinta Flora was first published in 1986 by USDA Forest Service-Intermountain Region, Ogden, Utah (in cooperation with USDA Forest Service-Ashley National Forest) as *Uinta Basin Flora* by Sherel Goodrich and Elizabeth Neese. The *Flora* was revised and updated in 2014 and in January 2016 as *Uinta Flora* by Sherel Goodrich and Allen Huber (these works are in the public domain). In May 2016, Steve Chadde, botanist and author of a number of plant identification books, prepared this new field edition of the *Flora*. Included are illustrations of rare plant species of the region by K. H. Thorne from *Illustrated Manual of Proposed Endangered and Threatened Plants of Utah* (Welsh 1979). Comments and corrections are appreciated and can be emailed to: steve@chadde.net.

AN ORCHARD INNOVATIONS BOOK

Printed in the United States of America

ISBN: 978-1951682309

Ver. 2.1 (4/15/2020)

CONTENTS

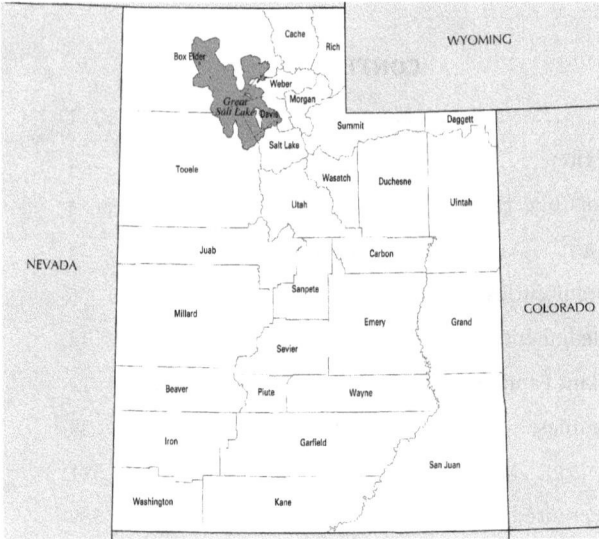

Above: Utah county map.

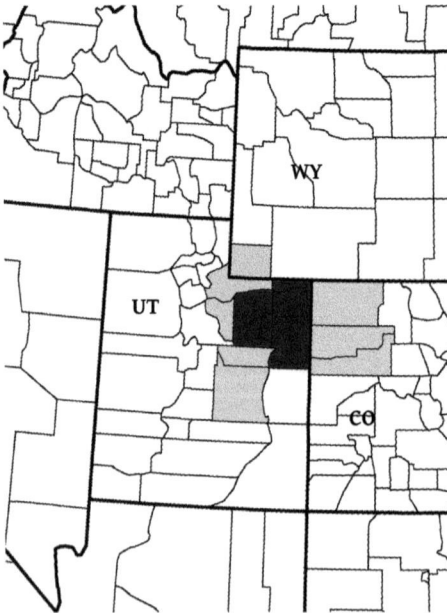

Above: General region included in *Uinta Flora*. Covered in Utah are all of Daggett, Duchesne, and Uintah Counties (dark shading), a large part of Wasatch Co. (Strawberry drainage), the portion of Summit Co. within the north slope of the Uinta Mountains and small parts of Emery, Grand, and Utah Counties (Tavaputs Plateau). In Wyoming, the portion of Uinta Co. within the Uinta Mtns is included. In Colorado, about the western half of Moffat and Rio Blanco Counties are included, as is the northwest corner of Garfield Co. (Baxter Pass-Douglas Pass area and upper Piceance drainage). See page 2 for more details.

PREFACE

This flora, containing about 1,700 species and about 400 subspecific taxa of vascular plants, is arranged alphabetically by family, genus, and species. It is intended as a field manual. To reduce the size of the volume, descriptions of species have been omitted except for those that are the only one of a genus, and those descriptions are often brief. Descriptions of families, genera, and species have been reduced to apply to plants as found within the boundaries of this flora. Also features listed in the keys are intended for use in this area only. Features listed in the keys are often not repeated in family, generic, or species descriptions. Expanded descriptions can be found in Harrington (1954) and *A Utah Flora* (AUF) (Welsh et at. 2008). Descriptions and illustrations can be found in *Intermountain Flora* (IMF) (Cronquist et al. (1972, 1977, 1984, 1989, 1994, 1997, and Holmgren et al. 2005 and 2012) and in *Flora of North America* (FNA) (Flora of North America Editorial Committee, eds). *The Plants Database* (TPD) [USDA, NRCS, 2014 (http//plants.usda.gov)] was frequently used as a source of plant distribution.

This work is intended to complement the works listed above by providing a manual specific for the Uinta Basin-Uinta Mountain area that is practical to carry into the field. Because descriptions have been omitted, some leads in the keys have been expanded to include several features. The reader should not feel obligated to wade through all of an expanded lead. However, since the *Uinta Basin Flora* (Goodrich & Neese 1986) was published, it has been our experience that students using that flora felt compelled to wade through expanded leads even when one feature would have been adequate. Thus many leads have been shorted in this treatment in hopes of speeding up the keying process. For expanded leads that remain we still recommend the feature that is most applicable to a particular specimen should be sought out. If a specimen has fruit but lacks flowers, then the features of the fruit, leaves, or other part of the plant should be used to key the specimen.

Because this book is intended for field use, we have attempted to incorporate, as often as possible, key features that can be seen without a dissecting scope. Features frequently used include size of plants, type of root, and presence or lack of rhizomes, bulbs, or other subterranean parts. All too often the lone specimen brought in from the field for identification is an odd one or incomplete. Examples include albino-flowered specimens of a *Penstemon* that mostly has blue flowers, or an *Allium* without the bulb. Advantages of keying specimens in the field include the abundance of material to work with and the opportunity to dig again for subterranean parts first missed.

This manual is intended for field identification, and with some experience and with the aid of a hand lens is often possible. However, in some cases and especially in some groups (for example, *Agrostis, Poa,* and *Carex*) a dissecting scope might be necessary. When collecting specimens that are to be identified at the office or herbarium, one should look at a number of plants to determine the range of variation of features of the plants, and dig thoroughly enough to determine the type of root and presence or absence of rhizomes, bulbs, or other underground parts. If the specimens are to be deposited in a herbarium, notes of variation in these features are valuable if included on the herbarium label.

In natural resource studies, one often encounters plants within study plots that do not have flowers or fruit. For this reason, vegetative features are frequently included in the keys. However, vegetative features can be highly variable, and in many cases such features were not found with diagnostic value. By keying a vegetative specimen a few ways through the keys, one might come to a few logical taxa for which the specimen can be checked against herbarium specimens or descriptions and illustrations in other manuals. The 2014 edition of the *Flora* included a DVD illustrating most of the plants treated in this manual; copies may be available from the USDA Forest Service, Intermountain Region office in Ogden.

INTRODUCTION

Our area includes the Uinta Mountains and the Uinta Basin bounded on the west by the divide of the Strawberry drainage and the west end of the Uinta Mtns, and on the south by the breaks of the West and East Tavaputs Plateaus. The northern boundary follows the Wyoming state line with Utah and Colorado, but it includes that part of Wyoming in the Uinta Mountains including Hickey Mtn. The eastern boundary south of the White River follows the Grand Hogback. Thus the Piceance Basin and Cathedral Bluffs are included, but the White River drainage to the east of these areas is not. The eastern boundary north of the White River roughly follows the Danforth Hills to Maybell and the Yampa River. North of the Yampa River the boundary roughly follows 108° longitude to the Colorado-Wyoming boundary.

It is anticipated that the principal users of this work will include personnel of the Ashley National Forest, Wasatch-Cache National Forest, National Park Service, and the Vernal, Utah, and Craig, Colorado, Districts of the Bureau of Land Management. Thus, all of the Ashley National Forest in Utah, the Mountain View and Evanston Districts of the Wasatch-Cache National Forest, Dinosaur National Monument, and Vernal District of the Bureau of Land Management are included. About the western half of the Little Snake and White River Resource Areas of the Craig District of the Bureau of Land Management are also included.

Covered in Utah are all of Daggett, Duchesne, and Uintah Counties, a large part of Wasatch Co. (Strawberry drainage), the portion of Summit Co. within the north slope of the Uinta Mountains and small parts of Emery, Grand, and Utah Counties (Tavaputs Plateau). The portion of Uinta Co., Wyoming, within the Uinta Mtns is included. In Colorado about the western half of Moffat and Rio Blanco Counties are included, as is the northwest corner of Garfield Co. (Baxter Pass-Douglas Pass area and upper Piceance drainage). In this work the area described above is referred to as "our area."

"Our plants" are the vascular plants growing in our area, and "our specimens" are the herbarium specimens collected from this area by all botanists. This usage is concise and commonly used in other botanical works including Arnow et al. 1980, IMF, Hitchcock et al. 1961, 1964, 1969, and Harrington 1954.

The taxa included in this work and information about distribution, habitat, and elevation are based on herbarium specimens. This information is not complete. Every field season of this study produced additional taxa and extensions in distribution and elevation. Additional taxa are expected, and distributional information will continue to expand with additional field work. Distributional and habitat information are given as an aid to identification and understanding of plant taxa of our area. This information also indicates where additional collections are needed. Distribution listed here is limited to our area. General distribution is readily available in TPD and is listed in AUF and IMF. Abundance is also indicated for many taxa. However, this is a rather loosely estimated value for many species.

Following the scientific names, synonyms and misapplied names are listed in parentheses and brackets. The term "misapplied" indicates a plant name of a valid taxon mistakenly applied to a plant for which a different name is correct. Misapplied names are usually included if they were used for plants in one or more lists or manuals cited in this work. Some taxa are known to our area by one or few collections. These collections are usually cited in the text. In some cases the acronym of the herbarium where the specimen is deposited is listed. Following is a list of acronyms of herbaria referred to in this work.

BRY Herbarium of Brigham Young University, Provo, Utah

CM Herbarium of Carnegie Museum of Natural History, Pittsburgh, Pennsylvania

COLO Herbarium of University of Colorado, Boulder, Colorado

CS Herbarium of Colorado State University, Fort Collins, Colorado

DINO* National Park Service Herbarium, Dinosaur National Monument

NY New York Botanical Garden, Bronx, New York

MO Missouri Botanical Garden

NRCS* Natural Resource Conservation Service, Roosevelt, Utah

OGDF Forest Service Herbarium, Ogden, Utah (moved to BRY)

PH Herbarium, Academy of Natural Sciences of Philadelphia, Philadelphia

RM Rocky Mountain Herbarium, University of Wyoming, Laramie, Wyoming

SSLP Herbarium of Shrub Science Laboratory, Intermountain Research Station, Provo

USUUB Utah State University Uintah Basin, Vernal Utah.

UI Uinta Herbarium, Bureau of Land Management, Vernal, Utah

UT Garrett Herbarium, University of Utah, Salt Lake City, Utah

UTC Intermountain Herbarium, Utah State University, Logan, Utah

* Acronym not included in *Index Herbariorum* (Holmgren, P. K., and N. H. Holmgren. 1998 [continuously updated]. *Index Herbariorum: A global directory of public herbaria and associated staff.* New York Botanical Garden's Virtual Herbarium. http://sweetgum.nybg.org/ih/)

Native plants are the focus of this flora. Cultivated plants that infrequently escape and rarely if ever persist are usually not included. Introduced weedy plants and especially those that have become naturalized are included. Nomenclature, with some exceptions, follows AUF and IMF. Nomenclature for Brassicaceae generally follows FNA (7; 224-746). This text has been near completion for several years while waiting for preparation of power point presentations that show photographs of plants included herein. Many changes have been made in families and genera in recent years. As a matter of convenience in avoiding major revision of the text, many of these changes are not followed.

Sources for names including those found in synonymy include: Holmgren and Reveal (1966), Harrington (1954), Graham (1937), Arnow et al. (1980), Goodrich et al. (1981), Hitchcock and Cronquist (1973), Holmgren (1962), Welsh

(1957), Bradley (1950), Flowers et al. (1960), Beidleman (1957), and Potter et al. (1983). Munz (1968) and AUF are followed in the method of listing synonyms. Synonyms found in old studies in files at offices of Ashley National Forest are also included. Many of these are listed in Tidestrom (1925).

Sources for common names include Plummer et al. (1977), Hitchcock and Cronquist (1973), Arnow et al. (1980), AUF and TPD. The abbreviations q.v. (which see), ssp. (subspecies), and var. (variety) are from Stern (1966). The abbreviations of north (n.), east (e.), south (s.), west (w.), northeast (ne.), northwest (nw.), southeast (se.), southwest (sw.), central (c.), Co. (Co.), Counties (Cos.), mountain (mt.), and mountains (mtns.) are those used by Hitchcock and Cronquist (1973). We have followed Harrington (1954) in using the metric system in keys and descriptions of plants and the English system for listing elevation and distance.

HISTORY OF COLLECTIONS

The first botanical collections taken from the Uinta Basin were made by the Fremont expeditions in 1844 and 1845 during which type specimens of *Gilia stenothyrsa, Penstemon fremontii,* and *P. pachyphyllus* were collected. Graham (1937) conducted the first extensive botanical study in the Uinta Basin. He spent the summers of 1931, 1933, and 1935 in our area and made about 4,200 collections that are deposited at the Carnegie Museum, Pittsburgh, Pennsylvania. Graham's treatment of plant communities, climate, geology, and other features of our area is superb. He listed over 1,100 specific and subspecific plant taxa, and his study has been a valuable source of information to this current work. Critical specimens from Graham's collections were obtained on loan for examination in this study.

Prior to Graham a number of early botanists collected at the margins of or sparingly within our area. Among these were John Wesley Powell, Lester Frank Ward, Sereno Watson, Marcus E. Jones, Leslie N. Gooding, L. H. Pammel, and George E. Osterhout. Graham (1937) discussed these and other early collectors.

Bradley (1950) worked on the vascular flora of Moffat Co. Collections from this project are deposited at the University of Colorado and Park Service Herbarium at Dinosaur National Monument. She also discussed other collectors of Moffat Co. Welsh (1957) provided a checklist of 360 specific and subspecific plant taxa that is based on collections from the Utah part of Dinosaur National Monument. His collections are deposited at the Stanley L. Welsh Herbarium at Brigham Young University.

Beidleman (1957) assembled a checklist of flora and fauna of Dinosaur National Monument. A list of plants of Dinosaur National Monument is maintained by the National Park Service (Headquarters at Dinosaur, Colorado). A list of plants of the Ouray National Wildlife Refuge is maintained by the U. S. Fish and Wildlife Service at Refuge Headquarters. Holmgren (1962) studied and collected vascular plants of Dinosaur National Monument and along the Green River from Flaming Gorge to Split Mountain. Woodbury et al. (1959) and Flowers (1960) conducted studies on lands that were subsequently covered by waters of Flaming Gorge Reservoir, and Flowers et al. (1960) assembled a checklist of plants for that area. Lewis (1970) described alpine rangelands of the Uinta Mountains and listed about 335 taxa for that area. Huber (1995) compared floras of limestone and quartz-rich sandstones of the Uinta Mountains, and he made numerous collections from our area. Brown (2006) provided a classification of alpine plant communities of the Uinta Mountains based on 308 plots scattered across the alpine of the Uinta Mountains Goodrich and others (1981) assembled a checklist of vascular plants of the Uinta Basin that has been a basis for a number of taxa included in this treatment. Vories (1974), Peterson and Baker (1982) and Baker (1983) studied plants and made collections in the Piceance Basin. Refsdal (1996) collected many specimens from the northern part of our area. Her specimens are deposited at RM and BRY.

The earliest collections deposited in Uinta Basin herbaria were made by Forest Service collectors, including John Bennett, Charles DeMoisy Jr., Selar Hutchings, Clyde Lambert, George Walkup, and K. E. Weight. Their specimens are now deposited at Utah State University Herbarium at Vernal (USUUB). Fredrick H. Hermann, Mont E. Lewis, and Duane Atwood made many collections in our area for the Forest Service.

Elizabeth Neese collected several thousand specimens from our area. A nearly complete set of her specimens is deposited at the Stanley L Welsh Herbarium at Brigham Young University, and many duplicates are deposited at the New York Botanical Garden, and in the Uinta Herbarium (UI) at the Vernal Field Office of Bureau of Land Management. In the previous edition of this work, the families Boraginaceae, Cactaceae, Ephedraceae, and Scrophulariaceae were written by Elizabeth Neese for the previous version of this work. Her work is carried forward in this edition with some changes and additions. The collections, family treatments, and enthusiasm of Elisabeth contributed greatly to this work.

Among those botanists whose collections are deposited at Brigham Young University are Duane Atwood, Bertrand Harrison, Stanley L. Welsh, Joseph Murdock, Jack Brotherson, Kaye Thorne, Ron Kass, Blain T. Welsh, Allen Huber, Jim Spencer, and James Reveal.

Patricia K. Holmgren, Noel H. Holmgren, Rupert C. Barneby, and Arthur Cronquist of the New York Botanical Garden have made numerous collections in preparation for the *Intermountain Flora*. Leila Shultz, Frank Smith, Kezia Snyder, and Betsy Neely have collected while working on various rare plant studies sponsored by the Bureau of Land Management. Lois Arnow and Beverly Albee (previous curators, UT) have collected in our area, and their excellent work in the central Wasatch Front (Arnow et al. 1980) is useful in our area and especially toward the west. Larry England made many collections, and he was instrumental in beginning the Uinta Herbarium (UI) kept by the Bureau of Land Management at Vernal, which now contains over 7,000 specimens.

William A. Weber (former curator, COLO) has been active in the Colorado part of our area, and he published a flora that included the Colorado part of our area (Weber 1987). Other collectors from Colorado include Harold D. Harrington (specimens at CS), Dieter H. Wilken (specimens at CS), Karen Wiley-Eberle, William L. Baker, Kimery Vories, Wayne Erickson, Walt A. Kelley, Robert Popp, and Joyce Walker. J. Scott Peterson has made many collections in both Colorado and Utah.

The National Park Service has sponsored and cooperated in a number of botanical projects in Dinosaur National Monument, and the Park Service keeps a herbarium, now housed at the Quarry, with principal collections from Ruth Ann Wolf and Kathleen Dever, M. MacLeod and W. MacLeod, Noel Holmgren, James Reveal, and Tom Jensen. Tamara Naumann has made many collections in the monument, and she has served as curator of the National Park Service Herbarium at the monument for many years.

Our field experience has been mostly in the Utah part of the area. However in recent years we have made numerous collections in Colorado. Within the area of this flora, we have made about 20,000 collections over a period of 40 years.

Lorain Squires started and maintained the herbarium at Utah State University, Vernal (USUUB) where he and his students deposited numerous specimens. The Ashley National Forest Herbarium was added to USUUB in 2001. This herbarium with about 8,000 collections has greatly increased in importance to this work in the past decade.

The two Vernal herbaria (UI and USUUB) coupled with the herbarium at Dinosaur National Monument (DINO) provide specimens for nearly all species listed in this work.

THE FLORA

With wide elevational differences (4,255 ft in Desolation Canyon to 13,528 ft on Kings Peak) and numerous geological formations, our area supports a diverse flora. The plants of lower and mid-elevations are typical of those of the Great Basin. Here desert shrub and sagebrush communities are dominated by shrubs of the Asteraceae and Chenopodiaceae families, and pygmy forests are made up of pinyon (*Pinus edulis*) and juniper (*Juniperus osteosperma*). *Redfieldia flexuosa, Spartina pectinata, Thermopsis rhombifolia,* and a few other species from the Plains reflect a slight influence from that flora. In sharp contrast, arctic and Rocky Mountain floras are strongly expressed in the Uinta Mountains where the genus *Carex* (the largest genus of our area) is represented by over 80 species. The flora of the Tavaputs Plateau has affinity to that of other parts of the Colorado Plateau and to the Great Basin. The diverse flora of Strawberry Valley shows influences from the Uinta and Wasatch Mountains, Colorado Plateaus and Great Basin. Some plants of the Wyoming flora are found in our area, especially in Daggett and Moffat Counties.

The floodplain of the Green River adds considerably to the diversity of the flora. Some species appear to be restricted in our area to this floodplain, and several others found elsewhere are most common on this floodplain.

There are about 50 species endemic to our area. Each endemic is usually confined to one or a few geological formations. Endemic plants have been the basis of a number of studies that have greatly increased information about the Uinta flora (Bio/West 1984; Neese and Smith 1982; Peterson and Baker 1982; Shultz and Mutz 1979; Shultz 1982; Welsh and Neese 1979a, b; Welsh 1981; Welsh and Atwood 2009).

PLANT COMMUNITIES

Moist alkaline lowlands support plant communities dominated by saltgrass (*Distichlis spicata*), greasewood (*Sarcobatus vermiculatus*), seepweed (*Suaeda* spp.), and alkali saccaton (*Sporobolus airoides*), threadleaf rubber rabbitbrush (*Chrysothamnus nauseosus* ssp. *consimilis*), and other halophytes. Along washes and other drainages, western goldenrod (*Euthamia occidentalis*) is locally common. With increasing wetness, rushes (*Juncus* spp.), sedges (*Carex* spp.), and bullrushes (*Scirpus* spp.) become abundant. Low elevation wetlands are often dominated by cattails (*Typha* spp.), bullrushes, and common reedgrass (Phragmites australis). The floodplains of the Green River and lower elevations of its tributaries support Fremont cottonwood (*Populus fremontii*) communities.

Dry lowlands support desert shrub communities that are dominated by low shrubs of the Chenopodiaceae and Asteraceae families. Some of the most common of these shrubs are shadscale (*Atriplex confertifolia*), Gardner saltbush (*A. gardneri*), bud sagebrush (*Artemisia spinescens*), winterfat (*Krascheninnikovia lanata*), and horsebrush (*Tetradymia* spp.). Mat saltbush (*Atriplex corrugata*) is common on geologic formations with high clay and silt content, principally the Mancos Shale, Duchesne River, and Uinta formations. Dune rubber rabbitbrush (*Chrysothamnus nauseosus* ssp. *turbinatus*) is locally common on aeolian sand. Indian ricegrass (*Oryzopsis hymenoides*), galleta (*Hilaria jamesii*), and blue gramma (*Bouteloua gracilis*) are among the common grasses of these communities. Forbs are numerous, but seldom is any one forb species abundant. One notable exception is small-leaf globemallow (*Sphaeralcea parvifolia*) which provides spectacular floral displays in desert shrub communities in years of favorable precipitation. Desert shrub communities are found from the lowest elevations of our area up to about 6,000 ft. They indicate an average annual precipitation of about 6 to 8 inches.

Greasewood forms dense thickets at lower elevations of drainages of the Tavaputs Plateau. With increasing elevation and probably decreasing salinity or alkalinity, it is associated with basin big sagebrush (*Artemisia tridentata* ssp. *tridentata*), and at about 7,000 ft greasewood is replaced by sagebrush and rubber rabbitbrush (*Chrysothamnus nauseosus*).

Pinyon and juniper communities form an elevational band across our area starting at about 6,000 ft. The band is wider and more continuous on the Tavaputs Plateau than it is on the Uinta Mountains. It commonly reaches to 8,000 ft on the Tavaputs Plateau, extending upward to about 8,500 ft on warm exposures. On the Uinta Mountains, it commonly reaches to about 7,000 ft extending to about 7,500 (rarely 8,000) ft on warm exposures. On the Tavaputs Plateau the band is made up of juniper at its lower elevations, a mixture of pinyon and juniper at mid-elevations, and pinyon with little or no juniper at its upper elevations. On the south slope of the Uinta Mountains, the band is mostly made up of pure stands of juniper with the pinyon being mostly limited to the upper elevations where the communities are restricted to warm exposures. On the north slope of the Uinta Mountains in Daggett Co. pinyon is rather well represented, but it is lacking toward the west. The elevational range of pinyon and juniper communities is apparently determined by thermal conditions. They seem to be limited on the lower end by severe cold air inversions that are common to our area in winter. Cold temperatures of long durations at higher elevations seem to be the determining factor at the upper end where these communities show a strong affinity for warm exposures. Low precipitation may be an additional limiting factor of some influence at the lower limit. Pinyon and juniper communities are found essentially on all soils and substrates available within the thermal zone to which they are confined.

Various kinds of sagebrush (*Artemisia* spp.) and rabbitbrush (*Chrysothamnus* spp.) and alderleaf mountain mahogany (*Cercocarpus montanus*) are shrubs commonly associated with pinyon and juniper. Indian ricegrass, Sandberg bluegrass (*Poa secunda*), muttongrass (*P. fendleriana*), and blue bunch wheatgrass

(*Elymus spicatus*) are common grasses, and on the Tavaputs Plateau, Salina wildrye or bullgrass (*Elymus salinus*) is common. Stemless hymenoxys (*Hymenoxys acaulis*), stemless goldenweed (*Haplopappus acaulis*), and scarlet gilia (*Ipomopsis aggregata*) are among the common forms of pinyon and juniper communities.

Sagebrush communities are dominated by *Artemisia* spp. with or without a codominant component of grasses. Black sagebrush (*A. nova*) communities are common from about 5,000 to 8,000 ft. They are typically formed on shallow, rocky soils, especially those high in carbonates. Black sagebrush is sometimes a component of desert shrub communities. Sandberg bluegrass is a common grass in these communities.

Wyoming big sagebrush (*Artemisia tridentata* var. *wyomingensis*) communities are found in and below the pinyon-juniper belt. Soils of these communities typically have a carbonate hardpan at about 12 to 20 inches below the surface. Associated grasses are often sparse where canopy cover of Wyoming big sagebrush is greater than 15%. They include needle-and-thread (*Stipa comata* var. *comata*), Indian ricegrass, and Sandberg bluegrass.

Mountain big sagebrush (*Artemisia tridentata* var. *pauciflora*) communities cover extensive areas from the upper edge of the pinyon-juniper belt up to about 10,000 ft. At upper elevations they are confined to warm aspects. They are typically found on soils with dark colored surface horizons and without a carbonate hardpan. The grass component is usually well developed. Among the common grasses are bluebunch wheatgrass, thickspike wheatgrass (*Elymus lanceolatus*), needle-and-thread (*Stipa comata* var. *intermedia*), muttongrass, Sandberg bluegrass, and sheep fescue (*Festuca ovina* var. *saximontana*). Mountain low rabbitbrush (*Chrysothamnus viscidiflorus* ssp. *lanceolatus*) is a common associated shrub. Bitterbrush (*Purshia tridentata*) is also a common shrub especially on the Uinta Mountains. Forbs are well represented in many mountain big sagebrush communities, among the common being arrowleaf balsamroot (*Balsamorhiza sagittata*) and Hooker balsamroot (*B. hooker*) and silver lupine (*Lupinus argenteus*).

Communities of basin big sagebrush are usually limited to alluvial soils of canyon bottoms and valleys from the lowest elevations of our area up to about 8,000 ft. In canyons of the Tavaputs Plateau, basin big sagebrush is often associated with greasewood at lower elevations and with rubber rabbitbrush at upper elevations. In valleys, basin big sagebrush is an indicator of rather deep soils of sandy loam texture, and much of the valley-land once dominated by this sagebrush is now in cultivation.

Mountain silver sagebrush (*Artemisia cana* var. *viscidula*) is codominant with grasses, sedges, and forbs in meadows or meadow-like communities where the water table is at or near the soil surface during some of the growing season. Such communities are particularly common in Strawberry Valley, Farm Creek Pass, and on Diamond Mtn. At the head of Cart Creek along Highway 44, silver sagebrush meadow communities are well developed on Browns Park Formation, but they are not known in the Uinta Mountains on formations of the Uinta Mountain Group. Apparently mountain silver sagebrush communities are lacking on the Tavaputs Plateau except at the far west end of the West Tavaputs Plateau from the Willow Creek-Beaver Creek area where this shrub is often associated with upland grasses in non-riparian settings.

Communities of snowfield or spike sagebrush (*Artemisia spiciformis*) are limited to the western part of our area. They are well developed along

Reservation Ridge between Cat Peak and Horse Ridge, near Red Creek Mtn., the east slope in Blind Stream drainage, Wolf Creek summit, and scattered locations on the north slope of the Uinta Mountains including the Whitney area. These communities are best developed on well-drained, leeward slopes where drifting snow accumulates. They are mostly found on soils of clay loam texture on limestone or other basic substrates at 9,000 to 10,000 ft. Although snowfield sagebrush is occasionally found on quartzitic sandstones of the Uinta Mountains, communities dominated by this shrub are infrequent in this setting.

Salina wildrye communities are best developed on the Tavaputs Plateau on Uinta and Green River formations. Salina wildrye forms nearly pure stands on steep raw slopes and colluvial fans as well as on lands of gentle relief. This plant is also associated with blue bunch wheatgrass, sagebrush, big wild buckwheat (*Eriogonum corymbosum*), pinyon, juniper, and with plants of mountain brush communities such as true mountain mahogany. Away from the Tavaputs Plateau, this grass is mostly scattered and is seldom found above 7,000 feet, but near Sheep Creek Gap and a few other places it forms stands on formations that tend to weather to badlands.

Mountain brush communities are dominated by species of *Cercocarpus, Amelanchier, Quercus gambelii, Prunus virginiana,* and a few other shrubs. These communities are most common from 7,000 to 8,500 ft. Grasses and forbs are mostly well represented in these communities except for those with a dense canopy of curlleaf mountain mahogany (*Cercocarpus ledifolius*). Alderleaf mountain mahogany forms rather extensive stands on slopes across the Uinta Mountains, and it is common across the Tavaputs Plateau. Communities of curlleaf mountain mahogany are well developed on the north slope of the Uinta Mountains, especially toward the east and on Diamond Mtn. There are sizable stands of this shrub in Lake Fork and Yellowstone Canyons, and it is common on Red Mtn. and vicinity north of Vernal, but it is rare elsewhere on the south slope of the Uinta Mountains. It is occasional toward the west on the West Tavaputs Plateau. Littleleaf mountain mahogany (*C. intricatus*) forms stands on outcrops of Navajo, Weber, and other sandstone formations about the flanks of the Uinta Mountains from about 6,000 to 7,000 ft. Gambel oak communities are well developed west of Blind Stream in the Uinta Mountains (but completely absent from the east part of these mountains) and west of Avintaquin on the West Tavaputs Plateau and again on the East Tavaputs Plateau. Communities dominated by serviceberry (*Amelanchier* spp.) are rather infrequent, but Saskatoon serviceberry (*A. alnifolia*) does form stands at a few places in the Uinta Mountains including John Starr Flat between Uinta and Yellowstone drainages.

Riparian communities are marked by an abundance of willows (*Salix* spp.), cottonwoods (*Populus* spp.), alder (*Alnus incana*), river birch (*Betula occidentalis*), and numerous sedges and grasses.

Ponderosa pine (*Pinus ponderosa*) communities are best developed on the north slope of the Uinta Mountains east of Birch Creek and lacking or essentially so west of Birch Creek. Ponderosa pine is found in the Provo River drainage at the west end of the Uinta Mountains. On the south slope of the Uinta Mountains, Ponderosa pine communities are best developed between the Lake Fork and Whiterocks drainages from about 7,000 to 8,000 ft. There are scattered stands elsewhere such as on Red Mtn. and Blue Mtn.

There is a sizable stand of ponderosa pine in the Cow Canyon-Timber Canyon area of the West Tavaputs Plateau, but elsewhere on this plateau ponderosa pine is either mostly lacking or is associated with other conifers. Ponderosa pine communities are generally aligned with thermal conditions similar to but cooler than those of pinyon-juniper communities, and they are best developed in the warmer elevations of the Uinta Mountains well away from cold air drainage. Ponderosa pine trees occasionally extend down into the Uinta Basin and sometimes survive for many years, but they are killed in winters with long and cold inversions. Bitterbrush is a common understory shrub, and greenleaf manzanita (*Arctostaphylos patula*) is common in the Lake Fork, Yellowstone, Dry Gulch, and Uinta drainages. Scattered ponderosa pine trees are found at lower elevations on the acidic Mowry Formation where precipitation would be expected to be below the tolerance of this tree.

Douglas-fir (*Pseudotsuga menziesii*) communities in the Uinta Mountains are mostly limited to limestone. Notable exceptions include Weber sandstone and the sandstones of Uinta Mtn. Group in Red Canyon and its tributaries. On the Tavaputs Plateau these communities are well developed below the elevations typically dominated by subalpine fir (*Abies lasiocarpa*). Toward its lower limits Douglas-fir is mostly restricted to slopes of northerly aspects. Elevational range is mostly from 7,500 to 9,000 ft. Oregon grape (*Mahonia repens*) and mountain snowberry (*Symphoricarpos oreophilus*) are common understory plants in both the Uinta Mountains and Tavaputs Plateau. Mallow ninebark (*Physocarpus malvaceus*) is a dominate understory shrub in some stands in the West Tavaputs Plateau where mountain hollyhock (*Iliamna rivularis*) becomes abundant after fire.

Aspen communities cover extensive areas on mountain sides and in canyons from about 7,500 to 9,500 ft. In the Uinta Mountains aspen stands found above 8,500 ft are more common toward the west on limestone. Isolated, small stands of unthrifty aspen are found in subalpine talus. Elk sedge (*Carex geyeri*) is sometimes a common understory plant especially in the Uinta Mountains where common juniper (*Juniperus communis*) is also abundant. In other places, mountain snowberry is the dominant understory plant, and in a few places mallow ninebark is common in the western Uinta Mountains.

Tall forb communities are often found in association with aspen communities in Strawberry Valley and from west of Rock Creek on the south slope of the Uinta Mountains including the upper West Fork Duchesne River drainage, the Whitney area of the north slope, and rather weakly developed from west of Indian Canyon on the West Tavaputs Plateau. Tall forb communities seem to be limited by the quartz-rich sandstones of the Uinta Mountains and by insufficient moisture on much of the Tavaputs Plateau. They are marked by an abundance of butterweed groundsel (*Senecio serra*), western coneflower (*Rudbeckia occidentalis*), Arizona bluebells (*Mertensia arizonica*), western sweet-cicely (*Osmorhiza occidentalis*), tall larkspur (*Delphinium occidentalis*) and other tall forbs. In wet places, California false-hellebore (*Veratrum californicum*) forms dense stands, and cow parsnip (*Heracleum lanatum*) is occasionally common.

Subalpine fir communities are well developed on concave slopes of cool exposures at the upper elevations of the Tavaputs Plateau and in the vicinity of the Strawberry Valley. On the Uinta Mountains these communities are found on limestone and other basic substrates. On quartzitic sandstones of the Uinta Mtn. Group, subalpine fir is found in association with lodgepole pine (*Pinus contorta*) and Engelmann spruce (*Picea engelmannii*), but it rarely dominates stands on this substrate.

Lodgepole pine communities cover extensive areas on the Uinta Mountains. These communities are closely associated with quartzitic sandstones of the Uinta Mtn. Group. This tree is lacking or nearly so on soils derived from limestone. The Blind Stream drainage provides a vivid example of absence or low presence of lodgepole pine on limestone. Lodgepole pine is not found on the Tavaputs Plateau where marly mudstones and calcareous sandstones of the Uinta and Green River Formations dominate.

Lodgepole pine often forms pure stands from about 9,000 to 10,000 ft. Subalpine fir is occasionally an associated tree, but it is excluded in vast areas especially where lodgepole pine occurs in pole or "dog-hair" thickets. At places in upper elevations, grouse whortleberry (*Vaccinium scoparium*) forms a ground cover while elk sedge forms a ground cover at lower elevations. But in many places the forest floor supports only a sparse cover of understory plants. However, where opened by fire or by logging, several successional species quickly increase. Ross's sedge (*Carex rossii*) is probably the most common seral species. Several other sedges, spike trisetum (*Trisetum spicatum*), strawberry (*Fragaria* spp.), sheep fescue, coast goldenrod (*Solidago simplex*), and low goldenrod (*Solidago multiradiata*) are among the many other seral plants.

Lodgepole pine and Engelmann spruce grow together from about 9,800 to 10,400 ft in the Uinta Mountains. Communities formed by these two trees often have an understory of grouse whortleberry. But as in lodgepole pine communities, understory plants are sometimes sparse. Subalpine fir is a common associate with lodgepole pine and Engelmann spruce at this elevation.

Engelmann spruce communities are found from about 10,400 ft to timberline (about 11,000 ft) in the Uinta Mountains. Near timberline these trees are commonly reduced to the krummholz condition. This tree is abundant on quartzite, limestone, and other substrates in the Uinta Mountains, but apparently it is replaced by blue spruce (*Picea pungens*) on the Tavaputs Plateau. Small stands of blue spruce are found in mesic places at upper elevations of the Tavaputs Plateau, and it is a common tree along drainages at lower to mid-elevations in the Uinta Mountains. Stands of blue spruce are also found on limestone uplands in the Brush Creek drainage.

Subalpine meadow communities are found in openings in the dense coniferous forests of the Uinta Mountains. In dry meadows, timber oatgrass (*Danthonia intermedia*) is abundant. In somewhat more moist places, tufted hairgrass (*Deschampsia caespitosa*) is abundant. And in very wet places, a number of sedges including water sedge (*Carex aquatilis*), Russet sedge (*C. saxatilis*), and beaked sedge (*C. utriculata*) are usually dominant. In boggy places, *C. limosa* and *Eleocharis quinqueflora* are common, and *Scirpus cespitosus* is locally abundant on hummocks. Numerous forbs and other grasses are found in these meadow communities. Some of the common forbs include elephant head (*Pedicularis groenlandica*), explorer gentian (*Gentiana calycosa*), fernleaf ligusticum (*Ligusticum tenuifolium*) marsh marigold (*Caltha leptosepala*), and American bistort (*Polygonum bistortoides*). Plain-leaf willow (*Salix planifolia*) is common, especially along drainages.

Alpine tundra communities cover extensive areas on the Uinta Mountains above timberline. These communities are typically dominated by grasses and sedges such as tufted hairgrass and Canada single-spike sedge (*Carex scirpoidea* var. *pseudoscirpoidea*). Alpine avens (*Geum rossii*) is essentially restricted to the alpine. However, within the alpine area this plant has wide ecological amplitude on rounded summits and low to moderate gradients of the Uinta Mountain Group. It is a dominant in turf communities with single spike sedge, kobresia, and tufted hairgrass and it is frequent in but of relatively low percent cover in fellfields with curly sedge and cushion plants. It is a colonizer in vegetated islands of stone-nets and stone-stripes and margins of snowbeds. Varileaf cinquefoil (*Potentilla diversifolia*) and alpine avens were the most frequent alpine forbs, and alpine avens was the forb of greatest cover in an alpine study in the Uinta Mountains (Brown 2006). Other forbs that are typical of these communities are alpine sagebrush (*Artemisia scopulorum*), American bistort (*Polygonum bistortoides*), moss campion (*Silene acaulis*), whitish gentian (*Gentiana algida*), and beautiful paintbrush (*Castilleja pulchella*).

ACKNOWLEDGMENTS

The Stanley L. Welsh Herbarium at Brigham Young University, with about 25,000 specimens from the Uinta Basin, has been a primary source of information for the Utah portion of our area. The Herbarium of Colorado State University has been a primary source of information for the Colorado portion of our area. Herbarium searches were also conducted at the Garrett Herbarium, Intermountain Herbarium, Bureau of Land Management at Vernal. The National Park Service provided access to the Herbarium at Dinosaur National Monument, and sent specimens of *Erigeron* to Art Cronquist for annotation.

Ashley National Forest Service Herbaria at Vernal, and two other Forest Service herbaria (OGDF and SSLP) were also used in preparation of this flora. Loans of Edward H. Graham's specimens were received twice from Carnegie Museum. Some specimens of other collectors cited by Graham (1937) were obtained on loan from Missouri Botanical Garden and Rocky Mtn. Herbarium. The cooperation of curators at each of the above herbaria is appreciated. Not only did Stanley L. Welsh (Curator, BRY) leave the herbarium open for this work, but he also gave helpful responses to numerous questions about the flora.

Dieter Wilken provided valuable information on the distribution in Colorado of a number of taxa. Karen Wiley-Eberle provided maps of the Colorado portion of our area and provided distribution information for some taxa. Beverly Albee and Lois Arnow (UT) responded to numerous questions about the flora. The enthusiastic long-term support and encouragement offered by personnel of the Forest Service, Bureau of Land Management, and Park Service is also appreciated.

KEY TO FAMILIES

1 Plants aquatic; leaves submerged, or some or all of them floating; inflorescence submerged or floating or sometimes aerial, but then usually not extending much above the water-surface . . . . **KEY 1**

1 Plants not aquatic, sometimes growing in water but then some or all of the leaves emergent and not floating; leaves sometimes much reduced but then the inflorescence well elevated above the surface of the water 2

2 Plants parasitic or saprophytic, growing on trees, or if not on trees then devoid of chlorophyll and not at all green; leaves obsolete or scale-like . . . . **KEY 2**

2 Plants not parasitic or saprophytic, with chlorophyll and almost always more or less green, or with well developed leaves. 3

3 Plants not producing flowers or seeds, reproducing by spores; aerial stems and scapes often lacking or else terete, hollow, and sheathing-jointed, or plants moss-like and evergreen (ferns, horsetails, and spikemosses) **KEY 3**

3 Plants producing flowers and seeds (the flowers sometimes small and inconspicuous or converted into bulblets), with aerial stems or scapes, or flowers sometimes sessile in a basal rosette of leaves, the stems sometimes hollow but not sheathing-jointed, rarely both moss-like and evergreen. . 4

4 Stems spiny and succulent; leaves lacking, or essentially so; flowers showy with several to numerous petals; stamens numerous (prickly pears and other cacti) . . . . **CACTACEAE**

4 Plants not as above in all features. 5

5 Plants vines or vine-like . . . . **KEY 4**

5 Plants not vines 6

6 Plants woody above ground level, trees and shrubs . . . . **KEYS 5A-5D**

6 Plants herbaceous above ground level. 7

7 Leaves strongly sheathing at the base and not opposite or flowers 3- or 6-merous or both; leaf blades mostly parallel-veined, simple, entire, or rarely basally lobed, sometimes lacking and leaves reduced to bladeless sheaths [class Liliopsida (Monocotyledons), rushes, sedges, grasses, lilies, irises, orchids, and similar plants] . . . . **KEY 6**

7 Leaves not sheathing, or if so then mostly opposite or compound, the blades often but not always pinnately, palmately or reticulate veined, simple or compound, entire, toothed or variously lobed or dissected, rarely lacking; flowers mostly 2-, 4-, or 5-merous or multiples there of [class Magnoliopsida (Dicotyledons)] . . . . **KEY 7**

— KEY 1 —
Plants aquatic

1 Plants 2-12 mm long, without stems, free floating, often forming a mass over the water surface; roots, if present, dangling from the lower surface of the leaf-like thallus . . . . **LEMNACEAE**

1 Plants not as above in all respects 2

2 Plants without evident stems or scapes; leaves appearing all basal; reproduction by spores; flowers and seeds lacking 3

2 Stems or scapes usually evident; leaves not as in the above two families in all features; plants producing flowers and seeds 4

3 Leaves linear, terete, similar to those of onions, the blades not petiolate . . . . **ISOETACEAE**

3 Leaves like those of a 4-leaved clover, the 4-foliate blade borne on a slender petiole . . . . **MARSILEACEAE**

4 At least the submerged leaves dissected into linear or filiform segments . 5

4 Leaves not dissected into linear or filiform segments, entire, toothed or lobed or trifoliate to pinnately compound 8

5 Flowers borne on naked pedicels or peduncles; corollas conspicuous; leaves twice or more dissected, either with stipulate bases or with small, buoyant bladders. 6

5 Flowers borne in the axils of leaves or sessile in a terminal spike, usually inconspicuous, corolla lacking or quickly deciduous; leaves variously dissected, not bearing bladders, not stipulate at the base 7

6 Corollas yellow, united, strongly 2-lipped; stamens 2; the peduncles usually conspicuously emergent, often with more than 1 flower; leaves 2-3 or more times pinnately dissected, bearing small, buoyant bladders, without stipulate bases . . . . **LENTIBULARIACEAE**

6 Corollas whitish, not 2-lipped, the 5 petals free; stamens 10 or more; pedicels usually submersed or nearly so with the solitary flowers near the water surface; leaves 2-3 or more times dichotomously, trichotomously or palmately dissected, with stipulate bases, without bladders (*Ranunculus*) . . . . **RANUNCULACEAE**

7 Submerged leaves once pinnately dissected, the segments entire; emersed leaves or bracts subentire or pectinately dissected . . . . . **HALORAGACEAE**

7 Leaves 1-2 (3) times palmately or dichotomously dissected, the segments finely toothed . . . . **CERATOPHYLLACEAE**

8 Leaves all basal, simple. 9

8 Leaves not all basal, or if apparently all basal then trifoliate. 11

9 Leaf blades 10-45 cm long, cordate, floating, on long submerged petioles; sepals several to many, the inner ones petal-like, bright yellow; flowers solitary . . . . **NYMPHAEACEAE**

9 Leaf blades smaller or flowers not solitary; sepals not as above 10

10 Scapes not over 4 cm long, shorter than the leaves, 1-several per plant; flowers solitary (*Limosella*) . . . . **SCROPHULARIACEAE**

10 Scapes over 4 cm long, shorter than or exceeding the leaves, 1-3 per plant but often solitary; flowers not solitary . . . . **ALISMATACEAE**

11 Leaves opposite or whorled 12

11 Leaves alternate or appearing basal. 18

12 Leaves 5-30 mm wide; flowers in axillary racemes; corolla blue, conspicuous (*Veronica*) . . . . **SCROPULARACEAE**

12 Leaves 0.5-4 mm wide; corolla lacking or inconspicuous 13

13 Leaves whorled with 4 or more per node 14

13 Leaves opposite, sometimes appearing whorled or 3 per node in *Elodea canadensis* 15

14 Stems emergent .... **HIPPURIDACEAE**

14 Plants submersed .... **NAJADACEAE**

15 Leaves 2-10 cm long, the stipules free of the blade and sheathing the stem; stamen 1 .... **ZANNICHELLIACEAE**

15 Leaves 0.2-3 cm long, stipules lacking or fused to the blade; stamens various 16

16 Flowers born on thread-like perianth tubes 4-30 cm long, mostly unisexual; leaves finely serrulate, submersed, all linear ..... **HYDROCHARITACEAE**

16 Flowers and fruit axillary and sessile or on pedicels to 6 mm long, often bisexual; Leaves entire, linear or not (see also leads 4 and 5 in Key 11).. 17

17 Leaves with short but definite connate-sheathing bases; flowers and fruit sessile or on pedicels to 6 mm long; stamens 4; petals 4, about 1.5 mm long (*Tillaea*) .... **CRASSULACEAE**

17 Leaves not connate-sheathing; flowers and fruit sessile; corolla lacking; stamens 1 or rarely more .... **CALLITRICHACEAE**

18 Leaves simple or entire... 19

18 Leaves either compound or not entire............................. 24

19 Corolla 3-16 mm long, deeply lobed .... **CAMPANULACEAE**

19 Perianth lacking or mostly less than 3 mm long, not lobed 20

20 Inflorescence a panicle; fruit not an achene (*Catabrosa* and possibly other genera) .... **POACEAE**

20 Flowers and fruit (achenes) sessile in a spike or in 1-several globose heads 21

21 Leaves pinnate-veined, not with sheathing bases, but with sheathing stipules; perianth pink to rose (*Polygonum*) .... **POLYGONACEAE**

21 Leaves parallel-veined, at least some with sheathing bases; perianth greenish.. 22

22 Lower leaves loosely sheathing, commonly over 30 cm long, longer than the upper sheathless leaves; stipules lacking; inflorescence of 1-several globose heads, the heads mostly over 1 cm in diameter .... **SPARGANIACEAE**

22 Lower leaves equal to or mostly smaller than the upper leaves, of various lengths; stipules sheathing; inflorescence a spike, sometimes head-like but then less than 1 cm thick 23

23 Ligules free or sheathing the stem, not adnate to the leaf blade, conspicuous but often soon shredding into fibrous threads; inflorescence borne on straight peduncles .... **POTAMOGETONACEAE**

23 Ligules lacking, the scarious margins of the sheaths prolonged on to the leaf blade as adnate small auricles; inflorescence sometimes enveloped in a leaf sheath at first but usually soon elevated to the water surface on a thread-like, elongated, peduncle that is sometimes coiled .. **RUPPIACEAE**

24 Leaves trifoliate, appearing all basal; leaflets entire or undulate-dentate, 5-10 cm long .... **MENYANTHACEAE**

24 Leaves not as above... 25

25 Leaves pinnately compound with 5-9 leaflets or seldom more, suborbicular to elliptic leaflets (*Nasturtium*) .... **BRASSICACEAE**

25 Leaves palmately lobed (*Ranunculus*) .... **RANUNCULACEAE**

— KEY 2 —
Plants parasitic or saprophytic

1   Plants parasitic on branches of coniferous trees, more or less green or yellow-green; leaves opposite, scale-like . . . . **VISCACEAE**

1   Plants not on branches of coniferous trees not green; leaves various 2

2   Stems twining on host plants, vine-like, seldom over 3 mm in diameter . . . . **CUSCUTACEAE**

2   Stems erect, not at all vine-like; sometimes over 3 mm in diameter 3

3   Plants scapose or with 1-3 bladeless sheaths enveloping the lower part of the stem, glabrous; flowers short pedicellate or nearly sessile; perianth of 6 similar free segments (*Corallorhiza*) . . . . **ORCHIDCACEAE**

3   Stems with few to numerous bract-like or scale-like leaves, these not sheathing, glandular-hairy; corolla united, with 4-5 lobes. 4

4   Plants 30-60 (100) cm tall (*Pterospora*) . . . . **MONOTROPACEAE**

4   Plants 3-20 (30) cm tall . . . . **OROBANCHACEAE**

— KEY 3 —
Plants reproducing by spores

1   Leaves compound, 2-100 cm long; spores borne on the leaves 2

1   Leaves lacking or simple and only 0.1-0.3 cm long; spores various. 4

2   Leaves palmate with 4 leaflets . . . . **MARSILEACEAE**

2   Leaves pinnate 3

3   Plants with a solitary evident aerial stem, the stem supporting one sterile leaf (lower) and one fertile leaf (terminal),; leaves 2-10 cm long, once compound, the young leaves not coiled at the apex . . **OPHIOGLOSSACEAE**

3   Plants without evident aerial stems; leaves more than 2, all basal, either longer than above or plants of dry rocky places, the young leaves coiled at the apex . . . . **POLYPODIACEAE**

4   Stems densely covered with spirally arranged overlapping scale-like leaves, creeping with ascending erect branches, the branches similarly covered with leaves; plants more or less forming mats, not extending over 10 cm above ground level; spores borne on sporophylls that are slightly different from the leaves . . . . **SELAGINELLACEAE**

4   Stems leafless (or leaves inconspicuous), erect, simple or sometimes with whorls of elongate branches, terete, hollow, sheathing-jointed, 10-150 cm tall; spores borne in a cone-like cap at the apex of stems . **EQUISETACEAE**

— KEY 4 —
Plants vines or vine-like

1   Leaves compound with 3 or more distinct leaflets. 2

1   Leaves simple, sometimes deeply cut but not truly compound 3

2   Leaves trifoliate to biternate (*Clematis*) . . . . **RANUNCULACEAE**

2   Leaves pinnate compound with 8-16 leaflets (*Vicia*) . . . . **FABACEAE**

3   Leaves palmately veined and usually palmately lobed, about as wide or wider than long, mostly 3-18 cm long 4

3 Leaves not palmately veined, not palmately lobed, often smaller than above
.... 6

4 Leaves opposite, deeply palmately divided; plants native (*Humulus*) ......
.... **CANNABACEAE**

4 Leaves alternate, not divided; plants introduced, cultivated, sometimes escaping. 5

5 Plants woody vines (*Vitis*) **VITACEAE** (Grape family, not treated in the text)

5 Plants not woody; cultivated, persisting, and escaping . **CUCURBITACEAE**

6 Leaves opposite (*Linnaea*) .... **CAPRIFOLIACEAE**

6 Leaves alternate 7

7 Leaves evergreen 3-12 mm wide, leathery; stems creeping on the ground
.... **ERICACEAE**

7 Leaves deciduous, not leathery; stems climbing. 8

8 Corolla white or pink without reflexed lobes (*Convolvulus arvensis*) ....
.... **CONVOLVULACEAE**

8 Corolla blue or purple with reflexed lobes (*Solanum dulcamara*) ....
.... **SOLANACEAE**

— KEY 5 —
Plants woody above ground level, shrubs and trees

1 Leaves needle-like or scale-like, evergreen, aromatic with pine-like odor 2

1 Leaves not needle-like or scale-like or if so then not evergreen, not aromatic with pine-like odor 3

2 Leaves needle-like; plants trees or shrub-like at timberline; seeds borne in cones; cones 2-15 (25) cm long, with several or many scales, brownish or purplish or blackish .... **PINACEAE**

2 Leaves scale-like or if needle-like then plants spreading shrubs; seeds born in globose, drupe-like cones; cones not over 1 cm long, without obvious scales, usually blue glaucous .... **CUPRESSACEAE**

3 Leaves compound, not scale-like .... **KEY 5A**

3 Leaves simple, sometimes scale-like or lacking 4

4 Some or all leaves and twigs opposite or whorled, not all basal ... **KEY 5B**

4 Leaves and twigs alternate or essentially basal .... **KEY 5C**

— KEY 5A —
Plants woody, leaves compound

1 Leaves palmately divided, the linear segments appearing as whorls of needle-like leaves, not over 2 cm long; plants not over 50 cm tall (*Leptodactylon*) .... **POLEMONIACEAE**

1 Leaves not as above.. 2

2 Leaves and twigs opposite; leaflets all petiolulate 3

2 Leaves and twigs alternate; leaflets sometimes opposite, mostly sessile or nearly so except for the terminal one 4

3 Leaves pinnately compound with 5-9 leaflets; leaflets finely serrate with more than 20 teeth (*Sambucus*) .... **CAPRIFOLIACEAE**

3   Leaves with 3-5 leaflets; leaflets entire or coarsely serrate or dentate with fewer than 20 teeth .... **ACERACEAE**

4   Leaves holly-like, evergreen, the leaflets with spiny-toothed margins; shrubs not over 30 cm tall, often of woods; flowers yellowish; fruit bluish, berry-like .... **BERBERIDACEAE**

4   Leaves not holly-like, the leaflets not with spiny-toothed margins; shrubs, flowers, and fruit various... 5

5   Leaflets 3, some usually over 2 cm long, not linear, stems not armed with bristles .... **ANACARDIACEAE**

5   Leaflets either more than 3 or less than 2 cm long and linear or nearly so, or stems armed with bristles....................................... 6

6   Leaves dissected into linear, or nearly linear, segments, seldom over 2 cm long, emitting a sagebrush-odor when crushed; plants 5-40 cm tall (*Artemisia*) .... **ASTERACEAE**

6   Leaves mostly pinnately compound with well marked leaflets, mostly over 2 cm long, or if dissected as above in *Potentilla fruticosa*, not emitting a sagebrush odor when crushed; plants mostly over 40 cm tall 7

7   Flowers sweetpea-type (papilionaceous), yellow; fruit a legume; leaves 4-10 cm long, pinnate with 8-12 leaflets (*Caragana*) .... **FABACEAE**

7   Flowers regular, yellow only in Potentilla and then leaves 1-5 cm long with 3-7 leaflets; fruit not a legume .... **ROSACEAE**

— KEY 5B —
Plants woody; some or all leaves and twigs opposite or whorled

1   Leaves inconspicuous or lacking; twigs mostly whorled, yellow-green, blue-green, or bright green .... **EPHEDRACEAE**

1   Leaves conspicuous; twigs not whorled, variously colored 2

2   Leaves palmately lobed or divided to the base 3

2   Leaves entire, toothed, or pinnately lobed 4

3   Trees or tall shrubs over 80 cm tall; leaves over 2 cm long, lobed .... **ACERACEAE**

3   Shrubs or sub-shrubs not over 80 cm tall; leaves not over 2 cm long, divided to the midrib, the linear segments sometimes appearing as whorls of needle-like leaves .... **POLEMONIACEAE**

4   Leaves whorled at least at the lower nodes; plants low sub-shrubs, mostly of woods (*Chimaphila*) .... **PYROLACEAE**

4   Leaves opposite, not whorled; plants low or tall 5

5   Corollas 2.5-4 cm long, united, lavender-pink, to blue-lavender; leaves serrate; plants rare, sub-shrubs to 30 cm tall, mostly of rocky places in Uinta Mtns. (*Penstemon*) .... **SCROPHULARIACEAE**

5   Corollas less than 2.5 cm long; leaves entire, minutely toothed, or lobed, or if crenate then plants over 30 cm tall............................... 6

6   Leaves rarely over 2.5 cm long or 2 cm wide, except on vigorous, young shoots; plants less than 1.5 m tall 7

6   Some leaves regularly over 3 cm long and/or over 2 cm wide, or plants over 1.5 m tall 11

7   Leaves scurfy and whitish or gray on both sides; plants not over 80 cm tall, of desert places below 6,300 ft (*Atriplex*) . . . . **CHENOPODIACEAE**

7   Leaves not as above; plants montane, or over 80 cm tall 8

8   Leaves evergreen; shrubs not over 20 cm tall, of if so then the leaves minutely toothed. 9

8   Leaves not evergreen; plants mostly over 20 cm tall; leaves not minutely toothed 10

9   Plants not over 20 cm tall; leaves entire, dark green above and pale beneath (*Kalmia*) . . . . **ERICACEAE**

9   Plants 20-60 cm tall; leaves minutely toothed, dark green on both sides . . . . **CELASTRACEAE**

10 Petals separate; fruit a woody capsule, not white . . . . . **HYDRANGEACEAE**

10 Corolla united; fruit berry-like, white (*Symphoricarpos*) . **CAPRIFOLIACEAE**

11 Bark of twigs bright red; leaves 5-9 cm long; flowers man in a flat-topped inflorescence . . . . **CORNACEAE**

11 Bark not red, of if so, leaves less than 5 cm long or else flowers few and borne in pairs. 12

12 Leaves (at least beneath) and young twigs covered with silvery-gray, brownish, or copper colored scurfy stellate hairs (*Shepherdia*) . . . . **ELAEAGNACEAE**

12 Leaves and twigs not as above. 13

13 Plants shrubs or small trees, 2-8 m tall, of rocky places near or along the Green River; fruit 1-seeded, winged; corolla lacking . . . . **OLEACEAE**

13 Plants shrubs, not over 2 m tall, not restricted as above; fruit with 1-several seeds, not winged; corolla present . . . . **CAPRIFOLIACEAE**

— KEY 5C —
Plants woody, leaves and twigs alternate or essentially basal

1   Flowers sessile, borne in heads (the heads of flowers subtended by and partly enveloped in an involucre of few to several bracts), yellow and with a pappus, or if not yellow and without a pappus then leaves with a sagebrush-like odor . . . . **ASTERACEAE**

1   Flowers not as above in all features; leaves not with a sagebrush-like odor 2

2   Flowers and fruit borne in aments; leaves not pinnately lobed; plants often along water courses and other places where the soil is kept wet for much of the growing season, but also in dry places especially at high elevations (alder, aspen, birch, cottonwoods, poplars, and willows). 3

2   Flowers and fruit not borne in aments; or if so (*Quercus*) then leaves pinnately lobed; habitat various. 4

3   Staminate and pistillate aments not borne on the same plant; leaves mostly more than 2.5 times longer than wide, or if wider then plants trees with whitish bark on young branches and sometimes on older branches and trunk . . . . **SALICACEAE**

3   Staminate and pistillate aments borne on the same plant; leaves mostly less than 2.5 times longer than wide; plants shrubs or small trees; bark of twigs,

branches, and sometimes older branches reddish or reddish-brown .... **BETULACEAE**

4   Some of the leaves lobed, parted, divided, or compound 5

4   Leaves entire or toothed, not lobed 11

5   Leaves pinnately lobed 6

5   Leaves palmately lobed 8

6   Plants half shrubs, less than 1 m tall; petals 4 (*Lepidium*) . **BRASSICACEAE**

6   Plants shrubs, commonly over 1 m tall; petals lacking or 5 7

7   Lobes of leaves entire or nearly so, broad and rounded; plants not armed with spines .... **FAGACEAE**

7   Lobes of leaves toothed; plants armed with spines (*Crataegus*) . **ROSACEAE**

8   Leaves sessile, less than 15 mm long, with linear or needle-like segments . ... **POLEMONIACEAE**

8   Leaves not as above.. 9

9   Corolla tubular and/or plants bearing spines or bristles; fruit a succulent berry .... **GROSSULARIACEAE**

9   Corolla of separate petals; plants without spines or bristles; fruit not succulent.. 10

10   Twigs with pungent odor when crushed .... **ANACARDIACEAE**

10   Twigs not as above .... **ROSACEAE**

11   Leaves toothed; the teeth sometimes small but then numerous and closely spaced all along the margins of the leaves 12

11   Leaves entire, or with a few irregularly spaced small teeth............ 17

12   Leaves entire on the lower 1/4 to 1/2, serrate or dentate above with rather coarse teeth .... **ROSACEAE**

12   Leaves about equally serrate from near the base to the apex (the teeth sometimes minute) 13

13   Shrubs less than 1.5 m tall 14

13   Shrubs or trees over 1.5 m tall when mature 15

14   Leaves with 1 prominent midrib, deciduous (*Vaccinium*) ..... **ERICACEAE**

14   Leaves with 3 prominent veins running parallel the margins, winter-persistent (*Ceanothus*) .... **RHAMNACEAE**

15   Flowers greenish; petals lacking; stamens 4-9 .... **ULMACEAE**

15   Petals present but sometimes small; stamens various................ 16

16   Flowers 1-3 in the axils of leaves; petals 4, about 1 mm long fruit with 2-4 seeds; plants known from Rio Blanco, Co. Colorado ...... **RHAMNACEAE**

16   Flowers more than 3 in racemes or corymbose cymes; petals 5, over 1 mm long; fruit with 1 seed; plants widespread .... **ROSACEAE**

17   Plants tall shrubs or trees, over 2 m tall when mature 18

17   Plants shrubs, rarely over 2 m tall .... **KEY 5D**

18   Leaves scale-like, 0.2-0.4 cm long, more or less appressed to and appearing more like a part of the twigs than as definite leaves .... **TAMARICACEAE**

18   Leaves not scale-like, 2-10 cm long................................ 19

19   Leaf blades oblique or cordate at the base, only about twice as long as wide, scabrous, conspicuously reticulate veined (*Celtis*) .... **ULMACEAE**

19 Leaf blades tapered or rounded at the base; often over twice as long as wide
.... 20

20 Leaves, young twigs, fruit and parts of the inflorescence covered with silvery-gray stellate scurfy pubescence; (*Elaeagnus*) ..... **ELAEAGNACEAE**

20 Plants not as above (*Rhamnus*) .... **RHAMNACEAE**

— KEY 5D —
Shrubs rarely over 2 m tall

1   Leaves rigid, sword-like, all basal, 10-30 (50) cm long, evergreen, the margins often with exfoliating curled string-like fibers; flowers showy, 3-6 cm long .... **AGAVACEAE**

1   Leaves not as above in all respects; flowers smaller 2

2   Leaves nearly all basal, 3-17 mm long; flowering stems scapose; plants matted, 2-12 cm tall, mostly growing in cracks of rock outcrops or in rocky places (*Petrophytum*) .... **ROSACEAE**

2   Plants not as above in all features................................... 3

3   Inflorescence umbel-like or dichotomously or trichotomously branched; leaves floccose or tomentose (*Eriogonum*) .... **POLYGONACEAE**

3   Inflorescence not as above 4

4   Leaves with 3 prominent veins that run the entire length of the blades, 1-2.5 cm long, not linear; plants montane (*Ceanothus*) ...... **RHAMNACEAE**

4   Leaves without 3 prominent veins 5

5   Leaves 7-17 mm long, 3-4 mm wide, oblanceolate, not scurfy; current year's twigs greenish; twigs turning into spines; petals 4-6 mm long, whitish; shrubs to 60 cm tall .... **CROSSOMATACEAE**

5   Leaves either larger than above or scurfy, or plants taller and petals lacking
.... 6

6   Flowers inconspicuous; corolla lacking .... **CHENOPODIACEAE**

6   Flowers conspicuous; corolla present................................ 7

7   Spinescent subshrubs 5-20 cm tall; flowers legume-like, bright pink-purple (*Polygala*) .... **POLYGALACEAE**

7   Plants not spinescent; flowers not legume-like 8

8   Plants 1-3 m tall, introduced, known from below 7,000 ft; leaves dull green or grayish on both sides (*Lycium*) .... **SOLANACEAE**

8   Plants commonly less than 1 m tall, native from above 7,000 ft on the Uinta Mtns.; leaves usually vivid green at least on one side .... **ERICACEAE**

— KEY 6 —
CLASS LILIOPSIDA (MONOCOTYLEDONS)
(See KEY 1 for aquatic plants of this group.)

1   Flowers without well developed petaloid tepals; perianth parts if present reduced and scale-like, chartaceous, or chaffy (cattails, burreeds, grasses, and grass-like plants) 2

1   Flowers with well-developed tepals, these more or less petal-like, sometimes small but not chartaceous 7

2 Inflorescence a terminal, cylindrical, dense, spike 15-40 cm long, the lower portion pistillate, brown at maturity, 1.5-3 cm thick, the upper portion staminate, pale or light brown; plants 1-3 (4) m tall, growing in water or mud (cattails) ....**TYPHACEAE**

2 Inflorescence not as above; plants mostly less than 1 m tall or inflorescence not a thick cattail-like spike 3

3 Flowers born in globose clusters, each subtended by 3-5 chaffy bracts, the staminate clusters borne above the pistillate clusters; fruit hardened and strongly beaked; plants not particularly ....**SPARGANIACEAE**

3 Flowers in spikes, racemes, panicles, or sometimes globose clusters, but then flowers bisexual or the clusters solitary and with both staminate and pistillate flowers; plants more or less grass-like 4

4 Leaves terete, basal or nearly basal; inflorescence a terminal, elongate, slender spike or spike-like raceme; flowers bisexual, with 6 perianth segments.. **JUNCAGINACEAE**

4 Plants rarely with both terete leaves and a spiked inflorescence, but if so then the flowers unisexual and perianth segments fewer than 6 or else reduced to bristles..5

5 Stems jointed with conspicuous often swollen solid nodes and mostly hollow internodes; flowers born in spikelets consisting of 1 or 2 glumes and 1 to several florets; perianth lacking; stamens and pistil subtended by 2 scales (lemma and palea) ....**POACEAE**

5 Stems not jointed, nodes lacking or inconspicuous and not swollen; internodes mostly not hollow; flowers various; perianth various........6

6 Flowers without a perianth or the perianth reduced to bristles, the staminate ones consisting of 3 stamens, the pistillate ones consisting of 1 pistil, each flower subtended by a single scale; fruit an achene ....
....**CYPERACEAE**

6 Perianth of 6, more or less similar chartaceous tepals; fruit a capsule with 3-several seeds ....**JUNCACEAE**

7 Stamens 6; ovary mostly superior...................................8

7 Stamens 1 or 3; ovary inferior.....................................10

8 Inflorescence with whorled branches; leaves all basal, the blades not linear, the petioles nearly as long or longer than the blade; pistils 6; plants growing in water or mud; tepals greenish or white ....**ALISMATACEAE**

8 Inflorescence without whorled branches, sometimes a terminal umbel; leaves not all basal, or if so then linear or sessile or with the petiole much shorter than the blade; pistil 1; habitat various; tepals various 9

9 Leaves not rigid, not sword-like, the edges smooth; plants caulescent or scapose ....**LILIACEAE**

9 Leaves rigid, sword-like, the edges usually with threads of exfoliating epidermis; plants scapose ....**AGAVACEAE**

10 Stamens 3, not united to the style; leaves strongly folded and flattened; stems flattened ....**IRIDACEAE**

10 Stamens 1 or 3, united to the style; leaves not strongly folded, stems not flattened ....**ORCHIDACEAE**

— KEY 7 —

1   Flowers sessile, in heads subtended by and partly enveloped in an involucre of separate bracts (or if the involucre united then prickly and forming a bur); ovary inferior 2

1   Flowers not both sessile and in heads subtended by free bracts (involucres if present not forming a bur), (note: a few taxa of *Eriogonum* might be keyed above, except with close observation they have pedicellate flowers and the ovary is superior). 7

2   Leaves all basal. 3

2   Leaves not all basal 4

3   Summit of scape enclosed in a sheathing bract up to 1.5 cm long . . . .
. . . . **PLUMBAGINACEAE**

3   Summit of scape not enclosed in a sheath . . . . **ASTERACEAE**

4   Leaves simple, opposite 5

4   Leaves compound or alternate . . . . **ASTERACEAE**

5   Plants with prickles (these often recurved) on the angles of stems, lower midrib of leaves and involucral bracts; leaves sessile, connate; heads brush-like, with each of the numerous flowers subtended by an awned chaffy bract that surpasses the flower; stamens 4; corollas 4-lobed, white to pale purple
. . . . **DIPSACEAE**

5   Plants not prickly; leaves various; heads not as above; stamens mostly 5; corollas various 6

6   Corollas all salverform, white or pink, showy; pappus lacking; leaves entire, petiolate; heads solitary on long peduncles . . . . **NYCTAGINACEAE**

6   Plants not as above in all features . . . . **ASTERACEAE**

7   Flowers sweet pea-type (the corolla papilionaceous), the uppermost petal (banner) the largest and more or less turned 90 degrees to the lower ones, the lateral petals (wings) equal and more or less enfolding the lower most petals (keel), which are united and strongly folded and more or less keel-shaped; stamens 10, 9 with united filaments forming a tube and the other 1 free, or all free; fruit a legume or a loment; leaves trifoliate, once pinnate, once palmate or rarely simple . . . . **FABACEAE**

7   Plants not as above in all features (flowers similar in Fumariaceae but then the leaves 2-3 times pinnate; petals or sepals other than 4, or if 4 then corolla united; corolla sometimes lacking but then stamens mostly other than 4, 6, or 8; fruit various 8

8   Petals 4, free, sometimes lacking but then the sepals 4 and free; sepals 4 or sometimes 2 and deciduous; stamens (4) 6 or 8, or if more numerous then plants keyed both ways (mustards, evening primroses, and others)   **KEY 8**

8   Petals or sepals other than 4, or if 4 then corolla united; corolla sometimes lacking but then stamens mostly other than 4, 6, or 8; fruit various 9

9   Plants with leafless stems except for a perfoliate leaf (or united pair of leaves) just below the inflorescence (*Claytonia perfoliata*) . . . **MONTIACEAE**

9   Plants not as above 10

10 Leaves all basal; flowers scapose or sometimes sessile among a basal rosette of leaves; the scapes sometimes with 1 or 2 (rarely more) entire or toothed

bracts or bract-like leaves that are much smaller and usually of different shape than the basal leaves .... **KEY 9**

10 Leaves not all basal; the flowering stems mostly with 3 or more leaves, sometimes with only one pair of compound or deeply dissected leaves . 11

11 Leaves compound .... **KEY 10**

11 Leaves not compound, sometimes deeply parted, or appearing divided and compound and then plants mostly keyed both ways.................. 12

12 Stems prostrate, usually radiating over the ground in several or all directions from a taproot, or sprawling or climbing; plants annual; corolla inconspicuous, 1-5 mm long, not over 3 mm wide, or lacking; leaves entire, sessile or nearly so or gradually tapered to an indistinct petiole, or distinctly petiolate in *Tiquilia* .... **KEY 11**

12 Plants ascending to erect, not prostrate, or if so then leaves not entire, or flowers larger, or plants perennial.................................. 13

13 Stems leafless, with a pair of opposite, persistent cotyledons 1-3 mm long near the base and a whorl of basally connate bracteate leaves forming an involucre just beneath the flower cluster; plants 1-4 cm tall, annuals (*Gymnosteris*) .... **POLEMONIACEAE**

13 Plants not as above 14

14 Leaves opposite or whorled, the upper ones sometimes alternate but then plants mostly keyed both ways .... **KEY 12A**

14 Leaves alternate, the lower 1 or 2 nodes of the stems sometimes with a pair of opposite leaves in *Plagiobothrys* and possibly other genera ..... **KEY 13**

— KEY 8 —
Petals 4; sepals mostly 4, mostly free

1 Leaves deeply lobed, divided, or compound 2

1 Leaves simple and entire or toothed 5

2 Leaves all basal; stamens more than 16; petals 1-3 cm long; plant alpine in the Uinta Mtns.................................... **PAPAVERACEAE**

2 Leaves not all basal; stamens less than 16; petals various.............. 3

3 Leaves trifoliate or palmate, with 3-7 leaflets; stamens 6-16, exserted beyond the petals .... **CLEOMACEAE**

3 Leaves not as above; stamens 6, not exerted beyond the petals 4

4 Corolla 12-18 mm long .... **FUMARIACEAE**

4 Corolla less than 10 mm long .... **BRASSICACEAE**

5 Leaves alternate 6

5 Leaves opposite or whorled....................................... 7

6 Stamens 4 or 8; sepals and petals borne above the apex of the ovary and fruit; styles slender and elongate, usually equaling or exceeding the petals; .... **ONAGRACEAE**

6 Stamens 6; sepals and petals borne at the base of the ovaries and fruit; styles short or obsolete, hardly distinct from the ovary or fruit; flowers without a floral tube .... **BRASSICACEAE**

7 Leaves 3-6 mm long, linear or linear-oblanceolate, entire, with short but definite connate-sheathing bases; stems 1-6 cm long, prostrate to erect;

plants of ephemeral pools and drying mud, rare (*Tillaea*) **CRASSULACEAE**

7   Leaves mostly longer than 6 mm 8

8   Flowers numerous in capitate clusters at the ends of scape-like peduncles (*Calyptridium*) .... **MONTIACEAE**

8   Flowers not in capitate clusters 9

9   Stamens 4-5 .... **GENTIANACAE**

9   Stamens 8 .... **ONAGRACEAE**

— KEY 9 —
Leaves all basal

1   Leaves compound or parted to divided 2

1   Leaves simple, entire, toothed, or lobed but not parted or divided 11

2   Leaves trifoliate, once pinnate, or once palmate; leaflets well defined, elliptic or obovate, entire, serrate, or dentate, but not lobed 3

2   Leaves either more than once compound or the primary divisions or leaflets again deeply lobed or dissected..................................... 4

3   Leaflets mostly sharply toothed; petals free, not fringed; plants seldom growing in water or bogs .... **ROSACEAE**

3   Leaflets entire or with rounded, poorly marked teeth; corolla united, the lobes densely fringed on the upper side; plants aquatic or in bogs .... **MENYANTHACEAE**

4   Flowers solitary and terminal; petals yellow; leaves mostly palmately parted, the primary divisions usually deeply lobed.................... 5

4   Flowers not solitary or else petals not yellow; leaves mostly pinnately compound or parted 6

5   Sepals 2, densely blackish hirsute; petals 10-30 mm long; fruit a capsule; scapes and leaves blackish hirsute .... **PAPAVERACEAE**

5   Sepals mostly 5, not blackish hirsute; petals 3-18 mm long; fruit an achene, the achenes numerous on an open receptacle; plants glabrous or pubescent but not with blackish hair .... **RANUNCULACEAE**

6   Petals pink; fruit with a beak, the beak 1-5 cm long, separating into 5 segments upon drying; plants winter annuals, often on disturbed ground (*Erodium*) .... **GERANIACEAE**

6   Petals not pink; fruit not beaked as above; plants perennial 7

7   Inflorescence an umbel; plants not above timberline, or if so then strongly aromatic, not glandular; flowers white, yellowish, or purplish . **APIACEAE**

7   Inflorescence not an umbel; plants sometimes of high elevation, not aromatic, usually glandular...................................... 8

8   Inflorescence a head-like or open cyme; petals separate, white or yellow; leaves pinnately compound with 20 or more sessile leaflets (*Ivesia*) .... **ROSACEAE**

8   Inflorescence a raceme or spike-like raceme; corolla blue, united, or lacking; leaves not as above in all respects 9

9   Corolla lacking; stamens 8-15; leaves ternately or biternately compound; some of the primary leaflets petiolulate; fruit of achenes (*Thalictrum alpinum*) .... **RANUNCULACEAE**

9 Corolla united, blue or bluish, sometimes with a yellowish tube; stamens 2 or 5; leaves pinnatifid; fruit a capsule 10

10 Stamens 2; corolla 4-7 mm long; plants known from above 9,000 ft on the Uinta Mtns. (*Synthyris*) .... **SCROPHULARIACEAE**

10 Stamens 5; corolla various; plants of lower elevations (*Gilia*) .... **POLEMONIACEAE**

11 Flowers hidden in the crowded leaves, inconspicuous; leaves 3-6 mm long, densely crowded with scarious stipules that are 1/2 to as long as the leaves; plants glabrous or slightly puberulent .... **CARYOPHYLLACEAE**

11 Flowers not hidden in crowded leaves, or if so (*Eriogonum*) then leaves without stipules and plants densely tomentose or floccose............ 12

12 Scapes lacking; flowers sessile in a basal rosette of leaves, the corolla bluish; functional stamens 4; plants puberulent (*Penstemon*) .... **SCROPHULARIACEAE**

12 Scapes present, or corolla not bluish and plants tomentose or floccose . 13

13 Flowers solitary and terminal, on simple scapes, often more or less showy 14

13 Flowers not solitary or rarely so in stunted plants 24

14 The upper 1 or 2 petals arched backwards; flowers irregular, usually nodding; leaf blades more or less toothed, sometimes cordate at the base .... **VIOLACEAE**

14 Petals not arched as above; flowers regular except in Scrophulariaceae; leaf blades various.. 15

15 Plants stoloniferous or at least freely rooting at the nodes of prostrate stems, of moist or wet places; petals yellow, free (*Ranunculus*) .... **RANUNCULACEAE**

15 Plants not stoloniferous; petals not yellow except in *Mimulus* and then united, mostly white, pinkish, or lavender 16

16 Leaves white tomentose beneath, glabrous and dark green above, regularly crenate, usually revolute, 1-3 cm long, 3-12 mm wide; styles numerous, elongate and plumose in fruit; plants alpine on the Uinta Mtns. (*Dryas*) .. .. **ROSACEAE**

16 Leaves glabrous to villous-hirsute but not tomentose, not crenate or mostly larger than above, not revolute; styles not plumose 17

17 Leaf blades 3-10 cm long, 1.5-5 cm wide, usually crenate or dentate at least in part, more or less cordate at the base; fruit a follicle, the follicles several (*Caltha*) .... **RANUNCULACEAE**

17 Leaf blades mostly smaller than above, mostly entire or lobed, not cordate at the base except in Parnassia; fruit not a follicle 18

18 Leaves linear or nearly so, about 1-3 mm wide, sessile or gradually tapered to an indistinct petiole 19

18 Leaves not linear, sometimes rather narrow but then mostly conspicuously narrowed to a petiole 20

19 Petals lacking or 5, to 1.5 mm long; plants annual; leaves not succulent; scapes without bracts (*Myosurus*) .... **RANUNCULACEAE**

19 Petals (5 7-18, 6-35 mm long; plants perennial; leaves more or less succulent; scapes sometimes with a pair or a whorl of bracts .... **MONTIACEAE**

20 Scapes mostly shorter than the leaves, 1-several in the basal rosette . . . 21

20 Scapes exceeding the leaves, mostly solitary 22

21 Leaf blades shorter than the petioles, 5-18 mm long, 1-7 mm wide, tapered to the petioles, these 1-8 cm long; flowers not over 3 mm long (*Limosella*) . . . . **SCROPHULARIACEAE**

21 Leaf blades equal or longer than the petioles, 15-50 mm long, to 20 mm wide; flowers 5-15 mm long (*Hesperochiron*) . . . . **HYDROPHYLLACEAE**

22 Corolla yellow, united; calyx united; leaves glabrous or glandular-villous, the blades elliptic or obovate, sessile or gradually tapered to a rather indistinct, short petiole (*Mimulus primuloides*) . . . . . **SCROPHULARIACEAE**

22 Corolla white; leaves glabrous, the blades mostly ovate or broader, with distinctly petioles about 1/3 as long to longer than the blade. 23

23 Styles lacking or nearly so; fertile stamens 5; flowers mostly erect; scapes 7-30 cm tall or taller; leaves not evergreen . . . . **SAXIFRAGACEAE**

23 Styles prominent, persistent on the fruit; fertile stamens 10; flowers often nodding; scapes 3-15 cm tall; leaves more or less evergreen (*Moneses*) . . . . **PYROLACEAE**

24 Plants stoloniferous, usually growing in mud; leaves crenulate-toothed; petals yellow, separate, 3-9 mm long; stamens more than 10; fruit of numerous achenes on an open receptacle (*Ranunculus*) **RANUNCULACEAE**

24 Plants not stoloniferous; leaves mostly entire except in Saxifragaceae; petals not yellow, perianth sometimes yellow in *Eriogonum* but then united; stamens 10 or fewer; fruit various 25

25 Leaves shallowly to conspicuously trilobate or palmately lobed and usually also toothed, or if not lobed then uniformly and conspicuously toothed . . . . **SAXIFRAGACEAE**

25 Leaves entire or toothed, but not lobed. 26

26 Inflorescence a spike or spike-like; flowers small, the petals lacking or to 3 mm long; leaf blades glabrous, sericeous to villose but not tomentose or floccose; scapes not glandular. 27

26 Inflorescence open to capitate, neither a spike nor spike-like; flowers as small as or larger than above; leaves and scapes various 29

27 Leaves 5-15 mm long, 1-5 mm wide; plants of rock faces or at least of rocky places (*Petrophytum*) . . . . **ROSACEAE**

27 Leaves longer and sometimes wider than above 28

28 Flowers solitary and terminal but the petals and sepals lacking or soon deciduous; receptacle elongating in fruit, with many achenes and simulating a spicate inflorescence; leaves linear (*Myosurus*) . . . . **RANUNCULACEAE**

28 Flowers numerous, in spikes; fruit a circumscissile capsule with 1-several seeds; leaves linear to broadly ovate . . . . **PLANTAGINACEAE**

29 Inflorescence a solitary simple raceme; petals 3-7 mm long; styles conspicuous, thick and persistent in fruit; stamens 10; leaves more or less evergreen; plants glabrous, 5-20 (40) cm tall (*Pyrola*) . . . . . . **PYROLACEAE**

29 Inflorescence not a solitary simple raceme; petals mostly either shorter or longer than above; styles various; stamens various; leaves not evergreen; plants glabrous or pubescent. 30

30 Stems with only 2 or 3 leaves, 2-5 cm tall; leaves only 1-3 mm wide (*Lewisia triphylla*) .... **MONTIACEAE**

30 Plants not with the above combination of features................... 31

31 Flowers solitary at the ends of rays of simple umbels, 15-28 mm long, and pink-lavender, and plants perennial, or only 3-5 mm long, and white and plants annual; stamens 5; corolla distinctly united or petals reflexed and anthers strongly exserted; fruit a capsule .... **PRIMULACEAE**

31 Flowers not as above, if in umbels, then the umbels compound and flowers not solitary, 2-10 mm long; plants annual or perennial; stamens 3, 9, or 10; fruit various 32

32 Scapes glandular pubescent; plants perennial, from the Uinta Mtns., 4-20 cm tall; flowers in a head-like cluster, or the inflorescence somewhat interrupted after anthesis, not subtended by an involucre; leaf blades deltoid to rhombic-deltoid, mostly toothed (crenate), rarely entire, ciliate .... **SAXIFRAGACEAE**

32 Plants not as above in all features; flowers few to several together in small involucres .... **POLYGONACEAE**

— KEY 10 —
Leaves compound

1 Leaves opposite or whorled.. 2

1 Leaves alternate.. 8

2 Stems prostrate, radiating in all directions from the taproot; leaves once pinnate with 4-8 pairs of opposite leaflets, the basal ones lacking or no longer than the stem leaves; flowers yellow; fruit with hard, sharp-pointed spines; plants annual, weedy .... **ZYGOPHYLLACEAE**

2 Plants not as above in all features.................................. 3

3 Leaves divided into filiform, sometimes needle-like segments; the segments sometimes radiating out from the stem and appearing as whorls of filiform leaves, not over 2 mm wide .... **POLEMONIACEAE**

3 Leaves not divided into filiform segments, the divisions well over 2 mm wide 4

4 Leaves once pinnatifid, the segments entire, some of the basal ones and sometimes the lowest pair of stem leaves simple and entire; stems with 2-6 pairs of leaves; flowers numerous; stamens 3 .... **VALERIANACEAE**

4 Leaves more than once compound and/or leaflets toothed to lobed, the basal ones not simple and entire, sometimes lacking; stems with 1 or 2 (more in Clematis) pairs of leaves; flowers solitary or up to about 10 per inflorescence; stamens more than 3................................ 5

5 Plants glabrous, rare; leaves ternate; leaflets with petiolules about as long as the blades, ternately or palmately lobed or cleft, or the terminal one ternate again; flowers 3-8 in a head-like terminal cluster; corollas 2-3 mm long, yellowish green .... **ADOXACEAE**

5 Plants pubescent at least in part; leaves not ternate or if so the leaflets sessile or nearly so; flowers not in head-like clusters; petals usually over 3 mm long, white, pink, or yellowish 6

6 Fruit with a stylar beak, the beak 1-5 (7) cm long, breaking into 5 twisted or coiled styles upon drying, the styles not plumose; stamens not more than 10 . . . . **GERANIACEAE**

6 Fruit of numerous achenes on a head-like receptacle; if the styles elongate then plumose; stamens more than 10. 7

7 Leaves once or occasionally twice pinnate; the basal leaves with blades mostly longer than the petioles, larger than the 1-2 pairs of stem leaves (*Geum triflorum*) . . . . **ROSACEAE**

7 Leaves either palmatifid with the basal blades mostly shorter than the petioles (*Anemone*), or 2-4 times pinnate with the middle stem leaves larger than the lower bract-like entire leaves (*Clematis*) . . . . . **RANUNCULACEAE**

8 Leaves trifoliate or once palmate 9

8 Leaves not trifoliate, not once palmate 12

9 Leaflets entire or obscurely toothed. 10

9 Leaflets with conspicuous teeth or lobes. 11

10 Leaves alternate but essentially basal; plants of ponds and bogs in the Uinta Mtns., also keyed in KEYS 1 and 8 . . . . **MENYANTHACEAE**

10 Leaves well distributed up the stem; plants of dry places, also keyed in KEY 8 with sepals and petals 4 each and free . . . . **CLEOMACEAE**

11 Stamens united into a column around the style . . . . **MALVACEAE**

11 Stamens free . . . . **ROSACEAE**

12 Plants glabrous fragile annuals, 2-10 cm tall, of shady places; flowers solitary on slender, axillary pedicels; petals 3, free, 1-3 mm long, about equal to or shorter than the 3 sepals; leaves once pinnatifid, the segments entire . . . .
. . . . **LIMNANTHACEAE**

12 Plants perennial or annual but not as above in all features 13

13 Inflorescence an umbel . . . . **APIACEAE**

13 Inflorescence not umbellate 14

14 Corolla united, not yellow 15

14 Corolla of separate petals, or yellow, or lacking. 17

15 Corolla irregular; stamens 2 or 4 . . . . **SCROPHULARIACEAE**

15 Corolla regular, 5-lobed; stamens 5 16

16 Style 2-lobed or 2-parted; stigmas entire; inflorescence exceeded by the leaves . . . . **HYDROPHYLLACEAE**

16 Style entire but the stigma 3-lobed; inflorescence exceeding the leaves . .
. . . . **POLEMONIACEAE**

17 Stamens 6; petals 4, the flowers somewhat like those of the legume family
. . . . **FUMARIACEAE**

17 Stamens mostly 8 or more; petals 5 or none; flowers not like those of the legume family. 18

18 Leaves once pinnate, the leaflets serrate or dentate (lobed in *Geum* and rarely in *Potentilla*); flowers with a saucer-like or campanulate floral cup, the sepals, petals, and stamens borne on the floral cup; sepals 5, alternating with 5 sepal-like bracteoles and thus sometimes appearing to be 10, persistent; petals yellow, not spurred; stamens numerous . . . . **ROSACEAE**

18 Leaves more than once compound or the leaflets deeply lobed to dissected and the primary lobes mostly lobed again; flowers without a floral cup (See also KEY 13 for plants with palmately cleft, parted, or divided leaves) . . . .
. . . . **RANUNCULACEAE**

— KEY 11 —
Stems prostrate, twining, or climbing

1 Leaves opposite or whorled; plants not dichotomously branched 2

1 Leaves alternate, sometimes congested at the nodes and appearing opposite or whorled, but then plants usually dichotomously branched 6

2 Leaves whorled, with curved or hooked hairs, ovaries and fruit with hooked hairs . . . . **RUBIACEAE**

2 Leaves opposite; plants without hooked hairs, mostly glabrous 3

3 Plants with milky juice, of dry places; leaves ovate or nearly so . . . .
. . . . **EUPHORBIACEAE**

3 Plants without milky juice, growing at margins of ponds and lakes, or aquatic but sometimes stranded and prostrate on mud; at least the lower leaves li—near or spatulate 4

4 Leaves with short but definite connate-sheathing bases, 3-6 mm long; flowers and fruit nearly sessile or on pedicels to 6 mm long (*Tillaea*) . . . . .
. . . . **CRASSULACEAE**

4 Leaves not connate-sheathing, sometimes over 6 mm long; flowers and fruit sessile. 5

5 Flowers with a perianth . . . . **ELATINACEAE**

5 Perianth lacking; flowers consisting of a naked stamen and a pistil . . . . . .
. . . . **CALLITRICHACEAE**

6 Perianth pink or lavender, conspicuous; leaves 1-3 cm long, (2) 5-10 mm wide; stems 10-70 cm long, mainly 2-4 mm thick; plants of low elevations often of alkaline or saline ground; capsules with 3-5 chambers . . . .
. . . . **AIZOACEAE**

6 Perianth not pink or lavender, inconspicuous or lacking; leaves linear or mostly shorter than above; stems not over 10 cm long or else not over 2 mm thick; plants mostly montane 7

7 Leaves with conspicuous, scarious stipules, the stipules more or less sheathing the stem . . . . **POLYGONACEAE**

7 Stipules lacking or inconspicuous. 8

8 Plants glabrous and succulent (sap easily pressed from the leaves and stems by gentle pressure between the thumb and fingers), weeds of gardens and other disturbed ground; petals free, yellow, or lacking (*Portulaca*) . . . .
. . . . **PORTULACEAE**

8 Plants pubescent, not especially succulent, sometimes weedy; corolla united or lacking, not yellow, sometimes concealed by the calyx 9

9 Corolla 1-5 mm long, united; plants usually densely hispid, or stems retrorsely hispid-prickly 10

9 Corolla lacking; plants not densely hispid; stems not retrorsely hispid-prickly. 11

10 Styles 2, united toward the base; leaves gradually tapered to the base, without a distinct petiole; corolla 2.5-5 mm long ... **HYDROPHYLLACEAE**

10 Style 1, or if 2 then leaves abruptly constricted to a distinct (but short) petiole with the blades broadly elliptic, ovate, or orbicular; corolla 1-3 mm long .... **BORAGINACEAE**

11 Styles 1, short; flowers not in numerous chaffy bracts ..... **URTICACEAE**

11 Styles 2; flowers hidden in numerous chaffy bracts ... **AMARANTHACEAE**

— KEY 12A —
Leaves simple and opposite or whorled

1 Plants stoloniferous, growing in water or mud; stems rooting at the nodes; leaves opposite or fascicled, mostly basal; petals yellow (*Ranunculus flammula*) .... **RANUNCULACEAE**

1 Plants not as above in all features. 2

2 Leaves whorled at least at some of the nodes 3

2 Leaves opposite 11

3 Leaves sharply toothed, more or less evergreen (*Chimaphila*) .... **PYROLACEAE**

3 Leaves not sharply toothed 4

4 Stems with a single whorl of leaves 5

4 Stems with more than one whorl of leaves 6

5 Perianth less than 4 mm long (*Eriogonum*) .... **POLYGONACEAE**

5 Perianth over 4 mm long .... **MONTIACEAE**

6 Basal leaves 25-50 cm long; lower stem leaves usually over 10 cm long and 1 cm wide; flowers greenish, mostly over 1 cm wide; plants (80) 100-200 cm tall .... **GENTIANACEAE**

6 Basal leaves lacking or smaller than above; stem leaves seldom over 10 cm long, not over 1 cm wide; plants mostly shorter than above 7

7 Flowers borne in umbels; plants with milky juice ..... **ASCLEPIADACEAE**

7 Flowers not in umbels; plants without milky juice 8

8 Corolla 2.5-25 mm long, united; leaves palmately dissected into filiform segments appearing as whorls of filiform leaves, or leaves entire and all clustered as an involucre just beneath the inflorescence, and stems naked except for two opposite cotyledons at the base .... **POLEMONIACEAE**

8 Corolla lacking, or 1-3 mm long, united or not; leaves not as above...... 9

9 Flowers borne in dense spike-like racemes; the racemes 6-20 mm long; plants glabrous, annual, rare........................ **POLYGALACEAE**

9 Flowers not in spike-like racemes. 10

10 Plants annual, glabrous except for the perianth, of dry open places below 6,500 ft; basal leaves spatulate, wider than the nearly linear stem leaves; stem leaves ascending to erect; flowers in small involucres, the involucres sessile or at the ends of filiform peduncles (*Eriogonum salsuginosum*)...... .... **POLYGONACEAE**

10 Plants perennial or herbage pubescent, of various habitats and elevations; basal leaves lacking; cauline leaves spreading; flowers various; fruit often pubescent .... **RUBIACEAE**

11 Stems trailing on the ground, slender, somewhat woody, with erect herbaceous branches, these mostly less than 10 cm tall, bearing 2-4 pairs of leaves and a long peduncle with a pair of descending to reflexed flowers with united corollas (*Linnaea*) .... **CAPRIFOLIACEAE**

11 Plants not as above in all features 12

12 Leaves toothed to parted 13

12 Leaves entire 24

13 Leaves 3-15 mm long; stems mostly prostrate; plants with milky juice .... **EUPHORBIACEAE**

13 Leaves over 15 mm long including the petiole 14

14 Flowers inconspicuous; petals lacking; leaves toothed to shallowly lobed but not parted; plants more or less weedy 15

14 Flowers more or less conspicuous, the corolla often united or partly so, if rather inconspicuous then the leaves usually parted to divided........ 16

15 Leaf blades uniformly serrate the entire length with more than 10 teeth per side .... **URTICACEAE**

15 Leaf blades irregularly toothed to sub-entire or hastate lobed, with fewer than 10 teeth per side (*Atriplex*) .... **CHENOPODIACEAE**

16 Leaves parted into 3 or more needle-like or filiform segments, the segments not over 2 mm wide .... **POLEMONIACEAE**

16 Leaves toothed to cleft, but if cleft then the segments not needle-like or filiform and over 2 mm wide 17

17 Leaves mostly basal, the blades palmately cleft, the basal ones born on long petioles that are mostly over 1/2 the height of the plant; stems mostly with one pair of leaves just below the inflorescence .... **GERANIACEAE**

17 Leaves not as above... 18

18 Flowers scarlet red, 2.5-3.5 cm long including a corolla-like floral tube; stamens 8 (*Epilobium*) .... **ONAGRACEAE**

18 Flowers not scarlet red; stamens not 8 19

19 Flowers 1-2 mm long, whitish, borne in racemes; sepals 2, spreading to reflexed, of the same color and texture as the petals; petals 2, notched; stamens 2; fruit covered with hooked hairs (*Circaea*) ...... **ONAGRACEAE**

19 Flowers and fruit not as above in all respects....................... 20

20 Petals united only at the base; styles 3, persistent in fruit; stamens 10 .... **ERICACEAE**

20 Petals united into a tubular corolla; styles 1; stamens less than 10...... 21

21 Inflorescence an umbel-like cyme; corollas 2-7 mm long, whitish; seeds tipped with plumose hair-like calyx bristles .... **VALERINACEAE**

21 Inflorescence not as above; corollas various; seeds not tipped with plumose calyx bristles....................................... 22

22 Flowers sessile or nearly so and solitary or verticillate in the axils of leaves or in a head-like or spike-like terminal inflorescence; stems more or less square in cross-section....................................... 23

22 Flowers pedicellate, axillary or terminal; stems not square in cross-section .... **SCROPHULARIACEAE**

23 Stamens 5 .... **VERBENACEAE**

23 Stamens 2 or 4 .... **LAMINACEAE**

24 Plants with milky juice.. 25

24 Plants without milky juice...................................... 27

25 Leaves 3-15 mm long; stems prostrate .... **EUPHORBIACEAE**

25 Leaves over 15 mm long; stems not prostrate....................... 26

26 Follicles not over 5 mm wide; flowers borne in dichotomously or more often
   trichotomously branched cymes .... **APOCYNACEAE**

26 Follicles often over 5 mm wide; flowers borne in simple or compound
   umbels, the umbels mostly with more than 3 rays .... **ASCLEPIADACEAE**

27 Stem leaves abruptly constricted to a distinct petiole 28

27 Stem leaves sessile or the blades gradually tapered to an indistinct petiole
   29

28 Flowers borne in the axils of leaves, inconspicuous; leaves glandular
   denticulate .... **ELATINACEAE**

28 Flowers not in the axils of leaves, conspicuous; leaves not glandular
   denticulate .... **NYCTAGINACEAE**

29 Flowers exceeded by the subtending leaves or bracts, axillary (terminal in
   Paronychia but then inconspicuous and hidden in crowded leaves) ......
   .... **KEY 12B**

29 Flowers exceeding the leaves or subtending bracts, axillary or terminal ..
   .... **KEY 12C**

— **KEY 12B** —
Flowers exceeded by the subtending leaves or bracts

1 Plants cushion-like caespitose perennials; leaves 3-6 mm long, densely
  crowded, with scarious stipules 1/2 to as long as the leaves; flowers
  inconspicuous; petals lacking or minute (*Paronychia*) **CARYOPHYLLACEAE**

1 Plants not as above; leaves mostly over 6 mm long.................... 2

2 Some or all of the flowers and bracts alternate; plants puberulent or
  glandular pubescent on the stems or in the inflorescence............. 3

2 Bracts and flowers all opposite or verticillate 4

3 Flowers subtended by bracts or bract-like leaves that are much smaller than
  the lower leaves; corolla 2-4 mm long, 4 lobed; stamens 2, usually exserted;
  fruit a somewhat flattened capsule, notched or obcordate at the apex
  (*Veronica*) .... **SCROPHULARIACEAE**

3 Flowers subtended by leaves that are only gradually reduced in size from
  the lower leaves; corolla 5-15 mm long, 5 lobed; stamens 5, included;
  capsules not flattened, not notched or obcordate at the apex (*Microsteris*)
  .... **POLEMONIACEAE**

4 Stems prostrate; leaves 3-15 mm long 5

4 Stems not prostrate or some leaves over 15 mm long................. 6

5 Plants with milky juice, of dry places, with short-petiolate leaves (also keyed
  above in lead 23) .... **EUPHORBIACEAE**

5 Plants not with milky juice, of wet places (see also lead 5 in KEY 11) .....
  .... **ELATINACEAE**

6 Plants puberulent throughout. 7

6 Plants glabrous, not aromatic; perianth various 8

7 Plants strongly aromatic; corolla united, 6-12 mm long, exceeding the calyx (*Hedeoma*) . . . . **LAMIACEAE**

7 Plants not aromatic; petals free, less than 4 mm long, shorter than the calyx (*Bergia*) . . . . **ELATINACEAE**

8 Plants perennials from rhizomes; leaves 4-25 mm long, crowded, the upper few alternate (*Glaux*) . . . . **PRIMULACEAE**

8 Plants annual (sometimes with numerous fibrous roots and appearing perennial) 9

9 Leaves 3-12 mm long usually with membranous stipules . . **ELATINACEAE**

9 Leaves (10) 20-40 mm long, entire; stipules lacking. 10

10 Leaves oblanceolate, spatulate or obovate, gradually tapering to a petiole-like base; flowers pink-lavender, 6-10 mm long . . . . **AIZOACEAE**

10 Leaves linear or nearly so, with auriculate-clasping bases; flowers inconspicuous, not over 4 mm long . . . . **LYTHRACEAE**

— KEY 12C —
Flowers exceeding the leaves or subtending bracts, axillary or terminal

1 Perianth of similar tepals, the segments free, pubescent, yellowish, about 2 mm long; flowers borne in small involucres, the involucres sessile or at the ends of filiform peduncles; some nodes of the stem usually with more than 2 leaves (*Eriogonum salsuginosum*) . . . . **POLYGONACEAE**

1 Perianth of sepals and petals or not yellow, mostly over 2 mm long; plants otherwise not as above in all features 2

2 Petals yellow, free; stamens 15 or more . . . . **HYPERICACEAE**

2 Petals not yellow or if so then united; stamens 2-12 3

3 Sepals apparently 2, free . . . . **MONTIACEAE**

3 Sepals 5 or calyx united 4

4 Petals free, (if petals lacking then sepals mostly free); stamens (5) 10-14 . 5

4 Corolla or corolla-like calyx united at least at the base, sepals free or united; stamens 2, 4, or 5 8

5 Plants 50-150 (200) cm tall; petals rose-purple (*Lythrum*) . . . **LYTHRACEAE**

5 Plants less than 50 cm tall or petals not rose-purple. 6

6 Stamens mostly 10 . . . . **CARYOPHYLLACEAE**

6 Stamens (4) 5 7

7 Petals and sepals 5-15 mm long . . . . **GENTIANACEAE**

7 Petals and sepals less than 5 mm long . . . . **CARYOPHYLLACEAE**

8 Flowers numerous in a solitary, terminal head-like cluster, the cluster closely subtended by an involucre of bracts; plants strongly aromatic, entering our area at the w. extremes (*Monardella*) . . . . **LAMIACEAE**

8 Flowers not in a solitary, terminal, head-like cluster; plants seldom aromatic, of various distribution 9

9   Flowers in umbel-like cymes, 2-7 mm long, whitish; stamens 3; seeds tipped
    with plumose, bristle-like pappus-hairs; leaves rarely all entire; plants
    rather malodorous .... **VALERIANACEAE**

9   Plants not as above in all features 10

10 Inflorescence open, often trichotomously branched, strongly glandular;
    stamens and style exserted from the corolla-like calyx, this subtended by a
    calyx-like involucre (*Mirabilis linearis*) .... **NYCTAGINACEAE**

10 Inflorescence not trichotomously branched, or if so then glabrous; stamens
    and style mostly included; true corolla and calyx present, not subtended by
    an involucre 11

11 Corolla irregular or if nearly regular then stamens 2 ....
    .... **SCROPHULARIACEAE**

11 Corolla regular; stamens 4 or 5 12

12 Stigmas 3-lobed; fruit a 3-chambered capsule .... **POLEMONIACEAE**

12 Stigmas entire, but styles sometimes 2 cleft; fruit a 2-chambered capsule .
    .... **GENTIANACEAE**

— KEY 13 —
Leaves simple, alternate

1   Perianth of a single whorl or the segments not strongly differentiated into
    sepals and petals (arbitrarily called tepals, the tepals sometimes petal-like),
    not over 4 mm long, inconspicuous or somewhat showy (especially when
    flowers are clustered together 2

1   Perianth of 2 whorls, the corolla usually showy, often over 4 mm long... 9

2   Leaves deeply, palmately lobed (*Trautvetteria*) .... **RANUNCULACEAE**

2   Leaves not deeply lobed 3

3   Leaf blades uniformly serrate on most of the margin; stamens 2, strongly
    exserted from the calyx; plants villous, of high elevations on the Uinta Mtns.
    (*Besseya*) .... **SCROPHULARIACEAE**

3   Leaf blades entire or occasionally toothed, or stamens not exserted and
    plants not villous nor of high elevations on the Uinta Mtns............ 4

4   Leaves with evident, usually membranous sheathing stipules, or if stipules
    lacking then the inflorescence umbel-like or cyme-like and borne on long
    naked peduncles and flowers few to several in small involucres ....
    .... **POLYGONACEAE**

4   Leaves without stipules as above; inflorescence not as above 5

5   Flowers more or less conspicuous; sepals more or less petal-like; plants
    perennial 6

5   Flowers mostly inconspicuous; sepals scale-like; plants annual, usually
    weedy... 7

6   Flowers in axils of lower and middle stem leaves; lower leaves often
    opposite; plants of moist saline or alkaline lowlands (*Glaux*) ....
    .... **PRIMULACEAE**

6   Flowers terminal and in axils of upper leaves; leaves all alternate; plants of
    dry places .... **SANTALACEAE**

7   Style 1, short; flowers in axillary clusters, subtended by several linear involucral bracts; leaves usually elongate-elliptic, with a slender petiole (*Parietaria*) .... **URTICACEAE**

7   Styles usually 2 or 3; flowers variously disposed, but not subtended by involucral bracts; leaves various.................................... 8

8   Flowers obscured among many, dry, scarious, persistent bracts; sepals scarious; leaves not scurfy, petiolate, entire, not linear .... **AMARANTHACEAE**

8   Flowers not obscured among many bracts, sometimes folded between 2 bracts; leaves often scurfy, sessile or petiolate, entire or toothed, sometimes linear .... **CHENOPODIACEAE**

9   Corolla united for 1/4 or more of its length; stamens 2, 4, or 5........ 10

9   Corolla of separate petals or petals united for less than 1/4 of the length (or stamens more than 5, and stems mostly woody toward the base) 20

10   Stems creeping, twining or climbing; from deep-seated, whitish rhizomes; corolla 1.5-2.5 cm long, white or pink; leaf blades often triangular or hastate, occasionally oblong-ovate .... **CONVOLVULACEAE**

10   Stems not as above, or if so corolla smaller 11

11   Corolla irregular 12

11   Corolla regular (radially symmetrical); stamens 5.................... 13

12   Stamens 5 (*Porterella* and *Downingia*) .... **CAMPANULACEAE**

12   Stamens 2 or 4 .... **SCROPHULARIACEAE**

13   Style 1; stigma 1, entire 16

13   Styles either 2 or 2-lobed or with 3-lobed stigmas 18

14   Corolla blue or reddish purple..................................... 15

14   Corolla not blue 16

15   Corolla not strongly divided into a tube and a limb; style terminal on an inferior ovary; fruit a capsule .... **CAMPANULACEAE**

15   Corolla strongly divided into a tube and a limb; style arising from below the base of 1-4 nutlets .... **BORAGINACEAE**

16   Corolla yellow.. 17

16   Corolla white and sometimes with bluish markings ..... **BORAGINACEAE**

17   Basal leaves 10-40 cm long, 2-13 cm wide; plants 40-120+ cm tall (*Verbascum*) .... **SCROPHULARIACEAE**

17   Leaves shorter or narrower than above; plants to 40 cm tall .... **BORAGINACEAE**

18   Style 1, stigmas entire; fruit a berry or a capsule; stamens sometimes exserted and united by the anthers into a column around the style .... **SOLANACEAE**

18   Styles 2, or 2-lobed, or with 3-1obed stigmas; fruit a capsule; stamens not united 19

19 Styles 2, or 2-lobed, stigmas entire; capsules with 1 (2) chamber(s) .... **HYDROPHYLLACEAE**

19   Style 1, entire, but with a 3-lobed stigma; capsules 3-chambered .... **POLEMONIACEAE**

20 Leaves lobed to divided. 21

20 Leaves entire or toothed but not lobed 24

21 Leaves pinnately lobed, harshly pubescent with minute, many-barbed hairs, rough to the touch; petals 5 or apparently 10, yellowish; stamens 10-numerous . . . . **LOASACEAE**

21 Leaves palmately lobed, cleft or divided, or trilobate, not pubescent as above 22

22 Stamens 5-10 . . . . **SAXIFRAGACEAE**

22 Stamens more than 10 23

23 Filaments of stamens united into a column around the styles; plants usually stellate-pubescent . . . . **MALVACEAE**

23 Filaments free; plants not stellate pubescent . . . . **RANUNCULACEAE**

24 Leaves sessile or the blades gradually tapered to a short or indistinct petiole, entire; basal leaves lacking or poorly developed or not much larger than the stem leaves. 25

24 At least some of the leaf blades narrowed to a distinct petiole (often abruptly so), usually also toothed; basal leaves often well developed and larger than the stem leaves 29

25 Flowers borne in umbels; plants with milky juice. . . . . . **EUPHORBIACEAE**

25 Flowers not in umbels; plants without milky juice 26

26 Flowers mostly solitary and terminal (*Saxifraga*) . . . . **SAXIFRAGACEAE**

26 Flowers not solitary and terminal. 27

27 Leaves pubescent with minute, barbed hairs, rough to the touch, brittle, and clinging to any porous or slightly porous object; stamens 10-numerous . . . . **LOASACEAE**

27 Leaves glabrous, often glaucous; stamens 5-10. 28

28 Leaves succulent; stamens 8-10; flowers nearly sessile . . . **CRASSULACEAE**

28 Leaves not especially succulent; stamens 5; flowers on conspicuous pedicels . . . . **LINACEAE**

29 Flowers in umbels; blades of basal leaves cordate at the base, uniformly serrate; at least some of the stem leaves usually compound (*Zizia*) . . . . **APIACEAE**

29 Flowers not in umbels; leaves various but all simple 30

30 Flowers bilaterally symmetrical (irregular), usually nodding, the upper 1 or 2 petals arched backwards at the tip, the lowest one spurred or saccate at the base; leaf blades sometimes cordate at the base . . . . **VIOLACEAE**

30 Flowers radially symmetrical (regular); petals not arched, not spurred, leaves not cordate at the base. 31

31 Leaves pubescent with minute, many-barbed hairs, rough to the touch, brittle, clinging to any porous or slightly porous object, mostly well distributed on the stems; stamens 10-numerous . . . . **LOASACEAE**

31 Leaves glabrous or nearly so, at least not with hairs as above, sometimes congested toward the base of the stem; stamens various. 32

32 Flowers yellow; petals separate; stamens more than 10; leaves not evergreen; fruit of achenes on an open receptacle (*Ranunculus*) . . . . **RANUNCULACEAE**

32 Flowers pink or rose-lavender to white; corolla more or less united; stamens 8-10; leaves usually more or less evergreen; fruit a capsule............33

33 Leaves well distributed along the stems, not crowded toward the base; flowers solitary in axils of leaves ....**ERICACEAE**

33 Leaves crowded toward the base of the stems, sometimes in rosettes; flowers in racemes (*Pyrola*) ....**PYROLACEAE**

Linnaea borealis L. — twinflower, page 208.

ACERACEAE Maple Family
Acer L. — Maple

Shrubs or trees, polygamous or dioecious; leaves opposite, simple or compound, palmately lobed to pinnately compound, often turning bright red in autumn; flowers in terminal or axillary racemes, corymbs or panicles; sepals (4) 5, free or united at the base; petals lacking or shorter than the calyx and inconspicuous; stamens 4-12, arising at the inner or outer edge of a lobed disk or the disk obsolete; pistil 1; ovary superior, usually 2-lobed and 2-chambered; styles 2; fruit a double samara, splitting into halves at maturity, each half with a wing.

---

1 Leaves compound 2
1 Leaves simple 3
2 Terminal leaflet on an extended rachis; leaflets mostly over 5 cm long; each half of the fruit 2-5 cm long . . . . **A. negundo**
2 Terminal leaflet sessile or nearly so; leaflets seldom to 5 cm long; each half of the fruit 1-2.5 cm long . . . . **A. glabrum**
3 Leaves sharply serrate; sinuses between the lobes and teeth mostly acute; shrubs or small trees . . . . **A. glabrum**
3 Leaves not serrate, if toothed then the teeth broad and rounded; sinuses between the lobes and teeth typically rounded; trees . . **A. grandidentatum**

Acer glabrum Torrey [*A. glabrum* var. *glabrum*; *A. glabrum* var. *tripartitum* (Nuttall) Pax] — Rocky Mtn. maple — Widespread and scattered; mountain slopes and canyons, avalanche paths; mostly 7,000-9,000 ft; April-June. Plants with parted, divided, or palmately compound leaves have been referred to var. *tripartitum*. Those with simple leaves have been referred to var. *glabrum*. Arnow and others (1980) reported the two forms to be sympatric and to intergrade completely. They are treated here in synonymy.

Acer grandidentatum Nuttall in Torrey & Gray — bigtooth maple — Rather infrequent at lower and mid-elevations of the s. slope of the Uinta Mtns., most common towards the w., specimens and sightings from Rhoades, N. Fork Duchesne, and Rock Creek drainages; one population with rather small leaves has been found as far e. as Whiterocks Canyon; reported by Graham (1937) for the bottom of Florence Canyon, Tavaputs Plateau; May-June.

Acer negundo L. (*A. interior* Britton; *A. interius* Britton; *Negundo aceroides* (L.) Moench — boxelder — Widespread; occasional to abundant along water courses or in depressions where run off accumulates; mostly below 7,500 ft, sometimes cultivated as a shade tree; May-June. Our trees are referable to var. *negundo*.

ADOXACEAE Moschatel Family
Adoxa L. — Muskroot; Moschatel

Adoxa moschatellina L. — moschatel — Perennial glabrous herbs from scaly or tuberous rootstocks, 5-15 cm tall; leaves basal and one opposite pair above mid-length of the stem, long petiolate, ternately compound, the leaflets 3-lobed or 3-cleft or the terminal one again ternate; flowers bisexual, in terminal head-like clusters, the clusters 6-8 mm wide, with 3-8 flowers; calyx 2-3 lobed; corolla united, 2-3 mm long, greenish, 4-5 lobed in terminal flowers, 5-6 lobed in lateral flowers; ovary inferior; styles 3-5-parted; fruit a greenish dry drupe with 3-5 nutlets. Listed by Tidestrom (1925) for the Uinta Mtns.

AGAVACEAE Agave Family
Yucca L. — Yucca

Stout rhizomatous subacaulescent shrubs from a semi-woody caudex or rhizomes; leaves saber-like, basal, thick and firm, lanceolate to nearly linear, fibrous margined, stiffly awl-pointed, the bases dilated; scape stout; inflorescence racemose, with papery bracts; flowers numerous, large, pendulous, the tepals thick and waxy when fresh, white, cream, or yellowish green; stamens 6, filaments fleshy; ovary superior, 3-loculed, with a 3-lobed stigma, ovoid-oblong in outline, woody in age.

1 Plants consisting of 1-few stems, these connected by short caudex-like rhizomes; leaves relatively wide, thick, and stiff, the marginal fibers freely exfoliating, coarse and curly; flowers usually greenish-yellow; plants not particularly of sandy places .... *Y. harrimaniae*

1 Plants consisting of clones (sometimes numbering many hundreds), individual stems solitary, rather widely spaced, connected by deep-seated horizontal rhizomes; leaves relatively narrow, thin, and flexible, the marginal fibers little exfoliating, mostly slender and nearly straight; flowers white; plants of sandy places .... *Y. sterilis*

Yucca harrimaniae Trelease — Harriman yucca — Locally common from e. Duchesne Co. and eastward in the Uinta Basin south of the Uinta Mtns.; dry slopes and foothills in desert shrub, Salina wildrye, and pinyon-juniper communities; specimens seen are from 4,800-7,000 ft; May-July.

Yucca sterilis S. L. Welsh & L. C. Higgins (*Y. harrimaniae* var. *sterilis* Neese & S. L. Welsh) — spreading yucca — Known from sandy sites at Walker Hollow south of Jensen, Cedar View northwest of Roosevelt, 25 miles south of Roosevelt, and along highway 246 between Gusher and Lapoint. This Uinta Basin endemic is strongly rhizomatous, clone-forming, apparently sterile, and has flaccid leaves that tend to lay on or near the soil surface.

AIZOACEAE Carpet-weed Family
Sesuvium L. — Seapurslane

Sesuvium verrucosum Rafinesque — verrucose seapurslane — Glabrous, more or less succulent, annual herbs; stems 10-70 cm long, prostrate or ascending, much branched; leaves opposite, spatulate to linear, the base narrow but clasping; flowers axillary, solitary, bisexual; sessile or on short pedicels; calyx tube turbinate, 5-lobed, the lobes 5-7 mm, scarious margined, short-horned near the apex; corolla lacking; stamens many; ovary half-inferior; fruit a circumscissile capsule with 3-5 chambers. The 7 specimens seen are from the edge of Pelican Lake and the floodplain of the Green River near Ouray.

ALISMATACEAE Water-Plantain Family

Herbs aquatic, emergent, or growing in wet mud of drying ponds, scapose; leaves erect and/or floating, with elongate, sheathing petioles, the blades with prominent parallel veins and transverse veinlets; inflorescence a series of (1) 2-8 (rarely more) verticils of simple or more often compound umbels or panicles; sepals 3; petals 3 or lacking; stamens (3) 6-many, included; pistils several to numerous, free, the ovary superior, the style often persistent as a beak on the fruit; fruit an achene, the achenes rather numerous, borne on a receptacle, the achenes and clusters of achenes similar to those in some of Ranunculaceae.

1 Blades of emergent leaves mostly with conspicuous sagittate or hastate lobes .... *Sagittaria*

1 Leaf blades neither sagittate nor hastate. 2

2 Stamens 12; achenes firm, with terminal beak, densely packed on the receptacle, not in a single row .... *Echinodorus*

2 Stamens (3) 6-9; achenes, rather papery, beakless or the rather fragile beak lateral, in a single row on the receptacle .... *Alisma*

Alisma L. — Water-plantain

Perennial herbs, aquatic, submerged or emergent or in mud of drying ponds, from rootstocks and fibrous roots; leaves all basal; inflorescence usually with 2-4 whorls of braches, the branches ending in umbels and often with a lateral whorl of verticillate flowers; flowers mostly bisexual, regular; sepals 3; persistent; petals 3, white, pinkish, or purplish, deciduous.

1 Inflorescence exceeding the leaves; emergent leaf blades broadly elliptic to oval, 2-20 cm wide, rather abruptly constricted to a distinct petiole; petals 3.5-6 mm long, white .... *A. plantago-aquatica*

1 Inflorescence shorter than or slightly exceeding the leaves; leaves sometimes all submerged and reduced to linear petioles; blades of emergent leaves narrowly elliptic or rarely oblong, mostly less than 3 cm wide, gradually narrowed to the petiole; petals 2-4 mm long, pinkish or whitish .... *A. gramineum*

Alisma gramineum Lejeune (*A. geyeri* Torrey) — narrowleaf water-plantain — Specimens seen are from the floodplain of the Green River near Ouray, Pelican Lake, ponds on Diamond Mtn. and Cottonwood wash nw. of Neola, listed for Irish Lake based on Porter 3683 (Bradley 1950); aquatic or in mud; 4,400-7,400 ft; July-Aug.

Alisma plantago-aquatica L. (*A. triviale* Pursh) — common water-plantain — Three specimens seen (Goodrich and Jepson 15960 and Goodrich 27130 are from Meadow Park in the Uinta Mtns., Daggett Co., Kass 4068 is from Lake Fork River near Altamont) to be expected elsewhere; ephemeral ponds at 6240-7,600 ft; June-Sept.

Echinodorus Rich — Bur Reed

Echinodorus berteroi (Sprengel) Fassett [*E. rostratus* (Nuttall) Engelmann] — upright burhead — Plants annual, 10-40 (60) cm tall; emergent leaves usually exceeded by the inflorescence, erect, not sheathing, the blades oval or cordate, strongly 5-9 nerved and reticulate-veined 2-8 (10) cm long, often about as wide, abruptly contracted to a petiole, the floating or submerged leaves narrower and lanceolate to elliptic; scapes 1-few, erect, ribbed, or angled; inflorescence of 1-several bracteate verticils, the lower ones often of a few peduncles each terminated by a bracteate umbel or umbel-like cluster of pedicellate flowers; the upper verticils simple and of few to several pedicellate flowers; internodes of the inflorescence reduced upward; sepals 3, persistent, reflexed, 2-4 (5) mm long; petals 3, 5-10 mm long, white to cream or greenish-white, the margin entire; stamens 12; achenes 1-4 mm long, with 2-winged ribs alternating with 3 non-winged ribs, the style persistent as a stout beak; fruiting head bur-like at maturity. The 2 specimens seen are from disturbed ground of wet bottomlands near Ouray; July-Aug.

Sagittaria L. — Arrowhead

Sagittaria cuneata Sheldon — wapato, arumleaf arrowhead — Plants perennial, aquatic or in mud of drying ponds, 20-100 (200) cm tall, arising from stout, tuber-bearing, spreading rhizomes; leaves shorter than or about equaling the inflorescence, blades of aerial leaves sagittate, the terminal lobe 1-10 cm long and 1-6 cm wide, lateral lobes widely divergent, 1-8 cm long and 1-3 cm wide; submerged leaves with lobes reduced and hastate or blades obsolete and leaves petiole-like the entire length; scapes few, erect; inflorescence of 2-7 mostly simple verticils of flowers on pedicels 0.5-6 cm long; lower flowers pistillate, upper flowers staminate, with some bisexual flowers in between, rarely all bisexual or all staminate; sepals 3, reflexed, 5-10 mm long, persistent; petals 3, 1-2 cm long, white; stamens 21 or more; achenes numerous, 2.5-4 mm long, in a globose head 1.5-3 cm in diameter; the style persistent as a beak on the achene. Specimens seen are from wet bottomlands at Vernal, a pond at Green Lakes, Daggett Co., and the floodplain of the Green River near Ouray, Stewart Lake, and Echo Park; 4,700-7,380 ft; July-Aug.

AMARANTHACEAE Amaranth Family

Amaranthus L. — Amaranth; Pigweed; Redroot

Annual herbs; leaves simple, alternate, entire or shallowly dentate to sinuate; flowers inconspicuous, regular, mostly crowded in dense racemes or spikes, subtended by 1 or more membranous or scarious, pointed, persistent bracts, bisexual or unisexual, the same plant often with a mixture of bisexual and unisexual flowers; bisexual and staminate flowers with 5 scarious or membranous sepals and 5 stamens, corolla lacking; pistillate flowers without a perianth; styles 3; fruit a 1-seeded, usually circumscissile, membranous or scarious capsule; seeds smooth, shiny, brown or blackish.

1   Flowers in terminal and axillary, simple to compound spikes; leaf blades usually over 3 cm long; plants branched or not . . . . **A. retroflexus**

1   Flowers in small axillary clusters; leaf blades mostly 1-3 cm long; plants
    much branched, prostrate to ascending 2

2   Capsules 1-1.7 mm long, 1-1.2 mm wide, subglobose, tardily dehiscent;
    pistillate sepals 0.8-1.5 mm long; herbage often papillate-puberulent at least
    on lower part of stems .... *A. albus*

2   Capsules 1.8-2.8 mm long, 1.5-1.6 mm wide, ovoid to ellipsoid, with
    circumscissile dehiscence; pistillate sepals 1.4-3.4 mm long; herbage
    glabrous or sparsely puberulent around the nodes .... *A. blitoides*

Amaranthus albus L. — pale amaranth, tumble pigweed — Roadsides and
cultivated areas, increasing after fire and other disturbance, various plant
communities; up to 7,600 (8,000) ft; July-Oct. Sepals 3; stamens 2-3; seeds 0.6-1
mm long.

Amaranthus blitoides S. Watson (*A. graecizans* L. misapplied) — mat amaranth,
prostrate pigweed, tumble pigweed — Cultivated, and other disturbed areas;
up to 8,200 (9, 000) ft; July-Oct. Sepals 4-5; stamens 3-5; seeds 1.3-1.7 mm long.

Amaranthus retroflexus L. — redroot pigweed, redroot amaranth — Roadsides,
gardens, and other cultivated areas, and waste places; up to about 7,000 ft;
July-Oct. Stamens 5; seeds about 1 mm long. Apparently the specimen reported
as Amaranthus powellii S. Watson by Graham (1937) for w. of Vernal, along a
ditch; Sept. belongs here. The similar *A. hybridus* L. (slim amaranth) might be
expected in the Uinta Basin area. FNA (4: 422) list *A. hybridus* for most western
states except Utah and Wyoming.

ANACARDIACEAE Sumac Family

Shrubs (in our area), trees or climbing plants, the sap usually acrid, resinous
or milky; leaves alternate, pinnately or palmately 3-foliate; flowers regular,
bisexual or polygamous, small, in axillary or terminal panicles; sepals usually
5 (4-6) with a glandular disk at the base; petals separate, usually 5 (4-6);
stamens the same number as the petals and alternate to the petals; ovary
superior, 1-chambered, with 1 ovule; styles usually 3; fruit drupe-like, but dry,
1-seeded.

1   Leaflets mostly less than 3 cm long, the terminal one sessile or on a short
    stalk much less than 1 cm long; fruit reddish, pubescent .... ***Rhus***

1   Leaflets over 3 cm long, the terminal one on a stalk usually over 1 cm long;
    fruit yellowish white, usually glabrous .... ***Toxicodendron***

Rhus L. — Sumac

Rhus aromatica Aiton (*R. trilobata* Nuttall) — skunkbush, skunkbush sumac —
Shrub 1-2 m tall; stems usually many from a large root crown; leaves alternate,
mostly compound with 3 sessile leaflets, rarely simple and crenate to shallowly
lobed or 1-3 cleft; flowers appearing before the leaves, small, densely clustered
at the ends of branches; petals yellow, 2-3 mm long; fruit 6-8 mm long,
subglobose, red or orange, densely puberulent with red, viscid hairs and some
long simple hairs, becoming glabrous in age. Widespread and common below
7,500 ft, abundant on the floodplains of the Green and White Rivers; April-

June. The plant is aromatic with a pungent, rather disagreeable odor that is particularly noticeable when the twigs are crushed or broken. Our plants with mostly trifoliate leaves belong to var. *trilobata* (Nuttall) A. Gray ex S. Watson. Var. *simplicifolia* (Greene) Cronquist with simple leaves is listed for south of our area in IMF (3A: 314) and AUF (37).

Toxicodendron Mill — Poison Ivy

Toxicodendron rydbergii (Small ex Rydberg) Greene [*Rhus radicans* L. var. *rydbergii* (Small) Rehder; *Toxicodendron radicans* L. var. *rydbergii* (Small) Erskine] — poison ivy — Shrub or subshrub to 1.5 m tall, dioecious; leaves alternate, congested toward the summit of the stem, compound with 3 (5) leaflets, the leaflets 3-20 cm long, 2-11 (17) cm wide, coarsely dentate to nearly entire, turning yellow, orange, or red in autumn; flowers appearing with the leaves, in compact or loose axillary panicles to about 5 cm long; sepals 5, about 1 mm long; petals 5, yellow-white, to 3 mm long; stamens 5; styles 3, fused at the base; fruit a subglobose drupe, 2.5-7 mm wide, white or cream to yellow, glabrous and smooth. Occasional; woods and canyons, specimens seen are from Split Mt., Dinosaur National Monument, and Uinta Mtns. from Cart Creek, Dry Fork, and Big Brush Cr. Canyons and at Dry Gulch; 4,800-7,850 ft; May-June.

APIACEAE (UMBELLIFERAE) Parsley Family; Carrot Family

Annual or perennial herbs, often aromatic, commonly hollow-stemmed; leaves alternate, basal or rarely opposite, once to many times compound, rarely simple, the petioles often sheathing; flowers in compound umbels, the rays elongate or sometimes reduced or nearly obsolete and then the umbel head-like; the umbel sometimes subtended by an involucre of separate or united bracts; the pedicels sometimes subtended by an involucel of separate or united bractlets; calyx of 5 minute teeth or lacking; petals 5, separate, mostly small; stamens 5, alternate with the petals; ovary inferior, 2-chambered, each chamber with a single ovule; styles 2, sometimes swollen at the base forming a stylopodium; fruit a schizocarp, splitting at maturity into 1 seeded mericarps and disclosing a wire-like or thread-like carpophore to which the mericarps are apically attached; carpophore sometimes lacking; mericarps typically 5-nerved or 5-ribbed, the ribs sometimes winged. Carrot (*Daucus carota* L.), celery (*Apium graveolens* L.), dill (*Anethum graveolens* L.), parsley (*Petroselinum crispum* (Miller) A. W. Hill are cultivated species that rarely escape in our area. They are excluded from the following treatment.

---

1 Leaves not all basal; the few to several peduncles mostly shorter than the leafy stem on which they are borne 2

1 Leaves all basal, the leaves sometimes whorled atop a pseudoscape, or the usually solitary peduncle longer than the short, leafy stem on which it is borne.. 3

2 Leaves simple to ternate or pinnate with leaflets mostly sessile .... **KEY 1**

2 At least the larger leaves more than once-compound; primary leaflets usually petiolulate .... **KEY 2**

3   Leaves ternate or biternate with 3-9 leaflets, or rarely a few simple, plants
     5-10 cm tall; petals white . . . . ***Orogenia***

3 . Leaves and leaflets not as above, or plants mostly taller and/or petals yellow
     . . . . **KEY 3**

— KEY 1 —

1   Basal leaves simple and cordate at the base, upper leaves sometimes ternate;
     leaves and leaflets seldom over 7 cm long or 3 cm wide . . . . ***Zizia***

1   Leaves pinnately compound, or if some simple or ternate then the leaves
     or leaflets over 7 cm long and over 3 cm wide 2

2   Leaflets 10-40 cm long, 10-35 cm wide; plants 1-3 m tall, tomentose; leaves
     pinnate, ternate, or the upper ones simple; the larger petals cleft, 3-10 mm
     long . . . . ***Heracleum***

2   Leaflets and plant smaller, not tomentose; leaves pinnate; petals not cleft
     3

3   Ultimate segments of leaves mostly entire and filiform, linear, or lanceolate,
     mostly over 20 times longer than wide . . . . ***Perideridia***

3   Ultimate segments of leaves toothed, mostly much broader than above. . 4

4   Involucre and involucels well developed, sometimes spreading or deflexed
     bracts and bractlets, the bracts 1-6, the bractlets (2) 4-12; fruit 1.5-3 mm
     long, the ribs not winged; plants of wet places, usually growing in water,
     from fibrous roots 5

4   Involucre lacking or infrequently of 1-2 bracts; involucels often lacking;
     fruit over 3 mm long or else winged; plants of various habitats, from a
     taproot or tuberous roots. 6

5   Stems often sprawling, sometimes stoloniferous; leaves with (3) 5-15
     opposite pairs of leaflets . . . . ***Berula***

5   Stems erect, not stoloniferous; leaves with 4-6 opposite pairs of leaflets . .
     . . . . ***Sium***

6   Umbels often more than 7 per plant; flowers yellow or red . . . . . ***Pastinaca***

6   Umbels often less than 7 per plant; flowers white or greenish-yellow. . . . 7

7   Inflorescence subtended by dilated, sheathing petioles; fruit 3-5 mm long
     . . . . ***Angelica***

7   Inflorescence seldom subtended by dilated, sheathing petioles; leaflets often
     hirtellous; fruit over 10 mm long . . . . ***Osmorhiza***

— KEY 2 —

1   At least some ultimate segments of leaflets over 2 cm long or over 1 cm
     wide, toothed or shallowly lobed 2

1   Ultimate segments less than 2 cm long, less than 1 cm wide, or if longer or
     wider then deeply lobed to pinnatifid 4

2   Involucels of about 6 bractlets 1-4 mm long; umbels 6-20 or more per stem,
     the rays 15-26, 1.5-4 cm long; fruit 2-4 mm long, the ribs corky but not
     winged . . . . ***Cicuta***

2   Involucels mostly lacking; umbels often fewer than 6 per stem and/or the
     rays either fewer or longer than above or both fewer and longer; fruit either
     longer or else winged 3

3   Fruit 3-5 mm long; inflorescence subtended by dilated sheaths; leaflets glabrous . . . . **Angelica**

3   Fruit 10-25 mm long; inflorescence seldom subtended by dilated sheaths; leaflets often hirtellous . . . . **Osmorhiza**

4   Stems usually purple-spotted, (50) 100-300 cm tall, usually much branched, mostly with 10-30 or more umbels, the umbels with 15-26 rays 1.5-4 cm long; involucres and involucels usually present . . . . **Conium**

4   Stems not purple-spotted, mostly less than 100 cm tall, mostly not much branched, with 3-9 (14) umbels, the rays usually either fewer or some longer than above; involucres mostly lacking; involucels various. 5

5   Fruit 3-4 mm long, the ribs filiform; plants biennial, introduced, rarely persisting in indigenous plant communities; involucels lacking or minute . . . . **Carum**

5   Fruit 4-8 mm long, the lateral and sometimes the dorsal ribs with small wings; plants perennial, native; involucels various 6

6   Involucels usually of 3-6 bractlets; fruit slightly compressed dorsally; root crown mostly simple, without long-persisting petiole-bases; plants rather rare, Daggett Co. and E. Tavaputs Plateau . . . . **Conioselinum**

6   Involucels lacking or of 1-3 bractlets; fruit terete or slightly compressed laterally; root crown simple or branched, usually with fibrous long-persisting petiole-bases; plants not of the E. Tavaputs Plateau . . . . **Ligusticum**

— KEY 3 —

1   Plants rather densely hirtellous-scabrous to villous 2

1   Plants glabrous or at most obscurely scabrous. 3

2   Rays of the umbel 1-6 mm long . . . . **Cymopterus alpinus**

2   Rays of the umbel longer . . . . **Lomatium**

3   At least some of the ultimate segments of leaves over 15 mm long and entire . . . . **Lomatium**

3   Ultimate segments of leaves less than 15 mm long or else not entire 4

4   Plants not strongly aromatic, from fibrous taproots with few if any long-persisting leaf bases 5

4   Plants strongly aromatic, from simple or more often branched, more or less woody caudices, these often clothed with long-persisting leaf-bases 6

5   Flowers white; plants from above 8,000 ft on the Uinta Mtns.   . **Ligusticum**

5   Flowers yellow or plants from below 8,000 ft . . . . **Cymopterus**

6   Calyx teeth lacking or to 0.3 mm long; ultimate segments of leaves 0.2-0.3 mm wide; fruit with lateral wing only . . . . **Lomatium grayi**

6   Calyx teeth 0.5-0.9 mm long; ultimate segments of leaves 0.5-1 (1.5) mm wide; fruit with dorsal and lateral wings . . . . . **Cymopterus** (*Pteryxia* group)

Angelica L. — Angelica

Perennial herbs from stout taproots, stems solitary; leaves 1-2 times pinnate or ternate, the ultimate segments broad and distinct, the lateral veins oriented toward the tips of the marginal teeth; petioles strongly sheathing; upper leaves often reduced to dilated sheaths; umbels 1-several, compound; involucre usually lacking; involucel lacking or of a few linear bractlets; calyx teeth mostly lacking; fruit strongly compressed parallel with the commissure, the lateral ribs with broad, thin to corky wings, dorsal ribs thread-like to corky-winged; carpophore divided to the base.

---

1   Stems over 100 cm tall; fruit 7-8 mm long at maturity; umbels globose; plants of the Piceance Basin .... *A. ampla*

1   Stems 30-100 cm tall, rarely taller; fruit 4-7 mm long; umbels rather flat-topped; plants of various distribution 2

2   Leaves once pinnate or partly bipinnate by some division in lower leaflets; some leaflets usually over 2 times as long as wide .... *A. pinnata*

2   Leaves twice or more pinnate, the leaflets less than 2 times as long as wide .... *A. roseana*

Angelica ampla A. Nelson — giant angelica — The one specimen seen (Erickson sn.) is from the Piceance Basin.

Angelica pinnata S. Watson — small-leaf angelica — Widespread; occasional in moist aspen or coniferous forest communities and along streams; 6,000-10,000 ft; June-Sept.

Angelica roseana L. Henderson — Rose angelica, talus angelica — Uinta Mtns., occasional or locally common in rocky places, often in boulder fields in the upper coniferous forest belt and above timberline; (9,300) 10,000-11,700 ft; July-Aug. Graham 8576 (CM!) reported as *A. lyallii* S. Watson (*A. arguta* Nuttall ex Torrey & A. Gray) by Graham (1937) belongs here.

Berula Hoffman — Berula

Berula erecta (Hudson) Coville — cutleaf waterparsnip — Plants perennial, often stoloniferous, with numerous fibrous roots; stems 20-80 cm long; submerged leaves much dissected into linear segments, aerial leaves once pinnate, dimorphic, the lower with 4-14 pairs of lanceolate or ovate serrate, crenate or lobed leaflets 1-7 cm long, 3-30 mm wide, the upper with 2-6 pairs of narrower, more deeply lobed, serrate leaflets; umbels with 6-15 (20) rays; involucre and involucel of narrow, often 3-nerved, more or less reflexed bracts and bractlets; petals white; fruit 1.5-2 mm long. Apparently widespread, seldom collected; emergent and in mud; not expected over 7,000 ft; July-Sept.

Carum L.

Carum carvi L. — caraway — Plants biennial, glabrous, 30-60 (100) cm tall, from a taproot; leaves 2-3 times pinnate and then often pinnatifid, with about 6-11 opposite or offset pairs of lateral primary leaflets; petioles to 15 cm long, the upper ones reduced and the blades sometimes sessile on a dilated sheath; leaf blades 5-16 cm long, oblong in outline; primary leaflets from less than 1/4 to about 1/2 as long as the leaf blade, sessile, the ultimate segments 2-8 (15) mm long, 0.5-2 mm wide, linear and entire or obovate and toothed to lobed; peduncles 4-12 cm long, usually subtended by a dilated sheath; umbels compound, 6-12 or more; involucre lacking or inconspicuous; rays 6-12 (14), 1.5-8 cm long; involucels lacking or of minute scarious teeth; pedicels (5) 8-20 mm long; petals white; filaments white, the anthers pale green or whitish; styles about 0.5-0.85 mm long; carpophore divided to the base; stylopodium low, conic; fruit 3-4 mm long, the ribs filiform. Introduced from Europe; specimens seen are from widely scattered locations; pastures, ditch banks, and occasionally in indigenous plant communities; 6500-8000 ft; June July.

Cicuta L. — Water Hemlock

Cicuta maculata L. [*C. douglasii* (de Candolle) Coulter & Rose misapplied; *C. occidentalis* Greene] — water hemlock — Perennial herbs 50-200 cm tall, from tuberous-thickened, chambered roots; leaves 1-3 times ternate-pinnate, the leaflets lanceolate, 3-10 cm long, 0.5-3.5 mm wide, sharply serrate; umbels 1-several; involucre lacking or of a few linear bracts; rays 15-28, 2-6 cm long; involucel of a few linear or elliptic bractlets, these 2-15 mm long, entire, toothed or occasionally lobed; petals white or greenish; fruit elliptic or subglobose, 2-4 mm long, the ribs corky-thickened. Widespread; occasional to common along ditches and natural waterways and other wet places; to about 7,000 ft; June-Aug. A violently poisonous plant. Our plants belong to var. *angustifolia* Hooker.

Conioselinum Hoffman — Hemlock-parsley

Conioselinum scopulorum (A. Gray) Coulter & Rose — Rocky Mtn. hemlock parsley — Plants perennial, 30-90 cm tall, from a taproot or cluster of fleshy roots; stems without marcescent material at the base; leaves 1-2 times pinnate or ternate-pinnate; involucral bracts lacking or narrow; rays 10-20; bractlets 2-8 mm long; flowers white; fruit 4-6 mm long, oval, dorsal ribs low, the lateral ribs slightly winged. Specimens seen are from the n. slope of the Uinta Mtns. and from the E. Tavaputs Plateau, from along streams and in aspen woods; (7,800) 8,500-9,000 ft; June-July. This is sometimes confused with *Ligusticum porteri* from which it differs by taproot or cluster of fleshy roots instead of fibrous caudex, without marcescent fibrous leaf bases, and with conspicuous bractlets.

Conium L. — Poison Hemlock

Conium maculatum L. — poison hemlock — Plants biennial, from a taproot; stems 50-200 (300) cm tall, hollow, purple spotted; lower petioles purple spotted; leaves 2-4 times pinnately dissected, the ultimate segments seldom over 1 cm long, incised; involucre of linear bracts; umbels numerous, with 15-28 rays; bractlets 2-15 mm long, linear or nearly ovate; petals white or greenish; fruit 2-4 mm long, the ribs corky-thickened and low. Introduced from Europe; to be expected throughout our area, but seldom collected; wetlands, water courses, often ditch banks; up to 8,000 ft; June-Aug. This plant is violently poisonous.

Cymopterus Rafinesque — Spring-parsley

Scapose to subscapose usually low perennials from fibrous often enlarged taproots or branched caudices; pseudoscape sometimes well developed; leaves mostly pinnately compound, often much dissected; umbels mostly solitary and terminal; involucre and involucel various; fruit strongly compressed parallel with the commissure, the lateral and most of the dorsal ribs winged;

carpophore divided to the base or obsolete. As treated here the genus includes Pteryxia Nuttall ex. Coulter & Rose. This group has the following features: Strongly aromatic perennials from branched caudices clothed with persistent bases of old petioles and peduncles; leaves basal or mostly so, pinnate compound; involucres lacking; bractlets of involucels linear or nearly so; calyx teeth present but small; carpophore divided to the base; fruit winged with some dorsal wings often reduced.

---

1 Stems from a branched caudex; leaves finely dissected with the ultimate segments mostly less than 2 mm wide; old leaf bases persistent and forming a thatch at the base of the plant 2

1 Stems from a simple or sparingly branched taproot or ultimate leaf segments wider than 2 mm; old leaf bases not persisting long enough to build up a thatch 5

2 Petioles 0.5-3 cm long; umbels 6-18 mm in diameter. 3

2 Petioles 2-14 cm long; umbels sometimes over 18 mm in diameter 4

3 Petals white; plants known from the rim of Ashley Gorge . . . . *C. evertii*

   Petals yellow; plants not restricted as above . . . . *C. alpinus*

4 Lowest pair of primary leaflets (1/4) 1/2-3/4 or more the length of the leaf blade, mostly 3-9 cm long; plants growing at 4,700-7,600 (8,400) ft . . . .
   . . . . *C. terebinthinus*

4 Lowest pair of primary leaflets 1/4 or less the length of the leaf blade, to 2.7 cm long; plants growing at (rarely 7,500) 9,000-12,000 ft *C. longilobus*

5 Involucels (and involucres if present) whitish or purplish with purple nerves, the bractlets mostly over 3 mm wide; petals white 6

5 Involucels not colored as above, the bractlets less than 3 mm wide; involucres lacking 8

6 Rays 0.3-1 cm long, rarely longer, not exserted beyond the always well developed involucre, obscured by the mature fruits; pedicels 1-4 mm long; fruit lacking a carpophore . . . . *C. purpurascens*

6 Rays 1-3.5 cm long, usually at least some exceeding the well developed to obsolete involucre through all stages of phenology, not obscured by the mature fruits; pedicels 5-12 mm long; fruit with well developed carpophore
   7

7 Bractlets connate to 1/3 the length, the free portion gradually expanded distally, obovate to spatulate, with mostly 3 veins arising from the base that are more or less parallel below, gradually flaring distally, equal or nearly so . . . . *C. constancei*

7 Bractlets connate for 1/3-2/3 the length, the free portion usually abruptly enlarged distally, broadly ovate to orbicular, with mostly 1 vein or with 1-2 pairs of shorter lateral veins that are parallel or divergent or branched .
   . . . . *C. bulbosus*

8 Petals white; plants from north of the Uinta Mtns. 9

8 Petals yellow or purplish 10

9 Dorsal wings of fruit smaller than the lateral, sometimes reduced to ribs .
   . . . . *C. lapidosus*

9 Dorsal wings of fruit about equal to the lateral . . . . *C. glomeratus*

10 Rays of umbels 1-13 mm long; pedicels lacking or to 2 mm long **C. glomeratus**

10 At least some of the rays of umbels usually over 13 mm long; pedicles in fruit 3-10 mm long. 11

11 An aerial pseudoscape usually rather quickly developing, (3.5) 5-24 cm long; leaf blades with 4-6 opposite or offset pairs of lateral primary leaflets; umbels sometimes nodding on recurved peduncles; plants mostly montane, above 7,000 ft . . . . **C. longipes**

11 Pseudoscape lacking or mostly subterranean, the aerial portion not over 3 cm long; leaf blades with 2-4 opposite pairs of lateral primary leaflets, or ternate; umbels not nodding 12

12 Leaves once pinnately compound with 2 opposite pairs of lateral primary leaflets, or a few ternate or rarely biternate, glaucous; confluent portions of the blades (3) 6-25 (40) mm wide; petals and stamens bright yellow when fresh, fading to cream or white in herbarium specimens; plants from 4,690-6,200 ft . . . . **C. duchesnensis**

12 Leaves ternate or 2-3 times pinnately compound with up to 4 opposite pairs of lateral primary leaflets, glaucous or not, the confluent portions mostly 1-4 mm wide or if wider then the leaves ternate; petals yellow or purple when fresh, if yellow then turning dark purple in herbarium specimens, the anthers remaining pale; plants of broad distribution . . . . **C. purpureus**

Cymopterus alpinus A. Gray [*Oreoxis alpina* (A. Gray) Coulter & Rose] — matted spring parsley — The few specimens seen are from the Uinta Mtns. on the n. slope near Hickerson Park, Round Park, and Coal Mine Hill and from Big Ridge on the s. slope; alpine and subalpine forb-grass, mt. big sagebrush, and lodgepole pine communities and limestone escarpments; 10,000-10,800 ft; July-Aug.

Cymopterus bulbosus A. Nelson — bulbose spring parsley — Widespread; locally common to abundant on clay soils of salt desert shrub communities; 4,700-5,800 ft; April-June. In many specimens the involucre is reduced or lacking (a condition not well documented for the species). Such a specimen is probably the basis for Graham's (1937) report of *C. planosus* (Osterhout) Mathias. See *C. purpurascens*.

Cymopterus constancei R. L. Hartman — Constance spring parsley — Listed for sw. Wyoming, e. Utah, and w. Colorado in grassland, sagebrush, pinyon-juniper, and ponderosa pine communities (Heil et al. 2013).

Cymopterus duchesnensis M. E. Jones — Uinta Basin spring parsley — Endemic, locally common from Myton to Raven Ridge and flank of Blue Mtn. in Rio Blanco Co., and disjunct in the Little Snake River drainage in Moffat Co. and in Daggett Co. near Manila, but by far most abundant in Uintah Co.; salt desert shrub communities and at the lower fringe of the juniper belt, on clay hills and slopes in Duchesne River, Uinta, Morrison, and other formations that weather to badlands; 4,700-5,600 (6,200) ft; April-June.

Cymopterus evertii Hartmann & Kirkpatrick — Evert spring parsley — Known from the rim of Ashley Gorge where disjunct from northwestern Wyoming; Douglas-fir/limber pine communities at 8,600 ft.

Cymopterus glomeratus (Nuttall) de Candolle [*C. acaulis* (Pursh) Rafinesque; (*C. fendleri* A. Gray)] Occasional, widespread; desert shrub, sagebrush, and juniper communities, mostly on sandy soil; 4,700-5,700 (7,200, Red Mt.) ft; April-June. There are two wholly intergrading vars. Plants with white petals and with peduncles mostly not exserted beyond the leaves belong to var. *acaulis* (Pursh) R. L. Hartman (plains spring parsley). Those with yellow petals and with peduncles mostly exserted beyond the leaves belong to var. *fendleri* (A. Gray) R. L. Hartman (chimaya). Plants of Daggett Co. (north of the Uinta Mtns.) clearly belong to var. *acaulis* which is widespread to the n. of our area and in the Plains, Plants from south of the Uinta Mtns. apparently all belong to var. *fendleri* which is common in the Uinta Basin and s. of the Tavaputs Plateau to Arizona and New Mexico. The yellow petals of var. *fendleri* rather quickly fade to cream or white making the distinction from var. *acaulis* difficult with age.

Cymopterus duchesnensis M. E. Jones — Uinta Basin spring parsley

Cymopterus lapidosus (M. E. Jones) M. E. Jones — Echo spring parsley — Entering our area near Hickey Mtn. on the North slope of the Uinta Mtns., locally common on semi-barrens of the Bridger Formation and Green River Formation in Uinta and Sweetwater Cos., Wyoming.

Cymopterus longilobus (Rydberg) W. A. Weber [*C. hendersonii* (Coulter & Rose) Cronquist misapplied (a plant limited to Idaho), *Pteryxia hendersonii* (Coulter & Rose) Mathias & Constance] Cronquist misapplied; *Pseudocymopterus hendersonii* Coulter & Rose misapplied] — mtn. parsley — Uinta Mtns.; occasional or locally common in talus, cliffs, rocky canyons, ridges, slopes, and alpine tundra and East Tavaputs Plateau from Cathedral Bluffs in shale-barrens (Neese 11958 and Smith 1801 BRY); (7,500 in canyons) 9,000-12,200 ft; June-Aug.

Cymopterus longipes S. Watson [*Aulospemum longipes* (S. Watson) Coulter and Rose] — long-stalk spring parsley — Widespread (no specimens seen from the E. Tavaputs Plateau; locally common in sagebrush, pinyon-juniper and mtn. brush communities; 7,000-9,000 ft; April-June.

Cymopterus purpurascens (A. Gray) M. E. Jones — widewing spring parsley — Scattered or locally common across our area; gravelly terraces, pediments, and hills in desert shrub, sagebrush, and pinyon-juniper communities; 4,700-7,000 ft and up to 9,000 ft on the Tavaputs Plateau; late Feb.-April. Similar to and often confused with *C. bulbosus,* but in addition to the differences given in the key, plants of *C. purpurascens* are more scattered than those of *C. bulbosus,* and they are found in sagebrush and pinyon-juniper communities or on slopes in salt desert shrub communities where soils are usually not so clayey, and they occur from the bottom of the Basin to 9,000 ft.

Cymopterus purpureus S. Watson — variable spring parsley — There are two more or less distinct vars. in our area.

1 Wings of fruit 5-8 mm long, to 2 mm wide; fruiting rays 5-8 (15), 0.2-2 (3) cm long; fruiting pedicels 1-5 (7) mm long; leaf blades 1-3.5 (4) cm long, mostly (not always) ternate, the leaflets with rounded lobes; plants glabrous, or more often scabrous; lower to mid-montane, of the Tavaputs Plateau .... **var. *rosei***
1 Wings of fruit 8-10 (12) mm long, (2) 2.5-4 mm wide; fruiting rays (8) 12-22, (2) 2.5-7 (9.5) cm long; fruiting pedicels 5-10 mm long; leaf blades commonly 3-9 (13) cm long, pinnately compound, rarely ternate, often with acute ultimate segments; plants mostly glabrous, rarely scabrous, of deserts and lower montane, widespread .... **var. *purpureus***

Var. *purpureus* — purple spring parsley — Widespread; occasional or common in desert shrub, sagebrush, and pinyon-juniper communities on a variety of geological formations, probably more common on sandy or sandy loam soils than on clayey soils; 4,700-6,000 ft or up to 7,600 ft on the Tavaputs Plateau where it intergrades with var. *rosei;* May-June.

Var. *rosei* (M. E. Jones) Goodrich (*C. rosei* M. E. Jones) — Rose spring parsley — Known from the Strawberry drainage from Willow Creek to Argyle Canyon, W. Tavaputs Plateau; Salina wildrye, and pinyon-juniper communities, steep, nearly barren, whitish, marly shale or limestone slopes; 6,200-7,000 ft; May-June.

Cymopterus terebinthinus (Hooker) Torrey & A. Gray (*Pteryxia terebinthina* Nuttall ex Torrey & A. Gray) — rock-parsley — Occasional across much of our area (no specimens seen from Duchesne Co.); colluvium and crevices of rocks, often with sandstone; 4,700-7,600 (8,400) ft; April-May. Our plants belong to var. *albiflorus* (Nuttall in Torrey & Gray) M. E. Jones.

Heracleum L. — Cow parsnip

Heracleum lanatum Michaux [*H. maximum* Bartram; *H. sphondylium* L. ssp. *lanatum* (Michaux) A. & D. Love; *H. s.* var. *lanatum* (Michaux) Dorn] — cow parsnip — Plants biennial or perennial from a taproot or fascicled fibrous roots, 1-3 m tall, usually conspicuously hairy at least in the inflorescence; larger leaves mostly ternate, the leaflets 10-75 cm wide, about as long, palmately lobed, coarsely toothed; umbels 10-20 cm wide; rays 15-30; involucre and involucels of 2-10 deciduous, lanceolate or linear bracts and bractlets, these 5-20 mm long; petals white, the outer ones enlarged and 2-lobed, 3-10 mm long; fruit often short-hairy, 6-12 mm long, 5-9 mm wide, flattened, the lateral wings well developed, the dorsal ribs filiform. Occasional to locally common from the Strawberry drainage to Rock Creek (rarely farther e.) in the Uinta Mtns. and to the Avintaquin drainage of the W. Tavaputs Plateau, and one specimen seen from the E. Tavaputs Plateau; seeps, springs, water courses, and aspen and tall forb communities; 7,200-9,600 ft; June-Aug.

Ligusticum L. — Ligusticum

Plants scapose to caulescent, perennial; root crown fibrous, surmounting a taproot; leaves ternate-pinnately compound; umbels one to several; involucres lacking or of 1 deciduous bract; involucels wanting or of 1-3 narrow bractlets; calyx teeth minute or evident; petals white; stylopodium low-conic; fruit oblong, slightly flattened or nearly round, the ribs narrowly winged, carpophore divided to the base.

1 Umbels mostly solitary, occasionally 2, rarely 3, never opposite; rays 0.5-3.6 cm long; petioles 1.2-13.5 cm long; leaf blades 3-19 cm long; plants 10-45 (64) cm tall, of the Uinta Mtns. .... *L. tenuifolium*

1 Umbels 2-5, the lateral ones frequently opposite or 3 per node; rays 2.5-6.5 (8) cm long; petioles 8-32 cm long; leaf blades (9) 12-30 cm long; plants (40) 60-100 cm tall, of various distribution 2

2 Ultimate segments of leaves mainly linear or narrowly triangular and widest at the base, rarely over 3 mm wide; plants of Wasatch and nw. Duchesne Cos. .... *L. filicinum*

2 Ultimate segments of leaves mainly narrowly to broadly elliptic, or at least tending to be wider toward the middle, to 8 mm wide, but as narrow as 1.5 mm; distribution mostly s. and e. of that listed for the above species .... *L. porteri*

Ligusticum filicinum S. Watson — fernleaf ligusticum, fernleaf licorice-root — Strawberry Valley to Red Creek Mt. and w. of Rock Creek in the Uinta Mtns.; occasional in woods and more commonly on open slopes and ridges; 7,700-10,200 ft; June-July.

Ligusticum porteri Coulter & Rose — Porter ligusticum — South slope of the Uinta Mtns. from Uinta Canyon eastward, and Tavaputs Plateau; aspen woods and parklands; 7,400-9,600 ft; June-July. Our plants belong to var. *porteri*.

Ligusticum tenuifolium S. Watson [*L. filicinum* S. Watson var. *tenuifolium* (S. Watson) Mathias & Const.] — slender-leafed ligusticum — Uinta Mtns.; occasional to common in dry and wet meadows, and along streams in moist woods; 8,000-11,200 ft; June-Aug. Hermann 5070 was listed as *Pseudocymopterus montanus* (A. Gray) Coulter & Rose by Graham (1937), but this specimen probably belongs here.

Lomatium Rafinesque

Plants perennial, acaulescent or caulescent, occasionally with a short pseudoscape, but this mostly lacking, glabrous or pubescent, from a slender taproot or from a thickened, woody, branched caudex, sometimes clothed at the base with marcescent material; stems simple or rarely branched and thus peduncles and umbels mostly solitary; leaves once or more pinnate, ternate, ternate-pinnately divided, or decompound, sheaths often dilated especially in lower leaves, the ultimate segments extremely variable; involucre lacking or inconspicuous; involucel mostly of separate or partly united bractlets, rarely wanting; rays few or many, spreading to ascending, the central ones often shorter and sterile; petals various; fruit flattened dorsally, linear to orbicular or obovate, dorsal ribs filiform or obsolete or occasionally with rudimentary wings at the base, lateral ribs winged. Closely related to the genus *Cymopterus*, separated in part from that genus by the absence of dorsal wings on the fruit; the strength of this feature is somewhat weakened by the reduced dorsal wings in some of *Cymopterus*.

1 At least some of the ultimate segments of leaves over 15 mm long, these less than 50 per leaf. 2

1 Ultimate segments of leaves less than 15 mm long, often over 50 per leaf 3

2 Plants from a thickened, woody, branched caudex, glabrous, strongly aromatic; marcescent material often clothing the caudex; leaves basal, the ultimate segments of leaves 0.3-5.5 cm long, 0.5-2 (4) mm wide . . . .
. . . . *L. graveolens*

2 Plants from a taproot or small caudex, puberulent, not strongly aromatic; marcescent material lacking or weakly persisting; leaves basal or sometimes 1-3 cauline; ultimate segments of leaves 1-12 cm long, 1-10 mm wide . . . .
. . . . *L. triternatum*

3 Larger mature leaves with blades (10) 15-30 cm long, ternate-pinnately compound, the larger ultimate segments 2-3 mm wide; plants (30) 50-130 cm tall; peduncles fistulose, (3) 4-6 (10) mm thick at the base. . . . .
. . . . *L. dissectum*

3 Blades of leaves 2-11 cm long, or if longer then either not ternate or with ultimate segments not over 1 mm wide; plants rarely over 50 cm tall; peduncles fistulose or not, often less than 4 mm thick. 4

4 Herbage pubescent 5

4 Herbage glabrous or at most scabrous. 7

5 Ovaries and fruit glabrous or occasionally somewhat scabrous; petals white or yellow; plants widespread. 6

5   Ovaries and young fruit rather densely pubescent, older fruit sometimes
    glabrous but often retaining some hirtellous hairs; petals yellow; plants
    rare, known from Daggett Co. .... *L. foeniculaceum*

6   Bractlets of the involucel about 10, the longer ones 4-10 mm long,
    pubescent; herbage more or less villous; the lowest pair of primary leaflets
    sessile or on petiolules to 1 cm long; mature fruit 9-12 (15) mm long; petals
    white .... *L. macrocarpum*

6   Bractlets of the involucre 1-5, 1-4.5 mm long, glabrous; herbage glabrate to
    puberulent; the lowest pair of primary leaflets on petiolules 1-3 cm long;
    mature fruit 5-8 (11) mm long; petals yellow or white ..... *L. juniperinum*

7   Fruit 3-6 mm wide; plants not strongly aromatic, from a fibrous taproot,
    the crown not clothed with long-persisting petiole bases; ultimate segments
    of leaves 0.7-1.5 mm wide; umbels with 3-12 rays; (rare glabrous specimens
    of a usually pubescent species) .... *L. juniperinum*

7   Fruit 6-8 mm wide; plants strongly aromatic, from branched caudices, these
    clothed with long-persisting petiole-bases; ultimate segments of leaves 0.2-
    0.3 mm wide; umbels with 10-26 rays .... *L. grayi*

Lomatium dissectum (Nuttall) Mathias & Const. — Indian parsley, fernleaf biscuitroot — Specimens seen are from the Uinta Mtns. and Blue Mt.; sagebrush
and mt. brush communities; 7,000-8,700 (10,400) ft; April-May. Our plants
belong to var. *multifidium* (Nuttall) Mathias and Constance.

Lomatium foeniculaceum (Nuttall) Coulter & Rose — desert parsley — The one
specimen seen (Flowers sn 28-30 July 1959 UT) is from Hideout Forest Camp at
about 5,900 ft, which is now below the high-water level of Flaming Gorge
Reservoir. Plants of our area belong to var. *macdougalii* (Coulter & Rose) Cronquist.

Lomatium graveolens (S. Watson) Dorn & R. L. Hartman [*Lomatium nuttallii* (A.
Gray) Macbride misapplied; *Cynomarathrum nuttallii* (A. Gray) Coulter & Rose)]
— stinking lomatium — Locally common; Strawberry Valley and east to the
Bear River and Rock Creek drainages in the Uinta Mtns. and to the Avintaquin
drainage, W. Tavaputs Plateau; locally common on limestone hills in the
spruce-fir belt, on sandy, exposed slopes and flats, and on shale barrens and
escarpments; 7,500-10,500 ft; June-Aug. This has often been confused with *L.
triternatum*, but it is different by the features given in the key. No specimen
was found at UT to support the Flowers and others (1960) listing for Hideout
Canyon (area now inundated by Flaming Gorge Reservoir, probably confused
there with *L. triternatum*). Our plants belong to var. *graveolens*.

Lomatium grayi Coulter & Rose [*Cogswellia grayi* (Coulter & Rose) Coulter &
Rose] — narrowleaf lomatium — Common on slopes along flanks of the Uinta
Mtns. in Daggett and Uintah Cos. and e. into Colorado in several plant
communities including sagebrush, pinyon-juniper, and mt. brush, often in
rocky places; 7,200-9,000 ft; March-June. Our plants belong to var. *grayi*. Brown
sn (BRY!) from the n. slope of Douglas Mt., reported as *L. leptocarpum* (Torrey &
A. Gray) Coulter & Rose (Bradley 1950), belongs here.

Lomatium juniperinum (M. E. Jones) Coulter & Rose (*Cogswellia juniperina* M. E.
Jones) — juniper lomatium, juniper biscuitroot — Strawberry Valley, Uinta
Mtns. to Round Top Mt., Moffat Co. and Tavaputs Plateau; occasional to
common in sagebrush, pinyon-juniper, and aspen communities; 7,200-9,400 ft;

April-July. Petal color varies from white to yellow. There is some correlation between geography and petal color that might support separating Uinta Basin plants into 2 taxa. On the n. slope of the Uinta Mtns. petals are yellow. On the Tavaputs Plateau and s. slope of the Uinta Mtns. petals are mostly white. However, the weight of this correlation is undermined by populations with yellow petals on the West Tavaputs Plateau and more seriously by populations of mixed flower color at the e. end of the Uinta Mtns. Plants with white petals are the basis for reports of *L. nevadense* (S. Watson) Coulter & Rose [*Cogswellia nevadensis* (S. Watson) M. E. Jones] for our area. Perhaps Uinta Basin plants could be included in an expanded concept of *L. nevadense.*

Lomatium macrocarpum (Nuttall ex Torrey & A. Gray) Coulter & Rose [*Cogswellia macrocarpa* (Hooker & Arnot)] — bigseed lomatium — Known from widely scattered locations from Hickey Mtn. to the Three Corners area north of the Uinta Mtns. and from Grouse Creek in Uintah Co. to e. of Rangely in Rio Blanco Co.; desert shrub and sagebrush-grass communities, and with scattered juniper; 5,400-8,400 ft; April-June. It is not uncommon to find populations with several to numerous vegetative specimens with few flowering or fruiting specimens.

Lomatium triternatum (Pursh) Coulter & Rose [*L. simplex* (Nuttall) Macbride; *Cogswellia platycarpa* (Torrey) M. E. Jones] — nineleaf biscuit root — Common on plateaus and flanks of the Uinta Mtns., Split Mt. and into Colorado, apparently lacking on the Tavaputs Plateau except at the far west in the Willow Creek drainage; 7,000-8,400 ft; April-June. Our plants belong to var. *triternatum.*

Orogenia S. Watson — Indian Potato

Orogenia linearifolia S. Watson — Indian potato — Plant perennial, scapose, from a globose or somewhat elongate tuber; scapes 1-several, 2-10 cm tall; leaves few, 1-3 times ternate or occasionally simple, the ultimate segments linear or narrowly lanceolate, 0.5-6 cm long, 1-8 mm wide; inflorescence compact, about 1-4 cm wide or wider in fruit; involucre lacking; rays 1-8, unequal, nearly obsolete or up to 5 cm long; involucels lacking or inconspicuous; pedicels not over 2 mm long; petals white; anthers purple; fruit 3-5 mm long, nearly as wide, the dorsal ribs narrow, the lateral ribs thickened. Specimens seen are from Strawberry Ridge and east end of the Uinta Mtns. including Diamond Mt., and Blue Mt., no specimens seen from Duchesne Co. or the Tavaputs Plateau, sagebrush-grass mt. brush, aspen, and ponderosa pine communities; March-June, usually flowering during or shortly after snowmelt. The roots reported used by Native Americans for food and to taste like potatoes. The small size of the root indicates an abundance of the roots would have to be gathered to supply a meal for a few people. The energy spent digging the roots might be about equal to the energy derived from eating them.

Osmorhiza Rafinesque — Sweet-cicely; Sweetroot

Perennial from elongate thick roots; leaves ternate or pinnate, the ultimate leaflets distinct; rays of umbels 3-14; involucre and involucels lacking or inconspicuous; calyx teeth obsolete; petals various; stylopodium conic; fruit narrowly cylindrical or club-shaped, black at maturity, glabrous or bristly, the ribs thread-like; carpophore bifid at the apex.

1 Fruit glabrous, the base generally obtuse; flowers yellow or greenish-white; leaflets generally 2-10 cm long, 0.5-5 cm wide, hirtellous to nearly glabrous on both sides; stems usually 2-several per root crown, 40-150 cm tall *O. occidentalis*

1 Fruit bristly pubescent, long pointed at the base into bristly tails; flowers white or greenish-white; leaflets hirtellous on veins and margins with translucent hairs; stems mostly 1-3 per root crown, 20-60 (80) cm tall . . . 2

2 Mature fruit including tails mostly 16-25 mm long, the apex concavely pointed into a beak 1-2 mm long; the most divergent rays spreading 30° to 65° from the peduncle; fruiting pedicels mostly ascending-spreading; blades of leaflets 1.5-6 (9) cm long, 1-4 cm wide; plants most common below 8,000 ft . . . . *O. berteroi*

2 Mature fruit including tails mostly 13-18 mm long, the apex convex and obtuse; the most divergent rays spreading 40° to 90° from the peduncle, some fruiting pedicels usually horizontally spreading to deflexed; blades of leaflets 1-4 (5.5) cm long, 1-3 cm wide; plants common above as well as below 8,000 ft . . . . *O. depauperata*

Osmorhiza berteroi de Candolle (*O. chilensis* Hooker & Arnot; *O. divaricata* Nuttall) — sweet-cicely — The few specimens seen are from riparian communities in the Uinta Mtns., abundant at Jones Hole with box elder; 5,570-8,100 ft; June-July.

Osmorhiza depauperata Phil. [*O. obtusa* (Coulter & Rose) Fernald] — bluntseed sweetroot — widespread and common, moist woods and streamsides; mostly above 8,000 ft and up to 10,600 ft; May-Aug.

Osmorhiza occidentalis (Nuttall ex Torrey & A. Gray) Torrey — western sweet-cicely — Strawberry Valley and e. in the Uinta Mtns. to Rock Creek; occasional to locally common in aspen-tall forb communities, and in parklands within the aspen and coniferous forest belts; 7,500-10,200 ft; June-July.

Pastinaca L. — Parsnip

Pastinaca sativa L. — parsnip — Plants biennial from a taproot, 30-120 cm tall, the stems often robust and thick; leaves oblong to ovate in outline, the basal ones to 50 cm long, once pinnate; leaflets distinct, lanceolate to ovate, 5-13 cm long, 2.5-10 cm wide, serrate, lobed or divided; umbels compound; rays 15-25, 2-10 cm long; involucre and involucels of a few filiform segments or lacking; petals yellow; stylopodium low-conic; fruit glabrous, 5-6 mm long, 3-5 mm wide, strongly compressed, the dorsal ribs filiform, the lateral ribs narrowly winged. Introduced from Europe; ditch banks, and other wet places at lower elevations; May-July.

Perideridia Reichenbach — Yampa

Perideridia montana (Blankenship) Dorn [*P. gairdneri* (Hooker & Arnot) Mathias misapplied] — yampa — Perennial herbs 30-80 cm tall, from fascicled, easily detached, tuberous roots; leaves ternate to pinnate, or the upper ones reduced and simple, the ultimate segments linear to narrowly lanceolate, to 15 cm long, 1-6 mm wide; umbels compound, the rays 10-30, 1.5-3.5 cm long; bracts of the involucre (4) 6-10, lanceolate to linear, 3-15 mm long, entire or incised, soon reflexed; involucels similar to the involucre but smaller; petals white; stylopodium low-conic; fruit glabrous, 2-3 mm long, ribbed but not winged. Known from Strawberry Valley, West Fork Duchesne River drainage, and from the north slope of the Uinta Mtns. south of Mtn. View, Wyoming; infrequent in meadow and sagebrush-grass communities; 7,500-7,700 ft; June-Aug. Our plants belong to ssp. *borealis* Chuang & Const.

Sium L. — Waterparsnip

Sium suave Walter — hemlock waterparsnip — Perennial, caulescent herbs from fibrous roots; leaves pinnate or occasionally partly bipinnate with 4-6 opposite pairs of sessile lateral leaflets, the lower petioles to 25 cm long, often septate, the upper ones smaller and sometimes reduced to dilated sheath; lower blades 14-32 cm long, the upper ones reduced; leaflets 2-8 (15) cm long, (1) 3-8 (20) mm wide, linear to lanceolate, sharply and uniformly serrate to pinnatifid with linear segments; peduncles 4-10 cm long; umbels 3-11 or more per stem; involucre of 1-6 separate, often reflexed bracts 2-9 mm long; rays 11-24, 1.5-3 cm long; involucels of (2) 5-12 separate bractlets 2-5 mm long; pedicels 2-8 mm long; petals and stamens white; styles about 1 mm long; fruit 2-3 mm long, the ribs prominent. The few specimens seen are from the Blacks Fork drainage of the north slope of the Uinta Mtns.; emergent in ponds or in mud of drying ponds. Specimens seen from the south slope that have been identified as this taxon belong to *Angelica pinnata.*

Zizia Koch.

Zizia aptera (A. Gray) Fernald — meadow zizia — Perennial, caulescent, glabrous herbs from fascicles of fleshy roots; leaves simple at least the basal) or ternate, serrate, the upper ones often lobed; umbels rather compact, usually solitary and terminal, sometimes also lateral, the rays to 3.5 cm long in fruit; flowers yellow; fruit 2-4 mm long, compressed laterally, glabrous, the ribs all filiform. Specimens seen are from Strawberry Valley and Mill Creek drainage, n. slope Uinta Mtns. (C. Refsdal 5004) in wet meadow and streamside communities at 7,500-7,800 ft; June-July.

APOCYNACEAE Dogbane Family

Plants perennial, with milky, acrid juice; leaves cauline, simple, opposite or alternate, entire; flowers bisexual, regular, mostly in corymbose cymes; calyx 5-lobed; corolla 5-lobed, united at the base, convolute in bud; stamens 5, attached to the corolla tube; ovaries superior, of 2 separate carpels, the carpels joined by their styles and with a common stigma; fruit of 2 follicles, sometimes only 1 developing.

1   Leaves opposite . . . .*Apocynum*
1   Leaves alternate . . . .*Amsonia*

Amsonia Walt.

Amsonia jonesii Woodson (*A. latifolia* M. E. Jones) — Jones amsonia, Jones bluestar — Herbs from thickened often woody roots, 20-50 cm tall, glabrous or pubescent; petioles about 4-5 mm long; leaf blades 3-6 cm long, ovate to lanceolate, glaucous; flowers many, densely crowded; calyx lobes 1-2 mm long; corolla white or nearly so, the tube 6-8 mm long, the lobes 3-6 mm long; follicles 5-10 cm long, terete. Occasional from near Roosevelt east to Dinosaur National Monument and near Gate Canyon of the West Tavaputs Plateau and east to Willow Creek; desert shrub, sagebrush, and pinyon-juniper communities, often on sandy or white shaly soils and gravelly pediments; 4,750-7,000 ft; late April-May. Specimens from near Roosevelt (Goodrich 27839) have pubescent leaves as found in *A. tomentosa* Torrey and Fremont, but the pods are not constricted between the seeds. These specimens are here considered to be part of *A. jonesii*.

Apocynum L. — Dogbane, Indian Hemp

Plants herbaceous or slightly woody at the base; corollas bearing 5 distinct appendages within, these adnate to the corolla tube and opposite the lobes; anthers united to the stigma; follicles terete, elongate; seeds with dense white hairs.

---

1   Corolla about 3 times longer than the calyx, usually 4-12 mm long, white or pinkish, the lobes spreading to recurved; follicles 6-15 cm long .... *A. androsaemifolium*

1   Corolla less than 2 times as long as the calyx, 2-5 mm long, white, the lobes erect or only slightly spreading; follicles 12-20 cm long............... 2

2   Leaves sessile or nearly so, cordate at the base; follicles less than 12 cm long .... *A. sibiricum*

2   Leaves with distinct petiole, narrowed at the base; follicles over 12 cm long .... *A. cannabinum*

Apocynum androsaemifolium L. (*A. scopulorum* Greene) — spreading dogbane — Strawberry Valley, Uinta Mtns. and E. Tavaputs Plateau; occasional to locally common, usually in or near aspen or coniferous woods; 7,200-9,000 ft; June-Aug. Our plants apparently all belong to var. *androsaemifolium*.

Apocynum cannabinum L. (*A. sibiricum* Jacquin; *A. hypericifolium* Aiton) — dogbane, Indian hemp — widespread but localized; forming patches (clones) along ditches, agricultural lands, and in riparian and marsh communities; 4,700-6,500 ft; June-Aug. This taxon freely intergrades with *A. androsaemifolium*, and hybrid segregates of these two have been called *A. floribundum* Greene and *A. medium* Greene.

Apocynum sibiricum Jacquin — Siberian dogbane — The one specimen seen (B. Welsh & Moore 238) is from Ouray National Wildlife Refuge, disturbed marsh community at 4,700 ft. In addition to features listed in the key, the flowers of *A. sibiricum* are greenish white or often pale yellow. Flowers are greenish white to cream but not yellow in *A. cannabinum*. Listed in synonymy under *A. cannabinum* in TPD.

ASCLEPIADACEAE Milkweed Family

Perennial herbs with milky juice; leaves simple, usually entire, alternate, opposite or whorled; flowers bisexual, regular, usually in umbels; calyx 5-

parted or divided; corolla of united petals, 5 lobed, the lobes often reflexed; stamens 5, inserted on the base of the corolla and closely united around the pistils into a tube that is fused above with the style column, these forming separate or united often hood-like appendages (hoods); pistils 2; ovary superior; fruit of paired follicles, only one usually developing; seeds numerous, each with a tuft of long silky hairs.

Asclepias L. — Milkweed

---

1   Leaves mostly less than 4 times longer than wide 2

1   Leaves mostly 5 or more times longer than wide 4

2   Plants decumbent, seldom over 20 cm tall; follicles 3-5 cm long . . . .
. . . . *A. cryptoceras*

2   Plants erect, 20-100 cm tall or taller; follicles 8-12 cm long 3

3   Follicles with soft subulate processes; hoods 10-13 mm long, attenuate-acuminate; plants widespread, more or less weedy along ditches and in cultivated places . . . . *A. speciosa*

3   Follicles smooth or nearly so; hoods 5-6 mm long, not attenuate-acuminate; plants known from the East Tavaputs Plateau, apparently rare . . . *A. hallii*

4   Umbel solitary and terminal on a long peduncle, conspicuously exceeding the leaves; leaves mostly alternate . . . . *A. asperula*

4   Umbels usually 2 or more, born on leafy stems; leaves various 5

5   Leaves less than 4 mm wide, mostly whorled, glabrous or nearly so . . . . . .
. . . . *A. subverticillata*

5   Leaves, at least some, over 4 mm wide, mostly alternate or opposite, often pubescent 6

6   Corolla white, greenish, or cream; follicles pendulous . . . . . *A. labriformis*

6   Corolla pink or occasionally white; follicles erect . . . . *A. incarnata*

Asclepias asperula (Decaisne) Woodson (*A. capricornu* Woodson) — spider milkweed — Apparently widespread but uncommon; the 6 specimens seen are from Daggett, Rio Blanco, and Uintah Cos.; pinyon-juniper communities and road-cuts; 5,000-7,000 ft; July-Aug. Plants of our area belong to var. *asperula*.

Asclepias cryptoceras S. Watson — Pallid milkweed — Occasional; widespread; desert shrub and pinyon-juniper communities, often on geologic formations that weather to badlands; 4,800-6,500 ft; May-June. Our plants belong to var. *cryptoceras*.

Asclepias hallii A. Gray — Hall milkweed — 2 specimens seen. One (Wiley-Eberle & England 0437) is from 2.5 mi sw. from Rio Blanco; talus slopes of Green River Shale at 7,200 ft; the other (Goodrich 25897) is from Seep Ridge Road at 6400 ft. July.

Asclepias incarnata L. — swamp milkweed — The one specimen (Riedel 124 BRY) seen is from wetlands near Jose Cabin on Cub Creek where there is small stand of this species. Our plants belong to ssp. *incarnata*.

Asclepias labriformis M. E. Jones — Utah milkweed — Ouray to Kennedy Wash and north to the intercept of the Green River and Highway 45; locally common along roadsides, washes, and floodplain of the Green River, mostly on sandy soil; 4,660-4,890 ft; June.

Asclepias speciosa Torrey — showy milkweed — Widespread; somewhat weedy, moist ground, especially along roadsides, ditches, fence lines, and disturbed riparian communities; up to about 7,000 ft and rarely to 8,480 ft; June-July.

Asclepias subverticillata (A. Gray) Vail — horsetail milkweed — The one specimen seen (Erickson 901) is from Piceance drainage from a ridge between Dry Fork and Hay Gulch; sandy, clay soil; 6,700 ft; Aug.

ASTERACEAE (COMPOSITAE) Sunflower Family

Plants annual, biennial or perennial, herbs or shrubs; leaves basal alternate, opposite, or whorled; flowers borne in heads subtended by an involucre of separate or united bracts; calyx modified to a pappus or none, the pappus of awns, scales, or capillary bristles, these simple or plumose; corollas of 2 types, one type (disk flowers) tubular, regular, and mostly 5-lobed, the other type (ray flowers or rays) flattened and strap-shaped and 2-5 toothed at the apex; stamens usually 5, inserted on the corolla, united by their anthers or sometimes by their filaments; ovary inferior, 1-celled, 1-ovuled; styles usually 2 branched; fruit an achene, the achene often bearing a pappus.

---

1 Plants with woody stems extending well above ground level; heads with disk flowers only (except in *Gutierrezia*) .... **KEY 1**

1 Plants not woody, sometimes with a woody caudex, but the woody portion not much above ground-level; heads with or without rays 2

2 Leaves all or nearly all basal, cauline leaves (if present) bract-like; heads often solitary .... **KEY 2**

2 Leaves not all basal; heads various 3

3 Some or all leaves opposite or whorled .... **KEY 3**

3 Stem leaves all alternate 4

4 Plants spiny or prickly .... **KEY 5**

4 Plants not spiny or prickly....................................... 5

5 Leaves compound or 2-3 times pinnatifid or palmatifid .... **KEY 4**

5 Leaves simple, or once ternate to once pinnatifid or palmatifid 6

6 Plants with milky juice; heads with rays only; pappus of capillary bristles except in *Cichorium* and *Microseris* .... **KEY 6**

6 Plants without milky juice; heads with some or all disk flowers; pappus various.. 7

7 Heads with ray and disk flowers .... **KEY 7**

7 Heads with disk flowers only or the rays inconspicuous .... **KEY 8**

— KEY 1 —
Plants with woody stems

1 Leaves aromatic with odor of sagebrush, lobed, to divided, entire only in *Artemisia cana;* pappus lacking; involucres 2-5 (7) mm high ..... ***Artemisia***

1 Leaves not with odor of sagebrush, entire and mostly linear or at least narrow except in *Brickellia microphylla;* pappus present, of capillary bristles or of scales in *Gutierrezia;* involucres sometimes larger than above 2

2 Involucral bracts striate; corollas greenish white or cream; plants glandular, strongly aromatic . . . . **Brickellia**

2 Involucral bracts not striate; corollas yellow; plants glandular or not. . . . 3

3 Pappus of scales; heads mostly with 1-4 rays; stems and leaves glabrous or at most scabrous, and somewhat viscid . . . . **Gutierrezia**

3 Pappus of capillary bristles; rays lacking; stems and leaves glabrous and viscid to densely pubescent 4

4 Involucral bracts 4-6, equal, in a single series, not imbricate; flowers 4 per head, or if more than 4 then plants usually spiny; achenes pubescent with long hairs . . . . **Tetradymia**

4 Involucral bracts usually more than 6, slightly to strongly imbricate; flowers usually more than 4 per head; plants not spiny; achenes glabrous or short-hairy. 5

5 Involucral bracts 10-13 mm long, all about equal, only slightly imbricate; heads 8-12 mm wide; plants known from Rock Creek and N. Fork Duchesne, Uinta Mtns. . . . . **Haplopappus macronema**

5 Involucral bracts mostly shorter, strongly imbricate, the outer shorter than the inner; heads narrower; plants widespread and abundant . . . .
. . . . **Chrysothamnus**

— KEY 2 —
Leaves all basal or nearly so; heads often solitary

1 Leaves exceeding the heads, pinnately divided, the margins translucent-white and irregularly toothed or lobed; plants about 2-4 cm tall . . . .
. . . . **Glyptopleura**

1 Heads exceeding the leaves or else leaves entire, or plants over 5 cm tall 2

2 Heads with rays only; plants with milky juice 3

3 Heads solitary and terminal. 4

4 Flowers orange . . . . **Agoseris aurantica**

4 Flowers yellow 5

5 Leaves laciniate 6

6 Plants mostly glabrous except sometimes on lower midrib of leaves . . . . .
. . . . **Taraxacum**

6 Plants pubescent . . . . **Agoseris**

5 Leaves entire or toothed but not laciniate . . . . **Agoseris**

3 Heads 2 to several per stem 7

7 Corollas exserted more than 1 mm beyond the involucres **Crepis runcinata**

7 Flowers not exerted more than 1 mm beyond the involucre . . . .
. . . . **Hieracium triste**

2 Heads with some or all disk flowers; plants without milky juice. 8

8 Leaves dissected or at least deeply cut. 9

9   Leaves once pinnatifid, at least some of the segments over 3 mm wide ...
.... ***Balsamorhiza***

9   Leaves not as above..10

10 Involucral bracts glabrous, the outer ones spreading to reflexed ....
.... ***Thelesperma***

10 Involucral bracts with hairs, all erect 11

11 Leaves ternately or palmately lobed or dissected .... ***Erigeron compositus***

11 Leaves twice pinnately dissected ....***Hymenopappus***

8   Leaves simple and entire 12

12 Heads with disc flowers only (rays if present minute) 13

13 Scapes lacking; heads sessile in the basal rosette of leaves ...***Parthenium***

13 Scapes at least 2 cm long or longer................................14

14 Plants 2-8 cm tall ....***Chamaechaenactis***

14 Plants 10-40 cm tall or taller ....***Enceliopsis nutans***

12 Heads with ray flowers..15

15 Rays not yellow ....***Erigeron***

15 Rays yellow..16

16 Pappus of capillary bristles 17

17 Involucral bracts in a single series; leaves usually not linear ***Senecio canus***

17 Involucral bracts in 2 or more series; leaves linear or nearly so ....
.... ***Haplopappus***

16 Pappus not of capillary bristles....................................18

18 Leaf blades less than 1 cm wide, without a distinct petiole .... ***Hymenoxys***

18 Leaf blades over 1 cm wide with well developed petioles that are abruptly
different from the blades 19

19 Leaf blades 2-6 cm long ....***Enceliopsis nudicaulis***

19 Leaf blades 10-40 cm long ....***Balsamorhiza***

— KEY 3 —
Some or all leaves opposite or whorled

1   At least the lower leaves whorled ....***Eupatorium***

1   Leaves opposite, at least the lower, the upper ones sometimes alternate . 2

2   Leaves, at least some, compound or deeply dissected..................3

2   Leaves simple, not dissected 6

3   Leaves mostly clustered on the lower 1/2 of the stem ....***Thelesperma***

3   Leaves somewhat equally distributed on lower and upper parts of the stem
....4

4   Leaves with 3-5 leaflets ....***Bidens frondosa***

4   Leaves more dissected than above 5

5   Involucres dotted with yellow glands, rays yellow, glandular dotted, about
equal to the disk flowers; pappus of scales that are cleft with 5-10 bristles
.... ***Dyssodia***

5   Involucres not glandular dotted, rays lacking; pappus lacking ..***Ambrosia***

6 Rays well developed, yellow. 7

6 Heads with disk flowers only. 13

7 Primary involucral bracts in a single series; pappus of numerous capillary
   bristles . . . . **Arnica**

7 Involucral bracts usually in more than 1 series; pappus not of capillary
   bristles. 8

8 Leaves serrate to incised, sessile or connate at the base, usually surpassing
   the heads . . . . **Bidens cernua**

8 Leaves entire or inconspicuously toothed or heads definitely surpassing the
   leaves. 9

9 Disk flowers dark purple, brown or black 10

9 Disk flowers yellow 11

10 Plants annual; rays 15-40 mm long . . . . **Helianthus**

10 Plants perennial; rays 10-13 mm long . . . . **Helianthella microcephala**

11 Stems with 3-5 pairs of leaves; leaves mostly all opposite, the lower ones
    12-50 cm long, with a midnerve and 1-2 pairs of lateral nerves prominent
    . . . . **Helianthella**

11 Stems usually with more leaves, the upper ones sometimes alternate, only
    the midnerve prominent or leaves less than 12 cm long 12

12 Rays 7-17 mm long; leaf blades 3-10 cm long; plants montane, usually on
    well drained soil, 25-70 (100) cm tall . . . . **Heliomeris**

12 Rays 20-30 mm long; leaf blades 5-16 cm long; plants mostly in lowland
    meadows and other wet places, 30-150 cm tall . . . . **Helianthus nuttallii**

13 Leaves sessile, serrate to incised, usually much surpassing the heads . . . .
    . . . . **Bidens cernua**

13 Leaves not sessile, or if so then entire 14

14 Pappus of capillary bristles; involucral bracts striate, over 7 mm long . . . .
    . . . . **Brickellia**

14 Pappus lacking; involucral bracts not striate, not over 5 mm long . . . . . **Iva**

— KEY 4 —
**Stem leaves alternate, compound or 2-3 times pinnatifid or palmatifid,
not spiny or prickly**

1 Heads with rays 2

1 Rays none 8

2 Rays yellow. 3

2 Rays white, pink or blue. 4

3 Rays (1) 1.5-3.5 cm long, tridentate at the apex; plants about 15-25 cm tall,
   perennial . . . . **Hymenoxys**

3 Rays 0.6-1.2 cm long, not tridentate; plants 25-100 cm tall, annual . . **Bahia**

4 Rays less than 4 mm long . . . . **Achillea**

4 Rays over 4 mm long. 5

5 Involucral bracts widely spreading at the tips; rays bluish . . . .
   . . . . **Machaeranthera**

5   Involucral bracts not spreading; rays white or bluish..................6

6   Heads solitary; pappus of capillary bristles ....*Erigeron compositus*

6   Inflorescence with more than 1 head; pappus none or a low crown......7

7   Achenes with 2 marginal and 1 ventral callous-thickened almost wing-like ribs; receptacle without chaffy bracts ....*Matricaria*

7   Achenes with about 10 obscure, turbiculate ribs; inner ½ of receptacle with persistent, chaffy that are about equal to the length of the flowers *Anthemis*

8   Pappus of capillary bristles; heads solitary ....*Erigeron compositus*

8   Pappus not of capillary bristles; inflorescence usually with more than 1 head 9

9   Involucres 5-16 mm high; pappus of hyaline scales...................10

9   Involucres to 4 (5) mm high; pappus lacking or minute 11

10 Flowers white or pink; pappus scales 4-10, 2-5 mm long; terminal segment of leaves less than 1 cm long ....*Chaenactis*

10 Flowers yellow; pappus scales 10-20, about 1-2 mm long; terminal segment of leaves sometimes over 1 cm long ....*Hymenopappus*

11 Flowers unisexual, the staminate borne in terminal spikes, the pistillate borne singly or in small clusters in the axils of leaves; fruit bur like, with spines at maturity ....*Ambrosia*

11 Flowers not as above; fruit without spines 12

12 Plants annual; receptacle strongly conical ....*Matricaria*

12 Plants perennial or biennial......................................13

13 Flowers yellow; leaf blades 6-15 cm long ....*Tanacetum*

13 Flowers not especially yellow or leaf blades often less than 6 cm long ....
     ....*Artemisia*

— KEY 5 —
Stem leaves alternate, simple or once pinnatifid;
plants spiny or prickly

1   Plants with milky juice; rays yellow; disk flowers none; involucral bracts not spiny; leaves sessile, often auriculate clasping 2

1   Plants without milky juice; rays bluish or lacking; disk flowers present ..3

2   Rays less than 6 mm long, 6-8 per head, withering and inconspicuous in the heat of the day; leaves prickly on the midrib ....*Lactuca serriola*

2   Rays over 6 mm long; over 10 per head, usually conspicuous in the day; leaves usually not prickly on the midrib ....*Sonchus*

3   Leaves not spiny; spines confined to the fruiting involucres; heads unisexual, the staminate ones in terminal spikes or racemes, the pistillate ones axillary ....*Ambrosia*

3   Leaves spiny or spinulose-margined; heads bisexual 4

4   Disk flowers yellow; conspicuous rays present in all but 1 species; leaves spinulose toothed but herbage otherwise spineless; plants seldom over 60 cm tall ....*Machaeranthera*

4   Disk flowers whitish, pinkish or purplish, rays lacking; leaves and sometimes stems and involucral bracts spiny; plants often over 60 cm tall
    5

5   Involucral bracts 2-6 (8) mm wide, the lower ones strongly reflexed in age
    ....*Carduus*

5   Involucral bracts mostly less than 3 mm wide erect to widely spreading but
    not reflexed 6

6   Pappus bristles not plumose ....*Onopordum*

6   Pappus bristles plumose ....*Cirsium*

— KEY 6 —

Stem leaves alternate, simple; plants with milky juice;
heads with ray flowers only

1   Leaves exceeding the heads, pinnately divided, the margins translucent
    white and irregularly toothed or lobed, plants 2-4 cm tall ... *Glyptopleura*

1   Plants not as above in all respects 2

2   Flowers yellow, orange, or white 3

2   Flowers pink, red, blue, or purple.....................................8

3   Involucral bracts 2.5-4 cm long in flower, 4-7 cm long in fruit . *Tragopogon*

3   Involucral bracts less than 2.5 cm long 4

4   Plants from creeping stems, known from alpine talus slopes of the Uinta
    Mtns.; flower heads more or less nested in the basal leaves ... *Crepis nana*

4   Plants not as above; head well exserted above the basal leaves.........5

5   Plants annual, of desert communities ....*Malacothrix*

5   Plants perennial, mostly of montane or moist areas and less commonly of
    desert communities 6

6   Involucre not over 1 cm high; rays 3-10 mm long ....*Hieracium*

6   Involucre 8-21 mm high; rays often over 10 mm long.................7

7   Herbage glabrous or slightly scurfy; pappus of scales with plumose bristles
    ....*Microseris*

7   Herbage pubescent at least in part except in *C. runcinata;* pappus of simple
    capillary bristles ....*Crepis*

8   Pappus of scales; heads sessile or short pedunculate, 1-3 in the axils of much
    reduced bract-like leaves, the rays strongly spreading ....*Cichorium*

8   Pappus of capillary bristles; heads usually on elongate peduncles, solitary,
    the rays erect or spreading 9

9   Involucres not over 1 cm high; lower leaves seldom over 5 cm long, upper
    leaves reduced and bract-like; rays 3-12 mm long...................10

9   Involucres 1-2.5 cm high; leaves various; rays usually 12-40 mm long...11

10 Pappus of plumose, capillary bristles ....*Stephanomeria*

10 Pappus of simple, capillary bristles ....*Prenanthella*

11 Leaves to 5 mm wide; heads borne singly at the ends of branches or stems;
    rays 5-10 per head, mostly 2.5-4 cm long ....*Lygodesmia*

11 Leaves 6-70 mm wide; inflorescence more or less racemose or paniculate;
    rays 18-50 per head, to 1.4 cm long ....*Lactuca tatarica*

— KEY 7 —
Stem leaves alternate, mostly simple; plants without milky juice; heads with ray and disc flowers

1 Disc flowers on a columnar receptacle.............................2

2 Leaves pinnatifid ....*Ratibida*

2 Leaves palmatifid ....*Rudbeckia*

1 Disc flowers on a flattened or rounded receptacle....................3

3 Rays blue, pink, white, or reddish; pappus lacking or of mostly capillary bristles..4

3 Rays yellow or orange 12

4 Pappus lacking; leaves pinnately lobed; rays white ......*Chrysanthemum*

4 Pappus well developed; leaves entire, or toothed with spinulose teeth and then the rays bluish..5

5 Involucres 2-4 mm high; plants annual, 40-100 cm tall; heads usually numerous, the rays inconspicuous ....*Conyza*

5 Involucres larger; plants perennial, or if annual then usually shorter than above and with rays well developed................................6

6 Involucre bracts mostly in a single series, usually numerous, mostly of the same color throughout, linear; rays over 20 and often 40-150 per head ...
   ....*Erigeron*

6 Involucral bracts imbricate (overlapping) in 2 or more series, sometimes bicolored, sometimes not linear; rays mostly less than 20 per head, but to 30 in some taxa...7

7 Plants seldom over 15 cm tall; involucral bracts scarious and ciliate, erose, or fimbriate at the margins; leaves not over 4 cm long, not over 4 mm wide; heads 1-3 per stem..8

7 Plants mostly over 15 cm tall; involucral bracts not ciliate or fimbriate at the margins; leaves often different from above; heads sometimes over 3 per stem...9

8 Basal leaves lacking; stem leaves several ....*Chaetopappa*

8 Basal rosette of leaves well developed, stems lacking or with few leaves ..
   ....*Townsendia*

9 At least some of the leaves spinulose toothed ....*Machaeranthera*

9 Leaves entire or if toothed the teeth not spinulose...................10

10 Plants from woody, branching caudices, of deserts; leaves villose or strigose
  11

10 Plants from rhizomes and fibrous roots or annual from a taproot; habitat various but often montane or moist or wet; leaves not villous ......*Aster*

11 Involucres 9-18 mm high, the bracts spinulose-acuminate ......*Xylorhiza*

11 Involucres 6-9 mm high, the bracts acute but not spinulose-acuminate ...
   ....*Erigeron pulcherrimus*

12 Pappus of capillary bristles 13

12 Pappus not of capillary bristles....................................19

13 Involucral bracts in single series, equal, or with a few much reduced outer ones, blackish at the tips in a few species; leaves entire, serrate, or pinnatifid
   ....*Senecio*

13 Involucral bracts in 2 or more series and imbricate (overlapping), not black at the tips; leaves entire or toothed in *Solidago* 14

14 Leaves pubescent on both sides with rather stiff, short, and appressed hairs (strigose); stems strigose throughout .... *Heterotheca*

14 Leaves glabrous at least on 1 side or merely puberulent; stems glabrous to puberulent 15

15 Ray flowers 5-14 mm long 16

15 Ray flowers 1-4 mm long 17

16 Outer bracts linear or linear-lanceolate .... *Haplopappus*

16 Outer bracts ovate or ovate-lanceolate .... ***Solidago parryi***

17 Rays 1-3 per head; leaves entire; stems not over 20 cm tall, usually several from a caudex .... ***Petradoria***

17 Rays more than 4 per head; leaves sometimes toothed; stems 10-100 cm tall, usually 1-few from taproots or rhizomes 18

18 Leaves resinous punctate .... ***Euthamia***

18 Leaves not resinous punctate .... ***Solidago***

19 Rays 2-3 mm long; leaves linear, entire, not over 7 mm wide 20

19 Rays longer; leaves various 21

20 Plants annual, tar-scented and malodorous; herbage with stalked greenish yellow glands .... ***Madia***

20 Plants perennial, sub-shrubs, not tar-scented .... ***Gutierrezia***

21 Some or all or the disk flowers reddish, purplish, brown, or black 22

21 Disk flowers yellow... 24

22 Rays 3 lobed or toothed at the apex, sometimes reddish at base; disk flowers reddish or purplish .... ***Gaillardia aristata***

22 Rays not 3 lobed at apex, yellow throughout; disk flowers brown or black 23

23 Plants perennial, not weedy; leaves entire; rays 10-13 mm long .... ***Helianthella microcephala***

23 Plants annual; more or less weedy; leaves entire or serrate; rays 15-40 mm long .... ***Helianthus***

24 Plants strongly resinous; some leaves usually serrate; involucral bracts more or less squarrose (strongly spreading or recurved) .... ***Grindelia***

24 Plants not strongly resinous; leaves entire or divided into linear segments; involucral bracts not squarrose 25

25 Leaves pinnately divided .... ***Gaillardia flava***

25 Leaves not as above, mostly entire................................ 26

26 Plants long-villose or floccose at least in part; leaves not over 10 cm long .... ***Psilostrophe***

26 Plants not pubescent as above or if so the larger leaves 10-30 cm long .. 27

27 Leaves with decurrent or clasping bases............................ 28

27 Leaves not decurrent, not clasping................................ 29

28 Leaves decurrent .... ***Helenium***

28 Leaves clasping but not decurrent .... ***Hymenoxys hoopesii***

29 Rays 6-14 mm long; involucres 7-11 mm high . . . . *Platyschkuhria*
29 Rays 18-45 mm long; involucres 20-40 mm high . . . . *Wyethia*

— KEY 8 —
**Stem leaves alternate, simple; plants without milky juice;
heads with disc flowers only or rays inconspicuous**

1 Pappus of capillary bristles 2
1 Pappus not of capillary bristles. 13
2 Leaves spinulose-toothed; strigose or glabrous; plants 5-20 (30) cm tall . .
    . . . . *Machaeranthera grindelioides*
2 Leaves not spinulose toothed 3
3 Plants more or less white or gray tomentose; leaves entire; inflorescence
    compact with the heads touching or close together, or solitary on stems not
    over 3 cm long 4
3 Plants not as above in all features. 6
4 Plants annual from taproots . . . . *Gnaphalium*
4 Plants perennial, usually from stolons or rhizomes 5
5 Lower involucral bracts strongly spreading or reflexed, pearly white above
    the middle; leaves usually green above and gray or white below Anaphalis
5 Involucral bracts erect, ascending, or slightly spreading, white or other
    color; leaves about the same color on both sides except in *A. marginata* . . .
    . . . . *Antennaria*
6 Heads including the flowers 2-4 mm high . . . . *Conyza*
6 Heads including flowers over 4 mm high 7
7 Disk flowers yellow, heads including flowers 5-12 mm high 8
7 Disk flowers white, cream, pink or purple, not hidden in the pappus; plants
    perennial; heads including the flowers often over 12 mm high 11
8 Involucral bracts imbricate in 3 or more series 9
8 Involucral bracts more or less in a single series. 10
9 Involucres 2-3 mm wide . . . . *Petradoria*
9 Involucres mostly over 3 mm wide . . . . *Aster*
10 Heads nodding; plants glabrous . . . . *Senecio pudicus*
10 Heads not nodding; plants sparsely to densely spreading hairy . . *Erigeron*
11 Corolla lobes over 3 mm long . . . . *Centaurea*
11 Corolla lobes less than 3 mm long. 12
12 Pappus bristles obviously plumose, the side cilia about 0.5 mm long; plants
    known from the E. Tavaputs Plateau . . . . *Kuhnia*
12 Pappus bristles not plumose; plants widespread in our area . . . . *Brickellia*
13 Some leaves over 12 cm wide; involucres covered with hooked bristles . . .
    . . . . *Arctium*
13 Leaves smaller; involucres without hooked bristles 14
14 Some or all of the leaves divided into long, linear segments; plants 1-2 m
    tall . . . . *Oxytenia*
14 Leaves not as above; plants mostly shorter. 15

15 Heads cone-like, 3-6 cm long .... *Rudbeckia occidentalis*

15 Heads less than 2 cm long 16

16 Plants tar scented and strongly malodorous, annuals; herbage with stalked yellow-green glands .... *Madia*

16 Plants sometimes aromatic but not tar scented, perennials or biennials, without glands as above 17

17 Plants sub-shrubs with entire, glabrous or puberulent leaves . *Gutierrezia*

17 Plants herbs; leaves often pubescent or toothed, lobed, or divided (entire and glabrous in *Artemisia dracunculus*) 18

18 Leaves finely serrate, otherwise uncut .... *Chrysanthemum*

18 Leaves entire or lobed to pinnatifid or palmatifid but not serrate 19

19 Corollas lacking; stamens naked in the involucres; heads unisexual, the staminate ones in terminal spikes or racemes, the pistillate ones axillary; involucres of the pistillate heads bur-like, with short spines .... *Ambrosia*

19 Corollas present; stamens included in the corolla or rarely lacking; heads mostly bisexual, not axillary or bur-like, without spines .... *Artemisia*

Achillea L. — Yarrow; Milfoil

Achillea millefolium L. (*A. lanulosa* Nuttall) — yarrow — Perennial, villous, aromatic herbs from slender rootstocks; stems 10-60 cm tall; leaves 3-10 cm long, finely dissected, the ultimate segments linear; inflorescence flat-topped; involucres 4-6 mm high; rays about 5, 2-3.5 mm long, mostly white, rarely pink; disk flowers 10-20. Widespread and common in many plant communities including sagebrush, mt. brush, riparian, meadow, aspen, coniferous forest, and alpine tundra; 6,000-11,100 ft and probably higher; June-Aug.

In FNA (19: 493) subspecific names are included in synonymy under an expanded view of *A. millefolium* where it is noted that at least 58 names have been applied to North American specimens. A note in IMF (5:134) includes "The intricate pattern of morphologic, cytologic, geographic, and ecologic variation within the species has frustrated all efforts to organize an infraspecific taxonomy on a circumboreal or even a strictly North American basis." Both FNA and IMF indicate trends in morphological differences are apparent, but these trends are difficult to translate to distinct taxa. If infraspecific status is applied, IMF (5: 134) indicates ssp. *lanulosa* (Nuttall) Piper would be applicable for our plants. Our native plants are included in var. *alpicola* (Rydberg) Garrett and var. occidentalis de Candolle in TPD. High elevation plants tend to have dark involucral bracts, fewer heads, and lower stature. These plants have been treated as var. *alpicola,* but they intergrade completely with plants of lower elevations (AUF 146). Introduced plants of this complex have been used as ornamentals and in reclamation seedings. These include plants with red, yellow, and white rays.

Agoseris Rafinesque — Mountain Dandelion

Perennial, mostly scapose herbs with milky juice; leaves all basal; heads solitary; involucral bracts in series, the outer shorter and broader than the inner, with ray flowers only; pappus of numerous, white, simple, capillary bristles; achenes 10-ribbed.

1 Achenes narrowed to a long slender beak, the beak equal or longer than the body; flowers yellow to orange, drying to red or purple .... **A. aurantiaca**
1 Beak of achenes obsolete or to about 1/2 as long as the body; flowers yellow ....**A. glauca**

Agoseris aurantiaca (Hooker) Greene [*A. arizonica* Greene; *A. gracilens* (A. Gray) — orange mtn. dandelion — Widespread but rather infrequent; sagebrush, aspen, coniferous forest, and parkland communities; 6,000-11,000 ft. June-Aug.

Agoseris glauca (Pursh) Rafinesque [*A. elata* (Nuttall) Greene misapplied (Graham 1937)] — pale mtn. dandelion — With 4 more or less intergrading vars.:

1 Plants glabrous or ciliate on the petioles and lower part of the leaves, 10-65 cm tall; leaves strongly acute or acuminate, entire or with a few teeth, not laciniate ....**var. glauca**
1 Plants more or less pubescent, if rather sparsely so then leaves laciniate, seldom over 25 cm tall...2
2 Leaves mostly laciniate, mostly lanceolate, mostly acute or acuminate ..
....**var. laciniata**
2 Leaves entire or sometimes weakly laciniate below, mostly oblanceolate or broader, more or less obtuse 3
3 Involucral bracts rather densely pubescent, not particularly purple .....
....**var. dasycephala**
3 Involucral bracts glabrous or pubescent at the base, strongly purple (anthrocyanic) throughout or at least in a central stripe . **var. cronquistii**

Var. cronquistii S. L. Welsh — purple mtn. dandelion — Occasional to common across the Uinta Mtns. 10,600-12,000 ft, mostly alpine and near alpine.

Var. dasycephala (Torrey & A. Gray) Jepson — short mtn. dandelion — The 13 specimens seen are from the Uinta Mtns.; alpine, krummholz, and coniferous forest parkland communities at 10,600-12,000 ft; July-Aug.

Var. glauca — wetland mtn. dandelion — Widespread; common in meadow and willow-streamside and other riparian communities, occasionally with sagebrush but then mostly near meadows; 5,400-10,300 ft; June-July.

Var. laciniata (D.C. Eaton) Smiley [*A. taraxacifolia* (Nuttall) D. Dietrich] — laciniate mtn. dandelion — Widespread; common in sagebrush, pinyon-juniper, mt. brush, and aspen communities; 5,800-9,000 ft; late May-Aug.

Ambrosia L. — Ragweed; Bur-sage

Annual or perennial herbs; leaves opposite or alternate, mostly lobed or dissected; heads unisexual or occasionally bisexual, the staminate heads in terminal spikes or racemes, the involucres saucer-shaped or hemispheric, of 5-12 partly united bracts, the anthers not united; pistillate heads below the staminate ones, axillary, 1-flowered, enclosed by a nut-like or bur-like involucre, the involucre with tubercles or spines near the apex; corollas lacking; pappus lacking.

1   Plants annual from taproots 2

1   Plants perennial from rhizomes 2

2   Lower stems and leaves with pustular-based, stiff, multicellular hairs; spines of burs in more than 1 series .... *A. acanthicarpa*

2   Lower stems without pustular-based hairs; spines of burs in one series ... . *A. artemisiifolia*

3   Leaves alternate, bicolored, the lower surface covered with appressed white hairs .... *A. tomentosa*

3   Leaves, at least the lower, opposite, not covered with appressed white hairs .... *A. psilostachya*

Ambrosia acanthicarpa Hooker [*Franseria acanthicarpa* (Hooker) Coville] — bur ragweed — Widespread; locally common; desert shrub, sagebrush, and piny-on-juniper communities, often on sandy soil on roadsides and other disturbed sites; 4,800-5,800 ft; mid Aug.-early Oct.

Ambrosia artemisiifolia L. — common ragweed — Duchesne Co. is included in the distribution of this species in AUF (149) but no specimen of this species was found at BRY in 2011.

Ambrosia psilostachya de Candolle — western ragweed — the one specimen seen (F. Smith & K. Snyder 2149 BRY!) is from a cottonwood/rabbitbrush community in Cowboy Canyon along the White River at 5,100 ft.

Ambrosia tomentosa Nuttall (*Franseria discolor* Nuttall; *F. tomentosa* A. Gray) — skeleton leaf ragweed — Occasional to locally common along the floodplain of the Green River from Dinosaur National Monument to s. of Ouray, specimens also seen from Blind Stream, Uinta Mtns., and Matt Warner Reservoir, Diamond Mt.; roadsides and other disturbed places; 4,700-8,950 ft; June-Aug. Holmgren and others 446 DINO!, reported as *Franseria linearis* Rydberg [*A. linearis* (Rydberg) Payne] by Holmgren (1962) belongs here.

Anaphalis D.C. — Pearly-everlasting

Anaphalis margaritacea (L.) Bentham & Hooker [*A. subalpina* (A. Gray) Rydberg] — pearly-everlasting — Perennial, tomentose or floccose herbs from rhizomes; stems erect, leafy, 25-60 cm tall; leaves alternate, 3-10 cm long, linear to narrowly lanceolate, usually green above, white-pubescent beneath; heads in corymbose clusters; involucres 6-7 mm high, of several series of imbricate bracts, the outer shorter, white, scarious, spreading or reflexed and persistent when dry; flowers all discoid, of two types, one type bisexual and sterile with undivided styles, the other type mostly pistillate with a few bisexual flowers in the center of the head; pappus of capillary bristles; anthers united; achenes glabrous. Uinta Mtns.; locally common in lodgepole pine, Engelmann spruce communities, occasionally in aspen woods, sometimes on open rocky ground including scoured boulder-armored channels of streams; 7,600-10,200 ft; July-Aug.

Antennaria Gaertner — Pussytoes

Perennial herbs, unisexual, tomentose; leaves alternate, and usually in a basal rosette, entire, usually narrow; heads in capitate or corymbose clusters;

involucral bracts imbricate in several series, scarious, the tips often whitish or pinkish; staminate heads with bisexual-appearing tubular flowers, and with clavate apical-flattened pappus bristles (except in *A. dimorpha*); pistillate heads with filiform corollas, and with a pappus of capillary bristles, the bristles united at the base and deciduous together; achenes usually glabrous. We have not seen specimens to support the listing by Goodrich and others (1981) of the more southern *A. rosulata* Rydberg. The report of that taxon for along the Yampa River (Potter and others 1983) is based on a vegetative specimen (Y 69!) of *Petrophytum caespitosum*.

---

1   Leaves 2-20 cm long, the larger ones mostly over 4 cm long, distinctly 3 (5) nerved 2

1   Leaves 0.3-3.5 cm long, with 0-1 distinct nerve (rarely with 3 indistinct nerves). 4

2   Stems mostly solitary from short to prolonged rootstocks; plants of wet meadows, bogs, stream sides; involucral bracts more or less tomentose toward the base . . . . *A. pulcherrima*

2   Plants somewhat caespitose, mostly from drier ground than above, the stems clustered on a caudex; involucral bracts various 3

3   Involucral bracts glabrous or nearly so, scarious and whitish to near the base; inflorescence compact to open corymbose; basal leaves 2-5 (8) cm long . . . . *A. luzuloides*

3   Involucral bracts tomentose in the lower 1/2, the lower portion not scarious; inflorescence compact cymose; basal leaves 2.5-19 cm long . . . . 4 . . . . *A. anaphaloides*

4   Heads solitary, sessile in the basal rosettes, or the stem seldom over 3 cm long; stolons lacking or short . . . . *A. dimorpha*

4   Heads more numerous; stems taller or plants with well-developed stolons 5

5   Involucral bracts white or pink; plants (5) 10-30 cm tall 6

5   Involucral bracts dirty brownish green, or blackish green, sometimes whitish at the tip; plants 2-10 (25) cm tall. 8

6   Involucres 7-11 mm high; dry pistillate corollas 5-8 mm long . *A. parvifolia*

6   Involucres less than 8 mm high; dry pistillate corollas 2.5-5 mm long. . . . 7

7   Involucral bracts white with a dark brown or black spot at the base of the scarious portion . . . . *A. corymbosa*

7   Involucral bracts white or pinkish, scarcely or not at all darkened below the scarious portion . . . . *A. microphylla*

8   At least the inner bracts usually abruptly acute to acuminate, the scarious terminal portion brownish green to blackish green throughout . . *A. media*

8   Involucral bracts mostly obtuse at the apex, the scarious terminal portion merely dirty tan or often whitish at the apex . . . . *A. umbrinella*

Antennaria anaphaloides Rydberg — pearly pussytoes — Uinta Mtns. from the Brush Creek drainage to Diamond Mtn. Plateau on the s. slope and Lodgepole Creek east to Goslin Mtn. and Round Mtn. on the n. slope; infrequent in sagebrush-grass, mt. brush, aspen, Douglas-fir, and spruce-aspen communities; 7,300-9,300 ft; June-mid-July.

Antennaria corymbosa E. Nelson — plains pussytoes — Uinta Mtns.; wet meadows, bogs, along streams, and wet areas in coniferous woods; 8,100-11,000 ft; mid June-Aug.

Antennaria dimorpha (Nuttall) Torrey & A. Gray (*A. rosulata* Rydberg misapplied) — low pussytoes — Widespread; desert shrub, sagebrush, pinyon-juniper, and mt. brush communities; 5,000-8,000 ft, one specimen (Harrison 11691) from among rocks at the outlet of Granddaddy Lake at 10,200 ft; April-June (July).

Antennaria luzuloides Torrey & A. Gray — rush pussytoes — Locally common w. of N. Fork Duchesne River in the Uinta Mtns. also on Diamond and Hoy Mtns. at the e. end of the Uinta Mtns.; forb-grass, sagebrush, mtn. brush, ponderosa pine, aspen, and spruce-fir communities, often in aspen and coniferous forest parklands; 7,700-9,900 ft; June-Aug.

Antennaria media Greene [*A. alpina* (L.) Gaertner var. *media* (Greene) Jepson] — alpine pussytoes, snowbed pussytoes — Uinta Mtns.; lodgepole pine and Engelmann spruce woods, rocky slopes, meadows, along streams, and locally abundant in alpine snowbeds and the dominant in an alpine pussytoes snowbed community type (Brown 2006); 10,200-11,500 ft; July-Aug. Apparently passing into *A. umbrinella.*

Antennaria microphylla Rydberg (*A. concinna* E. Nelson; *A. formosa* Greene; *A. rosea* Rydberg) — rose pussytoes — Widespread and common; sagebrush, pinyon-juniper, mt. brush, aspen, meadow, and coniferous forest communities; 6,000-10,600 ft; June-Aug. Plants with rose color involucral bracts have been referred to *A. rosea.* This feature seems variable within populations. FNA (19: 408) maintains *A. rosea* as a species based on pistillate verses dioecious plants with the acknowledgment that staminate plants are rare.

Antennaria parvifolia Nuttall (*A. aprica* Greene) — little leaf pussytoes — Occasional across the Uinta Mtns., flank of these mtns. to Douglas Mt. and upper elevations of the W. Tavaputs Plateau; coniferous woods, edge of meadows, sagebrush, and infrequently riparian communities; 6,500-10,000 ft; late May-July. Antennaria marginata Greene — sandstone pussytoes — is listed for Daggett and Uintah Cos. (AUF 152), but not included in our area in IMF (5: 374) or FNA (19: 405). The 3 specimens seen annotated as such are from lodgepole pine communities, and are here included in *A. parviflora.*

Antennaria pulcherrima (Hooker) Greene — showy pussytoes — Fens and wet meadows on the north slope of the Uinta Mtns. Known from the south slope of the Uinta Mtns. only in a calcareous bog in S. Fork Rock Creek at 9,200 ft; July-Aug.

Antennaria umbrinella Rydberg — umber pussytoes — Occasional or common across the Uinta Mtns., riparian, aspen, Engelmann spruce, Krummholz, alpine and subalpine meadow, snowbed and talus communities; 9,000-11,870 ft; June-Aug. In our opinion many specimens identified as *A. umbrinella* and *A. alpine* could go either way. Cronquist (IMF5: 376) noted *A. umbrinella* to pass freely into *A. microphylla* and *A. media.*

Anthemis L. — Chamomile; Dogfennel

Anthemis cotula L. — Mayweed, dogfennel, stinking chamomile — Annual, possible biennial herbs with rank odor; stems branching, leafy, 20-60 cm tall;

leaves alternate, 2-3 pinnatifid into linear or filiform segments, about 3-5 cm long; heads solitary at the ends of branches; involucres 4-6 mm high, the bracts in 2 series; pappus none; rays about 6-10 mm long, reflexed in age, white; disk flowers yellow; achenes 10-ribbed. Introduced from Europe. The one specimen seen is from a chained and seeded pinyon-juniper community on Brundage Ridge, W. Tavaputs Plateau where seed of this species was likely a contaminant in the seed mix. Mayweed did not persist at this site. Most specimens from our area that have been identified as *Anthemis cotula* belong to *Matricaria maritima.*

Arctium L. — Burdock

Arctium minus (Hill) Bernhard — Burdock — Biennial herbs 50-200 cm tall, with much branched stems; leaves alternate, long petiolate, the blade up to 50 cm long and 40 cm wide, thinly woolly or glabrous in age; heads in axillary raceme-like clusters, the involucral bracts numerous, linear, hooked at the apex, forming a bur; rays lacking, the disk flowers red-violet; pappus of numerous short rigid scale-like bristles. Introduced from Europe, an occasional weed of ditch banks and other moist places, up to about 7,200 ft, not particularly aggressive and usually not found in cultivated areas, although usually adjacent to such areas and particularly on abandoned farmlands; July-Oct.

Arnica L. — Arnica

Perennial herbs; herbage more or less pubescent with short or long crinkled hairs, the hairs often glandular; leaves simple, mostly opposite; heads solitary to several in cymose clusters; involucres hemispheric to turbinate, the bracts equal in length, in 1-2 series; receptacles naked; ray and disk flowers yellow, or rays lacking; pappus of capillary bristles.

1 Rays lacking, lower heads sometimes nodding ....*A. parryi*
1 Rays well developed, heads not nodding.............................2
2 Leaves 3-10 times longer than wide, sessile or gradually tapering to a more or less winged petiole, entire or obscurely toothed, with 3-7 conspicuous ribs or veins running parallel to the margins, hardly if at all netted veined, not cordate, large basal leaves often lacking, or these and the lower cauline ones often withering by anthesis 3
2 Leaves 1-3 times longer than wide, lower stem and basal (if present) leaves abruptly petiolate, or sometimes sessile, often conspicuously toothed, more or less netted veined, without prominent parallel veins, sometimes cordate, large basal leaves sometimes present on the rhizomes.................7
3 Stems with (4) 5-9 pairs of equally spaced leaves, the lower 1 or 2 pairs withering by anthesis, basal leaves lacking...........................
3 Stems with 2-4 pairs of leaves, the leaves equally spaced or crowded toward the base of the stem 4
4 Apex of involucral bracts with a tuft of pilose hairs on the inside ....
   ....*A. chamissonis*
4 Apex of involucral bracts not tufted pilose ....*A. longifolia*
5 Old leaf bases with dense tufts of brown hair in the axils ......*A. fulgens*

5   Old leaf bases not as above...5

6   Heads mostly with 9-18 rays; pappus subplumose, brownish or dirty white; leaves (2) 4-18 cm long, the lower ones more or less petiolate, occasionally connate sheathing; plants 15-65 cm tall ....***A. mollis***

6   Heads mostly with 6-10 rays; pappus barbellate, white; leaves 2-5 cm long, the lower ones sometimes reduced to bladeless sheaths; plants 10-25 cm tall ....***A. rydbergii***

7   Pappus brownish to dirty white, subplumose; lower stem leaves obtuse to cuneate basally ....***A. diversifolia***

7   Pappus bright white, barbellate; lower cauline leaves cordate, truncate or obtuse basally..8

8   Achenes glabrous at least at the base ....***A. latifolia***

8   Achenes uniformly hairy throughout ....***A. cordifolia***

Arnica chamissonis Lessing (*A. foliosa* Nuttall) — chamisso arnica — Strawberry Valley and Uinta Mtns.; along streams, meadows, and rarely in dry places; 7,600-10,000 ft and probably higher; June-Aug.

Arnica cordifolia Hooker (*A. pumila* Rydberg) With 2 vars.:

1   Plants mostly 2-6 dm tall; basal and lower cauline leaves slightly to deeply cordate, the blade mostly 4-12 cm long and 3-9 cm wide, coarsely and often irregularly toothed to sometimes entire; involucre usually long-hairy at the base ....**var. *cordifolia***

1   Plants mostly less than 2 dm tall; basal and lower cauline leaves only slightly or not at all cordate, commonly 2-5.5 cm long and 1.5-4 cm wide; involucre usually with fewer or no long hairs ....**var. *pumila***

Var. *cordifolia* — heartleaf arnica — Widespread, common to abundant in aspen and coniferous woods at 7200-11,000 ft.

Var. *pumila* (Rydberg) Maguire — dwarf arnica — apparently infrequent in coniferous woods and alpine talus slopes at 8,500-11,700 ft; June-Aug.

Arnica diversifolia Greene — varyleaf arnica — The few specimens seen are from the Uinta Mtns.; meadows, rocky ground, talus slopes, and open spruce forests; 10,800-11,000 ft; July-Aug.

Arnica fulgens Pursh — orange arnica — Locally common in the Greendale area, Daggett Co., Diamond Mt., Uintah Co., Cold Springs and Blue Mtns., Moffat Co., and Uinta Canyon and Burnt Mill Spring, Duchesne Co.; sagebrush communities; 7,400-8,300 ft; June. The closely related *A. sororia* Greene is reported for our area (Graham 1937) but we have seen no specimens.

Arnica latifolia Bongard — broadleaf arnica — The several specimens seen are from across the Uinta Mtns. with few of these from Uintah Co.; coniferous woods, openings in woods, meadows, and rocky places; (8,250) 9,500-11,200 ft; July-Aug. In addition to the features listed in the key, *A. latifolia* differs from *A. cordifolia* var. *cordifolia* by smaller rays, glabrous or short-hairy involucres, and non cordate or less cordate leaves with shorter petioles. However, these features more or less intergrade between the two species and especially when *A. cordifolia* var. *pumila* is included in the comparison. It appears that pubescence of achenes is perhaps the strongest diagnostic feature, and this likely fails some of the time.

Arnica longifolia D.C. Eaton in S. Watson — longleaf arnica — The 15 specimens seen are from Uinta Mtns. in Duchesne, Summit and west Daggett Cos., no specimens seen from Uintah Co.; along streams, in springs and seeps, talus and boulder fields usually with seepage, and moist cliffs; 9,200-11,200 ft; July-Aug.

Arnica mollis Hooker — hairy arnica — Common across the Uinta Mtns. no specimens seen from the Tavaputs Plateau; coniferous woods, moist and wet meadows, open rocky slopes, and talus; 8,600-11,000 (11,600) ft; July-early Sept.

Arnica parryi A. Gray — rayless arnica — The 6 specimens seen are from West Fork Duchesne and Provo River drainages east to Rock Creek in the Uinta Mtns.; aspen and coniferous woods, and openings in woods; 9,400-10,500 ft; mid July-Aug.

Arnica rydbergii Greene — alpine arnica — Uinta Mtns.; occasional in pine-spruce woods and open slopes; 10,000-11,000 ft; (June) July-Aug. Maguire 4317 UTC from Krummholz at 11,400 ft on Mt. Agassiz, Duchesne, Co. identified as *Arnica nevadensis* A. Gray belongs here. *Arnica nevadensis* is not included for our area in AUF, IMF (5:98), FNA (21: 374), or TPD.

Artemisia L. — Sagebrush; Wormwood; Mugwort; Sagewort

Annual or perennial herbs or shrubs, often aromatic; leaves alternate, entire, lobed or dissected; heads small, mostly in panicles, sometimes in racemes or spicate panicles; involucral bracts imbricate in 2-4 series, the inner ones scarious or scarious-margined; receptacles naked or hairy; marginal flowers pistillate and fertile or wanting, rays obsolete or lacking, the central disk flowers bisexual, fertile or sterile; pappus none. This is a complex group of plants in which boundaries between taxa are sometimes vague and sometimes totally obscured by hybridization. Shrubby plants often have ephemeral and persistent leaves, these often strikingly different in shape and size. This adds to the difficulty in distinguishing the taxa.

---

1 Plants shrubs or sub-shrubs, woody at least at the base 2
1 Plants herbs, annual to perennial, the above ground parts dying back to ground level each year (some species keyed both ways) 13
2 Leaves dissected, mostly with more than 5 linear or narrow segments; plants mostly 5-50 (60) cm tall 3
2 Leaves entire to palmately lobed or toothed, the lobes or teeth generally 2-5 per leaf; plants of various stature 6
3 Leaves distinctly once pinnatifid, green and nearly glabrous, 2-8 mm long; plants 5-20 cm tall . . . . *A. pygmaea*
3 Leaves not as above 4
4 Lower leaves lobed or entire, not dissected . . . . *A. pedatifida*
4 Lower and upper leaves dissected into narrow segments 5
5 Flowering stems shorter than or slightly surpassing vegetative stems; old flowering stalks developing into spines; plants flowering May-early June .
   . . . . *A. spinescens*
5 Flowering stems clearly surpassing vegetative stems; old flowing stalks not developing into spines; plants flowering Aug.-Sept. . . . . *A. frigida*

6   Plants 15-40 cm tall, the flowering stalks often 1/3-2/3 the height of the plant; leaves shallowly tridentate with pointed teeth, becoming entire upward, silvery canescent, 1-1.5 cm long; heads ascending, spreading or nodding; involucres not over 2 mm wide, the bracts villous throughout . . . . *A. bigelovii*

6   Plants not as above in all features. 7

7   Plants 10-40 (50) cm tall 8

7   Plants (20) 40-300 cm 11

8   Heads 3-5 mm wide, 6-11 flowered; plants flowering in mid June-July, approaching our area on the n. and e.; leaves deeply 3 lobed . *A. longiloba*

8   Heads about 3 mm wide, 2-6 flowered, plants flowering in late Aug.-Sept.; leaves 3-toothed or mostly shallowly lobed 9

9   Involucral bracts densely white pubescent. 10

9   Involucral bracts not densely white pubescent; plants widespread **A. nova**

10 Plants known from heavy clay soils of Duchesne and Morrison Formations blow 7,000 ft elevation . . . . *A. nova* var. *duchesnicola*

10 Plants known from Diamond Mtn. and Goslin Mtn. above 7,000 ft . . . . *A. arbuscula*

11 Summer-persistent leaves entire, killed by frost in fall and deciduous or persisting-dead in winter, but replaced by new leaves in spring; involucres mostly 4-5 mm high; plants sprouting from the base . . . . *A. cana*

11 Summer-persistent leaves tridentate (leaf-like bracts of the flowering stalks often entire), persisting alive through winter; involucres mostly smaller or larger than above; plants sprouting or not. 12

12 Heads 5-7 mm high, nearly as wide, 5-20 (120) per flowering stalk, with (8) 11-18 flowers, often sessile in the axils of bracts; inflorescence spiciform or narrow paniculate; plants sprouting from the base, montane in the w. part of our area, often in snow flush areas on soils with basic substrate . . . . *A. spiciformis*

12 Heads 2.5-4 mm high, commonly more numerous, with 2-6 (8) flowers; plants not sprouting from the base, different from above in one or more other features . . . . *A. tridentata*

13 Plants green, glabrous or sparingly pubescent, not aromatic; heads usually numerous, 2-3 mm high, 2-4 mm wide. 14

13 Plants gray or white from dense pubescence, or if sparingly pubescent then heads larger than above and/or plants aromatic, perennial 15

14 Leaves pinnatifid, the segments serrate for most of their length A. biennis

14 Leaves simple, entire or with 3-5 linear, entire segments . . *A. dracunculus*

15 Corollas bright yellow; plants 5-20 cm tall, of Moffat Co. 16

15 Corollas not bright yellow; plants widespread 17

16 Basal leaves deeply trifid or palmatifid with the larger segments often again trifid; inflorescence a globose head . . . . *A. capitata*

16 Basal leaves strongly to weakly tridentate or a few entire; inflorescence more or less compact-corymbose. . . . . *A. macarthuri*

17 Leaves 2 or 3 times pinnatifid or palmatifid 18

17 Leaves simple to once-pinnate 23

18 Receptacles beset with numerous long hairs between the flowers, these hairs dense and long enough to be visible without magnification 19

18 Receptacles not pubescent as above. 20

19 Plants from (10,000) 10,500-13,000 ft on the Uinta Mtns.; heads 5-20, in racemes; involucral bracts with black or dark brown scarious margins . . . . *A. scopulorum*

19 Plants known from 6,600-9,000 ft, heads usually more numerous, in panicles; involucral bracts without dark margins . . . . *A. frigida*

20 Plants from taproots or caudices, lacking rhizomes, usually with a basal tuft of persistent leaves larger or more divided than the stem leaves, of the Uinta Mtns.; leaves usually 2-3 times pinnatifid, the ultimate segments linear to lanceolate, 1-30 mm long or longer 21

20 Plants with rhizomes or slender-branched caudices, with or often without a basal tuft of leaves, and these often withering by anthesis when present; leaves variously lobed to pinnatifid but often not so divided as above, the ultimate segments seldom over 5 mm long. 22

21 Plants from Engelmann spruce and alpine communities; heads 4-5 mm high; inflorescence somewhat racemose . . . . *A. norvegica*

21 Plants known from sagebrush and ponderosa pine belts; heads 3-4 mm high; inflorescence paniculate . . . . *A. campestris*

22 Leaves green on the upper side, sparsely to densely tomentose beneath, the lower ones often larger than the upper ones, and persistent at anthesis, the primary divisions again divided and some leaves truly bipinnate . . . . *A. michauxiana*

22 Leaves more or less equally gray-white tomentose on both sides, the lower ones not much if any larger than the upper ones, usually withered and deciduous by anthesis, the primary segments merely lobed, toothed or entire and leaves not truly bipinnate . . . . *A. ludoviciana*

23 Leaves 1-3 cm long, usually fascicled, pinnately divided into linear or filiform segments about 0.5-1 mm wide . . . . *A. carruthii*

23 Leaves usually over 3 cm long, usually not fascicled, entire to pinnately divided, the segments 2-4 mm wide . . . . *A. ludoviciana*

Artemisia arbuscula Nuttall — low sagebrush — Locally abundant on Diamond Mtn., Goslin Mtn. and Cold Spring Mtn. and infrequent in the Pipe Cr.-Gorge Cr. area of the north slope of the Uinta Mtns., usually on soils with heavy clay layer within 12 inches of the surface.

Artemisia biennis Willdenow — biennial wormwood — Widespread, most common across the Tavaputs Plateau, occasional elsewhere; floodplains, drawdown basins of reservoirs, along drainages and ditches, mostly in moist or wet soils, tolerant of saline or alkaline conditions; 5,400-8,000 ft; mid Aug.-Sept.

Artemisia bigelovii A. Gray — Bigelow sagebrush — Across the E. and W. Tavaputs Plateaus and adjacent in the Uinta Basin from near Duchesne to the Green River and in Dinosaur National Monument; desert shrub, pinyon-juniper, and sagebrush communities; 4,900-6,000 ft. Occasionally Forming dominate stands within the desert shrub belt; mid July-Oct. The head have 1-3 central, bisexual flowers and 0-2 marginal, smaller flowers that lack stamens and sometimes have minute rays.

Artemisia campestris L. [*A. spithamaea* Pursh; *A. campestris* var. *petiolata* S. L. Welsh and Goodrich; in IMF (5: 156) our plants are included in ssp. *borealis* (Pallas) Hall & Clements var. *scouleriana* (Bess.) Cronquist. In FNA (19: 508) they are included in ssp. *pacifica* (Nuttall) Hall & Clements. *Oligosporus campestris* (L) Cassini] — field wormwood — Known from the Uinta Mtns. from Lake Fork Canyon, near Moon Lake (Goodrich 21096) and Death Valley, Daggett Co. and from Danforth Hills in Rio Blanco, Co.; July-Aug. Petiole length is highly variable within populations.

Artemisia cana Pursh [*Seriphidium canum* (Pursh) W. A. Weber] — Silver sagebrush — Two vars. as follows:

1 Leaves often over 5 cm long and 4 mm wide; plants known from below 7,800 ft elevation . . . . **var.** *cana*
1 Leaves often less than 5 mm long and 4 mm wide; plants of meadows and streambanks mostly above 7,800 ft elevation . . . . **var.** *viscidula*

Var. *cana* — plains silver sagebrush — Known from small populations in sandy soil at eastern Antelope Flat and along Lodgepole Creek in Daggett Co., Danforth Hills in Rio Blanco Co., Deep Creek and Farm Creek in Uintah Co.; 6,600-7,600 ft elevation; Sept.-Oct. These populations and one more from west of Flaming Gorge Reservoir in Sweetwater Co., Wyoming represent the known western extent of plains silver sagebrush. This plant is also known from the Yampa River drainage in Moffat Co. Most of the range of this plant is east of the Continental Divide. Our outliers might be from introductions.

Var. *viscidula* Osterhout [ssp. *viscidula* (Osterhout) Beetle] — mtn. silver sagebrush — Abundant in Strawberry Valley east to Red Creek Mtn and east on the West Tavaputs Plateau to the Beaver drainage, occasional or locally common on the Uinta Mtns. in the Blind Stream, Brush Creek, Cart Creek, Davenport Draw, Lambson Draw, and Jackson Draw drainages; dominant with grasses in meadows and along drainages, usually (but not always) on soils with slower drainage than typical of *A. tridentata* at 7,400-9,000 ft; mid Aug.-Sept. Although mtn. silver sagebrush is often considered a riparian plant, it commonly grows in well-drained uplands in the Red Creek Mtn. area and in the Willow Creek drainage south of Strawberry Valley and in Strawberry Valley.

Artemisia capitata (Nuttall) S. Garcia, Garnatje, McArthur, Pellicer, S. C. Sanderson, & Valles-Xirau [*Sphaeromeria capitatum* Nuttall; *Tanacetum capitatum* (Nuttall) Torrey & A. Gray] — rock tansy — Locally common along the rim of Vermillion Bluffs, Moffat Co.; sagebrush-saltbush-buckwheat community; 7,600 ft; June.

Artemisia carruthii Wood ex Carruth — Carruth sagewort — Approaching the Uinta Basin and perhaps entering it in Colorado.

Artemisia dracunculus L. [*A. dracunculoides* Pursh; *Oligosporus dracunculus* (L.) Poljakov] — tarragon — Widespread; sagebrush, rabbitbrush, desert shrub, pinyon-juniper, Douglas-fir-aspen, and grass-forb communities, disturbed marshland and open exposed ridges; 4,670-10,000 ft; mid July-Sept.

Artemisia frigida Willdenow — fringed sagebrush — Widespread; desert shrub, sagebrush, pinyon-juniper, mt. brush, and dry meadow communities, and windswept ridges; 4,900-9,000 ft; mid Aug.-Sept.

Artemisia longiloba (Osterhout) Beetle [*A. arbuscula* (Nuttall) spp. *longiloba* (Osterhout) L. M. Schultz] — early sagebrush, alkali sagebrush — This is a community dominant on clay soils that approaches our area in Sweetwater and Uinta Cos., Wyoming. Plotted for Uintah Co. in TPD.

Artemisia ludoviciana Nuttall (*A. mexicana* Willdenow) — Louisiana wormwood, prairie sage, white sagebrush — With 3 freely intergrading vars. as follows:

1 Leaves entire or some of them toothed or lobed; heads mostly with 6-21 flowers .... **var. *ludoviciana***
1 Leaves more or less deeply parted or divided; heads with 15-30 (45) flowers 2
2 Involucres 2.5-5 mm wide, 2.5-3.8 mm high .... **var. *incompta***
2 Involucres 4-7 mm wide, 3.5-4.2 mm high .... **var. *latiloba***

Var. *incompta* (Nuttall) Cronquist — Louisiana wormwood — Specimens seen are from w. of Avintaquin drainage, W. Tavaputs Plateau, and w. of Lake Fork, Uinta Mtns; coniferous forest, aspen, cottonwood, snowbank, and grass-forb communities; 7,500-11,000 ft; mid July-mid Sept. Much like and passing to *A. michauxiana.*

Var. *latiloba* Nuttall. Occasional, Uinta Mtns.; mtn. brush and montane communities.

Var. *ludoviciana* (*A. gnaphaloides* Nuttall) — Louisiana wormwood — Widespread in sagebrush, rabbitbrush, juniper, riparian, wet meadow, and ponderosa pine communities, roadsides and floodplains; 4,800-9,400 ft; late June-Sept.

Artemisia macarthuri S. Garcia, Garnatje, Pellicer, S. C. Sanderson, & Valles-Xirau (*Sphaeromeria argentea* Nuttall; *Tanacetum nuttallii* Torrey & A. Gray) — silver chicken sage — Known from w. summit of Lookout Mt. at 6,580 ft (Peterson & Wilkin 83-285, breaks of the Vermillion Bluffs (Goodrich & Huber 28119) and Sand Wash (Goodrich & Huber 28099) in Moffat Co., and adjacent to our area in Wyoming; desert shrub communities; May-June.

Artemisia michauxiana Besser in Hooker — Michaux wormwood — Uinta Mtns from Blind Stream and e. to the Whiterocks Drainage on the s. slope and Blacks Fork to Beaver Creek on the n. slope; Talus slopes, spruce forests and openings in spruce forests; 9,700-11,400 ft; mid July-Aug. Passing into *A. ludoviciana* var. *incompta.*

Artemisia norvegica Fries — boreal sagewort — Occasional to locally common in alpine basins and adjacent spruce forests of the central Uinta Mtns. at about 11,000 ft; mid July-Aug. Uinta Mtn. plants belong to var. *saxatilis* (Besser) Jepson (*A. n.* var. *piceetorum* S. L. Welsh & Goodrich). If plants called *A. norvegica* var. *saxatilis* were treated at the species level, they would take the name of *A. arctica* Lessing.

Artemisia nova A. Nelson (*Seriphidum novum* Nelson) — black sagebrush — With 2 vars.:

1 Mid and upper involucral bracts glabrous or sparingly pubescent ....
  .... **var. *nova***
1 Mid and some upper involucral bracts densely white-pubescent ....
  .... **var. *duchesnicola***

Var. *duchesnicola* S. L. Welsh and Goodrich — red-clay sagebrush — Known from the Duchesne River Formation from Tridell south to Gusher and east nearly to Asphalt Ridge and on Morrison Formation in Rough Draw and Six Mile Draw nw. of Rainbow Park; 5,600-5,950 ft.

Var. *nova* — black sagebrush — Widespread; common to dominate on flats, hills, mtn. slopes, and ridges, on shallow or rocky soil, often on calcareous substrate, sometimes a dominant understory plant in pinyon-juniper communities; 4,700-8,500 ft and to 9,000 ft on the Tavaputs Plateau; mid Aug.-early Oct.

Plants of *A. nova* var. *nova* of taller stature (over 50 cm tall) can be difficult to distinguish from *A. tridentata* ssp. *wyomingensis.* The following key might help.

1 Involucres to 4 mm high, about 3 mm wide, the bracts often curing to golden brown or reddish brown and persistent with these colors through much of the winter, commonly glabrous; persistent leaves broadly cuneate, 1-3 times longer than wide, sometimes with greenish, glandular dots conspicuous through the pubescence, aromatic with a rather spicy fragrance when crushed, rather sticky when rolled and crushed between the forefinger and thumb, the lobes 1-6 mm long, 1/5-1/3 as long as the entire leaf; cured flower-stalks reddish-brown and rather thinly pubescent
.... **A. nova var. nova**

1 Involucres to 3 mm high, about 2 (3) mm wide, the bracts sometimes curing to a golden brown but often fading and not so persistent through the winter, pubescent; persistent leaves cuneate, often about 3 times as long as wide, rarely with glandular dots showing through the dense pubescence, aromatic with a rather pungent odor when crushed, not particularly sticky when rolled and crushed between the forefinger and thumb, the lobes about 1 (2) mm long, seldom more than 1/10 as long as the entire leaf; cured flowering stalks mostly remaining white with dense pubescence ..
.... **A. tridentata ssp. wyomingensis**

Artemisia pedatifida Nuttall [*Oligosporus pedatifidus* (Nuttall) Poljakov] — birdfoot sagebrush — The 2 specimens seen are from Irish Canyon and confluence of Shell Creek and Hells Canyon of the Vermillion Creek drainage, Moffat Co.; sagebrush-grass and *Atriplex*-grass communities; 6,000-6,800 ft; May-June.

Artemisia pygmaea A. Gray [*Seriphidium pygmaeum* (A. Gray) W. A. Weber] — pygmy sagebrush — Mostly across the flank of the Tavaputs Plateau, rarely to the bottom of the Uinta Basin; desert shrub and pinyon-juniper communities, often on raw soils that are high in carbonates, often in association with relatively rare, endemic species; 5,300-6,000 ft; mid Aug.-early Oct.

Artemisia scopulorum A. Gray — alpine sagewort, Rocky Mt. sagewort — Widespread in the Uinta Mtns.; alpine tundra, pine-spruce woods, occasionally in meadows; more common in timber oatgrass, Ross avens, and Canada single-spike sedge communities with moderate snow cover in winter than in curly sedge-cushion plant communities or in deep snowbeds; (10,000) 10,500-13,000 ft; July-Aug. In an alpine plant community classification study in the Uinta Mtns., this was the 6th most frequently encountered species (Brown 2006).

Artemisia spiciformis Osterhout — spiked big sagebrush, snowfield sagebrush — Locally abundant to dominant, Tavaputs Plateau (Reservation Ridge), Red Creek Mt., and from Wolf Creek to Blind Stream on the s. slope of the Uinta Mtns. and in the Whitney area and other scattered locations to the Blacks Fork drainage on the n. slope of the Uinta Mtns.; moist slopes and snow flush areas, mostly on soils derived from basic substrate; 9,450-10,800 ft; Aug.-Sept.

A hybrid origin (*A. cana* × *A. tridentata*) is likely for this plant (Schultz 1983; Goodrich and others 1985). A combination (*A. tridentata* ssp. *spiciformis* Goodrich & McArthur) has been made. However, the habit (several or many small stems), large and many-flowered heads, and sprouting capability, are features that seem to align this plant to A. cana more than to *A. tridentata*. The persistent, tridentate leaves are about the only feature more aligned with *A. tridentata* than with *A. cana*. In habitat (well-drained soil), it might be considered closer to *A. tridentata*. However, the snowbed habitat of this plant often excludes *A. tridentata*, and the sprouting ability by which it persists in snowbank areas is a feature of *A. cana*. The heads are larger than those of either *A. cana* or *A. tridentata*. This feature is not intermediate between these two taxa. However, Garcia et al. (2011) includes this in the *A. tridentata* complex. With fewer heads per inflorescence (on the average), larger heads, earlier phenology, and different habitat than in either of the other two taxa, *A. spiciformis* seems just as well left as a species as treated by Shultz (2009) as it is subordinated to subspecific rank.

Artemisia spinescens D.C. Eaton in S. Watson (*Picrothamnus desertorum* Nuttall) — bud sagebrush, budsage — Widespread; occasional or locally dominant in desert shrub communities; 4,700-5,700 ft; May-early June. Bud sagebrush is included in *Artemisia* by Garcia et al. 2011.

Artemisia tridentata Nuttall (*Seriphidium tridentatum* Nuttall) — big sagebrush — With 3 intergrading ssp. Names are available at various infraspecific levels for each of the following taxa.

1 Vegetative twigs standing about the same height, giving a flat-topped appearance to the shrubs, the flowering stalks well exceeding and usually over 2 times as long as the subtending vegetative twigs; persistent leaves averaging 4 times longer than wide, with slightly pungent or pleasant camphor-like odor when crushed; plants in and above the pinyon-juniper belt . . . . ***ssp. vaseyana***

1 Vegetative twigs short and long; shrubs with an irregularly topped crown, the flowering stalks about equal to or less than 2 times as long as the vegetative twigs or leaves averaging less than 4 times as long as wide; leaves with pungent odor when crushed; plants mostly in and below the pinyon-juniper belt 2

2 Mature plants with a single, trunk-like, main stem, usually 1-2 (3) m tall; persistent leaves nearly linear or narrowly oblanceolate, hardly cuneate or fan shaped, mostly with nearly parallel margins; averaging 5.6 times longer than wide; panicle of each flowering stalk extremely variable but those of vigorous shrubs 5-8 cm wide or wider and 20-40 cm long, profusely flowered . . . . ***ssp. tridentata***

2 Mature plants with several main branches and mostly without a distinct, trunk-like, main stem, mostly less than 1 m tall; persistent leaves cuneate to broadly cuneate or fan shaped, averaging 3.1 times longer than wide; panicle of each flowering stalk various, but seldom over 5 cm wide or 15 cm long even on the most vigorous shrubs, and often small and relatively few flowered . . . . ***ssp. wyomingensis***

Ssp. tridentata (*A. t.* var. *tridentata*) — basin big sagebrush — Widespread; common to dominate with rubber rabbitbrush, basin wildrye, and sometimes greasewood on canyon bottoms, also along gullies, valley bottoms, ditch banks, and fence rows, usually on gravelly to fine sandy loam, deep alluvial soils, often where periodic deposition of alluvium is more rapid than in-place development of soil horizons; 4,700-8,000 ft; late Aug.-Sept.

Ssp. *vaseyana* (Rydberg) Beetle [*A .t.* var. *vaseyana* (Rydberg) B. Boivin] —
mtn. big sagebrush — Widespread; common to dominant in sagebrush-grass
communities, aspen-sagebrush ecotones, and upper elevations of the pinyon-
juniper belt, mtn. sides, ridges, canyons, and plateaus, usually on Mollisols
with dark colored surface horizons and without a carbonate hardpan; about
7,000-10,000 ft; late July-Sept. Plants of our area are referable to var. *pauciflora*
Winward & Goodrich. Vasey big sagebrush is a common name that has been
applied to plants of this complex. If separation of var. *pauciflora* is maintained
this name is better applied to plants of var. *vaseyana* that are known from n.
and w. of our area. Variety *pauciflora* is treated in synonymy by Shultz (2009).

Ssp. *wyomingensis* Beetle & Young [*A. t.* var. *wyomingensis* (Beetle & Young)
S. L. Welsh] — Wyoming big sagebrush — Widespread; common to dominant in
and below the pinyon-juniper belt in sagebrush and pinyon-juniper communities,
forming extensive stands on Aridisols with light-colored surface horizons and
with a carbonate hardpan at about 12-22 inches and following drainages out
into desert shrub dominated areas (4,600) 5,200-7,400 ft; late July-Sept. Wyoming
big sagebrush is a major component of winter diets of mule deer and pronghorn.

Aster L. — Aster

Annual or perennial herbs from taproots or rhizomes; leaves alternate, simple,
mostly entire; inflorescence usually corymbose or paniculate; involucral bracts
usually graduated in 3 or more series; receptacles naked, rays white, rose,
pink, lavender, or purple, or lacking, disk flowers mostly yellowish; pappus of
equal or subequal capillary bristles. We prefer divisions in *Aster* at section
level. This avoids proliferation of names, and we feel it provides a more user
friendly treatment.

---

1   Plants annual; rays reduced to a short tube or slightly if at all exceeding the
    pappus; plants mostly of moist saline or alkaline ground 2

1   Plants perennial; rays well developed, disk flowers not hidden in the
    pappus; habitat various 3

2   Involucral bracts linear, widest at or below the middle, tapering to an acute
    apex . . . . ***A. brachyactis***

2   Involucral bracts oblanceolate or spatulate-oblong, mostly widest above the
    middle; obtuse or abruptly acute or cuspidate . . . . ***A. frondosus***

3   Upper (inner) involucral bracts not green at the tip, often reddish or purple,
    with a distinct midvein; rays 5-15 (23) per head; basal leaves lacking; lower
    stem leaves smaller than middle stem leaves, often withering 4

3   Upper (inner) involucral bracts green and herbaceous at the tips or
    throughout, lacking a distinct midvein; rays (10) 15-50 per head; basal and
    lower stem leaves sometimes larger than those of mid-stem 6

4   Rays dark purple, 7-13 mm long; involucres 5-10 mm high, the bracts
    strongly graduated; stems (20) 30-50 cm tall; leaves 3-7 cm long, 3-14 mm
    wide . . . . ***A. perelegans***

4   Rays white, pink or pale lavender; involucres and stems various; leaves
    sometimes larger 5

5   Involucres 8-12 mm high, the bracts acute to acuminate; herbage green;

stems (20) 50-150 cm tall; lower leaves smaller than those of mid-stem; rays (13) 15-25 mm long .... *A. engelmannii*

5   Involucres 6-9 mm high, the bracts (at least the outer) obtuse or bluntly acute, herbage glaucous; stems 10-70 cm tall; lower leaves larger than those of mid-stem; rays 7-15 mm long .... *A. glaucodes*

6   Involucres and peduncles glandular................................. 7

6   Involucres and peduncles not glandular or inconspicuously so 8

7   Leaves 2-5 mm wide; 1.5-7 cm long, not especially clasping the stem; involucres 4-7 mm high; plants of saline or alkaline lowlands; rays pale to dark blue .... *A. pauciflorus*

7   Leaves 8-50 mm wide, 2.5-19 cm long, those of the stem strongly clasping; involucres 8-15 mm high; plants montane, rays dark blue or purple .....
.... *A. integrifolius*

8   Involucres 4-5 mm high, 4.5-6 mm wide, the bracts scabrous dorsally and ciliate; inflorescence sometimes secund; rays 3-8 mm long; stems and leaves rather densely and evenly hirsute or villose .... *A. pansus*

8   Involucres 4.5-12 mm high, 6-20 mm wide, the bracts glabrous or rays 5-15 (20) mm long; stems and leaves glabrous or unevenly pubescent........ 9

9   Leaves glaucous .... *A. glaucodes*

9   Leaves not glaucous... 10

10 Involucral bracts distinctly and strongly imbricate, most of the outer ones less than 1/2 as long as the inner ones, obtuse or nearly so .. *A. ascendens*

10 Involucral bracts loosely imbricate, most of the outer ones over 1/2 as long as the inner... 11

11 Stems mostly with (10) 15-30 or more heads, plants (40) 60-150 cm tall, rays white, pink, or blue; middle and upper stem leaves mostly less than 1 mm wide... 12

11 Stems usually with fewer than 10 heads; plants 10-50 (80) cm tall; rays blue or purple, or if pale lavender or white then the middle and sometimes upper stem leaves over 1 cm wide 13

12 Pubescence of stems in decurrent lines below the leaf bases; involucral bracts averaging less than 1 mm wide, distinctly acute or the inner ones long-tapered at the apex; rays mostly blue .... *A. hesperius*

12 Pubescence of stems more or less uniformly dispersed; involucral bracts averaging at least 1 mm wide, the outermost often obtuse and apiculate, the innermost abruptly acute; rays white, pinkish, or rarely bluish ......
.... *A. eatonii*

13 Pappus purplish or brownish; disk-corolla tube equaling or a little longer than the limb; plants known from the Blacks Fork Drainage on the north slope of the Uinta Mtns. .... *A. sibiricus*

13 Pappus white; disk-corolla tube shorter than the limb plants widespread .
.... 14

14 Middle stem leaves mostly over 1 cm wide, usually less than 7 times longer than wide, the upper ones not much reduced; involucral bracts sometimes over 2 mm wide .... *A. foliaceus*

14 Middle stem leaves mostly not over 1 cm wide, often over 7 times longer

than wide; upper leaves often much reduced: involucral bracts less than 2 mm wide .... *A. spatulatus*

Aster brachyactis Blake [*Symphyotrichum ciliatum* (Ledebour) Nesom] — rayless alkali aster, ciliate aster — Widespread; occasional or very locally common; lowland meadows, riparian communities, edge of ponds and lakes, often in alkaline or saline soil; 4,900-5,400 ft; mid Sept.-early Oct.

Aster ascendens Lindley in Hooker [*A. chilensis* Nees ssp. *adscendens* (Lindley) Cronquist; *Symphyotrichum ascendens* (Lindley) Nesom] — Pacific aster, everywhere aster — Widespread; common to abundant in many plant communities including riparian, sagebrush-rabbitbrush, Salina wildrye, coniferous and aspen forests, also along roadsides and ditches, weedy in lawns and gardens; up to 10,000 ft and occasionally higher; July-Oct. There is extreme variability between the high and low elevation plants, but all plants from our area are apparently referable to ssp. *ascendens* (Lindley) Cronquist (sometimes spelled *adscendens*).

Aster eatonii (A. Gray) Howell [*Symphyotrichum eatonii* (A. Gray) Nesom] — Eaton aster — The few specimens seen are from riparian communities in glacial and stream canyons of the Uinta Mtns. at 5,770-7,360 ft; Aug.-Sept.

Aster engelmannii (D.C. Eaton) A. Gray — Engelmann aster — The many specimens seen are from the W. Tavaputs Plateau, Strawberry Valley, and Uinta Mtns. from Rock Creek and west; mostly in coniferous woods, occasionally in aspen woods and open rocky slopes; 8,200-10,600 ft; mid July-Aug.

Aster foliaceus Lindley in de Candolle [*Symphyotrichum foliaceum* (Lindley) G. L. Nesom] — leafy bract aster — With 3 vars.:

1 Involucral bracts leaf-like, some 2-6 mm wide .... **var. canbyi**
1 Involucral bracts not leaf-like, mostly less than 2 mm wide............ 2
2 Plants mainly 5-25 cm tall, decumbent or ascending, alpine or nearly so; rays drying dark purple .... **var. apricus**
2 Plants 20-60 cm tall with more or less erect stems, montane and subalpine; rays blue-purple .... **var. parryi**

Var. apricus A. Gray — space aster — The lone specimens seen are from the Uinta Mtns from Bear River east to Henrys Fork drainage on the north slope and Lake Fork to the Uinta drainage on the south slope; Engelmann spruce parklands, krummholz spruce, and alpine communities at 10,200-11,600 ft.

Var. canbyi A. Gray — Canby aster — Occasional and locally common in the Currant Creek, W. Fork Duchesne, and Soapstone drainages; tall forb, willow-streamside, and spruce-fir communities; 9,000-9,600 ft; July-Aug.

Var. parryi (D.C. Eaton) A. Gray (*Aster parryi* A. Gray) — Parry aster — Specimens seen are from across the Uinta Mtns. and Hill Creek, Tavaputs Plateau, to be expected elsewhere; open woods, riparian and meadow communities; 7,200-10,700 ft; July-Aug. Similar to and freely hybridizing with *A. spatulatus* (IMF 5:292). However, var. *parryi* seldom has the long peduncles, dark blue-purple rays, nor the much reduced upper stem leaves of *A. spatulatus*.

Aster frondosus (Nuttall) Torrey & A. Gray [*Brachyactis frondosa* (Nuttall) A. Gray; *Symphyotrichum frondosum* (Nuttall) Nesom] — short-rayed alkali aster, leafy aster — The 3 specimens seen are from the floodplain of the Strawberry River (Neese 8462), a wet meadow in Sowers Canyon (Goodrich 24017), and the drawdown basin of Flaming Gorge Reservoir (Goodrich 27339) from 5900-7180 ft.

Aster glaucodes Blake [*Eurybia glauca* (Nuttall) Nesom] — blueleaf aster — Widespread; occasional to locally common in pinyon-juniper, mt. brush, Douglas-fir, aspen, ponderosa pine, and lodgepole pine communities, and talus and rocky slopes, and roadsides; 6,720-10,600 ft; Aug.-Sept.

Aster hesperius A. Gray (*Symphyotrichum hesperium* (A. Gray) Nesom; *S. lanceolatum* ssp. *hesperium* (A. Gray) Nesom var. *hesperium*] — Siskiyou aster, marsh aster — Occasional in canyons of the s. slope of the Uinta Mtns. and adjacent in the Uinta Basin from Duchesne to Vernal, one specimen seen from Nine Mile Canyon, W. Tavaputs Plateau, to be expected elsewhere; riparian communities including ditch banks; 5,400-8,160 ft; mid Aug.-Sept.

Aster integrifolius Nuttall [*Eurybia integrifolia* (Nuttall) G. L. Nesom] — thickstem aster — The several specimens seen are all from Wasatch Co.; tall forb, willow-streamside, and spruce-fir communities; 7,600-9,800 ft; late July-early Sept.

Aster pansus (Blake) Cronquist [*Symphyotrichum ericoides* var. *pansus* (S. F. Blake) G. L. Nesom; *A. ericoides* L. var. *pansus* (Blake) A. G. Jones; *A. falcatus* Lindley misapplied; *Virgulus ericoides* (L.) Reveal & Keener] — heath-leaved aster, tufted white prairie aster — The few specimens seen are from near Manila, near Vernal, glacial outwash of the Uinta and Whiterocks Rivers, and floodplain of the White River; streamside woods, meadows and roadside ditches; 5,400-6,240 ft; Aug,-Sept.

Aster pauciflorus A. Nelson & Macbride [*A. junciformis* Rydberg misapplied; *Almutaster pauciflorus* (Nutt.) A. Love & D. Love] — alkali marsh aster — Daggett, Duchesne, and Uintah Cos.; locally common in moist alkaline lowlands, roadside ditches, and riparian communities; 5,100-7,000 ft; Aug.-early Oct.

Aster perelegans A. Nelson & Macbride (*Eucephalus elegans* Nuttall) — elegant aster — Specimens seen are from the W. Tavaputs Plateau and from Layout Canyon near Strawberry Valley, western Uinta Mtns. as far east as the Rock Creek drainage, Daggett Co. on the north slope, and Round Top Mt. (MacLeod 52A DINO); mtn. big sagebrush, mt. brush, Douglas-fir, and aspen communities; 7,400-9,250 ft; late July-Aug.

Aster sibiricus L. [*Eurybia sibirica* (L.) G. L. Nesom] — Siberian aster — The 5 specimens seen are from alpine slopes, krummholz spruce-fir, and semi-barrens of Red Pine Shale in the East and West Forks of the Blacks Fork drainage, north slope of the Uinta Mtns. at 10,300-11,100 ft.

Aster spatulatus Lindley in Hooker [*A. occidentalis* (Nuttall) Torrey & A. Gray; *A. fremontii* A. Gray; *Symphyotrichum spatulatum* (Lindley) G. L. Nesom] — western mt. aster — Widespread in the Utah part of our area; mostly along streams or in meadows, sometimes in aspen and coniferous woods; 7,500-10,400 ft; mid July-Aug. Plants our area belong to var. *spatulatus*. See *A. foliaceus* var. *parryi*.

Bahia Lag. — Bahia

Bahia dissecta (A. Gray) Britton — ragleaf bahia — Annual herbs from a taproot, 25-100 cm tall; glandular in the inflorescence; leaves alternate, 2-3 times ternately divided into oblong or linear segments; involucres 5-7 mm high, the bracts in 2-3 series, the outer with 3 or more nerves; receptacles naked; rays 12-20, 7-9 mm long yellow; disk flowers yellow; pappus none. The specimens seen are from Eagle Creek, Dry Fork, and Whiterocks Canyons of the Uinta Mtns. and from the head of Willow Creek, E. Tavaputs Plateau; 7,500-8,600 ft; Aug.-Sept.

Balsamorhiza (Hooker) ex Nuttall — Balsamroot

1	Leaves entire	***B. sagittata***
1	Leaves pinnatifid	2
2	Leaves mostly 10-30 cm long, the lobes 1-5 cm long	***B. hookeri***
2	Leaves mostly 30-60 cm long, the lobes 5-12 cm long	***B. macrophylla***

Balsamorhiza hookeri Nuttall (*B. hirsuta* Nuttall, misapplied) — Hooker balsamroot — With 2 rather well marked vars. in our area as follows.

1 Involucres densely villose-tomentose or wooly at least at the base, not or sparsely glandular; scapes not or sparsely glandular, usually villose or tomentose at least above; leaves gray-green with more-or-less appressed hairs, those of the margins mostly appressed-ascending ... **var. *neglecta***

1 Involucres hispid and glandular; scapes densely glandular at least above with at least some glands stipitate, hispid; leaves green, hispid, ciliate hairs of the margins widely spreading .... **var. *hispidula***

Var. *hispidula* (Sharp) Cronquist (*B. hispidula* Sharp) — desert balsamroot — Specimens seen are from Cold Spring Mtn., mtn. big sagebrush communities and limestone outcrops; 8,300-8400 ft; May-June. In addition to features listed in the key, the leaves of this var. are commonly narrower than those of var. *neglecta*. This is the common var. of Utah, and it is reported for Utah in our area in AUF (172), but Utah specimens we have seen seem to fit better in var. neglecta.

Var. *neglecta* (Sharp) Cronquist — montane Hooker balsamroot — Common in the Uinta Mtns. including Blue Mtn. and Douglas Mtn. and extending somewhat out into valleys; desert shrub, sagebrush, mtn. brush, pinyon-juniper, and ponderosa pine communities; 5,500-8,600 ft; mid April-June. Occasionally hybridizing with *B. sagittata*. Our plants have villose or wooly involucres with plants of the var. from outside our area lacking this feature (IMF 5:22). Perhaps our plants could be recognized as another var. Seven vars. are listed in TPD. Four are treated in IMF (5: 22).

Balsamorhiza macrophylla Nuttall. — bigleaf balsamroot — Known from the west side of Kamas Valley and perhaps to be expected in the western Uinta Mtns. in the Provo and Weber River drainages.

Balsamorhiza sagittata (Pursh) Nuttall — arrowleaf balsamroot — Common in mtns., probably across our area, no specimens seen between Blind Stream and Whiterocks Canyon on the south slope of the Uinta Mtns.; sagebrush, pinyon-juniper and mt. brush communities; about 7,000-9,000 ft; May-June.

Bidens L. — Beggarticks

Plants annual sometimes appearing perennial, herbs; lower leaves opposite; involucral bracts in 2 definite series, the outer series often leaf-like and exceeding the head, the inner series smaller; rays (when present) yellow; disk flowers yellow; pappus of 2-4 retrorsely barbed awns.

1 Leaves simple, serrate, rarely incised, sessile, sometimes connate; rays lacking or to 15 mm long; heads often nodding .... *B. cernua*
1 Leaves pinnately divided or compound, with 3-5 divisions or leaflets, petiolate, the divisions or leaflets more or less petiolulate; rays lacking or to 4 mm long .... *B. frondosa*

Bidens cernua L. — nodding beggarticks — The 7 specimens seen are from or near the floodplains of the Green River and White River and from a wet alkaline meadow at Vernal; 4,655-5,380 ft; July-Sept.

Bidens frondosa L. — devils beggarticks — The 7 specimens seen from the floodplain of the Green River near Jensen and Ouray, Stewart lake, Pelican Lake, and a ditch bank in the town of Duchesne, at 4,655-5,515 ft; Aug.-Oct.

Brickellia Ell. — Brickellbush

Perennial herbs or shrubs, often aromatic; leaves simple, alternate or some opposite; heads variously disposed; involucral bracts imbricate in series, striate (with longitudinal lines), more or less chartaceous; receptacles naked; rays lacking; disk flowers white, pale greenish, or pinkish white; pappus of capillary bristles, these scabrous to subplumose.

1 Petioles 1-7 cm long; leaf blades 2-10 cm long, deltoid-ovate or deltoid-lanceolate, alternate or the lower opposite .... *B. grandiflora*
1 Leaves sessile or petioles not over 3 mm long, alternate; plants of deserts and lower montane 2
2 Leaves linear to elliptic-lanceolate, entire; heads 10-14 mm high, with about 40-50 flowers .... *B. oblongifolia*
2 Leaves ovate to oblong ovate, sometimes dentate; heads 7-9 mm high, with about 10-12 flowers .... *B. microphylla*

Brickellia grandiflora (Hooker) Nuttall [*Coleosanthus grandiflora* (Hooker) Kuntze] — tasselflower brickellbush — Uinta Mtns. and Echo Park; infrequent or very locally common; coniferous forests, riparian forests, and talus slopes; 5,500-9,600 ft; July-Aug. Heads usually nodding.

Brickellia microphylla (Nuttall) A. Gray [*B. scabra* (A. Gray) A. Nelson; *B. watsonii* B. L. Robinson misapplied; *Coleosanthus microphyllus* (Nuttall) Kuntze] — littleleaf brickellbush — Occasional across much of our area; desert shrub, sagebrush, pinyon-juniper, and dry mtn. brush communities, often in rocks or rocky places, sometimes on sandy soil; 5,000-7,100 ft; Aug.-mid Sept. Plants of our area belong to var. *scabra* A. Gray.

Brickellia oblongifolia Nuttall — Mohave brickellbush — Widespread; infrequent or very locally common in desert shrub, sagebrush, pinyon-juniper, and juniper-mt. brush communities; 4,800-6,700 ft; June-mid July. Our plants belong to var. *linifolia* (D. C. Eaton). B. L. Robinson.

Carduus L. — Musk thistle; Bristle-thistle

Carduus nutans L. — musk thistle — Robust, annual or biennial, weedy forbs 80-150 cm tall; leaves alternate, decurrent, pinnately lobed or deeply dentate, the lobes or teeth spinose; heads large, solitary on the branches; involucral bracts strongly imbricate in series, strongly spreading or reflexed in age, with a prominent midrib, this excurrent as a spine; receptacles bristly; rays none; disk flowers deeply 5-cleft, purple or sometimes pale; pappus of scabrous bristles. Introduced from Europe, adventive and weedy across much of our area; roadsides, ditches, fields, and invading many indigenous plant communities; up to about 7,500 (8,000) ft; June-Aug.

Centaurea L. — Knapweed

Annual or biennial or perennial herbs; leaves alternate; heads solitary to numerous toward the ends of branches, with disk flowers only; involucral bracts imbricate, with scarious erose to pectinate margins; receptacle bristly, the bristles capillary; flowers purple, blue or whitish (or yellow in *C. solstitalis* L. that might be expected in our area).

1　Leaves simple, entire or serrate; plants perennial from robust, horizontal root stocks . . . . **C. repens**

1　Leaves pinnatifid into linear segments, these occasionally again toothed or lobed; plants biennials or short lived perennials from a taproot and branched caudex 2

2　Involucre 3-6 mm wide, 8-10 mm high, the bracts fringed and spine-tipped with the spine 1.5-4 mm long, the apex pale . . . . **C. diffusa**

2　Involucre 10-13 mm wide, 10-15 mm high, the bracts fringed but not spine-tipped, the apex of middle and outer bracts dark-spotted . . . . **C. stoebe**

Centaurea diffusa Lamarck — diffuse knapweed — Introduced weedy species from the Mediterranean region. Apparently sparingly found in the Uinta Basin with known infestations being treated for eradication. However, this plant is abundant in western Utah and elsewhere. Continuing introductions of this pestiferous weed can be expected. Pappus wanting or very short.

Centaurea repens L. (*C. picris* Pallas, *Acroptilon repens* (L) de Candolle — Russian knapweed, hardheads — Introduced from Eurasia; weedy, becoming noxious in fields and pastures, abandoned farmland, along ditches, streams and roadsides, moist lowlands, riparian communities; 4,700-7,800 ft; late June-Aug. Pappus 6-11 mm long.

Centaurea stoebe L. (*C. biebersteinii* de Candolle; *C. maculosa* Lamarck misapplied) — spotted knapweed — Introduced from Europe, highly invasive and rapidly advancing along highways and other roads and high-use recreation areas at 5500-8400 ft; July-Aug. to be expected to advance across our area at lower and middle elevations. Pappus to 2 mm long, hidden in bristles of the receptacle. Plants introduced to our area belong to var. *micranthus* (Gugler) Hayek (FNA 19: 189).

Chaeactis de Candolle — Chaenactis; Dusty-maiden; False Yarrow

Annual, biennial, or perennial herbs from taproots; leaves alternate or mainly basal, pinnately dissected or entire; heads solitary to several in corymbose cymes, with disk flowers only; involucral bracts in 1-3 series; receptacle naked; flowers whitish or pinkish; pappus of 4-20 hyaline scales.

1 Plants annual or winter annual, lacking a basal rosette of leaves, of desert shrub and juniper communities; pappus of 4 scales ....*C. stevioides*
1 Plants biennial or perennial, with a well-developed basal rosette of leaves; distribution various; pappus of 10-16 scales..........................2
2 Plants 2-9 cm tall, perennial from a simple or branched caudex, alpine in the Uinta Mtns ....*C. alpina*
2 Plants 5-50 (60) cm tall, biennials or short-lived perennials from taproots, of many plant communities ....*C. douglasii*

Chaenactis alpina (A. Gray) M. E. Jones — alpine dusty-maiden — Alpine on the Uinta Mtns. in Duchesne and Summit Cos. in talus slopes and boulder fields; 11,200-11,600 ft; July-Aug.

Chaenactis douglasii (Hooker) Hooker & Arnot — Douglas dusty-maiden — Widespread; occasional in many plant communities including desert shrub, sagebrush, rabbitbrush, pinyon-juniper, mt. brush, Salina wildrye, ponderosa pine, Douglas-fir, aspen, and occasionally meadow; 4,800-9,500 ft; late May-mid Aug., depending on elevation.

Chaenactis stevioides Hooker & Arnot — annual dusty-maiden — Many specimens seen from e. Uintah Co. and adjacent Colorado, and se. Duchesne Co.; desert shrub and sagebrush communities; 4,800-5,000 ft; mid May-mid July.

Chaetopappa de Candolle (Leucelene Greene) — Heath aster

Chaetopappa ericoides (Torrey) G. L. Nesom [*Leucelene ericoides* (Torrey)] Greene; *Aster arenosus* Blake; *A. leucelene* Blake] — heath aster, rose-heath — Perennial herbs with branching caudices and rootstocks, 5-12 cm tall, the stems tufted; leaves 2-12 (20) mm long, 0.6-2 mm wide, linear, entire, hispid, ciliate, smaller and fewer upward on the stem; heads solitary on ends of slender branches; involucres 5-7 mm high, the bracts imbricate in about 3-7 series, the midribs green, the margins scarious; rays to 6 mm long, white; disk flowers yellow when fresh, whitish in age; pappus of capillary bristles. Widespread; locally common; desert shrub, sagebrush, pinyon-juniper, and mt. brush communities; 5,200-7,200 ft; mid May-July (Oct.).

Chamaechaenactis — Rydberg

Chamaechaenactis scaposa (Eastwood) Rydberg — fullstem — Perennial herbs 2-9 cm tall, from a pilose branching caudex, this clothed with old leaf bases, leaves all basal, simple, entire, oblanceolate to orbicular, the blades 4-18 mm long, 3-15 mm wide, pubescent; scapes long villous; heads solitary, with disk flowers only; involucre 7-17 mm high, the bracts about equal or the outer ones shorter, villous; receptacle naked; flowers whitish or pinkish; pappus of hyaline scales. Occasional across the Tavaputs Plateau and flank of the plateau; desert shrub, pygmy sage, juniper-mt. brush-greasebush, pinyon-juniper, and sagebrush-grass communities, often on whitish marl limestone; 5,200-7,900 ft; late April-June.

Chrysanthemum L.

Chrysanthemum leucanthemum L. (*Leucanthemum vulgare* Lamarck) — ox-eye daisy — Perennial herbs; leaves alternate, crenate and more-or-less lobed or cleft; involucral bracts in 2-4 series with brownish-scarious margins; pappus lacking or a short crown; achene glabrous. Introduced from Eurasia, specimens seen are from Strawberry Valley, Greendale and a roadside near Hacking Lake in the Uinta Mtns. at 10,300 ft (Neese 15114); July-Aug. The high elevation introduction at 10,300 ft apparently did not persist. However, the population at Greendale expanded rapidly until controlled with herbicide. This species is listed as a noxious weed in Utah. Additional introductions can be expected, and expansion of existing populations can also be expected.

Chrysopsis (Nuttall) Elliott — Hairy goldenaster

Chrysopsis villosa (Pursh) Nuttall [*Chrysopsis foliosa* Nuttall; *C. fulcrata* Greene; *C. hispida* (Hooker) de Candolle; *C. viscida* (A. Gray) Greene; *Heterotheca fulcrata* (Greene) Shinners; *H. horrida* (Rydberg) Harms; *H. villosa* (Pursh) Shinners; *H. viscida* (A. Gray) Harms] — hairy goldenaster — Perennial herbs; 10-60 cm tall; herbage strigose to canescent; leaves alternate, entire, 2-7 cm long, linear to obovate; heads solitary to several near ends of branches; involucral bracts imbricate, pubescent to glandular; receptacle naked and pitted; rays 6-15 (20) mm long, yellow; disk flowers yellow; pappus double, the outer series of short scales, the inner series of capillary bristles. Widespread; several plant communities including desert shrub, sagebrush, pinyon-juniper, mt. brush, ponderosa pine, and Douglas-fir; occasional or common to dominant in rocky places in burned areas within the pinyon-juniper belt; 5,000-9,000 ft; usually mid June-early Oct. with plants in full flower as late as 20 Nov. A highly variable taxon with several intergrading vars. none of which are discussed here. Dense patches of this plant appear to be highly competitive with cheatgrass in burned areas. Treatment of our plants under *Chrysopsis* rather than *Heterotheca* follows IMF (5: 238) where differences used to separate the 2 are considered of sectional rather than generic importance.

Chrysothamnus Nuttall — Rabbitbrush

Shrubs or sub-shrubs; leaves alternate, sessile or nearly so, linear to filiform, entire, commonly resinous and aromatic; inflorescence usually a panicle or cyme, occasionally racemose; heads small, the involucres cylindrical, the bracts imbricate in more or less distinct vertical rows; receptacle naked; rays none; disk flowers 4-7 to a head, all bisexual, yellow. Nomenclature in the genus is confusing as several of the taxa have been recognized at the species, subspecies, and varietal levels by different authors. This has added considerable difficulty to this morphologically difficult group of plants.

1   Stems covered with a felt-like tomentum, the hairs sometimes matted in resin and not conspicuous 2

1   Stems glabrous or pubescent but not as above 3

2   Involucral bracts attenuate, rather membranous; inflorescence mostly racemose, leafy . . . . **C. parryi**

2 Involucral bracts obtuse to acute, if attenuate then carinate and chartaceous; inflorescence mostly cymose, sparsely if at all leafy . . . .
. . . . *C. nauseosus*

3 Achenes glabrous; involucres 9-12 mm high, the bracts strongly aligned in vertical rows; plants 10-30 cm tall; young twigs scabrous or cinereous . . .
. . . . *C. depressus*

3 Achenes pubescent; involucres 4-8 mm high, the bracts variously aligned; young twigs glabrous or puberulent, greenish at first, but soon whitish and shining [Note: except for glabrous achenes, *C. vaseyi* (A. Gray) Greene will key here. This approaches our area near Soldier Summit, W. Tavaputs Plateau.] 4

4 Involucral bracts acuminate-cuspidate; leaves 1-2 mm wide, 1 nerved; corollas 4-4.5 mm long; plants mostly 10-35 cm tall . . . . *C. greenei*

4 Involucral bracts obtuse or acute; leaves 1-10 mm wide; corollas 4.5-7 mm long 5

5 Leaves 4-8 mm wide, not twisted, mostly 3 nerved; plants 80-240 cm tall, mostly growing along waterways on bottomland and roadsides, often in alkaline soil; style appendages long, 40-70 percent of the style branches; outer involucral bracts with a thick greenish or brownish spot near the apex
. . . . *C. linifolius*

5 Leaves 1-4 (6) mm wide, sometimes twisted; 1-3 (5) nerved; plants mostly 20-120 cm-tall, not restricted as above and often in desert shrub, foothill and montane communities; style appendages short, 30-45 percent of the style branches; involucral bracts not as above . . . . *C. viscidiflorus*

Chrysothamnus depressus Nuttall — dwarf rabbitbrush — Widespread; occasional in desert shrub, juniper, pinyon-juniper, Salina wildrye, and ponderosa pine communities at 6,500-7,900 ft; June-Sept.

Chrysothamnus greenei (A. Gray) Greene — desert rabbitbrush, needle-leaf rabbitbrush — Rather common from Currant Creek east into Colorado in the Uinta Basin and along the lower edge of the Tavaputs Plateau; mostly in desert shrub communities and also in pinyon-juniper communities; July-Sept. Often closely browsed in winter by ungulates.

Chrysothamnus linifolius Greene [*C. viscidiflorus* ssp. *linifolius* (Greene) Hall & Clements] — spreading rabbitbrush — Locally common to abundant across the bottom of the Uinta Basin and following the Green River and Yampa River in Daggett and Moffat Cos.; mostly along drainages and in moist bottomlands, often where alkaline, sometimes on roadsides up to 7,500 ft; Aug.-Sept.

Chrysothamnus nauseosus (Pallas ex Pursh) Britton [*Ericameria nauseosa* (Pallas ex Pursh) G. L. Nesom & Baird] — rubber rabbitbrush — A complex of highly variable plants; the following fairly distinct subspecific taxa have been recognized for our area.

1 Corolla lobes pubescent; plants mostly restricted to aeolian sand . . . .
. . . . **var. *turbinatus***

1 Corolla lobes glabrous 2

2 Ovaries and achenes glabrous . . . . **var. *leiospermus***

2 Ovaries and achenes pubescent 3

3 Plants 10-30 (50) cm tall 4

3 Plants mostly 50-200 cm tall. 6
4 Leaves flat, some 2 mm wide or wider, usually widely spreading . . . .
. . . . var. *uintahensis*
4 Leaves terete or nearly so, less than 2 mm wide, ascending spreading. . . 5
5 Leaves soon deciduous, greenish yellow . . . . var. *psilocarpus*
5 Leaves persistent, green . . . . var. *linearifolius*
6 Involucres glabrous, the outer bracts sometimes ciliate or scurfy 7
6 Involucres pubescent. 8
7 Leaves 1-3 mm wide, 4-6 cm long, mostly 3-5 nerved, mildly ill-scented .
. . . . var. *graveolens*
7 Leaves less than 1 mm wide, 2.5-5 cm long, 1 nerved, strongly ill-scented
. . . . var. *oreophilus*
8 Leaves 3-10 mm wide, 4-8 cm long; involucral bracts mostly obtuse; plants
restricted to the w. side of our area . . . . var. *salicifolius*
8 Leaves 0.5-3 mm wide, 2.5-4 cm long; involucral bracts mostly acute; plants
widespread 9
9 Corolla lobes 0.5-1 mm long; style appendages shorter than the stigmatic
portion . . . . var. *hololeucus*
9 Corolla lobes 1-2 mm long; style appendages longer than the stigmatic
portions 10
10 Corollas 6-8.5 mm long; involucres 7-9.5 mm tall; shrubs mostly 0.2-0.6 m
tall; plants approaching our area from the n. and e. . . . . . var. *nauseosus*
10 Corollas (8) 9-11 mm long; involucres mostly 9-11 mm tall; plants mostly
0.4-1.5 m tall, apparently widespread in our area . . . . var. *speciosus*

Var. *graveolens* (Nuttall) M. H. Hall [*C. nauseosus* var. *bigelovii* (A. Gray) Hall;
C. nauseosus ssp. *graveolens* (Nuttall) H. M. Hall; *C. graveolens* (Nuttall) Greene;
Ericameria nauseosa ssp. *nauseosa* var. *glabrata* (A. Gray) G. L. Nesom & Baird] —
green rubber rabbitbrush — Widespread and common to abundant mostly
along drainages and roadsides and in canyon bottoms on alluvial soil, usually
not as tolerant of alkaline conditions as var. oreophilus; up to about 7,500
(9,000) ft; Aug.-Oct.

Var. *hololeucus* (A. Gray) H. M. Hall [*C. n.* ssp. *hololeucus* (A. Gray) Hall &
Clements; *C. nauseosus* var. *gnaphaloides* (Greene) Hall; *Ericameria nauseosa* ssp.
nauseosa var. *hololeuca* (A. Gray) G. L. Nesom & Baird] — all white rubber
rabbitbrush — Widespread and occasional to common mostly along drainages,
roadsides, and canyon bottoms on deep alluvial soil; 6,000-8,000 ft; Aug.-Oct.
More common than the intergrading var. speciosus.

Var. *leiospermus* (A. Gray) H. M. Hall [*C. nauseosus* ssp. *leiospermus* (A. Gray)
Hall & Clements; *C. nauseosus* var. *abbreviatus* (M. E. Jones) Blake; *Ericameria
nauseosa* ssp. *consimila* var. *leiosperma* G. L. Nesom & Baird] — smoothseed
rubber rabbitbrush — Specimens seen are from face of Blue Mtn. in outcrops
of Weber Sandstone (Goodrich 25647), 4 mi ne. of Dinosaur, Moffat Co. (Goodrich
& Atwood 17983), rim of Red Canyon in Daggett Co. from bed rock of Uinta
Mtn. Group (Goodrich et al. 22832), Sheep Creek Canyon in Daggett Co.
(Goodrich28457), and (Neese 6628) from the roadside of Hwy. 40 at 30 mi w. of
Duchesne; 5,650-7,240 ft; Aug.-Oct. Involucral bracts glabrous and appear var-
nished; leaves thread-like, persistent. Some of our specimens might be aberrant
forms of var. oreophilus with glabrous achenes.

Var. *linearifolius* S. L. Welsh & S. Goodrich — White River rubber rabbitbrush

— Known from Wasatch and Green River Formations in the White River drainage in Rio Blanco Co, Colorado, locally abundant in desert shrub, greasewood, and Wyoming big sagebrush-greasewood communities at 5,650-6,300 ft. Highly selected by ungulates in winter.

Var. oreophilus (A. Nelson) H. M. Hall [*C. nauseosus* ssp. *consimilis* (Greene) Hall & Clements; *C. n.* var. *consimilis* (Greene) Hall; *C. n.* ssp. *pinifolius* (Greene) Hall & Clements; *Ericameria nauseosa* ssp. *consimila* var. *oreophila* (A. Nelson) G. L. Nesom & Baird] — threadleaf rubber rabbitbrush — widespread and common to abundant mostly along drainages and roadsides and in alkaline low lands; up to about 7,000 (9,000) ft; Aug.-Oct.

Var. psilocarpus S. F. Blake [*C. nauseosus* ssp. *psilocarpus* (S. F. Blake) L. C. Anderson; *Ericameria nauseosa* ssp. *nauseosa* var. *psilocarpa* (S. F. Blake) G. L. Nesom & Baird] — Hunting rubber rabbitbrush — Infrequent; one specimen is from Roosevelt Airport, the others are from the W. Tavaputs Plateau w. in the Indian Canyon and Current Creek drainages; sagebrush and Salina wildrye communities; 6,300-7,500 ft; Aug.-Sept.

Var. salicifolius (Rydberg) Hall [*C. nauseosus* ssp. *salicifolius* (Rydberg) Hall & Clements; *Ericameria nauseosa* ssp. *nauseosa* var. *salicifolia* (Rydberg) G. L. Nesom & Baird] — mtn. rubber rabbitbrush — Infrequent or locally common from Strawberry Valley e. to Lake Canyon of the W. Tavaputs Plateau and to Rock Creek of the Uinta Mtns.; mostly in canyons above 7,600 ft; Aug.-Oct.

Var. speciosus (Nuttall) H. M. Hall [*C. nauseosus* ssp. *speciosus* (Nuttall) Hall & Clements; *C. nauseosus* var. *albicaulis* (Nuttall) Rydberg; *Ericameria nauseosa* ssp. *nauseosa* var. *speciosa* (Nuttall) G. L. Nesom & Baird] — white rubber rabbitbrush — Widespread and occasional to common; mostly along drainages, roadsides, and canyon bottoms on deep alluvial soil; 6,000-8,000 ft; Aug.-Oct. Distinct from var. *hololeucus* to the north of the Uinta Basin, but wholly intergrading into that var. within this area.

Var. turbinatus (M. E. Jones) S. F. Blake [*C. nauseosus* ssp. *turbinatus* (M. E. Jones) Hall & Clements; *C. nauseosus* ssp. *consimilis* var. *turbinatus* (M. E. Jones) Blake; *Ericameria nauseosa* ssp. *consimila* var. *turbinata* (M. E. Jones) G. L. Nesom & Baird] — sand-dune rubber rabbitbrush — Locally common to abundant from near Bluebell, Duchesne Co. to near the Utah-Colorado line (and perhaps further e.) on aeolian sand from Duchesne River, Navajo Sandstone, and probably other formations; mostly below 6,000 ft; Aug.-Oct.

Var. uintahensis (L. C. Anderson) S. L. Welsh [*C. nauseosus* ssp. *uintahensis* L. C. Anderson; *Ericameria × uintahensis* (L. C. Anderson) G. L. Nesom & Baird] — Uinta rubber rabbitbrush — Endemic, locally common from Lapoint to Red Fleet Reservoir and Willow Creek, Tavaputs Plateau; on Duchesne River, Morrison, Dakota, Green River and other formations that tend to weather to badlands; desert shrub communities, and colonizing on road cuts; mostly below 6,000 ft; Aug.-Oct.

Chrysothamnus parryi (A. Gray) Greene — Parry rabbitbrush — This is a complex of variable plants. Three more or less distinct vars. have been recognized from our area. Annotations on specimens seen indicate common disagreement in identification of the vars. of *C. parryi*. Plants with upper leaves exceeding the inflorescence grow among those with upper leaves shorter than the inflorescence at some locations in our area. The value of recognizing var. *howardii* seems questionable.

1 Heads with 8-20 flowers .... **var. parryi**
1 Heads mostly with 5-7 flowers.................................... 2
2 Corolla lobes 0.6-1.2 mm long, averaging less than 1 mm; inflorescence more or less racemose .... **var. affinis**
2 Corolla lobes mainly 1.3-2.2 mm long, inflorescence more or less corymbose.. 3
3 Involucral bracts acuminate; leaf surfaces not glandular; flowers 8-10 mm long, pale yellow, the lobes mainly 1.8-2.2 mm long .... **var. howardii**
3 Involucral bracts attenuate; leaf surfaces glandular; flowers 9-12 mm long, yellow, the lobes mostly 1.3-1.7mm long .... **var. nevadensis**

Var. affinis (A. Nelson) Cronquist [*C. p.* ssp. *affinis* (A. Nelson) L. C. Anderson; Ericameria parryi var. affinis (A. Nelson) G. L. Nesom & Baird] — short-lobe Parry rabbitbrush — The 8 specimens seen are from widely scattered locations including tuffaceous outcrops in Browns Park of Daggett Co., East and West Tavaputs Plateaus with scattered pinyon-juniper on semi-barrens of Green River and Uinta Formations, rim of Ashley Gorge, and Dowd Hole and Beaver Creek of the n. slope of the Uinta Mtns.; pinyon-juniper, sagebrush, ponderosa pine-mahogany communities; 5,800-8,200 ft; July-Sept.

Var. nevadensis (A. Gray) Kittell in Tidestrom & Kittell [*C. p.* ssp. *attenuatus* (M. E. Jones) Hall & Clements; *C. p.* var. *attenuatus* (M. E. Jones) Kittell in Tidestrom & Kittell; Ericameria parryi var. nevadensis (A. Gray) G. L. Nesom & Baird] — short-leaf Parry rabbitbrush — Infrequent from Myton to Split Mtn and south to the flank of the Tavaputs Plateau, locally common to abundant from Myton (especially w. of Duchesne on the Duchesne River formation) to Currant Creek and Windy Ridge in Strawberry Valley; occasional at higher elevations on shale ridges of the W. Tavaputs Plateau; infrequent in gorges of the eastern Uinta Mtns.; pinyon-juniper, black sagebrush, big sagebrush-grass, sagebrush-Douglas-fir, and rarely riparian communities; 4,800-8,700 ft; July-Sept. A specimen (E. Neese & L. England 6649) from Antelope Flat, Daggett Co., appears to be of hybrid origin involving *C. nauseosus* and perhaps *C. parryi* var. *nevadensis*. Specimens with leaves overtopping the inflorescence are found growing with plants with leaves not exceeding the inflorescence in Dowd Hole and Sheep Creek Canyon in Daggett Co.

Var. howardii (Parry) Kittell in Tidestrom & Kittell [*C. p.* ssp. *howardii* (Parry) Hall & Clements; Ericameria parryi var. howardii (Parry) G. L. Nesom & Baird] — longbract Parry rabbitbrush — The 5 specimens seen are from widely scattered locations including East and West Tavaputs Plateaus, Dry Fork Canyon of the south slope of the Uinta Mtns. and n. flank of the Uinta Mtns from Sheep Creek to Bare Top Mtn.; pinyon-juniper, sagebrush, oak brush, and Salina wildrye communities; 6280-10,050 ft.

Var. parryi — Parry rabbitbrush — Indicated to be across much of our area in TPD, but indicated to be south of our part of Utah in AUF (187) and IMF (5: 224). The one specimen seen (Huber & Goodrich 5304) is from the Cathedral Creek drainage of the Piceance Basin at 8,580 ft.

Chrysothamnus viscidiflorus (Hooker) — Nuttall Low rubber rabbitbrush, little rabbitbrush — A complex of variable plants; 3 somewhat distinct subspecific taxa have been found in our area:

1 Leaves and upper stems puberulent; leaves 2-5 mm wide **var. lanceolatus**

1 Leaves and upper stems glabrous or the leaves merely ciliate-puberulent
   2
2 Leaves 0.5-1.5 mm wide; plants mainly 20-30 cm tall ... **var. *stenophyllus***
2 Leaves mostly 1-4 mm wide; plants commonly 30-100 cm tall ....
   .... **var. *viscidiflorus***

Var. *lanceolatus* (Nuttall) Greene [*C. lanceolatus* Nuttall; *C. v.* ssp. *lanceolatus* (Nuttall) attributed to Greene by Piper; *C. viscidiflorus* ssp. *elegans* (Greene) Hall & Clements; *C. viscidiflorus* var. *elegans* (Greene) Blake; *Ericameria viscidiflora* ssp. *lanceolata* (Nuttall) L. C. Anderson] — yellowbrush, mtn. low rabbitbrush — Common to abundant across, our area; many plant communities; 7,000-10,500 ft; July-Sept. Livestock use this plant quite heavily especially after flowering. Elk and deer readily use yellowbrush in winter.

Var. *stenophyllus* (A. Gray) H. M. Hall [*C. viscidiflorus* ssp. *stenophyllus* (A. Gray) Hall & Clements; *Ericameria viscidiflora* var. *stenophylla* (A. Gray) L. C. Anderson] — slenderleaf sticky rabbitbrush — Six specimens seen are from Tridell to Raven Ridge and s. to Bonanza and Rainbow, and one is from Grasshopper Island in the Green River in Daggett Co. This is included in var. *viscidiflorus* in IMF (5: 233).

Var. *viscidiflorus* [*C. viscidiflorus* ssp. *pumilus* (Nuttall) Hooker & Arnot; *C. viscidiflorus* var. *pumilus* (Nuttall) Jepson; *C. stenophyllus* (A. Gray) A. Gray; *C. viscidiflorus* ssp. *stenophyllus* (A. Gray) Hall & Clements; *C. viscidiflorus* var. *tortifolius* (A. Gray) Greene; *Ericameria viscidiflora* L. C. Anderson] — stickyleaf low rabbitbrush — widespread and common to abundant in desert shrub, juniper, and pinyon-juniper communities; 6,000-7,000 ft; July-Sept. Livestock and big game animals generally avoid browsing this plant.

Cichorium L. — Chicory

Cichorium intybus L. — chicory — Perennial from a deep taproot, 30-170 cm tall, glabrous or hirsute; lower leaves petiolate, 8-25 cm long, 1-7 cm wide, toothed or more often pinnatifid, upper leaves becoming reduced, sessile, entire, bract-like; heads sessile or short pedunculate, the branches of the inflorescence racemose; rays blue, rarely white, to 2 cm long; involucre 9-15 mm high; pappus of narrow, minute scales; achenes 2-3 mm long. Introduced from the Mediterranean Region, weedy and apparently increasing in our area; mostly on agricultural lands and roadsides at 5,000-5,800 ft.

Cirsium Mill — Thistle

Biennial or perennial, spiny, coarse herbs; leaves basal and alternate; involucral bracts in several series, subequal to imbricate, usually spine tipped; receptacle densely bristly; rays none; disk flowers pink, purple, reddish, or whitish; pappus of plumose bristles, those of the outer flowers sometimes barbellate; achenes glabrous, flattened or 4-angled, with 4-many nerves. This is a complex and difficult genus. Harrington (1954) stated that it was badly in need of revision in North America. It remains badly in need of revision, but the work in recent floras (Dorn 2001, AUF 189-197, IMF 5: 338-414, FNA 19: 95-164) has made it possible to separate most of the plants in our area. However, notes in FNA 19: 95-97 make clear that many problems remain to be worked out in North American *Cirsium* where species complexes appear to be evolutionary works in progress, and some of the species of the mountainous western part of

North America are frustrating polymorphic with much overlapping variability and intergradation of characters.

---

1   Plants perennial from creeping robust rhizomes, often forming dense patches or large clones; heads unisexual, comparatively small, the involucre 1-2 (2.5) cm high . . . . *C. arvense*

1   Plants biennial or perennial, seldom obviously rhizomatous, not forming patches 2

2   Upper surface of stem leaves with forward-pointing prickly hairs . . . .
. . . . *C. vulgare*

2   Leaves glabrous to woolly-villous but not with forward-pointing prickly hairs. 3

3   Inner involucral bracts conspicuously dilated, erose but not lacerate, or tan to silver and contrasting with the outer bracts; heads generally sessile or nearly so in a terminal cluster, the cluster of heads closely subtended by a cluster of leaves; plants of meadows or other moist places, sometimes stemless and the heads sessile in the basal rosette of leaves, or with stems to 1 m tall . . . . *C. scariosum*

3   Inner involucral bracts not especially dilated nor colored as above; heads various; plants of various habitats; stems well developed 4

4   Plants greenish, at least the upper surface of the leaves glabrous or only thinly pubescent, stems often glabrous or thinly pubescent; outer involucral bracts sometimes pinnate-spinose; heads sometimes rather congested, sometimes closely subtended by bract-like leaves of the same texture and vestiture as the lower involucral bracts 5

4   Plants white or gray; leaves and stems densely pubescent; involucral bracts not pinnate-spinose; heads usually not much congested, sometimes solitary on the ends of branches, often not closely subtended by bract-like leaves, or if so then the bract-like leaves usually of different texture and vestiture than the involucral bracts 8

5   Plants known from 5,400-8,000 ft in n. Uintah and Daggett Cos.; leaves finely divided and more or less tripinnatified, the divisions about l-2 (3) mm wide, green and glabrous on both sides or sparingly tomentose along the midrib beneath; stems winged along the internodes by spiny decurrent leaf bases; involucral bracts sparingly tomentose along the margins; heads not in a terminal cluster . . . . *C. ownbeyi*

5   Plants known from higher elevations; leaves once or twice pinnatifid, some of the divisions regularly over 2 mm wide, glabrous or pubescent beneath; stems not particularly winged; involucral bracts various; heads various . 6

6   Plants rather common above 9,000 ft on the Uinta Mtns.; leaves about 15-25 (35) mm wide, the sinuses 5-10 (15) mm deep 7

6   Plants of the E. Tavaputs Plateau; some leaves usually over 30 mm wide, the larger sinuses 15-25 mm deep or deeper; involucral bracts various . . . . . .
. . . . *C. clavatum*

7   Involucral bracts copiously villous with 3-4 mm long multicellular hairs, not particularly pinnate-spinose; corollas yellowish white; heads more or less sessile in a compact terminal cluster, the cluster often closely subtended by leaves . . . . *C. murdockii*

7   Involucral bracts merely white-tomentose or rarely with short multicellular hairs, at least the outer ones copiously pinnate-spinose; corollas mainly pink or rose; heads sessile and clustered as above or remote and distinctly pedunculate . . . . *C. eatonii*

8   Leaves white tomentose beneath and green and thinly pubescent or glabrate above . . . . *C. pulcherrimum*

8   Leaves white pubescent on both sides or greenish on both sides 9

9   Stems winged-decurrent; involucral bracts usually with cobwebby hairs that appear to connect alternate bracts and cross over the bract between them . . . . *C. subniveum*

9   Stems not winged-decurrent; involucral bracts not as above 10

10 Involucres 15-20 mm high, 15-25 mm wide; stems and leaves usually densely white pubescent; plants endemic . . . . *C. barnebyi*

10 Involucres 19-34 mm high, some over 20 mm high in nearly all specimens, 15-45 mm wide; plants gray-white pubescent, but sometimes not so densely as above, widespread 11

11 Heads often turbinate to sub-cylindrical, the involucre 15-45 mm wide, the bracts seldom with a thickened glutinous, dorsal ridge, at least the inner ones narrowly wedge-shaped, minutely scabrous, often suffused with bright red or light purple toward the tips; corolla lobes 9-17 mm long . . . .
. . . . *C. arizonicum*

11 Heads campanulate, sometimes cordate in outline at the base, the involucre 20-60 mm wide, the bracts often with a thickened, glutinous, dorsal ridge, the inner ones various but not particularly narrowly wedge-shaped, not scabrous, usually brownish or grayish, dark brown or dark red upon drying; corolla lobes 6-9 mm long . . . . *C. undulatum*

Cirsium arizonicum [(A. Gray) Petrak] *C. calcareum* (M. E. Jones) Wooton & Standley; *C. pulchellum* (Greene) Wooton & Standley; *C. calcareum* var. *pulchellum* (Greene) S. L. Welsh] — Arizona thistle — Occasional across the Tavaputs Plateau and 2 specimens seen from south slope of Uinta Mtns from Rock Creek and Mosby Creek.; rubber rabbitbrush-Great Basin wildrye, Salina wildrye, sagebrush, mt. brush, ponderosa pine, and Douglas-fir communities; 5, 600-8,800 ft; mid July-Aug. Distinct from similar species with corolla lobes 9-17 mm long. A highly variable complex (FNA 19: 141-144). Most of our specimens likely belong to var. *bipinnatum* (Eastwood) D. J. Keil.

Cirsium arvense (L.) Scopoli — Canada thistle — Introduced from Eurasia; sometimes a serious weed in agricultural areas, forming patches or clones on ditch banks, and in pastures, fields, and riparian communities where frequent in beaver dams; 4,700-7,500 ft, and introduced (in contaminated hay) at recreation, logging, and livestock camps in mountains up to about 10,000; July-Aug.

Cirsium barnebyi S. L. Welsh & Neese in S. L. Welsh — Barneby thistle — Endemic to the Uinta Basin where rather widespread and locally common on the East and West Tavaputs Plateaus and one specimen from Lookout Mtn. in Moffat Co.; nearly barren, white, shale or marl limestone hills and slopes of Green River and Uinta Formations. Grading into *C. arizonicum,* but with corolla lobes 3-7 mm long.

Cirsium clavatum (M. E. Jones) Petrak — Fish Lake thistle — The 9 specimens seen are from the East Tavaputs Plateau from semi-barrens in desert shrub, subalpine fir, Douglas fir, aspen-fir-aspen parkland, and big sagebrush-snowberry communities at 7,000-8,600 ft; July. Annotations on these specimens indicate difficulty distinguishing this taxon from *C. eatonii* as they have been annotated as *C. scopulorum*. Morphology of some plants of southeastern Utah indicates hybridization between *C. clavatum* and *C. eatonii* (FNA 19: 127).

Cirsium eatonii (A. Gray) Robinson [*C. polyphyllum* (Rydberg) Petrak] — Eaton thistle — Uinta Mtns.; most specimens seen are mostly from w. of Rock Creek.; common to abundant on rocky ground and slopes at and above the upper edge of spruce woodlands at 10,600-11,500 ft, but with small-headed, pale-flowered plants at 7,800-8,250 in ponderosa pine and lodgepole pine communities in the Sheep Creek and Carter Creek drainages, Daggett Co.; July-Aug. Our plants apparently all belong to var. eatonii. However, var. *eriocepalum* (A. Gray) D. J. Keil [*C. scopulorum* (Greene) Cockerell ex Daniels with densely tomentose involucres might be found in our area.

Cirsium murdockii (S. L. Welsh) Cronquist — Uinta Mtn. thistle — Specimens seen are from e. of Rock Creek in the Uinta Mtns.; lodgepole pine and spruce forests, lake margins, krummholz, and rocky ground near and above timberline, and common on roadsides and in timber harvest areas; 9,800-11,400 (12,000) ft; July-Aug.

Cirsium ownbeyi S. L. Welsh — Ownbey thistle — Specimens seen are from Crouse Canyon and Browns Park (Daggett Co.), breaks and base of Blue Mtn. (Uintah Co.), and Cross Mtn. and Yampa River Canyon (Moffat Co.); juniper, sagebrush, ponderosa pine, lodgepole pine, and riparian communities, sometimes growing in crevices of bed rock and in talus; 5,400-8,000 ft; May-Aug. In addition to locations listed above, specimens were also seen from lower Green River Basin of Sweetwater Co., Wyoming. One specimen was seen from the House Range in Millard Co. Utah (Kass 1019 BRY). The specimen from Millard Co. indicates considerable potential for additional locations for this species.

Cirsium pulcherrimum (Rydberg) K. Schumann — pretty thistle — Mostly known from the Uinta Mountains within our area but also the East and West Tavaputs Plateau in sagebrush, mtn. brush, aspen, and mixed coniferous forest communities at 7500-9000 ft; July-Aug. Corolla lobes 5-7 mm long.

Cirsium scariosum Nuttall [*C. coloradense* (Rydberg) Cockerell; *C. drummondii* Torrey & A. Gray misapplied; *C. drummondii* var. *acaulescens* (A. Gray) Macbride; *C. foliosum* Torrey & A. Gray misapplied; *C. tioganus* Congdon; *C. tioganum* var. *coloradoense* Dorn] — elk thistle, meadow thistle — Widespread and common; moist and dry-meadows, and occasionally in aspen-conifer, and sagebrush communities but then usually adjacent to meadows. The tall phase is more common from 5,300-8,900 ft; June-Aug. The stemless phase is more common from 9,000-10,000 ft. Both phases occur together in some populations.

Cirsium subnivium Rydberg — snowy thistle — The few specimens seen are from roadsides in Daggett Co. within the mtn. big sagebrush and ponderosa pine belts. With corolla lobes 3.5-7 mm long this is similar to *C. undulatum*. Included in *C. inamoenum* (Green) D. J. Keil in FNA.

Cirsium undulatum (Nuttall) Sprengel A. — wavyleaf thistle — Occasional; widespread from Strawberry Valley across the Uinta Mtns. and West and East Tavaputs Plateaus and with a few specimens from the bottom of the Uinta

Basin; desert shrub, black sagebrush, Wyoming big sagebrush, mtn. big sagebrush, pinyon-juniper, Salina wildrye, ponderosa pine, Douglas-fir, limber pine, and disturbed riparian communities; 4,900-9,000 ft; June-July. Graham 6250, 9087, and 9180 (CMI), referred to *C. megacephalum* (A. Gray) Cockerell (Graham 1937), belong here. Two vars. listed in AUF (196) for our area:

1 Heads mainly less than 2.5 cm wide, (1) 3-10 per stem .... **var. *tracyi***
1 Heads, at least the largest, over 2.5 cm wide, commonly 1-3 per stem ...
....**var. *undulatum***

Var. *tracyi* (Rydberg) S. L. Welsh [C. tracyi (Rydberg) Petrak] — Tracy thistle — Summit and Uintah Cos.

Var. *undulatum* — wavyleaf thistle — Daggett, Duchesne, Summit, and Uintah Cos.

Cirsium vulgare (Savi) Tenor (*C. lanceolatum* Hill misapplied) — bull thistle — Introduced from Eurasia; widespread and weedy in many plant communities, and on agricultural lands, often on disturbed ground; up to about 9,000 ft; July-Aug. This species increases rapidly and abundantly following fire, logging, and other activities, and usually decreases to low presence within a few years. Although the abundant increase of this plant following disturbance might invoke an urgent sense for control measures, a little patience will likely achieve the same results and save a lot of money.

Conyza Less.

Conyza canadensis (L.) Cronquist (*Erigeron canadensis* L.) — Canadian horseweed — Annual herbs from taproots, 40-100 cm tall; stems usually much branched; leaves alternate, simple, 2-10 cm long, 2-8 mm wide, linear to oblanceolate, entire or ciliate-serrate, often deciduous by late anthesis; heads numerous, in panicles, small, the involucres 2-4 mm high, 3-7 mm wide, the bracts more or less imbricate, the receptacle naked; rays numerous, inconspicuous, shorter than the pappus, white or purplish; disk flowers yellowish; pappus of capillary bristles. Widespread in many plant communities at 5,400-7,500 ft; July-Aug.

Crepis L. — Hawksbeard

Perennial herbs with milky juice; leaves basal or alternate; heads few to numerous in corymbose or paniculate clusters; involucral bracts in 1 or 2 series, the outer ones much reduced; receptacle naked; rays yellow; disk flowers none; pappus of capillary bristles. *Crepis nana* is included in *Askellia* W. A. Weber, and our other species are included in *Psilochenia* Nuttall in Weber (1987).

---

1 Leaves and stems glabrous or hispid; leaves entire to pinnatifid, mostly basal..2
1 Leaves and stems more or less tomentose at least in part; leaves pinnatifid, some cauline..3
2 Heads more or less nested in the leaves; plants mostly less than 10 cm tall, of alpine talus slopes ....***C. nana***
2 Heads well exserted above the leaves; plants mostly of meadows, not alpine ....***C. runcinata***

3   Involucres glabrous or nearly so, the outer bracts sometimes with some pubescence; heads (5) 20-100 per stem, each with 5-10 flowers . . . . *C. acuminata*

3   Involucres tomentose and sometimes hispid; heads 2-30 (60) per stem, each with 7-60 flowers 4

4   Terminal leaf lobe more than 5 cm long . . . . *C. atribarba*

4   Terminal lobe less than 5 cm long; achenes usually yellowish or brownish 5

5   Involucres more than twice longer than wide, with 7-8 (12) inner bracts; flowers generally 7-12 (16) in each head . . . . *C. intermedia*

5   Involucres less than twice longer than wide, with 8-14 inner bracts; flowers generally 10-40 in each head 6

6   Involucre and/or stems conspicuously setose, the seta not glandular . . . . *C. modocensis*

6   Involucre and stems not or sparingly setose, if setose then the seta gland-tipped . . . . *C. occidentalis*

Crepis acuminata Nuttall — tapertip hawksbeard — Widespread and mostly occasional in sagebrush, rabbitbrush, Salina wildrye, mt. brush, ponderosa pine, aspen, and Douglas-fir communities at 7,600-9,400 ft; June-July.

Crepis atribarba A. Heller — slender hawksbeard — Specimens seen are from Daggett Co. Little Brush Creek drainage and eastward in Uintah Co. and from Beaver Canyon and Slab Canyon drainages of the West Tavaputs Plateau in Wasatch Co., no specimens seen from Duchesne Co. or western Uintah Co.; forb-grass, pinyon-juniper, fringed sagebrush, mtn. big sagebrush, mtn. brush, aspen parkland, and ponderosa pine communities at 6,800-9,640 ft; June-Aug. Also spelled *C. atrabarba.*

Crepis intermedia A. Gray — Gray hawksbeard — Widespread; infrequent in desert shrub, sagebrush, pinyon-juniper, and mtn. brush communities; 5,600-9,300 ft; June. This taxon represents a series of intermediate hybrids involving *C. acuminata* and *C. modocensis* and *C. occidentalis.*

Crepis modocensis Greene (*C. scopulorum* Coville) — low hawksbeard — Common in Daggett Co from Sheep Creek and eastward, occasional in Uintah Co from Taylor Mtn eastward into Colorado, no specimens seen from Duchesne, Summit, or Wasatch Cos.; desert shrub, black sagebrush, Wyoming big sagebrush-grass, mtn. big sagebrush-grass, pinyon-juniper, and mtn. brush communities; 6,000-8,500 ft; May-June (July).

Crepis nana Richardson in Franklin — dwarf hawksbeard — Uinta Mtns., infrequent on alpine talus slopes where rock fragments are gravel size and creep down slope. The root system is highly adapted to talus-creep. Our plants belong to var. *nana.*

Crepis occidentalis Nuttall — western hawksbeard — Widespread and occasional across much of our area (no specimens seen from Summit or Wasatch Cos.; desert shrub, black sagebrush, big sagebrush, pinyon-juniper, and rabbitbrush communities; 5,100-8,900 ft; late May-mid July. Our plants belong to var. *occidentalis.*

Crepis runcinata (James) Torrey & A. Gray — dandelion hawksbeard, meadow hawksbeard — With 2 vars. in our flora:

1 Involucres hispid with black hairs . . . . **var. *hispidulosa***
1 Involucres glabrous or merely puberulent . . . . **var. *glauca***

Var. *glauca* (Nuttall) S. L. Welsh (C. glauca Nuttall) Widespread and occasional to common in seeps, meadows, pastures, along ditches, and other waterways, often in alkaline places; 5,300-7,000 ft; June.

Var. *hispidulosa* Howell ex Rydberg. Occasional; widespread in the Uinta Mtns. and adjacent to these mtns. and Strawberry Valley; non-alkaline, moist, wet, or boggy meadows, and riparian communities; 6,400-8,700 ft; July-Aug. Some specimens from our area have been mistakenly identified as *Hieracium*.

Dyssodia Cavanilles

Dyssodia papposa (Ventenat) A. S. Hitchcock — pappose glandweed — Plants annual, 15-40 cm tall, glabrous or sparingly puberulent; lower leaves opposite, the upper ones alternate, pinnatifid into 11-15 lobes, these sometimes again lobed; involucral bracts 6-12, 6-10 mm long, oblanceolate, with yellowish oil glands; rays 8 or fewer, little longer than the involucre, yellow-orange; disk flowers dull yellow; pappus of about 20 scales, each dissected into 5-10 bristles that might be mistaken for capillary bristles; achenes 8-35 mm long. This somewhat weedy species is likely introduced from the Great Plains. The few specimens seen are from gardens and roadsides below 6,000 ft.

Enceliopsis (A. Gray) A. Nelson

Perennial herbs; leaves all basal, the blades spatulate, lanceolate, oblanceolate, ovate or orbicular; heads solitary; involucral bracts in 2-3 series; receptacle chaffy, the scales clasping the achenes; flowers yellow; pappus of 2 awns, with or without small scales between them, or none.

---

1 Heads with disk flowers only, often nodding . . . . *E. **nutans***
1 Heads with ray and disk flowers, not nodding . . . . *E. **nudicaulis***

Enceliopsis nudicaulis (A. Gray) A. Nelson — false sunflower — The several specimens seen are all from e. Uintah Co., plotted in Moffat Co. in TPD; desert shrub and juniper communities; 4,800-6,500 ft; May-mid June. Graham's (1937) listing of *E. argophylla* (D.C. Eaton) A. Nelson is likely based on a specimen of *E. nudicaulis*. Herbage more or less tomentose; plants from caudices. Our plants belong to var. *nudicaulis*.

Enceliopsis nutans (Eastwood) A. Nelson — noddinghead — Several specimens-seen from Uintah Co. and one from Duchesne Co.; desert shrub and juniper communities; 4,700-5,700 ft; late April-mid June. Herbage pilose-hirsutulous; plants from tuberous roots.

Erigeron L. — Daisy; Fleabane

Annual or perennial herbs; leaves alternate, rarely all basal; heads solitary to several; involucral bracts narrow, often equal, in a single series, but sometimes imbricate mostly herbaceous throughout; receptacles naked; rays white, pinkish, bluish, or purple, mostly numerous, sometimes lacking; disk flowers yellow; pappus of capillary bristles.

1   Lower leaves deeply palmately lobed or divided into linear segments . . . .
.... *E. compositus*

1   Leaves entire to toothed or shallowly lobed 2

2   Rays lacking or inconspicuous not over 5 mm long, mostly erect. 3

2   Rays conspicuous, spreading, usually over 5 mm long 4

3   Involucral bracts more or less imbricate, the inner ones usually long
attenuate, often glandular; cauline leaves narrowly lanceolate to broader,
rarely linear; plants known from Lake Fork and Uinta drainages of the Uinta
Mtns. . . . . *E. acris*

3   Involucral bracts mostly in a single series, not glandular; cauline leaves
more or less linear; plants widespread . . . . *E. lonchophyllus*

4   Stem leaves well developed, only slightly reduced in size and number
upward on the stems, not linear; heads often more than 1 per stem 5

4   Stem leaves often much reduced in size and number upward on the stem
or lacking, often linear or narrowly oblong, the basal ones often persisting
and functional after flowering time 10

5   Rays 2-4 mm wide; achenes with 4-7 nerved . . . . *E. peregrinus*

5   Rays 1-2 mm wide; achenes with 1-4 nerves 6

6   Hairs of the involucres with black crosswalls near the base; rays white;
plants of the n. slope of the Uinta Mtns. . . . . *E. coulteri*

6   Hairs of the involucres not as above; rays white, pink, or blue 7

7   Stem leaves glabrous; lower part of stems glabrous   . . . . *E. eximius*

7   Stems leaves pubescent at least on the margins; stems commonly hairy . 8

8   Plants with more or less well developed caudex . . . . *E. speciosus*

8   Pants fibrous rooted without or with weakly developed caudex. 9

9   Hair of stems appressed; plants common in our area . . . . *E. glabellus*

9   Hair of stems spreading; plants apparently rare in our area . . . .
. . . . *E. formosissimus*

10 Leaves basal or nearly so, stem leaves, if present, mostly not extended above
the basal tuft; plants mostly 2-10 cm tall. 11

10 Leaves not all basal; stems mostly with 1 or more leaves borne above the
basal tuft 17

11 Leaves commonly glabrous on the lower surface. 12

11 Leaves mostly pubescent on the lower surface. 13

12 Involucral bracts villose with some of the hairs with purple cross walls;
plants known from Lake Fork Mountain . . . . *E. radicatus*

12 Involucral bracts strigose, the hairs without purple cross walls . . . .
. . . . *E. nematophyllus*

13 Involucral bracts about equal; involucres and peduncles villose with
contorted multiple-cellular hairs; plants known from Daggett Co. *E. nanus*

13 Involucral bracts imbricate 14

14 Achenes ciliate on the 2 marginal nerves, otherwise glabrous . . . .
. . . . *E. compactus*

14 Achenes more uniformly pubescent. 15

15 Pubescence of the stem widely spreading to somewhat retrorse ....
.... *E. caespitosus*

15 Pubescence of the stem mostly ascending to appressed............... 16

16 Leaves narrowly but conspicuously oblanceolate; involucral bracts not so uniformly linear, sometimes with rather broad hyaline margins, sometimes purplish, the inner ones 1-1.5 mm wide .... *E. untermannii*

16 Leaves linear or nearly so; involucral bracts rather uniformly linear, the inner ones about 1 mm wide .... *E. nematophyllus*

17 Rays 2-4 mm wide; plants silvery pubescent; xylorhiza-like with imbricate involucral bracts and well developed woody caudices .... *E. pulcherrimus*

17 Rays less than 2 mm wide; plants different from above in 1 or more features
.... 18

18 Stems slender, rather flexuous, at least some usually trailing along the ground with leaves oriented upward; flowering stems more upright, usually with fewer leaves, with a single head .... *E. flagellaris*

18 Stems slender to rigid but not trailing............................. 19

19 Pubescence of stems widely spreading 20

19 Pubescence of stems mostly appressed or ascending-spreading........ 25

20 Plants annual or biennial with simple taproot; a few of the lower leaves sometimes lobed .... *E. divergens*

20 Plants perennial... 21

21 Involucres densely villous or wooly-villous with long, multiple-cellular hairs; leaves villous or villous-ciliate .... *E. simplex*

21 Pubescence not as above 22

22 Plants with simple or sparingly branched subrhizomatous caudex; heads 1-6 per stem .... *E. formosissimus*

22 Plants commonly with branched caudex............................ 23

23 Basal leaves with the midnerve and usually 2 lateral nerves prominent ..
.... *E. caespitosus*

23 Basal leaves with 0-1 conspicuous nerves 24

24 Rays exceeding the involucre and disk flowers by 7-12 mm, white to blue; caudex often with a noticeable thatch of fibrous old leaf bases; pubescence of stems widely spreading .... *E. pumilus*

24 Rays exceeding the involucre and disk flowers by about 5-7 mm, white, rarely bluish; plants without much buildup of old leaf bases; pubescence of stems mostly ascending-spreading; pubescence of petioles and lower part of leaves widely spreading .... *E. engelmannii*

25 Basal leaves oblanceolate or broader, the blade well defined, usually more or less abruptly contracted to a petiole............................ 26

25 Basal leaves linear or rather narrowly oblanceolate, the blade, if distinguishable, gradually tapering to a petiole...................... 29

26 Leaves essentially glabrous .... *E. leiomerus*

26 Leaves more or less hairy... 27

27 Heads 1-13 (15) per stem; involucres10-20 mm wide .... *E. glabellus*

27 Heads 1-3 per stem 28

28 Rays blue to purple; leaves acute apically .... *E. tener*

28 Rays mostly white; leaves rounded apically .... *E. wilkenii*

29 Stems erect or strongly ascending, the upper half with bract-like leaves shorter than the internodes; lower leaves usually withered by anthesis; involucres 3-5 mm high; plants known from Chandler Canyon of Desolation Canyon drainage .... *E. utahensis*

29 Stems decumbent to erect, often with leaves nearly equal to or longer than the internodes; lower leaves mostly as persistent as the upper leaves; plant of various distribution 30

30 Basal leaves (0.5) 1-2.3 mm wide; plants of Moffat and Uintah Cos. .... *E. nematophylloides*

30 Some of the basal leaves usually over 2.3 mm wide.................. 31

31 Base of leaves with widely spreading hairs, the blades with 0-1 conspicuous nerves; plants not purplish, often without a conspicuous tuft of enlarged basal leaves; leaves of the stem often about as long, or longer than those of the base; rays white .... *E. engelmannii*

31 Leaves without widely spreading hairs, the blades with 1-3 conspicuous nerves; plants often purplish at the base, usually with a few enlarged basal leaves; stems decumbent-ascending, sometimes with much reduced leaves above; rays white to blue .... *E. eatonii*

Erigeron acris L. — bitter fleabane — The 9 specimens seen are from the Hades, Blind Stream, Swasey Hole, Yellowstone, and Uinta drainages of the s. slope and Henrys Fork and Beaver Creek drainages of the n. Slope of the Uinta Mtns.; spruce-fir-lodgepole pine communities and at timberline, usually on sandy or rocky ground; 8,100-11,500 ft; June-July. Our plants belong to var. *debilis* A. Gray.

Erigeron caespitosus Nuttall [*E. goodrichii* S. L. Welsh] — tufted daisy — Common; known from Strawberry Valley, Uinta Mtns., Cold Spr. Mtn.; sagebrush, dry aspen, and alpine communities, often on rocky, dry ground; 8,600-13,000 ft; June-July. Small plants with narrow leaves have been included in *E. goodrichii*. However, a continuum in size, leaf shape, and other features is found in the complex. The small phase is mostly found at high elevations. However, random plants of lower elevations have the smaller features of the alpine plants. Also some high elevation plants have wide leaves and other features of typical *E. caespitosus*. The widely spreading and sometimes retrorse hairs of the stem is a common feature in large and small plants.

Erigeron compactus Blake — mound daisy — Specimens seen are from near Myton, Duchesne Co. (Neese & B. Welsh 7214 BRY), s. of Duchesne and e. of Indian Canyon (Atwood 32184), Jesse Ewing Canyon, Daggett Co. (Neese & Peterson 5544 BRY), Vermillion Creek (Peterson & Wiley-Eberly 83-79 CS) and from near summit of Lookout Mt. (Peterson & Wilken 83-289; Goodrich and Huber 28116); desert shrub and Utah juniper communities and soft, white, tuffaceous, eroding mounds; 5,300-6,850 ft; May-June. Our plants belong to var. *consimilis* (Cronquist) Blake.

Erigeron compositus Pursh — fern-leaf daisy — Common; Uinta Mtns., Strawberry Valley, and Kleins Hill near the Three Corners area; sagebrush, coniferous forest, and alpine communities, usually on rocky ground; 7,800-13,000 ft; June-July.

Erigeron coulter Porter — larger mtn. daisy — North slope of the Uinta Mtns. in aspen and coniferous forests.

Erigeron divergens Torrey & A. Gray — spreading daisy — Widespread but no specimens seen from the W. Tavaputs Plateau; occasional in desert shrub, sagebrush, aspen, ponderosa pine, and lodgepole pine communities, also on disturbed ground of riparian communities; 4,950-8,800 ft; June-July (Sept. with sufficient fall moisture).

Erigeron eatonii A. Gray — Eaton daisy — Widespread; occasional to frequent in pinyon-juniper, mtn. brush, mtn. big sagebrush, lodgepole pine, Engelmann spruce, and alpine; (5,500) 7,100-11,000 ft; June-Aug. High elevation plants often have narrow leaves with only 1 instead of the 3 conspicuous nerves common in lower elevation plants. Our plants belong to var. *eatonii*.

Erigeron engelmannii A. Nelson — Engelmann daisy — Widespread; occasional in desert shrub, greasewood, pinyon-juniper, and sagebrush communities 5,000-7,200 ft; may-June. Based on Wilken 13058, Erigeron ochroleucus Nuttall has been reported for our area. However, this is a plant of the Plains from w. of the Continental Divide (Cronquist 1947), and specimens of our area identified as such likely belong with *E. engelmannii*.

Erigeron eximius Greene (*E. superbus* Greene ex. Rydberg) — spruce-fir daisy — Uinta Mtns.; occasional and locally common in aspen, coniferous forest, streamside, and meadow communities; 7,400-9,800 ft; July-Aug.

Erigeron flagellaris A. Gray — trailing daisy — Widespread; common in sagebrush, pinyon-juniper, mt. brush, aspen, coniferous forest, meadow, and streamside-willow communities; 7,000-9,600 ft; late May-July.

Erigeron formosissimus Greene (*E. viscidus* Rydberg) — beautiful daisy — Reported for Duchesne and Summit Counties (AUF 210), Daggett Co. (TPD), Uinta Mountains (IMF 5: 314). Listed by Graham (1937) for near Moon Lake; meadow, aspen, and mtn. brush communities. A note by Cronquist (IMF 5: 314) that "Diverse forms of *E. formosissimus* approach *E. glabellus* var. *pubescens*, *E. eximus*, and *E. ursinus*," indicates odd specimens of these species could be mistaken for *E. formosissimus*.

Erigeron glabellus Nuttall — Smooth daisy — Strawberry Valley, Uinta Mtns., and pastures and meadows at the n. part of the Uinta Basin; 5,400-7,900 ft; June. Our plants belong to var. glabellus.

Erigeron leiomerus A. Gray — rockslide daisy, glaber daisy — Occasional across the Uinta Mtns. in coniferous woods, rocky slopes, ledges, cliffs, talus, fell fields, and alpine tundra; 9,200-12,000 ft; July-Aug.

Erigeron lonchophyllus Hooker — longleaf daisy — Widespread but seldom collected; riparian communities, wet ground of seeps, wet meadows, and bogs; 5,200-7,200 (9,300) ft; mid June-July.

Erigeron nanus Nuttall — dwarf daisy — The 5 specimens seen are from the north flank of the Uinta Mountains on Phil Pico Mt. and in Sols Canyon area, Daggett Co.; open ridges and forb-grass communities at 9,300-9,400 ft; July.

Erigeron nematophylloides S. L. Welsh & A. Huber — linear-leaf daisy — Weber Sandstone in the vicinity of Split Mtn. and Madison Limestone in Irish Canyon and on Limestone Mtn.; endemic to Moffat and Uintah Cos.; rock debris and face and cracks of rock outcrops at 5,300-7,380 ft.

Erigeron nematophyllus Rydberg — needleleaf daisy — Locally common; Cross

Mt., Three Corners area, Goslin Mt. to Round Top Mt. also Tavaputs Plateau
from Atachee Ridge to Nutters Ridge in desert shrub, pinyon-juniper and mt.
brush communities at (5,800) 7,200-9,400 ft, and limestone plateaus and ridges
of the Blind Stream area of the Uinta Mtns. in alpine and Krummholz
communities to 11,200 ft; mostly on rocky ground or soil of poor substrate;
May-July. With small stature (mostly 1-10 cm) and high elevation habitat, the
Uinta Mtn. plants have been confused with *E. radicatus*, but the hairs of the
leaves and involucres are different. The high elevation plants of the Uinta
Mtns. have also been confused with *E. caespitosus*. However the mostly ascending
and appressed hairs of the stem set this apart from *E. caespitosus*. With more
work the Uinta Mtn. plants might be assigned elsewhere or described as
something new. However, no new taxon is proposed here. The report of *E.
filifolius* for Daggett Co. (AUF 210) is perhaps based on specimens of this
species.

Erigeron peregrinus (Pursh) Greene [*E. peregrinus* ssp. *callianthemus* var. *scaposus*
(Torrey & A. Gray) Cronquist; *E. salsuginosus* (Richardson) A. Gray] — peregrine
daisy, subalpine daisy — Uinta Mtns.; common in streamside meadows and
aspen and moist coniferous woods, and occasionally in rocky places near or
above timberline; 7,500-11,600 ft; late June-Sept. Uinta Mtn. plants have been
referred to as var. *callianthemus* Cronquist. A mostly alpine, often subscapose
phase that is mostly shorter than 20 cm tall with relatively ample, apically
rounded or obtuse basal leaves and very much smaller cauline leaves is
referable to var. *scaposa* (Torrey & A. Gray) Cronquist. However, this phase
passes freely into the larger phase.

Erigeron pulcherrimus A. Heller — basin daisy — Uinta Basin and lower
elevations of the Tavaputs Plateau in desert shrub, and pinyon-juniper communities, often on formations that weather to badlands; 4,900-6,500 ft; late
May-June. Reports of *E. argentatus* A. Gray (Graham 1937) and *E. utahensis* A.
Gray (Holmgren 1962) are based on specimens of *E. pulcherrimus*.

Erigeron pumilus Nuttall [*E. concinnus* (Hooker & Arnot) Torrey & A. Gray] —
Vernal daisy, shaggy daisy — Widespread; occasional to common in desert
shrub, sagebrush, pinyon-juniper, and mtn. brush communities often on stony
hills and ridges and gravel or cobble terraces and benches; 4,700-7,900 ft; late
April-June. Our plants belong to ssp. *concinnoides* Cronquist var. *concinnus*
(Hooker & Arnot) Dorn. Some plants have been referred to ssp. *intermedius*
Cronquist. Macleod 8A DINO, apparently the basis of reports of *E. vetensis*
Rydberg, belong here.

Erigeron radicatus Hooker (*E. huberi* S. L. Welsh & N. D. Atwood) — taproot
daisy — Known from Lake Fork Mtn., Uinta Mountains where locally common
among limestone gravel and cobbles at 10,600-10,900 ft; July-Aug. Disjunct
from central Idaho and Sublette Co, Wyoming and northward to Alberta and
Saskatchewan.

Erigeron simplex Greene — one-flower daisy — Uinta Mtns.; occasional to
common in alpine tundra, fell fields, rock strips, talus, and upper Engelmann
spruce communities; 10,600-13,000 ft; July-Aug. The closely related *E.
melanocephalus* Rydberg might be expected on the Uinta Mtns. The cross-walls
of involucral hairs of our plants are white or sometimes a few are light reddish
or light purplish or rarely dark, but none of our numerous specimens show
the consistently dark purple cross-walls of *E. melanocephalus* from the LaSal

Mtns. of se. Utah or from high mtns. of Colorado.

Erigeron speciosus (Lindley) D.C. — Oregon fleabane, showy fleabane — With 4 freely intergrading and often sympatric vars. in our area:

1 Leaves ciliate, glabrous on the surface; stems mostly glabrous . . . . **var. *speciosus***

1 Leaves glandular or pubescent on one or both sides; stems hairy or glandular or both 2

2 Involucral bracts merely glandular, rarely also somewhat spreading-hairy; upper leaves often ovate . . . . **var. *macranthus***

2 Involucral bracts glandular and commonly also spreading-hairy; upper leaves lance-attenuate. 3

3 Leaves and usually the stems with spreading hairs but not glandular or the upper few leaves slightly glandular; involucres spreading hairy and more or less glandular . . . . **var. *mollis***

3 Leaves and stems glandular or glandular-scabrous, sometimes also sparsely long-hairy . . . . **var. *uintahensis***

Var. *macranthus* (Nuttall) Cronquist (*E. macranthus* Nuttall) — aspen daisy — Common; Tavaputs Plateau, mostly e. of the Avintaquin drainage; sagebrush, mt. brush, aspen, and Douglas-fir communities; 8,200-10,050 ft; July-Aug.

Var. *mollis* (A. Gray) S. L. Welsh (*E. subtrinervis* Rydberg) — three nerve daisy — Occasional to common; Strawberry Valley, e. to Rock Creek in the Uinta Mtns., and E. Tavaputs Plateau; open slopes and openings in coniferous woods, and dry meadows; 7,200-10,000 ft; July-Aug.

Var. *speciosus* — aspen daisy — Occasional; Tavaputs Plateau, and Uinta Mtns. from Rock Creek and west, apparently replaced by var. *uintahensis* to the e. in the Uinta Mtns.; sagebrush, mt. brush, aspen, spruce, fir, riparian, and meadow, communities; 7,500-10,000 ft; July-Aug.

Var. *uintahensis* (Cronquist) S. L. Welsh (*E. uintahensis* Cronquist) — Uinta daisy — Occasional; Strawberry Valley and Uinta Mtns.; sagebrush, aspen, ponderosa pine, lodgepole pine, and riparian communities; 7,300-9,000 ft; July-Aug.

Erigeron tener (A. Gray) A. Gray — thin daisy — The one specimen seen (Goodrich 21439) is from Smith and Morehouse Canyon of the north slope of the Uinta Mtns., from limestone ledges at 8050 ft.

Erigeron ursinus D.C. Eaton. — bear daisy — Common; Uinta Mtns., W. Fork Duchesne drainage, and on Strawberry Peak of the W. Tavaputs Plateau; dry meadows especially at the edge of coniferous woods, streamside meadows, open ridge tops, sometimes aspen and coniferous woods, rarely in sagebrush communities and then usually near meadows; (7,500) 8,500-11,000 ft; June-Aug.

Erigeron untermannii S. L. Welsh & Goodrich — marly ridge daisy — Endemic, locally common from Lake Canyon e. to Wild Horse Ridge, W. Tavaputs Plateau; mostly on open, windswept, marly ridge tops with Salina wildrye and on adjacent slopes with pinyon-juniper, scattered limber pine and Douglas-fir communities at 7,000-9,300 ft; May-June. Plants of the Wasatch Plateau included in *E. carringtoniae* S. L. Welsh in AUF (208) are included in *E. untermannii* in FNA (20:284).

Erigeron utahensis — Utah fleabane — Known in our area from a single collection (N. D. Atwood & P. Fontane 32466) from Chandler Canyon about 1 mile up from the Green River. The specimen belongs to var. *sparsifolius* (Eastwood) Cronquist. This species is rather common to the s. of our area.

Erigeron wilkenii O'Kane — Wilken fleabane — Known from alluvium and colluvium at the base of sandstone walls in the canyons of the Yampa River, Dinosaur National Monument at 4,950-7,570 ft. Rays commonly dry bluish or pinkish but apparently they are all white when fresh. Dry specimens might be difficult to distinguish from *E. tener*.

Eupatorium L. — Joe-Pye-weed

Eupatorium maculatum L. — spotted Joe-pye-weed — Perennial herbs 50-150 cm tall, more or less purple-spotted; leaves opposite or at least the lower ones usually whorled, short-petiolate or nearly sessile, simple, serrate or crenate serrate; inflorescence flat-topped; involucres about 7-9 mm high, the bracts in 4 or 5 series, striate and pinkish; receptacle naked; rays none; disk flowers purple, pink, or rose; pappus of capillary, scabrous bristles. The few specimens seen are from ditch banks and riparian communities near the towns of Lapoint, Tridell and Whiterocks; 5,500-6,000 ft; Aug.-early Oct.

Euthamia (Nuttall) Elliott

Euthamia occidentalis Nuttall [*Solidago occidentalis* (Nuttall) Torrey & A. Gray] — western goldenrod — plants 50-100 cm tall or taller; leaves resinous punctate, linear or narrowly elliptic-linear, not at netted-veined, the midrib flanked by 2 (rarely more) parallel, lateral veins, these sometimes faint or lacking; inflorescence leafy-bracteate, not at all secund; heads in small glomerules, with yellow ray and disk flowers; pappus of capillary bristles. Widespread; locally abundant in seeps, alluvium of rivers, along gullies, ditches, edges of ponds, and in wet bottomlands, often where alkaline or saline; 4,700-6,200 ft; late July-mid Oct. The report of *Selloa glutinous* Sprengel for along the Green and Yampa Rivers (Potter and others 1983) is based on specimens belonging here. Differing from tall species of *Solidago* in the not at all secund inflorescence.

Gaillardia Fougeroux — Blanket flower; Gaillardia

Perennial herbs; leaves alternate; heads solitary on stem or branches; involucral bracts in 2-3 series, ovate or lanceolate, strongly reflexed in fruit; receptacles with subulate or bristle-like chaffy bracts; rays yellow to reddish-purple; pappus of lanceolate, scarious, awned scales, the awn twice as long as the body.

---

1   Disk flowers dark purple; rays yellow or reddish-purple at the base ......
     .... *G. aristata*
1   Disk and ray flowers yellow.. 2
2   Leaves pinnately lobed or parted .... *G. flava*
2   Leaves entire or less commonly toothed .... *G. spathulata*

Gaillardia aristata Pursh — common perennial blanket flower — Eastern Uinta Mtns. from Dry Fork and Carter Creek and e. to Blue Mt.; uncommon or occasional in mtn. big sagebrush, ponderosa pine, aspen, and lodgepole pine communities; 7,000-9,400 ft; June-Sept. Flowers and others 195 (UT!), reported as *G. gracilis* A. Nelson (*G. pinnatifida* Torrey) (Flowers and others 1960) for Hideout Canyon belongs here.

Gaillardia flava Rydberg — yellowflower — Entering the Uinta Basin area in Desolation Canyon; cottonwood riparian, and shrub communities at 4,100-4,200 ft.

Gaillardia spathulata A. Gray. — basin blanket flower — The 2 specimens seen are from the Walker Hollow drainage on the east side of the Green River, Uintah Co., desert shrub communities.

Gaillardia flava Rydberg — yellowflower

Glyptopleura D.C. Eaton

Glyptopleura marginata D.C. Eaton. — crustweed — Depressed winter-annual forbs, seldom over 3 cm tall, with milky juice; leaves in a compact rosette, pinnatifid, the margins toothed, white-crustaceous and scarious, 2-5 cm long; heads sessile in the rosette or short pedunculate; involucre about 10-12 mm high, the bracts in 2 series, the inner scarious-margined, the outer crustaceous-margined and pinnatifid above and lacerate-fringed near the base; receptacle naked, rays inconspicuous, hardly if at all exceeding the involucre; disk flowers none; pappus of capillary bristles. Three specimens seen are from the mouth of Pariette Draw (Neese 4422), Willow Creek (Atwood 15178), and 4 miles s. of Ouray (L. Shultz & J. Shultz 3367) in desert shrub communities; 4,700-4,900 ft; May-June. This low growing plant might be overlooked, and perhaps it is more common in our area than the few specimens indicate. However, the crustose-margined leaves are an eye-catching feature that indicates serious botanists would see it and make collections.

Gnaphalium L. — Cudweed

Annual or biennial herbs from taproots, more or less woolly throughout; leaves alternate, simple, narrow, entire; heads in glomerules terminating the stems and branches; involucral bracts imbricate in several series, scarious; receptacles naked, usually pitted; rays none; disk flowers inconspicuous; pappus of capillary bristles.

---

1   Stems (10) 20-80 cm tall, rarely much branched; inflorescence 2-3 cm wide or wider. 2

1   Stems 5-20 cm tall, usually much branched; inflorescence narrower than above 3

2   Stems and both surfaces of leaves woolly throughout, not viscid . . . . ***G. stramineum***

2   Stems and upper surface of leaves strongly viscid with multicellular glandular hairs, the lower surface (especially of lower leaves) often woolly . . . . ***G. macounii***

3   Involucral bracts generally brown with a whitish tip; plants loosely tomentose, native; leaves linear to oblanceolate or oblong . . . . ***G. palustre***

3   Involucral bracts generally greenish or brownish to the tip; plants appressed tomentose, introduced; leaves linear or narrowly oblanceolate . . . . ***G. exilifolium***

Gnaphalium exilifolium A. Nelson [*Filaginella uliginosa* (L.) Opiz] — slender-leaf cudweed — Widespread, but no specimens seen from west of the Uinta drainage on the south slope of the Uinta Mtns; floodplain of the Green River, pond-margins, mud flats of draw-down basins, recent burns, clear-cuts in lodgepole pine forests, ephemeral pools, drained beaver ponds, ephemerally moist ground at watering places for livestock, 4,725-9,950 ft. June-Sept.

Gnaphalium macounii Greene (*G. viscosum* Humboldt, Bonpland, & Kunth misapplied) — viscid cudweed — Specimens seen are from the Uinta Mtns. from Dry Fork Canyon and east on the south slope and Green Lakes and east on the north slope; ponderosa pine and lodgepole pine communities, in ephemeral pools and clearcuts and in well drained soil of forests and rock outcrops at 7,900-9,800; July-Sept. Inflorescence of a single cluster as in *G. stramineum* or with few to several clusters, these sometimes on branches up to 12 cm long.

Gnaphalium palustre Nuttall [*Filaginella palustris* (Nuttall) Holub] — lowland cudweed — Widespread but no specimens seen from the Tavaputs Plateau; ephemeral pools, margins of ponds and reservoirs, sandbars, ditch banks, clearcuts in coniferous forests, dry meadows, disturbed or open ground of riparian areas including drained beaver ponds; 4,660-9,800 ft; June-Sept.

Gnaphalium stramineum H.B.K (*G. chilense* Sprengel) — cotton-batting cudweed — Specimens seen are from along the Green and Yampa Rivers in Daggett and Moffat Cos., and from Dripping Springs and near Brownie Lake in Daggett Co., outwash plains of the Uinta and Whiterocks drainages, and at a seep in the Grouse Creek drainage; riparian, pasture, and hillside-seep, burned pinyon-juniper and lodgepole pine communities, 5,460-8,600 ft; July-Aug. In addition to features of the key, this differs from *G. macounii* with the inflorescence mostly a single cluster of heads with branches not over 2 cm long.

Grindelia Willdenow — Gumweed; Resin-weed

Grindelia squarrosa (Pursh) Dunal — curlycup gumweed — Strongly resinous, biennial or short-lived perennial herbs from taproots; stems 1-several, 10-70 (120) cm tall, green to straw-colored; leaves alternate, 2-8 cm long, entire to toothed; heads terminal on stems and branches; involucres 7-9 mm high, the bracts narrow, imbricate in several series, often strongly recurved, the outer ones usually reflexed from the base; receptacles naked; rays yellow, 7-15 mm long; disk flowers yellow; pappus of 2-3 (6) deciduous awns. Widespread; common; more or less weedy along roads, and in neglected pastures and fields; up to about 7,500 ft; July-Oct. Variety *serrulata* (Rydberg) Steyermarck (with upper cauline leaves 2-4 times longer than wide, and var. squarrosa with upper cauline leaves 5-8 times longer than wide) are found in our area. Wolf and Dever 5268 DINO! reported as *G. aphanactis* Rydberg (Welsh 1957), belongs here. Other taxa may be present, but none are apparent in the specimens seen.

Gutierrezia Lag. — Snakeweed

Plants shrubs, and sub-shrubs woody only at the base, 10-70 cm tall; leaves alternate, entire, linear, or linear-filiform, 1-5 cm long, 1-2 mm wide, glabrous or puberulent; heads small, in terminal corymbs; rays 3-10 per head but inconspicuous, yellow; disk flowers yellow; pappus of scales, present at least on the disk flowers.

---

1   Heads usually solitary; involucres 5-7.5 mm long, 2-5 mm wide, turbinate to cylindrical; disk flowers 5-23 .... ***G. pomariensis***

1   Heads usually clustered; involucres usually less than 5 mm long and 2 mm wide; turbinate; disk flowers 3-8 .... ***G. sarothrae***

Gutierrezia pomariensis (S. L. Welsh) S. L. Welsh [*G. sarothrae* var. *pomariensis* S. L. Welsh; *Xanthocephalum sarothrae* var. *pomariense* (S. L. Welsh) S. L. Welsh] — Uinta Basin snakeweed — Endemic, occasional to common in Uintah Co. and rather infrequently west to Mt. Home in Duchesne Co.; desert shrub, juniper, and sagebrush communities; 4,700-5,700 ft; mid July-early Oct.

Gutierrezia sarothrae (Pursh) Britton & Rusby — broom *snakeweed* — [*Xanthocephalum sarothrae* (Pursh) Shinners] Widespread; common and locally abundant; desert shrub, sagebrush, pinyon-juniper, Salina wildrye, and Douglas-fir communities; 4,700-9,100 ft; mid Aug.-early Oct. Reports of *G. microcephala* (D.C.) A. Gray, a plant s. and w. of the Uinta Basin, are based on plants of *G. sarothrae*.

Haplopappus Cassini — Goldenweed

Perennial herbs and shrubs; leaves alternate, entire or less commonly tooted; heads solitary or several; involucral bracts in 2 or more series but not in vertical rows; heads with ray and disc flowers or disc flowers only; flowers yellow; pappus of many, capillary bristles. Sometimes spelled Aplopappus in older manuals

---

1 Plants shrubs; heads with disc flowers only .... **H. macronema**

1 Plants herbaceous or woody only at the base; heads with disc and ray flowers.. 2

2 Plants glabrous or scabrous, from woody, branched caudices........... 3

2 Plants pubescent 4

3 Involucral bracts acute .... **H. acaulis**

3 Involucral bracts rounded at the apex .... **H. armerioides**

4 Stems commonly with 2 or more heads (sometimes only 1 in small plants); involucres 5-10 mm high; plants known from below 8,100 ft **H. lanceolatus**

4 Stems commonly with 1 head (rarely 2); plants known from above 8,100 ft 5

5 Achenes 4-6 mm long; involucre 9-15 mm high; outer involucral bracts broadly lanceolate to ovate, to 3 mm wide, often more densely pubescent on the margin than on the back and thus conspicuously ciliate with flattened, more or less straight, multicellular hairs; leaves typically entire and glabrous to sparsely pubescent and then much less pubescent than the stem .... **H. clementis**

5 Achenes 1-4 mm long; involucre 6-10 (12) mm high; outer involucral bracts narrowly lanceolate to nearly linear, to 1.5 mm wide, about equally pubescent on the margins and backs, the hairs fine and often tangled; 1 or more of the leaves of a plant usually with a few small teeth, often pubescent with tangled hairs and about equally pubescent as the stem or more so .. .. **H. uniflorus**

Haplopappus acaulis (Nuttall) A. Gray [*Stenotus acaulis* (Nuttall) Nuttall] — stemless goldenweed — Wide spread, common, desert shrub, sagebrush, and pinyon-juniper communities, often on rocky ground and geologic strata that weather to badlands; 5,600-7,600 ft; May-June.

Haplopappus armerioides (Nuttall) A. Gray (*Stenotus armerioides* Nuttall) — thrifty goldenweed — Widespread, locally common or abundant; desert shrub, sagebrush, and pinyon-juniper communities, often on rocky ground and geologic strata that weather to badlands Plants from white semi-barrens of the Green River Formation that are mainly 3-8 cm tall with leaves 1-3 mm wide are referable to var. *gramineus* S. L. Welsh & F. J. Smith. Taller plants with wider leaves are referable to var. *armerioides*.

Haplopappus clementis (Rydberg) S. F. Blake (*Pyrrocoma clementis* Rydberg) — tranquil goldenweed — Uinta Mtns. dry parts of montane and subalpine meadows, edges of coniferous forests and with krummholz Engelmann spruce at 8,960-10,700 ft; July-Sept. In AUF (227) plants with dentate as well as entire leaves are included in *P. clementis*. IMF (1994: 198) describes the leaves of P. clementis to be strictly entire. Plants of the Uinta Mtns. with dentate leaves have other features of *H. uniflorus* and are include there in this treatment.

Haplopappus lanceolatus (Hooker) Torrey & Gray [*Pyrrocoma lanceolata* (Hooker) Greene — lanceleaf goldenweed — The 5 specimens seen are from the floodplain of the Duchesne River near Tabiona, north and west of Roosevelt, Sowers Canyon (W. Tavaputs Plateau) and an outwash plane of Beaver Creek on the n. slope of the Uinta Mtns.; meadows, and floodplains at 5,200-8,020 ft; July-Aug. Perhaps reports of *H. racemosus* (Nuttall) Torrey for our area are based on odd specimens of *H. lanceolatus*.

Haplopappus macronema A. Gray [*Macronema discoideum* Nuttall; *Ericameria discoidea* (Nuttall) G. L. Nesom] — whitestem goldenbush — Uinta Mtns. on the limestone plateau between North Fork Duchesne and Rock Creek and in North Fork Duchesne Canyon above Castle Rocks; open slopes and with scattered coniferous trees; 9,500-10,800 ft; July-Sept.

Haplopappus uniflorus (Hooker) Torrey & A. Gray — one-flower goldenweed — Uinta Mtns. in "parks" on the n. slope from Henrys Fork east to Half Moon Park and Little Brush Cr drainage and Taylor Mtn. of the s. slope at 8,430-9,475 ft. Likely more closely related to *H. lanceolatus* than to *H. clematis*. However, in our area this has been more often confused with *H. clementis* than with *H. lanceolatus*.

Helenium L. — Sneezeweed

Helenium autumnale L. (*H. montanum* Nuttall) — common sneezeweed — Perennial herbs; leaves alternate, with decurrent bases, simple, entire or toothed; heads corymbose; involucral bracts in 2-3 series, spreading or reflexed in age; receptacles naked or sometimes with a few chaffy, scales between the ray and disk flowers; ray and disk flowers well developed, yellow to yellow-orange; pappus of 5-8 acuminate or awn-tipped scales. The 8 specimens seen are all from the floodplain of the Green River from Little Hole in-Daggett Co. to Ouray, Uintah Co., also likely along the Yampa River; 4,700-6,600 ft; July-Sept. Our plants belong to var. *montanum* (Nuttall) Fernald. Orange sneezeweed (*Helenium hoopesii*) long included in this genus is treated in *Hymenoxys*.

Helianthella Torrey & A. Gray

Perennial herbs from stout taproots and woody caudices; leaves simple, entire, often with 3-5 prominent nerves, at least those of the lower stem opposite; 'heads solitary or few at the ends of branches; involucral bracts more or less imbricate; receptacles chaffy; ray and disk flowers well developed, the rays yellow; pappus of a pair of awns or teeth, usually with a crown of scales present also; achenes pubescent, sometimes flattened with winged margins.

1   Rays 10-13 mm long; disk flowers dark purple or blackish; plants 20-60 cm
    tall . . . . **H. microcephala**
1   Rays 20-30 mm long; disk flowers yellow; plants 40-150 cm tall 2
2   Leaves mostly with 5 prominent nerves, the basal ones large, to 50 cm long;
    involucral bracts ciliate; heads nodding . . . . **H. quinquenervis**
2   Leaves mostly with 3 prominent nerves, basal leaves small or lacking;
    involucral bracts not conspicuously ciliate; heads erect . . . . **H. uniflora**

Helianthella microcephala (A. Gray) A. Gray — smallhead helianthella — South
flank of the Uinta Mtns., Diamond Mt., and probably to Blue Mt.; most common
in juniper and mtn. brush communities, sometimes with sagebrush, rarely in
desert shrub communities; 5,400-7,600 ft; late July-Sept. Heads with 8-10 rays.

Helianthella quinquenervis (Hooker) A. Gray — five-nerve helianthella — Uinta
Mtns. and Strawberry Valley; occasional; tall forb, aspen, spruce-fir, and
sagebrush communities and open grassy slopes; 7,600-10,400 ft; July-Aug.

Helianthella uniflora (Nuttall) Torrey & A. Gray — one-flower helianthella —
Widespread; occasional or locally abundant; sagebrush, forb-grass, tall forb,
and aspen communities, often in leeward snowbeds just under the crest of
ridges; 7,200-9,600 ft; mid June-mid Aug.

Helianthus L. — Sunflower

Annual or perennial herbs; leaves simple, entire, at least the lower ones
usually opposite, the upper ones often alternate; heads large, solitary or few at
the ends of branches; involucral bracts imbricate in several series; receptacles
chaffy; rays well developed, yellow; pappus of 2-few readily deciduous scales
or awns; achenes more or less 4-angled, not at all winged.

1   Disk flowers yellow; plants perennial, usually of wet places; leaves usually
    3-5 times longer than wide . . . . **H. nuttallii**
1   Disk flowers dark purple, reddish brown, or blackish; plants annual, often
    on disturbed ground or dry places; leaves about 1-3 times longer than wide
    2
2   Central bracts of the receptacle inconspicuously pubescent at the tip;
    involucral bracts ovate, abruptly contracted above the middle . **H. annuus**
2   Central bracts of the receptacle conspicuously white-bearded at the tip;
    Involucral bracts lanceolate, not abruptly contracted above the middle . .
    . . . . **H. petiolaris**

Helianthus annuus L. (*H. aridus* Rydberg; *H. lenticularis* Douglas ex Lindley) —
common sunflower, wild sunflower — Widespread; weedy in various places,
often on disturbed ground including roadsides, ditches, fields, gardens,
sometimes in indigenous plant communities; up to 7,000 (8,000) ft; (June) July-
Sept. Additional contrast with H. petiolaris: disk of the head often over 3 cm
wide; leaves 4-20 cm long, about 3-15 cm wide, ovate, rarely ovate lanceolate,
the lower ones often cordate at the base. Small specimens of *H. annuus* are
commonly mistaken for *H. petiolaris*.

Helianthus nuttallii Torrey & A. Gray — perennial sunflower, wetland sunflower — Widespread; along ditches and streams, wet lowlands and riparian communities; 5,200-6,400 ft; mid June-Sept.

Helianthus petiolaris Nuttall — prairie sunflower — Widespread (most of the specimens seen are from Uintah Co.); desert shrub, sagebrush, and juniper communities, and roadsides; 4,700-6,000 ft; (mid May) July-Sept. Additional contrast with *H. annuus:* disk seldom over 3 cm wide; leaf blades mostly 2-6 cm long, 1-3.5 cm wide, narrowly lanceolate to deltoid-ovate, mostly cuneate at base; involucral bracts less commonly ciliate or with fewer ciliate hairs.

Heliomeris Nuttall — Goldeneye

Heliomeris multiflora Nuttall [*Viguiera multiflora* (Nuttall) Blake] Cronquist (IMF 5:24) includes this in *Viguiera.* — goldeneye — Perennial herbs, 20-100 cm tall, branched; leaves lanceolate or linear lanceolate, entire or serrate, short petiolate, the lower ones opposite, the upper ones alternate; heads in loose panicles; involucral bracts in 2-3 series; receptacles chaffy; rays 7-17 mm long, yellow; disk flowers yellow or brownish; pappus lacking. With 2 vars.:

1 Leaves averaging over 8 mm wide; plants mostly above the pinyon-juniper belt .... **var. *multiflora***
1 Leaves averaging less than 8 mm wide; plants in and below the pinyon-juniper belt .... **var. *nevadensis***

Var. *multiflora* — showy goldeneye — widespread but relatively rare in the e. Uinta Mtns.; occasional to common; aspen, sagebrush, coniferous forest, mt. brush, tall forb, and meadow communities; 6,000-9,800 ft; late July-early Oct.

Var. *nevadensis* (A. Nelson) Yates [*Viguiera multiflora* var. *nevadensis* (A. Nelson) S. F. Blake] — Nevada goldeneye — Specimens seen are from the Red Wash area from Utah juniper and black sagebrush communities at 5,600-5,800 ft. Average plant height (mostly less than 40 cm tall) is less than that of var. *multiflora*.

Hieracium L. — Hawkweed

Perennial herbs with milky juice; leaves basal or alternate, entire to dentate; heads various; involucral bracts in 1-2 primary series, these sometimes subtended by shorter ones; heads with rays only; pappus of brownish or sordid-white capillary bristles; achenes fusiform, 10-15 ribbed, not beaked.

---

1 Leaves mostly all basal, the cauline ones if present much reduced and bract-like; flowers yellow .... **H. triste**
1 Stems with at least 1-2 leaves, if the stem leaves reduced upwards then flowers white 2
2 Flowers yellow; lower leaves about equal to or smaller than the middle stem leaves .... **H. scouleri**
2 Flowers white; basal leaves larger than those of the middle stem .... **H. albiflorum**

Hieracium albiflorum Hooker [*Chlorocrepis albiflora* (Hooker) W. A. Weber] — white hawkweed — Widespread in the Uinta Mtns. where infrequent or locally

common; lodgepole pine and spruce woods; 8,200-11,000 ft; mid July-early
Sept. Involucres 6-11 mm high with flowers exserted about 3 mm beyond the
involucre.

Hieracium scouleri Hooker (*H. cynoglossoides* Arv.-Touv.) — houndstongue
hawkweed — Occasional from Strawberry Valley to N. Fork Duchesne drainage
on the s. slope of the Uinta Mtns. and Red Mountain of the Dahlgren drainage
on the north slope; aspen and open coniferous woods, and open slopes in forb-
grass communities; 7,500-9,800 ft; July-Aug. Our plants apparently belong to
var. *griseum* (Rydberg) A. Nelson. Involucres 9-15 mm high with flowers
exserted about 3-6 mm beyond the involucre.

Hieracium triste Willdenow ex Sprengel [*Chlorocrepis tristis* (Willdenow) Love &
Love] — slender hawkweed — Uinta Mtns. where occasional or common in
coniferous woods and less common in alpine communities; 10,000-11,000 ft;
July-Aug. Uinta Mtn. plants belong to var. *gracile* (Hooker) A. Gray. Involucres
6-8 mm high with ray flowers exserted about 1 mm beyond the involucre.

Hymenopappus L'Heritier — Hymenopappus

Hymenopappus filifolius Hooker — fineleaf hymenopappus — Perennial herbs,
10-50 cm tall, glabrate to white-floccose; leaves alternate, sometimes mostly
basal, usually twice-pinnatifid into filiform segments; involucres 5-10 cm high,
the bracts about 6-12 in 1-2 nearly equal series; receptacles naked; rays
lacking; disk flowers cream or yellow; pappus of 10-20 minute hyaline scales,
sometimes obscured by the long hairs of the achene. With 3 more or less intergrading vars. in our area:

1 Basal leaf axils without dense tomentum, stems and leaves glabrous or
  sparsely tomentose; stems 5-30 (45) cm tall, with 0-1 (2) leaves; corolla 3-
  5 mm long; achenes 4-6 mm long .... **var. *nudipes***
1 Basal leaf axils with dense tomentum; stems and leaves sometimes densely
  tomentose; stem with 0-7 leaves 2
    2 Corollas 2-3 mm long; flowers 10-30 (averaging 20) per head; involucral
      bracts 3-7 mm long; stem leaves 0-3; pappus 0.1-1 mm long; plants 5-20 cm
      tall .... **var. *luteus***
    2 Corollas 2.5-7 mm long; flowers (15) 20-70 per head; involucral bracts (5)
      6-14 mm long; pappus 1-3 mm long; plants various .... **var. *cinereus***

Var. *nudipes* (Maguire) Turner [*H. filifolius* var. *alpestris* (Maguire) Shinners
— glabrate hymenopappus — The few specimens seen are from Fruitland and
e. to the Argyle-Nine Mile drainage on the W. Tavaputs Plateau; sagebrush,
pinyon-juniper, and aspen communities; 6,800-9,000 ft; June-Aug.

Var. *cinereus* (Rydberg) I. M. Johnston. [*H. arenosus* A. Heller; *H. cinereus*
Rydberg; *H. filifolius* var. *lugens* (Greene) Jepson; *H. filifolius* var. *megacephalus*
Turner possibly misapplied; *H. lugens* Greene] — fineleaf hymenopappus —
Widespread; occasional to rather common in desert shrub, sagebrush, pinyon-
juniper, and ponderosa pine-bunchgrass communities; 5,100-6,000 (8,000) ft;
late May-June. Plants from below 5,500 ft appear to have slightly longer
flowers and achenes than those above 6,000 ft of the W. Tavaputs Plateau.
Those with larger heads from lower elevations have been referred to as *H.
filifolius* var. *megacephalus*. Such plants were referred to *H. eriopodus* A. Nelson
by Graham (1937), but this is a taxon from s. of the Uinta Basin. Those with
smaller flowers and of higher elevations are more typical of var. *cinereus*. In-

tergradation greatly weakens the value of distinction between var. *cinereus* and var. *megacephalus* (AUF 235). However, var. *megacephalus* is recognized as a distinct taxon in IMF (5:132) and FNA (21:313), but the Uinta Basin is not included in range of var. *megacephalus* in IMF (5:132). It appears best to follow Turner (1956) and AUF (235) and include our large headed plants in var. *cinereus.*

Var. *luteus* (Nuttall) Turner (*H. luteus* Nuttall) — fineleaf hymenopappus — Most specimens seen range from Bald Mtn. Range and Hickey Mtn. on the north slope of the Uinta Mtns. eastward to Browns Park and to the Little Snake River, Moffat Co. and adjacent to our area in Sweetwater Co., Wyoming with few specimens seen from the south flank of the Uinta Mtns.; desert shrub and juniper communities; 5,000-8,720 ft; late May-June. Over much of its range, var. *luteus* is distinct and easily recognized. However, the features intergrade completely with those of var. cinereus in Moffat Co., Colorado (Turner 1956). In addition to the features given in the key, the heads are usually more numerous and the leaves are often more finely dissected than the other vars. of our area.

Hymenoxys Cassini — Actinea; Hymenoxys; Woollybase

Caespitose perennial herbs from taproots and simple or branched caudices; herbage glabrate to densely floccose or woolly; often with dense tufts of long hairs in the axils of the basal leaves; leaves basal or alternate; heads solitary on the stem or branches; involucral bracts in 2 or 3 series; receptacles naked; ray and disk flowers well developed, yellow; pappus of 5-8 scales, achenes 5-angled, pubescent.

---

1   Leaves all basal, simple, linear or nearly so, entire 2
1   Leaves basal and cauline, mostly divided into linear segments 4
2   Heads sessile in the basal rosettes; involucral bracts recurved, thickened and reddish at the apex; plants of Blue Mt. and vicinity . . . . . . *H. lapidicola*
2   Heads mostly scapose; involucral bracts not recurved; plants widespread 3
3   Involucral bracts with broad scarious margins . . . . *H. torreyana*
3   Involucral bracts without broad scarious margins . . . . *H. acaulis*
4   Rays 26-34 mm long; leaves 2-3 times divided; plants mostly growing above 10,000 ft. . . . . *H. grandiflora*
4   Rays 8-23 mm long; leaves once divided into 3-7 segments, occasionally simple; plants mostly growing below 10,000 ft . . . . *H. richardsonii*

Hymenoxys acaulis (Pursh) Parker [*Actinea acaulis* Sprengel; *Tetraneuris acaulis* (Pursh) Greene] — stemless woolly base — Additional contrast with *H. torreyana:* backs of rays and involucral bracts with or without glandular dots; leaves glandular dotted or not, more or less densely pubescent with long hairs. With 4 vars. with overlapping features as follows:

1   Leaves linear and green, mainly 1-2 cm long; heads sessile among tufts of leaves or on scapes to about 2 cm long; plants compactly caespitose . . . .
 . . . . **var. *nana***
1   Leaves not as above in all respects, if linear and green then mostly over 2

cm long; heads varying from sessile to long pedunculate 2
2 Leaves linear or linear-oblanceolate, conspicuously glandular punctate, sparingly long-hairy to glabrous, typically green . . . . **var. *arizonica***
2 Leaves narrowly to broadly oblanceolate, inconspicuously glandular-punctate, merely punctate, or epunctate 3
3 Leaves epunctate or nearly so, glabrous or less commonly silky-hairy . . . . **var. *epunctata***
3 Leaves punctate, silky-hairy, or less commonly glabrous . **var. *caespitosa***

Var. *arizonica* (Greene) K. L. Parker [*Tetraneuris arizonica* Greene; *T. acaulis* (Pursh) E. L. Green var. *arizonica* (Greene) K. F. Parker] — Arizona woollybase — Widespread and occasional to common in desert shrub, sagebrush, pinyon-juniper, ponderosa pine, and Douglas fir communities from 5,600-10,500 ft.

Var. *caespitosa* (A. Nelson) K. L. Parker (*Tetraneuris acaulis* (Pursh) Greene var. *caespitosa* A. Nelson) — lanate woollybase — Widespread in desert shrub, sagebrush, pinyon-juniper, bunchgrass, and alpine communities often on windswept ridges at 4,750-11,300 ft.

Var. *epunctata* (A. Nelson) Cronquist [*Tetraneuris epunctata* A. Nelson; *T. acaulis* (Pursh) Greene var. *epunctata* (A. Nelson) Kartesz & Gandhi] — epunctate woollybase — Widespread at least in the Utah part of the area; desert shrub, sagebrush, pinyon-juniper, mtn. brush, ponderosa pine, spruce-fir, and alpine communities from 5,200 -12,500 ft. Apparently the most common var. at high elevations in the Uinta Mtns.

Var. *nana* S. L. Welsh [*Tetraneuris acaulis* (Pursh) Greene var. *nana* (S. L. Welsh) Kartesz & Gandhi] — low woollybase — West Tavaputs Plateau where known from Wire Fence Canyon of the Sowers Canyon drainage at 7,240 ft (Franklin 6444) and head of Left Fork Antelope Canyon at 8315 ft (Huber and Goodrich 4864); marly semi-barrens. This var. is placed in synonymy under var. *arizonica* in FNA (21: 453).

Hymenoxys grandiflora (Torrey & A. Gray) Parker [*Actinea grandiflora* (Torrey & A. Gray) Kuntze; *Rydbergia grandiflora* (Torrey & A. Gray) Greene; *Tetraneuris grandiflora* (Torrey & A. Gray) Parker] — graylocks — Uinta Mtns.; occasional and locally common; open slopes and ridges, mostly above timberline; 10,400-12,000 ft; July-Aug.

Hymenoxys hoopesii (A. Gray) Bierner [*Helenium hoopesii* A. Gray; *Dugaldia hoopesii* (A. Gray) Rydberg; *Picradenia helenioides* Rydberg] — orange sneezeweed — Specimens seen are from Strawberry Valley, near Mirror Lake, Uinta Mtns., Farm Creek Mtn., e. of Tabiona, and W. Tavaputs Plateau west of Horse Ridge; sagebrush-grass, aspen, and sometimes coniferous forest communities; 8,200-10,200 ft; July-Aug.

Hymenoxys lapidicola S. L. Welsh & Neese — rock hymenoxys — Endemic, Blue Mtn. Cliff Ridge and vicinity, pinyon-juniper and ponderosa pine-manzanita communities, on Weber Sandstone; 6,000-8,100 ft; May-June.

Hymenoxys richardsonii (Hooker) Cockerell [*Actinea richardsonii* (Hooker) Kuntze; *Picradenia richardsonii* Hooker] — Colorado rubberweed, pinnatifid woollybase — Widespread, particularly common in Daggett Co.; sagebrush, pinyon-juniper, ponderosa pine, and occasionally streamside communities; 5,000-9,500 ft; late May-Aug.

Hymenoxys torreyana (Nuttall) Parker [*Actinea torreyana* (Nuttall) Macbride; *Hymenoxys depressa* (Torrey & A. Gray) S. L. Welsh & Reveal; *Tetraneuris torreyana*

(Nuttall) Greene] — Torrey woollybase — The 12 specimens seen are from 1-4 mi s. of Manila, Jesse Ewing Canyon, and Phil Pico Mt. in Daggett Co., Douglas Mt., Moffat Co., and from the E. Tavaputs Plateau in the Seep Ridge area, Uintah Co. and one specimen from 5 mi ne. of Lapoint; pinyon-juniper communities and open ridges; 6,200-9,400 ft; May. Additional contrast with *H. acaulis:* backs of rays and involucral bracts glandular with yellow crystalline dots; leaves densely glandular dotted, glabrous or with scattered, long, more or less flexuous hairs.

Iva L.

Herbs from taproot or rhizomes, with leafy stems; lower leaves opposite, the upper ones often alternate; heads rather inconspicuous, with disk flowers only in cup-like involucres; pappus lacking.

---

1 Leaf blades 6-15 cm long, coarsely serrate or incised; petiole about as long as the blade . . . . *I. xanthifolia*

1 Leaf blades 1-3 cm long, entire, sessile . . . . *I. axillaris*

Iva axillaris Pursh — stinking povertyweed — Widespread and common; floodplains of rivers and streams, pastures, roadsides, ditch banks, often where alkaline or saline; 4,700-6,800 ft; June-Oct. Plants 10-50 cm tall, perennial, from creeping, horizontal rhizomes; heads solitary in axils of leaves. Other than as follows our plants belong to var. axillaris. Small leaved plants of Mancos Shale in the vicinity of Blue Hill in Moffat Co. are referable to *I. axillaris* var. *parvifolia* S. L. Welsh & A. Huber — small-leaved povertyweed.

Iva xanthifolia Nuttall [*Cyclachaena xanthifolia* (Nuttall) Fresenius] — marshelder — Widespread; occasional along ditches, roadsides, and in pastures and alkaline and saline riparian communities; about 5,000-6,600 ft; July-Sept. Plants 70-120 cm tall, annual from a taproot; heads in axillary and terminal panicles.

Kuhnia L. — False Boneset

Kuhnia eupatorioides L [*K. chlorolepis* Wooton & Standley; *K. rosmarinifolia* Ventenat var. *chlorolepis* (Wooton & Standley) Blake; *Brickellia eupatorioides* (L) Shinners; *B. rosmarinifolia* (Ventenat) W. A. Weber] — false bonset — Perennial herbs from a taproot, 30-70 cm tall; leaves alternate, entire, or rarely with a basal pair of teeth, sessile or nearly so, linear or narrowly lanceolate; heads solitary on the ends of branches or in loose clusters; involucres 8-12 mm high, the bracts imbricate in a few series, striate and chartaceous; receptacle naked; rays lacking; disk flowers whitish to purplish; pappus of 10-20 plumose bristles, the side cilia about 0.5 mm long; achenes columnar, 10-20 ribbed. The few specimens seen are from Bogart, Sweetwater, and Bull Canyons of the E. Tavaputs Plateau; gravely alluvium on wash bottoms, rabbitbrush communities; 6,200-6,700 ft; Aug. Plants of the Uinta Basin belong to var. *chlorolepis* (Wooton & Standley) Cronquist.

Lactuca L. — Lettuce

Annual or perennial herbs with milky juice, 30-100 cm tall; leaves alternate; heads paniculate, with rays only; involucres narrowly cylindrical, the bracts in 3 or more series; receptacles naked; pappus of capillary bristles; achenes strongly flattened, beaked.

---

1   Leaves entire, dentate or some pinnatifid but not spinulose; flowers blue .
   .... *L. tatarica*
1   Leaves pinnatifid, the lobes spinulose-toothed; flowers-yellow . *L. serriola*

Lactuca serriola L. — prickly lettuce — Introduced from Europe; widespread and weedy; roadsides, ditches, around ponds and lakes, lowland meadows, cultivated ground, and in indigenous plant communities, becoming abundant following fire in pinyon-juniper communities; 4,700-7,000 ft; late July-early Oct. Plants annual or biennial, from a taproot; midrib of leaves usually prickly hairy beneath; beak of achenes longer than the body; flowers opening by night, quickly withering by day and inconspicuous. Despite the weedy nature of this plant, it is readily eaten by wild and domestic ungulates.

Lactuca tatarica (L.) C. A. Meyer [*L. oblongifolia* Nuttall; *L. pulchella* (Pursh) de Candolle] — blue lettuce — Widespread; along ditches, gullies and streams, riparian communities, and wet lowland meadows; 4,900-7,500 ft; July-Aug. Our plants belong to ssp. *pulchella* (Pursh) Stebbins. Plants perennial, from spreading rootstocks; leaves glabrous or glaucous without prickly hairs; beak of achenes less than 1/2 as long as the body.

Lygodesmia D. Don — Skeleton plant; Rushpink

Lygodesmia grandiflora (Nuttall) Torrey & A. Gray [*L. juncea* (Pursh) D. Don misapplied] — showy rushpink — Perennial herbs from deep-seated rootstocks, with milky juice, 10-40 cm tall; leaves alternate, entire, linear and grass-like, often crowded, ascending; heads solitary at the ends of stems and branches, 18-25 mm high; involucral bracts few, the inner series much longer than the outer; receptacles naked; rays 2.5-4 cm long, rose or pink; disk flowers none; pappus of simple capillary bristles. Widespread; infrequent to occasional in desert shrub, sagebrush, pinyon-juniper, Salina wildrye, and mt. brush communities; 4,800-8,300 ft; late May-July.

Machaeranthera Nees — Tansyaster; Aster

Annual or perennial herbs; leaves alternate, simple, spinulose toothed; heads variously disposed; involucral bracts imbricate in a few series; pappus of capillary bristles. Our 3 species are in 3 different genera in FNA. Treatment at subgenus or section level maintains these in *Machaeranthera*.

---

1   Heads with disk flowers only .... *M. grindelioides*
1   Heads with bluish, purplish, or rarely white rays 2
2   Leaves pinnatifid .... *M. tanacetifolia*
2   Leaves toothed or entire .... *M. canescens*

Machaeranthera canescens (Pursh) A. Gray [*M. bigelovii* (A. Gray) Greene (*Aster bigelovii* A. Gray) misapplied; *M. leucanthemifolia* (Greene) Greene; *M. linearis* Greene misapplied; *Aplopappus leucanthemifolius* Greene; *Aster canescens* Pursh;

Aster leucanthemifolius Green; *Aster rubrotinctus* Blake (*M. rubricaulis* Rydberg) misapplied; *Dieteria canescens* (Pursh) Nuttall] — hoary aster — Widespread; desert shrub, sagebrush, pinyon-juniper, mtn. brush, aspen, open coniferous forest, and parkland communities, also roadsides and other areas of disturbance; 4,700-9,600 ft; late June-Oct. Apparently our plants belong to var. *canescens*. Some of our low-elevation, biennial plants with numerous bract-like leaves in the inflorescence and stipitate glands approach var. *leucanthemifolia* (Greene) S. L. Welsh. However, plants of that taxon are apparently limited to the s. and w. of our area.

Machaeranthera grindelioides (Nuttall) Shinners [*Haplopappus nuttallii* Torrey & A. Gray; *H. nuttallii* Torrey & A. Gray; *Xanthisma grindelioides* (Nuttall) D. R. Morgan & R. L. Hartman] — discoid aster — Widespread; occasional in desert shrub, black sagebrush, juniper-greasebush, pinyon-juniper, mt. brush, and Salina wildrye-Douglas-fir communities, often on shallow rocky soils or on poor substrates, sometimes on exposed windswept ridges; 4,800-10,400 ft; June-Aug. Our plants belong to var. *grindelioides*.

Machaeranthera tanacetifolia (H.B.K) Nees — tansyleaf aster — Known from roadsides and about buildings at Fort Duchesne. Possibly introduced from the Wind River area of Wyoming or from southern Utah. To be expected to spread to other areas within the Uinta Basin.

Madia Molina — Tarweed

Madia glomerata Hooker — cluster tarweed — Annual, malodorous herbs, 5-30 (60) cm tall; leaves alternate, entire, linear, 2-6 cm long; heads glomerate near ends of branches; involucres about 6-8 mm long, 3-5 mm wide, the bracts in a single series, the margins involute and enclosing the disk flowers; rays 2-5, inconspicuous, 3-lobed, yellowish; disk flowers about 8-12, yellowish; pappus lacking. Except for apparent introductions at E. Mckee Draw, Diamond Mtn., and at Farm Creek Mt. (Snake John Spring) all the specimens seen are from w. of N. Fork, Duchesne River, in the Uinta Mtns. and west to Strawberry Valley where the plant is fairly common; openings in aspen and coniferous woods, forb-grass communities, often along roads or in other areas of disturbance; 7,500-9,000 ft; July-Aug.

Malacothrix de Candolle — Desert dandelion

Annual herbs with milky juice; leaves alternate and basal, mostly pinnatifid; heads solitary to several, principal involucral bracts in 1-2 series, with several shorter outer ones; receptacles naked or bristly; rays yellow; disk flowers none; pappus of capillary bristles; achenes 5-10 ribbed.

1   Middle and upper cauline leaves sagittate or auriculate at the base; achenes 1.7-3 mm long at maturity; pappus bristles all quickly deciduous . . . .
. . . . *M. sonchoides*

1   Middle and upper cauline leaves with strap-shaped base; achenes 3-4 mm long at maturity; most of pappus quickly deciduous, but 1 or more bristles sometimes persistent . . . . *M. torreyi*

Malacothrix sonchoides (Nuttall) Torrey & A. Gray — sowthistle desert dandelion — Widespread and occasional in desert shrub communities at 5,500-5,780 ft; May-June. Without mature achenes, the plants are difficult to distinguish from those of *M. torreyi*. In addition to the features given in the key, the involucral bracts are acute in *M. sonchoides* and often long acuminate in *M. torreyi*. Reported to be mostly glabrous while *M. torreyi* is reported to often have gland-tipped hairs. However, these feature do not seem consistently correlated with features of the leaves listed in the key.

Malacothrix torreyi A. Gray — Torrey desert dandelion — Widespread and occasional in desert shrub and Wyoming big sagebrush communities; 4,800-6,000 (7,500) ft; May-June. See *M. sonchoides*.

Matricaria L. — Mayweed

Annual herbs, branched from the base or simple, glabrous or pubescent; leaves rather finely dissected, heads solitary to many, paniculate; involucres saucer shaped, the bracts with hyaline margins; receptacle conical; rays lacking; disk corollas 4-lobed; pappus a very low crown.

---

1   Rays lacking .... *M. matricarioides*
1   Rays white, 12-25 mm long .... *M. maritima*

Matricaria maritima L. [*M. chamomilla* L; *Tripleurosperma perforatum* (Merat) M. Lainz], — scentless chamomile, false mayweed — Introduced from Europe; widespread and weedy, as of yet very localized; specimens seen are from near Sheep Creek Lake, Daggett Co., Blacks Fork River, Summit Co., Strawberry Valley and head of West Fork Duchesne River in Wasatch Co; 7,650-10,000 ft; June-Sept. To be expected to spread across moderate elevations of our area. Plants of this taxon have been generally misidentified as *Anthemis cotula*.

Matricaria matricarioides (Lessing) Porter [*Matricaria discoidea* de Candolle; *Chamomilla suaveolens* (Pursh) Rydberg; *Lepidotheca suaveolens* Nuttall] — pineapple-weed, disk mayweed — Introduced from Europe, specimens seen are from scattered locations in the Uinta Mtns. and Strawberry Valley, weedy along roadsides and other disturbed places, apparently spreading rapidly in our area; 7,600-8,200 ft; June-Aug. The discussion in IMF (5: 138) is followed here in using *M. matricarioides* as the name for this plant.

Microseris D. Don — Microseris

Microseris nutans (Geyer) Schultz-Bip. — nodding microseris — Perennial herbs with milky juice, 10-40 (70) cm tall; leaves basally disposed but hardly in a basal rosette, alternate, linear, entire or nearly so, 10-30 cm long, 1-20 mm wide; heads solitary on scapose peduncles, with rays only; involucres 1-2 cm high, the bracts in about 2 series; flowers yellow; pappus of 15-20 narrow scales 1-3 mm long, each with a terminal, subplumose awn, the awns giving the pappus the appearance of capillary bristles; achenes with 8-10 ribs, not beaked. Specimens seen are from Diamond Mt. Plateau (where fairly common), Blue Mt. and Daggett Co., plotted for Rio Blanco and Moffet Cos in TPD, and reported (Graham 1937) for E. Tavaputs Plateau; sagebrush, ponderosa pine and meadow communities 7,600-8,200 ft; June.

Onopordum L. — Thistle

Onopordum acanthium L. — Scotch thistle, cotton thistle — Coarse, branching, strongly spiny, sparsely to densely tomentose, biennial herbs 50-150 (300) cm tall; leaves of basal rosettes 50-60 cm long, 2-20 cm wide or wider, pinnately lobed and serrate-dentate, leaves pinnatifid, strongly decurrent and forming wings the full length of the stem internodes; involucres 25-35 mm high, 30-65 mm wide, the bracts spine-tipped, the spines 3-5 mm long, yellowish, spreading; receptacle fleshy, honeycombed; corollas reddish-purple to pink; pappus bristles barbellate; achenes glabrous, 4- or 5-ribbed. Introduced from Europe, spreading in much of the United States, recently found at several locations in the Uinta Basin and apparently held in check by an aggressive weed control program. Additional introductions can be expected. Similar to *Carduus nutans* but distinguished by the white tomentose herbage, erect or ascending heads, and features listed in the key.

Oxytenia Nuttall — Copperweed

Oxytenia acerosa Nuttall — copperweed — Plants perennial, woody at the base, 1-2 m tall; stems slender and leafy or sometimes leafless and rush-like, grayish-strigose; and often canescent especially above; leaves alternate, pinnately parted into 3-5 filiform segments or the upper ones sometimes entire; heads numerous in panicles, each with 10-20, white, staminate disk flowers and 5 marginal pistillate flowers, the pistillate ones lacking corollas; involucres about 4-5 mm wide, with 5 bracts; pappus none or of an obsolete scale; achenes long villose. Known from Nine Mile Canyon, Tabyago Canyon (Atwood 32,257, Desolation Canyon (Graham 9973) and from 5 mi w. of Roosevelt (Stoddart sn UTC) and (Hamblin sn USUUB; 5,000-6,000 ft. Although the site w. of Roosevelt is disjunct from the main body of the species, the specimens clearly belong here.

Parthenium L. — Feverfew

Parthenium ligulatum (M. E. Jones) Barneby [*P. alpinum* (Nuttall) Torrey & A. Gray var. *ligulatum* M. E. Jones; *Bolophyta ligulata* (M. E. Jones) W. A. Weber] — low feverfew — Pulvinate caespitose, perennial herbs, 1-2 (3 cm tall; leaves in basal rosettes, 0.3-2 cm long, oblanceolate or spatulate, densely strigose; heads sessile in the leaves, involucres about 5 mm high, the bracts broadly rounded; receptacles with firm scales, these partly enveloping the disk flowers; ray and disk flowers white, the rays 1-2 mm long, apparently attached to the apex of the inner involucral bracts; pappus none or obsolete. occasional or rather common, but inconspicuous; Tavaputs Plateau and flank of the plateau, disjunct in Browns Park area.; desert shrub, pygmy sagebrush, juniper-greasebush-mt. brush, and pinyon-juniper communities, often on whitish, marly semi-barrens; 5,600-7,000 ft; May-mid June.

Petradoria Greene — Rock goldenrod

Petradoria pumila (Nuttall) Greene (*Solidago petradoria* Blake) — rock goldenrod — Plants perennial, tufted from woody caudices, 10-20 cm tall, the aerial stems herbaceous, glabrous, more or less resinous; leaves simple, alternate, entire, linear-oblanceolate, often 3 nerved, reduced in size and number upward on the stem; inflorescence a flat-topped cyme; involucres about 5-6 mm high, the bracts more or less aligned in rows, greenish or brownish at the apex; flowers few, yellow, the rays 1-3, sometimes lacking, usually inconspicuous; pappus of capillary bristles; achenes glabrous. Widespread; occasional or common, and locally abundant; desert shrub, sagebrush, and pinyon-juniper communities, often on shallow rocky soil; 4,700-8,500 ft, possibly higher on windswept ridges; late June-Sept.

Platyschkuhria (A. Gray) Rydberg

Platyschkuhria integrifolia (A. Gray) Rydberg [*Bahia integrifolia* (A. Gray) Macbride; *B. nudicaulis* A. Gray] — basindaisy — Plants perennial from a woody caudex, 10-40 cm tall, puberulent and more or less glandular above; leaves alternate or nearly all basal, entire, simple, the basal ones 2-7 cm long, ovate to lanceolate, 3-nerved, the stem leaves much reduced; heads 1-few, the involucres 7-10 mm high, the bracts about equal, in about 2 rows; ray and disk flowers yellow, the rays 8-20, 7-10 mm long; pappus of 8-14 scales; achenes sparsely hairy. With 2 vars. in our area:

1 Involucral bracts obtuse to acute or acuminate; peduncles, involucres, and upper stems with numerous stipitate glandular hairs ... **var. *desertorum***
1 Involucral bracts caudate-attenuate; peduncles, involucres, and upper stem with no or few, sessile or short-stipitate glands .... **var. *ourolepis***

Var. *desertorum* (M. E. Jones) Ellison — desert basindaisy — The single specimen seen (Neese 5776) is from the Badlands Cliffs, Duchesne Co., from a pinyon-juniper community at 6,750 ft; June.

Var. *ourolepis* (Blake) Ellison (Bahia ourolepis Blake) — basindaisy — Common in Uintah Co., apparently limited and localized in Rio Blanco Co.; desert shrub, juniper, and sagebrush communities; 4,700-6,000 ft; late May early July. Uinta Basin plants have been mistakenly referred to var. *oblongifolia* (A. Gray) Ellison [*P. oblongifolia* (A. Gray) Rydberg; *Bahia oblongifolia* (A. Gray) A. Gray]. Occasional specimens with caudate-attenuate involucral bracts have abundant stipitate glandular hairs (S. Goodrich 28554).

Prenanthella Rydberg

Prenanthella exigua (A. Gray) Rydberg (*Lygodesmia exigua* A. Gray) — prenanthella — Plants annual, 6-20 cm tall, much branched, with milky juice; leaves alternate, toothed to pinnatifid, the lower ones 1-3 cm long, withered by anthesis, the upper ones reduced and bract-like; heads at the ends of panicle branches; involucres 4-5 high, the bracts in 2 series, the outer ones 1-few and much reduced, the inner ones 4-5; rays pink or about 7 mm long; disk flowers none; pappus of simple or barbellate, bright white, capillary bristles. The 2 specimens seen (Neese & Neese 7557 & 7588) are from near Randlett and Pelican Lake, both from desert shrub communities; 4,800-4,900 ft; June.

Psilostrophe de Candolle — Paperflower

Psilostrophe bakeri Greene — Baker paperflower — Plants perennial with caudices, 10-20 cm tall, long-villous or floccose; leaves alternate, 5-10 cm long, spatulate or oblanceolate, entire, or rarely more or less lobed; flowers yellow, rays 4-6, 3-lobed, 8-15 mm long, becoming papery in age and persistent on the achenes; pappus of 4-6 scales, about 1/3-1/2 as long as the disk corollas. The only specimen seen (Brown s.n. CS, collected in 1938) is from Sunbeam, Moffat Co. This plant is otherwise known from well s. of our area. The Sunbeam specimen appears to be a waif.

Ratibida Rafinesque

Ratibida columnifera Nuttall) Wooton & Standley — prairie coneflower — Perennial herb; leaves pinnatified; heads with disk flowers on an elongate, columnar receptacle; rays reflexed in age, mostly yellow sometimes purple; pappus an awn-tooth. Known from Bitter Creek where likely planted, and cultivated in flower gardens. This is a native of the plains to the east of Utah.

Rudbeckia L. — Coneflower

Tall herbaceous perennial herbs; leaves alternate; heads terminal on the stems or branches, long-pedunculate; involucral bracts in 2 or more series, herbaceous to foliaceous; receptacles conic or cylindrical in fruit, chaffy; pappus a short crown, a low border, or lacking; achenes 4-angled.

1 Leaves entire; disk flowers dark purplish-brown or blackish, rays lacking; heads cylindrical in fruit .... *R. occidentalis*
1 Leaves pinnately divided into 3-7 ovate or lanceolate segments, these cleft and coarsely serrate, disks and rays yellowish; heads ovoid in fruit ......
.... *R. laciniata*

Rudbeckia laciniata L. — cutleaf coneflower — Known in our area from the Piceance Basin (Goodrich 23151). Plants of our area belong to var. *ampla* (A. Nelson) Cronquist.

Rudbeckia occidentalis Nuttall — western coneflower — Strawberry Valley and e. to N. Fork Duchesne drainage on the s. slope of the Uinta Mtns. and Bear River drainage on the n. slope, and south to the Willow Creek drainage of West Tavaputs Plateau; occasional and locally abundant in tall forb, aspen, and streamside communities; 7,500-9,000 ft; July-Aug.

Senecio L. — Groundsel; Ragwort

Perennial herbs; leaves alternate, rarely nearly all basal; heads various; involucral bracts equal in a single series, sometimes with a few, much reduced outer ones; receptacles naked; flowers yellowish; pappus of capillary bristles.

1 Rays lacking 2
1 Rays well developed, exceeding the involucre and disk flowers 3
2 heads not nodding; leaves pinnatifid .... *S. vulgaris*
2 Rays lacking; head nodding; leaves entire or shallowly dentate . *S. pu*dicus
3 Leaves only slightly reduced in size or number upward on the stem, glabrous or nearly so, serrate, dentate, lobed, or pinnatifid, occasionally entire, those of the lower stem sometimes withering and deciduous by late anthesis; basal leaves usually lacking. 4
3 Leaves progressively reduced in size and number upward on the stem, if

only slightly so then densely pubescent, entire to parted, those of the lower stem usually persistent; basal leaves mostly well developed, persistent . 11

4   Plants 5-15 (20) cm tall, sometimes sprawling, of rocky places, mostly near or above timberline on the Uinta Mtns.; stems usually several from slender caudices, the caudex branches often elongate and rhizome-like; leaves not over 5 cm long .... *S. fremontii*

4   Plants mostly taller and otherwise different from above in 1 or more ways; leaves over-5 cm long 5

5   Leaves filiform, entire, or parted with filiform segments; plants growing below 7,500 ft 6

5   Leaves not as above; plants often growing above 7,500 ft.............. 7

6   Plants tomentose; main involucre bracts 13-21 .... *S. douglasii*

6   Plants glabrous; main involucre bracts 8-13 .... *S. spartioides*

7   At least some (usually most) of the leaves deeply lobed to parted 8

7   Leaves serrate, dentate, or occasionally nearly entire 9

8   Involucral bracts black-tipped .... *S. eremophilus*

8   Involucral bracts not black-tipped .... *S. multilobatus*

9   Leaves widest beyond the middle; plants 20-50 cm tall; involucres 8-15 mm wide .... *S. crassulus*

9   Leaves widest below the middle; plants often 50-150 cm tall; involucres mostly narrower than above 10

10 Leaves widest at the base, triangular, the widest ones often over 3 cm wide .... *S. triangularis*

10 Leaves widest between the base and the middle, lanceolate; seldom over 3 cm wide .... *S. serra*

11 At least some of the stem leaves pinnatifid or deeply sinuate-toothed, if only shallowly so then often auriculate clasping; basal leaves toothed, lobed or parted 12

11 Leaves all entire, serrate, or cuneate, not lobed or pinnatifid.......... 17

12 Basal and stem leaves (except sometimes the uppermost) pinnatifid .... *S. multilobatus*

12 At least some of the basal leaves toothed or lobed, not pinnatifid 13

13 Heads 1-6; plants 2-15 cm tall .... *S. werneriifolius*

13 Heads usually more, plants mostly taller............................. 14

14 Rays orange or yellow-orange; plants from Strawberry Valley to Rock Creek, Uinta Mtns. (Note: this and the following 3 taxa grade into each other) .. .. *S. crocatus*

14 Rays yellow, distribution various 15

15 Lower and middle cauline leaves as long or longer than the basal leaves, usually with large auriculate bases; plants known from Rock Creek, Uinta Mtns. .... *S. dimorphophyllus*

15 Lower and middle cauline leaves reduced, not with bases large and auriculate-clasping 16

16 Stems mostly solitary; plants of moist soil of meadows and along streams; involucral bracts 6-9 mm long .... *S. pauperculus*

16 Stems solitary, but more commonly few to several; plants mostly of well-drained soil; involucral bracts 5-7.5 mm long .... *S. streptanthifolius*

17 Plants glabrous, 70-100 cm tall, growing in marshy or boggy places below 8,500 ft; involucres 3-6 mm wide; leaves entire or minutely denticulate, the larger ones 15-40 cm long .... *S. hydrophyllus*

17 Plants pubescent at least in part when young, if glabrate when mature, then differing from the above in 1 or more other ways.................... 18

18 Involucres 8-12 mm wide; leaves glabrous, mostly dentate or serrate, often not much reduced in size or number upward on the stem .... *S. crassulus*

18 Involucres mostly narrower and/or leaves pubescent at least when young, the upper ones usually much reduced............................. 19

19 Involucres 3-4 (5) mm wide; leaves permanently tomentose throughout, entire to dentate, the larger ones 10-20 cm long including the petiole; plants (25) 30-60 cm tall, the stems often tomentose .... *S. atratus*

19 Involucres mostly wider or plants otherwise different from above 20

20 Plants mostly 20-80 cm tall, of wet poorly drained sites, from short rhizomes or simple caudices; stems mostly solitary; stem leaves 4-30 cm long, the basal ones if present not noticeably larger than those of the lower stem 21

21 Leaves 4-10 (15) cm long including the petioles, entire, occasionally dentate, the surfaces often glabrous at maturity, the axils and lower petioles and stem with some translucent, crinkly, multicellular hairs with flattened segments, the alternating segments often twisted 90°; plants with a short button-like caudex, growing in moist or dry places ...... *S. integerrimus*

21 Largest leaves 13-30 cm long, including the petioles, with minute callus-tipped, dentate teeth, the surface more or less permanently finely pubescent, the hairs not noticeably flattened and twisted; plants from short stout rhizomes, of moist to wet places .... *S. sphaerocephalus*

20 Plants 2-20 (30) cm tall, of dry places, from prolonged rootstocks and branching caudices; stem leaves seldom over 5 cm long, the basal ones usually noticeably larger than the lower cauline ones 22

22 Plants densely white tomentose even in age, 5-30 cm tall, rarely taller; heads sometimes more than 6; stem leaves usually well developed ..... *S. canus*

22 Plants tomentose at first, glabrous or glabrate in age, 2-10 (15) cm tall; heads 1-6; stem leaves small or lacking .... *S. werneriifolius*

Senecio atratus Greene — black groundsel — Occasional in the western Uinta Mtns. and sparingly eastward to East Fork Whiterocks; rocky open ground, talus, and spruce-pine woods; 9,600-10,600 ft; late June-early Sept.

Senecio canus Hooker [*S. harbourii* Rydberg; *S. purshianus* Nuttall; *Packera cana* (Hooker) W. A. Weber & A. Love] — woolly groundsel — Widespread; sagebrush, pinyon-juniper, open coniferous forest, krummholz, and alpine communities; 7,800-11,500 ft; June-Aug. See *S. streptanthifolius*.

Senecio crassulus Rydberg — thickleaf groundsel — Specimens seen are from the Blind Stream-Log Hollow-Rock Creek area, Uinta Mtns.; Engelmann spruce woods, openings in woods, and open rocky slopes; 9,700-10,600 ft; mid July-Aug.

Senecio crocatus Rydberg [*S. pseudaureus* Rydberg misapplied; *Packera crocata* (Rydberg) W. A. Weber & A. Love] — saffron groundsel — Locally common from Strawberry Valley to Rock Creek and sparingly to Uinta Canyon, Uinta Mtns.; willow-streamside, aspen, meadow, and Engelmann spruce communities; 7,600-10,000 ft; mid July-Aug. See *S. streptanthifolius.*

Senecio dimorphophyllus Greene [*Packera dimorphophylla* (Greene) W. A. Weber & A. Love] — varyleaf groundsel — The few specimens seen from Blind Stream and Rock Creek drainages, Uinta Mtns.; riparian and spruce-fir communities at 7,500-10,700 ft belong to var. *dimorphophyllus.* One specimen from Slab Canyon, W. Tavaputs Plateau, aspen-fir parkland at 8,200 ft (Goodrich 23159) belongs to var. *intermedius* T. M. Barkley; late June-Aug. See *S. streptanthifolius.*

Senecio eremophilus Richardson in Franklin — King groundsel — Widespread; aspen, riparian, forb-grass, and meadow communities, sometimes along roads and in clear-cuts in coniferous forests; 7,200-10,000 ft; late July-Sept. Our plants belong to var. *kingii* (Rydberg) Greenman (*S. ambrosioides* Rydberg).

Senecio flaccidus Lessing [*Senecio douglasii* de Candolle var. *longilobus* (Bentham) L. D. Benson] — threadleaf groundsel, threadleaf ragwort — The 1 record seen (Welsh et al. 9434) is from the floodplain of the Duchesne River about 6 mi e. of Myton; salt grass-greasewood alkali flat. Plants of the Uinta Basin belong to var. *flaccidus.*

Senecio fremontii Torrey & A. Gray — talus groundsel — Uinta Mtns.; coniferous woods and locally common in talus; 9,950-13,000 ft; mid July-Aug. Plants of the Uinta Mtns. belong to var. fremontii. Occasional specimens with somewhat lobed leaves approach var. *inexpectans* Cronquist that is limited to the LaSal Mtns of se. Utah.

Senecio hydrophyllus Nuttall — water groundsel — Seldom collected, but probably widespread; wet places; 4,800-7,800 ft; July-Sept.

Senecio integerrimus Nuttall (*S. hookeri* Torrey & A. Gray) — western groundsel — Widespread; common; sagebrush-grass, aspen, streamside, and meadow communities, and rocky ground in pine-spruce woods; 7,200-11,000 ft; late May-early Aug. Our materials belong to var. *exaltatus* (Nuttall) Cronquist.

Senecio multilobatus Torrey & A. Gray [*S. millelobatus* Rydberg, a species from far s. of the Uinta Basin, misapplied; *S. uintahensis* Greenman; *Packera multilobata* (Torrey & A. Gray) W. A. Weber & A. Love] — lobe-leaf groundsel — Widespread; common; sagebrush, pinyon-juniper, mt. brush, Salina wildrye, aspen, and coniferous forest communities, and rocky slopes and burned areas; 5,700-11,200 ft; May-Aug. depending on elevation. See *S. streptanthifolius.*

Senecio pauperculus Michaux [*Packera paupercula* (Michaux) A. Love & D. Love] — balsam groundsel — Apparently rare; known from meadow and streamside communities, Uinta Mtns., Daggett and Uintah Cos.; 8,600-9,000 ft; July-Aug. See *S. streptanthifolius.*

Senecio pudicus Greene [*Ligularia pudica* (Greene) W. A. Weber] — rayless groundsel, rayless ragwort — Three specimens seen are from Mt. Bartles, and Lake Canyon drainage, W. Tavaputs Plateau. A fourth specimen is from the Cart Creek drainage, Uinta Mtns. where the species was not found on a second visit; open ridge tops, rocky canyons, spruce-fir, and Douglas-fir communities; 8,000-10,100 ft; July-early Oct.

Senecio serra Hooker (*S. admirabilis* Greene) — butterweed groundsel, tall groundsel, saw groundsel — Common to abundant in Strawberry Valley and eastward on the W. Tavaputs Plateau to the Avintaquin drainage and to Rock Creek and sparingly to Uinta Canyon and Black Fork drainage in the Uinta Mtns.; aspen, tall forb, and riparian communities, coniferous woods and parklands; 7,400-10,000 ft; late June-Aug.

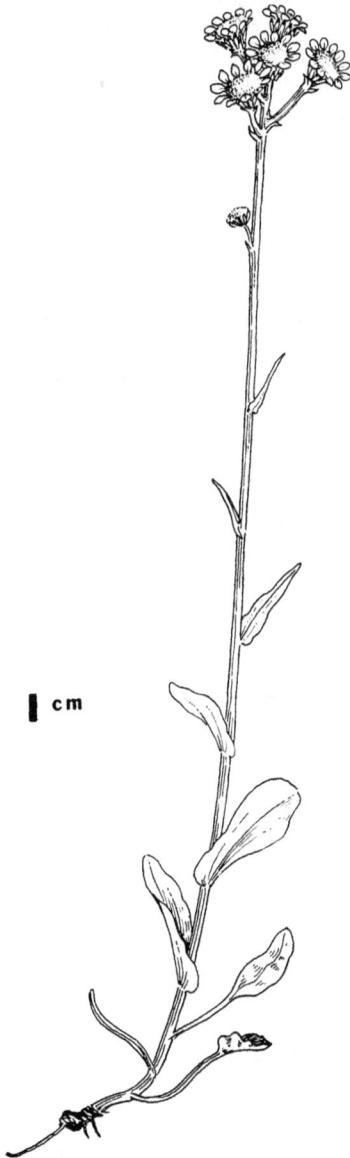

cm

Senecio dimorphophyllus Greene — varyleaf groundsel

Senecio spartioides Torrey & A. Gray — broom groundsel — Occasional at scattered locations across the s. flank of the Uinta Mtns. and adjacent in the Uinta Basin; sagebrush, pinyon-juniper, mt. brush, and ponderosa pine communities, usually in sandy soil; 5,800-7,400 ft; Aug.-early Oct. *S. spartioides* var. *multicapitatus* (Greenman) S. L. Welsh has been reported for Uintah Co. and some of our specimens have leaves that are divided into linear segments and have inflorescences with up to 60 heads. These plants could belong to var. *multicapitatus,* but some or most of the blades are 2 mm wide or wider, a feature referable to var. *spartioides.* We are unable to clearly distinguish 2 taxa in our material, and refer all plants of the Uinta Basin to var. *spartioides.*

Senecio sphaerocephalus Greene — roundhead groundsel, wet meadow groundsel — Strawberry Valley and Uinta Mtns.; occasional and locally common in meadow, willow-streamside, and other riparian communities; 7,600-10,500 ft; July-early Sept.

Senecio streptanthifolius Greene [*S. rubricaulis* Greene; *S. cymbalarioides* Nuttall; *Packera streptanthifolia* (Greene) W. A. Weber & A. Love] — manyface groundsel, Rocky Mtn. groundsel — Many specimens seen are from the Uinta Mtns. and Strawberry Valley, to be expected across our area; aspen, and coniferous woods, meadows, and rocky slopes at and somewhat above timberline; 7,600-11,200 ft; June-Aug. Forming intermediates with *S. multilobatus, S. dimorphophyllus, S. pauperculus,* and *S. canus.* Some specimens are not clearly assignable to a definite taxon. *S. multilobatus* and *S. canus* as well as *S. werneriifolius* also form intermediates. This further complicates separation of plants in the group. *Senecio crocatus* is also a part of the above complex and is close to and possibly passes into *S. dimorphophyllus* and *S. pauperculus.* References to *S. cymbalarioides* Nuttall (a plant known from well n. of our area) have been based on plants of *S. streptanthifolius* and its allies. *S. L. Welsh 487,* referred to as *S. fendleri* A. Gray (Welsh 1957; Holmgren 1962), belongs here.

Senecio triangularis Hooker — arrowleaf groundsel — Uinta Mtns.; locally common to abundant along streams, in coniferous woods, sometimes with aspen and willows; 7,600-11,000 ft; (mid June) July-early Sept.

Senecio vulgaris L. — common groundsel — Introduced from Europe, weedy, mostly about cities and towns; rarely collected in our area.

Senecio werneriaefolius A. Gray [*Packera werneriifolia* (A. Gray) W. A. Weber & A. Love] — montane groundsel — The 3 specimens seen are from the south rim of South Fork Rock Creek, from krummholz spruce and spruce parkland communities at 10,500 ft. July-Aug. Plants of this taxon inter-grade freely into *S. canus.* Our plants belong to var. *werneriaefolius.*

Solidago L. — Goldenrod

Perennial herbs; leaves alternate, simple, entire or toothed; heads usually small, few to numerous, variously arranged on the branches; involucral bracts imbricate or subequal, chartaceous, often green-tipped (herbaceous in S. parryi); receptacle naked; rays yellow, small, sometimes inconspicuous and not much exceeding the disk flowers; disk flowers yellow; pappus of capillary bristles.

1   Involucres 5-12 (15) mm wide; inflorescence with (1) 3-20 heads, rarely more; plants mostly 2-40 cm tall, basal leaves usually well developed and larger than those of the stem 2

1   Involucres 3-4 (5) mm wide; inflorescence mostly with more than 20 heads; plants (15) 30-120 (200) cm tall; basal leaves various, sometimes lacking or withering and deciduous by late anthesis 4

2   Involucral bracts herbaceous, somewhat foliaceous, the larger ones 8-11 mm long, 2-5 mm wide .... *S. parryi*

2   Involucral bracts chartaceous, smaller than above, seldom over 6 mm long 3

3   Margins of petioles ciliate with crinkled multicellular hairs, these usually much different from other hairs of the leaves; stems sometimes also with crinkled, multicellular hairs, rarely red-purple .... *S. multiradiata*

3   Margins of petioles not ciliate as above; stems glabrous or sparsely scabrous, sometimes glandular, often red-purple .... *S. simplex*

4   Stems glabrous below the inflorescence, sometimes glandular or sparingly scabrous; heads not secund except sometimes in *S. missouriensis* 5

4   Stems puberulent for some distance below the inflorescence and sometimes to near the base; heads more or less secund........................... 6

5   Plants from simple or short-branched caudices (mostly keyed above in lead 3); leaves obovate oblanceolate, often crenate toward the apex; heads not secund, the inflorescence racemose, not at all flat-topped ..... *S. simplex*

5   Plants from rhizomes or caudices with slender elongate branches, or the branches occasionally short as above; leaves oblanceolate to narrowly oblanceolate, mostly entire; heads sometimes secund, in a rounded to flat-topped inflorescence .... *S. missouriensis*

6   Leaves densely and evenly puberulent (use 10-20X magnification); plants from simple or short branched caudices; basal leaves usually persisting at late anthesis .... *S. nana*

6   Leaves glabrous or the margins and often the midribs more densely pubescent than the surfaces; plants from rhizomes or caudices with slender elongate branches or the branches occasionally short as above; basal leaves lacking or often deciduous by late anthesis 7

7   Leaves mostly elliptic-lanceolate, often serrate, those of the lower 1/4 of the stem withering by anthesis; rays 10-17; inner involucral bracts tapering from near the base; plants 30-100 (200) cm tall, of moist or wet places; inflorescence 2-20 cm wide .... *S. canadensis*

7   Leaves mostly oblanceolate, usually not serrate, those of the lower stem usually persisting; rays 5-9; inner involucral bracts slightly widened toward the apex; plants 20-60 (100) cm tall, mostly of well-drained soil; inflorescence 2-8 cm wide .... *S. velutina*

Solidago canadensis L. (*S. altissima* L.; *S. elongata* Nuttall; *S. lepidus* de Candolle) — Canada goldenrod — Widespread; locally common to abundant in riparian and wet meadow communities, and along ditches and gullies; 4,800-7,600 ft; late July-Sept. Our plants belong to var. *salebrosa* (Piper) M. E. Jones.

Solidago missouriensis Nuttall — Missouri goldenrod — Widespread; occasional in riparian, lodgepole pine, and meadow communities, and sandy alluvium of rivers; 5,500-8,800 ft; July-Aug.

Solidago multiradiata Aiton (*S. ciliosa* Greene) — low goldenrod, mtn. goldenrod — Uinta Mtns.; common in meadows, coniferous woods, rocky slopes including talus, and lower alpine communities; 8,600-11,000 ft; July-Aug. Our plants belong to var. *scopulorum* A. Gray.

Solidago nana Nuttall — baby goldenrod — Widespread; occasional; alluvium along rivers, cracks of rocks, seeps, and sagebrush, pinyon-juniper, and ponderosa pine communities; 4,800-8,200 ft; June-Sept.

Solidago parryi (A. Gray) Greene [*Haplopappus parryi* A. Gray; *Oreochrysum parryi* (A. Gray) Rydberg] — parry goldenrod — Widespread in mountains; aspen and coniferous woods and openings, open slopes in grass-forb communities, talus or rocky slopes, and occasionally above timberline; 8,500-11,600 ft; late July-Aug.

Solidago simplex H.B.K. (*S. spathulata* de Candolle) — coast goldenrod — Occasional to common in the Uinta Mtns. and apparently rare on the E. Tavaputs Plateau in ponderosa pine, aspen, lodgepole pine, spruce-fir, and lower alpine communities; July-Aug. Two intergrading vars. can be detected in plants of our area:

1 Inflorescence short and compact, branches or peduncles seldom over 1 cm long; plants 3-15 cm tall, from about 10,000-11,500 ft on the Uinta Mtns. [*S. decumbens* Greene; *S. spathulata* var. *nana* (A. Gray) Cronquist)] . . . .
. . . . **var. nana** (A. Gray) G. S. Ringius
1 Inflorescence more or less elongate, some of the lower branches often 1-5 (8) cm long; plants from about 8,800-9,800 ft on the Uinta Mtns. and E. Tavaputs Plateau [*S. spathulata* var. *neomexicana* (A. Gray) Cronquist] . . . .
. . . . **var. simplex**

Solidago velutina de Candolle (*S. sparsiflora* A. Gray; *S. trinervata* Greene) — slender goldenrod — Widespread; locally common in sagebrush, pinyon-juniper, mt. brush, aspen, coniferous forest, and meadow communities, sometimes in rocky places; mid July-Aug. Sometimes confused with *S. canadensis* but usually smaller, more montane and/or of drier and rockier habitats. More like *S. nana* in stature and general appearance than like *S. canadensis,* and perhaps grading somewhat toward *S. nana* in pubescence of leaves, but mostly noticeably different.

Sonchus L. — Sowthistle

Annual or perennial herbs with milky juice; leaves alternate, sometimes also basal, usually auriculate clasping, mostly toothed to pinnatifid with prickly margins or the teeth spine-tipped, rarely entire; involucres bell-shaped, the bracts in 3-5 series; receptacles naked; rays yellow; disk flowers none; pappus of capillary bristles.

1 Plants perennial from stout rootstocks, 40-200 cm tall; heads including the rays usually 3-5 cm wide in flower, the involucre mostly 14-22 mm high in fruit; rays 8-15 mm long, bright yellow-orange; pappus 10-14 mm long . . .
. . . . *S. arvensis*

1   Plants annual from taproots, these often robust and appearing more than annual, 10-100 cm tall; heads mostly 1-3 cm wide, the involucre mostly 9-14 mm high in fruit; rays to 7 mm long, yellow; pappus to 9 mm long. . . . 2

2   Achenes with prominent smooth ribs; leaves with rounded auriculate lobes, the margins sharply and narrowly toothed and sometimes lobed . *S. asper*

2   Achenes striate to weakly ribbed, scabrous or cross-wrinkled; auriculate lobes of leaves rounded but sharply acute; the margins sharply and broadly toothed, or merely toothed and lyrate pinnatifid . . . . *S. oleraceus*

Sonchus arvensis L. With 2 ssp. as follows:

  1   Involucres and peduncles bearing coarse stipitate glands . . . *ssp. arvensis*
  1   Involucres and peduncles not stipitate glandular . . . . *ssp. uliginosus*

Ssp. *arvensis* — field sowthistle — Adventive from Europe, weedy, to be expected across our area; the 4 specimens are from Duchesne Co. near Mt. Home, Neola, Talmage, and Red Creek, to be expected elsewhere; along drainages; 6,600-7,000 ft; June-Sept.

Ssp. *uliginosus* (Bieberstein) Nyman (*S. uliginosus* Bieberstein; *S. arvensis* var. *glabrescens* Gunther, Grabner & Wimmer) — meadow sowthistle — Adventive from Europe, weedy, widespread, the common sowthistle of our area; riparian-wet meadow, saltgrass-bullrush, and common reed communities, along roadside ditches, edge of ponds, seeps, and floodplains; 4,700-5,600 ft and probably to 7,000 ft; late June-early Oct.

Sonchus asper (L.) Hill — prickly sowthistle — Introduced from Europe, weedy, to be expected across our area, but seldom collected; edge of ponds, along ditches, roadsides, and waste places, usually on moist or wet ground; 4,900-7,000 ft; June-July.

Sonchus oleraceus L. — common sowthistle — Introduced from Europe, weedy, to be expected across our area, but seldom collected; disturbed mesic areas as in S. asper; June-Aug.

Stephanomeria Nuttall — Wirelettuce

Herbs with milky juice, more or less rush-like; leaves alternate, linear to oblong, the upper ones often scale-like; heads with rays only, borne at the ends of branches; involucres cylindrical or oblong, the principal bracts few, about equal, usually with a few outer, much reduced bracts; receptacles naked; rays pinkish, about 3-8 per head, rather small, quickly withering; pappus of 12-20 bristles, these plumose at least above.

---

1   Plants annual; involucres about 7 mm high; pappus bristles plumose on the upper 1/2 or 2/3, simple below; rays 3-5 mm long . . . . *S. exigua*

1   Plants perennial; involucres mostly 9-16 mm high; pappus bristles plumose to the base; rays 4-12 mm long 2

2   Plants 10-20 cm tall; leaves runcinate-pinnatifid; achenes with ridges pitted and tuberculate; rays mainly 8-12 mm long . . . . *S. runcinata*

2   Plants 20-50 cm tall; leaves various; achenes with smooth ridges; rays 4-10 mm long . . . . *S. tenuifolia*

Stephanomeria exigua Nuttall [*Ptiloria exigua* (Nuttall) Greene] — annual wirelettuce — Uintah and Moffat Cos.; occasional in desert shrub and juniper communities; 4,700-5,000 ft; July-Sept.

Stephanomeria runcinata Nuttall — desert wirelettuce — Widespread but not plotted for Rio Blanco Co. in TPD; occasional in desert shrub, juniper and mtn. brush communities; 4,200-8,600 ft; July-Sept.

Stephanomeria tenuifolia (Torrey) Hall [*Ptiloria tenuifolia* (Torrey) Rafinesque] — slender wirelettuce — Apparently widespread, seldom collected, with 2 vars.:

1 Involucres 10-16 mm high, the bracts attenuate; basal leaves, at least some, bipinnatifid; stems solitary or 2 from a rhizome-like creeping root . . . . .
. . . . **var. *uintaensis***
1 Involucres 8-11.2 mm high, the bracts hardly attenuate; basal leaves not bipinnatifid; stems usually several from a branched caudex . . . .
. . . . **var. *tenuifolia***

Var. *tenuifolia* — slender wirelettuce — Apparently widespread, seldom collected; desert shrub, sagebrush, pinyon-juniper, and lodgepole pine-Engelmann spruce communities; 5,000-9,700 ft; July-Sept.

Var. *uintaensis* Goodrich & S. L. Welsh — Uinta wirelettuce — Known from Brownie Canyon, Uinta Mtns.; ponderosa pine-Rocky Mtn. juniper communities at about 8,000 ft; Aug.-Sept.

Tanacetum L.

Aromatic, perennial, rhizomatous herbs; leaves alternate; heads numerous, corymbose, with disk flowers only or with both rays and disks; flowers perfect, yellow; involucres hemispheric with bracts in 2 or 3 series and more or less imbricate; receptacle low-convex, naked; anthers entire at the base; pappus a minute crown; achenes 5-angled.

1 Leaves finely serrate . . . . *T. balsamita*
1 Leaves 1-3 times pinnatified with coarsely serrate leaflets . . . . . *T. vulgare*

Tanacetum balsamita L. — costmary — Strigose herbs 50-100 (120) cm tall; leaves 6-15 cm long, the lower ones with petioles, the upper sessile or nearly so; involucres about 4-5 mm high. Introduced from Europe, cultivated, persisting, and apparently rarely escaping. The one specimen seen (Goodrich 21499) is from Vernal. This specimen and all other plants observed in the Uinta Basin are about older homes. Apparently preference for this plant as a cultivated species has declined.

Tanacetum vulgare L. — Tansy — Glabrous or sparingly tomentose herbs, 30-100 (150) cm tall; leaves 6-15 cm long, sessile or nearly so; involucres about 4-5 mm high. Introduced from Europe, rather commonly cultivated, persisting, and apparently rarely escaping.

Taraxacum Weber — Dandelion

Perennial plants from taproots, with milky juice; leaves in a basal rosette, pinnatifid to toothed, rarely entire; heads solitary on a hollow, more or less translucent, succulent scape, with rays only; involucral bracts in a short outer series and a much longer inner series; receptacles naked; rays yellow, 5-toothed at apex; pappus of simple capillary bristles; achenes ribbed and nerved, strongly beaked.

---

1 Introduced weedy plants, widespread from 4,700 to about 10,500 ft; involucres 12-18 mm high, the outer bracts strongly spreading or more often reflexed; leaves toothed to divided, often with a large terminal lobe
.... *T. officinale*

1 Native plants, mostly above 10,500 ft; involucres shorter or the outer bracts mostly erect or only slightly spreading; leaves various 2

2 Involucres 7-10 mm high (in our plants), the inner bracts seldom corniculate; leaves divided to nearly entire, 2-7 mm wide; plants 2-6 cm tall (reported to 15 cm tall); mature achenes blackish, sometimes reddish at the apex .... *T. lyratum*

2 Involucres 11-17 mm high, the inner bracts often corniculate; leaves entire, toothed or shallowly lobed, commonly 8-17 mm wide; plants 4-10 cm tall; mature achenes greenish or light Brown .... *T. ceratophorum*

Taraxacum ceratophorum (Ledebour) de Candolle — horned dandelion — The 20 specimens seen are from across the Uinta Mtns.; talus slopes, Red Pine Shale barrens, open dry slopes, upper Engelmann spruce belt and alpine, 10,000-12,600 ft; July-Aug.

Taraxacum lyratum (Ledebour) de Candolle [*T. scopulorum* (A. Gray) Rydberg] — dwarf alpine dandelion — The 15 specimens seen are from the Uinta Mtns.; Engelmann spruce and alpine rocky slopes and talus communities; 10,160-13,000 ft; July-Aug.

Taraxacum officinale Weber in F. H. Wiggers (*Leontodon taraxacum* L.) — common dandelion — Introduced from Eurasia, weedy and extremely variable; fields, lawns, gardens, roadsides, and in many plant communities; 4,700-10,300 ft and perhaps higher; April-Oct. Over 1,000 species have been described in the genus Taraxacum, but conservative treatments recognize fewer than 60. Taxonomic problems are numerous. Hybridization, polyploidy, and apomixis are all common complicating factors. Some of the specimens seen have reddish or reddish-brown achenes and somewhat corniculate inner involucral bracts, and thus might be referred to *T. laevigatum* (Willdenow) de Candolle (*T. erythrospermum* Andrezejowiski in Besser). However, they also have features of *T. officinale,* and we are not able to distinguish more than 1 taxon in our specimens of weedy dandelions. The 2 indigenous taxa listed above have been included in an expanded *T. officinale* by some authors.

Tetradymia de Candolle — Horsebrush

Shrubs with stems more or less tomentose at least when young; leaves alternate, solitary or fascicled, entire, the primary leaves sometimes modified into spines; heads axillary or clustered at the tips of the branches; involucral bracts 4-6, in a single row, all about equal; rays lacking; disk flowers yellow; pappus of numerous, whitish, capillary bristles; achenes densely pubescent, usually with long hairs.

1   Twigs armed with spines 2
1   Twigs unarmed. 3
2   Twigs densely, uniformly, and permanently tomentose, the tomentum of older twigs often peppered with minute blackish flakes and appearing blackish beneath; spines recurved, 5-10 (15) mm long; involucral bracts 5-6 . . . . *T. spinosa*
2   At least some of the older twigs tomentose in lines alternating with glabrous lines, or glabrate in age; spines straight or nearly so, 5-25 mm long; involucral bracts 4 . . . . *T. nuttallii*
3   Leaves densely gray- or white-pubescent, 1-5 mm wide, 6-30 mm long, linear-oblanceolate or linear elliptic; involucral bracts tomentose . . . .
. . . . *T. canescens*
3   Leaves glabrous or scattered-pubescent, mostly greenish, not over 2 mm wide, 3-8 mm long, linear or filiform; involucral bracts glabrous or rather thinly pubescent . . . . *T. glabrata*

Tetradymia canescens de Candolle A. — gray horsebrush — Widespread; occasional in sagebrush-grass, pinyon-juniper, mt. brush, Salina wildrye-Douglas-fir, and aspen-coniferous forest communities; 6,800-9,800 ft; June-Aug.

Tetradymia glabrata A. Gray — littleleaf horsebrush — The 2 specimens seen (Neese 4970, Goodrich 27590) are from Big Pack Mt.; desert shrub communities; 5,500 ft; May-June.

Tetradymia nuttallii Torrey & A. Gray — Nuttall horsebrush — Widespread; desert shrub and juniper communities; 4,800-6,000 ft; June-mid July.

Tetradymia spinosa Torrey & A. Gray — cotton-thorn horsebrush — Widespread; desert shrub and juniper communities; 4,800 6,200 ft; late May-early July. Different from other species of *Tetradymia* of our area by pubescence of achenes nearly concealing the pappus.

Thelesperma Lessing — Greenthread

Thelesperma subnudum A. Gray Plants perennial herbs from rather woody roots, subscapose; leaves clustered toward the base of the stem, opposite, pinnately or bipinnately divided into linear or broadly linear segments; heads solitary on long peduncles, the involucral bracts in about 2 rows, the outer about 1/4-1/2 as long as the inner, the inner ones 7-10 mm long, united at the base; receptacles chaffy; rays yellow or lacking; disk flowers yellow; pappus a crown with 4-5 teeth or lacking.

With 4 vars. as follows. All 4 and other taxa from outside our area are included in an expanded view of a single taxon under *T. subnudum* in FNA (21: 201). Considering the small and intergrading differences in the group, inclusion of the complex in one taxon seems reasonable. It seems the only dependable difference for the latter 3 taxa is geographic isolation.

1   Plants with solitary or few stems arising from a simple or sparingly branched root, 10-40 cm tall, glabrous; heads most often (not always) with ray flowers . . . . **var. subnudum**
1   Plants caespitose with stems clustered on a branched caudex, mostly less than 12 cm tall, often pubescent on petioles, heads with disk flowers only
. . . . 2

2 Plants of the Tavaputs Plateau in Utah; outer involucral bract 5.3-6.5 × 1.2-
   2.5 mm; inner involucral bracts 7.7-10.8 × 2.8-4 mm .. **var. *maliterrimum***
2 Plants of Wyoming; outer involucral bract 3.6-5 × x1.0-1.4 mm; inner
   involucral bracts 6.5-9 × 1.8-3.5 mm 3
3 Leaves hairy throughout; plants of Hickey Mtn. area, Uinta Co., Wyoming
   .... **var. *pubescens***
3 Leaves hairy on the petioles only or rarely on the lower blade; plants from
   near Green River, Sweetwater Co., Wyoming .... **var. *caespitosum***

Var. *caespitosum* (Dorn) S. L. Welsh [*T. caespitosum* Dorn; *T. pubescens* var. *caespitosum* (Dorn) C. J. Hansen] — Green River greenthread — Locally common to abundant on Green River Formation near the town of Green River in Sweetwater Co., Wyoming in desert shrub communities and on semi-barrens at about 6,200 ft.

Var. *maliterrimum* S. L. Welsh & N. D. Atwood — Tavaputs greenthread — Locally common on semi-barren, eroding slopes and ridges of Green River For-mation in the Badlands Cliffs and other areas of the Tavaputs Plateau at 6,000-8,800 ft; July-Aug. Leaves glabrous or nearly so as in var. caespitosum or pubescent as in var. pubescens. Although there is a trend for larger involucres in plants of the Tavaputs Plateau, the whole of the Tavaputs population includes plants with heads as small as those from Wyoming.

Var. *pubescens* (Dorn) S. L. Welsh (*T. pubescens* Dorn) — Uinta greenthread — Known from fluvial deposits in the Hickey Mtn. area on the n. flank of the Uinta Mtns. in Uinta Co, Wyoming; mtn. brush and windswept grass-forb com-munities at about 8,500-8,900 ft; July-Aug.

Var. *subnudum* — Navajo-tea greenthread — Infrequent to locally common on the W. Tavaputs Plateau and its flank, and some distance (Blue Bench, Windy Ridge, and Duchesne River valley nw. of Duchesne) into the Uinta Basin, apparently rare on the E. Tavaputs Plateau and in the desert shrub belt east of the Green River; desert shrub, pinyon-juniper, and sagebrush communities, barrens, often on white, marly mudstone; 5,300-5,980 ft; June-Sept.

Townsendia Hooker — Townsendia, Townsend Daisy

Plants herbs; leaves basal or alternate, entire, mostly narrow and broadest above the middle; heads variously disposed; involucral bracts imbricate, appressed, mostly with scarious margins; receptacles naked; rays white to violet; disk flowers yellow; pappus of scabrous bristles, that of the rays sometimes shorter than that of the disks; achenes of disk flowers with 2-forked or glochidiate hairs.

1 Plants annual, biennial, or short-lived perennial, not pulvinate caespitose,
   the stems often conspicuous, commonly over 1 cm long, canescent, or
   strigose 2
1 Plants perennial, more or less pulvinate caespitose; heads sessile or nearly
   so in the basal rosettes of leaves; stems rarely over 1 cm long except in
   vigorous juvenile plants.. 3
2 Stems reddish, densely to sparsely strigose .... **T. *strigosa***
2 Stems grayish-white, densely canescent .... **T. *incana***

3   Involucral bracts fimbriate-ciliate, tending to taper into a fimbriate-ciliate appendage, linear to narrowly lanceolate, acute to acuminate, the inner ones about 10 mm long; leaves linear or narrowly oblanceolate, (10) 15-20 (25) mm long, uniformly and densely strigose on both sides; pubescence persistent on the marcescent leaf bases .... *T. hookeri*

3   Involucral bracts broadly lanceolate to oblanceolate, rounded to broadly acute, not acuminate, the inner ones 5-8 (9) mm long; leaves linear to oblanceolate, generally shorter than above and/or glabrous to sparingly pubescent 4

4   Leaves rather densely and uniformly gray-white canescent on both sides, 5-10 (15) mm long, linear or narrowly oblanceolate; rays minutely densely glandular on the upper (inside) surface; plants common from 5,600-9,000 ft on the Tavaputs Plateau .... *T. mensana*

4   Leaves glabrous or rather sparsely to densely pubescent, more or less greenish, some either longer than above or else distinctly oblanceolate to spatulate; rays generally not glandular or sparsely so; distribution various 5

5   Plants common, widespread, from 4,700-7,000 ft; leaves uniformly and rather densely gray-white canescent; heads sometimes on short but conspicuous stems; achenes pubescent; see lead 2 above .... *T. incana*

5   Plants less common, restricted, from 7,000-10,600 ft; leaves green, glabrous or sparingly strigose; heads sessile or nearly so in the basal rosettes; achenes various 6

6   Leaves narrowly oblanceolate or nearly linear; involucral bracts linear-lanceolate to narrowly elliptic; plants known from the w. end of the Uinta Mtns.; achenes pubescent .... *T. leptotes*

6   Leaves spatulate to narrowly oblanceolate; involucral bracts elliptic to broadly elliptic; plants known from the W. Tavaputs Plateau and e. end of the Uinta Mtns.; achenes glabrous or pubescent .... *T. montana*

Townsendia hookeri Beaman — Hooker townsendia — The specimens seen are from widely scattered locations from Pigeon Basin, Clay Basin, Uinta Mtns., Strawberry Valley, and West Tavaputs Plateau, none seen from the East Tavaputs Plateau; big sagebrush, fringed sagebrush, mt. brush, and ponderosa pine communities; 7,000-10,200 ft; May-June.

Townsendia incana Nuttall — silvery townsendia — Widespread; common in desert shrub, sagebrush, and pinyon-juniper communities; 4,700-7,000 ft; late April-June. Involucral bracts strigose, 3-10 mm long, 1-3.5 mm wide

Townsendia leptotes (A. Gray) Osterhout — limestone townsendia — The lone specimens seen are from Rhoades Canyon and east to Brush Creek on the south slope of the Uinta Mtns., and Provo River drainage east to Blacks Fork on the north slope all from limestone gravelly ground; 9,000-10,800 ft; June-July.

Townsendia mensana M. E. Jones — Tavaputs townsendia — Endemic to the Tavaputs Plateau and adjacent in the Basin in Duchesne Co., occasional to McCook Ridge, Uintah Co., and perhaps in Colorado; desert shrub, pinyon-juniper, Douglas-fir, and sagebrush-Salina wildrye communities, often on bare shale or marl limestone slopes and ridges; 5,600-9,000 ft; late April-June.

m m

c m

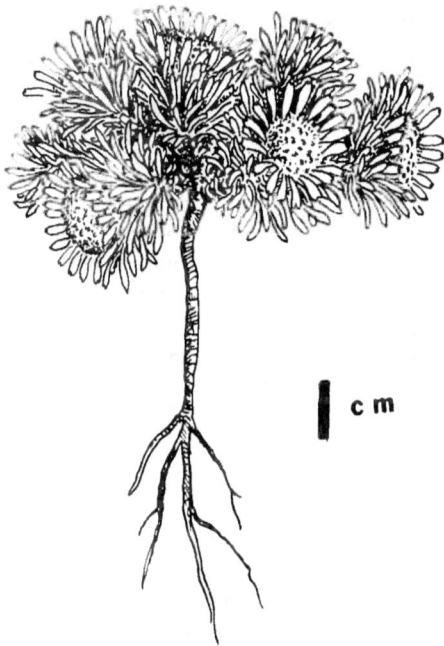

Townsendia mensana M. E. Jones — Tavaputs townsendia

Townsendia montana M. E. Jones — mtn. townsendia — With 2 vars.:

1 Leaves spatulate, rounded at the tip; plants from calcareous places of the
  W. Tavaputs Plateau . . . . **var. *caelilinensis*.**
1 Leaves oblanceolate to narrowly spatulate, some usually pointed at the tip;
  plants from the e. end of the Uinta Mountains . . . . **var. *montana***

Var. *caelilinensis* S. L. Welsh (*T. minima* Eastwood misapplied) — skyline
townsendia — The 20 specimens seen are from the W. Tavaputs Plateau, from
Argyle Canyon, Sowers Canyon, Indian Canyon, and Willow Creek drainages;
pinyon-juniper, Douglas-fir, and bristlecone pine communities, and on nearly
barren slopes and ridges of Green River and Uinta Formations; 7,000-9,200 ft;
June-July.

Var. montana (*T. goodrichii* S. L. Welsh & N. D. Atwood) — mtn. townsendia — Known from Blacks Fork and east to Daggett Co. on the north slope of the Uinta Mtns., and from the Whiterocks drainage east to Dyer Mtn on the south slope.; gravelly limestone; 8,300-11,100 ft; July. Some specimens seem to intergrade into *T. leptotes*.

Townsendia strigosa Nuttall [*T. strigosa* var. *prolix* (M. E. Jones) S. L. Welsh] — strigosa townsendia — Widespread; occasional in desert shrub and sagebrush communities; 4,700-7,000 ft; May-June. Involucral bracts glabrate to moderately strigose, 3-7 mm long, 1.2-1.9 mm wide.

Tragopogon L. — Goatsbeard; Salsify

Mostly biennial herbs from fleshy taproots with milky juice; leaves alternate, elongate, linear, solitary entire, strongly nerved; heads with rays only, solitary on long, hollow peduncles, the peduncles sometimes swollen just below the head; involucral bracts in a single row, about equal, united at the base; pappus of plumose bristles; achenes ribbed, long beaked, the achenes and pappus bristles forming a large feathery ball-shaped head, to 10 cm or more in diameter at maturity, the achenes easily broken from the head and carried long distances by wind.

---

1   Flowers purple or rose; leaf-axils glabrous .... *T. porrifolius*
1   Flowers yellow; leaf-axils with tufts of woolly hair .... *T. dubius*

Tragopogon dubius Scopoli — yellow salsify — Introduced from Europe, widespread; many plant communities both in areas of disturbance and in native plant communities; 4,700-8,000 ft; May-Sept. This plant is often encountered in study plots on the Ashley National Forest where only 1 or 2 plants are present. It has been present in our area for many years without evidence of displacing native plants.

Tragopogon porrifolius L. — vegetable-oyster salsify — Introduced from Europe, no specimens seen from our area, but to be expected.

Wyethia Nuttall — Wyethia; Mules-ears

Perennial herbs from stout taproots; leaves alternate, mostly entire; heads large, solitary or few; involucral bracts in 2-4 rows, the outer often leaf-like; receptacles chaffy; ray and disk flowers yellow; pappus a crown of unequal, laciniate, persistent scales, these sometimes awned; achenes 6-15 mm long.

---

1   Leaves 0.3-1.7 cm wide .... *W. scabra*
1   Larger leaves 5-15 cm wide .... *W. amplexicaulis*

Wyethia amplexicaulis (Nuttall) Nuttall — mules-ears — Strawberry Valley, Cold Spring Mtn. (Harrington 2127), and 2 mi n. of Rio Blanco (Harrington 4565); forb-grass and aspen communities and open slopes; June-July A few plants (apparently roadside waifs) have been found at Greendale Junction in Daggett Co. These have been eradicated. References to *W. arizonica* A. Gray for the Uinta Basin area are based on specimens belonging here.

Wyethia scabra Hooker — badlands wyethia — Widespread but mostly in scattered populations in desert shrub and juniper communities, mostly in sandy places or in sandstone; 5,400-5,600 (7,200) ft; June. England 232 UI, reported as *Verbesina encelioides* (Cavanilles) Bentham in A. Gray (Goodrich and others 1981), belongs here.

Xanthium L. — Cocklebur

Xanthium strumarium L. (*X. italicum* Moretti; *X. saccharatum* Wallroth) — cocklebur — Weedy, annual herbs, 20-80 cm tall or taller; leaves alternate, the petioles slender, about as long as the blades, the blades to 15 cm long, ovate, often cordate, more or less sinuately lobed or dentate, scabrous; heads unisexual; staminate heads borne above the pistillate, with separate involucral bracts, in 1-3 series, the flowers all discoid; pistillate heads bur-like, the bracts united, these 2-beaked at the apex and bearing stiff, hooked prickles, the corollas lacking; pappus none. Widespread weed of fields, ditch banks, gardens, roadsides, drawdown basins of reservoirs, lakes and ponds, and along streams, increasing with disturbance; 4,700-7,500 ft; Aug.-Sept.

Xylorhiza Nuttall — Desert Daisy

Perennial herbs (sometimes slightly woody at the base) from branching, woody caudices; leaves alternate, narrowly oblanceolate, villous, entire or undulate, heads solitary on peduncles; involucral bracts in 1-2 series, spinulose, acuminate, short villous on the back; receptacles naked; rays white; pappus of barbellate, capillary bristles; achenes pubescent.

---

1 Rays 15-25 mm long; involucres 13-16 mm high; stems leafy only near the base, the peduncles 5-15 cm long . . . . **X. venusta**

1 Rays 8-12 mm long; involucres 8-12 mm high; stems leafy to near the top, the peduncles about 1-6 cm long . . . . **X. glabriuscula**

Xylorhiza glabriuscula Nuttall [*Machaeranthera glabriuscula* (Nuttall) Cronquist & Keck.; *Aster glabriuscula* (Nuttall) Torrey & A. Gray; *Aster xylorhiza* Torrey & A. Gray] — smooth woodyaster — Known from Mancos (Hillard) Shale within a 5 mile radius of Manila, Daggett Co. and from Lookout Mtn, and Vermillion Bluffs in Moffat Co. at 6,200-7,500 ft. The species is common to the n. in Wyoming.

Xylorhiza venusta (M. E. Jones) A. Heller [*Aster venustus* M. E. Jones; *Machaeranthera venusta* (M. E. Jones) Cronquist & Keck.] — Cisco woodyaster, charming woodyaster — Uintah and Daggett Cos. and e. into Colorado; common and locally abundant in desert shrub communities or barrens in the desert shrub belt, rarely in sagebrush and pinyon-juniper communities; 4,800-6,200 ft; late April-mid June.

BERBERIDACEAE Barberry Family
Mahonia Nuttall— Barberry

Mahonia repens (Lindley) G. Don [*Berberis repens* Lindley; *Odostemon aquifolium* (Pursh) Rydberg misapplied] — Oregon grape — Low shrubs usually not over

30 cm tall; leaves pinnately compound with 3-7 leaflets; leaflets holly-like, 3-7 cm long, thick and evergreen, often turning red in autumn or winter, the margins toothed, each tooth usually ending in a sharp bristle; flowers in dense racemes to 8 cm long; sepals 6 in 2 series of 3 each, petal-like, the outer series greenish-yellow and 2-3 mm long, the inner bright yellow and 5-8 mm long; petals 6 in 2 series; stamens 6, opposite the petals; fruit a glaucous, blue-black berry 5-9 mm long. Common throughout our mountains, perhaps most common under aspen and coniferous trees, but also with sagebrush and in other shrub communities; 7,000-9,500 ft; May-June. *Mahonia fremontii* (Torrey) Fedde (*Berberis fremontii* Torrey) approaches the s. slopes of the E. Tavaputs Plateau. It is distinguished from *M. repens* by taller stature (1-3 m tall), and by having leaflets less than 3 cm long with fewer than 5 teeth on each side.

BETULACEAE Birch Family

Monecious shrubs and trees; leaves deciduous, alternate, simple, serrate; staminate flowers numerous in spreading or drooping aments (catkins), the perianth parts inconspicuous; pistillate flowers in cone-like aments, without a perianth, each bract (scale-like in *Betula,* or woody in *Alnus*) of the catkin subtending and obscuring 2-3 flowers; stamens 2 (*Betula*) or 4 (*Alnus*); ovary inferior or superior; fruit a 1-seeded nut or samara.

---

1   Leaves double-serrate, sinuate or lobed, the lobes or larger teeth with smaller teeth; twigs without glandular dots; scales of the pistillate aments woody and persistent . . . . ***Alnus***

1   Leaves more or less single-serrate with uniform sized teeth; twigs with glandular dots; scales of the pistillate aments thin and deciduous . . ***Betula***

Alnus Mill — Alder

Alnus incana (L.) Moench (*A. tenuifolia* Nuttall) — mtn. alder — Shrubs or small trees mostly not over 7 m tall; bark thin, gray or reddish; twigs pubescent when young; leaves 2-9 (12) cm long, 1.5-6.5 (8) cm wide; flowers developing before the leaves; staminate catkins in clusters of 2-4, linear, 2-7 (10) cm long; pistillate catkins in clusters of 3-9, sessile or pedunculate, 1-1.5 cm long; fruit a thin margined nutlet. widespread; occasional to abundant along water courses, most common in canyons but also in basins at lower elevations; (5,400) 6,000-8,000 ft; April-June. Our plants belong to var. *occidentalis* (Dippel) C. L. Hitchcock.

Betula L. — Birch

With features of the family.

---

1   Leaf blades usually not over 2 cm long, about as wide as long; shrubs mostly less than 2 m tall, above 8,500 ft . . . . ***B. glandulosa***

1   Leaf blades 2-5 (7) cm long, usually slightly longer than wide; shrubs often over 2 m tall; mostly below 8,500 ft . . . . ***B. occidentalis***

Betula glandulosa Michaux — bog birch, scrub birch — Occasional in the Uinta Mtns.; streamside meadows, boggy meadows, and well-drained soil in snowbank areas; 9,000-11,000 ft; July-Aug.

Betula occidentalis Hooker (*B. fontinalis* Sargent) — river birch, water birch, red birch — Widespread; most common along banks of rivers of our major mtn. drainages, sometimes around seeps and springs far removed from other water sources; not expected over 9,000 ft; May-June.

BORAGINACEAE Borage Family

Pubescent often bristly-setose or brtisly-hispid herbs (sometimes glabrous in *Heliotropium* and *Mertensia*); leaves simple, entire usually alternate; inflorescence a modified cyme, the branches frequently unilateral and coiled, usually elongating as the lower flowers mature; calyx 5-lobed or parted; corolla regular, rotate, salverform, or tubular-campanulate, 5-lobed, usually crested at the narrow throat with 5 rounded pouch-like appendages (fornices); stamens 5, inserted on the corolla tube; ovary 4-lobed, usually deeply so with the style gynobasic and arising from between the lobes, breaking at maturity into 4 bony, 1-seeded nutlets.

---

1 Corolla dull reddish-purple or maroon; fruit 12-15 mm broad **Cynoglossum**

1 Corolla blue, lavender, white, or yellow; fruit less than 8 mm broad 2

2 Calyx much enlarging in fruit, the segments 2-lobed, flattened at maturity, bract-like, to 2 cm wide; flowers blue-purple, tubular-campanulate, 2-3 mm long, solitary or few in the upper axils; stems 4-angled, bearing retrorse prickles; plants weak sprawling, rough-hispid annuals . . . . ***Asperugo***

2 Not as above 3

3 Dwarf, cushion-like, silvery-villous perennial to about 6 cm tall, of tundra habitats above 10,000 ft; corolla bright sky-blue, rotate . . . . . . ***Eritrichium***

3 Distribution various but not limited to tundra habitat, if alpine then taller, neither cushion forming nor silky villous 4

4 Style branched, each branch ending in a capitate stigma; plants with taproots, prostrate, divaricately and dichotomously branched annuals of sandy places; leaves with strongly impressed veins, ovate, petiolate, the blades 3-7 mm long, the margins revolute; flowers small, clustered in the axils, pinkish-lavender . . . . ***Tiquilia***

4 Styles simple, with a single stigma; other characteristics not as above . . . 5

5 Plants glabrous, glaucous, somewhat fleshy, of wet saline of alkaline places below 6100 ft; flowers white with a violet-purple eye in the throat . . . . . .
. . . . ***Heliotropium***

5 Plants pubescent or of higher elevations and flowers blue 6

6 Margins of nutlets with barbed prickles that cling tenaciously to clothing and fur. 7

6 Nutlets smooth or roughened, but not with barbed "stick-tight" prickles 8

7 Plants usually slender annuals, rarely biennial; pedicels erect in fruit; flowering branches bracteate; nutlet scar elongate, basal; gynobase narrowly pyramidal, elongate . . . . ***Lappula***

7   Plants biennial to perennial; pedicels reflexed in fruit; flowering branches without bracts subtending the flowers; nutlet scar almost round, median on the nutlets inner face; gynobase low and broad .... *Hackelia*

8   Corolla orange, yellow, or yellowish green 9

8   Corolla blue or white (throat appendages may be yellow) 12

9   Corolla tube 1.5-4.0 cm long .... *Lithospermum incisum*

9   Corolla tube 1.2 cm or less long..................................... 10

10 Plants annual; flowers yellow-orange, known from Strawberry Valley ... . *Amsinckia*

10 Plants perennial, widespread 11

11 Nutlets attached at their base to a flat gynobase, at maturity glossy, hard and pebble-like, grayish-white, not concealed by the calyx lobes......... .... *Lithospermum*

11 Nutlets attached at their ventral face to a conical gynobase, the attachment scar elongate or narrowly triangular; nutlets concealed by the calyx lobes .... *Cryptantha*

12 Plants biennial to perennial....................................... 13

12 Plants slender annuals 14

13 Corolla blue (rarely pink); herbage glabrous or finely pubescent, never setose or hispid .... *Mertensia*

13 Corolla white or yellowish; herbage usually bristly-hispid and with pustulate-based setose hairs, also often densely strigose ..... *Cryptantha*

14 Nutlets attached at their base to a flat gynobase, smooth and glossy on both sides, flattened-ovoid, the margins acute, the inner face not keeled or grooved; corolla blue .... *Myosotis*

14 Nutlets attached laterally to a conical gynobase and with a groove or keel running the length of the ventral face; corolla white; distribution various 15

15 Ventral face of nutlet with a longitudinal elevated keel or ridge; plants of drying mud of ephemeral pools and reservoir, lake, and stream margins . ... *Plagiobothrys*

15 Ventral face of nutlet with a longitudinal, impressed groove or slit; plants of mostly dry sites .... *Cryptantha*

Amsinckia Lehmann — Fiddleneck

Amsinckia menziesii (Lehmann) Nelson & Macbride (*A. retrorsa* Suksdorf) — fiddleneck — Erect, bristly, usually weedy annuals; inflorescence of 1-sided scorpioid cymes, the branches elongating with age; corolla yellow-orange, the throat open, fornices obsolete; nutlets coarsely roughened or tessellate, attached near the base of the ventral face to a short-pyramidal gynobase; cotyledons deeply lobed. Known in our area by a single specimen (L. C. Higgins 13946) from Bryant's Fork campground, w. side of Strawberry Reservoir, Wasatch Co., from a sagebrush-aspen community.

Asperugo L. — Catch-weed; Madwort

Asperugo procumbens L. — common catchweed — Nutlets obliquely compressed, ovate, attached above the middle, enveloped in the calyx; other characters as given in the key. A sparingly introduced weed of fairly moist places, 3 specimens seen are from the Three Corner area in Daggett Co. near a heavily used spring (Neese and Peterson 5619), Rock Creek Huber (5000), and Mosby Canyon Corral (Goodrich 25994), 7,300-8,060 ft.

Cryptantha Lehmann — Cryptantha

Plants annual, perennial, or biennial; herbage pubescent, usually harshly hispid or setose; corolla usually white, sometimes yellow, the fornices usually yellow; nutlets smooth to variously sculptured, 1-4 maturing, attached most of their length to an elongate, narrowly pyramidal gynobase; style slender, included or exserted; stigma capitate.

Our species comprise 2 distinct groups: the small-flowered annuals (Section Krynitzkia) and the showy-flowered perennials (Section Oreocarya). These sections were treated at the level of genus by Weber (1987). Characters of the nutlets such as surface texture, size, number maturing, and size and shape of attachment scar are generally consistent for each species and serve as useful diagnostic features. They are especially important in the annual species, since the annuals are often vegetatively similar. A dissecting scope is often required for adequate evaluation of nutlet characters; also, these characters are not always evident except in fully mature fruit, a time of phenology when the plants are seldom noticed or collected. The following keys emphasize vegetative characters whenever possible to facilitate identification of immature specimens.

---

1 Slender annuals from taproots; basal leaf rosette absent; corolla inconspicuous, the limb not more than 2.5 mm wide (Section Krynitzkia) . . . . **KEY 1**

1 Biennials or perennials from stout caudices; basal leaves well-developed; corolla showy, the limb 4-14 mm wide (Section Oreocarya) . . . . **KEY 2**

— KEY 1 —

1 Nutlets with conspicuous white marginal wings, these irregularly toothed along the margin, at maturity the wings partially exposed between the calyx lobes; fruiting calyx 4-5(8) mm long . . . . ***C. pterocarya***

1 Nutlets not winged, concealed by the calyx 2

2 Calyx circumscissile, 2-3 mm long, at maturity fracturing above the midpoint, the green apical portion falling away as a unit from the paler scarious cupulate base; plants diminutive, branched from the base, forming small cushions 1-5 cm high; inflorescence congested, bracteate throughout, not elongating in age . . . . ***C. circumscissa***

2 Calyx not circumscissile; plants mostly taller than 5 cm; inflorescence not bracteate throughout (some of the lowermost flowers sometimes subtended by leafy bracts). 3

3 Calyx becoming stiffly recurved, asymmetrical, bent at base, 2.5-3.5 mm long at fruiting; stems strigose, the hairs all or nearly all closely appressed . . . . ***C. recurvata***

3 Calyx symmetrical, not bent or curved at base; stems hispid with stiffly spreading hairs, sometimes also with appressed hairs 4

4   Cymes borne in terminal, more or less globose clusters at the end of slender peduncles, congested, scarcely elongating in age; only 1 nutlet maturing per flower, smooth and shining; calyx densely pubescent with soft, straight, white hairs, with few or no pustulate based coarse bristle-hairs, 2-3 mm long in fruit . . . . **C. gracilis**

4   Cymes elongating in fruit, at least a few of the mature lower flowers remote from the apex; all 4 nutlets maturing in most of the flowers; calyx bristly with coarse, stiffly spreading hairs. 5

5   Nutlets roughened, the surface sparsely tuberculate, the scar open for ½ or more of the length; fruiting calyx 4-7 mm long 6

5   Nutlets all smooth or obscurely granular, usually glossy, the scar closed from near the base. 7

6   Nutlets ovate, all alike (homomorphic), the scar closed expect at the base . . . . **C. ambigua**

6   Nutlets lanceolate, heteromorphic, the one nearest the stem slightly larger, often more firmly attached, the scar open to near the middle   **C. kelseyana**

7   Nutlets asymmetric, the line of attachment and scar nearer one margin, not median on the ventral face; inflorescence becoming elongate, usually of 2- or 4-paired slender erect branches held well above the leaves; fruiting calyx 2.5-4 mm long . . . . **C. affinis**

7   Nutlets symmetrical, the scar equal distance from the edges; inflorescence usually more freely branched or the branches spreading, or not greatly exceeding the leaves 8

8   Nutlets with margins definitely angled; fruiting calyx 2-4 mm long . . . . **C. watsonii**

8   Nutlets with margins rounded; fruiting calyx 4-8 mm long. 9

9   Nutlets lanceolate, the basal scar acutely angled; plants of very sandy places . . . . **C. fendleri**

9   Nutlets broadly ovate, the basal scar obtusely angled; plants seldom of very sandy places . . . . **C. torreyana**

— ALTERNATE KEY 1 —

1   Fruiting calyx 4-8 mm long 2

1   Fruiting calyx 2-4 mm long 4

2   Nutlets winged . . . . **C. pterocarya**

2   Nutlets not winged 3

3   Nutlets roughened **C. ambigua** and **C. kelseyana**; see lead 6 of previous key

3   Nutlets smooth . . . . **C. fendleri** and **C. torreyana**; see lead 9 of previous key

4   Calyx circumscissile, at maturity fracturing above the midpoint, the green apical portion falling away as a unit from the paler scarious cupulate base . . . . **C. circumscissa**

4   Calyx not circumscissile 5

5   Calyx becoming stiffly recurved, asymmetrical, bent at base, 2.5-3.5 mm long at fruiting; stems strigose, the hairs all or nearly all closely appressed . . . . **C. recurvata**

5   Calyx symmetrical, not bent or curved at base; stems hispid with stiffly spreading hairs, sometimes also with appressed hairs 6

6   Cymes borne in terminal, more or less globose clusters at the end of slender peduncles, congested, scarcely elongating in age; nutlets solitary, only 1 maturing per flower, smooth and shining; calyx densely pubescent with soft, straight, white hairs, with few or no pustulate based coarse bristle-hairs also see lead 4 of previous key . . . . *C. gracilis*

6   Cymes elongating in fruit, at least a few of the mature lower flowers remote from the apex; all 4 nutlets maturing in most of the flowers; calyx bristly with coarse, stiffly spreading hairs . . . .
. . . . *C. affinis* and *C. watsonii*; see leads 7 and 8 of previous key

— KEY 2 —

1   Corolla tube mostly longer than the calyx; style exceeding mature nutlets by (2.3) 4-8 mm, usually equaling or exceeding the calyx in all stages of phenology. 2

1   Corolla tube about equaling the calyx; style exceeding the mature nutlets by 2 mm or less, conspicuously shorter than the calyx in all stages of phenology. 6

2   Corolla bright yellow; nutlets at maturity lanceolate, smooth and glossy, 1 (rarely 2 or 3) maturing . . . . *C. flava*

2   Corolla white (rarely light yellow in *C. flavoculata* but then nutlets coarsely roughened, ovate, all 4 usually maturing). 3

3   Corolla limb campanulate, never flat-opening. 4

3   Corolla limb flat-opening. 5

4   Plants biennial or (rarely) short-lived perennial from a simple taproot or sparingly branched caudex; corolla tube, when withered, filiform, 0.2-0.5 mm in diameter, evidently swollen at base near nutlets and at summit near anthers; pubescence whitish or obscurely yellowish; nutlets 1.6-1.8 mm long, finely roughened dorsally; widespread . . . . *C. rollinsii*

4   Plants perennial from much-branched caudices, forming large mounds in age; corolla tube when withered 1-1.5 mm in diameter, the sides parallel, not swollen at base nor summit; pubescence usually prominently yellowish, harshly bristly; nutlets 2.1-2.3 mm long, smooth dorsally; known only from barren shale knolls of the E. Tavaputs Plateau . . . . *C. barnebyi*

5   Plants not over 15 cm tall, uncommon; leaves often folded, not over 4 mm wide; corolla fornices when fresh low, rounded, as broad as long; nutlet scar closed . . . . *C. paradoxa*

5   Plants often over 15 cm tall, widespread, common; some leaves usually well over 4 mm wide; corolla fornices elongate, evidently longer than wide; nutlet scar open . . . . *C. flavoculata*

6   Pustulate hairs absent or inconspicuous on upper side of active leaves; the upper side closely silky strigose, with a satiny sheen 7

6   Pustulate hairs usually rather prominent on upper side of active leaves, non-pustulate hairs often rather harsh, the leaves without a satiny sheen
. . . . 8

7   Plants long-lived, caespitose with a much-branched caudex, forming

mounds; leaves silky strigose on both sides; hairs of lower stem appressed; nutlets muricate .... *C. breviflora*

7   Plants short-lived, from a simple taproot or sparingly branched caudex, never forming-mounds; leaves usually bristly-hairy on lower side with pustulate-based hairs; hairs of lower stem widely spreading; nutlets wrinkled .... *C. sericea*

8   Limb of corolla 11-15 mm wide; style exceeding mature nutlet by 1.5-2 mm, conspicuously widened near base, tapering to tip; plants known from exposed white shale barrens of the E. Tavaputs Plateau ...... *C. grahamii*

8   Limb of corolla not over 10 mm wide; style exceeding mature nutlet by 1 mm or less; plants not restricted as above 9

9   Plants strict, erect, 10-40 cm tall, biennial or short-lived perennials from simple taproots or sparingly branched caudices; pubescence harshly setose; nutlets glossy, sparingly wrinkled on dorsal surface, smooth on ventral surface; known from the n. and s. flanks of the Uinta Mtns. In Daggett and e. Uintah Cos. and in near-adjacent Colorado .... *C. stricta*

9   Plants caespitose, or not distributed as above, seldom over 15 cm tall; nutlets roughened and dull both dorsally and ventrally 10

10 Plants short-lived perennials, not especially caespitose, known from Argyle Canyon of W. Tavaputs Plateau; leaves sometimes over 4 cm long; calyx 4-5 mm long at anthesis; flowers white when dry; pubescence of calyx whitish, setose-hirsute; nutlets bowed, the margins not in contact, the scar open .
    .... *C. mensana*

10 Plants long-lived perennials, sometimes caespitose mound-forming; leaves not over-4 cm long; calyx 2-4 mm long at anthesis; flowers white or yellowish when dry; pubescence of calyx often yellowish, sub-tomentose; nutlets not bowed, their margins in contact

11 Plants known from Daggett Co. on Phil Pico Mt. and in Browns Park, and in adjacent Moffat Co.; mound plants, forming dense low clumps; stems short, the above-ground portion of plants 5-15 cm tall; calyx short, about 4 mm long at fruiting; corolla white; nutlets wrinkled, the projections rounded
    .... *C. caespitosa*

11 Plants widely distributed in the Basin but not from Daggett or Moffat Cos.; plants caespitose but not forming mats or mounds, 5-30 cm tall; calyx at fruiting usually 5-8 mm long; corolla light yellow fading yellowish-brown; nutlets muricate-rugulose, the projections often acute .... *C. humilis*

Cryptantha affinis (A. Gray) Greene — ally cryptantha — Our few specimens are from near Strawberry Reservoir and from n. of Vernal; disturbed places in sagebrush, bitterbrush and aspen communities; 6,000-7,700 ft; June-July. Listed for Daggett Co. (AUF 58) perhaps based on Williams 473 which has been annotated to *C. watsonii.*

Cryptantha ambigua (A. Gray) Greene — ambiguous cryptantha — Uncommon; the few specimens seen are from Strawberry Valley to Diamond Mt. and reported for Zenobia Peak (Bradley 1950); desert shrub, sagebrush-grass, and Douglas-fir communities; 4,840-7,800 ft; April-July . The subtle nutlet characters given in the key seem to be the only consistent means of separating this from *C. kelseyana.* See note under that species.

Cryptantha barnebyi I. M. Johnston — Barneby cryptantha — Narrowly endemic, locally abundant; restricted to domed or gently sloping white shale barrens of the Green River Formation between Hill Creek and the Colorado border, mostly in the pinyon-juniper belt, usually growing with one or more of the following: *Chamaechaenactis scaposa, Cirsium barnebyi, Eriogonum ephedroides, Hymenoxys acaulis,* and *Machaeranthera grindelioides;* 5,600-7,200 ft; May-June.

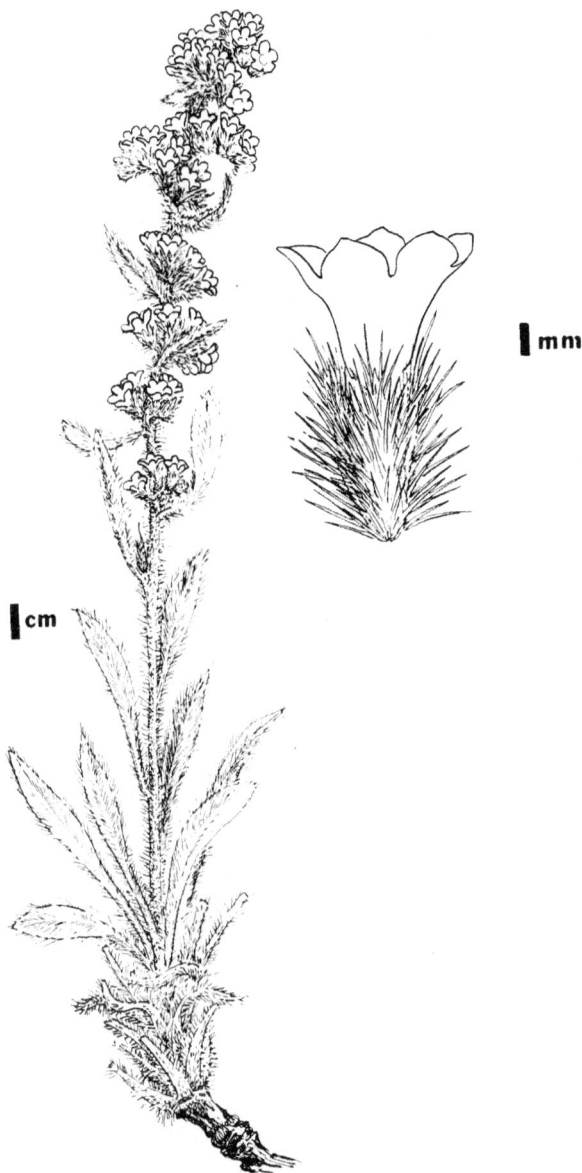

Cryptantha barnebyi I. M. Johnston — Barneby cryptantha

Cryptantha breviflora (Osterhout) Payson — small-flower cryptantha — Endemic, sporadic across the Uinta Basin from Current Creek near Fruitland to the Colorado border; occasional in mixed desert shrub and juniper communities on sparsely vegetated, dry, eroding knolls and badland slopes in several geologic formations, especially common on the Duchesne River Formation, commonly colonizing on roadcuts; 4,800-7,400 ft; late April-May.

Cryptantha caespitosa (A. Nelson) Payson — tufted cryptantha — Uncommon; known from Brown's Park in Daggett and Moffat Cos., where it grows on barren sandy or tuffaceous ridges and knolls in the pinyon-juniper belt, and the summit of Phil Pico Mt. w. of Manila, on barren ridges with limber pine, and Lookout Mtn. and Vermillion Bluffs, Moffat Co., also in the Bridger Formation near Lone Tree, Wyoming; 5,500-8,500 ft; June-July.

Cryptantha circumscissa (Hooker & Arnot) I. M. Johnston — cushion cryptantha — Our few specimens are from near Altamont and Ioka in Duchesne Co. and in the vicinity of Vernal, Jensen, and Dry Fork Settlement in Uintah Co., and one from Clay Basin in Daggett Co.; usually in pinyon-juniper and sagebrush communities, in sand or sandstone crevices; 5,400-7,600 ft; May-July.

Cryptantha fendleri (A. Gray) Greene [*C. pattersonii* (gray) Greene] — Fendler cryptantha — Common along the s. flank of the Uinta Mtns., also found at Snake John Wash, Raven Ridge, se. of Vernal, and in Moffat Co., one specimen is from junction of Strawberry River and Willow Creek, and one specimen is from Sheep Creek in Daggett Co.; desert shrub, sagebrush-grassland, rabbitbrush, pinyon-juniper, mt. brush, and ponderosa pine communities and along roadsides, usually on sand dunes or in other sandy places; (5,200) 5,600-9,000 ft; June-Sept.

Cryptantha flava (A. Nelson) Payson — yellow cryptantha — Widespread; especially common in sandy soil and on sandstone outcrops in pinyon-juniper, rabbit brush-Indian rice grass, sagebrush, galleta-black sagebrush, bitterbrush, and curl leaf mt. mahogany communities; 4,700-7,200 ft; May-July.

Cryptantha flavoculata (A. Nelson) Payson — roughened cryptantha — Probably our most widespread and common perennial cryptantha; salt desert shrub, mixed desert shrub, pinyon-juniper, sagebrush, and mt. brush communities, usually in less sandy places than *C. flava;* from valley bottoms to about 9,000 ft; May-July. *Cryptantha flavoculata* is a variable taxon. Future study may justify recognition of infraspecific taxa, but little if any correlation is apparent in the geographic and morphological variation of our specimens.

Cryptantha gracilis Osterhout — slender cryptantha — Common in the e. end of our area in Uintah, Daggett, Moffat, and Rio Blanco Cos.; apparently less common westward; sagebrush, pinyon-juniper, and ponderosa pine communities; 5,000-6,500 (8,200) ft; May-July.

Cryptantha grahamii I. M. Johnston — Graham cryptantha — Endemic, Gate Canyon in se. Duchesne Co. eastward to the Willow Creek-Buck Canyon area; locally common in salt desert shrub and pinyon-juniper communities, restricted to sparsely vegetated shale terraces, benches, gentle talus slopes, and knolls of Green River Formation; 4,750-6,750 ft; May-June. This distinctive species grows on similar habitats with much the same distribution as *C. barnebyi,* but the species are quite different, and they do not grow together.

Cryptantha humilis (A. Gray) Payson [*C. nana* (Eastwood) Payson] — low cryptantha — Scattered across Duchesne, Uintah, and Rio Blanco Cos.; infrequent in salt desert and mixed desert shrub, and pinyon-juniper communities, usually on dry, sparsely vegetated ridges and slopes, usually on soils derived from Duchesne River, Uinta, and Green River Formations; 4,800-7,800 ft; late April-June. Our plants belong to var. *nana* (Eastwood) L. C. Higgins.

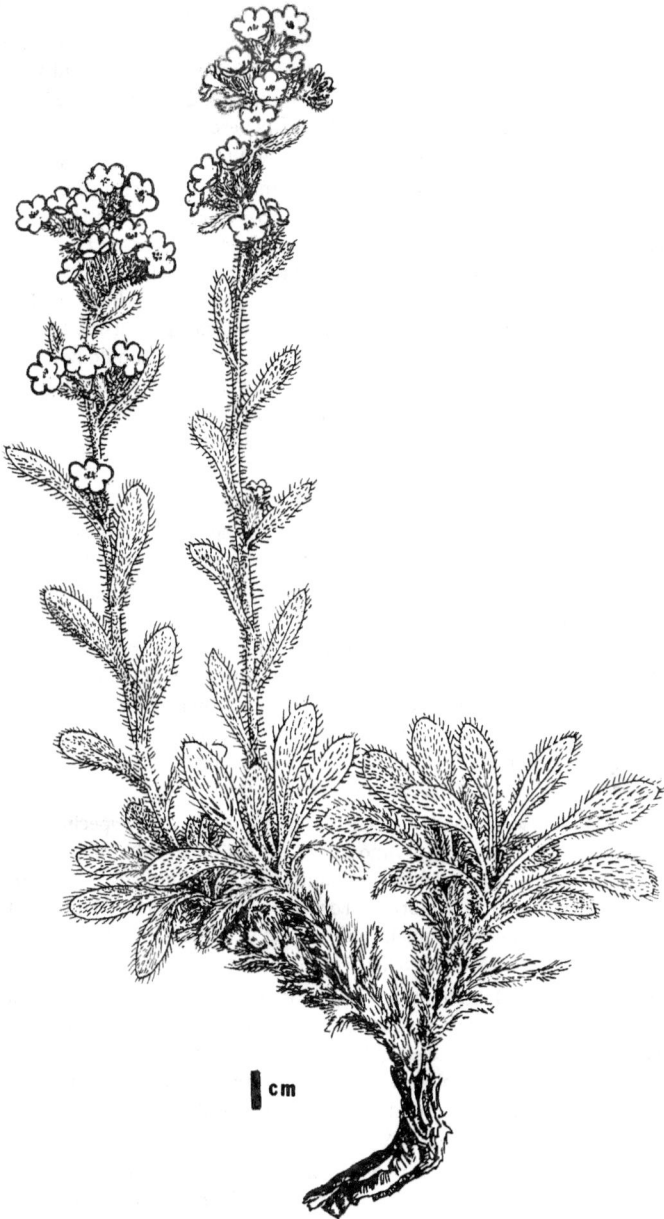

cm

Cryptantha grahamii I. M. Johnston — Graham cryptantha

Cryptantha kelseyana Greene — Kelsey cryptantha — Common and locally abundant in Uintah and Daggett Cos. and e. into Colorado, apparently rare in Duchesne Co., and absent from the w. end of the Basin; usually in desert shrub communities with sagebrush, greasewood, horsebrush, shadscale, and hopsage, usually in sandy soil; 4,800-8,200 ft;-May-Aug. Calyx lobes linear with the midrib thickened and with pustular based bristle-like hairs. The larger nutlet is reported to be less turbiculate than the smaller ones. However, there appears to be little difference in this feature between the larger and smaller nutlets in many specimens from our area. Graham 7872 (CM!), referred to as *C. scoparia* A. Nelson (Graham 1937) likely belongs here.

Cryptantha mensana (M. E. Jones) Payson — table cryptantha — Specimens seen are from Argyle Canyon (Neese & England 5861), Wire Fence Canyon of Sowers Canyon (Huber & Wedig 3670), and Desolation Canyon (Atwood & Evenden 24359, 24370, 24401); desert shrub, pinyon-juniper, and mtn. brush communities; 4,700-7,500 ft; May-June. Additional specimens from just outside our area are from the top of Soldier Creek drainage in Carbon Co. Note: a similar species, *C. bakeri* Payson, has been collected just outside the Uinta Basin near the Carbon-Duchesne Co. line due n. of Price. It would key here or to *C. sericea,* and is characterized by nutlets that are more sharply wrinkled and with a closed, not open scar: It is to be expected on limestone or limey shale of the Green River Formation at about 7,200 ft in sagebrush communities.

Cryptantha paradoxa (A. Nelson) Payson — Paradox cryptantha — Mostly on gravelly margins of abandoned, old pediments and river terraces above the Green and Duchesne Rivers in Duchesne and Uintah Cos. where locally common (one specimen from near Steinaker Reservoir); desert shrub and juniper communities; 4,700-5,600 ft; April-May.

Cryptantha pterocarya (Torrey) Greene — winged cryptantha — Uintah Co. from Steinaker-Red Fleet areas. to Desolation Canyon, and Rio Blanco Co., one specimen from Teepee Mtns. in Daggett Co.; usually in sandy soil and crevices of sandstone outcrops in juniper-mtn. mahogany communities; 4,800-6,700 ft; May-July. Graham (1937) listed *C. scoparia* A. Nelson for the Uinta Basin based on a specimen (Graham 7894 CM!) of *C. pterocarya.*

Cryptantha recurvata Coville — recurved cryptantha — The 2 specimens seen are from Moffat Co. 13 mi e. of Dinosaur (Dunn & Gallian 17731 BRY) roadside and adjacent sagebrush-Bromus community and Dinosaur Ledge (Quarry), Dinosaur National Monument, Morrison Formation (Weber 5445 DINO). Distribution of this species is generally much to the s. and w. of the Uinta Basin. Perhaps merely introduced in Moffat Co. (IMF 4: 264).

Cryptantha rollinsii I. M. Johnston — Rollins cryptantha — Endemic, common from Starvation Reservoir east across the West and East Tavaputs Plateaus, uncommon near Raven Ridge and near Highway 121 between Lapoint and Maeser; desert shrub and pinyon-juniper communities, usually on semi-barren land of shaly or silty-clay slopes; 4,800-6,800 ft; late May-early Aug.

Cryptantha sericea (A. Gray) Payson — silky cryptantha — Widespread and common except apparently rare or lacking on the s. slope of the Uinta Mtns. w. of the Brush Creek drainage; many plant communities, but mostly in dry places of the pinyon-juniper belt on eroding soils; 5,400-9,000 ft; May-July.

Cryptantha stricta (Osterhout) Payson — strict cryptantha — Locally common in Moffat and Daggett Cos. and Diamond Mtn, Blue Mtn. and Raven Ridge, Uintah Co.; desert shrub, juniper, sagebrush-grass, and mt. brush-limber pine communities, usually on rocky ridges or on barrens; 6,600-9,400 ft; June-July. Anomalous specimens (Neese & Fullmer 11715, Smith and others 1716) from the sw. rim of the Blue Mt. plateau are tentatively placed here. They have morphological features of *C. rugulosa* (Payson) Payson, a species well outside the Uinta Basin in Nevada and w. Utah.

Cryptantha torreyana (A. Gray) Greene — Torrey cryptantha — Four of the 5 specimens seen are from Strawberry Valley area, one is from Dry Fork Canyon above Vernal; sagebrush communities and roadcuts at 7,600-8,300 ft;

Cryptantha watsonii (A. Gray) Greene — Watson cryptantha — Specimens seen are from Daggett Co., Whiterocks Canyon and eastward to Blue Mtn. in Uintah and Moffat Cos. The one specimen from Duchesne Co. (Nutter's Ridge) is from a chained and seeded area; desert shrub, sagebrush-grassland, pinyon-juniper, and ponderosa pine communities; (5,300) 6,600-8,200 ft; June-Aug.

Cynoglossum L. — Houndstongue

Cynoglossum officinale L. — common houndstongue — Taprooted, single-stemmed, villous biennial; leaves elliptic to oblong or lanceolate, the lower ones petiolate; inflorescence branches elongating, strongly coiled only when young; flowers racemose, calyx lobes broad, blunt, in fruit enlarging to 8 mm long and loosely spreading or reflexed; corolla short-salverform or salver-form-campanulate, quickly deciduous, 6-9 mm wide, the prominent fornices broadly rounded; nutlets subglobose, flattened dorsiventrally, evenly covered with short barbed prickles, attached to style at tips after splitting from base. Introduced, locally abundant along roadsides and other disturbed places, common and locally abundant on the Tavaputs Plateau, and uncommon in the Uinta Mountains. Most common on canyon bottoms at mid-elevations.

Eritrichium Schrader

Eritrichium nanum (Villars) Schrader (*E. argenteum* W. F. Wight; *E. elongatum* Rydberg) — Alpine forget-me-not — Plants sub-acaulescent or with flowering stems to 1 dm tall; leaves in basal tufts and on lower portions of flowering stems, elliptic, to about 15 mm long; flowers born in cymose clusters; corolla sky-blue, the tube and fornices usually white or yellowish, the limb 4-8 mm wide; nutlets 1-4, glabrous, the dorsal surface crowned with a lacerate submarginal flange. Occasional above timberline in the Uinta Mtns. on rocky slopes and ridges; July-Aug. Our plants belong to var. *elongatum* (Rydberg) Cronquist.

Hackelia Opiz — Stickseed; Wild Forget-me-not; Tick-weed

Biennial to perennial taprooted herbs; leaves basal and cauline, reduced upward, becoming bract-like near the base of the inflorescence; inflorescence branches elongating in age, bractless except near base; corolla rotate to short-salverform, blue, or white with blue markings, the fornices usually white; nutlets with a marginal ring of barbed or glochidiate prickles, these dilated and flattened toward the base.

1   Flowers white with blue markings near the center; fornices short-hairy; nutlets with numerous small intramarginal prickles .... *H. patens*

1   Flowers blue (sometimes pinkish when young); fornices not pubescent; nutlets with no or a few intramarginal prickles....................... 2

2   Plants biennial, from taproots; lower leaves often withering by or soon after anthesis; intramarginal face of all or almost all nutlets with no prickles .. .. *H. floribunda*

2   Plants long-lived perennials from branched caudices; lower leaves seldom withered by anthesis; intramarginal face of nutlets with a few minute barbed prickles .... *H. micrantha*

Hackelia floribunda (Lehmann) I. M. Johnston [*H. leptophylla* (Rydberg) I. M. Johnston] — western stickseed, western tick-weed — Common and locally abundant from Strawberry Valley and e. to the Rock creek drainage in the Uinta Mtns. and e. to Indian Canyon on the W. Tavaputs Plateau and less common nw. of Vernal in the Dry Fork area and on the E. Tavaputs Plateau in the Piceance Basin; tall forb communities and aspen and fir parklands, usually with aspen and sagebrush, often in disturbed areas and tolerant of high levels of pocket gopher activity; 7,200-9,000 ft; mid May-Aug. Goodrich et al. (1981) listed *H. gracillentia* (Eastwood) I. M. Johnston] for the Uinta Basin based on a specimen [England 219 (UI!)] belonging to *A. floribunda.*

Hackelia micrantha (Eastwood) Gentry [*H. jessicae* (McGregor) Brand] — Jessica stickseed, Jessie tick-weed — Known from Strawberry Valley and e. in the Uinta Mtns. to North Fork Duchesne drainage on the s. slope and Bear River drainage on the n. slope, and e. on the Tavaputs Plateau to Argyle Canyon; 6,200-10,000 ft; June-Aug, Similar to *H. floribunda,* immature specimens often difficult to place, but the flowers tend to be a little larger and more showy, and not so much associated with disturbance.

Hackelia patens (Nuttall) I. M. Johnston — common stickseed, white tick-weed — Occasional from Strawberry Valley and e. in the Uinta Mtns. to North Fork Duchesne drainage and e. on the Tavaputs Plateau to Horse Ridge/Reservation Ridge; sagebrush, mtn. brush, aspen, and spruce-fir parkland communities; 6,000-9,600 ft; June-July. More common westward in the Wasatch Mtns.

Heliotropium L. — Heliotrope

Heliotropium curassavicum L. (*H. spathulatum* Rydberg; *H. xerophilum* Cockerell) — salt heliotrope, quail-plant — inflorescence dichotomously branched, the spike-like branches densely flowered, 1-sided, coiled at the tip; ovary scarcely lobed; stigma and obsolescent style terminated on the summit of the ovary, the style neither gynobasic nor persistent, the stigma with a broad flange-like disk at the base; nutlets with a marginal corky or bony ridge, the scar asymmetrically placed near the summit of the ventral face. The few specimens seen are from the Ouray Wildlife Refuge, clay soil at margin of a disturbed marsh community (B. Welsh & G. Moore 189), from Linwood Bridge (Flowers 127 UT), and from Moffat Co. at Irish Lakes (Peterson & Kennedy 83-386). Also cited for the Green River near Quarry (Graham 1937). Our plants belong to var. *obovatum* de Candolle.

Lappula Moench — Stickseed

Taprooted annuals (usually) or biennials; herbage puberulent, hirsute, or partly strigose, not setose; stems mostly erect, simple, or variously branched; inflorescence branches spreading-ascending, racemose; corolla blue or white, small and relatively inconspicuous, short-funnelform, the throat closed by fornices.

1   Nutlets ringed with 2-3 rows of slender distinct prickles; corolla 2.5-4 mm wide .... *L. squarrosa*

1   Nutlets ringed with 1 row of prickles, these distinct or (usually) weakly to strongly fused, swollen and cup-like at the base; corolla averaging smaller, usually about 2 mm wide .... *L. occidentalis*

Lappula occidentalis (S. Watson) Greene [*L. redowskii* (Horneman) Greene; *L. texana* (Scheele) Britton] — desert stickseed — Widespread and common to abundant especially following disturbance; many plant communities in dry to moderately mesic places; 4,900-8,300 ft; late April-July. The phase with the prickles seated on and confluent with a swollen cupulate collar may be recognized as var. *cupulata* (A. Gray) L. C. Higgins.

Lappula squarrosa (Ritzius) Dumortier (*L. echinata* Gilibert; *L. myosotis* Moench) — European stickseed — Introduced from Eurasia, the 3 specimens seen are from Daggett Co. at 12 mi s. of Manila, Uintah Co. between Dry Fork and Tridell, and Garfield Co. at Douglas Pass at 7,400-8,400 ft.

Lithospermum L. — Stoneseed; Gromwell

Perennial herbs, the pubescence not bristly; roots red or purple-staining; flowers axillary or in short, few-flowered, terminal clusters, the inflorescence not noticeably coiled; fornices absent or inconspicuous; nutlets ovate, hard and glossy and appearing enameled, sparingly pitted, keeled centrally, 1-4 maturing.

1   Corolla (when developing) yellow, conspicuous, the tube mostly 2-4 cm long, the limb about 1.5 cm wide, the lobes erose; flowers of 2 kinds, the early season ones large, showy, long-styled, seldom setting seed, the later ones remaining closed, short-styled, highly fertile; nutlets mostly 3-4 mm long
    .... *L. incisum*

1   Corolla light yellowish-green, inconspicuous, about 0.5 cm long; flowers homomorphic and homostylic; nutlets mostly 4.5-7 mm long .. *L. ruderale*

Lithospermum incisum Lehmann — fringed stoneseed, showy stoneseed, plains stoneseed, trumpet puccoon — Rock Creek and eastward on foot slopes of the Uinta Mtns., Tavaputs Plateau, Piceance Basin, and Daggett Co.; infrequent in rocky, gravelly, or sandy places in desert shrub, sagebrush-grassland, and pinyon-juniper communities; 4,800-8,200 ft; May-June (inconspicuous flowers are produced in late summer and fall).

Lithospermum ruderale Douglas ex Lehmann (*L. lanceolatum* Rydberg) — western stoneseed — Widespread; occasional in pinyon-juniper, mtn. brush, bitterbrush, mtn. big sagebrush-grassland, and open ponderosa pine communities; 7,000-8,100 ft; May-June.

Mertensia Roth — Bluebells

Caulescent leafy herbs; flowers borne in the upper axils and at tips of stems in inconspicuously coiled modified cymes, usually somewhat nodding or pendulous; corolla tubular near the base, flaring distally to a lobed campanulate limb; nutlets ovoid, roughened, attached laterally. There is some disagreement in treatment of some of the shorter species of the genus in AUF, FNA, and INF. The following treatment is adapted from AUF.

1 Plants relatively robust and tall (greater than 4 dm in well-developed stems), summer-blooming; stem leaves relatively broad, generally over 1.5 cm wide, with well-developed lateral veins 2

1 Plants relatively short (almost always less than 4 dm tall), spring-blooming; stem leaves relatively narrow, mostly less than 1.5 cm wide; lateral veins of stem leaves absent or inconspicuous. 3

2 Calyx 1/3 or less as long as the corolla tube in most flowers, the lobes blunt; anthers mostly less than 2 mm long; plants of wide distribution in the Uinta Mtns. . . . . **M. ciliata**

2 Calyx mostly 1/2 or more as long as the corolla tube, the lobes acute; anthers mostly longer than 2 mm; plants of Wasatch and w. Duchesne Cos., and of the Piceance Basin of Colorado . . . . **M. arizonica**

3 Style 2.5 mm or less, included in the corolla tube; anthers virtually sessile, included within the tube; flowers mostly erect, not strongly nodding . . . .
   . . . . **M. brevistyla**

3 Style 4 mm or more long, exceeding or about equaling the corolla tube; stamens shortly but evidently exserted from the tube, the filaments more or less equaling the anthers; flowers often noticeably nodding 4

4 Corolla tube glabrous within . . . . **M. oblongifolia**

4 Corolla tube pubescent within at least near the base 5

5 Lower leaf surface glabrous; upper leaf surface evenly and densely strigose with the hairs mostly directed toward the leaf-margins; plants mostly of moderate elevations in mtns. . . . . **M. fusiformis**

5 Leaves glabrous to pubescent on one or both surfaces, the upper and lower surfaces not markedly different, or the hairs pointing toward the leaf tips not strongly oriented toward the margins; plants alpine and of moderate elevations . . . . **M. lanceolata**

Mertensia arizonica Greene [*M. a.* var. *grahamii* L. O. Williams; *M. a.* var. *leonardii* (Rydberg) I. M. Johnston] — aspen bluebells, Arizona bluebells — Locally abundant from Strawberry Valley to the North Fork Duchesne drainage on the s. slope of the Uinta Mtns. and to the Blacks Fork drainage on the n. slope, and east to the Avintaquin drainage on the W. Tavaputs Plateau, also Piceance Basin in Rio Blanco Co. (Kelley & Riefler 82-124); mostly in aspen, spruce, aspen-parklands, and sagebrush communities, usually associated with tall forbs on deep loamy soil, not known from the quartz-rich sandstones of the Uinta Mtn. Group where M. ciliata is common; 7,600-9,600 ft; June-Aug.

Mertensia brevistyla S. Watson — short style bluebells — The few specimens seen are from Strawberry Valley and west end of the Uinta Mtns., listed for Carter Creek, Daggett Co. (Flowers and others 1960), but a specimen from this location was not found in Utah Herbaria by Albee et al. (1988).; dry to moist soil, often on rocky slopes and ridges, in mtn. big sagebrush, aspen, and spruce-fir communities; 7,600-9,500 ft; May-Aug, often one of the first plants in bloom.

Mertensia ciliata (M. E. Jones) G. Don. — streamside bluebells — Common across the Uinta Mtns. (but uncommon w. of Duchesne Co., where *M. arizonica* is abundant); spruce-lodgepole pine, spruce, willow, and wet meadow communities, often in shady places along, rocky streams, and above treeline in talus below melting snow banks and along melt-water streams; 8,000-12,500 ft; June-Sept.

Mertensia fusiformis Greene — spindle-root bluebells — Common in the Uinta Mtns., occasional on the E. Tavaputs Plateau, apparently absent from Strawberry Valley and the W. Tavaputs Plateau; mostly in mtn. big sagebrush-grass communities, but occurring as well in open spruce, fir, and lodgepole pine woodlands, and pinyon-juniper, mt. brush, and riparian communities; 6,000-10,650 ft; May-Aug. A few specimens seen from the upper limits of the elevational range of *M. fusiformis* are scarcely separable from *M. viridis,* and might be of hybrid origin.

Mertensia lanceolata (Pursh) de Candolle — lanceleaf bluebells — With 2 vars.:

1 Leaves glabrous on both surfaces, upper surface often with pustules without hair; calyx lobes mostly glabrous except for the ciliate margins; plants mostly below 9,200 ft . . . . **var. *coriacea***
1 Leaves pubescent on at least one surface; calyx lobes strigose on the back; plants mostly above 10,400 ft . . . . **var. *nivalis***

Var. *coriacea* (A. Nelson) L. C. Higgins & S. L. Welsh (*M. coriacea* var. *dilatata* A. Nelson; *M. viridis* var. *dilatata* L. O. Williams) — smooth-leaf bluebells — Specimens seen are from Daggett, Summit, and Uintah Cos. mostly on the north slope of the Uinta Mtns. in pinyon-juniper, mtn. big sagebrush, mtn. brush, ponderosa pine, and Douglas-fir communities at 6,100-9,200 ft

Var. *nivalis* (S. Watson) L. C. Higgins [*M. lanceolata* var. *viridis* A. Nelson; *M. viridis* var. *viridis* (A. Nelson) A. Nelson; *M. bakeri* Greene; *M. viridis* var. *cana* L. O. Williams] — snowy bluebells, alpine bluebells — Uinta Mtns.; common in alpine and near-alpine communities of ridges, boulder fields, rock stripes, and open slopes with Geum rossii, sedges, and grasses; 10,400-12,700 ft.

Mertensia oblongifolia (Nuttall) G. Don. — oblong-leaf bluebells — Specimens seen are from Yampa Plateau of Blue Mt. and from Dinosaur National Monument on soil derived from the Weber Formation; pinyon-juniper, ponderosa, and Vasey sagebrush communities; 8,150 ft; May-June.

Myosotis L. — Forget-me-not

Plants annual or short-lived perennial; leaves alternate, entire; flowers borne in naked, false racemes that elongate in age or a few in axils of leaves; calyx 5-lobed; corolla blue (ours), 5-lobed; nutlets 4, smooth and shiny.

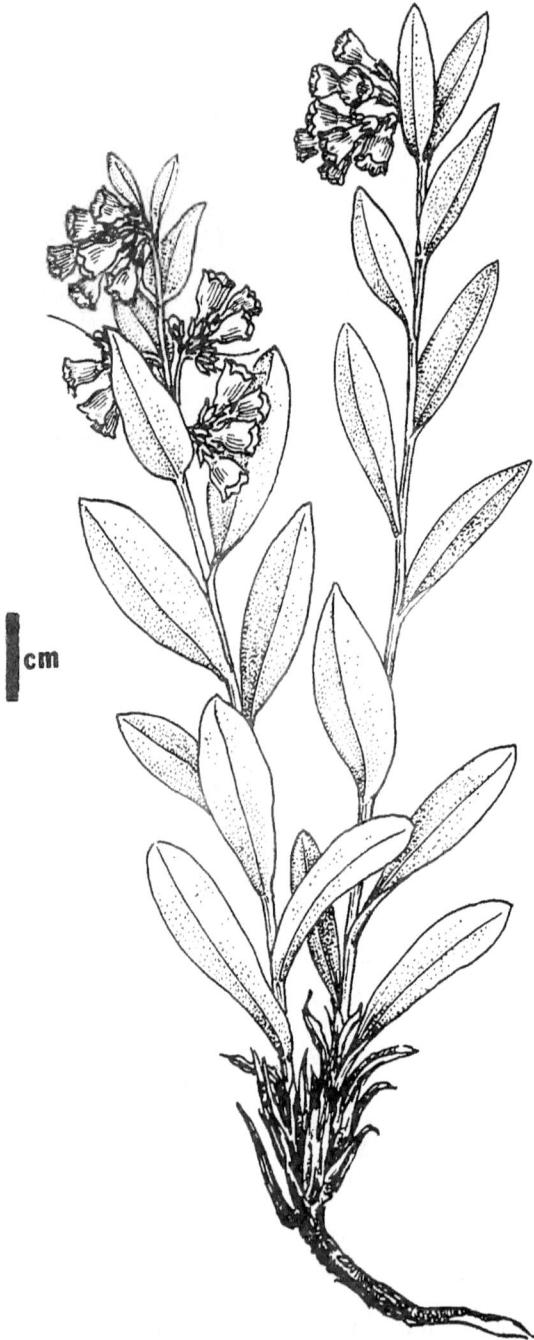

Cymopterus duchesnensis M. E. Jones — Uinta

1   Calyx with appressed hairs; corolla limb 2-5 mm wide .... **M. laxa**
1   Calyx with spreading hairs, some hairs uncinate; corolla limb 1-2 mm wide
    .... **M. micrantha**

Myosotis laxa Lehmann — lax forget-me-not — Two specimens seen (Goodrich 22956, Kass 4078) are from a ditch bank and riparian community near Altamont at 6,240-6,730 ft. Native to North America, but likely introduced in our area.

Myosotis micrantha Pallas ex Lehmann — small-flowered forget-me-not — Specimens seen are from Greendale, Daggett Co. and Burnt Mill Spring, Duchesne Co. from roadside, meadow, and ponderosa pine communities; native of Eurasia. Additional introductions to our area can be expected.

Plagiobothrys F. & M. — Popcorn-flower

Plagiobothrys scouleri (Hooker & Arnot) I. M. Johnston [*P. cognatus* (Greene) I. M. Johnston; *P. nelsonii* (Greene) Johnston; *P. scopulorum* Greene; *Allocarya nitans* Greene] — Scouler popcorn-flower — Low annual, the stems 1-many, usually spreading to erect or sometimes prostrate, to about 2 dm long; herbage more or less strigose; lower leaves opposite, the bases connate; flowers white, small, about 2 mm wide; nutlets rough-tuberculate, attached laterally near the base. Locally common in Strawberry Valley and across the Uinta Mtns., apparently uncommon on the E. Tavaputs Plateau; drying mud or crevices of sandstone in fresh water habitats along streams, reservoir margins, and potholes, and in montane wet meadows, in Vasey sagebrush, silver sage, sedge-willow, lodgepole, and ponderosa communities; 7,000-9,800 ft; June-Aug. Our plants belong to var. *penicillatus* (Greene) Cronquist.

Tiquilia Persoon

Tiquilia nuttallii (Hooker) A. Richardson (*Coldenia nuttallii* Hooker) — Nuttall tiquilia — Taprooted, prostrate annual, the dichotomously divaricate branches forming a flat, open mat; herbage hirsute, prickly-hispid in the inflorescence; flowers sessile, the corolla tubular-funnelform, about 3 mm long; nutlets smooth, lance-ovate, about 1 mm long. Specimens seen are from near Ouray and the Bonanza-Red Wash area of Uintah Co., also near Pariette Draw and w. of Roosevelt, Duchesne Co.; uncommon or locally frequent in sandy places of salt desert shrub communities; 4,800-5150 ft; May-Sept.

BRASSICACEAE (CRUCIFERAE) Mustard Family

Plants annual or perennial, glabrous or often pubescent with simple, forked, or stellate hairs; leaves alternate or basal, simple to compound, the stem leaves sometimes sessile and auriculate; inflorescence mostly a raceme, spike, or corymb; flowers bisexual, regular or nearly so; sepals 4; petals 4, rarely lacking; stamens 6, the outer 2 shorter than and inserted lower than the other 4, rarely 4 or 2; ovary superior, usually 2-loculed, with a thin papery partition (septum), the valves usually separating from the partition in fruit; fruit (pods) linear or nearly so and several times longer than wide (silique) or mostly not linear and 1-3 times as long as wide (silicle), terete or square or compressed parallel to or contrary to the septum, seeds 1-several per locule. A large and complex family in which generic lines are as difficult or more so to determine than are species lines.

---

1   Fruit 2-9 mm long, or if longer then about as wide as long and either bladdery inflated and with a style 3.5-7 mm long or strongly flattened with thin broad wings, mostly not linear; petals lacking or 1-5 (7.5) mm long, or if longer then ovaries with distinct slender styles, and plants densely covered with appressed many-rayed stellate hairs (*Physaria*), white or yellow (or sometimes pink to lavender only in *Smelowskia* and *Noccaea*) .. . . **KEY 1**

1   Fruit (8) 10-100 (130) mm long, linear, not inflated, the style not over 3 mm long (fruit sometimes beaked and the beak simulating a style); petals often 5-28 mm long, but shorter in several genera, white, yellow, pink or purplish; ovaries,without distinct slender styles; plants mostly not appressed stellate pubescent . . . . **KEY 2**

— KEY 1 —

1   Upper leaves perfoliate clasping, simple, nearly orbicular, about 3.5 mm long; lower leaves compound, more or less finely dissected . . . . *Lepidium perfoliatum*

1   Upper leaves not perfoliate clasping; lower leaves simple to compound. . 2

2   Petals strongly bilobed, white; leaves simple 3

2   Petals not strongly bilobed, white or yellow; leaves various 4

3   Plants 30-100 cm tall, leaves not all basal . . . . *Berteroa*

3   Plants less than 25 cm tall . . . . *Draba verna*

4   Styles 2-7 mm long; petals 4-12.5 mm long, yellow; plants covered with dense, appressed many-rayed stellate hairs; leaves not strictly basal, not auriculate clasping . . . . *Physaria*

4   Styles lacking or to 2.5 mm long; petals either less than 4 mm long or not yellow or plants scapose or not pubescent as above 5

5   Stem leaves deeply lobed, pinnatifid or dissected into fine segments 6

5   Stem leaves lacking or entire or toothed 10

6   Petals white to lavender; leaves with fewer than 10 lobes 7

6   Petals yellow; at least some of the leaves usually with more than 10 lobes or segments 9

7   Fruit about as wide as long. 8

7   Fruit 2 or more times as long wide, not notched at apex . . . . . . *Smelowskia*

8   Fruit often notched at apex . . . . *Lepidium*

8   Fruit not notched at apex . . . . *Hornungia*

9   Leaves twice or more pinnatifid or dissected into fine segments . . . . *Descurainia*

9   Leaves lobed or once pinnatifid . . . . *Rorippa*

10 Stem leaves not auriculate clasping. 11

10 Stem leaves auriculate clasping 17

11 Fruit retuse or notched at the apex, usually about as wide as long. 12

11 Fruit rounded or pointed but not notched 13

12 Petals yellow, plants annual, appressed-stellate pubescent ...... *Alyssum*

12 Petals white, plants glabrous or mostly with simple hairs ...... *Lepidium*

13 Fruit with a stout beak 1.7-2.6 mm long, length of the beak about equal to the body .... *Euclidium*

13 Fruit not beaked. 14

14 Petals yellow .... *Draba*

14 Petals white 15

15 Fruit pubescent; plants scapose or nearly so .... *Draba*

15 Fruit glabrous; plants with stem leaves 16

16 Fruit flattened, with only 1 seed per locule; perennial, or if annual then fruit narrowly winged .... *Lepidium*

16 Fruit not strongly flattened, not winged, with 2-10 seeds per locule; plants annual, glabrous .... *Hornungia*

17 Plants pubescent with forked or branched hairs, introduced annual weeds; stem leaves sessile and auriculate; pedicels of fruit 6-18 mm long 18

17 Plants glabrous or with simple hairs only, annual or perennial; stem leaves various; pedicels of fruit various but often shorter than above 19

18 Fruit obcordate, flattened; petals white or pinkish, 2-4 mm long; leaves of the basal rosette deeply lobed to pinnatifid .... *Capsella*

18 Fruit obovoid, slightly inflated, not at all flattened; petals cream to white, 3-5 mm long; leaves of basal rosette lacking or withered by anthesis, entire or obscurely toothed .... *Camelina*

19 Petals yellow; plants 35-100 cm tall, introduced weeds, stem leaves entire or nearly so, the lower ones 3.5-15 cm long; fruit flattened, 10-18 mm long, 4-7 mm wide, 1-seeded, the mature pedicels reflexed .... *Isatis*

19 Petals not yellow 20

20 Plants glabrous and glaucous 21

20 Plants pubescent, introduced weeds with fruit not over 6 mm long, annual or perennial, not high montane or alpine 22

21 Fruit 10-17 mm long; plants annual, introduced .... *Thlaspi*

21 Fruit 3-8 mm long; plants perennial, native .... *Noccaea*

22 Plants annual, from taproots; fruit 5-6 mm long, longer than wide; upper cauline leaves lanceolate; fruiting racemes 5-10 cm long or more .... *Lepidium campestre*

22 Plants perennial, from aggressive rhizomes; fruit less than 5 mm long, wider than long; upper cauline leaves ovate; fruiting racemes 2-5 cm long .... *Cardaria*

— KEY 2 —

1 Plants pubescent at least in part (note basal leaves) with forked, branched, or stellate hairs, sometimes nearly glabrous or with simple hairs in *Draba* but then plants scapose or nearly scapose with only 1-2 reduced stem leaves; leaves entire to toothed, not auriculate except in *Boechera* and *Transberingia* 2

1　Plants glabrous or with simple hairs only, not scapose (except in *Parrya* and then differing from *Draba* with petals pink or lavender and 16-23 mm long, and with fruits 2-6 cm long); stem leaves often auriculate. 14

2　Leaves pinnatifid or pinnately compound, dissected into rather fine segments . . . . *Descurainia*

2　Leaves entire to toothed but not dissected 3

3　Fruit 2-4 times longer than wide, or if narrower then plants scapose, 4-20 mm long, 2-5 mm wide; petals yellow or white, 1.7-5 (7) mm long; plants annual or perennial, with dendritic hairs or scapose or nearly scapose with only 1-2 reduced stem leaves, 2-20 (40) cm tall . . . . *Draba*

3　Fruit mostly over 8 times longer than wide, (8) 12-120 mm long, 1-2 mm wide, or if to 3 mm wide then mostly well over 20 mm long; petals various; plants various but not scapose 4

4　Petals yellow; fruit terete to square in cross-section; pubescence of malpighian hairs . . . . *Erysimum*

4　Petals white, pink or lavender; fruit flattened, terete or square in cross-section; pubescence not of malpighian hairs 5

5　Petals 15-25 mm long; leaves petiolate or subsessile to auriculate, 2-20 cm long, 6-40 mm wide; fruit terete or square in cross-section, 31-100 mm long; plants introduced, cultivated, persisting and sometimes escaping *Hesperis*

5　Petals less than 15 mm long or leaves not over 6 mm wide; plants native or introduced 6

6　Pedicels of fruit 1-2 mm long; fruit nearly terete, 33-66 mm long; petals pink or lavender, 6.2-9.5 mm long; plants annual, introduced, weedy . . . .
　. . . . . *Malcolmia*

6　Pedicels of fruit 3-24 mm long; fruit usually flattened, except in *Transberingia*, (8) 12-120 mm long; petals various; plants native, not or hardly weedy 7

7　Fruit 8-14 mm long; plants annual, 3-30 cm tall; stem leaves not auriculate
　. . . . . *Arabidopsis*

7　Fruit mostly over 14 mm long 8

8　Fruit erect or nearly so, the pedicels ascending and forming a 30°-40° angle with (and holding the pods away from) the axis of the raceme; racemes not dense; plants biennial (see discussion under *Transberingia virgata* for further comparison with biennial species of *Boechera* with erect pods) . . . .
　. . . . . *Transberingia*

8　Fruit erect to pendulous, if erect then the pedicels also erect or nearly so, not holding the pods away from the axis of the racemes, the racemes usually rather dense 9

9　Fruit spreading to reflexed . . . . *Boechera* (*Arabis*)

9　Fruit erect or strongly ascending, pedicels erect or spreading 10

10　Plants strong perennials with branched caudices . . . . *Boechera* (*Arabis*)

10　Plants biennial or short lived perennial from a mostly unbranched taproot
　. 11

11　Fruit 1.3 -3 cm long; stem leaves not auriculate; plants of Strawberry Valley
　. . . . . *Arabis nuttallii*

11 Fruit 3-10 cm long; stem leaves auriculate or not; plants widespread . . . 12

12 Plants perennial; basal leaves 2-12 mm wide, usually not forming a flat rosette; seeds winged, the wing at the tip of the seed about 1 mm wide; petals 6.5-10.5 mm long . . . . ***Boechera stricta*** (*Arabis drummondii*)

12 Plants biennial, basal leaves 3-50 mm wide, often forming a basal rosette that lies flat on the ground; seeds wingless (note: the superficially similar *Transberingia virgata* may key here. See note under that taxon) 13

13 Stigmas expanded, 0.8-1.1 mm broad, wider than the style base; fruit not strongly compressed; outer sepals not gibbous at the base; petals cream to rarely pinkish, 4.7-7 mm long; stem leaves 15-90 (120) mm long, 4-35 mm wide . . . . ***Turritis*** (*Arabis glabra*)

13 Stigmas not obviously expanded, 0.3-0.6 mm broad, not much wider than the style base; fruit strongly compressed; outer sepals gibbous at the base; petals white to pink, 3.2-9 mm long; stem leaves 6-50 mm long, 2-20 mm wide . . . . ***Arabis hirsuta***

14 Fruit not over 12 mm long; leaves pinnatifid; petals yellow; plants rhizomatous, of wet places . . . . ***Rorippa sinuata***

14 At least the longer fruits regularly over 12 mm long; leaves, petals, and plants various. 15

15 Petals yellow . . . . **KEY 3**

15 Petals white or lavender 16

16 Some of the stem leaves compound with 3 or more leaflets. 17

17 Fruit erect, 15-30 mm long; pedicels 3-10 mm long; plants annual . . . .
. . . . ***Cardamine***

17 Fruit spreading, 10-18 (25) mm long; pedicels 5-13 (20) mm long; plants perennial from rhizomatous or stolons, submerged or emergent . . . .
. . . . ***Nasturtium***

16 Stem leaves simple, entire, toothed or lobed but not compound, plants of dry to wet places but not aquatic 18

18 Leaves glabrous, usually glaucous, not both strongly petiolate and cordate-ovate; stems glabrous except sometimes at the very base . . . . **KEY 4**

18 Leaves pubescent, sometimes nearly glabrous except for the ciliate hairs of basal leaves; stems often pubescent, (plants sometimes glabrous in Cardamine but then the leaves strongly petiolate and cordate-ovate) . . . .
. . . . **KEY 5**

— KEY 3 —

1 Stamens exserted, the filaments about twice as long as the petals; petals 10-19 mm long; fruits 30-80 mm long with a stipe 10-25 mm long; racemes 10-50 cm long or longer, many-flowered; plants perennial, 25-120 (150) cm tall .
. . . . ***Stanleya***

1 Stamens about equal to or shorter than the petals; fruit pedicellate but not stipitate; racemes various but often shorter than above; plants various. . 2

2 Leaves compound, pinnatifid, or if only lobed or toothed then 5-20 cm long, and 3-10 cm wide; herbage glabrous or hirsute 3

2 Leaves entire or toothed or sometimes the lower ones pinnatifid in

Schoenocrambe, but then not over 2.5 cm wide; herbage glabrous and usually glaucous 6

3   Stem leaves pinnatifid with about (5) 8-20 primary lobes or divisions, the primary leaf divisions often narrow; lower part of stem and parts of lower leaves strongly hirsute .... *Sisymbrium*

3   Stem leaves toothed, lobed or compound, but usually with only 3-5 primary lobes or leaflets, most of the primary lobes or leaflets broad and rounded; stems and leaves glabrous or hirsute 4

4   Fruit not at all beaked, (15) 20-50 mm long; pedicels only 2-4 mm long; petals 4-6 mm long; plants biennials, of wet places, meadows and woods, not weedy .... *Barbarea*

4   Fruit constricted to a seedless beak, the beak 5-15 mm long, or if only 1-5 mm long then the fruit only 10-25 mm long; pedicels 2-6 (20) mm long; petals 5-15 mm long; plants annuals, often weedy..................... 5

5   Valves of fruit with 1 vein .... *Brassica*

5   Values of fruit with 3-7 veins .... *Sinapis*

6   Cauline leaves sessile and strongly auriculate, entire, the basal ones also entire; plants annual, 20-50 (70) cm tall; fruit 7-10 (13) cm long . *Conringia*

6   Cauline leaves not auriculate, the basal ones lacking or entire to pinnatifid; plants perennial; fruit 1-7.5 cm long .... *Sisymbrium*

— KEY 4 —

1   Stem leaves sessile and strongly auriculate 2

1   Stem leaves sessile or petiolate but not auriculate 4

2   Leaves dentate at the truncate apex; fruit flattened, 5-8.5 cm long, 3-5.8 mm wide; petals chestnut brown; sepals sometimes pubescent at the apex; plants perennial .... *Streptanthus*

2   Leaves entire and/or fruit not over 2 mm wide; petals cream, pinkish or purplish but not at all reddish or purplish brown; sepals glabrous; plants annual or short-lived perennial 3

3   Fruit 7-10 (13) cm long; petals white or cream, 6.2-12 mm long; leaves entire .... *Conringia*

3   Fruit 1-7.5 cm long, if over 7 cm then petals pink or purple and 11-14.5 mm long; leaves entire or toothed to sinuate dentate .... *Thelypodiopsis*

4   Stems usually strongly inflated and hollow; calyx hirsute; fruit 7-14 cm long; pedicels 1-4 mm long, sometimes hirsute; petals brownish-purple; leaves of the basal rosette often pinnatifid; plants perennial .... *Caulanthus*

4   Stems not inflated; calyx glabrous; fruit 1-6 cm long; pedicels glabrous; petals white, pink or purple; leaves entire or toothed 5

5   Fruit reflexed-descending; plants annual or winter annual; leaves 1.5-8.5 cm long, 1-12 mm wide .... *Streptanthella*

5   Fruit spreading to ascending; plants biennial to perennial; leaves various. 6

6   Plants perennial, 10-30 cm tall; basal leaves, if present, not larger than the stem leaves,0.9-3 (4.5) cm long, 8-20 mm wide, entire .... *Hesperidanthus*

6   Plants biennial, (15) 50-300 cm tall; basal leaves often larger than stem

leaves, over 3 cm long and up to 30 cm long and to 10 cm wide, entire, toothed or pinnatifid .... *Thelypodium*

— KEY 5 —

1 Leaf blades cordate-ovate or broader, mostly glabrous, borne on slender petioles; petioles of lower leaves 2-5 times longer than the blades; petals white; plants of wet places in mtns., not stipitate glandular .... *Cardamine cordifolia*

1 Leaves narrower than cordate-ovate, sessile or plants stipitate glandular; petioles much shorter than above; petals sometimes white but more often pink or purple 2

2 Plants stipitate-glandular; stem leaves lacking or petiolate, not auriculate; petals 9-23 mm long, pink to purple................................. 3

2 Plants not stipitate-glandular; stem leaves sessile, often auriculate; petals various.. 4

3 Plants scapose, alpine on the Uinta Mtns., perennial; petals 16-23 mm long; fruit 3-3.5 mm wide, flattened, not beaked .... *Parrya*

3 Plants not scapose, not alpine, annual, weedy; petals 9-12.5 mm long; fruit about 2 mm wide, terete, constricted to a seedless beak, the beak 8-22 mm long .... *Chorispora*

4 At least some leaves ciliate .... *Arabis*

4 Leaves not ciliate ....

5 Stem leaves sessile and strongly auriculate .... *Thelypodiopsis*

5 Stem leaves petiolate or sessile but not auriculate .... *Thelypodium*

Alyssum L. — Alyssum

Annual, stellate pubescent herbs; leaves alternate, simple, entire, not auriculate; petals yellow; fruit a sessile silicle, broadly elliptic to oval in outline, flattened parallel to the septum; seeds 1 or 2 per locule.

1 Fruit glabrous or nearly so; styles 0.5-0.8 mm long; sepals deciduous soon after anthesis .... *A. desertorum*

1 Fruit stellate-pubescent; styles various; sepals persistent until the fruit is nearly mature.. 2

2 Fruit 4-5 mm broad; styles 0.8-1.2 mm long .... *A. simplex*

2 Fruit about 3-4 mm wide; styles 0.3-0.6 mm long .... *A. alyssioides*

Alyssum alyssioides (L.) L. — pale alyssum, yellow alyssum, madwort — Introduced from Europe, more or less weedy on disturbed ground, The 4 specimens seen are from widely scattered locations in desert shrub, sagebrush, pinyon-juniper, and mt. brush communities at 6,100-7,920 ft; May-June. Number of specimens seen indicates pale alyssum is less common our area than is *A. desertorum*.

Alyssum desertorum Stapf — desert alyssum — Recently Introduced from Eurasia, more or less weedy on disturbed ground, becoming widespread, and to be expected to spread across our area; specimens seen are from desert shrub, sagebrush, pinyon-juniper, and mtn. brush communities; 6,400-8,320 ft. April-June.

Alyssum simplex Rudolphi [*A. parviflorum* M. Bieberstein, *A. minus* (L.) Rothmaler] — small alyssum — Introduced from Eurasia, the few specimens seen are from widely scattered locations including: Snake John Ridge, Uintah Co., Taylor Flat Bridge and Cedar Springs, Daggett Co., and Dudley Bluffs, Rio Blanco Co. at 5,450-6,525 ft. Apparently relatively recently introduced. To be expected to spread across low elevations of our area in floodplain, pinyon-juniper, sagebrush, and other communities.

Arabidopsis Heynhold

Plants annual, glabrous or with simple or forked hairs; leaves mostly basal, stem leaves not auriculate; pedicels divaricate ascending; petals white; fruit linear, spreading-ascending to nearly erect, glabrous.

Arabidopsis thaliana (L.) Heynhold — mouse-ear cress — The 5 specimens seen are from Bare Top Mtn, Green Lakes-Red Canyon, and Mustang areas of Daggett Co.; mtn. big sagebrush and ponderosa pine communities; 6,550-7,600 ft; April-May.

Arabis L. — Rockcress

Plants biennial or perennial, glabrous or pubescent with simple, branched, or stellate hairs; leaves simple, entire, dentate, serrate, or sinuate, sessile and usually auriculate; flowers in racemes; pedicels erect to pendulous; sepals deciduous; petals white, cream, pink, lavender, or purple; fruit a silique, many times longer than wide. Only 2 of our species (*A. hirsuta* and *A. nuttallii*) remain in the genus Arabis in IMF and FNA where *A. glabra* is included in *Turritis* and the others are included in *Boechera*.

---

1   Fruit 1.3 -3 cm long; stem leaves not auriculate . . . . *A. nuttallii*

1   Fruit 3-10 cm long; stem leaves auriculate; plants widespread . . *A. hirsuta*

Arabis hirsuta (L.) Scopoli — hairy rockcress — (*A. rupestris* Nuttall) Infrequent or rather rare; apparently widespread; sagebrush, silver sagebrush-grass, mt. brush, aspen, subalpine meadow, and spruce-fir communities; 6,230-10,000 ft; June-Aug. Our plants belong to var. *glabrata* Torrey & A. Gray [*A. h.* var. *pycnocarpa* (Hopkins) Rollins]. In FNA (7: 260-261) *A. hirsuta* is excluded from N. America and plants likely for our area are treated as *A. pycnocarpa* M. Hopkins and *A. eschscholtziana* Andrezejowiski in C. F. von Ledebour.

Arabis nuttallii B. L. Robinson in A. Gray — Nuttall rockcress — Infrequent or locally common in Strawberry Valley; silver sagebrush and meadow communities; 7,600 ft; June-July.

Barbarea R. Br. — Wintercress

Plants glabrous or sparsely hirsute on basal leaves, biennial or rarely annual from taproots; basal leaves lyrate-pinnatifid to pinnately compound; stem leaves smaller and less divided than the basal ones, sometimes entire, auriculate-clasping; sepals yellowish; petals yellow; fruit a silique, many times longer than wide.

1 Mature stylar beak 1.3-3 mm long; fruit 0.5-1.1 mm thick ..... **B. vulgaris**
1 Mature stylar beak 0.5-1.1 mm long; fruit 1.1-1.5 mm thick .. **B. orthoceras**

Barbarea orthoceras Ledebour — American wintercress, erectpod wintercress — Specimens seen are from Strawberry Valley, Uinta Mtns. and W. Tavaputs Plateau; 7,000-10,700 ft; June-July. Upper leaves pinnatifid to dentate, the venation distinctly pinnate.

Barbarea vulgaris R. Br. — common wintercress, yellow-rocket — The one specimen seen (N. Holmgren & P. Holmgren 13469 is from Nine Mile Canyon Road 1.3 mi s. of Highway 40 from along a ditch at 5250 ft elev. Introduced from Eurasia. Upper leaves broadly ovate to suborbicular and coarsely dentate, the venation palmate-pinnate.

Berteroa de Candolle — Berteroa

Berteroa incana (L.) de Candolle Plants annual, 30-100 cm tall, stellate pubescent; leaves simple, entire, the basal ones 3-5 cm long, slender-petiolate, the cauline reduced upward, sessile; racemes many-flowered; petals 4-6 mm long, conspicuously bilobed, white; fruit 5-7 mm long, about 2-3 mm wide; styles 2-3 mm long. Introduced from Europe; the one specimen seen (Goodrich 21986) is from Deep Creek, Uinta Mtns., Daggett Co.; roadside in ponderosa pine community; 7,970 ft. Subsequent visits to the site indicate this plant did not persist at this location.

Boechera A. Love & D. Love — American Rockcress

Plants perennial, glabrous or pubescent with simple, branched, or stellate hairs; leaves simple, entire, dentate, serrate, or sinuate, sessile, those of the stem usually auriculate; flowers in racemes; pedicels erect to pendulous; sepals deciduous; petals white, cream, pink, lavender, or purple; fruit a silique, many times longer than wide. This is a difficult genus. One should not expect success in initial efforts to differentiate taxa. Study of the genus and familiarity with several taxa will contribute greatly to use of the key. Traditionally all of our species have been include in the genus *Arabis*. As treated in FNA and IMF only 2 of our species (*A. hirsuta* and *A. nuttallii*) remain in the genus *Arabis*, and *A. glabra* is included in *Turritis*. The others are included in *Boechera*. This classification is followed here.

1 Plants from a branched caudex, sometimes mat-forming, 5-30 (50) cm tall; stems usually several; basal leaves 5-25 (30) mm long 2
1 Plants from a taproot or simple to sparingly branched caudex, not at all mat-forming, (8) 15-80 (100) cm tall; some leaves usually over 25 mm ... 7
2 Basal leaves glabrous or with scattered hairs......................... 3
2 Basal leaves densely pubescent..................................... 4
3 Fruits strictly erect .... **B. lyallii**
3 Fruits ascending to spreading .... **B. microphylla**
4 Racemes secund; fruiting pedicels 2-6 mm long .... **B. lemmonii**
4 Racemes not particularly secund; fruiting pedicels 5-13 mm long 5

5   Basal leaves green .... *B. microphylla*

5   Basal leaves densely gray-green pubescent........................... 6

6   Plants forming mats; petals 6-8.5 (11) mm long .... *B. fernaldiana*

6   Plants not forming mats; petals 4-6 (7.5) mm long .... *B. gracillentia*

7   Fruit densely pubescent to glabrate; lower leaves smaller than the main-stem leaves; plants sometimes slightly woody above ground level, of dry places, mostly below 6,000 ft; petals about 1-2 cm long, white, pink, or lavender .... *B. pulchra*

7   Fruit glabrous; lower leaves larger than the stem leaves; plants herbaceous above ground level, from (5,600) 6,000-11,000 ft, or higher; petals usually less than 1 cm long 8

8   Pedicels and fruit erect; stems and leaves glabrous or hirsute with simple, forked, or malpighian hairs; leaves not uniformly ciliate; petals white to cream or sometimes pinkish .... *B. stricta*

8   Pedicels or fruit or both spreading to descending; stems and leaves often with branched, dendritic (or stellate?) hairs or lower leaves rather uniformly ciliate with simple or forked hairs; petals pink, lavender, or purple, rarely white 9

9   Lower leaves hirsute-ciliate with simple or once-forked hairs; stems and leaf surfaces glabrous or with simple or forked hairs; pedicels glabrous. 10

9   Lower leaves not ciliate exclusively with simple or once-forked hairs; lower part of stems and leaf surfaces of at least lower leaves with forked, branched, or dendritic hairs; pedicels glabrous or pubescent.......... 11

10   Stems solitary (14) 20-60 cm tall, not arising from between the basal rosette and a tuft of secondary leaves; stem leaves auriculate, (0.5) 1-3.5 (4) cm long, 3-7 mm wide; petals 5-8 mm long .... *B. fendleri*

10   Stems usually several to many, 10-30 cm tall, arising from between the basal-rosette and a tuft of secondary leaves; stem leaves auriculate or not, 5-10 cm long, 1-4 mm wide; petals 4-6.5 mm long .... *B. pendulina*

11   Stems usually 3 or more, arising from between the basal rosette and a secondary tuft of leaves .... *B. gracillentia*

11   Stems solitary or few; tuft of secondary leaves lacking or poorly developed 12

12   Pedicels and fruit ascending; fruit straight, not curved; stems usually pubescent with 3-forked or malpighian hairs, with 10 or more well-developed, often overlapping leaves; plants seldom collected *B. divaricata*

12   Pedicels and fruit spreading, descending, or reflexed; fruit often curved; hairs of stems sometimes more than 3-branched; stem leaves various .. 13

13   Leaves and lower part of stem grayish, densely appressed-pubescent with minute dendritic hairs; basal leaves linear-oblanceolate, always entire; plants rarely collected in our area .... *B. lignifera*

13   Lower leaves and lower part of stem with spreading hairs; basal leaves linear-oblanceolate to broader, often toothed 14

14   Fruiting pedicels hirsute; fruit strongly curved, widely spreading, 6-12 cm long; petals 7-14 mm long, 2-5 mm wide .... *B. sparsiflora*

14   Fruiting pedicels glabrous to pubescent but not hirsute, curved to sharply deflexed; fruit straight or curved, 2-7 cm long; petals 4-9 mm long..... 15

15 Fruiting pedicles abruptly reflexed; fruit more or less appressed to the rachis .... **B. retrofracta**
15 Fruiting pedicles arched and gently recurred; fruit not appressed to the rachis .... **B. pendulocarpa**

Boechera divaricarpa (A. Nelson) A. Love & D. Love [*Arabis divaricarpa* A. Nelson; *A. holboellii* var. *brachycarpa* (Torrey & Gray) S. L. Welsh] — spreading pod rockcress — Scattered locations; to be expected in sagebrush, pinyon-juniper, mtn. brush, ponderosa pine, aspen, lodgepole pine, and spruce-fir communities at 5,000-9,200 ft; May-July. The concept of *B. divaricarpa* in FNA (7:274) encompasses apomictic triploid populations containing three distinct genomes, one each derived from *B. retrofracta*, *B. sparsiflora*, and *B. stricta*. It appears this could be a confusing mess for field botanists as the orientation of fruit of *B. retrofracta* is the complete opposite of that of *B. stricta*.

Boechera fendleri (S. Watson) W. A. Weber [*Arabis fendleri* (S. Watson) Greene] — Fendler rockcress — The 5 specimens seen are from the W. Tavaputs Plateau. To be expected across our area in pinyon-juniper, Douglas-fir-snowberry, and forb-grass communities; 7,000-10,300 ft; June-July.

Boechera fernaldiana (Rollins) W. A. Weber [*Arabis fernaldiana* Rollins; *A. vivariensis* S. L. Welsh] — Fernald rockcress — Specimens seen are from Jones Hole and Dinosaur National Monument, in Uintah Co. and w. Moffat Co.; piny-on-juniper, and mtn. mahogany communities, often in sandy soil; 5,500-7,280 ft; May-June. Uinta Basin plants are far disjunct from the main body of the species in Nevada, and they belong to ssp. *vivariensis* (S. L. Welsh) Windham and Al-Shehbaz.

Boechera gracilentia (Greene) Windham & Al-Shehbaz [*Arabis selbyi* Rydberg; *Boechera selbyi* (Rydberg) W. A. Weber; *B. selbyi* var. *thorneae* (S. L. Welsh) N. H. Holmgren; *Arabis perennans* S. Watson [*Boechera perennans* (S. Watson) W. A. Weber] misapplied; *Arabis perennans* var. *thorneae* S. L. Welsh] — slender rockcress — Common and widespread in desert shrub, sagebrush, pinyon-juniper, Salina wildrye, mt. brush, and Douglas-fir communities; 5,600-8,900 ft; April-June. In FNA (7: 396) *Boechera perennans* (*Arabis perennans*) is treated as a species with strongly dentate basal leaves mostly of warm deserts s. of our area and *B. gracilentia* is listed as scattered across the Colorado Plateau. In IMF (2B: 373-374) *B. perennans* is listed for se. Wyoming and sw. Colorado and most of Utah as well as California to New Mexico and Texas. Some of our plants from the W. Tavaputs Plateau have some basal leaves that are weakly dentate and generally less densely pubescent and a little wider than specimens with all entire leaves. These plants might be considered *B. perennans*. *Boechera crandallii* (Robinson) W. A. Weber (*Arabis crandallii* Robinson) (a plant of the Gunnison Basin, Colorado FNA 7:372) has been reported for our area based on specimens belonging to *B. gracilentia*.

Boechera lemmonii (S. Watson) W. A. Weber (*Arabis lemmonii* S. Watson) — Lemmon rockcress — Widespread across the alpine of the Uinta Mtns.; fellfields, rock strips, talus, Red Pine Shale barrens, often in snowbeds; (10,100) 11,000-13,000 ft; July-Aug.

Boechera lignifera (A. Nelson) W. A. Weber (*Arabis lignifera* Nelson) — woody rockcress — Probably widespread; the one specimen seen (Brotherson 982) is from 10 mi n. of Altonah; sagebrush community, and Harrington 2078 reported for Cold Springs Mt. (Bradley 1950), reported to be common on stabilized dunes (FNA 7:388); May-June. Perhaps some specimens from our area that have been keyed here belong to *B. pendulocarpa.*

Boechera lyallii (S. Watson) Dorn (*Arabis lyallii* S. Watson) — alpine rockcress — Occasional; Uinta Mtns.; alpine tundra, talus, boulder fields, moraines, and krummholz; 10,000-11,500 ft and probably higher; June-Aug.

Boechera microphylla (Nuttall) Dorn (*Arabis microphylla* Nuttall in Torrey & A. Gray) — littleleaf rockcress — The 6 specimens seen are from the Uinta Mtns., Browns Park, and E. Tavaputs Plateau; sagebrush, pinyon-juniper, and Douglas-fir communities; 6,000-8,600 ft; May-June.

Boechera pendulina (Greene) W. A. Weber [*Arabis pendulina* Greene; *A. demissa* Greene var. *russeola* Rollins; *A. pendulina* var. *russeola* (Rollins) Rollins; *Boechera demissa* (Greene) W. A. Weber var. *pendulina* (Greene) N. H. Holmgren; *B. demissa* (Greene) var. *russeola* (Rollins) N. H. Holmgren] — low rockcress — Occasional; widespread; desert shrub, sagebrush, juniper, and ponderosa pine communities; 5,700-8,600 ft; late April-June. In FNA (7: 391) *Boechera demissa* (*Arabis demissa*) is treated as a synonym of *B. oxylobula* (Greene) W. A Weber that is limited to central Colorado outside our area.

Boechera pendulocarpa (A. Nelson) Windham & Al-Shehbaz [*Arabis pendulocarpa* A. Nelson; *A. holboellii* Hornemann var. *pendulocarpa* (A. Nelson) Rollins] — pendulous rockcress — Reported (AUF 297) for Daggett, Summit, and Uintah Cos. in sagebrush, pinyon-juniper, mtn. brush, aspen, and spruce-fir communities.

Boechera pinetorum (Tiedstrom) Windham & Al-Shehbaz [*Arabis holboellii* (Hornemann) var. *pinetorum* (Tiedstrom) Rollins] — Holboell rockcress — Common, widespread in desert shrub, sagebrush, pinyon-juniper, mtn. brush, aspen, and coniferous forest communities at (5,400) 6,000-9,350 ft; May-July.

Boechera pulchra (M. E. Jones ex Watson) W. A. Weber (*Arabis pulchra* M. E. Jones; *A. formosa* Greene) — beauty rockcress — Our most commonly collected rockcress of lower elevations; widespread; desert shrub, sagebrush, and juniper communities; 4,700-6,000 ft; May-mid June. In FNA (7: 375) plants with fruits glabrous proximally are treated at species level as *Boechera duchesnensis* (Rollins) Windham and the remainder of our plants as *Boechera formosa* (Greene) Windham & Al-Shehbaz with petals 8-18 mm long. In IMF (2B: 382) and AUF (300) var. *gracilis* (M. E. Jones) Dorn and var. *pallens* (M. E. Jones) Dorn are recognized for our area with the former having lavender to purple petals 8-11(12) mm long and the latter having white or less commonly lavender petals (10) 11-18 mm long.

Boechera retrofracta (Graham) A. Love & D. Love [*Arabis exilis* A. Nelson; *Arabis holboellii* (Hornemann) var. *retrofracta* (Graham) Rydberg; *A. holboellii* var. *secunda* (Howell) Jepson; *Boechera holboellii* var. *secunda* (Howell) Dorn; *B. exilis* (A. Nelson) Dorn] — deflexed rockcress — Widespread; sagebrush pinyon-juniper, mtn. brush, ponderosa pine, aspen, and lodgepole pine communities at 5,000-9,400 ft; Apr.-Aug. depending on elevation. *Boechera retrofracta* forms hybrids with at least 12 other species (FNA 7: 402-403).

Boechera sparsiflora (Nuttall) Dorn (*Arabis sparsiflora* Nuttall in Torrey & A. Gray) — sicklepod rockcress — Reported (AUF 301) for Daggett, Duchesne, and Uintah Cos. in sagebrush, mtn. brush, pinyon-juniper, ponderosa pine, and aspen communities. In IMF (2B: 362) 2 vars. are keyed as follows with a note that the suite of characters do not always correlate.

1 Basal leaves entire, narrowly oblanceolate; pedicels divaricately ascending, sparsely pubescent or glabrous: **var. *sparsiflora*** (a note in AUF (301) indicates most specimens from our area belong to this var.)

1 Basal leaves dentate, rarely entire, oblanceolate or broader; pedicels horizontally to somewhat descending, hirsute: **var. *subvillosa*** (S. Watson) Dorn. In FNA (7: 393) this var. is treated as a synonym of *Boechera pauciflora* (Nuttall) Windham & Al-Shehbaz where the petals this taxon are listed as 5-8 mm long and 0.5-2 mm wide compared to 7-13 mm long and 2-5 mm wide for *B. sparsiflora*. Also where the fruits of this taxon are reported to be horizontal compared to ascending to divaricate-ascending or rarely horizontal in *B. sparsiflora*.

cm

Boechera pendulina (Greene) W. A. Weber — low rockcress

Boechera stricta (Graham) Al-Shehbaz (*Arabis drummondii* A. Gray) — Drummond rockcress — Common; widespread; sagebrush, forb-grass, aspen, aspen-alder, coniferous forest, and alpine communities; 7,600-11,000 (12,500) ft; June-Aug. Hybridizing with at least 15 other species of the genus (FNA 7: 408). Commonly recorded in study plots but always with low percent cover.

Brassica L. — Mustard

Plants annual, glabrous or hirsute; leaves alternate and basal; the basal ones often lyrate-pinnatifid, the cauline ones reduced upward, petiolate or sessile to auriculate; flowers in racemes; sepals deciduous; petals yellow; fruit a silique, terete or nearly so, sometimes torulose, the beak 1-3 nerved.

---

1   Stem leaves sessile, auriculate-clasping, glaucous, entire or nearly so; pedicels of fruit 7-20 mm long; fruit 30-70 mm long, the beak 8-15 mm long; petals 6-10 mm long . . . . **B. rapa**

1   Stem leaves petiolate and not auriculate, or if (rarely) seeming so, then falsely petiolate above the clasping base; petals 5-15 mm long 2

2   Fruit 10-25 mm long, erect and more or less appressed against the branches of the inflorescence, the beak 1-5 mm long . . . . **B. nigra**

2   Fruit 20-50 mm long, spreading to erect but not appressed, the beak 7-15 mm long 3

3   Fruiting valves with 1 vein; fruiting pedicels 4-10 (17) mm long . **B. juncea**

3   Fruiting valves with 3-7 veins; fruiting pedicels 2-7 mm long . . . .
    . . . . **Sinapis arvense**

Brassica juncea (L.) Czernjaew — Indian mustard — Introduced from Asia; the one specimen seen (Goodrich 19807) is from Timothy Creek between Yellowstone and Uinta Canyons from disturbed ground at a campsite at 8,080 ft; June-Aug.

Brassica nigra (L.) Koch in Roehling — black mustard — Introduced from Europe, more or less weedy; the one specimen seen (Goodrich 20968) is from Hanna, to be expected across our area at moderate and lower elevations in fields, pastures, ditch banks, and roadsides.

Brassica rapa L. (*B. campestris* L.) — field mustard, turnip, common bird rape — Introduced from Europe, cultivated and more or less weedy; gardens, fields, and roadsides, only 2 specimens seen, but to be expected to be widespread.

Camelina Crantz — False Flax

Camelina microcarpa Andrezejowiski in de Candolle — hairy false flax, littlepod false flax — Plants annual, 8-80 cm tall with simple or forked to stellate hairs; leaves alternate, simple, entire or obscurely toothed, about 1-8 cm long, at least the upper ones auriculate; sepals 2-2.7 mm long, often reddish, more or less villous; petals white or cream; fruit a silicle, moderately inflated, glabrous, 5-7 mm long. Introduced from Asia, widespread and more or less weedy with disturbance, but also in desert shrub, sagebrush, and pinyon-juniper communities; up to 8,000 ft; May-June.

Capsella Medicus — Shepherds-purse

Capsella bursa-pastoris (L.) Medicus — shepherds-purse — Plants annual, 10-50 cm tall, stellate pubescent and hirsute; basal leaves lyrate-pinnatifid to merely toothed; stem leaves much reduced upward, auriculate; sepals 1.2-2.5 mm long, pubescent or glabrous; petals white; fruit a silicle, 4.5-8 mm long, strongly flattened, obcordate. Introduced from Europe, widespread and more or less weedy with disturbance; up to about 8,000 ft; May-June.

Cardamine L. — Bittercress; Toothwort

Plants glabrous or pubescent with simple hairs; leaves alternate, sometimes also in basal rosettes; flowers mostly racemose; petals white to pinkish; fruit a silique, many times longer than wide.

---

1 Leaves all simple, cordate-ovate to orbicular, the basal ones on long petioles 2-5 times longer than the blades; petals 7-12 mm long; pedicels mostly 10-20 mm long; fruit 20-35 mm long; plants perennial from rhizomes . . . . *C. cordifolia*

1 Leaves pinnatifid or pinnately compound; fruit 15-30 mm long 2

2 Petals 3.7-6 mm long; fruiting pedicels 10-20 mm long; plants perennial from rhizomes . . . . *C. breweri*

2 Petals 2-3.5 mm long; fruiting pedicels 3-10 (12) mm long; plants annual or biennial from taproots 3

3 Principal cauline leaves all pinnately compound with distinctly petiolulate leaflets; fruit 1-1.6 mm thick; seeds 10-24 per fruit, 1.2-1.7 mm long . . . . *C. oligosperma*

3 Principal cauline leaves pinnatifid, at least distally, with the lobes decurrent on the rachis; fruit 0.8-1.1 mm thick; seeds 20-40 per fruit, 0.7-1.1 mm long . . . . *C. pensylvanica*

Cardamine breweri S. Watson — Brewer bittercress — The 3 specimens seen are all from the n. slope of the Uinta Mtns. from a gravel bar on the Blacks Fork River (S. L. Welsh et al. 9190), moist area in recently burned Engelmann spruce-lodgepole pine forest, West Fork Blacks Fork (Goodrich 27141), and riparian community at Cottonwood Creek (Goodrich and R. Zobell 26257); 8,640-9,400 ft.; July-Aug.

Cardamine cordifolia A. Gray (*C. infausta* Greene) — heartleaf bittercress, large mt. bittercress — Strawberry Valley and eastward in the Uinta Mtns. to the Rock Creek drainage and eastward on the West Tavaputs Plateau to the Avintaquin drainage; along streams usually in woods; 7,000-10,500 ft;-June-Aug.

Cardamine oligosperma Nuttall ex Torrey & A. Gray — little western bittercress — The 9 specimens seen are from wet places scattered across the Uinta Mountains in West Fork Blacks Fork, North Fork Ashley Creek, Dry Fork, Lake Fork, East Fork Whiterocks, Beaver Creek of Carter Creek drainage, Head of Smiths Fork, and China Meadows at 8,350-11,050 ft.

Cardamine pensylvanica Muhlenberg in Willldenow — Pennsylvania bittercress — The 25 specimens seen are from the Uinta Mtns. from Rock Creek to Red

Springs (Highway 191) on the south slope, and Daggett Co. on the north slope, no specimens seen from Summit Co., but to be expected there; riparian communities, often growing about seeps and springs and along spring-fed streams and often in shade of conifers and willows; 7,200-10,480 ft; May-June. Annotations seen on specimens of *C. pensylvanica* and *C. oligosperma* indicate these taxa are often confused and difficult to distinguish.

Cardaria Desvaux — Whitetop; Hoary cress

Plants perennial from deep-seated rhizomes, pubescent; leaves alternate, sinuate-dentate, auriculate-clasping; flowers in paniculate racemes; petals white; fruit a silicle. Introduced weedy plants included on noxious weed lists of our area.

---

1   Fruit puberulent . . . . *C. pubesecens*

1   Fruit glabrous. 2

2   Fruit cordate to reniform . . . . *C. draba*

2   Fruit globose or nearly so . . . . *C. chalepensis*

Cardaria chalepensis (L.) Handel-Mazzetti (*Lepidium chalepense* L.) — lens-podded hoarycress, whitetop — listed for Daggett Co. in AUF (306). treated as *C. draba* var. *repens* (Schrenk) O. E. Shultz in IMF (2B:244-245). Reported for wetter soils than *C. draba* and *C. pubescens* (Arnow et al. 1980).

Cardaria draba (L.) Desvaux (*Lepidium draba* L.) — heart-podded hoarycress, whitetop — Introduced from Eurasia, widespread; weedy in fields, ditch banks, roadsides, favored by disturbance but also invading native plant communities.

Cardaria pubescens (C. A. Meyer.) Jarmolenko (*Lepidium appelianum* Al-Shehbaz) — globe-podded hoary cress, hairy whitetop — Introduced from Eurasia, weedy, the 4 specimens seen are from Daggett Co.; riparian and sagebrush-grass, and ponderosa pine communities, usually associated with disturbance; 5,500-8,230 ft; May-July. To be expected elsewhere in our area.

Caulanthus S. Watson — Wild Cabbage; Caulanthus

Caulanthus crassicaulis (Torrey) S. Watson — thickstemmed wild cabbage, spindlestem — Short-lived perennial (biennial) from a taproot; (20) 30-90 cm tall, with a single unbranched, usually strongly inflated stem; basal rosette of entire or more often pinnatifid leaves 3-12 (17) cm long; stem leaves reduced, entire, narrow; pedicels 1-4 mm long; sepals hirsute, brownish-purple; petals dull-purplish, 10-14 mm long; fruit 7-14 cm long. Widespread; occasional in desert shrub, sagebrush, and pinyon-juniper communities; up to 7,200 ft; May-June. The sepals vary from more or less hirsute (var. *crassicaulis,* plants widespread) to glabrous or nearly so (var. *glaber* M. E. Jones, known from Duchesne Co.).

Chorispora R. Brown

Chorispora tenella (Pallas) de Candolle — musk mustard, blue mustard, purple mustard — Annual, malodorous herb, 20-45 cm tall, stipitate-glandular throughout and often hirsute at least at the base; leaves 5-85 mm long, 1-28 mm wide, entire, dentate or pinnatifid; petals 9-12.5 mm long, pink to lavender; fruit 30-45 mm long, with a seedless beak 8-22 mm long. Introduced from Asia, widespread, weedy; roadsides and other disturbed ground where it becomes abundant enough to blanket the ground, occasionally in indigenous plant communities; up to 8,500 ft or perhaps higher; March-June. This plant has increased many-fold in recent years in our area where it continues to expand in range. It is displacing native plant communities on canyon bottoms of the Tavaputs Plateau. Conversion from native dominance to displacement by musk mustard can take as little as 3 years in some areas.

Conringia Heister & Fabricius

Conringia orientalis (L.) Dumortier — hares-ear mustard — Annual herb, 20-50 (70) cm tall, glabrous and usually glaucous; leaves 2-12 cm long, 1-6 cm wide, those of the stem strongly auriculate-clasping; petals 6-12 mm long, white to cream; fruit 7-10 (12.3) cm long. Introduced from Europe; the 3 specimens seen are from Rio Blanco Co., also reported for Moffat Co. (Bradley 1950); desert shrub communities on clayey soils, roadsides; 5,500-6,000 ft; May-June.

Descurainia Webb & Berth. — Tansy-mustard

Annual or biennial herbs, stellate-pubescent, stipitate-glandular, or glabrate; leaves 1-3 times pinnately compound or pinnatifid; flowers racemose; petals yellow; fruit 3 to many times longer than wide, terete or nearly so.

---

1   Fruits 14-30 mm long, rarely shorter; sepals 1.6-2.8 mm long; upper as well as lower leaves 2-3 times pinnate; plants mostly of disturbed sites, mostly below 7,000 ft . . . . ***D. sophia***

1   Fruits 2.2-18 mm long; sepals 0.8-2 mm long; upper leaves once-pinnate; distribution various. 2

2   Fruits 2.2-4.7 mm long; plants of aspen and coniferous forest belts . . . . ***D. californica***

2   Fruits 4-20 mm long 3

3   Fruiting pedicels strictly ascending, nearly appressed to the rachis, 2-6.5 mm long; fruit pointed; plants commonly above the pinyon-juniper belt . . . . ***D. incana***

3   Fruiting pedicles divaricately ascending to widely spreading or slightly reflexed, 1.5-23 mm long; fruit rounded or pointed; plants most common in and below the pinyon-juniper belt. 4

4   Fruit clavate (wider apically) . . . . ***D. pinnata***

4   Fruit linear . . . . ***D. incisa***

Descurainia californica (A. Gray) O. E. Schulz — California tansy-mustard — Occasional or locally common, widespread; Strawberry Valley, Tavaputs Plateau, Uinta Mtns. and Middle Mt. in the Three Corners area; mtn. big sagebrush-snowberry, aspen, Douglas-fir, and spruce fir communities; 7,000-10,100 ft; late June-Aug.

Descurainia incana (Bernhardi ex Fischer & C. A. Meyer) Dorn [*D. incana* var. *brevipes* (Nuttall) S. L. Welsh; *D. incana* ssp. *procera* (Greene) Kartesz & Gandhi;

D. richardsonii var. *macrosperma* O. E. Schulz] — mtn. tansy-mustard — Daggett, Duchesne, Summit, and Wasatch Cos. in aspen, lodgepole pine, spruce-fir, and alpine communities mostly above 9,000 ft; June-Aug. *Descurainia incana* is treated as a species without vars. or spp. with a long list of synonyms in FNA (7: 522) and in IMF (2B: 410). This species is listed with 3 vars. in AUF (309) and with 4 ssp in TPD. Hybrids between *D. incana* and other species of *Descurainia* are likely (FNA 7: 522)

Descurainia incisa (Engelmann) Britton [*D. richardsonii* (Sweet) O. E. Schulz; *D. serrata* (Greene) O. E. Schulz] — Richardson tansy-mustard — Widespread; occasional in sagebrush, sedge-willow, tall forb, aspen, lodgepole pine, and spruce-fir communities; 7,000-10,500 ft; May-Sept. In FNA (7: 523) 2 ssp. are recognized. In TPD 3 ssp. were listed in the past. However, as of 25 Oct. 2012 TPD included all of *D. incisa* in *D. incana* ssp. *incisa* (Engelmann ex A Gray) Kartesz & Gandhi. In IMF (2B: 412) 4 vars. are recognized. In AUF (309) 3 vars. are recognized including *D. incana* as var. *brevipes*). Synonymy listed in IMF (2B: 410-411) covers about ¾ of a page. The variation is likely a result of hybridization with all species of the genus with overlapping ranges (FNA 7: 523). This and *D. pinnata* are closely related, and detailed study might show need to combine these taxa (IMF 2B: 412). We prefer not to get any deeper into this entanglement. Thus the species is listed without keys to vars. or ssp.

Descurainia pinnata (Walter) Britton [*D. brachycarpa* (Richardson) O. E. Schulz; *D. halictorum* (Cockerell) O. E. Schulz; *D. longipedicellata* (Fournier) O. E. Schulz] — western tansy-mustard — Occasional to common; widespread; desert shrub, sagebrush, pinyon-juniper, and rarely aspen communities; 4,700-7,000 (8,000) ft; May-Sept. In AUF (310) 7 vars. are recognized with some of these included in D. incisa in other treatments. In IMF (2B: 414) 3 vars. are recognized. In FNA (7: 527) and TPD 4 and 10 ssp. are recognized respectively. We prefer not to get any deeper into this entanglement. Thus the species is listed without keys to vars. or ssp.

Descurainia sophia (L.) Webb ex Prantl — flixweed — Introduced from Eurasia, widespread; more or less weedy with disturbance; up to about 7,000 ft and probably higher; May-Aug.

Draba L. — Whitlow-grass; Draba

Annual or perennial, mostly small herbs, nearly glabrous to densely pubescent with branched to stellate hairs; leaves simple and entire to toothed; flowers racemose; petals white or yellow; fruit orbicular to linear.

1   Plants annuals or short-lived perennials from taproots; styles obsolete or to 0.2 mm long 2

1   Plants perennial, mostly from branching caudices: styles mainly 0.2-1.2 mm long 9

2   Petals white; leaves all basal or nearly so 3

2   Petals yellow, or cream, sometimes fading white when dried; leaves various 5

3   Petals deeply bilobed; fruits glabrous, not crowded toward the ends of the scapes; pedicels 2-12 mm long . . . . **D. verna**

3 Petals not bilobed; fruits pubescent, crowded toward the ends of scapes; pedicels 1-3 mm long 4

4 Upper part of the peduncle, axis of raceme, pedicels, and fruit pubescent .... *D. cuneifolia*

4 Upper part of the peduncle, axis of raceme, and pedicels glabrous; fruit pubescent .... *D. reptans*

5 Fruit glabrous 6

5 Fruit pubescent 7

6 Leaves glabrous or ciliate with simple or rarely forked hairs; stems glabrous; plants of the lodgepole pine-Engelmann spruce belt of the Uinta Mtns. .. .. *D. crassifolia*

6 At least some of the leaves pubescent at least on the lower surface with stalked,-2-4-rayed hairs; stems usually pubescent below; plants of broader distribution .... *D. albertina*

7 Petals (3.5) 4-6 mm long; some hairs of the fruit usually branched; fruit sometimes twisted; style 0.5-1.5 mm long .... *D. aurea*

7 Petals 1.5-3.5 mm long; hairs of the fruit simple; fruit plain; style obsolete or to 0.1 mm long 8

8 Fruiting pedicels 2-9 mm long, mostly less than twice as long as the fruit . ... *D. rectifructa*

8 Some of the fruiting pedicels usually over 9 mm long, commonly over twice as long as the fruit .... *D. nemorosa*

9 Plants scapose or nearly so, the stem sometimes with 1 or 2 reduced leaves; leaves 3-15 mm long 10

9 Plants not scapose, the stems mainly with 3-several leaves 5-40 (80) mm long 15

10 Leaves ciliate with simple or rarely forked hairs, the upper and sometimes the lower surfaces glabrous; scapes often glabrous; fruit and pedicels glabrous .... *D. densifolia*

10 Leaves not ciliate with simple hairs, usually pubescent on both surfaces with branched to stellate hairs; scapes often pubescent 11

11 Stems usually with 1-3 reduced bract-like leaves; racemes apparently not over 1 cm long in fruit .... *D. inexpectata*

11 Leaves basal or nearly so; racemes often exceeding 1 cm when in fruit . 12

12 Petals white; fruit linear or narrow-elliptic, 5-14 mm long, 1-2 mm wide, about 3-7 times longer than wide; pedicels glabrous; plants from above timberline .... *D. lonchocarpa*

12 Petals yellow; fruit not linear, 3-8 mm long, mostly 2-5.5 mm wide, not more than twice as long as wide; pedicels glabrous or stellate pubescent..... 13

13 Leaves 1-2 mm wide, the surface pubescent with pectinate hairs that are more or less parallel to the leaf axis; fruit not strongly flattened 14

13 Leaves 2-4 mm wide, densely cinereous with simple, forked or branched hairs; fruit flattened, densely pubescent with forked, branched or stellate hairs; plants usually soboliferous .... *D. ventosa*

14 Fruit with pectinate hairs; petals 4-6 mm long; pedicles 7-14 mm long; plants mostly of pinyon-juniper, sagebrush and ponderosa pine communities .. .. *D. pectinipila*

14 Fruit with simple hairs or glabrous; petals 2.5-4 mm long; pedicels 2-7 (12) mm long; plants mostly of subalpine and alpine communities . . . . ***D. oligosperma***

15 Fruit and ovaries glabrous; leaves 20-80 mm long, the surfaces mostly glabrous, the margins ciliate with simple hairs; plants rare . . . . . ***D. crassa***

15 Fruit and ovaries pubescent; leaves 5-40 mm long, the plants rather common 16

16 Petals white; leaves 1-4 mm wide, the basal ones pubescent with branched to stellate hairs; plants at or above timberline on the Uinta Mtns. . ***D. cana***

16 Petals yellow; leaves 2-13 mm wide, with a mixture of forked, branched, and stellate hairs; plants from lower montane to above timberline. 17

17 Fruiting pedicels ascending to divaricately ascending; fruit usually twisted; petals 3.5-5 mm long; plants widespread across our area . . . . ***D. aurea***

17 Fruiting pedicels widely spreading, sometimes slightly descending; fruit not twisted; petals 2.8-3.8 mm long; plants rare . . . . ***D. brachystylis***

Draba albertina Greene (*D. nitida* Greene; *D. stenoloba* Ledebour misapplied) — Alaska draba, slender draba, Alaska whitlow grass — Specimens seen are from Current Creek, Uinta Mtns., and W. Tavaputs Plateau; meadows, woods, along streams, and in sagebrush-grass communities; 7,200-10,900 ft; June-Aug. Reports of *D. fladnizensis* Wulfen for the Uinta Mtns. are perhaps based on specimens of this taxon.

Draba aurea M. Vahl in Hornemann — golden draba — Uinta Mtns. and Mt Bartles area of W. Tavaputs Plateau; occasional in sagebrush, snowberry, aspen, various coniferous forest, streamside, meadow, krummholz, and alpine communities; 7,000-12,500 ft; June-Aug.

Draba brachystylis Rydberg — Wasatch draba — The one specimen seen (Goodrich 18784) is from Swift Creek of the North Fork Duchesne drainage from limestone cliffs at 9,800 ft with Douglas-fir and limber pine.

Draba cana Rydberg (*D. lanceolata* Royle misapplied) — lanceleaf draba — Occasional across the Uinta Mtns. from peaks of the Provo River drainage to Leidy Peak; mostly alpine and krummholz and occasionally upper Engelmann spruce communities; 10,600-12,000 ft; July-Aug.

Draba crassa Rydberg (*D. spectabilis* Greene misapplied) — thick draba — The 4 specimens seen are from the Uinta Mtns. near Ostler Lake in the east fork of Stillwater Fork of the Bear River, Bald Mtn. between East Fork Blacks Fork and Smiths Fork, and near Gilbert Peak and Kings Peak in the Uinta drainage from 10,680-12,450 ft, alpine communities including snowbeds and Red Pine Shale barrens; July-Aug.

Draba crassifolia R. Graham — hairy draba, Rocky Mt. draba — The 28 specimens seen are from Rock Creek and Blacks Fork east to Leidy Peak, Uinta Mtns.; Engelmann spruce, mixed conifer, and alpine communities from (9,600) 10,400-12,400 ft; June-Aug. Grading into *D. albertina* in pubescence and habitat.

Draba cuneifolia Nuttall in Torrey & A. Gray — wedgeleaf draba — The 8 specimens seen are from the south slope of the Uinta Mtns. from Pole Creek to Red Mtn., and Raven Ridge, Rainbow-Dragon area and Coyote Basin; desert shrub, juniper, sagebrush, and mtn. brush communities; 5,830-8,085 ft; April-June. Our plants belong to var. *cuneifolia*, and those from the Uinta Basin are clearly more closely related to *D. reptans* than to plants from outside the Uinta Basin that have passed under the name of *D. cuneifolia* var. *platycarpa* (Torrey & A. Gray) S. Watson by the scape-like peduncles and narrow fruits. From a local view it would seem best to reduce *D. cuneifolia* to synonymy or varietal status under *D. reptans*. However, *D. reptans* is an Old World plant, and outside the Uinta Basin, plants of *D. cuneifolia* do have elongate racemes. More study is indicated, and a new combination is not proposed here.

Draba globosa Payson [*D. densifolia* Nuttall in Torrey & A. Gray misapplied; *D. densifolia* var. *globosa* (Payson) S. L. Welsh; *D. apiculata* C. L. Hitchcock; *D. densifolia* var. *apiculata* (C. L. Hitchcock) S. L. Welsh] — rockcress draba — The 35 specimens seen span the alpine of the Uinta Mtns. from the peaks of the Provo River drainage to Leidy Peak; alpine communities often in long-persistent snowbeds; July-Aug.

Draba inexpectata S. L. Welsh — Uinta Mtns. draba — The 3 known collections of this Uinta Mtn. endemic (Franklin 6293, 6328, 6331) are from near Kletting Peak, Hayden Peak and near Tamarack Lake, Uinta Mountains in Summit Co. from spruce-fir, krummholz, and fellfield communities at 14,400-12,240 ft.

Draba lonchocarpa Rydberg (*D. nivalis* var. *elongata* S. Watson) — lancefruit draba — Occasional across the Uinta Mtns. from Bald Mtn. to Leidy Peak at 9,400-12,600 ft in subalpine and alpine communities; July-Aug. Plants with elliptic to linear fruit less than 7 mm long belong to var. *exigua* O. E. Schultz. Those with linear to narrow elliptic fruit mostly over 10 mm long belong to var. *lonchocarpa*. The report by Graham (1937) of *D. uncinalis* Rydberg (*D. sobolifera* Rydberg) is based on a specimen (Hermann 5001 MO!) that belongs to *D. lonchocarpa*.

Draba nemorosa L. — woods draba — Strawberry Valley, Uinta Mtns., Blue Mt., and Yampa Plateau, no specimens seen from e. of Willow Creek of the W. Tavaputs Plateau but to be expected there; occasional in pinyon-juniper, mt. brush, and ponderosa pine communities; 5,800-8,200 ft; May-June.

Draba oligosperma Hooker — doublecomb draba, few-seeded draba — Occasional across the Uinta Mtns. and Diamond Peak in Moffat Co. in a variety of plant communities, but mostly alpine and wind-swept slopes below timberline; 8,240-12,500 ft; June-Aug.

Draba pectinipila Rollins [*D. juniperina* Dorn; *D. oligosperma* var. *juniperina* (Dorn) S. L. Welsh; *D. oligosperma* var. *pectinipila* (Rollins) C. L. Hitchcock] — juniper draba, Dorn draba — Common and locally abundant on the north flank of the Uinta Mtns. in Daggett Co. and into Moffat Co. and apparently rather restricted in distribution on the south slope of the Uinta Mountains in ne. Uintah Co.; pinyon-juniper, mtn. brush, sagebrush, ponderosa pine, and Douglas-fir communities; 6,800-7,600 ft; May-June. We follow FNA (7: 328) in listing this taxon as *D. pectinipila*.

Draba rectifructa C. L. Hitchcock — mtn. draba — Of the 10 specimens seen, 7 are from Strawberry Valley and south slope of the Uinta Mtns. (N. Fork Duchesne, Rock Creek and Yellowstone drainages), 1 is from the north slope of the Uinta Mtns (Bear River drainage), 1 is from the E. Tavaputs Plateau (Hill Creek), and 1 is from W. Tavaputs Plateau (Strawberry Peak); sagebrush, aspen, and spruce-fir communities; 7,600-10,200 ft; June-Aug. Like *D. nemorosa* in pubescence of stems and leaves, and like *D. albertina* in the tendency to have a small but-rather well-developed basal rosette.

Draba reptans (Lamarck) Fernald (*D. micrantha* Nuttall) — Carolina draba., Carolina whitlow grass — The 16 specimens seen are from south slope of the Uinta Mtns. from Petty Mtn. to Brush Creek and Uinta Basin from Tridell east into Colorado; desert shrub, sagebrush, mtn. brush, and juniper communities; 5,100-8,270 ft; May-June.

Draba ventosa A. Gray — Wind River draba — The 4 specimens seen are from alpine talus slopes of the Uinta Mtns. from West Fork Beaver Creek, Henrys Fork, Bald Mtn. (Blacks Fork drainage), and Lake Basin of the Rock Creek drainage from 10,400-12,980 ft.; July-Aug.

Draba verna L. — spring draba — Five specimens seen are from pinyon-juniper, sagebrush, and ponderosa pine communities in Daggett Co. and one is from Blue Mtn. in Moffat Co.; 6100-7,770 ft.

Erysimum L. — Wallflower

Plants annual, biennial, or perennial, pubescent with 2-3 (4)-rayed hairs; leaves entire to toothed, linear, narrow-elliptic, lanceolate or oblanceolate; petals yellow; fruit linear, many times longer than wide, square to nearly terete in cross-section.

---

1 Petals (10) 12-20 mm long; fruit (17) 20-115 mm long, the styles mostly 1.5-3 mm long; pedicels 3-17 mm long .... *E. capitatum*

1 Petals 3-12 mm long; fruit 12-50 mm long, the styles 0.8-1.5 mm long.... 2

2 Sepals 1.8-3.3 mm long; petals 3-6 mm long .... *E. cheiranthoides*

2 Sepals 3.5-7 (8) mm long; petals 5-12 mm long....................... 3

3 Pedicles 4-15 mm long; fruit ascending to erect .... *E. inconspicuum*

3 Pedicles 2-5 mm long; the fruit spreading or curved-ascending at maturity .... *E. repandum*

Erysimum capitatum (Douglas ex Hooker) Greene [*E. asperum* (Nuttall) de Candolle misapplied; *E. wheeleri* Roth] — pretty wallflower, prairie rocket — Widespread; occasional in desert shrub, sagebrush, pinyon-juniper, mt. brush, aspen-fir, Douglas-fir, Engelmann spruce, and meadow communities; 5,000-11,300 ft; May-June.

Erysimum cheiranthoides L. — treacle mustard, wormseed wallflower — The 6 specimens seen are from Strawberry Valley to Wolf Creek; willow-streamside, meadow, silver sagebrush-grass, and aspen communities; 7,140-8,600 ft; July-Aug.

Erysimum inconspicuum (S. Watson) MacMillan (*E. parviflorum* Nuttall) — small-flower rocket, small wallflower — The 9 specimens seen are from Strawberry Valley, W. Tavaputs Plateau, and Farm Creek Pass, Uinta Mtns. and Blue Mt.; sagebrush, snowbank, and meadow communities; 7,500-9,000 ft; June-Aug. The fruits of *E. inconspicuum* are (1.5) 3-5 (7) cm long and those of *E. cheiranthoides* are 1-3 (4) cm long. Although 3 cm marks a common difference, the overlap is frequent enough to limit the value of fruit-length as a dependable key feature.

Erysimum repandum L. — spreading wallflower — Introduced from Europe. The 2 specimens seen (Goodrich 24771& Goodrich 27512) are from disturbed sites at Tridell and road to Island Park on Duchesne River Formation and

Mancos Shale at about 5,400 ft. This introduced, European, weedy annual can be expected to spread across low elevations of our area.

Euclidium R. Br.

Euclidium syriacum (L.) R. Brown. — Syria-weed — Plants 4-50 cm tall, annual from a taproot, pubescent with forked hairs; leaves mainly cauline, alternate, not auriculate, subsessile or petiolate, 0.7-6.5 mm long, 2-17 mm wide; sepals tinged with purple; petals 0.8-1.1 mm long, white; fruit (silicles) 2.8-4 mm long including a beak 1.2-1.5 mm long, pubescent with simple or forked hairs. Known from a roadside about 4 miles north of Lapoint, White River valley upstream from Rangely, Clay Basin, Diamond Gulch, and Grouse Creek; 5,900-7,230 ft. This ugly Eurasian weed was recently introduced to the Uinta Basin. It can be expected to spread along road sides and other disturbed areas as has Chorispora tenella.

Hesperis L.

Hesperis matronalis L. — sweet rocket — Perennial herbs with simple and forked hairs, 50-130 cm tall; leaves lanceolate to ovate-lanceolate, serrate-dentate to entire, 5-20 cm long, the lower ones long-petiolate, the upper ones often nearly sessile and sometimes auriculate; flowers fragrant; petals purple or white to rose, 18-25 mm long; fruit 40-100 cm long, linear. Introduced from Europe, cultivated as a garden flower, persisting and escaping. The one specimen seen (Neese & Peterson 5821) is from a roadside 1 mi n. of Tridell where it likely did not persist.

Hesperidanthus (B. L. Robinson) Rydberg

Plants perennial from semi woody caudices, glabrous or with simple hairs; leaves linear, lanceolate or oblanceolate, entire or the lower ones somewhat pinnatifid; flowers in racemes; fruit terete, linear, many times longer than wide.

---

1   Fruit 1-2 cm long; leaves petiolate . . . . *H. suffrutescens*
1   Fruit 4.5-7 cm long; middle and upper leaves sessile . . . . *H. argillacea*

Hesperidanthus argillaceus (S. L. Welsh & N. D. Atwood) Al-Shehbaz [*Schoenocrambe argillacea* (S. L. Welsh & Atwood) Rollins; *Thelypodiopsis argillacea* S. L. Welsh & Atwood] — clay reed-mustard — Endemic, known from mouth of Sand Wash, Wild Horse Bench, and Big Pack Mt., with a range of about 15 miles wide on the E. Tavaputs Plateau; Uinta and Green River Formations, desert shrub communities; 4,900-5,600 ft; May-June.

Hesperidanthus suffrutescens (Rollins) Al-Shehbaz [*Schoenocrambe suffrutescens* (Rollins) S. L. Welsh & Chatterley [*Glaucocarpum suffrutescens* (Rollins) Rollins; *Thelypodium suffrutescens* Rollins] — shrubby reed-mustard — Plants 10-25 cm tall; leaves entire, 7-25 mm long, 3-10 mm wide, elliptic, lanceolate or oblanceolate, not auriculate; pedicels 3-12 mm long; sepals 4-6 mm long, yellowish or greenish; petals 9-11 mm long, pale yellow; fruit 10-20 mm long, linear. Endemic, known only from the vicinity of Big Pack Mt. in Hill Creek and

Agency Draw drainages; desert shrub and juniper-mahogany communities on Green River Formation; 5,400-6,500 ft; May-early June.

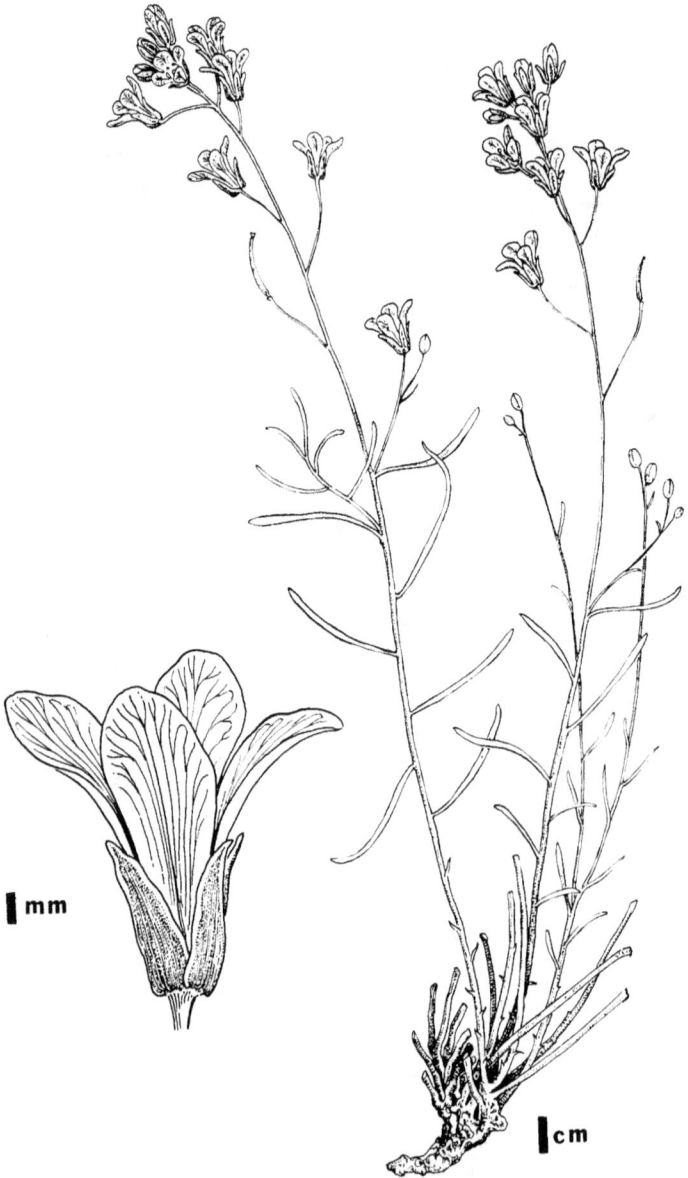

Hesperidanthus argillaceus (S. L. Welsh & N. D. Atwood) Al-Shehbaz
clay reed-mustard

Hesperidanthus suffrutescens (Rollins) Al-Shehbaz — shrubby reed-mustard

Hornungia R. Brown — Hutchinsia

Hornungia procumbens (L.) Hayek [This species has long been called *Hutchinsia procumbens* (L.) Desvaux, but the name *Hutchinsia* is illegitimate (IMF 2B: 266)]. — slenderweed — Annual glabrous herb, 5-30 cm tall; stems slender, mostly branched, erect or prostrate; leaves ovate to nearly linear, 5-30 mm long, 1-13 mm wide, entire to pinnatifid; sepals about 1 mm long, greenish or purplish; petals about 1 mm long, white; fruit 2.5-4 mm long, elliptic to obovate. Specimens seen are from Sheep Creek Canyon, Daggett Co.; near Island Park, Tridell, and Gorge Draw in Uintah Co., and near Roosevelt in Duchesne Co.; streamside, greasewood, and sagebrush communities; 5,200-6,000 ft; May.

Isatis L. — Woad

Isatis tinctoria L. — dyers woad — Plants biennials or short-lived perennials from robust taproots, glabrous or with long simple hairs at the base; leaves simple, alternate, the basal ones petiolate, 3.5-18 cm long, 0.8-4 cm wide, entire to inconspicuously crenulate, ciliate to pilose with simple hairs, the cauline leaves sessile, strongly auriculate, gradually reduced upward; pedicels 4.5-9 mm long, reflexed; petals 3-4.2 mm long, yellow; fruit 10-18 mm long, 4-7 mm wide, strongly flattened, winged, oblong to oblanceolate, black at maturity, 1-seeded. Introduced (from Europe) and known from several widely scattered locations, a troublesome weed to be expected to invade roadsides and other disturbed sites as well as indigenous plant communities from low elevations up to about 9,000 ft; May-July. This is abundant along the Wasatch Front, and new introductions of this weed from that area and other areas can be expected to be frequent as our area is a frequent destination for recreationists from infested areas.

Lepidium L. — Peppergrass; Pepperweed

Annual or perennial herbs or low shrubs (ours), glabrous or with simple hairs; leaves simple, entire to tripinnatifid; flowers in racemes; petals white (yellow in *L. perfoliatum*); fruit a silicle, usually less than twice as long as wide, flattened contrary to the partition.

1	Stem leaves clasping the stem 2
1	Stem leaves petiolate or sessile but not clasping the stem 3
2	Leaves all entire or toothed, the stem leaves auriculate-clasping; petals white, 2-2.5 mm long . . . . **L. campestre**
2	Lower leaves pinnatifid into fine segments; upper leaves entire or dentate and falsely perfoliate clasping; petals yellow, 1-2 mm long . **L. perfoliatum**
3	Plants strictly annuals, from taproots; fruit more or less with small wings, notched at the apex; styles obsolete or not over 0.2 mm long, arising from between and shorter than the notch of the fruit 4
3	Plants biennial or perennial; fruit not or only slightly winged, not notched at the apex or else the style exceeding the notch; styles 0.2-1.2 mm long. 7
4	Pedicels flattened, about twice as wide as thick, glabrous or puberulent on the lower side; fruit pubescent . . . . **L. lasiocarpum**
4	Pedicels either nearly terete or else puberulent on the upper side; fruit glabrous except sometimes in *L. ramosissimum* 5

5 Fruits oblong to obovate, broadest above the middle; petals lacking, or mostly shorter than the sepals .... *L. densiflorum*

5 Fruits elliptic-rotund to orbicular, broadest about the middle; petals various 6

6 Petals lacking-or shorter than the sepals; fruits often ciliate or uniformly pubescent, elliptic or nearly so; plants commonly much branched .... *L. ramosissimum*

6 Petals usually longer than the sepals; fruits glabrous, elliptic-rotund to nearly orbicular; plants usually sparingly branched, widespread in our area .... *L. virginicum*

7 Leaves (at least some) deeply lobed to pinnatifid..................... 8

7 Leaves entire or serrate 10

8 Plants with woody branches well above the base; plants long-lived perennials, commonly 30-60 cm tall .... *L. huberi*

8 Plants not woody much above the base, branches mostly not woody; plants biennials or perennials 9

9 Plants mostly 60-120 cm tall .... *L. alyssoides*

9 Plants mostly less than 60 cm tall .... *L. montanum*

10 Plants over 35 cm tall, arising from deep-seated robust rootstocks, introduced, weedy, along water courses, roadsides, and in fields......... .... *L. latifolium*

10 Plants less than 30 cm tall 11

11 Leaves linear or oblong-linear, 1-3.5 mm wide; petals 3.5-4.2 mm long; plants known only from the vicinity of Indian Canyon, Duchesne Co. .... *L. barnebyanum*

11 Leaves oblanceolate to elliptic, 3-12 mm wide; petals 2.7-3.1 mm long; plants not known from Duchesne Co. .... *L. integrifolium*

Lepidium alyssoides A. Gray [*L. montanum* var. *spathulatum* (B. L. Robinson) C. L. Hitchcock misapplied] — tall pepperplant, mesa pepperwort — Mostly E. Tavaputs Plateau, one specimen seen from Moffat Co.; wash bottoms, stream-sides, and desert shrub, sagebrush, and pinyon-juniper communities, common in burned pinyon-juniper communities; 5,400-7,400 ft; mid Aug.-Sept. Uinta Basin plants belong to var. eastwoodiae (Wooton) Rollins.

Lepidium barnebyanum Reveal — ridgecress — Endemic, known only from the vicinity of lower Indian Canyon within 2-4 miles south of Duchesne, W. Tavaputs Plateau; pinyon-juniper communities on whitish, marly mudstone of the Uinta Formation; 6,200-6,500 ft; May-June.

Lepidium campestre (L.) R. Brown in Aiton (*Thlaspi campestre* L.) — fieldcress — Introduced from Europe, more or less weedy; specimens seen are from Daggett and Uintah Cos. but to be expected to be widespread.

Lepidium densiflorum Schrader — common peppergrass, densecress — With 2 more or less distinct vars.:

1 Pedicels subterete, pubescent all around; fruit averaging less than 3 mm long; plants commonly of roadsides and other disturbed places, over a wide elevational range .... **var. *densiflorum***

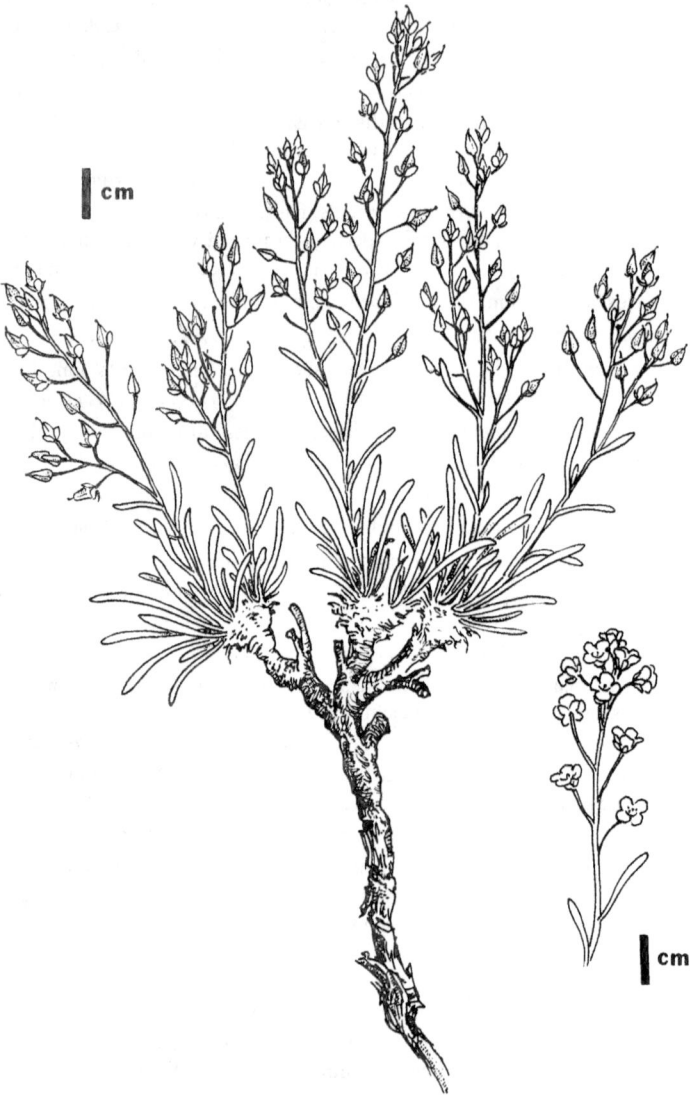

Lepidium barnebyanum Reveal — ridgecress

1 Pedicels flattened, especially on the upper side, tending to be glabrous on the lower side; fruit averaging at least 3 mm long; plants often in indigenous plant communities at low elevations .... **var. *ramosum***

Var. *densiflorum* Perhaps introduced, widespread; often weedy along roadsides, in sheep bedgrounds, and other disturbed places; 7,000-10,000 ft, to be expected at lower elevations; June-Aug.

Var. *ramosum* (A. Nelson) Thellung Apparently native, widespread; desert shrub, greasewood, sagebrush, cottonwood, and ponderosa pine communities; 4,800-6,200 (7,600) ft; May-June.

Lepidium huberi S. L. Welsh & Goodrich — Huber pepperplant — Locally common in Brush Creek and Ashley Creek drainages of the Uinta Mtns. and adjacent Red Mtn. in mtn. brush and less commonly in pinyon-juniper and ponderosa pine communities. Also known from Rio Blanco and Garfield Counties, Colorado on the E. Tavaputs Plateau. One specimen (R. & K. Rollins 8387) from s. of Hamilton, Moffat Co. Colorado also appears to belong here. This endemic plant appears quite secure by its rather wide range and variable habitat that includes steep rocky canyons. It also occurs as waifs in sediments washed out of canyons.

Lepidium integrifolium Nuttall in Torrey & A. Gray — entire-leaf pepperplant — Rare; moist meadows at lower elevations including the outwash plain of Henrys Fork at Lonetree, Wyoming, and listed for Uintah Co. (AUF 325).

Lepidium lasiocarpum Nuttall in Torrey & A. Gray — hispidcress — Harper sn is from the White River area, K. Thorne & J. Thorne 6184 is from Red Wash; desert shrub communities at 5000 ft. Brotherson 809 is from Dinosaur National Monument, Mowry Shale. Our plants belong to var. *georginum* (Rydberg) C. L. Hitchcock. Except for the pubescent fruit, this hardly different from *L. densiflorum* and *L. virginicum*.

Lepidium latifolium L. — broadleaf pepperweed, giant whitetop, tall whitetop — Introduced from Eurasia; weedy along water courses, roadsides, fields, and other moist places; up to about 7,200 ft; May-Sept. This rather recently introduced weed probably entered our area from the n. along the Green River. It has spread rapidly especially along the floodplain of the Green River, and it has become one of the most noxious weeds of our area.

Lepidium montanum Nuttall in Torrey & A. Gray — mtn. pepperplant — A variable complex with little agreement among North American authors as to characters, number of infraspecific taxa, and their synonymies (FNA 7: 587). With 4 more or less distinct vars. in our area as follows:

1 Stems glabrous 2
1 Stems puberulent; plants uncommon. 3
2 Silicles 2.2–2.4 mm long, 1.8–2.1 mm wide; plants known from Irish Canyon, Moffat Co. . . . . **var. *diffusum***
2 Silicles averaging larger than above; plants widespread and common . . . . **var. *jonesii***
3 Plants long lived perennials with few to several stems arising from a branched woody base . . . . **var. *wyomingense***
3 Plants annual, biennial, or short lived perennial from a simple taproot or less commonly from a branched base . . . . **var. *montanum***

Var. *diffusum* S. L. Welsh & S. Goodrich — Irish Canyon pepperplant — Known from Madison Limestone in Irish Canyon, Moffat Co. at about 6,500 ft.

Var. *jonesii* (Rydberg) C. L. Hitchcock [*L. montanum* var. *alyssoides* (A. Gray) M. E. Jones] widespread; common in desert shrub, sagebrush, pinyon-juniper, and cottonwood-skunkbush communities; 4,700-7,400 ft; May-June (July).

Var. *montanum* The 2 specimens seen (S. L. Welsh & C. Higgins 6204; Garrett 8302) are from Duchesne Co.; pinyon juniper communities; May-June.

Var. *wyomingense* (C. L. Hitchcock) C. L. Hitchcock (*L. montanum* var. *soliarborense* S. L. Welsh) — Wyoming pepperplant — Rim of Vermillion Bluffs e. of Lookout Mtn. at 7390 ft. (S. Goodrich & A. Huber 28122), occasional in the badlands of Bridger Formation at Hickey Mtn. and perhaps to be expected in Daggett Co.

Lepidium perfoliatum L. — clasping peppergrass — Introduced from Europe, widespread; weedy especially on disturbed ground, not expected above 8,000 ft; May-June.

Lepidium ramosissimum A. Nelson — branched pepperwort — Reported for Moffat Co. (Bradley 1950), mapped for Rio Blanco Co. but not Moffat Co. in PDB, and approaching our area in western Wyoming (Dorn 2001); in intermountain parks, oak, and aspen, often strongly asymmetrical with the main stem widely spreading and the racemes erect along one side (Weber 1987).

Lepidium virginicum L. (*L. medium* Greene) — Virginia cress, poor-man's pepper — The few specimens seen are from the s. slope and flank of the Uinta Mtns., West Fork Duchesne drainage, and from The Timber Canyon drainage near the mouth of Cow Hollow and Rough Hollow, West Tavaputs Plateau; various plant communities, often on disturbed ground; 6,800-8,500 ft; May-July. Our plants belong to var. *pubescens* (Greene) Thellung

Malcolmia (L.) R. Brown — Malcolmia

Malcolmia africana R. Brown in Aiton — African mustard — Plants annual, decumbent to erect, 3-40 cm tall, pubescent with forked or 3-rayed hairs; leaves 1.2-9 cm long, 0.3-2.3 cm wide, oblanceolate to elliptic, sinuate-dentate, petiolate or sessile, not auriculate; petals 6-10 mm long, pink to lavender; fruit 33-66 mm long, 1-2 mm wide. Introduced from Africa; widespread and more or less weedy on disturbed ground and invading some indigenous plant communities; up to 7,400 (8,600) ft; April-June. Sometimes confused with *Chorispora tenella.* Invasive but less so than *C. tenella,* and somewhat more likely to be in indigenous communities than in vacant lots and roadsides.

Nasturtium R. Brown — Watercress

Nasturtium officinale R. Brown in Aiton [*Rorippa nasturtium-aquaticum* (L.) Hayek] — watercress — Perennial herbs with rhizome-like stolons; stems 3-10 cm long, glabrous; leaves 1-10 cm long, simple to pinnately compound; petals 3-5 mm long, white; fruit 10-18 (25) mm long, about 2 mm wide. Locally widespread and locally common to abundant; submersed or emergent in springs, seeps, ditches, and sluggish streams; 4,800-8,000 ft; May-Oct.

Noccaea Moench

Noccaea montana (L.) F.K. Meyer (*Thlaspi montanum* L.; *T. alpestre* L. misapplied; *T. fendleri* A. Gray; *T. glaucum* A. Nelson) — wild candytuft — Perennial, from a branched caudex, 5-25 cm tall; herbage glabrous; basal leaves with petiole 5-25 mm long and blade5-20 mm long; cauline leaves auriculate clasping, 7-10 (20) mm long; fruit obovate to obcordate, winged or wingless, not notched at the apex, 3-8 mm long, 1.5-5 mm wide; style 0.5-2.5 mm long. Widespread; occasional in sagebrush, mt. brush, aspen, meadow, and Douglas-fir communities, most common in subalpine forests and alpine areas of the Uinta Mtns.; 7,300-

12,500 ft; May-Aug. depending on the elevation. Our plants belong to var. *montana* [*N. fendleri* ssp. *glauca* (A. Nelson) Al-Shehbaz & M. Koch].

Parrya R. Brown

Parrya nudicaulis (L.) Regel (*P. platycarpa* Rydberg. *P. rydbergii* Botschantzev) — Uinta parrya, naked-stemmed parrya — Perennial, scapose, stipitate-glandular herbs from caudices clothed with persistent leaf-bases, 7-12 cm tall; leaves 3-10 cm long, 0.6-2 cm wide, oblanceolate to elliptic, coarsely dentate to incised; flowers 3-10, in racemes, the petals 16-23 cm long, pink to lavender; fruit 25-47 mm long, 3-3.5 mm wide, strongly flattened parallel to the septum. The 8one specimens seen are from across the alpine belt of the Uinta Mtns. in rock-fields and talus, edges of stone strips, and small patches of turf; 11,000-12,400 ft; June (July). Uinta Mtn. and Wyoming plants belong to var. *rydbergii* (Botschantzev) N. H. Holmgren. This is recognized at species level in FNA (7: 513) and AUF (330). Although several features are listed in the key in FNA (7: 512) for separation of *P. nudicaulis* and *P. rydbergii,* these features all overlap. Var. *nudicaulis* is known from British Columbia and northward.

Physaria (Nuttall) A. Gray — Twinpod; Double-bladderpod

Plants perennial, grayish, whitish, or silvery with dense appressed-stellate hairs; leaves mainly basal, the stem leaves much reduced, simple; petals yellow; fruit bladdery-inflated, often broader than long, strongly indented at apex and usually at the base.

---

1 Fruit deeply notched apically and sometimes at the base, strongly inflated
   2

1 Fruit not notched apically nor basally, not or somewhat inflated 6

2 Stems 1.5-4 (5.5) cm long; fruit 4.8-6 mm long; style 1.5-4 mm long; basal leaves 0.5-2.7 cm long . . . . ***P. condensata***

2 Stems 4-25 (30) cm long; fruit (5) 6.5-16 mm long; style 4-6 (9) mm long; basal leaves (2.5) 4-11 cm long. 3

3 Fruit with widely spreading upper part, obcordate, not notched basally . .
   . . . . ***P. obcordata***

3 Fruit not obcordate, usually notched basally 4

4 Fruiting pedicels recurved; silicles pendent . . . . ***P. floribunda***

4 Fruiting pedicels spreading to divaricately ascending, straight or slightly sigmoid; silicles more or less erect 5

5 Basal leaves repand to coarsely dentate; plants uncommon . . . ***P. grahamii***

5 Leaves entire, rarely dentate or repand; plants common . . . . . ***P. acutifolia***

6 Leaves linear or if spatulate then gradually tapered to a petiole, the blades seldom over 4 mm wide 7

6 Leaves spatulate or broader, rather abruptly contracted to a petiole, the blades commonly over 4 mm wide (except in *P. subumbellata*). 11

7 Plants 2-10 (12) cm tall, mostly densely caespitose and sometimes mound forming, from a slightly to much branched caudex; styles 1-4 mm long; fruit pointed at the tip, ovate-lanceolate; basal leaves arising with the stems, 1-4 cm long. 8

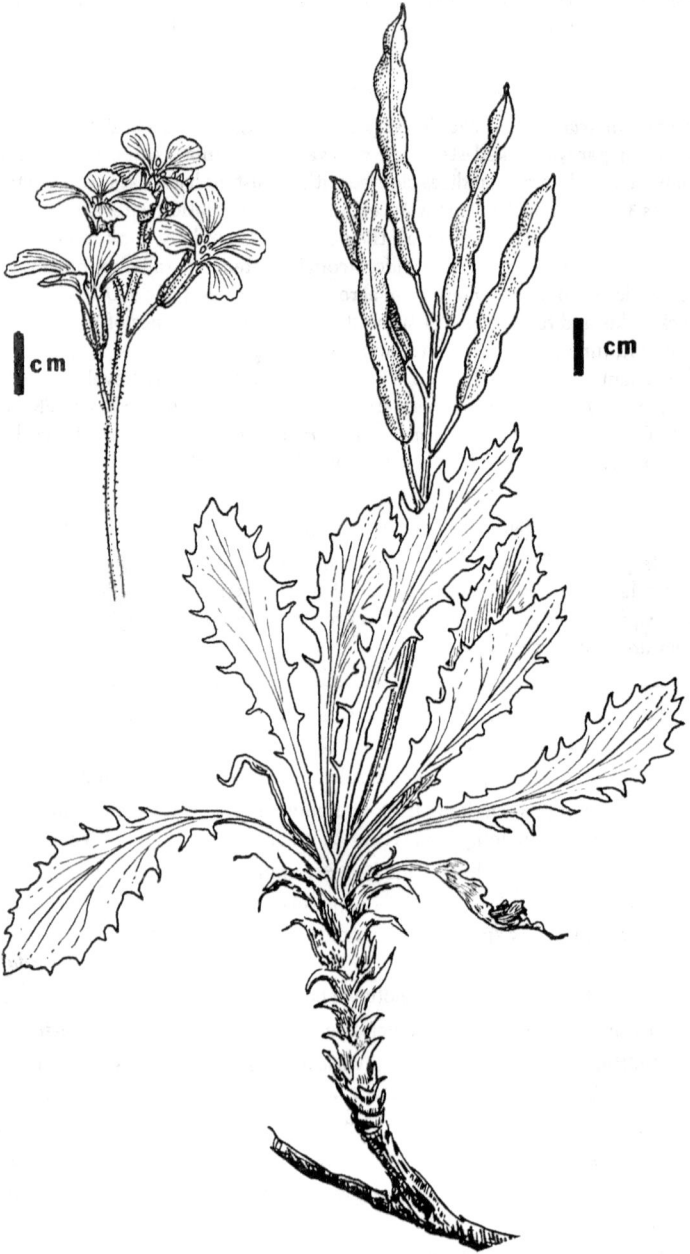

Parrya nudicaulis (L.) Regel — Uinta parrya, naked-stemmed parrya

7   Plants often over 10 cm tall, not caespitose, from a taproot and simple or slightly branched rootcrown; styles 2-7 mm long; fruit rounded, orbicular to ellipsoid; stems erect or more or often decumbent, often with a basal tuft of leaves arising separately from the stems; basal leaves 1-10 cm long . . 10

8   Styles 1-1.5 mm long; silicles flattened on margins and apex; plants of the Piceance Basin . . . . *P. congesta*

8   Styles 1.5-4 mm long; silicles flattened at apex only 9

9   Plants 2-10 cm tall; inflorescence conspicuously exceeding the basal tuft of leaves . . . . *P. parvula*

9   Plants 1-2.5 cm tall; inflorescence included, or nearly so, in the basal tuft of leaves . . . . *P. nelsonii*

10 Pedicels generally recurved or arched downward in fruit or less commonly almost straight; leaves mostly linear, rarely over 4 mm wide; plants widespread . . . . *P. ludoviciana*

10 Pedicels more or less S-shaped; basal leaves often narrowly spatulate, the blades sometimes over 4 mm wide; plants uncommon, known from the E. Tavaputs Plateau . . . . *P. rectipes*

11 Plants mostly caespitose, sometimes mound-forming, from a branched caudex or this simple in young plants, 1-7 (10) cm tall; styles 1.5-4 mm long; fruit pointed at the tip; basal leaf blades 1-5 mm wide   . . . *P. subumbellata*

11 Plants from taproots and somewhat enlarged simple rootcrowns, rarely with slightly branched caudices, the stems sometimes over 7 cm long; styles 2-7 mm long; fruit rounded apically; basal leaf blades often over 5 mm wide 12

12 Seeds 2 per locule; plants of the Piceance and Cathedral drainages, Rio Blanco Co.; stems 10-30 cm tall; basal leaves distinctly differentiated into a comparatively broad blade and a short petiole . . . . *P. parviflora*

12 Seeds 3-8 per locule; distribution otherwise and stems mostly less than 15 cm long, or else the basal leaves gradually tapered to a petiole 13

13 Leaves mostly gradually tapered to a petiole; stems ascending, erect, or decumbent at the base, often with 6 or more leaves, 4-40 cm long; fruit not obcordate; plants of the E. and W. Tavaputs Plateaus . . . . *P. rectipes*

13 Leaves abruptly tapered to a petiole; at least the outer stems often strongly spreading to prostrate, seldom with more than 6 leaves, 2-10 (15) cm long 14

14 Fruit obcordate in outline, sometimes subglabrous; plants of Strawberry Valley and W. Tavaputs Plateau . . . . *P. hemiphysaria*

14 Fruit orbicular, pubescent, slightly flattened contrary to the septum, not obcordate or retuse at the apex; plants of Uinta Mtns. . . . . *P. kingii*

Physaria acutifolia Rydberg [*P. a.* var. *purpurea* S. L. Welsh & Reveal; *P. australis* (Payson) Rollins; *P. didymocarpa* (Hooker) A. Gray var. *australis* Payson; *P. stylosa* Rollins (*P. acutifolia* var. *stylosa* (Rollins) S. L. Welsh] — common twinpod — widespread; common in desert shrub, sagebrush, pinyon-juniper, mt. brush, Salina wildrye-Douglas-fir, and ponderosa pine communities; 4,700-8,900 ft, and occasional on talus slopes and alpine communities up to 11,300 ft; April-June (July). Occasional specimens of the W. Tavaputs Plateau with repand leaves grow in populations of plants with mostly entire leaves. Such specimens show intergradation into *P. grahamii.*

Physaria condensata Rollins — tufted twinpod — Lincoln, Sublette, and Uinta, Cos, Wyoming (the type from 3 mi w. of Fort Bridger); barren clay slopes in sagebrush and shadscale country. Perhaps to be expected in Daggett Co. (IMF 3B: 204) and to be expected in the Bridger Formation between Mtn. View and Lonetree.

Physaria congesta (Rollins) O'Kane & Al-Shehbaz (*Lesquerella congesta* Rollins) — Dudly Bluffs bladderpod — Endemic; Piceance Basin, Rio Blanco Co.; shale barrens of Green River Formation.

Physaria floribunda Rydberg — floribund twinpod — In AUF this is limited in Utah to Grand Co. from s. of our area based on a single specimen (N. D. Atwood 22284). In IMF (3B: 304) this is excluded from Utah altogether with a note of potential for Utah south of our area. In TPD this is excluded from Moffat and Rio Blanco Cos. of Colorado, but Duchesne and Uintah Cos. in Utah are included. These disagreements are vivid examples of the difficult taxonomy of this genus, and perhaps examples of splitting of taxa based on minutia that leads to confusion.

Physaria grahamii Morton in Graham [*P. repanda* Rollins; *P. acutifolia* var. *repanda* (Rollins) S. L. Welsh)] — Graham twinpod — Endemic, rare or perhaps occasional; W. Tavaputs Plateau, from Indian Canyon to Sand Wash and s. to Range Creek; desert shrub, pinyon-juniper, mt. brush, and aspen-Douglas-fir communities; 6,000-9,000 ft; May-June. A poorly understood taxon; more collections are needed to better understand this plant. Perhaps not distinct from *P. acutifolia.* Plants with leaves entire, repand, and dentate are found together in some populations.

Physaria hemiphysaria (Maguire) O'Kane & Al-Shehbaz (*Lesquerella hemiphysaria* Maguire) — skyline bladderpod — The 19 specimens seen are from Strawberry Valley and across the W. Tavaputs Plateau to Bruin Point and Mt. Bartles, no specimens seen from the East Tavaputs Plateau; sagebrush-snowberry, Salina wildrye-sagebrush, rubber rabbitbrush, Douglas-fir-sagebrush, and spruce fir communities, often in semi-barren outcrops of Green River Formation; 7,250-10,200 ft; (May) June-Aug. Plants of our area with styles 1.8-3 mm long, mature pedicels 2-4 (5.5) mm long; silicles glabrous or sparsely pubescent belong to var. *lucens* (S. L. Welsh & Reveal) O'Kane & Al-Shehbaz — Tavaputs bladderpod.

Physaria kingii (S. Watson) O'Kane & Al-Shehbaz (*Lesquerella utahensis* Rydberg) — King bladderpod — Occasional in the Uinta Mtns. where it is mostly on limestone; sagebrush-grass, forb-grass, spruce-fir, Engelmann spruce, and krummholz communities; 8,400-10,800 ft; June-Aug. Our plants belong to ssp. *utahensis* (Rydberg) O'Kane — Utah bladderpod

Physaria ludoviciana (Nuttall) O'Kane & Al-Shehbaz [*Lesquerella ludoviciana* (Nuttall) S. Watson; *L. intermedia* (S. Watson) A. Heller misapplied] — silver bladderpod — Occasional or common across the n. and s. flanks and lower slopes of the Uinta Mtns., from the Uinta drainage to Blue Mt. and Moffat Co., apparently uncommon across the Uinta Basin in e. Uintah Co. and across the E. Tavaputs Plateau, not known from the W. Tavaputs Plateau; desert shrub, sagebrush, pinyon-juniper, and ponderosa pine-manzanita communities; May-June.

Physaria nelsonii O'Kane & Al-Shehbaz [*Lesquerella condensate* A. Nelson; *L. alpina* var. *condensate* C. L. Hitchcock; *L. alpina* ssp *condensate* (A. Nelson) Rollins & Shaw) not *Physaria condensate* Rollins] — dense bladderpod — Approaching our area in Sweetwater and Uinta Cos., Wyoming.

Physaria obcordata Rollins — Piceance twinpod — Known from semi-barrens of Green River Formation in the Piceance Basin. 5,500-7,550 ft; May-June.

Physaria parviflora (Rollins) O'Kane & Al-Shehbaz (*Lesquerella parviflora* Rollins) — Piceance bladderpod — Occasional; Cathedral and Piceance drainages; mt. brush communities and shale barrens of the Parachute Creek Member of the Green River Formation; 6,250-8,515 ft; June-July.

Physaria parvula (Greene) O'Kane & Al-Shehbaz [*Lesquerella alpina* ssp. *parvula* (Greene) Rollins & Shaw] — small bladderpod — Common across the north and south flanks of the Uinta Mountains and e. through Moffat Co and at isolated stations on Tabby Mtn. and Red Creek Mtn.; 6,000-10,400 ft; sagebrush, juniper-mahogany, pinyon-juniper, mtn. brush, and Douglas-fir communities and semi-barren, wind-swept ridges on a variety of substrates; late April-June depending on elevation.

Physaria rectipes (Wooton & Standley) O'Kane & Al-Shehbaz (*Lesquerella rectipes* Wooton & Standley) — Colorado Plateau bladderpod — Cronquist 11496 is from Cliff Ridge. The few other specimens seen are from the East Tavaputs Plateau from Sweetwater Canyon east to near the Utah-Colorado line, also reported for the W. Tavaputs Plateau (Rollins and Shaw 1973) in Carbon Co.; pinyon-juniper communities and barren, shaly slopes; 5,900-8,400 ft; June-July. Reports of *P. montana* (A. Gray) Greene [*L. montana* (A. Gray) S. Watson] for our area might be based on specimens of *L. rectipes,* to which it is most closely related (Rollins and Shaw 1973).

Physaria subumbellata (Rollins) O'Kane Y Al-Shehbaz — Rollins bladderpod — Widespread in shadscale, pinyon-juniper, and sagebrush communities from 5,400-8,000.

Rorippa Scopoli — Cress; Yellowcress

Plants annual, biennial, or perennial, glabrous or pubescent with simple hairs; leaves pinnatifid, toothed or the upper ones subentire, petiolate to sessile, sometimes auriculate; flowers in racemes, the petals yellow, fading white or pinkish; fruit orbicular to sub linear, but short (rarely over 1 cm long).

1 Fruit strigose . . . . ***R. calycina***

1 Fruit glabrous or minutely papillose, not strigose 2

2 Plants perennial, from rhizomes, lacking basal rosettes even when young, rarely collected, known from below 6,800 ft; petals 3.5-5.5 mm long; leaves rather uniformly sinuate . . . . ***R. sinuata***

2 Plants from taproots, with basal rosettes when young, occasional to common, from low to high elevations; petals about 1-2 mm long; leaves various. 3

3 Stems erect or somewhat decumbent, mostly 30-60 (100) cm tall, 4-12 mm in diameter; fruit 1-2 times longer than wide, the pedicels equal to or longer than the fruit (including the style); inflorescence becoming 10-15 cm wide; plants from 5,400-7,000 (8,000) ft . . . . ***R. palustris***

3   Stems erect to prostrate, often less than 30 cm long, about 1-4 mm thick; fruit mostly 2-5 times longer than wide, the pedicels shorter than or equaling the fruit; inflorescence less than 10 cm wide; plants of various elevations 4

4   Fruit minutely papillose; stems prostrate to decumbent; plants known from 4,700-5,400 ft, uncommon . . . . **R. tenerrima**

4   Fruit glabrous; stems prostrate to erect; plants known from (7,000) 8,000-11,000 ft, common . . . . **R. curvipes**

Rorippa calycina (Engelmann) Rydberg — persistent sepal yellowcress — Known from along Flaming Gorge Reservoir 1.5 mi below Blacks Fork in Wyoming (B. E. Nelson eat al 35245), perhaps to be expected along the reservoir in our area.

Rorippa curvipes Greene [*R. alpina* S. Watson; *R. obtusa* (Nuttall) Britton var. *alpina* S. Watson; *Nasturtium obtusum* Nuttall var. *alpinum* S. Watson] — common yellowcress — Common in Strawberry Valley and across the Uinta Mtns., apparently rare at the extreme west of the W. Tavaputs Plateau; wet places, often in mud at the edge of ponds and lakes and along streams; (7,000) 8,000-11,000 ft; June-Aug. Three vars. [var. *alpina* (S. Watson) Stuckey; var. *curvipes* Greene; and var. *truncata* (Jepson) Rollins (var. *integra* (Rydberg) Stuckey)] have been collected from our area. See AUF (338-339) for features used for separation. Var. *alpina* [treated at species level in IMF (3B: 340) and FNA (7: 496)] is the common phase above 9,800 ft on the Uinta Mtns.

Rorippa palustris (L.) Besser [*Rorippa islandica* (Oeder ex Murray) Borbas misapplied; *Nasturtium palustre* (L.) de Candolle] — marsh yellowcress, hispid yellowcress — Six specimens seen from Whiterocks-Tridell-Lapoint area, Rock Creek, floodplain of the Green River, and the W. Tavaputs Plateau at 5,400-8,000 ft with stems more or less hispid belong to var. *hispida* (Desvaux) Rydberg [*R. hispida* (Desvaux) Britton]. One specimen from the bottomlands of the Duchesne River near Bridgeland at 5,500 ft and another from Blacks Fork, Uinta Mtns. at 8,740 ft with glabrous stems are referable to var. *glabra* (0. E. Schulz) Taylor and McBride. The 2 glabrous specimens indicate random distribution within the range of pubescent plants. This species is also plotted for upper elevations of the E. Tavaputs Plateau by Albee et al. (1988).

Rorippa sinuata (Nuttall) A. S. Hitchcock (*Nasturtium sinuatum* Nuttall) — spreading yellowcress — The 4 specimens seen are from the floodplain of the Green River on or near Ouray National Wildlife Refuge and a dry pond at Twelve Mile Wash near Hwy. 121; disturbed marsh communities and pond margins; 4,655-6,760 ft; June-Sept. Graham (1937) lists Graham 6113 from between Green River and Quarry.

Rorippa sphaerocarpa (A. Gray) Britton — round pod yellowcress — The one specimen seen (Goodrich 22920) is from Uinta Canyon from a riparian community at 7,360 ft.

Rorippa tenerrima Greene — low yellowcress — The 7 specimens seen are from the floodplain of the Green River at Browns Park, Stewart Lake near Jensen, Leota Bottoms near Ouray, draw-down basin of Steinaker Reservoir; 4,655-5,550 ft, and a spring near Bear Wallow of the Lake Fork Drainage at 8750 ft; Aug.-Oct.

Sinapis L. — Charlock

Similar to *Brassica* but fruiting valves with 3-7 nerves instead of 1.

Sinapis arvensis L. [*Brassica arvensis* Rabenhorst; *B. kaber* (de Candolle) L. C. Wheeler] — charlock, wild mustard — Introduced from Europe, more or less weedy; only 4 specimens seen, but to be expected across our area at lower and mid-elevations in fields, roadsides, ditch banks, and recently disturbed areas.

Sisymbrium L. — Tumblemustard

Plants annual or rarely biennial, glabrous or hirsute, from taproots; lower leaves pinnatifid, the upper ones sometimes entire; flowers in racemes, petals yellow; fruit linear, terete or nearly square in cross-section, many times longer than wide.

---

1   Plants glabrous, perennial with rhizomes . . . . ***S. linifolium***

1   Plants pubescent, at least in part, annual with taproot 2

2   Fruit 5-9 cm long; lower leaves strongly pinnatifid, upper leaves linear and entire . . . . ***S. altissimum***

2   Fruit 2-3.5 cm long; lower leaves pinnatifid or lobed, upper leaves similar to the lower ones . . . . ***S. loeselii***

Sisymbrium altissimum L. — Jim Hill mustard, tumble mustard — Introduced from Europe, widespread; roadsides and other disturbed sites where it is especially abundant in years with abundant spring rains, dominant and forming dense, tangled patches following fire in pinyon-juniper communities where it decreases with time and with increase of cheatgrass, also in sagebrush, and mt. brush communities up to 7,500 ft; May-June. Response to fire and to variation in seasonal precipitation indicates this species forms seed-banks in the soil with germination of dormant seeds triggered by fire and weather events.

Sisymbrium linifolium (Nuttall) Nuttall in Torrey & A. Gray [*Schoenocrambe linifolia* (Nuttall) Greene — flaxleafed plainsmustard — Occasional; widespread; desert shrub, sagebrush, pinyon-juniper, and mt. brush communities; 4,700-8,300 ft; late April-July.

Sisymbrium loeselii L. — Loesel tumble mustard — Introduced from Eurasia and recently found in great abundance in an old corral along the Seep Ridge Road at 5655 ft (Goodrich and Newberry 27370), along roads and in fields between Randlett and Ouray, and along much of the White River valley bottom between Rangely and White River City. This invasive species can be expected to spread across lower elevations of the Uinta Basin.

Smelowskia C. A. Meyer — Smelowskia

Smelowskia calycina C. A. Meyer in Ledebour (*S. lineariloba* Rydberg) — alpine smelowskia., Siberian smelowskia — Plants perennial, caespitose, from branched caudices, 4-20 cm tall, pubescent with simple and branched hairs; leaves pinnatifid, 5-50 mm long, 3-16 mm wide, the upper ones sometimes smaller; flowers mostly in racemes; petals 3-8 mm long, white, cream, pink, or lavender; fruit 5-9 mm long, linear or nearly so. Common; Uinta Mtns.; open lodgepole pine and Engelmann spruce woods, krummholz, dry and moist meadows, moraine, talus, fell fields, rock stripes, and alpine tundra; 9,700-13,000 ft; June-Aug. Our plants belong to var. *americana* (Regel & Herder) Drury & Rollins. North American plants are treated at the species level as *S. americana* Rydberg in FNA (7: 672) with readily deciduous calyces rather than persistent calyces of central Asian plants.

Stanleya Nuttall — Prince's plume; Stanleya

Plants perennial, from taproots or caudices, glabrous or with simple hairs; flowers in racemes, yellow, the stamens strongly exserted beyond the petals; fruit long-stipitate, linear, many times longer than wide.

1   Middle and upper cauline leaves sessile, auriculate, entire . . . **S. *viridiflora***
1   Leaves not sessile, not auriculate 2
2   Upper and lower leaves deeply pinnate or pinnatifid; sepals and petals 5-10 mm long; fruit contorted, the body 2.5-4 mm long, the stipe 6-11 mm long . . . . **S. *bipinnata***
2   At least some upper leaves usually entire; sepals and petals (8) 10-16 mm long; fruit straight or gradually arched, the body 4-8 cm long, the stipe 12-21 mm long . . . . **S. *pinnata***

Stanleya bipinnata Greene [*S. pinnata* var. *bipinnata* (Greene) Rollins] — bipinnate prince's plume — Known in our area from Lookout Peak in Moffat Co., along roads and along the rim of the peak at 7,640-8,120 ft elevation. Listed for "Uinta" Co., Utah in FNA (7: 696).

Stanleya pinnata (Pursh) Britton — prince's plume, desert prince's plume — With 2 vars. as follows. Plants with all entire leaves are sometimes found with plants with some leaves pinnatified (Goodrich 28874A & 28874B). Plants with pinnatified leaves were found in a population n. of Vernal one year where only entire leaves were found the next year. The distinction of 2 vars. is weak at best. This plant is an indicator of seleniferous soils.

1   Leaves all entire or toothed, or the lower ones coarsely toothed to lobed, the cauline leaves ovate or lanceolate . . . . **var. *integrifolia***
1   At least the lower leaves pinnately divided, the cauline leaves lanceolate, sometimes narrowly so . . . . **var. *pinnata***

Var. *integrifolia* (James) Rollins (*S. integrifolia* James in Torrey) Occasional; Daggett and Uintah Cos., and e. into Colorado; desert shrub and juniper communities; 5,500-7,000 ft; May-July.

Var. *pinnata* (*S. arcuata* Rydberg) Apparently less common then var. *integrifolia;* widespread; desert shrub and pinyon-juniper communities; 4,600-7,500 ft; June-Aug.

Stanleya viridiflora Nuttall — green prince's plume — The several specimens seen are from widely scattered locations in Daggett (Sheep Creek Canyon), Duchesne, Summit (Hole In The Rock), Uintah, and w. Moffat Cos.; desert shrub, pinyon-juniper, black sagebrush, big sagebrush, and alderleaf mtn. mahogany communities, sometimes growing in crevices of cliffs and rock outcrops, and semi-barrens of geologic formations that weather to badlands; 5,590-7,300 ft; July-Aug. Racemes commonly more than ¼ the height of the plant.

Streptanthella Rydberg — Streptanthella

Streptanthella longirostris (S. Watson) Rydberg — beaked streptanthella — Annual herbs from taproots, 10-50 cm tall, without basal rosettes; leaves 1.5-8.5 cm long, 0.1-1.2 cm wide, oblanceolate to elliptic or nearly linear, entire to sinuate-dentate, the lower ones deciduous by flowering time; petals 5-8 mm long, white with purplish veins; fruit 30-60 mm long, 1.5-2 mm wide, reflexed or descending, narrowed to a 3-7 mm long beak. Widespread; occasional or locally common in desert shrub, sagebrush, and pinyon-juniper communities, often (but not always) on sandy soils; 4,700-7,100 ft; May-mid June.

Streptanthus Nuttall — Streptanthus

Streptanthus cordatus Nuttall in Torrey & A. Gray — heartleaf twistflower or jewelflower — Perennial, glabrous, glaucous herbs from taproots, 18-60 (80) cm tall; basal leaves 1.5-8 (15) cm long, 0.5-2 cm wide, obovoid to oblanceolate, dentate at the apex; stem leaves sessile and strongly auriculate; petals 10-15 mm long, purple to chestnut-brown; fruit 50-85 mm long, 3.5-8 mm wide. Widespread; almost always associated with pinyon or juniper, rarely in desert shrub or ponderosa pine communities; 5,200-7,800 ft; May-June. Our plants belong to var. *cordatus*.

Thelypodiopsis Rydberg

Plants annual, biennial, or short-lived perennial, glabrous or pubescent with simple hairs, usually glaucous; at least some of the stem leaves strongly articulate; petals white, pink, or lavender; fruit many times longer than wide.

---

1   Stigmas expanded and deeply bilobed; petals obovate, 3-6 mm wide, 11-14.5 mm long, constricted to a broad claw; lower leaves 3-15 mm wide, entire or dentate or irregularly toothed to lobed; fruit 45-75 mm long; plants known from 4,800-7,200 ft elevation .... *T. elegans*

1   Stigmas narrow, not deeply bilobed; petals oblanceolate, 1-4 mm wide, 5-15 mm long, tapering to a narrow claw; lower leaves 10-50 mm wide, entire; fruit 10-65 mm long; plants known from 7,500-8,600 ft elevation ....
.... *T. sagittata*

Thelypodiopsis elegans (M. E. Jones) Rydberg [*T. wyomingensis* (A. Nelson) Rydberg; *Sisymbrium elegans* (M. E. Jones) Payson] — elegant thelypody — Widespread; occasional to locally common in desert shrub and pinyon-juniper communities; 4,800-7,200 ft; late April-mid June.

Thelypodiopsis sagittata (Nuttall) O. E. Schulz [*Thelypodium sagittatum* (Nuttall) Endlicher; *T. paniculatum* A. Nelson misapplied] — arrowleaf thelypody — The few specimens seen are from Duchesne and Wasatch Cos., from Farm Creek Pass of the Uinta Mtns., Strawberry Valley, and Willow Creek of the W. Tavaputs Plateau, reported for Uintah Co. (AUF 345); wet meadows, silver sagebrush, and riparian communities; 7,500-8,600 ft; June-Aug. Harrington 3712 reported for Blue Mt., 7,500 ft (Bradley 1950) was apparently annotated elsewhere prior to 1985. Uinta Basin plants belong to var. *sagittata*.

Thelypodium Endlicher — Thelypody

Annual, biennial or short-lived perennial herbs from taproots, glabrous or pubescent with simple hairs; leaves entire to pinnatifid; flowers in racemes; petals white, pink, or lavender; fruit linear, many times longer than wide, short stipitate.

---

1   Cauline leaves sessile, entire to dentate; terminal branch of the inflorescence about equal to or shorter than one or more of the lateral branches; racemes densely flowered and commonly with more than 30 flowers; plants 50-300 cm tall, from along waterways and floodplains . . . .
. . . . *T. integrifolium*
1   Lower cauline leaves petiolate, sometimes pinnatifid; terminal branch of the inflorescence usually well exceeding the lateral ones; racemes rather loosely flowered, commonly with fewer than 30 flowers; plants 15-150 cm tall or taller. . . . . *T. laxiflorum*

Thelypodium integrifolium (Nuttall) Endlicher [*T. rhomboideum* Greene var. *gracilipes* (Robinson) Payson] — tall thelypody — Infrequent; widespread; along water ways, meadows, and seeps, tolerant of alkali; 4,875-7,350 ft; July-Oct. With 2 vars. as follows:

1   Racemes densely congested, the one of the main stem 1-4.5 (6) cm long; Browns Park of Daggett Co. (C. H. Refsdal 6373, N. H. Holmgren & P. K. Holmgren 13675) . . . . **var. *integrifolium***
1   Racemes moderately congested, the one of the main stem (4.5) 8-30 cm long; Daggett, Duchesne, and Uintah Cos. . . . . . . **var. *gracilipes*** B. L. Rob.

Thelypodium laxiflorum Al-Shehbaz — slate thelypody — S. L. Welsh & K. Taylor 15148 and E. Neese & L. England 6149 are from the W. Tavaputs Plateau, Carbon Co. S. Goodrich 24443 is from Bogart Canyon of the E. Tavaputs Plateau, Grand Co., and A. Huber 861B is from Big Brush Creek Gorge, Uinta Mtns.; mtn. brush, ponderosa pine-Gambel oak, spruce-fir communities; 7,350-9,000 ft; June-Aug. *Thelypodium wrightii* A. Gray [*Stanleyella wrightii* (A. Gray) Rydberg] was listed for Post Canyon at 8,000 ft by Graham (1937) based on Graham 9922. Plants of this taxon will key as *T. laxiflorum* in this flora. AUF (347) and IMF (2B: 210) exclude the Uinta Basin from the range of *T. wrightii*. The Graham specimen likely belongs to *T. laxiflorum*.

Thlaspi L. — Pennycress; Stinkweed

Thlaspi arvense L. — field pennycress, fanweed — Plants annual 10-70 cm tall, glabrous and usually glaucous; stem leaves 1-8 cm long, 2-25 mm wide, auriculate, simple, entire, dentate, or lobed, the upper ones sessile and auriculate; pedicels 5-12 mm long; petals 3-4.5 mm long, white or sometimes pinkish or lavender; fruit 10-17 mm long, 7-12 mm wide, strongly flattened and winged. Introduced from Europe, infrequent, more or less weedy; roadsides, disturbed ground, desert shrub, pinyon-juniper, aspen, and Douglas-fir communities; up to 9,000 ft; June-July.

Transberingia Al-Shehbaz & O'Kane

Transberingia virgata (Nuttall) N. H. Holmgren [*Halimolobos virgata* (Nuttall) Schulz in Engelmann; *Transberingia bursifolia* ssp. *virgata* (Nuttall) Al-Shehbaz & O'Kane] — strict weed, twiggy fissurewort — Biennial plants but occasionally flowering the first year, 10-40 cm tall, pubescent with mixed, simple, forked and branched hairs at least toward the base; basal leaves 3-6 cm long, 0.5-1.8 cm wide denticulate to shallowly lobed, rarely entire; cauline leaves reduced upward, at least the upper ones sessile and auriculate; pedicels 7-11 mm long, ascending at a 30°-40° angle from the axis of the raceme; petals 4-4.5 mm long, white, the veins often purplish; pods erect, parallel or nearly so with the axis of the raceme, 15-40 mm long, 1-1.5 mm wide, rounded to nearly square in cross-section. The 4 specimens seen are from Trout Creek Peak, Wasatch Co., Bigelow Bench between Ft. Bridger and Evanston, Uinta Co., Wyoming, and Phil Pico Mt. and Sheep Creek Canyon, Daggett Co.; windswept ridges, fringed sagebrush, sagebrush, and forb-grass communities; 6,300-9,350 ft; July-Aug.

Turritis glabra and *Arabis hirsuta* also with erect pods and biennial habit, are sometimes mistaken for this taxon. The following key may help to distinguish these taxa. In addition the mature pods of *Transberingia* are terete or nearly square in cross-section. Those of *Arabis* are moderately to strongly compressed.

1 Pedicels diverging at a 30°-40° angle from the rachis, the pods erect, their angle abruptly changed from the angle of the pedicels; stems often branched .... ***Transberingia***
1 Pedicels and fruit erect or nearly so, the fruit not assuming a strongly different angle from the pedicels; stems mostly simple at least in unbroken or ungrazed specimens .... ***Arabis hirsuta*** and ***Turritis glaber***

Turritis L — Tower mustard

Turritis glabra L [*Arabis glabra* (L.) Bernhardi] — tower mustard — Widespread; Infrequent in pinyon-juniper, aspen and spruce-fir communities, and along streams; 7,000-8,000 ft; June-July. Our plants belong to var. *glabra*.

CACTACEAE Cactus Family

Perennial, succulent, usually spiny plants with globose, cylindrical, or flattened stems; leaves lacking at least at maturity; areoles axillary (regardless of apparent position), evenly spaced, bearing spines, branches, flowers, or (in *Opuntia*) fine short-barbed bristles (glochids); stamens numerous; ovary inferior; fruit a many-seeded, dry or fleshy berry.

1 Stems jointed, the joints more or less flattened; areoles bearing numerous minute glochids as well as spines .... ***Opuntia***
1 Stems hemispheric or cylindrical, not jointed; areoles with spines but without glochids 2
2 Stems ribbed, the tubercles longitudinally confluent at the base and thus aligned in rows... 3
2 Stems not ribbed, the tubercles not confluent at the base 4
3 Flowers bright red; stems cylindrical, usually few to numerous in compact

clusters; flowers borne laterally, the buds breaking through the epidermis
above an areole; spines never hooked .... *Echinocereus*

3   Flowers pinkish to rose-purple; stems subglobose or ovoid (rarely in old
     individuals short-cylindrical), usually solitary (sometimes few-branched
     due to injury of terminal bud); flowers sub-terminal, borne at the stem apex;
     spines hooked or not .... *Sclerocactus*

4   Upper side of tubercles with a longitudinal groove; petaloids usually rose-
     pink; plants rare .... *Coryphantha*

4   Upper side of tubercles not grooved; petaloids usually whitish, yellow, or
     peach, occasionally pinkish; plants common .... *Pediocactus*

Coryphantha (Engelmann) Lemaire — Pincushion Cactus; Ball Cactus

Coryphantha vivipara (Nuttall) Britton & Rose [*C. neomexicana* (Engelmann)
Britton & Rose probably misapplied; *Mammillaria vivipara* (Nuttall) Haworth] —
Nuttall pincushion — Plants 2-5 cm long, and broad, the stems depressed-
globose to ovoid, solitary or clustered; tubercles separate and spirally arranged;
central spines 4, rather prominent, radial spines 12-20, slender, the spines
mostly 1-2 cm long, straight; flowers borne near the summit of the stem on the
tubercle at the base of a longitudinal felty groove, pink-purple or rose, about 4
cm wide; fruit green, ellipsoid, 1-2 cm long.

Reported for Theodore (Duchesne), benches of the Uinta Mtns., 8,000 ft
based on a collection by M. E. Jones of 13 May 1908 (L. D. Benson 1982) and for
a bench w. of Green River, n. of Sand Wash, 4,500 ft based on Graham 7908, collected 28 May 1933 (Graham 1937). Plants of our area would belong to var.
vivipara. Our area is not included in the distribution of this species as given in
IMF (2A: 685-687). This strongly indicates The M. E. Jones collection from
benches of the Uinta Mtns. likely belongs to *Pediocactus simpsonii* and the
Graham specimen from Sand Wash likely belongs to one of the endemic
species of *Sclerocactus.*

Echinocereus Engelmann — Hedgehog Cactus; Strawberry Cactus

Echinocereus mojavensis (Engelmann & J. M. Bigelow) Rumpler [*E. coccineus*
Engelmann; *E. triglochidiatus* var. *mojavensis* (Engelmann & Bigelow) L. D.
Benson; *E. octacanthus* (Muhlenberg) Britton & Rose misapplied] — scarlet
claretcup — Stems few to numerous in mounds, mostly 8-15 cm long, 4-7-cm
thick; ribs 9 or 10, the tubercles not prominent; central spines 1-3; radial
spines 5-9, shorter than the central one; flowers narrowly vase-shaped, the
petaloids scarlet, greenish toward the base. Uncommon; specimens seen are
scattered across the Basin from near Duchesne ne. across the foot-slopes of
the Uinta Mtns. to the Green River in Browns Park (Swallow Canyon overlook,
Neese 5664A — this apparently the most northerly known location for the
species), E. Tavaputs Plateau including the Piceance Basin; dry, often rocky or
sandy places; June.

Opuntia Miller — Pricklypear

Stems jointed, the joints flattened (sometimes nearly terete in *O. fragilis*),
never ribbed; leaves small, fleshy, scale-like, caducous; areoles with glochids
as well as spines (rarely nearly spineless); flowers borne in areoles of previous

year's growth, yellow, peach, pink, rose, or purple, tube of hypanthium short, cup-shaped; ovary spiny or not; fruit (in ours) dry; seeds light colored, flattened. Note: The pricklypears of our area form a complexly intergrading group; morphological characters traditionally treated as diagnostic (spine number, shape, color, and distribution, pad shape, nature of the areole, and so forth) seem little-correlated, the populations demonstrating a perplexing array of combinations.

1   Largest joints 2-8 cm long, 1.5-3.5 cm wide, relatively thick, sometimes nearly terete, readily detached; plants mat-forming (transitional to the next) . . . . *O. fragilis*

1   Largest joints mostly 7-15 cm long, 4-12 cm broad, flattened, not readily detached; plants mat-forming or not . . . . *O. polyacantha*

Opuntia fragilis (Nuttall) Haworth — brittle pricklypear — Duchesne, Uintah, and Moffat Cos.; occasional to common, occurring in many habitats but usually in dry, open, often rocky or sandy places in the juniper and mt. brush belts, sometimes under ponderosa pine. Plants of our area are rarely seen in flower apparently due to a very short flowering period. Lisa Boyd checked a population in Rock Creek weekly and found plants in flower for only 2-3 days.

Opuntia polyacantha Haworth — plains pricklypear — Common, sometimes locally abundant, widespread; salt desert shrub, Wyoming big sagebrush, and juniper communities, usually in dry open rocky or sandy places; 4,700-7,000 ft (and probably higher); June-July.

Note: the following names have been variously applied to the *O. polyacantha* complex in the Uinta Basin:

O. erinacea Engelmann var. *utahensis* (Engelmann) L. D. Benson

O. hystricina Engelmann & Bigelow (Graham 8169; Vernal-Manila road n. of Vernal).

O. juniperina Britton & Rose.

O. phaeacantha Engelmann (Graham 913; Red Wash above Island Park).

O. polyacantha var. *juniperina* (Britton & Rose) L. D. Benson

O. rhodantha Schumann

O. rutila Nuttall in Torrey & A. Gray

Pediocactus Britton & Rose — Hedgehog Cactus

Pediocactus simpsonii (Engelmann) Britton & Rose (*Echinocactus simpsonii* Engelmann) — mtn. ball cactus — Plants subglobose, usually solitary, to 15 cm high; tubercles spirally arranged; areoles woolly, at least when young; spines straight; flowers borne at one side of the areoles near the tubercle apex, small, often numerous, in a ring near the summit of the stem; fruit dry. Specimens seen are from near Starvation Reservoir and the Uinta River drainage e. to Diamond Mt., also near Manila, Daggett Co., and on the Tavaputs Plateau; salt desert shrub, mixed desert shrub, sagebrush-grasslands, juniper, and mt. brush communities, usually on gravelly benches or in sandy rocky places; 5,500-7,700 ft; April-June. Graham 9122 (CM!), reported as *Coryphantha neomexicana* (Engelmann) Britton & Rose (Graham 1937) belongs here.

Sclerocactus Britton & Rose — Fishhook Cactus

This treatment follows IMF (2A: 669-676).

---

1 Radial spines 8-17; spines often obscuring the stem, some spines commonly over 32 mm long, the central hooked spine mostly longer than the other spines . . . . *S. parviflorus*

1 Radial spines 6-10 (rarely more); spines not, or only slightly to moderately, obscuring the stem; spines mostly less than 32 mm long 2

2 Abaxial central spine usually 1-2 mm long, black, curved to hooked (if longer then hooked or curved as a result of introgression); lateral central and radial spines not well differentiated; flowers campanulate; inner tepals purple-pink . . . . *S. brevispinus*

2 Adaxial central spine 14-28 mm long, straight to curved, less often hooked; lateral central and radial spines differentiated or not; flowers campanulate to funnelform; inner tepals pale pine to dark pink . . . . *S. wetlandicus*

Sclerocactus brevispinus Heil & Porter [*S. parviflorus* var. *roseus* (Clover) L. D. Benson; *S. whipplei* var. *glaucus* (J. A. Purpus) S. L. Welsh] — shortspine cactus, Pariette Draw cactus — Endemic to the Uinta Basin. Plants mostly found on the Wagonhound member of the Uinta Formation in Pariette Draw-Castle Cliff area in se. Duchesne Co. This and two other populations of different but intergrading morphology have been mapped and discussed (Bureau of Land Management 1985). The central spine is very short (to 2 mm), stout, black, and mostly hooked, and the juvenile stage persists for several years after the initiation of flowering. These plants loose turgidity and shrink below the ground level after flowering. Graham 8839 (CM), questionably referred to *Utahia sileri* (Engelmann) Britton & Rose by Graham (1937), belongs here. Graham 7941a, reported by Graham (1937) as *Neolloydia texensis* Britton & Rose for a bench on the w. side of the Green River n. of the mouth of Sand Wash, probably belongs here.

Sclerocactus parviflorus Clover & Jotter [*S. whipplei* var. *roseus* (Clover) L. D. Benson] — Devil's-claw cactus — Listed for Uintah and Duchesne Cos. (IMF 2A: 671; AUF: 89 as *S. whipplei* var. *roseus*). The variability found in *S. parviflorus* throughout its range nearly encompasses that of all 3 species of *Sclerocactus* listed in this flora.

Sclerocactus wetlandicus Hochstatter [*S. glaucus* ssp. *wetlandicus* (Hochstatter) Luthy; *S. whipplei* var. *ilseae* (Hochstatter) S. L. Welsh] — Ouray cactus, Uinta Basin hookless cactus — Endemic to the Uinta Basin. Occasional and locally common on old terraces associated with the Green River from about 8 mi above Ouray to Minnie Maud Creek, also w. to near the Duchesne Co. line on the slopes above Nine Mile Creek; gravelly terrace and bluff margins, sloping gravelly, pediments, and gravel-littered draws, usually where clay or silty clay underlies the stony surface, in shadscale and mixed desert shrub communities; 4,700-5,800 ft; May-June. At least a few individuals in most populations possess moderately to strongly hooked spines. *S. glaucus* (K. Schumann) L. D. Benson is maintained by some workers to be confined to Colorado. Others consider the Utah and Colorado plants of this complex to be the same.

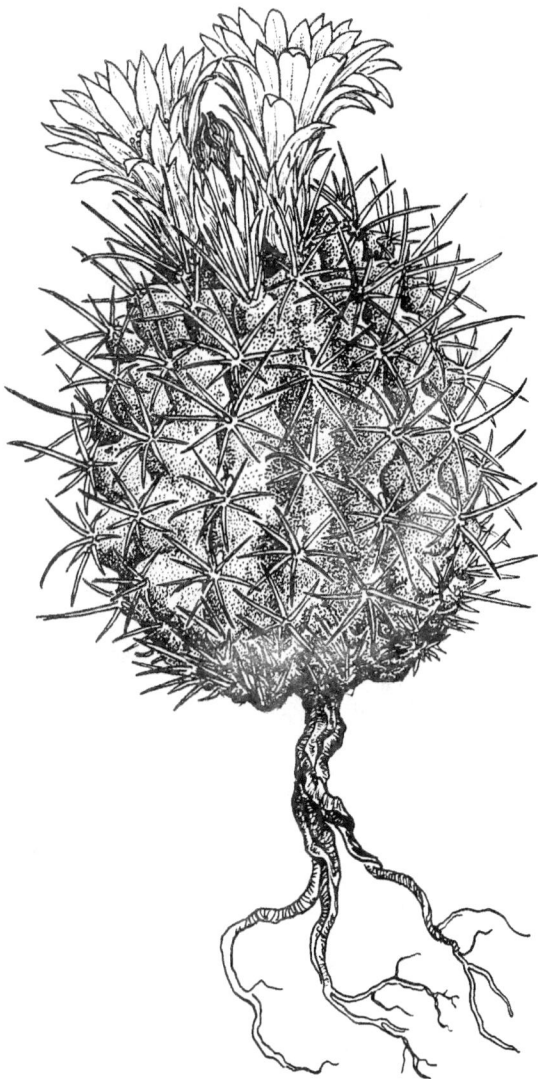

Sclerocactus wetlandicus Hochstatter
Ouray cactus, Uinta Basin hookless cactus

CALLITRICHACEAE Water-starwort Family

Callitriche L. — Water-starwort

Aquatic, submerged or emergent, slender-stemmed, inconspicuous, perennial herbs, sometimes stranded on mud of drying pools; leaves opposite or the floating ones tufted at the ends of stems; flowers inconspicuous, solitary or 2-3 in the axils of leaves, mostly unisexual, consisting of a single pistil or 1 (rarely more) stamens; ovary separating at maturity into four 1-seeded achene-like fruits. This family is now included in the Plantaginaceae.

1   Leaves all linear, 1 nerved, all submerged; floral bracts lacking; fruit not wing-margined, the faces with minute, pit-like, irregularly distributed markings .... *C. hermaphroditica*
1   Upper (floating) leaves usually broader; floral bracts usually present; fruit usually slightly wing-margined at the top. 2
2   Fruit wing-margined at the top, rectangular or at least not wider above the middle except for the wings, 1/5-1/3 longer than broad, the tiny pit-like markings of the faces in rather regular vertical lines .... *C. palustris*
2   Fruit not wing-margined, obcordate or at least widest above the middle, about as wide as long, the tiny pit-like markings of the faces irregularly distributed .... *C. heterophylla*

Callitriche hermaphroditica L. — secret water-starwort — The 2 specimens seen are from Blue Mt. (Neese & Snider 11907) and Diamond Mt. (Neese 1 832); aquatic in ponds and lakes; 7,050-7,800 ft; June-Sept.

Callitriche heterophylla Pursh ex Darby — larger water-starwort — The 4 specimens seen are from the Uinta Mtns.; pools, ponds, and lakes; 7,600-10,330 ft; July-Sept.

Callitriche palustris L. (*C. verna* L.) — vernal water-starwort — The several specimens seen are all from the Uinta Mtns.; pools, ponds, lakes, and perhaps in slow streams; 7,100-10,500 ft; July-Sept.

CAMPANULACEAE Harebell Family

Annual or perennial herbs with simple, alternate, estipulate leaves; flowers bisexual; sepals 5; corolla united, 5-lobed; stamens 5; pistils 1; ovary inferior, with 2-5 lobes; style 1; fruit a capsule.

1   Leaves petiolate; corolla regular .... *Campanula*
1   Leaves sessile; corolla irregular. 2
2   Flowers on slender pedicels; corollas with yellow spots near the base ....
     .... *Porterella*
2   Flowers sessile at the apex of elongate hypanthium tubes; corollas with yellow and purplish spots near the base .... *Downingia*

Campanula L. — Bellflower; Harebell

Perennial herbs from rootstocks; leaves entire to toothed; corolla mostly bell-shaped; stamens 5, alternate with the corolla lobes; style 1 with 3-5 stigmas; fruit a many-seeded capsule.

1   Plants mostly over 50 cm tall; middle and upper stem leaves lanceolate, distinctly toothed; flowers usually many .... *C. rapunculoides*

1 Plants rarely to 50 cm tall; middle and upper stem leaves linear or nearly
   so, mostly entire 2
2 Corolla 7-10 (12) mm long, narrowly bell-shaped; anthers 1.5-2.5 mm long;
   flowers solitary; capsules erect, opening by pores near the summit; lower
   leaves sessile or the petioles not sharply distinct from the blades, gradually
   expanding into the blade .... *C. uniflora*
2 Corolla 12-20 mm long, about as wide as long; anthers 4-6.5 mm long;
   flowers 1-several; capsules nodding, opening by pores near the base; lower
   leaves often early deciduous, the petioles often as long or longer than the
   blades, often abruptly contracted from the blade .... *C. rotundifolia*

Campanula rapunculoides L. — creeping bellflower — Cultivated and escaping,
native of Eurasia; the 3specimens seen are from a roadside at Neola from a
flower garden in Roosevelt and in Brownie Canyon, Uinta Mtns. in an aspen-
conifer community where the species did not persist; May-Oct.

Campanula rotundifolia L. (*C. petiolata* A. de Candolle) — Scotch bluebell — The
common bellflower of the Uinta Mtns.; many plant communities; 7,500-11,500
ft; June-Aug.

Campanula uniflora L. — arctic harebell — Occasional; Uinta Mtns.; above timberline; July-Aug.

Downingia Torrey

Downingia laeta (Greene) Greene — Great Basin calicoflower — Plants annual,
sometimes rooting at the lower nodes, glabrous; stems soft; leaves entire, 0.5-
2 cm long, 0.5-2 mm wide; flowers sessile in axils of bracts, born on long hypanthium tubes that might appear to be pedicels; calyx segments 2-7 mm long,
linear or nearly so; corolla 3-5 (7) mm long, white, pink, pale blue, or lilac,
divided into a lower and upper lip, the lower lip with 3 lobes and the upper lip
with 2 lobes, the lower lobes with larger yellow spots alternating with smaller
pink or purple spots; capsules (mature hypanthium) 1.8-3.5 cm long. The one
specimen seen (Goodrich 28158) is from an ephemeral pool at Greendale, n.
slope Uinta Mtns. at 7,240 ft elevation.

Porterella Torrey

Porterella carnosula (Hooker & Arnot) Torrey in Hayden — fleshy porterella —
Plants annual, rooting at the lower nodes, glabrous, somewhat fleshy, mostly
3-15 cm tall; leaves sessile, 4-20 mm long, linear-subulate to elliptic; sepals 2-6
(9) mm long; corolla blue with white or yellow center, 6-10 cm long, with 2-
lips, the lower lip 3-lobed, the upper lip with 2 erect lobes. Mostly in water or
mud recently exposed by drying of ponds and streams, known from the north
slope of the Uinta Mountains in the Blacks Fork drainage.

CANNABACEAE Endlicher Hemp Family
Humulus L. Hop

Humulus lupulus L. (*H. americanus* Nuttall) — American hop — Perennial,
herbaceous, dioecious vines; stems to 10 m long, scabrous with stiff, recurved
hairs; leaves opposite, palmately 3-5 (7) lobed and coarsely toothed, those of

the inflorescence alternate; flowers small, greenish, not showy, the staminate ones in loose panicles from upper leaf-axils, the sepals and stamens 5, the many pistillate flowers in pairs under a large persistent bract, the many bracts forming a large cone-like hop. Widespread but infrequent; usually growing on bushes along streams or in other moist places; not expected much over 7,500 ft; July-Aug. The glands on the bracts of the hops secrete lupulin that is used for flavor in brewing of beer, and the plant is cultivated (but not in the Uinta Basin area). Plants of our area belong to var. *neomexicanus* Nelson & Cockerell.

CAPRIFOLIACEAE Honeysuckle Family

Trees, shrubs, and woody vines; leaves opposite, simple or compound; flowers bisexual; sepals 4-5, mostly reduced to small teeth or obsolete; corolla, rotate to salverform, (4) 5-lobed; stamens (4) 5, arising from the corolla tube, alternate with the lobes; pistil 1; ovary inferior, 1-3 or 5 chambered; style 1 or obsolete; stigmas 1-3 or 5; fruit a berry or drupe.

1 Plants woody, prostrate-creeping, vine-like, the aerial stems not over 15 cm tall . . . . *Linnaea*

1 Plants upright shrubs 2

2 Leaves pinnate compound with 5-9, large, serrate leaflets . . . . . *Sambucus*

2 Leaves simple 3

3 Fruit a red or black berry; flowers and fruit borne in pairs on slender peduncles over 1 cm long . . . . *Lonicera*

3 Fruit a white or greenish drupe; flowers and fruit not in distinct pairs, peduncles not as above . . . . *Symphoricarpos*

Linnaea L. — Twinflower

Linnaea borealis L. — twinflower — Creeping vine-like shrub; stems prostrate to 1 m long; aerial (flowering) stems not over 15 cm tall; leaves simple, ovate to orbicular, 8-20 mm long, entire or a few teeth above the middle; flowers in pairs, nodding, borne on naked peduncles 3-12 cm long; corolla 8-15 mm long, white to pink; fruit small, dry, 1-seeded. Scattered across the Uinta Mtns. in coniferous forests at 8,600-9,700 ft; July-Aug.

Lonicera L. — Honeysuckle; Twinberry

Shrubs; leaves simple, opposite, mostly entire; flowers in pairs, borne on slender, elongate peduncles from axils of upper leaves; corolla cylindrical to bell-shaped, 4 (5) -lobed or 2-lipped with the upper lip 4-lobed, the tube pouched near the base; fruit a fleshy, several-seeded berry.

1 Fruit red; corollas 15-20 mm long, pale yellow, subtended by narrow, green bracts; leaves 2-6 cm long, pale blue-green . . . . *L. utahensis*

1 Fruit black; corollas 12-15 mm long, yellow, subtended by conspicuous broad, reddish-purple or blackish bracts; leaves 5-15 cm long, green . . . . . .
. . . . *L. involucrata*

Lonicera involucrata (Richardson) Banks ex Sprengel — black twinberry —
Widespread; occasional in canyons, mountains, and into valleys, along streams
and other moist places; 5,000-10,000 ft; May-Aug. The red-purple bracts are
rather easily mistaken for flowers. Our plants belong to var. *involucrata.*

Lonicera utahensis S. Watson — Utah honeysuckle — specimens seen are from
the Avintaquin drainage and w. on the W. Tavaputs Plateau and from Rock
Creek and w. on the Uinta Mtns.; mostly on basic substrates, aspen, fir, and
spruce communities; June-July.

Sambucus L. — Elderberry

Shrubs; leaves opposite, pinnately or bipinnately compound, the leaflets
serrate; flowers small and numerous, usually in compound umbel-like cymes;
white to cream, rotate or nearly obsolete; sepals inconspicuous; corolla nearly
so, the tube short, the limb horizontally flaring; stamens (4) 5; style 5 stigmas;
fruit a juicy drupe.

---

1 Fruit glaucous blue, blackish beneath the bloom; inflorescence flat-topped,
   often over 15 cm wide; plants 2-5 m tall, mostly growing below 8,000 ft ..
   .... *S. caerulea*
1 Fruit red or yellowish; inflorescence rounded or pyramidal, to about 8 cm
   wide; plants 0.5-2.5 m tall, mostly growing above 8,000 ft .... *S. racemosa*

Sambucus cerulea Rafinesque [the specific name has often been spelled *S.
caerulea* or *S. caerulea.* The original spelling of Rafinesque is followed here; *S.
nigra* ssp. *cerulea* (Rafinesque) R. Bolli] — blue elderberry — Widespread but
apparently quite limited on the E. Tavaputs Plateau; infrequent in sagebrush,
pinyon-juniper, mt. brush, and ponderosa pine communities; 7,000-8,000
(8,800) ft; June-Aug.

Sambucus racemosa L. (*S. melanocarpa* A. Gray; *S. microbothrys* Rydberg; *S.
pubens* Michaux) — red elderberry — Common to locally abundant toward the
west end of the W. Tavaputs Plateau, more scattered and occasional in the
Uinta Mtns often in aspen openings and adventive on coniferous forest areas
that have been opened up by fire or timber harvest; 8,000-10,500 ft; June-Aug.
Our plants belong to var. *racemosa* [var. *microbothrys* (Rydberg) Kearney &
Peebles] based on the red or yellowish fruits and mostly glabrous leaves. Apparently var. *melanocarpa* (A. Gray) McMinn with black fruits does not enter
the Uinta Basin.

Symphoricarpos Duhamel — Snowberry

Shrubs with exfoliating bark; leaves simple, opposite, entire to lobed, glabrous
or pubescent; flowers small, pink or white, in axillary or terminal clusters, or
solitary in upper axils; calyx teeth 4 or 5; corolla regular, campanulate,
funnelform or salverform, 4-5 lobed; stamens 4-5; ovary 4-celled; style 1; fruit
a 2-seeded berry-like drupe.

---

1 Corolla tube glabrous inside, 11-13 mm long, salverform, the lobes 1/5-1/3
   as long as the tube .... *S. longiflorus*

1 Corolla tube pubescent inside, 7-13 mm long, wider than above, campanulate or funnelform, the lobes at least 1/3 as long as the tube ... 2

2 Corolla short-campanulate, the lobes about as long as the tube or longer; leaves 3-10 cm long; petioles 4-10 mm long; plants mostly below 7,000 ft .
.... *S. occidentalis*

2 Corolla tubular-funnelform, the lobes 1/3-1/2 as long as the tube; leaves commonly 1-3 cm long, those of sterile shoots sometimes larger; petioles mostly less than 5 mm long; plants above 6,500 ft .... *S. oreophilus*

Symphoricarpos longiflorus A. Gray — long-flower snowberry — The few specimens seen are from 4-5 mi s. of Manila in the Sheep Creek area at the n. flank of the Uinta Mtns. and along the rocky canyons of the Green River from Red Canyon to near Split Mt.; rocky places, Welsh 426 is from Harpers Corner at 7,400 ft; May-June.

Symphoricarpos occidentalis Hooker — western snowberry — Apparently widespread but scattered; forming clones along ditches, streams and on floodplains of rivers; 5,000-7,000 ft; May-July. Flowers and others (1960) reference to *S. albus* (L.) Blake is based on specimens (UT!) belonging here.

Symphoricarpos oreophilus A. Gray (*S. tetonensis* A. Nelson; *S. utahensis* Rydberg; *S. vaccinioides* Rydberg; *S. rotundifolius* A. Gray misapplied) — mtn. snowberry — Widespread; common to abundant in mtn. big sagebrush, pinyon-juniper, mt. brush, aspen, and fir communities; 7,000-9,500 ft; May-July. Our plants are more or less referable to var. *utahensis* (Rydberg) A. Nelson.

CARYOPHYLLACEAE Pink Family

Annual or perennial herbs, sometimes suffrutescent; stems mostly with swollen nodes; leaves opposite, simple, entire, often linear to lanceolate; flowers mostly cymose, sometimes solitary, terminal, or axillary, complete or apetalous, usually bisexual, usually 5 (4)-merous; petals white or sometimes pinkish or reddish; stamens equal to or twice as many as the petals; fruit a utricle or capsule, the capsule sessile or elevated on a stalk (carpophore).

1 Leaves with prominent scarious stipules; fruit various; stamens 5 or 10 .. 2

1 Leaves without scarious stipules; fruit a many-seeded, 1-chambered capsule with free central placentation; stamens 10 3

2 Plants perennial, pulvinate caespitose, not glandular; flowers sessile, included among the dense leaves; leaves 3-6 mm long, the scarious stipules 1/2 to as long as the leaves; petals lacking or minute; fruit a 1-seeded utricle; stamens 5; style 1 .... **Paronychia**

2 Plants annual or winter annual, often glandular; flowers not sessile, about equaling or exceeding the leaves; leaves 5-25 (40) mm long; stipules mostly less than 1/2 as long as the leaves; petals about equal or a little shorter than the sepals; fruit a capsule; stamens various; styles 3 .... **Spergularia**

3 Calyx united, the tube equal or longer than the lobes. 4

3 Sepals separate to the base or nearly so 9

4 Calyx 2-10 mm long. 5

4 Calyx 10-25 mm long. 6

5 Calyx 4-10 mm long; petals 6-12 mm long .... *Silene*

5 Calyx about 2 mm long; petals 2-3.5 mm long .... *Gypsophila*

6 Calyx, pedicels, and upper parts of stems with glandular hairs ..... *Silene*

6 Plants mostly glabrous 7

7 Calyx urn-shaped; flowers in open cymes; plants annual or perennial ....
.... *Vaccaria*

7 Calyx not urn-shaped; flowers in densely crowded clusters, sessile or nearly so (the pedicels to 6 mm); plants perennial........................ 8

8 Each flower subtended by a pair of linear, elongate bracts, these equal or exceed the calyx; leaves 1-2 cm wide, the pairs 4-10 per stem and connate for 2-4 mm; calyx about 40-nerved .... *Dianthus*

8 Flowers not closely subtended by bracts, the bracts much shorter than the calyx; leaves 1.5-4 cm wide, the pairs up to 20 per stem, not connate; calyx about 20-nerved .... *Saponaria*

9 Leaves less than 2 mm wide, needle-like 10

9 Some leaves usually 2 mm wide or wider, or at least not needle-like.... 14

10 Sepals 1.5-2.5 mm long .... *Sagina*

10 Sepals 3-9 mm long 11

11 Leaves 1.5-10 (15) mm long 12

11 Some leaves usually over 15 mm long .... *Eremogone*

12 Plants glabrous or hirtellous but not glandular...................... 13

12 Plants glandular .... *Minuartia*

13 Sepals 5-8 mm long; plants seldom found above the pinyon-juniper belt ..
.... *Eremogone hookeri*

13 Sepals 3-5 (5.5 in fruit) mm long; plants subalpine and alpine ....
.... *Minuartia macrantha*

14 Petals with 3 or more teeth at the apex .... *Holosteum*

14 Petals entire or 2-lobed 15

15 Petals entire; stems retrosely pubescent with non glandular hairs 16

15 Petals shallowly to deeply bilobed 17

16 Petals 2 or more times longer than sepals .... *Moehringia*

16 Petals shorter than or equal to the sepals .... *Arenaria*

17 Stems glabrous or pubescent but not glandular .... *Stellaria*

17 Stems usually with glandular hairs at least above.................... 18

18 Styles mostly 5; capsules curved near the tip, opening by 10 usually revolute-margined teeth .... *Cerastium*

18 Styles mostly 3; capsules not curved, opening by 6 (8) teeth ....
.... *Pseudostellaria*

Arenaria L. — Sandwort

With features of the family and as listed in the key.

Arenaria lanuginosa (Michaux) Rohrbach — sprawling sandwort — Perennial herbs from a branched caudex, the caudex sometimes becoming rhizomes or stolons; stems erect when small and becoming procumbent and unable to support themselves when elongated; herbage puberulent, the stems retrosely so; leaves all cauline, ciliate on margins and midvein beneath, sessile, 5-23(28) mm long, (1) 1.5-5 (9) mm wide; flowers solitary or in few-flowered cymes in the widely branched upper ½ of the plant; petals white, 2-5 mm long, usually shorter than the sepals, sometimes lacking, obtuse or rounded at the apex; stamens 10; styles 3; fruit a capsule opening by 6 apical teeth. The one specimen seen (Goodrich 23398) is from a burned ponderosa pine community in Uinta Canyon at 7,900 ft. Our plants belong to var. *saxicola* (A. Gray) Zarucchi, R. L. Hartman & Rabeler.

Cerastium L. — Chickweed

With features of the family and as listed in the key.

1　Petals distinctly longer than the sepals, (5.5) 6-12 mm long 2

1　Petals shorter than to slightly longer than the sepals, 4-6 (7) mm long. . . 3

2　Middle and lower primary leaves of the flowering stems often with fascicles of secondary leaves in the axils; plants with creeping rhizomes, commonly from below 9,200 ft . . . . *C. arvense*

2　Axillary fascicles of leaves lacking or present in lower nodes only; plants rarely with rhizomes, commonly from above 9,200 ft . . . . *C. beeringianum*

3　Plants perennial, usually decumbent or sprawling and forming loose mats, often rooting at the lower nodes . . . . *C. fontanum*

3　Plants annual; stems ascending to erect, not matted, not rooting at the nodes . . . . *C. brachypodum*

Cerastium arvense L. (*C. oreophilum* Greene) — field chickweed, starry chickweed — Specimens seen are from Strawberry Valley, Uinta Mtns., Blue Mt., and Yampa Plateau; occasional in sagebrush, pinyon-juniper, riparian, and ponderosa, pine communities, and rock outcrops and locally abundant in dry meadows; 7,000-9,000 ft; June. Our plant belong to the variable var. *strictum* (Gaudin) W. D. J. Koch.

Cerastium beeringianum C. & S. (*C. buffumae* A. Nelson) — Bering chickweed, alpine chickweed — Scattered or locally common across the Uinta Mtns.; around seeps and springs, along streams, and in moist meadows; 9,200-12,800 ft; June-Aug. Our plants belong to ssp. *earlei* (Rydberg) Hulten.

Cerastium brachypodum (Engelmann ex A. Gray) B. L. Robinson [*C. nutans* Rafinesque misapplied] — nodding chickweed — The 3 specimens seen are from Matt Warner Reservoir on Diamond Mt., Meadow Park in Daggett Co., and Christmas Meadows in Summit Co. from riparian communities and ephemeral pools; 7,600-8,800 ft; June-July.

Cerastium fontanum Baumgarten (*C. vulgatum* L.) — mouse-ear chickweed — Introduced from Europe; the 20 specimens seen are from across the Uinta Mtns from riparian communities, wet and dry meadows, campgrounds, lodgepole pine and ponderosa pine woods; 7,120-9,000 ft; June-Aug. Our plants belong to var. *vulgare* (Hartman) M. B. Wyse Jacks.

Dianthus L. — Pink; Carnation

Dianthus barbatus L. — sweet William — Perennial herbs, 20-60 cm tall; basal leaves several, oblong-lanceolate, 1-2 cm wide; cauline leaves lanceolate, mostly 4-9 cm long; calyx about 15 mm long; petals white, pink, or red, the narrow claws about equaling the calyx, the expanded blade 6-10 mm long, erose-dentate; capsule on a 3-4 mm long carpophore. Native of the Old World, planted (especially about older homes), persisting and occasionally escaping. No specimens seen.

Eremogone Fenzl — Sandwort

With features of the family and as listed in the key. Retained in Arenaria in TPD as of 2014 and AUF (102).

---

1   Leaves 2-5.5 (6.5) mm long; plants 1-4 (7) cm tall . . . . *A. hookeri*

1   At least some leaves usually over 6 mm long; plants (5) 8-25 (30) cm tall . 2

2   Flowers in head-like clusters . . . . *E. congesta*

2   Flowers in open cymes 3

3   Sepals 2.2-5 (5.4) mm long . . . . *E. kingii*

3   Sepals (4) 4.5-8 cm long 4

4   Short-shoot leaves 0.4-1.8 cm long, rigid and pungent; sepals narrowly acute to acuminated; styles 1-2 mm long . . . . *E. eastwoodiae*

4   Short-shoot leaves 2.5-5 cm long, relatively soft and flexible; sepals obtuse to acute; styles 3.4-6 mm long . . . . *E. loisiae*

Eremogone congesta (Nuttall) Ikonnikov (*Arenaria burkei* Howell; *A. congesta* Nuttall in T. & G.) — ballhead sandwort — Strawberry Valley, Uinta Mtns. and E. Tavaputs Plateau; occasional or common in sagebrush, pinyon-juniper, grass-forb, dry meadow, mt. brush, ponderosa pine, aspen, Douglas-fir, open pine-spruce, and alpine tundra communities; 7,000-11,500 ft; June-Aug. Our plants belong to var. *congesta.*

Eremogone eastwoodiae (Rydberg) Ikonnikov [*Arenaria fendleri* var. *eastwoodiae* (Rydberg) S. L. Welsh] — Eastwood sandwort — Widespread; occasional to common in desert shrub, sagebrush, pinyon-juniper, and ponderosa pine communities, often in rocky places; 5,000-7,500 (8,500) ft; May-June. Most of our plants are referable to var. *eastwoodiae.*

Eremogone hookeri (Nuttall) W. A. Weber (*Arenaria hookeri* Nuttall) — Hooker sandwort — Widespread; occasional and locally common in sagebrush, pinyon-juniper, mt. brush, and windswept grass-forb communities, often on nearly barren ground; 5,700-8,400 (9,000) ft; late May-early July. Our plants belong to var. *hookeri.*

Eremogone kingii (S. Watson) Ikonnikov [*Arenaria kingii* (S. Watson) M. E. Jones; *A. uintahensis* A. Nelson; *A. kingii* var. *uintahensis* (A. Nelson) C. L. Hitchcock; *A. fendleri* var. *glabrescens* S. Watson] — King sandwort — var. *glabrescens* (Watson) Dorn is known from sw. Wyoming (Dorn 2001) and n. Utah (IMF 2A:432) and perhaps in our area.

Eremogone loisiae N. H. Holmgren & P. K. Holmgren — Lois sandwort — Uinta Co., Wyoming and n. Utah but not listed for the Uinta Basin (IMF 2: 432). Perhaps to be found in our area.

Gypsophila L. — Babysbreath

1	Inflorescence glandular	*G. scorzonerifolia*
1	Inflorescence glabrous	*G. paniculata*

Gypsophila paniculata L. — babysbreath — Seen in abundance in the town of Dinosaur, Colorado in 2012 (Goodrich 28425). This is available at nurseries, and it is a common component of floral arrangements. This species is likely to be found elsewhere in our area.

Gypsophila scorzonerifolia Seringe in DG. — pink babysbreath — Introduced from Eurasia, specimens seen are from Lapoint and Vernal. One specimen seen was from a roadside population of many plants in Vernal where search in later years indicates the species did not persist. Repeated escapes of the species can be expected.

Holosteum L.

Holosteum umbellatum L. — holosteum — Plants annual; basal leaves petiolate; cauline leaves sessile and more or less connate, 6-20 mm long, glandular ciliate, entire; inflorescence umbellate with 3-10 flowers; pedicels erect at flowering and soon deflexed; sepals 3.8-5.3 mm long, lanceolate, with hyaline margins; petals about equal to or longer than the sepals, white or pinkish; capsules cylindrical, longer than the sepals, with 6, reflexed teeth. The one specimen seen (Goodrich 27356) is from Bare Top Mtn. in Daggett Co. from 7,600 ft elevation, from a mtn. big sagebrush-grass community. To be expected elsewhere.

Minuartia L. — Sandwort

Annual and perennial forbs from taproots and branched caudices and/or rhizomes, sometimes mat forming; leaves linear or subulate (in our species); petals entire or nearly so, about equal to or a little longer than the sepals; stamens 10; styles 3 (4); fruit a capsule with 1-25 seeds.

1	Plants glabrous or hirtellous but not glandular	*M. macrantha*
1	Plants glandular	2
2	Stems brittle, generally shattering at the nodes; leaves mostly pungent, 3-nerved; capsules shorter than the acuminate to pungent sepals; seeds about 1.5 mm long, papillate in concentric rows	*M. nuttallii*
2	Stems not brittle, not shattering; leaves mostly obtuse or only slightly mucronate,1-3 nerved; capsules often exceeding the obtuse to acute sepals; seeds mostly less than 1 mm long, lightly reticulate to tuberculate	2
3	Sepals 4-5 mm long, the tip mostly obtuse, more or less erose, slightly incurved and hooded, purplish; stems sprawling, clothed with current and

marcescent leaves, with 1 (rarely 2-3) flowers; plant more mat-forming than cushion-forming, from a rather thick taproot . . . . **M. obtusiloba**

3   Sepals 3-4 mm long, obtuse to acute, not erose, not hooded; stems more upright,-usually not clothed with leaves, with (1) 2-5 (7) flowers; plants usually forming small cushions, from a slender taproot . . . . **M. rubella**

Minuartia macrantha (Rydberg) House [misapplied names include: *Arenaria rossii* R. Brown in Richards, *Minuartia rossii* (R. Br. in Richards) Graebner. With 2 vars.:

1   Petals 3-6.5 mm long, longer than the sepals . . . . **var. macrantha**
1   Petals 2.5-3.4 (3.7) mm long, shorter than the sepals . . . . **var. filiorum**

Var. macrantha [*Arenaria macrantha* (Rydberg) A. Nelson ex Coulter & Rose — large-flower stichwort — Two specimens reported for talus slopes of the Uinta Mtns. in Daggett and Uintah Cos. (AUF 104).

Var. filiorum (Maguire) N. H. Holmgren & P. K. Holmgren [*Arenaria filiorum* Maguire; *A. rubella* var. *filiorum* (Maguire) S. L. Welsh; *Minuartia filiorum* (Maguire) McNeill] — threadbranch stichwort — The 12 specimens seen are from the Uinta Mtns. from Blind Stream-S. Fork Rock Creek area, Dry Ridge, and Buck Ridge in Duchesne Co., Brush Creek Cave in Uintah Co., Blacks Fork-Smiths Fork area of the north slope in Summit Co., and Birch Creek in Daggett Co.; distribution of this var. demonstrates a strong affinity for limestone; Douglas fir, spruce-fir (often krummholz), and open alpine tundra communities; (8,640) 10,500-11,100 ft; July-Aug. This var. is included in *A. rubella* in AUF (105). Collections from Buck Ridge include glabrous plants (A. Huber 4882) and glandular pubescent plants (A. Huber 4886). Other collections from our area also indicate *M. macrantha* and *M. rubella* are not geographically separated. Perhaps inclusion in *M. rubella* is as plausible as inclusion in *M. macrantha*.

Minuartia nuttallii (Pax) Briquet (*Arenaria nuttallii* Pax) — Nuttall sandwort — Three specimens seen (Peterson 83305 from Lookout Mt., Moffat Co.; 7,800 ft,; Goodrich & Atwood 16165 from Rock Creek-Brown Duck divide, Uinta Mtns.; fine limestone talus at 11,300 ft; Huber 1892 from Rock Creek Peak w. of Moon Lake on limestone ridges at 11,200 ft); June-Aug.

Minuartia obtusiloba (Rydberg) House [*Arenaria obtusiloba* (Rydberg) Fernald; *A. sajanensis* Willdenow ex Schlechtendal misapplied; *Lidia obtusiloba* (Rydberg) Love & Love] — arctic sandwort — Occasional to common across the Uinta Mtns. with minor presence in many of the alpine plant communities listed by Brown (2006) including fellfield and snowbed communities but lacking or rare in moist or wet meadow communities; 10,500-12,000 ft; July-Aug.

Minuartia rubella (Wahlenberg) Graebner in Ascherson & Graebner [*Arenaria rubella* (Wahlenberg) Smith, *A. propinqua* Richardson] — boreal sandwort, reddish sandwort — Occasional across the Uinta Mtns.; mostly on rocky alpine tundra or meadows and talus; 10,000-11,500 ft but occasionally down to 7,600 ft in ledges and rock outcrops and windswept ridges; June-Aug. Goodrich 28237 from the n. flank of the Uinta Mtns, with glabrous leaves and stipitate glandular upper stems and sepals is here included in *M. rubella*.

Moehringia L.

Moehringia lateriflora (L.) Fenzl (*Arenaria lateriflora* L.) — bluntleaf sandwort — Seldom collected but widespread in the Uinta Mtns. along streams, and in moist woods and meadows; 7,200-8,620 ft; June-Aug.

Paronychia Miller — Nailwort; Whitlow-wort

Perennial, densely caespitose, low herbs from woody caudices; aerial stems short or essentially lacking, often hidden in the crowded leaves, hirtellous; leaves sessile, crowded, 3-6 mm long, puberulent, with scarious stipules about as long as the leaves; flowers terminal and solitary or rarely paired, small and often inconspicuous, with scarious bracts; sepals 5, concave or hooded at the awn-tipped apex; petals lacking or minute; stamens 5; styles partly united; fruit an ovoid or globose utricle included in the persistent calyx.

1   Leaves not linear or seldom so, nerveless; plants at or above timberline on
    the Uinta Mtns.; flowers more or less hidden in the leaves, sometimes sessile
    or nearly so in the basal rosettes .... *P. pulvinata*
1   Leaves linear, often with a prominent midrib; plants from well below
    timberline; flowers usually borne on definite but short stems *P. sessiliflora*

Paronychia pulvinata A. Gray — Rocky Mtn. nailwort — Occasional; Uinta Mtns., at or above timberline, most common on convex slopes and ridges where wind keeps snow depth less than 1 ft deep for much of the winter where it is associated with other cushion plants and curly sedge, less common at margins of snowbeds; 10,800-13,000 ft; July-Aug.; 11,100-13,000 ft; Jul-Aug.

Paronychia sessiliflora Nuttall — creeping nailwort — Occasional; widespread; desert shrub, sagebrush, pinyon-juniper, and mt. brush communities, and exposed ridges and slopes, often on rocky ground or semi-barrens; 5,800-8,400 ft; July-Aug.

Pseudostellaria Pax — Sticky Starwort

Pseudostellaria jamesiana (Torrey) W. A. Weber & R. L. Hartman [*Alsine curtisii* Rydberg; *A. jamesiana* (Torrey) A. Heller; *Stellaria jamesiana* Torrey]. — sticky starwort — Widespread in the Uinta Mtns., occasional to common and locally abundant following fire; apparently less common and more restricted on the Tavaputs Plateau; sagebrush, pinyon-juniper, mt. brush, aspen, fir, lodgepole pine, and Engelmann spruce communities; 6,800-10,600 ft; June-Aug. Sticky starwort is sometimes abundant following fire. This is likely a function of high percent survival of tuber-like roots.

Sagina L. — Pearlwort

Sagina saginoides (L.) Britton — arctic pearlwort — Biennial or tufted to matted perennial, sometimes flowering the first season and appearing annual; stems numerous, 0.3-5 cm long, ascending to procumbent; basal leaves 3-20 mm long, 0.2-0.5 mm wide; stem leaves 3-6 (10) mm long, occasionally with secondary leaves in the axils; flowers usually solitary and terminal, a second flower occasionally in the upper leaf axil, on filiform pedicels 0.3-2.5 cm long;

sepals 1.5-2.5 mm long; petals lacking or about equal to the sepals, white; capsule 3.5-5 mm long, about twice as long as sepals at maturity. Specimens seen are from Strawberry Valley and the Uinta Mtns.; along streams, around seeps and springs, in woods, meadows, and alpine tundra; 8,100-13,000 ft; June-Aug.

Saponaria L. — Soapwort

Saponaria officinalis L. — bouncing-bet — Rhizomatous perennial herbs, 30-90 cm tall; stems simple; leaves lanceolate to oblanceolate, mostly 4-10 (12) cm long; calyx 15-20 mm long at anthesis, up to 25 mm long in fruit, often deeply cleft in one or more places; petals white to pink, the narrow claw slightly exceeding the calyx, the blade 10-15 mm long, shallowly retuse; capsule on a short carpophore. Introduced from Europe, rather showy, widely planted as an ornamental (especially about older homes), persisting and occasionally escaping. Two specimens seen from abandoned homesteads at Dry Fork Settlement, and Whiterocks, another from disturbed ground in a rabbitbrush community in the mouth of Uintah Canyon, also grown as an ornamental in Vernal where mostly at older homes; not expected much above 7,500 ft; July-Sept.

Silene L. — Campion; Wild Pink; Silene

With features of the family and as listed in the key.

---

1 Calyx 4.5-11 mm long 2
1 Calyx 11-17 mm long. 4
2 Plants pulvinate caespitose perennials, forming cushions or mats, with branched caudices and woody taproots, 2-5 (10) cm tall; leaves linear, 0.4-1 (2) cm long, 1-2 mm wide; flowers solitary, often unisexual, the petals lavender, pink, to (rarely) white; calyx often purplish . . . . *S. acaulis*
2 Plants annual, or if perennial then from rhizomes, 5-50 (80) cm tall; leaves (0.3) 2-6 (10) cm-long, 2-25 mm wide; flowers seldom solitary, bisexual, the petals white, pink, or purple 3
3 Plants annual, 5-50 (80) cm tall; stems erect, simple below; calyx tube glabrous; petals white, pink, or purple . . . . *S. antirrhina*
3 Plants perennial from rhizomes, 5-20 (40) cm tall; stems decumbent, often branched below; calyx tube glandular; petals white . . . . *S. menziesii*
4 Plants 20-60 cm tall; inflorescence with more than 1 flower *S. drummondii*
4 Plants 5-15 (20) cm tall, inflorescence a single terminal flower 5
5 Leaves mostly glabrous except for the ciliate margins; seeds broadly winged, 1.2-1.7 mm wide; anthers 0.6-07 mm long . . . . *S. hitchguirei*
5 Basal leaves retrosely purberulent at least above and ciliate; seeds not winged, 0.8-1.1 mm wide; anthers 1-1.2 mm long . . . . *S. kingii*

Silene acaulis L. — moss campion — Uinta Mtns.; common to abundant on alpine, convex slopes and ridges where wind keeps snow depth less than 1 ft deep for much of the winter where it is associated with other cushion plants and curly sedge, less common at margins of snowbeds; 10,800-13,000 ft; July-Aug.

Silene antirrhina L. — annual catchfly — Infrequent or at least seldom collected, apparently widespread, mtn. brush, sagebrush, and pinyon-juniper communities at 5,700-7,870 ft; June.

Silene drummondii Hooker [*Lychnis drummondii* (Hooker) S. Watson] — Drummond campion — Widespread and frequent but of minor cover in Vasey sagebrush, mt. brush, ponderosa pine, Douglas-fir, aspen, lodgepole pine, Engelmann spruce, and meadow communities at 7,200-11,200 ft; July-Aug. Rather frequently found in study plots on the Ashley National Forest where often represented by only 1 or 2 plants in 1/10 acre plots.

Silene hitchguirei Bocquet [*Lychnis montana* Watson; *L. apetala* var. *montana* (S. Watson) C. L. Hitchcock; *Silene uralensis* (Rupreht) Bocquet ssp. *montana* (S. Watson) McNeill] — mtn. campion — Across much of the alpine of the Uinta Mtns. Specimens at USUUB indicate *S. hitchguirei* is much less common than *S. kingii,* and more likely to be found on talus slopes and in snowbeds than in fellfields. *Silene uralensis* is listed for Utah in FNA. This species is not listed for our area in IMF (2A: 452) where more study is recommended to determine the relationship of *S. hitchguirei* and *S. kingii* to *S. uralensis.*

Silene kingii (S. Watson) Bocquet [*Lychnis kingii* S. Watson; *L. apetala* var. *kingii* (S. Watson) S. L. Welsh] — King campion, King catchfly — Across much of the alpine of the Uinta Mtns. where most commonly associated with curly sedge and other wind-swept communities. Concerning *S. hitchguirei* and *S. kingii,* Welsh (AUF: 109) noted that pubescence and seed differences seem to be the most useful features to separate the two with features of flower buds and petal length of questionable value, but pubescence varies considerably and the two not only occur together within populations but are occasionally found mounted on the same herbarium sheet.

Silene menziesii Hooker — Douglas-fir campion — Widespread; infrequent or occasional in mt. brush, aspen, and Douglas-fir communities and open rocky slopes; 7,000-10,400 ft; June-Aug.

Spergularia (Persoon) J. & C. Presl — Sandspurry

Annual to perennial herbs, glabrous to stipitate-glandular; leaves linear, mucronate (in ours); flowers in open leafy-bracteate terminal cymes.

1 Leaves mostly fascicled in the axils; plants mostly above 7,000 ft; stamens 9-10; stems decumbent to erect .... ***S. rubra***

1 Leaves not fascicled or occasionally some nodes with 1 or 2 axillary leaves; plants of alkaline areas below 7,000 ft; stamens 2-5; stems prostrate .....
   .... ***S. marina***

Spergularia marina (L.) Grisebach — salt sandspurry — Introduced from Europe or possibly native; the 3 specimens seen (B. Welsh & G. Moore 219; England 315 UI; Goodrich 27337) are from near Jensen and Leota Bottoms near Ouray, sandbars of the Green River, and draw-down basin of Flaming Gorge Reservoir; to 6040 ft; Aug-Sept.

Spergularia rubra (L.) J. & C. Presl — red sandspurry — Introduced from Europe; specimens seen are from Strawberry Valley, Uinta Mtns, and Cold

Spring Mtn.; in roads along roadsides and in recently burned coniferous forest communities; 7,800-9,900 ft; July-Sept.

Stellaria L. — Starwort; Chickweed

Stems 4-angled but this feature obscured in pressed specimens, otherwise with features of the family and as listed in the key. *Moehringia lateriflora* and *Arenaria lanuginosa* are similar in habit and habitat to some of the taxa listed below. However, they differ by retrorse hairs on at least part of the stem.

---

1   At least the lower leaves with petioles; leaf blades 2-15 mm long. 2

1   Leaves sessile 3

2   Flowers solitary in axils of leaves; sepals glabrous; petioles to 2 mm long .
. . . . *S. obtusa*

2   Flowers in terminal cymes with few or many flowers; sepals pubescent; petioles 2-12 mm long . . . . *S. media*

3   Petals equal to or longer than the sepals, always present 4

3   Petals shorter than the sepals, sometimes lacking 6

4   Sepals 3.5-6 mm long; petals (3) 3.5-7 mm long, the cleft about ½ or less the length of the petal; capsules 4-7 mm long, dark purple . . . . *S. longipes*

4   Sepals 2-3.6 (4) mm long; petals 2-5 mm long, the cleft ¾ or more the length of the petal; capsules 2.5-4 mm long. 5

5   Leaves at midstem 15-55 mm long; stem-angles and leaf margins papillate-scabrous (at 10x); inflorescence terminal with 2-several flowers; stems 10-55 cm long, not matted . . . . *S. longifolia*

5   Leaves at midstem 5.5-15 mm long; herbage glabrous; at least some of the flowers solitary in axils of leaves; stems 3-13 cm long, more or less matted
. . . . *S. crassifolia*

6   Bracts of inflorescence 1-2.5 mm long, scarious; styles 0.15-0.25 mm long; inflorescence umbellate . . . . *S. umbellata*

6   Bracts of inflorescence 5-20 mm long, wholly herbaceous; styles 0.7-2 mm long; flowers solitary in axils of leaves and in few-flowered terminal cymes
. . . . 7

7   Mature capsules 3.4-6.5 mm long, about twice as long as wide; leaves 10-58 mm long; petals (0) 1.5-23 mm long . . . . *S. borealis*

7   Mature capsules 2.7-4.2 mm long, about as wide as long; leaves 6-20 (26) mm long; petals 0-1.5 mm long . . . . *S. calycantha*

Stellaria borealis Bigelow [*Alsine borealis* (Bigelow) Britton] — boreal starwort — Both vars. *borealis* and *sitchana* (Steudel) Fernald [*S. calycantha* var. *sitchana* (Steudel) Fernald; *S. calycantha* var. *bongardiana* (Fernald) Fernald] are listed for northern Utah in IMF (2A: 412). Specimens from our area that have been identified as *S. crispa* C. & S. [*Alsine crispa* (C. & S.) Holt.] most likely belong here. Herbage glabrous or finely papillate on stem ridges and leaf margins, and leaves sometimes ciliate at the base.

Stellaria calycantha (Ledebour) Bongard — northern starwort — Widespread across the Uinta Mtns., but no specimens seen east of Whiterocks Canyon on the s. slope or east of Sheep Creek Park on the n. slope; along streams, around seeps and springs, and in peat bogs, in various plant communities; 7,100-10,300 ft; June-Aug. Herbage glabrous or pilose with the hairs concentrated at the nodes and leaf bases.

Stellaria crassifolia Ehrhart — thick-leaved starwort — The 6 specimens seen are from the n. slope of the Uinta Mtns. from wet meadows and about springs in the Sheep Creek drainage at 8,650-9000 ft. and north side of Bald Mtn., Smiths Fork drainage at 11,050-11,110 ft. Herbage glabrous.

Stellaria longifolia Muhlenberg ex Willdenow — long-leaved starwort — The 4 specimens seen are from the s. slope of the Uinta Mtns. at 7,360-10,560 ft. Herbage appearing glabrous, but the stems angles and leaf margins often papillate-scabrid.

Stellaria longipes Goldie [*S. longipes* var. *altocaulis* (Hulten) C. L. Hitchcock; *S. longipes* var. *monantha* (Hulten) S. L. Welsh; *Alsine laeta* (Richards) Rydberg; *A. longipes* (Goldie) Coville] — longstalk starwort — Strawberry Valley, Uinta Mtns., and Yampa Plateau; streambanks, wet meadows, margins of ponds and lakes, and on well-drained rocky slopes; (6,000) 7,000-11,800 ft; June-Aug. Herbage glabrous or leaves sometimes ciliate near the base.

Stellaria media (L.) Villars (*Alsine media* L.) — common chickweed — Listed for Duchesne Co. in AUF (116). Stems with single line of fine, woolly hairs along each internode; the petioles often ciliate, the blades usually glabrous.

Stellaria obtusa Engelmann — blunt starwort — The one specimen seen (Huber 4896) is from a spring on Petty Mtn. at 10,170 ft. Listed for Uinta Mtns., Summit Co. in IMF (2A: 414). Herbage glabrous or the leaves sometimes ciliate on the winged petioles and base of the blade.

Stellaria umbellata Turczaninow ex Karel & Kirilov (*Alsine baicalensis* Coville) — umbellate starwort — Strawberry Valley and Uinta Mtns.; streambanks, around springs, margins of lakes, and rocky ground, in aspen, fir, lodgepole pine, and spruce communities and perhaps above timberline; 9,000-11,400 ft; July-Aug. Graham 6544, 6559, and 8447 (CM!), referred to as *Alsine longifolia* (Muhlenberg) Britton (*Stellaria longifolia* Muhlenberg) belong here. Herbage glabrous; inflorescence an umbellate-cyme, with dichotomous or trichotomous branches, the branches spreading as much as 130°-180° from each other; pedicles recurved; sepals 2-3 mm long at anthesis.

Vaccaria Medicus — Cowcockle; Cowherb

Vaccaria hispanica (Miller) Rauschert [*V. pyramidata* Medicus; *V. segetalis* (Necker) Garcke ex Ascherson; *Saponaria vaccaria* L.] — cowcockle, cow soapwort — Annual glabrous herbs 15-80 cm tall; leaves 3-8 cm long, 5-40 mm wide, sessile, sometimes cordate-clasping; flowers usually numerous in an open flat-topped panicle composed of leafy-bracteate cymes; calyx tube greenish, inflated, keeled or winged on the nerves at fruiting; petals 15-20 mm long, usually exceeding the calyx by 3-10 mm, deep pink to red, clawed, retuse; styles 2 (3); capsules enclosed in the calyx. Introduced from Europe, more or less weedy, no specimens seen from the Uinta Basin but plotted for Rio Blanco Co. in TPD.

CELASTRACEAE Staff-tree Family

Shrubs; leaves simple; flowers regular, usually bisexual, small, usually inconspicuous; calyx deeply 4-5 parted; petals 4-5, separate; stamens as many or twice as many as the petals, inserted on or below the margins of a disk; ovary superior, 2-5 celled; style short or lacking; stigma 2-5-lobed; fruit a capsule.

Paxistima Rafinesque
(orthographic variants: *Pachistima* and *Pachystima*)

Paxistima myrsinites (Pursh) Rafinesque — mtn. lover — Low shrubs with stems growing along the ground or arising to 50 cm tall; leaves evergreen, opposite, serrulate at least above the middle, 5-40 mm long, oval to oblanceolate, glossy above, paler beneath; flowers bisexual, sessile or nearly so in the axils of leaves, the sepals 4, less than 1 mm long, the petals 4, about twice as long as the sepals, deep red to brownish-red; capsules 4-5 mm long, 2-chambered, each chamber with 2 dark brown seeds. Widespread and common in the Uinta Mtns. and occasional or infrequent on the Tavaputs Plateau, mostly in aspen and coniferous forests, sometimes along streams or on open slopes, and marginally alpine; 7,000-10,400 ft; June-Aug. Mountain lover has been used as an indicator of a moist Douglas-fir habitat type in Utah (Mauk and Henderson 1984).

CERATOPHYLLACEAE Hornwort Family

Ceratophyllum L. — Hornwort

Ceratophyllum demersum L. — common hornwort — Submerged, aquatic, rootless perennial herbs; stems slender, 20-100 cm long; leaves verticillate, 5-12 in each whorl, dichotomously forked into linear or filiform segments, these often minutely toothed; flowers solitary in axils of leaves, unisexual, both the staminate and the pistillate with a perianth or involucre of 8-15 greenish segments; stamens 10-16, the filaments short; fruit a hardened achene, the achene with a persistent beak-like style, smooth or with -2 horn-like appendages at the base. No specimens seen, but to be expected.

CHENOPODIACEAE Goosefoot Family

Plants herbaceous or shrubby, annual or perennial; leaves alternate or opposite, sometimes succulent or fleshy; flowers borne in small glomerules in the axils of bracts or leaves, in spikes, panicles, or cymes, bisexual or unisexual, the perianth of a single set, regular or nearly so, lobed or parted with 5 (2-6) lobes or reduced to a single scale or sometimes lacking in pistillate flowers; stamens generally equal to the perianth lobes and opposite them; ovary superior or sometimes adnate to the perianth, 1-chambered, with l ovule; fruit an utricle, indehiscent or irregularly rupturing.

1 Plants perennial, more or less woody at least at the base. 2
1 Plants annual, herbaceous 11
2 Leaves glabrous or glaucous, fleshy-succulent, linear to subterete 3
2 Leaves either densely scurfy or otherwise pubescent, whitish or grayish, sometimes wider than above, succulent or not, but hardly fleshy 4

3 Shrubs more or less spiny, 0.3-3 m tall, woody throughout except for twigs
   of the season .... **Sarcobatus**

3 Plants not spiny, 0.1-0.8 m tall, woody only at the base .... **Suaeda**

4 Leaves and twigs glabrate, hirsute or sericeous to densely tomentose, not
   scurfy; leaves linear to nearly filiform; plants mostly suffrutescent
   subshrubs 5

4 Plants scurfy, the hairs inflated and collapsing when dry and leaving a
   grayish or whitish mealy coating on leaves and twigs; leaves linear or
   broader; plants suffrutescent subshrubs or shrubs with woody, rigid stems
   well above ground level 7

5 Twigs and leaves stellate-tomentose; leaves linear, slightly to strongly
   revolute, hardly terete, not subulate, the midnerve conspicuous beneath .
   .... **Krascheninnikovia**

5 Twigs not tomentose or if so not with stellate hairs; leaves glabrate,
   sericeous or hirsute, terete or nearly so or subulate, the midnerve
   sometimes not conspicuous.. 6

6 Current year's twigs thinly to densely tomentose; perianth 4-merous,
   wingless in fruit; leaves filiform-subulate, rigid, appearing pungent,
   fascicled, rather thinly hirsute with widely spreading hairs; inflorescence
   branched .... **Camphorosma**

6 Current year's twigs sericeous but not tomentose; perianth 5-merous,
   winged in fruit; leaves linear but not subulate, glabrate or sericeous, the
   hairs appressed or ascending but not widely spreading .... **Bassia**

7 Plants spiny; leaves oblanceolate to orbicular 8

7 Plants not spiny 9

8 Leaves scurfy pubescent when young with some of the hairs usually forked
   or stellate, glabrate or glabrous and green when mature, oblanceolate; bark
   of twigs exfoliating in long, whitish strips; bracts of fruit wholly united into
   a sac; anthers attached near the base, erect, parallel to each other . **Grayia**

8 Leaves permanently and densely gray-white scurfy with simple hairs,
   mostly obovate, ovate, or orbicular; bark not exfoliating in strips; bracts of
   fruit free toward the apex; anthers attached near the middle, the lower 1/2
   widely divergent .... **Atriplex confertifolia**

9 Leaves linear, 3-10 times longer than wide, or if broader then some of the
   lower leaves and/or branches opposite or sub opposite and stems not rigid;
   bracts of fruit dorsally compressed, sometimes tuberculate on the faces with
   wart-like appendages, margins of bracts free towards the apex, often
   toothed .... **Atriplex**

9 Leaves mostly less than 3 times as long as wide, alternate; stems rigid; bracts
   of fruit compressed or not, united to the apex, smooth on the faces or ribbed
   but not tuberculate, the margins entire 10

10 Fruiting bracts 5-12 mm long, 6-15 mm wide, glabrous at maturity,
   thickened and spongy within; leaves 5-30 mm long, 2-12 mm wide, scurfy
   when young, some or most of the hairs forked or branched, glabrate in age;
   bark of twigs exfoliating in long, whitish strips .... **Grayia**

10 Fruiting bracts 5-6 mm long, about 4-8 mm wide, scurfy at maturity, papery,
   not thickened; leaves 10-70 mm long, (4) 10-45 mm wide, scurfy even in age,
   the inflated hairs simple; bark of twigs seldom as above .... **Zuckia**

11 Leaves opposite, scale-like, not over 2 mm long, fleshy, embedded and barely discernible in the terete fleshy stem; some or most of the branches of the inflorescence opposite; flowers sunken in the axis of spikes; plants known from about Pelican Lake and near Manila . . . . *Salicornia*

11 Leaves alternate, some usually over 2 mm long, not embedded in the stem; branches of the inflorescence alternate, or if opposite then the flowers not sunken in the axis of spikes; plants of various distribution 12

12 Leaves linear, entire 13

12 Leaves neither terete nor linear, entire, toothed or lobed 20

13 Plants pilose, sericeous, or stellate at least in part 14

13 Plants glabrous or scurfy with inflated hairs that collapse upon drying and leave a grayish whitish mealy coating on the leaves and sometimes the stems 15

14 Pubescence of branched hairs at least in part; flowers mostly solitary in axils of scarious margined bracts; fruit 3.5-4.5 mm long, strongly flattened, the margins winged . . . . *Corispermum*

14 Pubescence of simple hairs, flowers various; fruit not flattened, smaller than above, the margins not winged . . . . *Bassia*

15 Plants dichotomously branched, rare in our area; the outer branches capillary or nearly so; leaves 2-6 (13) mm long, not terete*Monolepis pusilla*

15 Plants not branched as above, leaves mostly over 10 mm long or else terete 16

16 Plants scurfy, whitish or grayish, not especially succulent . . . . *Chenopodium leptophyllum*

16 Plants glabrous, green, succulent when young (except in *Corispermum*) . 17

17 Leaves 5-10 mm long, tipped with a single bristle-like mucronate hair; seed ringed by a 3-4 mm wide horizontal wing . . . . *Halogeton*

17 Leaves often over 10 mm long, not tipped with a bristle-like hair; seed various. 18

18 Leaves 3-6 cm long, filiform; plants much branched, succulent when young but becoming spiny tumbleweeds upon drying; seed ringed by a horizontal wing . . . . *Salsola*

18 Leaves sometimes shorter than above; plants not becoming spiny tumbleweeds; seed not ringed by a horizontal wing 19

19 Fruit strongly flattened, 3.5-4.5 mm long, about as wide, the margins winged, solitary in axils of scarious-margined bracts; plants of sandy ground . . . . *Corispermum*

19 Fruit not flattened, smaller than above, the margins not winged, solitary or in clusters; plants usually of moist or wet alkaline ground . . . . *Suaeda*

20 Perianth with a horizontal wing in fruit; plants villous-tomentose when young, glabrate in age . . . . *Cycloloma*

20 Perianth without a horizontal wing; plants not villous-tomentose 21

21 Plants greenish, 4-15 (30) cm tall; most of the leaves with a pair of hastate lobes, otherwise entire, 1-8 (11) mm wide; calyx with 1 (rarely 2-3) lobes; flowers and fruit not enclosed in bracts . . . . *Monolepis nuttalliana*

21 Plants scurfy and whitish or grayish at least in part, or if greenish then often taller than above or leaves not as above; calyx with 5 lobes or else the flowers and fruit enclosed in bracts.................................22

22 Flowers bisexual, the perianth regular, with (3-4) 5 lobes; leaves all alternate; seeds mostly brownish or blackish and shiny .... *Chenopodium*

22 Flowers unisexual, the pistillate naked or with greatly reduced perianth and subtended by and enveloped in 2 bracts, the bracts enlarging as the flowers and fruit mature; perianth of staminate flowers with 3-5 lobes; leaves alternate or opposite at some of the lower nodes; seeds not brownish, blackish, nor shiny ....*Atriplex*

Atriplex L. — Orach; Saltbush

Annual or perennial, monecious or dioecious herbs and shrubs, often scaly, scurfy, or mealy; leaves alternate or some of the lower ones opposite; flowers bisexual or unisexual, in axillary clusters, terminal spikes, or panicles; staminate flowers without bracts, the perianth, inconspicuous, 3- to 5-parted; stamens 3-5; pistillate flowers lacking a definite perianth, subtended by 2 bracts, these dorsally compressed and united at the base but free toward the apex; stigmas 2; fruit an utricle, the pericarp usually free; seeds erect or rarely horizontal. Vegetative features are used throughout the keys, but positive identification is not always possible without mature fruit.

1 Plants perennial shrubs or subshrubs; leaves entire...................2

1 Plants annual herbs; leaves entire or toothed 5

2 Fruiting bracts 4, conspicuously wing-like, 6-15 mm long, 4-8 (10) mm wide entire, undulate, or toothed; plants 40-200 cm tall, with erect and spreading, rigid stems and twigs; leaf blades mostly linear to narrowly oblong, 1.5-5 cm long, 2-8 mm wide ....*A. canescens*

2 Fruiting bracts 2, mostly smaller than above, sometimes rather inconspicuous; stems not rigid, or if so (as in *A. confertifolia*) then the leaves ovate to elliptic 3

3 Twigs rigid except for those of the current season, these becoming spine-like late in the season or in the second year; leaves 1-2 cm long, broadly ovate, obovate or elliptic, alternate; plants 20-100 cm tall; fruiting bracts 6-12 mm long, nearly as wide, the margins entire, the faces smooth ....
....*A. confertifolia*

3 Twigs not rigid, not turning into spines; leaves often linear or oblong, or if broader then some of the lower ones usually opposite or sub opposite; fruiting bracts smaller than above, the margins toothed, the faces smooth or tuberculate with wart-like appendages 4

4 Leaves 2-10 (15) mm long, 1-4 (5) mm wide, often densely crowded; plants more or less mat-forming, (2) 5-10 cm tall ....*A. corrugata*

4 Leaves 15-50 mm long, (2) 5-25 mm wide; plants sometimes with low spreading stems but hardly mat-forming, (15) 20-50 cm tall ...*A. gardneri*

5 Fruiting bracts truncate at the apex or the faces heavily set with wart-like tubercles, or the leaves sessile and entire; seeds all alike, either black or brown; plants native, mostly of dry alkaline flats and hills in desert shrub communities, sometimes along roads or other disturbed places, of various

stature but often 10-40 cm tall; leaves entire or at most wavy-toothed, hardly if at all hastate, mostly much smaller than in *A. hortensis,* mostly gray-scurfy, alternate or rarely the lower 2 opposite 6

5 Fruiting bracts ovate to orbicular, 3-12 mm wide, entire, and with smooth faces, or more or less triangular with the tips pointed and the faces smooth or slightly tuberculate; leaves not both sessile and entire; seeds of 2 types (black and brown); plants introduced, mostly of roadsides, ditch banks, fence lines, gardens, fields, and other disturbed areas, mostly 30-200 cm tall; leaves sinuate-dentate or mostly hastate (entire in *A. hortensis* but then the blades 5-20 cm long and 2-10 cm wide), more or less greenish and only slightly gray-scurfy except in *A. rosea* lower part of stems sometimes with 1-3 or more pairs of opposite leaves or branches. 11

6 Leaves (except sometimes the lower ones) linear, nearly filiform, about 1-2 (3) mm wide, seldom over 2 cm long . . . . *A. wolfii*

6 Leaves not linear, at least some over 2 mm wide 7

7 Leaves sessile, lanceolate or elliptic, rarely ovate, acute to acuminate, entire, mostly 1-3 cm long, 3-10 mm wide; plants low and spreading, 10-30 cm tall, rare at the e. edge of our area; small glomerules of staminate flowers borne above the pistillate flowers . . . . *A. suckleyi*

7 Leaves either petiolate or of different shape; plants various, either unisexual or with staminate and pistillate flowers mixed 8

8 Plants unisexual; leaves with 3 prominent veins; fruiting bracts blunt at the apex, the faces with tubercles . . . . *A. powellii*

8 Plants bisexual; leaves seldom with 3 prominent veins; fruiting bracts various. 9

9 Fruiting bracts of 2 kinds, those of the lower axils sessile to subsessile, truncate, without crested appendages, those of the upper axils long-stalked, sometimes reflexed, with crested or horny appendages toward the apex; leaves somewhat cordate . . . . *A. saccaria*

9 Fruiting bracts all alike; leaves not cordate 10

10 Fruiting bracts sessile, truncate, entire except for 3 minute teeth at the apex, the faces smooth . . . . *A. truncata*

10 Fruiting bracts sometimes stalked, dentate well below the apex, the faces sometimes with crested appendages . . . . *A. argentea*

11 Fruiting bracts ovate to orbicular, entire, smooth-faced 12

12 Fruiting bracts 6-12 mm wide, 1-veined at the base; leaf blades 5-20 cm long, 2-10 cm wide, ovate-triangular to broadly lanceolate; plants 50-200 cm tall . . . . *A. hortensis*

12 Fruiting bracts 3-5 mm wide, 5-veined at the base; leaf blades (1) 2-8 cm long, 0.5-6 cm wide, triangular-hastate; plants (20) 30-80 cm tall or taller . . . . *A. micrantha*

11 Fruiting bracts more or less triangular with the tips pointed, often toothed, the faces smooth or slightly tuberculate 13

13 Leaves mostly sinuate-dentate, alternate . . . . *A. rosea*

13 Leaves hardly if at all sinuate-dentate, at least the lower ones opposite . . . . *A. dioica*

Atriplex argentea Nuttall — silverscale — Widespread; occasional in desert shrub communities and floodplains with Sarcobatus; 4,680-6,600 ft perhaps higher; July-Oct.

Atriplex canescens (Pursh) Nuttall — fourwing saltbush — widespread; occasional to common in desert shrub, and pinyon-juniper communities, and locally abundant on wind-sand and in alluvial bottoms; up to about 7,500 (8,000) ft; June-Aug. Fourwing saltbush crosses with *A. confertifolia* and members of the *A. gardneri* complex. A population of an apparent hybrid is found in Antelope Canyon of the West Tavaputs Plateau (Goodrich 27059). Hybrid populations are more common in the Great Basin.

Atriplex confertifolia (Torrey & Fremont) S. Watson — shadscale — Widespread; common to dominant in desert shrub communities, occasional in Wyoming big sagebrush communities and extending upward into pinyon-juniper communities on the Tavaputs Plateau; seldom above 7,000 ft; May-June.

Atriplex corrugata S. Watson — mat saltbush — Common to dominant on clay of Mancos Shale, Duchesne River, and other formations that weather to badlands from 4,700-5,400 ft; April-early June. Some larger specimens with larger leaves are transitional to *A. gardneri* var. *cuneata.*

Atriplex dioica Rafinesque — thickleaf orach — The 5 specimens seen are from roadsides, ditches, and moist saline sites on the floodplains of the Strawberry River, Duchesne River, and Green River (Stewart Lake), and Tom Patterson Canyon of the East Tavaputs Plateau from 4,900-5,900 ft elevation.

Atriplex gardneri (Moquin-Tandon) Dietrich (*A. nuttallii* S. Watson). There are 3 more or less distinct vars.:

1 Leaves all alternate, linear or narrowly oblong or oblanceolate; fruiting bracts widest at or below the middle, the faces smooth; stems mostly erect
   . . . . **var. *utahensis***
1 Some of the lower leaves and branches usually opposite or subopposite; leaves wider than in var. utahensis or fruiting bracts widest at or above the middle; stems often decumbent; plants not restricted as above 2
2 Plants n. of the Uinta Mtns.; leaves usually 4-12 mm wide, greenish, sometimes widest above the middle; fruiting bracts 2-5 mm wide, the faces smooth or with tubercles less than 1 mm long . . . . **var. *gardneri***
2 Plants s. of the Uinta Mtns.; leaves 10-25 mm wide, gray-green, usually widest below the middle; fruiting bracts 5-9 mm wide, the faces heavily tuberculate with wart-like appendages, these sometimes over 1 mm long
   . . . . **var. *cuneata***

Var. *cuneata* A. Nelson [*A. cuneata* A. Nelson; *A. nuttallii* S. Watson var. *cuneata* (A. Nelson) Hall & Clements; *A. oblanceolata* Rydberg] — Castle Valley saltbush, Castle Valley clover — Common to dominant; desert shrub communities across the Uinta Basin, typically on valley fill washed in from silty or clayey formations that weather to badlands, but also on hills; 4,700-5,700 (6,000) ft; May-June. Perhaps intergrading into *A. corrugata* (q.v.).

Var. *gardneri* [*A. nuttallii* S. Watson var. *gardneri* (Moquin-Tandon) H. &-C.] — Gardner saltbush — Common to dominant; desert shrub communities n. of the Uinta Mtns.; up to about 6,200 ft; May-June. See Arnow and others (1980) for a discussion of the spelling of *A. gardneri.*

Var. *utahensis* (M. E. Jones) Dorn [*A. tridentata* Kuntze; *A. gardneri* var. *tridentata* (Kuntze) Macbride; *A. nuttallii* var. *utahensis* M. E. Jones; *A. nuttallii* S.

Watson var. *tridentata* (Kuntze) Hall & Clements] — basin saltbush, threetooth saltbush — Locally common on the floodplains of the Green River, White River and associated river systems, and at isolated locations elsewhere including a gravel pit between Lapoint and Maeser where apparently included in a reclamation planting; not expected over 6,500 ft; July-Sept. Basin saltbush also appears to have been seeded along highways in southwestern Wyoming.

Atriplex hortensis L. — garden orach — Introduced from Eurasia; the 6 specimens seen are from Rangely, Brush Creek near Jensen, Tridell, Nine Mile Canyon and Cowboy Canyon (White River); roadsides, ditches, fence lines, and disturbed riparian communities; 5,100,-7,000 ft; June-Sept. This plant does not appear to spread and persist and as well as does *A. micrantha.*

Atriplex micrantha Ledebour (*A. heterosperma* Bunge) — Russian atriplex, vary-seed orach — Introduced from Eurasia; widespread and common along ditches, roadsides, moist alkaline lowlands and floodplains; 4,600-7,800 ft; Aug.-Oct. This appears to have been recently introduced into our area where it spread rapidly.

Atriplex powellii S. Watson — Powell orach, Powell saltweed — widespread; occasional to common in desert shrub, greasewood, and pinyon-juniper communities, often on Mancos Shale, Duchesne River and other formations that weather to badlands; 4,700-6,000 ft; July-Oct. This is our most common, native, annual *Atriplex.*

Atriplex rosea L. — red orach, tumbling orach — Introduced from Eurasia, more or less weedy; occasional in mudflats of floodplains, along roadside, and in other disturbed sites in various plant communities up to 8,000 ft; July-Oct.

Atriplex saccaria S. Watson — stalked orach, medusa-head orach — Three of the 7 specimens seen are from the Pariette area, the other 4 from Red Fleet, Island Park area (Rainbow Draw and Six Mile Draw), and Willow Creek, E. Tavaputs Plateau; desert shrub communities, and semi-barrens of Morrison Formation; 5,000-6,000 ft; June-Aug. Our plants belong to var. *saccaria* [var. *ca-put-medusae* (Eastwood) S. L. Welsh].

Atriplex suckleyi (Torrey) Rydberg [*A. dioica* (Nuttall) Macbride; *Endolepis dioica* (Nuttall) Standley] — rillscale — The few specimens seen are from Little Snake drainage, Moffat Co. and just n. of Rio Blanco Lake, Rio Blanco Co. on clay slopes and knolls with desert shrubs; 6,000-6,700 ft.

Atriplex truncata (Torrey) A. Gray — wedgescale orach — The one specimen seen (Refsdal & McNight 6389) is from Browns Park, desert shrub community with scattered junipers at 5,400 ft. One specimen reported (Graham 1933) for the west side of the Green River near the Quarry; July-Sept. Much like and hardly distinct from *A. argentea.*

Atriplex wolfii S. Watson (*A. tenuissima* A. Nelson) — slender orach — Infrequent; 4 of the 7 specimens seen are from 0.5-4 mi w. of Duchesne, the other 3 from near Red Fleet and east to the Utah/Colorado line, Nielson 45 from near Massadona is listed in Bradley (1950); desert shrub, greasewood, and sagebrush communities; 5,300-7,000 ft; July-Oct. Uinta Basin plants belong to var. *tenuissima* (A. Nelson) S. L. Welsh.

Bassia Allioni — Bassia; Molly; Summer Cypress

Annual herbs or perennial subshrubs, occasionally glabrous but more often sericeous at least in part; leaves mostly alternate, entire; flowers mostly bisexual or some pistillate, solitary or a few in axils of leaves; perianth 6-lobed, persistent; stamens 3-5; stigmas 2 or 3; pericarp free from the seed.

1   Plants annual herbs, usually much branched, 30-100 cm tall; leaves 2-7 cm long, linear to lanceolate, short petiolate 2

1   Plants perennial subshrubs; stems usually simple at least below the inflorescence; leaves 0.5-3 cm long, narrowly linear, sessile 3

2   Calyx in fruit with hooked bristles . . . . **B. hyssopifolia**

2   Calyx without hooked bristles . . . . **B. scoparia**

3   Plants branched in the inflorescence, 10-75 cm tall, introduced; stems often reddish in age; fruit not encompassed by a horizontal wing . . **B. prostrata**

3   Plants not branched in the inflorescence, 10-30 (40) cm tall, native; stems seldom reddish; fruit encompassed by a horizontal wing . . . . **B. americana**

Bassia americana (S. Watson) A. J. Scott [*Kochia americana* S. Watson; *K. vestita* (S. Watson) Rydberg; Neokochia americana (S. Watson) G. L. Chu & S. C. Sanderson] — gray molly, green molly, red sage — Widespread; scattered to locally common in desert shrub communities and with scattered juniper; up to about 6,000 ft; June-Sept.

Bassia hyssopifolia (Pallas) Kuntze [*Echinopsilon hyssopifolium* (Pallas) Moquin-Tandon] — fivehook bassia, smotherweed — Widespread; moist alkaline lowlands, roadsides, desert shrub, and pinyon-juniper communities; up to 7,000 ft; Aug.-Sept. Similar to and possibly hybridizing with *B. scoparia* (Arnow and others 1980). Number of specimens seen indicate this is not nearly so common in our area as is *B. scoparia.*

Bassia prostrata (L.) A. J. Scott [*Kochia prostrata* (L.) Schrader] — forage kochia — Native of Eurasia, recently introduced from Russia, used in reclamation seedings on mine and oil shale spoils, gravel pits, and other disturbed areas, perhaps escaping.

Bassia scoparia (L.) A. J. Scott [*Kochia scoparia* (L.) Schrader; *K. iranica* Bornmuller misapplied; *K. sleversiana* (Pallas) Meyer] — kochia weed, Belvedere summer cypress — Introduced from Eurasia, weedy, widespread; common to abundant in fields, gardens, fence-lines, roadsides, and waste places; up to about 7,000 ft; Aug.-Oct. This is a tumble weed that breaks at the base and blows in autumn winds and stacks up against fences and buildings.

Camphorosma L.

Camphorosma monspeliaca L. — Mediterranean camphorfume — Plants perennial, woody at the base, usually several-stemmed and more or less caespitose, aromatic with odor of camphor but perhaps only weakly so, 10-60 cm tall; current years stems tomentose; leaves 3-10 mm long, linear-subulate, rigid and appearing (but not) pungent, spreading to recurved, fascicled; flowers in dense spikes; perianth 3-3.5 mm long, glabrous or pubescent, the 2 longer lateral teeth herbaceous and recurved in fruit, the 2 middle lobes shorter and scarious; seed glandular. Native of Eurasia, recently introduced from Russia, and planted in experimental plots s. of White River and near Watson and Rainbow, possibly escaping.

Chenopodium L. — Goosefoot; Pigweed; Lambsquarters

Annual herbs; leaves alternate, glabrous or grayish or whitish scurfy (farinose), entire to toothed or hastate, often reddish or purplish in age; flowers bisexual or some pistillate, in terminal or axillary crowded spikes or panicles; perianth herbaceous or slightly fleshy, 2-5 lobed; stamens 1-5; styles usually lacking; stigmas 2-5; seed erect or horizontal in the pericarp.

---

1 Leaves lobed, toothed, or at least with wavy margins, sometimes also with a pair of hastate lobes 2

1 Leaves entire or with a pair of hastate lobes near the base, but entire above the lobes 7

2 Mature leaves green, not or sparsely farinose 3

2 Mature leaves densely farinose at least on the lower side 5

3 Flowers in axillary and terminal open, cymose panicles . . . . *C. simplex*

3 Flowers in glomerules; leaves not cordate 4

4 Flower glomerules subglobose, (3) 4-7 (15) mm in diameter, axillary and in a terminal, mostly leafless, elongate spike; leaves more or less strongly hastate . . . . *C. capitatum*

4 Flower glomerules 1-4.5 mm in diameter, in axillary clusters with the terminal spike of clusters not or slightly elevated above the leaves; leaves not or weakly hastate . . . . *C. rubrum*

5 Mature stems prostrate to spreading, 7-25 (40) cm long; principal leaves 0.8-2.5 (3) cm long, 0.5-1.5 (2 cm) wide . . . . *C. glaucum*

5 Mature stems mostly erect 20-70 (200) cm tall; principal leaves sometimes larger than those of C. glaucum 6

6 Inflorescence crowded, the glomerules individually indistinct; median keel of sepal lobes weakly developed; seed coat smooth or faintly cellular-pitted, uniformly colored . . . . *C. album*

6 Inflorescence comparatively open, the glomerules individually distinct; median keel of sepal lobes prominent; seed coat distinctly honeycombed, the area around the style base yellow . . . . *C. berlandieri*

7 Leaves linear to narrowly lanceolate, entire, those of the main stem 1-4 (6.5) mm wide 8

7 Leaves lanceolate to ovate, deltate, often hastate-angled, -toothed, or -lobed, those of the main stem (3) 4-40 mm wide 10

8 Plant branching throughout with branches much shorter than the main stem; leaves of the main stem linear and with 1 conspicuous vein at 10x . . . . *C. leptophyllum*

8 Plant more strongly branched below than above with lower branches nearly as long as the main stem; leaves of the main stem often 3-nerved from near the base, narrowly lanceolate or narrowly oblanceolate 9

9 Calyx segments opening to expose the fruit at maturity; seed coat remaining adherent to the seed; plants ill-scented, mostly tall and slender with ascending branches . . . . *C. hians*

9 Calyx segments remaining closed and concealing the fruit at maturity; seed coat loose, readily separating from the seed; plants not ill-scented, bushy

with divaricately spreading branches . . . . **C. desiccatum**

10 Leaves ovate to deltate-ovate, hastate lobed or toothed . . . . . . **C. fremontii**

10 Leaves lanceolate to narrowly ovate, sometimes weakly hastate 11

11 Blades of the main stem leaves 1.5-3.5 cm long, broadly lanceolate to ovate, less than 3 times as long as wide; plants montane and submontane, often in mesic places . . . . **C. atrovirens**

11 Blades of the main stem leaves 2-5 cm long, lanceolate, at least 3 times as long as wide; plants of valley bottoms and foothills in drier places . . . . **C. pratericola**

Chenopodium album L. (*C. berlandieri* Moquin-Tandon) — lambsquarters, pigweed — Widespread; common in several plant communities, more or less weedy and increasing with disturbance; 5,000-6,800 ft; June-Sept. Rare plants with sculptured seeds (alveolate-reticulate or reticulate) belong to var. *berlandieri* (Moquin-Tandon) Mackenzie & Bush (*C. acerifolium* Andrezejowiski). Our common phase with smooth or faintly striate seeds is referable to var. *album*.

Chenopodium atrovirens Rydberg [*C. fremontii* var. *atrovirens* (Rydberg) Fosberg] — mtn. goosefoot — Strawberry Valley, Tavaputs Plateau, Uinta Mtns., and Douglas Mt.; occasional to common in roadside, sagebrush, aspen, Douglas-fir, and subalpine communities; 7,700-9,500 ft; July-Sept. See discussion under *C. fremontii*.

Chenopodium capitatum (L.) Asch. [*C. chenopodioides* (L.) Aellen misapplied; *C. overi* Aellen; *Blitum capitatum* L.]. With two vars.:

1 Flower clusters often 6-12 mm wide, the calyx becoming red and fleshy at maturity; plants mostly of gravelly streams of the north slope of the Uinta Mtns. . . . . **var. capitatum**

1 Flower clusters commonly less than 6 (to 8) mm wide, the calyx not fleshy, though sometimes reddish at maturity; plants common and widespread . . . . **var. parvicapitatum**

Var. capitatum — strawberry-spinach — Gravelly streams of the north slope of the Uinta Mtns. and one specimen from a burned aspen stand in Pole Creek Basin, Uinta drainage.

Var. parvicapitatum S. L. Welsh — smallhead chenopod — Widespread; occasional; sagebrush, pinyon-juniper, mt. brush, aspen, and spruce-fir communities and along streams and roadsides, often with disturbance; 7,000-9,000 ft; late June-Sept.

Chenopodium desiccatum A. Nelson (*C. pratericola* Rydberg) — desert goosefoot — The 7 specimens seen are from widely scattered locations; desert shrub, sagebrush, pinyon-juniper, and riparian communities and roadsides; 4,750-7,800 ft; late June-Sept. See discussion under *C. fremontii*.

Chenopodium fremontii S. Watson [*C. incanum* (S. Watson) A. Heller] — Fremont goosefoot — Widespread; common in desert shrub, sagebrush, pinyon-juniper, mt. brush, and Douglas-fir communities, floodplains and other wet places, often along roadsides or other disturbed places, becoming abundant and dominant for a year or two after fire; 5,200-8,700 ft; July-Sept. Plants mostly less than 25 cm tall that branch from the base with the curved ascending branches subequal to the main stem and with leaves more or less white scurfy beneath belong to var. *incanum* S. Watson. Specimens seen of this variety are from below 7,500 ft. Plants 5-80 cm tall with lower branches lacking or shorter than the main stem and with leaves glabrous to scurfy are referable to var. *fremontii*. Plants of *C. fremontii* and *C. leptophyllum* are strikingly different, but through *C. atrovirens* and *C. desiccatum* a continuum is formed. Mostly specimens can be assigned to one of the above taxa, but intermediate leaf forms are not uncommon.

Chenopodium glaucum L. (*C. salinum* Standley) — oakleaf goosefoot — widespread; occasional to locally common; mostly along moist or ephemerally moist alkaline drainages and floodplains, and margins of ponds, lakes, and reservoirs, and seeps, rarely dry ground; up to 7,000 (8,000) ft; July-Oct. Our plants are referable to var. *salinum* (Standley) J. Boivin.

Chenopodium hians Standley — hians goosefoot — The 2 specimens seen are from Horse Canyon, E. Tavaputs Plateau at 7,220 ft with sagebrush, oak, juniper and Douglas-fir (Thorne 15133) and Sheep Creek, Daggett Co. at 6120 ft in a Populus-Salix community (Welsh & Moore 18691).

Chenopodium leptophyllum Nuttall in S. Watson — slim leaf goosefoot — widespread; occasional to common and becoming abundant following fire or other disturbance, in many plant communities at about 5,200-9,000 t; July-Sept. See discussion under *C. fremontii*.

Chenopodium rubrum L. Widespread; locally common to abundant on mud flats, margins of lakes and ponds, and floodplains of streams, often increasing with disturbance; 4,900-6,500 ft; Aug.-Oct. Plants with erect stems 10-90 cm tall and leaves 2.5-7.5 cm long belong to var. rubrum — red goosefoot. Plants with prostrate, sprawling, or ascending stems 3-20 cm long and leaves 1.2-2.5 (3.7) cm long belong to var. humile (Hooker) S. Watson — low goosefoot. In draw down basins of reservoirs it is common to find plants with the growth form of var. rubrum more abundant at the upper edge of the basin and plants with the growth form of var. humile more abundant toward the lower part of the basin where release from water cover is relatively short. Plants of the 2 growth forms grow together.

Chenopodium simplex Torrey (Rafinesque). (*C. hybridum* L. var. *simplex* Torrey; *C. gigantospermum* Aellen) — mapleleaf goosefoot — Collected from and reported as common at Echo Park (N. Holmgren 458 DINO), and collected from Hideout Forest Camp at 5,800 ft (Flowers 150 UT) that was later inundated by Flaming Gorge Reservoir. Goodrich & Huber 26943 is from Wild Horse Ridge (W. Tavaputs Plateau) at 8,300 ft from under Douglas-fir where cattle had bedded.

Corispermum L. — Bugseed; Tickseed

Corispermum americanum (Nuttall) Nuttall (*C. hyssopifolium* L. misapplied; *C. nitidum* Kitaibel misapplied; *C. villosum* Rydberg) — American bugseed — Annual herbs, 15-60 cm tall, much branched; leaves simple, entire, linear, 1-7 cm long, 1-3 mm wide; inflorescence a densely flowered spike; flowers bisexual, in the axils of leaf-like, scarious-margined, ovate to lanceolate, usually imbricate bracts 1-2.4 mm long; fruit 2-4.5 mm long, conspicuously winged. The 3 specimens seen are from Red Creek in Daggett Co (Atwood 29026). and Brush Creek-Red Fleet area in Uintah Co (Neese & Chatterley 9894 & England 1063); Listed for Duchesne Co. (AUF 134), to be expected elsewhere in sagebrush-rabbitbrush and juniper communities in sandy soil; 5,200-6,000 and perhaps to 6400 ft; Aug.-Oct.

Cycloloma Moquin-Tandon — Ringweed; Winged Pigweed

Cycloloma atriplicifolium (Sprengel) Coulter — tumble ringweed — Plants annual, diffusely branched, 10-40 cm tall, villous-tomentose when young, glabrate in age; stems striate; leaves 1-6 cm long, alternate, petiolate, early deciduous, coarsely and irregularly sinuate-dentate, the teeth or lobes mucronate; inflorescence paniculate, with spicate branches, usually turning a deep red in age; flowers unisexual or bisexual or both in the same plant; perianth 5-lobed, the lobes inflexed, carinate, developing a membranous, horizontal wing that surrounds the fruit; stamens 5; stigmas 3; fruit depressed-globose; pericarp free; seed horizontal. The 2 specimens seen (Goodrich 22006, 27325) are from a sandy floodplain of the Green River at the Jensen Bridge.

Grayia Hooker & Arnot — Hopsage

Grayia spinosa (Hooker) Moquin-Tandon in de Candolle [*Atriplex spinosa* (Hooker) Collotzi] — spiny hopsage — Dioecious (monoecious) shrubs 50-150 cm tall; branches with whitish bark exfoliating in long strips, pubescent with scurfy and stellate hairs when young, often spiny in age; leaves alternate, 5-30 mm long, 2-12 mm wide, linear-oblanceolate, spatulate, or obovate, entire, tapering to a short petiole, scurfy pubescent when young but glabrate in age; flowers mostly unisexual, the staminate usually with a 4-lobed perianth enclosing the stamens; pistillate flowers in short spicate inflorescences, enclosed in 2 united, dorsally flattened bracts with lateral margins winged; wings of the fruiting bracts thickened and spongy within, 6-15 mm wide, greenish or straw-colored and turning bright red in age; seeds vertical. Locally common to abundant; widespread; desert shrub communities, often on aeolian sand; up to about 6,500 ft; May-June and with fruiting through July.

Halogeton C. A. Meyer — Halogeton

Halogeton glomeratus (Bieberstein) C. A. Meyer in Ledebour — halogeton — Annual, succulent, fleshy herbs, 8-30 cm tall, the stems branched, glabrous except tomentose in leaf-axils; leaves 5-10 mm long, about 1-3 mm wide, alternate, linear, or narrowly oblong, terete, entire, sessile, with a single hairlike bristle at the apex; flowers bisexual or some pistillate, in axillary clusters; perianth segments 5; stamens 5 or 3; stigmas 2; fruiting encompassed by a scarious, horizontal, 3-4 mm wide wing. Introduced from Eurasia with the first U.S. specimen collected near Wells, Nevada, in 1934. By 1954 it had spread across the Great Basin and Colorado Basin in Utah, Colorado, and Wyoming, and by 1980 it had spread to the 11 western states. Also plotted for Nebraska and South Dakota in TPD. Widespread in our area; mostly along roads and other areas of disturbance and spreading into desert shrub communities where it has capacity to displace native plants; up to about 7,500 ft; July-Sept.

Krascheninnikovia Gueldenstaedt — Winterfat; White Sage; Winter Sage

Krascheninnikovia lanata (Pursh) Meeuse & Smit [*Diotis lanata* Pursh; *Eurotia lanata* (Pursh) Moquin-Tandon; *Ceratoides lanata* (Pursh) J. T. Howell] — winterfat, white sage — Suffrutescent sub shrubs or occasionally the stems woody well above the base, 15-100 cm tall; twigs of the season densely stellate tomentose and also with longer straight hairs; leaves simple, alternate, entire, linear to narrowly lanceolate, sessile or short-petiolate, the blades 1-4 cm long, stellate-tomentose, the margins revolute, the midnerve conspicuous beneath; flowers unisexual, in axillary clusters or terminal in a spike-like inflorescence; staminate flowers with a 4-parted perianth; stamens 4; pistillate flowers lacking a perianth, but with 2 united pilose bracts; stigmas 2; fruit covered with white pilose hairs. Widespread; common to dominant in desert shrub communities, occasional in sagebrush and pinyon-juniper communities and on windswept ridges up to about 8,000 (9,000?) ft; June-Sept. Our plants belong to var. *lanata.*

Monolepis Schrader — Povertyweed

Annual herbs with branching stems; leaves alternate; flowers unisexual or bisexual; perianth segments (0) 1-3, inconspicuous, bract-like; stamens. 1; stigmas 2, cleft to the base; fruit laterally compressed.

---

1 Leaves bract-like, 0.2-0.6 (1.3) cm long, entire, oblong to obovate; plants dichotomously branched, rarely collected; flowers 1-5 per axil, borne on slender pedicles; perianth segments 1-3 .... **M. pusilla**

1 Leaves not at all bract-like, 1-7 cm long, triangular to lanceolate, hastate-lobed; plants not dichotomously branched, common; flowers in dense axillary clusters and in terminal interrupted spikes, sessile or nearly so; perianth segment 1 .... **M. nuttalliana**

Monolepis nuttalliana (Schultes) Greene — Nuttall povertyweed — Widespread; occasional to common in desert shrub communities and infrequent to occasional in sagebrush, pinyon-juniper, mtn. brush, ponderosa pine, aspen, Douglas-fir, and lodgepole communities; 5,000-10,400 ft; May-Sept.

Monolepis pusilla Torrey [*Micromonolepis pusilla* (Torrey ex S. Watson) Ulbrich] — dwarf povertyweed — Widespread, rarely collected, the 4 specimens seen are from 10.5 mi south of Ouray on the bottom of a dried up draw where locally abundant, Clay Basin, and Bridger Formation se. of Mtn. View, also reported by Graham (1937) for near Dinosaur Quarry and mouth of Ashley Creek; greasewood and desert shrub communities; 4,900-6,725 ft. Also collected from the draw-down basin of Flaming Gorge Reservoir in Sweetwater Co., Wyoming.

Salicornia L. — Glasswort; Pickleweed; Samphire

Salicornia rubra A. Nelson [*S. europaea* ssp *rubra* (A. nelson) Breitung] — marshfire pickleweed — Halophytic, glabrous, annual, herbs 5-30 cm tall, from slender taproots; stems fleshy, erect or ascending, commonly branched, the whole plant often reddish at maturity; leaves opposite, scale-like, triangular, to about 1.5 mm long; flowers bisexual or some unisexual, borne in spikes; spikes usually several to numerous, with internodes 2-4 mm long; perianth scale-like, sunken in depressions of the axis of the spikes, consisting of 4 basally united segments, enclosing the fruit; seed vertical, retrorsely pubescent. Specimens seen are from the margin of Pelican Lake, floodplain of the Green River near Ouray, White River east of Rangely, and near Manila; 4,800-6200 ft; Aug.-Sept.

Salsola L. — Russian Thistle; Tumbleweed

Annual herbs much branched and succulent when young, becoming rigid, spiny tumbleweeds in age; stems spreading or ascending; leaves alternate, entire, filiform, or linear-filiform, acute; flowers bisexual, solitary or few in axils of leaves, subtended by 2 bracts; perianth 4-5 parted; fruit encompassed by a horizontal, scarious, dorsal wing.

---

1 Fruits lacking prominent wings; bracts appressed at maturity or somewhat recurved; inflorescence elongate-spicate and usually 15-20 cm long . . . . .
. . . . **S. collina**

1 Fruits with prominent horizontal wings; bracts strongly spreading and usually recurved; inflorescence variable in length, sometimes less than 15 cm long 2

2 Apices of perianth segments long-acuminate and more or less fused into a slender columnar beak; fruiting perianth (including wings) 7-12 mm in diameter with the wings mainly 1-2 mm long . . . . **S. paulsenii**

2 Apices of perianth segments obtuse or weakly acuminate, sometimes reflexed, not forming a fused columnar beak; fruiting perianth (including wings) 4-10 mm in diameter with the wings mainly 3-4 mm long. **S. tragus**

Salsola collina Pallas — slender Russian thistle or tumbleweed — Collected from at the high-water line of Flaming Gorge Reservoir (Goodrich 23256) and along a road north of Lapoint (Goodrich 26522). To be expected across our area along roadsides and other disturbed places. Adventive weed from Asia, apparently of recent introduction.

Salsola paulsenii Litvinov in Isvet — barbwire Russian-thistle — To be expected in similar habitat as *S. tragus*. Apparently more recently introduced and less common than *S. tragus*. In addition to the feature of the key, this is different from *S. tragus* by the apices of the perianth segments forming a long-acuminate columnar beak.

Salsola tragus L. [*S. iberica* Sennen & Pau; *S. kali* L. misapplied; *S. pestifer* A. Nelson] — Russian-thistle, tumbleweed — Introduced from the Old World, widespread; scattered to locally abundant on disturbed ground; up to 8,000 ft; July-Oct.

Sarcobatus Nees — Greasewood

Sarcobatus vermiculatus (Hooker) Torrey — black greasewood — Spiny, much branched shrubs with rigid stems, (0.3) 1-2 (3) m tall, from deep taproots and branched root crowns; bark of older stems dull brown or gray, often pitted; twigs of the season green and succulent at first, turning white and spiny at the end of the season; leaves alternate or occasionally opposite, 1-4 cm long, about 1-3 mm wide, linear, entire, slightly flattened to nearly terete, fleshy-succulent, green, glabrous; the flowers unisexual, the staminate in cone-like or catkin-like spikes, these 5-25 mm long, the pistillate ones solitary or 2 in axils of leaves; fruit encompassed by a horizontal, scarious wing 6-12 mm wide. Widespread; common to dominant in saline or alkaline lowlands, forming dense thickets on deep alluvial soils of canyon bottoms of the Tavaputs Plateau; up to about 7,000 ft where commonly 2-3 m tall; May-June. Our plants belong to var. *vermiculatus.* Treated in Sarcobataceae in IMF 2A: 572.

Suaeda Forsskal ex Scopoli — Seepweed

Leaves alternate, succulent, terete or nearly so, entire, glabrous or glaucous; flowers unisexual or bisexual, solitary or clustered in axils of upper leaves or bracts; perianth 5-lobed or 5-parted; stamens 5; styles usually 2; fruit included in but usually free from the pericarp.

1 Plants perennial, somewhat woody at the base .... **S. nigra**

1 Plants annual, from slender taproots................................ 2

2 Mature calyx lobes with horn-like projections on the back and some with well-developed transverse wings at the face; flower borne in axils of bracts; plants native .... **S. calceoliformis**

2 Mature calyx lobes rounded on the back, lacking horn-like projections or transverse wings; flowers borne on the apex of bract petioles; plants introduced from Asia, known from along the Blacks Fork River in Wyoming (IMF 2A: 536 based on Fertig & Glennon 19011) and perhaps to be expected in our area (pin-leaf seepweed) .... **S. linifolia** Pallas

Suaeda calceoliformis (Hooker) Moquin-Tandon [*S. maritima* (L.) Dumortier; *S. occidentalis* S. Watson; *S. depressa* (Pursh) S. Watson misapplied] — slender seepweed — widespread; scattered to locally abundant in moist to marshy, alkaline lowlands, roadsides, edges of ponds and reservoirs; 4,700-6,000 (7,000) ft; Aug.-Oct. Perianth lobes unequal, with corniculate appendages. Plants of this species range from 5-80 cm tall with habit ranging from small, unbranched simple stems to much-branched, robust specimens.

Suaeda nigra (Rafinesque) J. F. Macbride [*S. diffusa* S. Watson; *S. fruticosa* (L.) Forsskal misapplied; *S. intermedia* S. Watson; *S. moquinii* (Torrey) Greene; *S. torreyana* S. Watson; *Dondia torreyana* (S. Watson) Standley] — bush seepweed — Widespread; occasional to common in moist lowlands, and dry slopes, tolerant of and perhaps dependent on alkaline or saline conditions; not expected over 6,500 ft; July-Sept. Perianth lobes equal, without corniculate appendages. Perhaps not much different from *S. vera* G. F. Gmelin (*S. fruticosa* Forsskal) of the Old World (Arnow and others (1980).

Zuckia Standley

Zuckia brandegeei (A. Gray) S. L. Welsh & Stutz [*Atriplex brandegeei* (A. Gray) Collotzi; *Grayia brandegeei* A. Gray] — siltbush — Monoecious or more often dioecious shrubs 10-80 cm tall, not thorny; herbage pubescent with inflated but readily collapsing hairs (scurfy); leaves alternate, subsessile or with a short petiole, 8-70 mm long, (0.4) 10-45 mm wide, elliptic, ovate, obovate, or oblanceolate, entire or rarely hastate lobed; staminate flowers 2-5 in clusters in axils of bracts, the perianth with 4 or 5 lobes, not enclosed in bracts; pistillate flowers 1-several in the axils of bracts, each enveloped by small, united bracts. Plants of what have been referred to as *Grayia brandegeei* and *Zuckia arizonica* are vegetatively similar, and plants of *Z. arizonica* are scattered through much of the range of *G. brandegeei* from n. Arizona to the Uinta Basin,

Utah. Although the features of the fruit are strikingly different, staminate flowers seem to be alike. The treatment of Welsh and Stutz (Welsh 1984b) is followed here in treating the two taxa as vars. of a single species, with the reservation that the two might well be treated at the species level especially if features of the staminate flowers are found to contrast, and also with the reservation that *Zuckia* could as well be included in *Grayia*. The differences in the seeds are similar to those found in *Chenopodium,* with seeds vertical in some taxa and horizontal in others. The 2 taxa are as follows:

1 Fruiting bracts forming an oval or rounded (in cross-section), apically depressed sac 2-4 (5) mm wide, this 6-ribbed, 2 of the ribs slightly larger, united except for a small orifice at the retuse apex; seeds horizontal . . . . **var.** *arizonica*
1 Fruiting bracts thin and flat, 4-8 mm wide, completely united, not ribbed, winged; seeds vertical . . . . **var.** *plummeri*

Var. *arizonica* (Standley) S. L. Welsh (*Z. arizonica* Standley) — Arizona siltbush — The 5 specimens seen are from near Bonanza, 5 mi s. of Ouray, and Big Pack Mtn.; desert shrub communities: 4,800-5,620 ft elevation.

Var. *plummeri* (Stutz & Sanderson) Dorn — Plummer siltbush — Widespread but no specimens seen from Duchesne Co. and westward; desert shrub communities and up to the lower edge of the juniper belt, often on formations that weather to badlands including Duchesne River, Green River, Uinta, and Morrison; 4,800-6,700 ft; May-June with fruit through Sept.

CLEOMACEAE (CAPPARIDACEAE) Cleome or Spider-flower Family

Annual herbs, often ill-scented; leaves alternate, ternate or palmately compound with 3-7 leaflets, or upper ones sometimes simple, the leaflets usually entire or serrulate; inflorescence a raceme; flowers bisexual; petals 4, usually clawed; stamens 6-16 (ours) exserted; ovary superior; fruit a 2-valved capsule, the 2 valves separating at maturity.

1 Capsules not over 6 mm long, usually wider than long; leaflets 0.7-2.5 cm long; stamens 6, the filaments about 5 mm long; plants 5-20 (30) cm tall . . . . ***Cleomella***
1 Capsules 10-55 mm long, longer than wide; leaflets 2-5 (8) cm long; stamens 6-16, the filaments 7-15 mm long; plants 20-80 (100) cm tall 2
2 Capsules sessile or short stipitate, the stipe 1-3 mm long; stamens usually 12-16, the filaments 7-10 mm long; plants viscid-villose, clammy **Polanisia**
2 Capsules stipitate, the stipe 10-20 mm long; stamens 6, the filaments 10-15 mm long; plants glabrous or nearly so . . . . ***Cleome***

Cleome L. — Spiderflower; Beeplant

With features of the family.

1 Petals yellow, 5-8 mm long; at least the lower leaves with 5-7 leaflets . . . . ***C. lutea***
1 Petals purple, 8-12 mm long; leaves trifoliate . . . . ***C. serrulata***

Cleome lutea Hooker — Yellow beeplant — Widespread (no specimens seen from Daggett or Summit Cos.); occasional to common; desert shrub, and sage-brush-juniper communities, roadsides and floodplains, sometimes in sandy soils, often where soil is disturbed; 4,700-6,000 ft; May-Aug.

Cleome serrulata Pursh — Rocky Mtn. Beeplant — Widespread; occasional and locally common in sagebrush, basin wildrye-rubber rabbitbrush, mt. brush, and aspen communities, floodplains and roadsides, often on disturbed soils; 4,650-7,400 ft; mid June-Sept.

Cleomella de Candolle

Tap rooted annuals (ours); herbage glabrous or puberulent (ours); leaves trifoliate; leaflets entire; flowers borne in racemes; petals 4, yellow; stamens 6, exserted beyond the petals; fruit a silicle borne on a stipe.

---

1   Plants with a dominant central stem; inflorescence in flower and fruit extending conspicuously above the foliage .... *C. hillmanii*
1   Plants diffusely branched, usually without a obvious central axis; inflorescence in flower and fruit barely extending above the foliage .....
.... *C. palmeriana*

Cleomella hillmanii A. Nelson — desert stinkweed, disjunct stinkweed — fruit rhombic, about 5-6 mm long, 7-9 mm wide, with horned-like lateral appendages. Specimens seen are from the Morrison Formation near Island Park, Uintah Co; 5,400 ft. 6,070 Plants of the Uinta Basin belong to var. *goodrichii* (S. L. Welsh) P. K. Holmgren. Graham 6124 reported by Graham (1937) to have been collected in the mixed desert shrub zone between Green River and the Quarry; 4,700 ft likely belongs here.

   Cleomella palmeriana M. E. Jones — Palmer cleomella — Leaves and flowers similar to those of *C. hillmanii* but with branched stems and fruit without horned appendages or these smaller. The one specimen seen is from 6 mi south of Randlett on west side of Duchesne River in Uintah Co. (N. D. Atwood 2653 BRY).

Polanisia Rafinesque — Clammyweed

Polanisia dodecandra (L.) de Candolle (*P. trachysperma* Torrey & A. Gray) — clammyweed, western clammyweed — Plants 20-28 cm tall, viscid, clammy, strongly ill-scented; leaves trifoliate or rarely simple; sepals purple-tinged, deciduous; petals 8-12 mm long, yellowish-white; stamens about twice as long as the petals; capsules 3-5.5 cm long, 5-7 mm wide. Several specimens seen from Bonanza to s. of Ouray in the vicinity of the Green and White Rivers; 3 specimens seen from Jessie Ewing Canyon, Browns Park area, Daggett Co. and reported for Hideout Canyon, Daggett Co. (Flowers and others 1960) but no specimens found at UT or elsewhere to verify this report; desert shrub, rabbit-brush, and sagebrush communities, gravelly washes, and tuffaceous outcrops; 5,000-5,800 ft; July-early Sept. Our plants belong to ssp. *dodecandra* var. *trachysperma* (Torrey & A. Gray) J. Iltis.

CONVOLVULACEAE Morning-glory Family

Perennial herbaceous vines from extensive creeping rhizomes; stems creeping or climbing by twining; leaves alternate, petiolate; pedicels or peduncles solitary in axils, with 1-2 flowers, these 5-merous except for the 2-chambered pistil; calyx segments free or united at the base; corolla united, funnelform to campanulate, entire to 5-lobed, pleated, white, creamy, pink or purplish; fruit an ovoid to globose capsule. In addition to the natives listed below, morning glory [*Ipomoea purpurea* (L.) Roth] is cultivated in our area. This is distinguished from the natives by globose rather than linear stigmas.

---

1 Corollas 4-6 cm long; calyx 10-17 mm long; bracts 15-30 mm long, immediately subtending and enveloping the calyx, broadly cordate-ovate, obtuse to acute; pedicels 3.5-12 cm long, with a solitary flower; stigma lobes oblong, blunt apically, distinct from the style .... *Calystegia*

1 Corollas 1.5-3 cm long; calyx 3-5 mm long; bracts 1.5-3 mm long, often remote from the calyx, narrow, acute; pedicels or peduncles 0.5-3 cm long, with 1-2 flowers; stigma lobes linear or filiform, acute apically, continuous with the style .... *Convolvulus*

Calystegia R. Br. — Morning-glory; Bindweed

Calystegia sepium (L.) R. Brown [*Convolvulus sepium* L.; *C. americanus* (Sims) Greene misapplied] — hedge false bindweed — Two specimens seen. Graham 9787 CM! is from the mouth of Ashley Creek, 4,750 ft. Atwood & Evenden 24404 is from Desolation Canyon from a riparian community. Plants of the Intermountain Region belong to var. *angulata* (Brummitt) N. Holmgren.

Convolvulus L. — Bindweed; Morning-glory

Convolvulus arvensis L. — small or field bindweed, orchard morning glory — Introduced from Europe; fields, gardens, ditch banks, roadsides, and other disturbed places; a troublesome weed that is nearly impossible to eradicate because of its extensive, deep rhizomes; not expected much over 9,000 ft; June-Sept. Bindweed is much less of a problem in wildland communities than it is in lawns, gardens, and fields of crops.

CORNACEAE Dogwood Family

Cornus L. — Dogwood

Cornus sericea L. [*C. stolonifera* Michaux; *C. instolonea* A. Nelson; *Swida sericea* (L.) Holub] — red-osier dogwood — Shrub to 4 m tall; bark bright red or reddish-purple; leaves opposite, entire, 5-10 (15) cm long, 2-5 (7) cm wide, or larger on young vegetative twigs, ovate to elliptic or oblong; flowers in terminal, flat-topped to slightly rounded clusters 3-6 cm across; sepals 4, minute to obsolete; petals 4, white, 2-4 mm long, arising with the stamens at the base of a disk; stamens 4, alternate with the petals; ovary inferior; style 1; fruit a white or bluish drupe, 7-9 mm in diameter. Widespread; occasional or locally common; usually along streams or other places where the soil is kept wet for most of the growing season; up to about 9,000 ft; May-July.

CRASSULACEAE Stonecrop Family

Annual or perennial, usually succulent herbs; sepals and petals 4 or 5, free or connate at the base; fruit of follicles.

---

1 Plants annual, usually of ephemeral pools or drying mud; leaves opposite, 3-6 mm long, linear or linear-oblanceolate, with short but definitely sheathing bases; stems weak, often prostrate, sometimes rooting at the nodes, 1-6 cm long; flowers inconspicuous, whitish . . . . *Tillaea*

1 Plants perennial, of dry or wet places; leaves alternate, linear or orbicular to obovate, not connate sheathing; stems stouter than above, ascending to erect; flowers conspicuous, not whitish . . . . *Sedum*

Sedum L. — Stonecrop

Plants perennial, succulent; leaves alternate (ours), fleshy, terete or flat, often closely imbricate; flowers bisexual, 4-5-parted; petals as many as the sepals, separate or united at the base; stamens twice as many as the petals, the filaments adnate to the perianth at the base; carpels 3-5, separate or united at the base; fruit of 1-chambered follicles dehiscent along the ventral side.

---

1 Flowers pinkish or reddish; leaves usually over 16 mm long; stems 10-35 cm tall; plants of moist or wet places . . . . *S. rhodanthum*

1 Flowers yellow; stems 5-20 cm tall; plants usually of dry places 2

2 Leaves of flowering stems opposite at least in part, orbicular to obovate, rounded at the apex, to 8 mm long, about as wide as long . . . . *S. debile*

2 Leaves of flowering stems alternate, narrowly lanceolate to occasionally ovate, acute at the-apex, 3-20 mm long . . . . *S. lanceolatum*

Sedum debile S. Watson — weakstem stonecrop — Known from a few locations in Wasatch Co. and to as far east as Rock Creek in Duchesne Co. and Provo River Drainage in Summit Co. from 9,000-11,500 ft; Aug.

Sedum lanceolatum Torrey [*S. stenopetalum* Pursh misapplied; *Amerosedum lanceolatum* (Torrey) Love & Love] — Lanceleaf stonecrop — widespread; common in many plant communities; from about 7,000 ft to above timberline, often in dry, rocky places; June-Aug.

Sedum rhodanthum A. Gray [*Clementsia rhodantha* (A. Gray) Rose] — redpod stonecrop — Uinta Mtns.; occasional to common along streams and wet meadows; 9,700-11,800 ft; July-Aug.

Tillaea L. — Pygmy-weed

Tillaea aquatica L. — Pygmyweed — Plants as described in the key and flowers single in axils, usually 4-merous, greenish, shortly pedicellate with the pedicels to 6 mm long in fruit; calyx 0.5-1 mm long, united at the base; petals about 1.5 mm long, whitish; stamens 4, shorter than the petals; fruit of follicles less than 2 mm long, purplish. The 1 record seen (Neese 13963) is from Little Davenport Draw, Diamond Mt.; in shallow flowing water of a wet meadow; 6,890 ft; June-Aug.

CROSSOMATACEAE Crossosoma Family

Glossopetalon A. Gray — Greasebush

Glossopetalon spinescens A. Gray [*Glossopetalon meionandra* Koehne; *Forsellesia meionandra* (Koehne) A. Heller; *F. spinescens* (A. Gray) Greene]. Low shrubs not over 60 cm tall; twigs terminating in spines, young twigs greenish or grayish-yellow; leaves 7-17 mm long, 3-4 mm wide; flowers solitary, axillary; sepals about 3 mm long; petals 4-6 mm long, greenish-white or cream; follicle about 4 mm long, 3 mm wide. Widespread; locally common in desert shrub, sagebrush, pinyon-juniper, and mtn. brush communities at 4,900-7,600 ft; April-June. Plants of our area belong to var. *mionandrum* (Koehne) Trelease in A. Gray — Utah greasebush — Often moderately to heavily hedged by wild ungulates despite the spiny twigs.

CUCURBITACEAE Gourd Family

Echinocystis Torrey & A. Gray — Balsam Apple; Wild mock cucumber

Echinocystis lobata (Michaux) Torrey & A. Gray — wild cucumber — Annual climbing vines with branched tendrils; leaves alternate, simple, palmately lobed, with 3-7 triangular lobes, scabrous, the blades about 3-8 cm long, cordate at the base; staminate flowers numerous, racemose or paniculate, pistillate flowers solitary or few from the same axils as the staminate; sepals 5-6; corolla united, deeply 5-6 parted, the lobes 3-6 mm long, rotate, pale yellowish green or whitish; stamens 2-3, the anthers more or less united; fruit an ovoid pepo, 1- or 2-celled, covered with numerous soft spines, 3-5 cm in diameter, spongy and fibrous within, pendulous, becoming papery and rupturing apically at maturity. The few specimens seen are from Vernal, Whiterocks, and Dry Fork Settlement; ditch banks and fence-rows; 5,600-6,400 ft; Aug.

CUPRESSACEAE Cypress Family

Juniperus L. — Juniper

Trees or shrubs, dioecious, occasionally monoecious, evergreen, aromatic; wood resinous; leaves opposite or whorled, scale-like or needle-like, often appressed to and clothing the twigs and smaller branches; staminate cones solitary or clustered, rounded, 3-6 mm in diameter; pistillate cones berry-like, blue, black, or reddish brown, covered with a waxy bloom; seeds 1-4.

1   More or less prostrate or spreading-ascending shrubs, seldom over 1.5 m tall; leaves needle-like, not appressed to the branches . . . . *J. communis*

1   Ascending or erect shrubs or trees, mostly over 1.5 m tall; leaves on mature growth scale-like, appressed to the branches. 2

2   Fruit 4-8 mm in diameter; trunks usually straight and undivided, to 13 m tall; bark platy in age; heartwood red or purple-red; leafy twigs relatively slender, the ultimate ones mostly less than 1 mm wide . . . . . *J. scopulorum*

2   Fruit (6) 8-14 mm in diameter; trunks usually branched near the base and/or twisted, to 6.5 m tall; bark shredding in long strips in age; heartwood white; leafy twigs relatively stout, the ultimate ones mostly more than 1 mm thick . . . . *J. osteosperma*

Juniperus communis L. (*J. sibirica* Brugsdorf) — common juniper, prostrate juniper — Widespread; most common in aspen and Douglas-fir communities where sometimes forming a rather dense undergrowth, occasionally as solitary or scattered plants in sagebrush and other communities; 7,000-11,000 ft. Our plants belong to var. *depressa* Pursh.

Juniperus osteosperma (Torrey) Little [*J. utahensis* (Engelmann) Lemmon] — Utah juniper, bone seed juniper, white cedar — This juniper (alone or in association with pinyon) forms the extensive pygmy forests or woodlands (often referred to locally as "the cedars") that flank the Uinta and other mtns. and the Tavaputs Plateau at about (5,600) 6,000-7,500 ft. Due to its abundance, this juniper has been the principal source of "cedar posts" that have been used in great numbers in thousands of miles of fence and in livestock shelters. The wood is quite rot-resistant in the ground, considerably more so (according to local ranchers) than that of the following species. See *J. scopulorum.*

Juniperus scopulorum Sargent [*J. virginicum* L. var. *scopulorum* (Sargent) Lemmon] — Rocky Mt. juniper, Rocky Mt. red cedar — Widespread; most common in canyons and on cool exposures or at higher elevations than the preceding species, often scattered and not forming extensive stands, although sometimes in small stands in Vasey sagebrush, mt. brush, ponderosa pine, Douglas-fir, and other communities; 7,000-9,000 ft. Also distinguished from *J. osteosperma* by entire and opposite leaves rather than minutely toothed and sometimes whorled leaves (Arnow and others 1980).

CUSCUTACEAE Dodder Family

Cuscuta L. — Dodder

Parasitic annual or perennial herbs, lacking chlorophyll and whitish or yellowish, the ephemeral root systems quickly withering and the twining stems not attached to the ground, but rather attached to their host plants by intrusive haustoria; leaves scale-like, obsolete, alternate; flowers small, inconspicuous, regular, bisexual; calyx persistent, 4-5 lobed to divided, shorter or longer than the corolla; corolla 4-5 lobed, narrowly tubular to campanulate or urceolate, mostly white; stamens inserted just below the sinuses of the corolla; fringed or fimbriate-margined scale-like appendages (infrastaminal scales) usually attached near the base of the corolla; pistil 1, superior, with 2 separate styles; fruit a membranous globose or depressed-globose capsule. The following key was adapted from IMF (4: 78-80). Some of our species were included in the genus *Grammica* Loureiro by Weber (1984).

---

1   Stigma appearing as a slender, tapering-cylindrical continuation of its style, several times longer than thick; introduced plants; mainly on legumes, especially alfalfa . . . . **C. approximata**

1   Stigma capitate, appearing as a small expanded knob terminating the style, not longer than thick; native species on a wide range of hosts 2

2   Infrastaminal scales wanting or much reduced, sometimes forming faint winged or toothed ridges; flowers sessile; corolla lobes acuminate; plants probably rare or infrequent, no specimens seen **C. occidentalis** Millspaugh

2   Infrastaminal scales well developed, generally more or less fringed 3

3   Corolla lobes obtuse or rounded; flowers (3) 4 (5)-merous, sessile or nearly so; plants probably infrequent, no specimens seen .... *C. cephalanthi*

3   Corolla lobes acute; flowers mostly 5-merous, subsessile to pedicellate . . 4

4   Capsule ovoid, usually 1-seeded; plants often on desert shrubs .... *C. denticulata*

4   Capsule globose, usually 2-4 seeded; plants seldom on desert shrubs 5

5   Corolla often papillose-nerved; capsules slightly crested (evidently thickened around the styles that appear to be set in a shallow pit surrounded by a collar) .... *C. indecora*

5   Corolla not papillose-nerved; capsules not evidently thickened around the styles .... *C. pentagona*

Cuscuta approximata Bab. [*C. planifolia* Ten. misapplied (Graham 1937)] — slender dodder — To be expected throughout our area, but only 3 specimens seen. These from Tridell and Browns Park at 5,400-5,700 ft; parasitic on alfalfa, kochia weed, and desert shrubs.

Cuscuta cephalanthi Engelmann — slenderflower dodder — Listed for Uintah Co. in AUF (351).

Cuscuta denticulata Engelmann — smalltooth dodder — Rarely collected, apparently widespread; the few specimens seen are from 5,200-7,600 ft elevation; often on desert shrubs, especially *Artemisia* and *Chrysothamnus,* but apparently not on Chenopodiaceous shrubs.

Cuscuta indecora Choisy — plain dodder, bigseed alfalfa dodder — The one specimen (B. Welsh & Moore 179) from near Ouray was parasitic on *Salsola.*

Cuscuta pentagona Engelmann [*C. glabrior* (Engelmann) Yuncker] — field dodder, bushclover dodder — The 5 specimens seen are from Uintah Co and mostly along the Green River from 4,700-5310 ft elevation. Parasitic on *Ambrosia, Bidens, Convolvulus, Grindelia, Salsola, Solanum* and likely other plants. Our plants belong to var. *glabrior* (Engelmann) Gandhi, R. D. Thomas, & S. L. Hatch.

CYPERACEAE Sedge Family

Grass-like, mostly perennial herbs; stems usually solid, not jointed as in grasses, often triangular in cross-section or terete; leaves mostly 3 ranked, the sheaths mostly closed, the blades elongate, narrow or reduced or even lacking; inflorescence variable; flowers unisexual in 2 of our genera and bisexual in the others, sessile, subtended by a scale, arranged in spikes or spikelets; perianth lacking or of 1-many bristles; stamens 1-3; ovary superior; style 1 with 2-3 (4) stigmas; fruit an achene, lenticular (flattened or lens shaped in cross-section) or trigonous (triangular in cross-section), the lenticular fruits with 2 stigmas and the trigonous ones with 3 stigmas. This is a large family. The genus *Carex* with about 70 species is the largest genus of our area. Differences are small between the taxa, and mature plants are needed to identify specimens. Identification without a dissecting scope or good lens is not practical in many taxa.

1   Achene enclosed or folded in a closed or open sac or small bract as well as subtended by scale; perianth bristles lacking; flowers unisexual 2

1   Achene exposed, not enveloped by nor folded in a sac or bract, merely
    subtended by a scale and usually by perianth bristles; flowers bisexual . . 3

2   Achene completely enclosed in a sac (perigynium), this closed to the tip;
    attachment of the style to the achene concealed by the perigynium, with
    only the stigmas exserted . . . . **Carex**

2   Upper part of the achene usually exposed, the subtending bract with
    unsealed margins and exposing the attachment of the style to the achene
    as well as the stigmas . . . . **Kobresia**

3   Bristles subtending the achene numerous, much elongate and giving the
    inflorescence the appearance of a tuft of cotton; plants subalpine and alpine
    on the Uinta Mtns. . . . . **Eriophorum**

3   Bristles not as above 4

4   Inflorescence a solitary spike 5

4   Inflorescence of few or many spikes. 6

5   Leaf sheaths entire; scales entire; perianth bristles minutely retrorsely
    barbellate . . . . **Eleocharis**

5   Uppermost sheath of at least some culms with a bristle-like blade 4-6 mm
    long; lowest scale with a blunt awn 1-3 mm long; perianth bristles smooth
    . . . . **Scirpus cespitosus**

6   Scales spirally arranged; perianth bristles present; plants perennial except
    in S. supinus . . . . **Scirpus**

6   Scales arranged in 2 vertical ranks; perianth bristles lacking; plants annual
    (note: perennial species might be expected in our area) . . . . **Cyperus**

Carex L. — Sedge

Monoecious or dioecious perennials; stems tufted to arising singly from
creeping rhizomes; inflorescence of congested and short to remote and elongate
spikes; spikes staminate, pistillate, androgynous (staminate flowers above the
pistillate ones) or gynaecandrous (pistillate flowers above the staminate ones)
each spike with few or many unisexual sessile or nearly sessile flowers;
perianth lacking; staminate flowers reduced to 2 or 3 stamens in the axil of a
scale, the pistillate ones reduced to a pistil, this enclosed in a scale-like
perigynium (plural, perigynia) with only the stigmas exserted from the apex,
each one borne in the axils of a scale (a small bract).

This is a delightful genus to work with. Many of the taxa have sharp
boundaries even though the differences may be small. Hybridization in the
genus as well as the rest of the family is rare. However, the taxa of Section
Ovales are challenging and do present some taxonomic problems.

1   Spikes solitary, the perigynia attached directly to the rachis . . . . . . **KEY 1**

1   Spikes more than 1 (sometimes densely congested into a head that
    resembles a solitary spike), the perigynia attached to a rachilla and the
    rachilla attached to the rachis. 2

2   Spikes all sessile, aggregated into a head-like or spike-like inflorescence,
    the terminal one not staminate and the lower ones commonly not all
    pistillate except in unisexual specimens 3

2   At least the lower spike pedunculate, or if sessile then well separated from the upper ones and/or the terminal one staminate; lower spikes mostly all pistillate 6

3   Stigmas 3; achenes trigonous; inflorescence with 3-5 spikes, often subtended by a leaf-like bract; terminal spike gynaecandrous, the lateral ones mostly all pistillate; pistillate scales black or blackish purple (Section Atratae) . . . . **KEY 5**

3   Stigmas 2; achenes lenticular; inflorescence commonly with more than 5 spikes, or else the pistillate scales paler than above, not subtended by a leaf-like bract (except in *C. athrostachya*); spikes mostly all androgynous, all gynaecandrous, or all unisexual 4

4   Spikes androgynous, or if unisexual then plants with well-developed creeping rhizomes . . . . **KEY 2**

4   Spikes gynaecandrous; plants caespitose; rhizomes lacking or short. 5

5   Perigynia round-margined, not winged, not conspicuously flattened, mostly less than 3.5 mm long; scales pale green to brown; inflorescence commonly less than 2 cm long and/or less than 1 cm wide; plants mostly of wet places . . . . **KEY 3**

5   Perigynia wing-margined, often conspicuously flattened, (2.5) 3.5-7.5 (8) mm long; scales commonly brownish to dark brown, often with green midrib; inflorescence often longer or wider than above; plants of dry or wet places (Section Ovales) . . . . **KEY 4**

6   Terminal spike gynaecandrous, the lateral ones all pistillate or gynaecandrous with few staminate flowers; pistillate scales dark brown-purple to black; plants tufted, without rhizomes or these short (except in *C. buxbaumii*) . . . . **KEY 5**

6   Terminal spike and sometimes 1 or more lateral spikes staminate or androgynous (the terminal one occasionally gynaecandrous in a few taxa but then the scales pale green, or if brown then the culms from slender rhizomes) 7

7   Spikes all androgynous; staminate flowers few, inconspicuous, and exceeded by the by the upper perigynia; lower pistillate scales leaf-like, exceeding the perigynia, commonly 1-7 cm long; achenes about 3 mm long . . . . *C. backii*

7   Spikes not as above 8

8   Stigmas 2; achenes lenticular; scales often black or black-purple or with blackish lines flanking a greenish or pale mid-stripe, often contrasting with the greenish or stramineous perigynia . . . . **KEY 6**

8   Stigmas 3; achenes trigonous; scales greenish or brownish, or if blackish then usually about the same color as the perigynia (except in *C. raynoldsii* of well drained soil) . . . . **KEY 7**

— KEY 1 —

1   Spikes unisexual; scales dark brown-purple to black 2

1   Spikes androgynous; scales green to dark brown. 4

2   Perigynia glabrous; plants of Strawberry Valley and vicinity, unusual specimens of a species usually with more than one spike . . . . *C. idahoa*

11 Spikes with only 1 or 2 perigynia 5-6 mm long, the staminate portion 1-2 (3) cm long .... *C. geyeri*

11 Perigynia either more numerous or smaller than above; staminate portion of spike usually less than 1 cm long 12

12 Rachilla exserted beyond the orifice of the perigynium; perigynium soon strongly reflexed, these and the scales deciduous at maturity, greenish or pale brown; plants rare .... *C. microglochin*

12 Rachilla lacking or included in the perigynia; perigynia ascending and persistent, or if strongly spreading and deciduous then these and the scales black or blackish-purple at least in part; plants various............... 13

13 Leaves (at least some) exceeding the spikes, often strongly curved to completely curled at the tip; rachilla lacking; plants alpine ... *C. rupestris*

13 Leaves mostly exceeded by the spikes, straight or at least not curled; rachilla about as long as the achene (except in *C. nigricans*) 14

14 Rachilla lacking; perigynia narrowed to a substipitate base, ascending at first, these and the scales widely spreading and deciduous in age .... *C. nigricans*

14 Rachilla about as long as the achene; perigynia sessile or inconspicuously stipitate, these and the scales ascending and persistent............... 15

15 Perigynia firm, mostly 1-6 per spike; achenes filling the perigynia; plants occasional or locally common .... *C. obtusata*

15 Perigynia thin, mostly more than 6 per spike; achenes various; plants rare (Section Inflatae) 16

16 Perigynia 4-7.5 mm long, mostly over 1.8 mm wide; spike 6-12 mm wide, 1.2-2.5 times as long as wide; achenes much smaller than the perigynia; plants of alpine, well-drained sites .... *C. breweri*

16 Perigynia 2.5-3 (4) mm long, to 1.5 mm wide; spike commonly 4-6 mm wide, 2.5-3 times as long as wide; achenes about as large as the perigynia; plants of subalpine wet places .... *C. subnigricans*

— KEY 2 —

1 Culms arising singly or few together from creeping rhizomes 2

1 Culms caespitose; rhizomes lacking or short 8

2 Spikes mostly well separated from each other, with 1-3 staminate and pistillate flowers, each about 5 mm long; scales and perigynia green .... *C. disperma*

2 Spikes congested into a head-like or continuous spike-like inflorescence, mostly with more flowers and longer than above; scales and perigynia brownish or blackish... 3

3 Spikes closely aggregated into a compact, nearly globose head, none distinguishable without teasing the head apart; plants mostly 5-25 cm tall .... *C. foetida*

3 Spikes not so closely aggregated, the lower ones and sometimes the upper readily distinguishable without teasing the head apart; plants often taller or of lower elevations, variously distributed 4

4 Perigynia winged, deeply bidentate, serrulate to below the middle; lateral

spikes androgynous or staminate, the upper one often wholly pistillate; plants 19-36 cm tall; rhizomes light brown .... *C. siccata*

4 Perigynia not winged or scarcely so, bidentate or not, if serrulate then mostly so only on the upper half; spikes androgynous or unisexual 5

5 Perigynia 1.7-2.6 mm long, yellow-green or yellow-brown, thick walled and firm especially toward the base, the beaks 0.2-0.5 mm long, slightly winged and serrulate at the confluence of the beak and the body only; plants usually of boggy meadows .... *C. simulata*

5 Either the perigynia or their beaks longer than above, the walls rather thin, mostly brownish, wings and not serrulate or inconspicuous so except in *C. praegracilis* and then usually extending onto the body; plants of dry or wet places... 6

6 Rhizomes averaging 2-4 mm wide, blackish or dark brown; lower leaves reduced to bladeless sheaths, blackish or dark brown; plants 10-70 cm tall, often (not always) of wet places; perigynia slightly winged, serrulate distally and sometimes to the middle .... *C. praegracilis*

6 Rhizomes less than 2 mm wide, brownish or tan; lower leaves with blades, or if reduced to bladeless sheaths these light brown or greenish; plants 8-28 cm tall, mostly of well-drained soil; perigynia wingless, not serrulate or inconspicuously so 7

7 Plants unisexual; beak of perigynia 1-1.5 mm long; inflorescence 1.5-5.5 cm long, with the most conspicuous coloration coming from the hyaline margins of the scales .... *C. douglasii*

7 Plants bisexual, the spikes androgynous, the staminate portion usually conspicuous; inflorescence 0.8-2.0 cm long, with the most conspicuous coloration coming from the brown or dark brown body of the scales .... *C. duriuscula*

8 Inflorescence simple, ovoid or ovoid-oblong, or if linear then plants of well-drained soil; stature various 9

8 Inflorescence compound (at least some of the lower spikes borne on branched rachillas), usually oblong to linear; plants mostly of wet places, (20) 30-120 cm tall, rare 12

9 Perigynia tapered to a beak, not or inconspicuously serrulate; plants mostly of wet places .... *C. neurophora*

9 Perigynia abruptly contracted to a beak or conspicuously serrulate; plants of well-drained soil (section Bracteosae)........................... 10

10 Perigynia entire or serrulate on the beak only, somewhat rounded on the margins, with the marginal nerves displaced onto the ventral surface, pale green or straw-colored, about 3-7 per spike, the beak not (or only slightly) bidentate .... *C. vallicola*

10 Perigynia serrulate on the beak and usually onto the upper 1/2 of the slightly winged-margins of the body, with the marginal nerves on the wing-edges or slightly displaced onto the ventral surface, green to dark brown, about 4-12 per spike, the beak bidentate........................... 11

11 Spikes closely congested into an ovoid or ovoid-oblong head, rarely the lower ones noticeably separated, the internodes obscure .... *C. hoodii*

11 Spikes loosely congested into an oblong to linear head or spike-like inflorescence, the lower ones generally noticeably separated, the lowest internode conspicuous, generally 2-7 mm long .... *C. occidentalis*

12 Sheaths conspicuously cross-rugose ventrally; perigynia very gradually tapering to a slender beak .... *C. stipata*

12 Sheaths not cross-rugose ventrally; perigynia more abruptly tapered to the beak 13

13 Leaf sheath copper-colored at the mouth, often red-dotted; inflorescence 3-8 cm long; perigynia 2.75-4 mm long, nearly concealed by the scales; leaves 3-5 mm wide; plants known from the Black Fork in the Uinta Mtns.
    .... *C. cusickii* Mackenzie

13 Leaf sheath not copper-colored at the mouth, not red-dotted; inflorescence 2-3.5-(5) cm long, not interrupted or but slightly so; perigynia 2-3 mm long, not concealed by the scales; leaves 1-2.5 (3) mm wide; plants known from lower Uinta Canyon, Uinta Mtns. .... *C. diandra*

— KEY 3 —

1 Perigynia widely spreading at maturity, the lower ones sometimes reflexed (section Stellulatae). 2

1 Perigynia appressed 3

2 Perigynium beak 1/4-1/3 (1/2) the length of the body, up to ca 1 mm long, inconspicuously bidentate with broad short teeth; perigynia mostly 2.2-3.2 mm long .... *C. interior*

2 Perigynium beak 1/2 or more the length of the body, conspicuously bidentate, the teeth narrow; perigynia mostly 2.8-3.5 (4) mm long ....
    .... *C. echinata*

3 Perigynia and scales black or dark brown at maturity; inflorescence an ovoid head, 0.8-1.4 cm long, the spikes closely congested and hardly distinguishable without teasing the head apart .... *C. illota*

3 Perigynia greenish or light brown and/or the inflorescence more open with the spikes distinguishable (section Heleonastes, in part) 4

4 Pistillate scales dark brown, equal to the body of the perigynium but often surpassed by the beak; perigynia 2.4-3.4 mm long; spikes 2-4, all approximate; plants rare, alpine on the Uinta Mtns. .... *C. bipartita*

4 Pistillate scales greenish to dark brown, shorter than the body of the perigynium; perigynia 1.7-2.5 mm long; spikes 4-8, the lower ones sometimes well separated; plants various........................... 5

5 Lower spikes 4-8 mm long, equal or shorter than the internode of the rachis (to 8 times shorter), with 5-10 (15) perigynia; perigynia somewhat spreading, smooth or nearly so except for the serrulate beak, thin and easily ruptured distally, the ventral surface nerveless or nearly so, the dorsal suture conspicuous the length of the beak and encroaching onto the body; scales tending to be pale and hyaline except for the greenish midstripe ..
    .... *C. brunnescens*

5 Lower spikes sometimes larger, usually equal to or longer than the internodes of the rachis, with (10) 15-32 perigynia; perigynia appressed usually granular roughened, firm-walled and not easily ruptured, the

ventral surface conspicuously nerved, the dorsal suture hardly if at all encroaching onto the body, or if so then the scales brown to dark brown 6

6 Pistillate scales brown to dark brown, the hyaline margins narrow to broad; perigynia slightly granular-roughened, not or sparingly serrulate distally, the nerves often reddish or tinged with brownish red, the dorsal suture conspicuous through the length of the beak and encroaching onto the body; plants 10-31 cm tall . . . . *C. praeceptorum*

6 Pistillate scales greenish or pale, occasional with light brown markings, the hyaline margins usually broad; perigynia conspicuous granular-roughened and serrulate distally (at 20 times or greater magnification), the nerves greenish or pale, the dorsal suture evident only on the beak if at all; plants 18-50 cm tall . . . . *C. canescens*

— KEY 4 —

1 Lowest bract and sometimes 1 or more upper bracts of at least some of the inflorescences as long or longer than the inflorescence . . . *C. athrostachya*

1 Bracts lacking or shorter than the inflorescence or rarely equaling it . . . 2

2 Pistillate scales as long and wide as the perigynia or nearly so; inflorescence pale green to stramineous or light reddish brown, rarely bicolored and/or tending to be spicate with the internodes of the rachis conspicuous, with (1) 3-6 (7) spikes 3

2 Pistillate scales shorter and narrower than the perigynia; inflorescence various colored but often dark brown to blackish or bicolored with the perigynia lighter than the scales; tending to be capitate, with the internodes of the rachis concealed by the congested spikes; with various number of spikes 7

3 Perigynia 2.9-4.3 mm long, 1.0-1.5 mm wide; plants 11-28 cm tall, edge of ponds and lakes, known from the western part of the Uinta Mtns. . . . . *C. leporinella*

3 Perigynia 4-8 mm long, 1.2-2.8 mm wide; plants of various stature places, distribution various. 4

4 Perigynia with flattened, winged, and serrulate, often ill-defined beak; plants commonly with conspicuous rhizomes with short internodes, apparently rare . . . . *C. tahoensis*

4 Perigynia with a terete entire beak; plants rarely with conspicuous rhizomes 5

5 Perigynia (5.7) 6-7.5 (8) mm long; plants lower montane to subalpine . . . . *C. petasata*

5 Perigynia 4-6 mm long; plants various. 6

6 Spikes mostly all 2 or more times longer than the internodes of the rachis; inflorescence rather strict, 1.4-3.0 cm long; plants mostly 10-30 cm tall, subalpine and more commonly alpine . . . . *C. phaeocephala*

6 At least the lower 2-3 spikes only 1-2 times longer than the internodes of the rachis; inflorescence rather flexuous, sometimes over 3 cm long; plants mostly 40-80 cm tall, lower montane to subalpine . . . . *C. praticola*

7 Perigynium with a flattened, winged, serrulate, often ill-defined beak . . . 8

7 Perigynia with a narrow, more or less terete beak, this not serrulate in the distal portion 12

8 Perigynia 2.3-3.7 mm long, 1.0-1.2 mm wide; plants known from wet places in Uintah Co. . . . . *C. bebbii*

8 Perigynia 4.5-8 mm long, 1.6-3.5 mm wide; plants usually of well-drained soil, of various distribution 9

9 Perigynia plano-convex; scales reddish brown with broad hyaline margins; plants known from the n. slope and e. end of the Uinta Mtns. . . . .
   . . . . *C. multicostata*

9 Perigynia strongly flattened except where distended by the achene; scales often brownish with green markings; plants of various distribution 10

10 Inflorescence spicate, the first 2 internodes collectively (8) 10-18 mm long; plants of the Tavaputs Plateau . . . . *C. tahoensis*

10 Inflorescence capitate, the first 2 internodes collectively only 4-7 (9) mm long; distribution various. 11

11 Perigynia 2.5-3.5 mm wide, commonly over 5.6 mm long, up to 45 or more per spike; spikes 7-12 mm wide; anthers 2.5-3.7 mm long; plants (15) 30-90 cm tall, of rather broad distribution . . . . *C. egglestonii*

11 Perigynia 1.6-2.5 mm wide, not over 5.6 mm long, to about 20 per spike; spikes 5-7 mm wide; anthers 1.5-2 mm long; plants 24-45 cm tall, known from w. of Rock Creek . . . . *C. straminiformis*

12 Perigynia 2.5-3.2 (3.5) mm long; inflorescence 0.7-1.8 cm long: perigynia sharp edged but hardly wing-margined, not serrulate, 2.5-3.0 mm long; inflorescence dark brown to blackish; plants common, mostly above 9,000 ft. . . . . *C. illota*

12 Perigynia 3.2-7.1 mm long; inflorescence sometimes longer (the following taxa, except possibly *13* form a complex that might be considered at a subspecific level under *C. macloviana* Urville) 13

13 Perigynia more or less strongly flattened except where distended by the relatively small achenes or else the scales dark brown to blackish; plants generally more common and widespread 14

13 Perigynia more or less plano-convex, the cavity nearly filled by the plump achene; scales greenish, light brown or reddish brown; plants rather rare and somewhat restricted 16

14 The longer perigynia 5.2-7.1 mm long, 4-6.4 times longer than wide; spikes commonly averaging over 10 mm long, the lower 1 or 2 regularly 11-13 (15) mm long, often elliptic, smooth with the perigynia appressed; inflorescence often cuneate at the base . . . . *C. ebenea*

14 Perigynia either shorter than above or else only 1.6-4 times as long as wide; spikes averaging less than 10 mm long, the lower 1 or 2 mostly less than 11 mm long, usually ovate, with the perigynia more or less spreading; inflorescence usually truncate at the base 15

15 Perigynia 1.3-2.2 mm wide, the body green or brownish in age, the beak often darker brown; scales brownish; inflorescence often conspicuously bicolored from the green or paler perigynia contrasting with the brownish scales . . . . *C. microptera*

15 Perigynia (1.75) 2.2-2.8 mm wide, usually averaging over 2.2 mm wide, the

body brown to dark brown, sometimes the winged margins greenish; scales brown to dark brown or blackish; inflorescence usually not conspicuously bicolored . . . . *C. haydeniana*

16 Pistillate scales equaling or scarcely shorter than the appressed perigynia, at least some with broad, shining, white-hyaline margins 0.1-0.3 mm wide . . . . *C. multicostata*

16 Pistillate scales conspicuously shorter than the spreading perigynia, generally lacking white hyaline margins or these usually less than 0.1 mm wide . . . . *C. pachystachya*

— KEY 5 —

1 Spikes sessile, congested into a head, the internodes of the rachis obsolete or at least short and hidden by the closely aggregated spikes 2

1 Spikes sessile or pedunculate, at least the lowest spike separated from the others by a conspicuous rachis internode 3

2 Perigynia 1.1-1.8 mm wide, usually somewhat inflated, without flattened margins or these generally not conspicuous, the width nearly filled by the mature achene, papillate at least apically (this most conspicuous along the margins at 20x or greater magnification) . . . . *C. nelsonii*

2 Perigynia 2-3 mm wide, strongly flattened with conspicuous broad margins, conspicuously wider than the mature achene, more or less glossy on the faces, glabrous or sometimes ciliolate apically . . . . *C. nova*

3 Lowest bract with a closed sheath 7-45 mm long; perigynia ciliolate-serrulate; spikes all on peduncles (section Ferrugineae) . . . . . *C. fuliginosa*

3 Lowest bract not sheathing or if so the sheath closed for less than 5 mm; perigynia glabrous papillate; spikes various 4

4 Pistillate scales conspicuously aristate, the awn 1-2 mm long; culms arising singly or few together from long rhizomes; leaves all on the culms, the lower ones reduced to bladeless sheaths; plants rare, of wet places . . . . *C. buxbaumii*

4 Pistillate scales acute to acuminate but not aristate; some leaves basal . . 5

5 Culms arising singly or few together from long creeping rhizomes, these covered with a yellowish felt-like tomentum; lateral spikes spreading or drooping, on slender peduncles; scales equaling or longer than the perigynia, not blackish or dark purple (section Limosae) . . . *C. paupercula*

5 Culms tufted and rhizomes lacking or short, if spikes spreading or drooping then the scales shorter than the perigynia, often blackish or dark purple 6

6 Spikes not over 1.3 cm long; perigynia 2-2.8 mm long, green at maturity and strongly contrasting with the dark scales . . . . *C. norvegica*

6 At least the largest spike over 1.3 cm long and/or the perigynia over 2.8 mm long, these sometimes as dark as the scales at least distally 7

7 Upper spikes not particularly more crowded than the lower ones, all sessile or nearly so; inflorescence not over 1 cm wide; perigynia 1.9-3.6 mm long . . . . *C. utahensis*

7 Upper spikes generally more crowded than the lower ones, the lowest one commonly on a conspicuous peduncle; inflorescence sometimes over 1 cm wide; perigynia 2.7-5 mm long 8

8 Perigynia light green, rarely with red markings, strongly contrasting with the dark scales; lateral spikes ascending, spreading, or drooping at maturity, each usually on a slender peduncle, at least some in each inflorescence commonly gynaecandrous .... *C. bella*

8 Perigynia olive green or commonly marked with dark red, purple or black, often about dark as the scales distally; lateral spikes ascending to erect, the upper ones commonly sessile or short-pedunculate, all pistillate or infrequently gynaecandrous in *C. heteroneura*. 9

9 Perigynia 2.7-3.5 mm long, 1.6-2.1 mm wide, consistently papillate, sometimes completely black or black-purple at maturity; lowest rachis internode 0.2-1 cm long; lowest spike sessile or on a peduncle to 0.3 cm long, the upper ones sessile or nearly so; plants 10-28 cm tall, rather rare, at or above timberline .... *C. albonigra*

9 Perigynia mostly over 3.5 mm long or over 2.1 mm wide, rarely papillate, sometimes partly or wholly greenish at maturity; lowest internode of the rachis ca 0.3-7 cm long; lowest spike sessile or on a peduncle to 4.5 cm long; plants (20) 25-100 cm tall, mostly below timberline 10

10 Perigynia not flattened, lanceolate or narrowly elliptic, olive green, becoming yellow-brown at maturity, seldom suffused with dark color except at the beak, not much wider than the mature achene; plants known from the Rock Creek drainage .... *C. atrosquama*

10 Perigynia usually flattened, broadly elliptic to ovate or obovate, usually conspicuously suffused with dark purple or black, wider than the achene; plants widespread .... *C. heteroneura*

— KEY 6 —

1 Scales pale green; lowest peduncle enveloped at the base by a closed sheath 2-6 mm long, often originating near the base of or on the lower 1/2 of the culm .... *C. aurea*

1 Scales black or blackish purple, sometimes with a green or paler midstripe; lowest peduncle not enveloped in a closed sheath, or if so the sheath mostly less than 2 mm long, mostly originating from the upper 1/2 of the culm . 2

2 Perigynia about as dark as the scales at maturity, slightly inflated; stigmas 2 or 3; style continuous with, persistent on, and of the same firm texture as the achene, strongly bent at maturity .... *C. saxatilis*

2 Perigynia greenish or at least paler than and strongly contrasting with the dark scales except in C. scopulorum, not inflated; stigmas 2; style jointed to, deciduous from, and of softer texture than the achene, straight (section Acutae) 3

3 Flowering culms all or mostly arising laterally and not enveloped at the base by the previous year's tufts of leaves; inner side of sheaths translucent and breaking up at maturity; plants forming large tussocks 0.5-2 m and larger in diameter, known from along the Green and White Rivers ......
.... *C. emoryi*

3 Flowering culms arising from the center of the previous year's tufts of leaves and surrounded at the base with dried leaves of the previous year; sheaths and distribution various; plants not as above in all respects..... 4

4   Leaves 1-2.6 mm wide; plants caespitose with numerous fibrous roots, the larger roots covered with yellowish or yellow-brown felt-like hairs; pistillate spikes 3-4 mm wide; perigynia more or less conspicuously stipitate, rather quickly deciduous after maturity, faintly nerved on both sides C. lenticularis

4   Some of the leaves over 2.6 mm wide or the perigynia not nerved except on the margins; plants rhizomatous, the culms arising singly or few together; fibrous roots often rather few, if with felt-like hairs then these whitish; perigynia not or hardly stipitate. 5

5   Perigynia nerved on both sides as well as on the margins, the beak (0.2) 0.4-0.6 mm long and bidentate and/or ciliolate; midrib of scales conspicuous throughout the length of the scale, at least some in each inflorescence usually excurrent into a mucro 0.5-1 (2) mm long; plants widespread in valleys and to mid-montane . . . . *C. nebrascensis*

5   Perigynia nerveless except on the margins, the beak not over 0.3 mm long, entire or obliquely cleft, not ciliolate; midrib of scales often inconspicuous toward the apex, not excurrent; plants low- to high-montane 6

6   Lowest bract shorter than the inflorescence; mature perigynia usually about as dark as the scales at least where exposed beyond the scale; pistillate spikes 5-10 mm wide . . . . *C. scopulorum*

6   Lowest bract usually exceeding the inflorescence; mature perigynia greenish or suffused with red or reddish brown, but not so dark as the scales; pistillate spikes 3-5 mm wide . . . . *C. aquatilis*

— KEY 7 —

1   Perigynia pubescent 2

1   Perigynia glabrous. 5

2   Pistillate spikes with only 1-4 perigynia, the lower ones originating from near the base of the plant; plants 5-35 cm tall, densely caespitose, without rhizomes, usually growing on dry or at least well-drained soil; staminate spikes 5-12 mm long . . . . *C. rossii*

2   Pistillate spikes with more than 4 perigynia, originating from the upper half of the culm; plants often taller, rhizomatous or not, mostly of moist or wet places; staminate spikes (at least some) commonly over 12 mm long 3

3   Perigynia densely pubescent throughout; staminate spikes (1) 2-3 (4) . . . 4

3   Perigynia sparsely pubescent . . . . *C. utahensis*

4   Leaves rolled or folded at least the distal 3/4, often appearing terete, 1-1.5 (2) mm wide where rolled or folded; plants known from ponds and marshes in subalpine moraine toward the e. end of the Uinta Mtns. . . *C. lasiocarpa*

4   Leaves flat or nearly so, the larger ones commonly 2-6 mm wide; plants widespread . . . . *C. lanuginosa*

5   Style continuous with, persistent on, and of the same firm texture as the achene, not withering; staminate spikes more than 1 and/or usually 2.1-8.5 cm long; larger pistillate spikes various but commonly (1) 2-5 (10.5) cm long; perigynia usually 5-7 (10) mm long (sometimes smaller in *C. saxatilis* but then plants keyed both ways), more or less inflated 6

5 Style jointed to, deciduous from, and of softer texture than the achene, withering; staminate spikes mostly 1, 0.3-2.1 cm long; pistillate spikes 0.5-2 (3) cm long; perigynia 1.9-5 mm long, not inflated 10

6 Leaf sheaths pubescent; plants known from along the White River in Colorado . . . . *C. atherodes*

6 Leaf sheaths glabrous; plants of various distribution (Section Vesicariae) 7

7 Pistillate scales narrowed to a serrulate-ciliolate awn; plants known from the floodplain of the Green River . . . . *C. hystricina*

7 Pistillate scales not awned; plants of higher elevations 8

8 Pistillate scales soon purple-black or black except at the often white hyaline acute tip; perigynia with the 2 marginal nerves evident, otherwise nerveless or faintly nerved dorsally, only slightly inflated, rather abruptly tapered to a short entire or inconspicuously bidentate beak, 3.2-5.2 mm long, commonly turning as dark, or nearly so, as the scale where exposed beyond it; stigmas 2 or 3; achenes lenticular or trigonous . . . . *C. saxatilis*

8 Scales green, stramineous, reddish brown, or if (rarely) dark brown then usually acuminate or caudate-acuminate; perigynia conspicuously nerved dorsally, conspicuously inflated, abruptly or gradually tapered to a conspicuously bidentate beak, 4-10 mm long, about the color of the scales, not blackish; styles 3; achenes trigonous. 9

9 Perigynia spreading often strongly so at maturity, the ellipsoid to subglobose body more or less abruptly contracted to a conspicuous beak; plants with robust, long-creeping rhizomes, common . . . . *C. utriculata*

9 Perigynia appressed or slightly ascending, the lanceolate to lance-ovate body gradually tapering to the often poorly defined beak; plants tufted, with rather short rhizomes . . . . *C. vesicaria*

10 Peduncle of the lowest and often the upper pistillate spikes enveloped in a closed sheath 0.4-2 (4) cm long or longer 11

10 Peduncles not enveloped in a sheath or this closed for less than 0.4 cm . 14

11 Plants with slender, creeping rhizomes. 12

11 Plants caespitose; rhizomes lacking or short 13

12 Perigynia blue glaucous, not golden in age; spikes borne on upper ¼ of culms . . . . *C. livida*

12 Perigynia not glaucous, turning golden in age; lowest spike often borne on the lower half of the culm . . . . *C. aurea*

13 Pistillate spikes sessile or on erect or ascending peduncles, all crowded or the lowest one sometimes separated; inflorescence 2-5 (10) cm long; staminate spike 0.7-2.1 cm long . . . . *C. oederi*

13 Pistillate spikes borne on capillary, spreading or drooping peduncles, remote; inflorescence 5-30 cm long; staminate spike 0.3-0.9 cm long . . . . *C. capillaris*

14 Pistillate spikes 4-6 mm long, sessile; perigynia widely spreading, greenish; scales greenish or pale; plants caespitose . . . . *C. interior*

14 Pistillate spikes longer and/or at least one on a conspicuous peduncle; perigynia and/or scales brownish, blackish, or black-purple 15

15 Pistillate spikes borne on slender, spreading to drooping peduncles; perigynia pale green; scales greenish brown to purple-brown; culms arising

singly or few together from long, creeping rhizomes; fibrous roots covered with a yellowish felt-like tomentum 16

15 Pistillate spikes sessile or on erect peduncles, or if the lower peduncles spreading then the perigynia and/or the scales often blackish or black-purple at least in part; culms tufted or arising singly or few together; fibrous roots not covered with a felt-like tomentum 17

16 Terminal spike 9-12 mm long, occasionally bisexual; lateral spikes often with a few staminate flowers at the base, never at the apex; leaves mostly with well-developed blades .... *C. magellanica*

16 Terminal spike (10) 15-21 mm long, staminate; lateral spikes sometimes with a few staminate flowers at the apex, never at the base; lower leaves usually bladeless .... *C. limosa*

17 Perigynia 1.9-3 (3.6) mm long; pistillate spikes 4-5 mm wide, commonly all sessile or nearly so .... *C. utahensis*

17 Perigynia (3) 3.3-5.2 mm long; pistillate spikes 5-8 mm wide, at least the lowest one commonly on a conspicuous peduncle.................... 18

18 Perigynia green or olive green, strongly contrasting with the dark scales; plants from 7,200-10,400 ft, of well-drained soil, tufted; rhizomes lacking or short .... *C. raynoldsii*

18 Mature perigynia as dark or nearly so at least in part as the blackish or black-purple scales; plants of higher elevations or mostly of wet places, commonly with short or long rhizomes............................ 19

19 Perigynia slightly inflated, not flattened; leaf blades 1.5-3 mm wide; styles continuous with and persistent on the achene, contorted in age; stigmas 2 or 3; staminate spikes 1-2 (3); plants mostly of wet places ..... *C. saxatilis*

19 Perigynia flattened, not inflated; leaf blades 2-6 mm wide; styles jointed with and deciduous from the achene, straight; stigmas 3; staminate spikes solitary; plants commonly on well-drained soil .... *C. paysonis*

Carex albonigra Mack — black-and-white-scaled sedge — The few specimens seen are all from above 11,000 ft on the Uinta Mtns.; July-Aug. Intermediate between *C. nova* and *C. atrata* and perhaps intergrading into both of these.

Carex aquatilis Wahlenberg — water sedge — Widespread in the Uinta Mtns. from (7,000) 9,500 ft to above timberline; riparian communities, most abundant in wet and boggy meadows where often associated with plane leaf willow; June-Aug. This is the dominant sedge in wet areas of alpine, morainal basins where cold air drainage is sufficient that frost is a potential on any night during summer. Water sedge is the dominant in an alpine plant community type in the Uinta Mtns. (Brown 2006). Easily confused with *C. lenticularis* and *C. nebrascensis*, but consistently different from them. Reports of *C. bigelowii* Torrey (*C. concolor* Mackenzie) are apparently based on specimens of *C. aquatilis* with short bracts.

Carex atherodes Sprengel — awned sedge — Apparently known from wetlands along the White River in Rio Blanco Co.; July-Sept.

Carex athrostachya Olney — slenderbeak sedge — Uinta Mtns.; occasional in moist or wet places and abundant in the drawdown basin at Moon Lake and other reservoirs; 7,500-9,500 ft; June-Sept. Distinguished from similar species by the long lower bracts, but this character is variable, and not in every flowering culm do the bracts exceed the inflorescence. Specimens with reduced bracts resemble *C. microptera,* but the perigynia are narrower than those of *C. microptera.*

Carex atrosquama (Mackenzie) Kelso (*C. atrosquama* Mackenzie) — dark-scale sedge — Known in our area from the head of Blind Stream and S. Fork of Rock Creek in the Uinta Mtns.; rare in open spruce woods and dry alpine tundra; July-Aug.

Carex aurea Nuttall — golden sedge — Common across the Uinta Mtns., occasional elsewhere; along streams, around seeps, springs, and other wet places; about 7,000-11,000 ft; June-Aug. *Carex hassei* Bailey is a similar taxon of questionable status that has been separated from *C. aurea* by greenish rather than golden perigynia and appressed rather than spreading scales. Some of our specimens resemble *C. hassei,* but these still seem to belong to *C. aurea.*

Carex backii F. Boott (*C. cordillerana* Sarella & B. A. Ford misapplied) — long-scale sedge — The one specimen seen is from the mouth of Carter Creek just above Flaming Gorge Reservoir.

Carex bebbii Olney ex Fernald — Bebb *sedge* — (*C. subfusca* Boott in S. Watson misapplied) The few specimens seen-are from Whiterocks Canyon, and White-rocks-Bench; wet places; 6,000-7,400 ft; June-July.

Carex bella Bailey — beautiful sedge — Uinta Mtns. and Bruin Point, Tavaputs Plateau; infrequent in lodgepole pine, spruce-fir, and alpine communities at 8,000 ft to above timberline; usually on well drained soil, sometimes showing a slight increase on areas cleared of timber; June-Aug.

Carex bipartita Allioni — two-parted sedge — The few specimens seen (including Maguire 14497 UTC; Lewis 648, 648a BRY) are from Gilbert Bench, Gilbert Peak, Bald Mtn. Smiths Fork, and head of W. Fork Blacks Fork, Uinta Mtns.; talus slopes and wet ground at seeps, along streams; 11,000-12,100 ft; .July-Aug.

Carex brunnescens (Persoon) Poiret — brownish sedge — Uinta Mtns.; infrequent in wet places at 7,300-9,500 ft; June-Aug.

Carex buxbaumii Wahlenberg — Buxbaum sedge — Known from a few, small, scattered populations in the Uinta Mtns. in wet or marshy meadows at 7,600-9,800 ft; June-Aug.

Carex canescens L. — silvery sedge — Common across the Uinta Mtns. along drainages and edges of ponds and lakes, from canyon bottoms to near timberline, sometimes an early successional species on sandy bottoms of dry beaver ponds; June-Aug.

Carex capillaris L. — hair-like sedge — Known from shady seeps or springs or open meadows in Whiterocks, Lake Fork, Rock Creek, and Sheep Creek drainages, Uinta Mtns.; 7,300-9,300 ft; June-Aug.

Carex capitata L. — capitate sedge — Uinta Mtns.; the few specimens seen are from wet meadows and snowbeds at the head of the Uinta drainage (Painter Basin and Gilbert Bench) at 12,000-12,400 ft; July-Aug. This is the dominant in a minor alpine plant community type in the Uinta Mtns. (Brown 2006).

Carex curatorum Stacey — canyon-seep sedge — Known from Ely Creek & Con-fluence of Ely and Jones Hole Creek, seeps in sandy ledges and hanging gardens; 5,600-5,640 ft (R. Harms 196 & T. Naumann 340 DINO!).

Carex cusickii Mackenzie ex Piper & Beattie — Cusick sedge — The 2 specimens seen are from meadows in the Blacks Fork drainage, n. slope Uinta Mtns. at 9,060-9,280 ft.

Carex diandra Schrank — lesser panicled sedge — The one specimen seen (Goodrich 19882 USUUB) is from a boggy meadow in Uinta Canyon at 7,500 ft.

Carex dioica L. [*C. gynocrates* Wormskjold ex Drejer] — yellow bog sedge — Uinta Mtns.; specimens seen are from Blacks Fork, Mill City Creek, Birch Creek, and Hickerson Park on the n. slope and S. Fork Rock Creek, Mill Park at the head of Hells Canyon, and S. Fork Ashley Creek on the s. slope; wet and boggy meadows in sunny and shady places; 8,900-10,260 ft; July-Aug.

Carex disperma Dewey — softleaved sedge — Uinta Mtns.; locally common in bogs and seeps, often in shade; from about 7,500 ft to near timberline; June-Aug.

Carex douglasii W. Boott — Douglas sedge — Widespread locally common to abundant; dry, often alkaline places, sagebrush communities and in meadows that are ephemerally moist, sometimes in compact soil of roads and around livestock watering places; 4,800-9,000 ft; May-Aug.

Carex duriuscula C. A. Meyer (*C. eleocharis* Bailey; *C. stenophylla* Wahlenberg misapplied) — needleleaf sedge — Widespread; occasional to common; usually associated with sagebrush and pinyon-juniper and rarely on windswept ridges; 7,000-10,720 ft; June-Aug.

Carex ebenea Rydberg — ebony sedge — Uinta Mtns.; occasional to common; edges of meadows, along streams, in open coniferous woods, and on well-drained slopes, mostly in the lodgepole pine-Engelmann spruce belt but occasionally above timberline; June-Aug. See *C. microptera*.

Carex echinata Murray (*C. angustior* Mackenzie; *C. muricata* L.) — narrowleaf sedge, boreal sedge — Uinta Mtns.; locally abundant in boggy meadows and seeps at (7,600) 9,500-10,400 ft; June-Aug. Similar to *C. interior*.

Carex egglestonii Mackenzie — Eggleston sedge — Infrequent across much of the Uinta Mtns. and common from Rock Creek and w. and on the W. Tavaputs Plateau; open slopes and openings in woods; 8,000 ft to timberline. Small specimens may be confused with *C. straminiformis*, but the perigynia are quite distinct, and the spikes of small plants are still larger than those of *C. straminiformis*.

Carex elynoides Holm. — blackroot sedge, kobresia-like sedge — Uinta Mtns.; common to dominant on dry, exposed ridges and slopes at subalpine and alpine elevations and rarely on exposed ridges at the upper limits of the mtn. big sagebrush belt including Coal Mine Hill on the n. slope; July-Aug. Of greatest extent and abundance in the Blind Stream-Rock Creek area on limestone where it is the dominate in an alpine community type (Brown 2006). Similar to and easily confused with *Kobresia myosuroides*, which grows in the same habitat. However, *Kobresia* myosuroides is common along the crest of the Uinta Mountains, and *Carex elynoides* is rather uncommon along the crest and more common at the lower limits of the alpine area of these mountains. The following key may help to distinguish plants of the 2 taxa:

1 Inflorescence of a solitary androgynous spike, the staminate flowers above and not mixed with the pistillate flowers; perigynium with sealed margins, enclosing the achene .... ***C. elynoides***
1 Inflorescence (spike) consisting of small, few-flowered spikelets, the

terminal spikelet staminate, the lateral spikelets mostly androgynous and thus staminate and pistillate flowers commonly mixed throughout most of the inflorescence; perigynia with unsealed margins, the achene exposed at least distally . . . . **Kobresia myosuroides**

Carex emoryi Dewey — riverbank tussock sedge, Emory sedge — Reported for low elevations of the White River drainage in Colorado, forming large tussocks along the Green River from Jensen to Ouray but producing relatively few flowering stems; May-June with perigynia dehiscent by July. Reported to have flat leaves. However, the lower part of leaves are strongly folded.

Carex engelmannii Bailey — Engelmann sedge — Uinta Mtns.; specimens seen are from Gilbert Bench, the Blind Stream Plateau between Rock Creek and N. Fork Duchesne of the s. slope, and Middle Fork Blacks Fork on the n. slope; well-drained snowbeds; 10,900-12,400 ft; Aug. Included in *C. breweri* F. Boott in IMF (6: 108). This is the dominant in a minor alpine plant community type in the Uinta Mtns. (Brown 2006).

Carex filifolia Nuttall — threadleaf sedge — Known from Daggett, Uintah, and Moffat Cos.; sagebrush, pinyon-juniper, and windswept grass-forb communities; 6,000-7,900 ft; May-June.

Carex foetida Allioni (*C. vernacula* L. H. Bailey) — foetid sedge — apparently rare in the Uinta Mtns. where known from a subalpine wet meadow, at 9,800 ft at Lost Lake, Wasatch Co. (Hayward 9948) and near Gilbert Peak at 12,100 ft (Goodrich 22760). Listed as *C. vernacula* in FNA (23:311).

Carex fuliginosa Schkuhr (*C. misandra* R. Brown) — few-flowered, shortleaf sedge — The few specimens seen are from alpine communities in the vicinity of Gilbert Peak, Uinta Mtns. at 11,500-12,800 ft; July-Aug.

Carex geyeri F. Boott — elk sedge — Common to dominant understory species in aspen, Douglas-fir, and lodgepole pine-communities of the Uinta Mtns. up to about 9,000 ft and infrequent to 10,000 ft, and Gamble oak communities of the E. Tavaputs Plateau, and occasionally in Douglas-fir and sagebrush-grass communities on the W. Tavaputs Plateau in and w. of the Avintaquin drainage; June-Aug. Plants remain green throughout the winter even when covered with considerable snow.

Carex haydeniana Olney — cloud sedge — Uinta Mtns. common on dry or moist but well-drained sites above and somewhat below timberline, dominant in late persisting, semi-barren, alpine snowbeds (Brown 2006) where in some years there might be little or no release from snow cover. June-Aug. See discussion under *C. microptera*.

Carex heteronerua W. Boott (*C. atrata* L. misapplied) — blackened sedge — Uinta Mtns.; occasional to common in moist openings in coniferous forests, moist meadows, along streams, and occasionally above timberline; 7,500-12,700 ft; July-Sept. Our plants belong to var. *erecta* (Mackenzie) F. Hermann. Rare specimens from the Uinta Mtns. have the scales slightly longer than the perigynia. These might be intermediate to var. *chalciolepis* (T. H. Holm.) F. J. Hermann (*C. chalciolepis* T. H. Holm).

Carex hoodii Boott — hood sedge — Known from w. of Lake Fork on the s. slope of the Uinta Mtns. and across the n. slope and in and w. of the Avintaquin drainage on the W. Tavaputs Plateau; occasional or locally common in open aspen and coniferous woods and open grassy slopes from 8,000 ft to timberline;

June-Aug. Wolf & Dever 5213 DINO, reported by Bradley (1950) for Round Top Mt., belongs C. occidentalis.

Carex hystericina Muhlenberg ex Willdenow — bottlebrush sedge, porcupine sedge — The 3 specimens seen are from Jones Hole (A. Holmgren & N. Holmgren 14251 UTC & A. Holmgren & T. Jensen 13980 DINO) and Desolation Canyon (A. Ross sn UTC and Atwood & Evenden 24395 BRY) along or near the Green River; 4,400-5,200 ft; June.

Carex idahoa L. H. Bailey — Idaho sedge — Strawberry Valley and Current Creek; silver sagebrush and moist meadows; 7,600-9720 ft; July-Aug.

Carex illota Bailey — sheep sedge, small-headed sedge — Uinta Mtns.; locally abundant; boggy meadows, along streams and around shores of ponds and "pot holes" in alpine moraine in lodgepole pine and Engelmann spruce forests; (7,500) 9,500-11,500 ft; July-Aug. This is the dominant in a minor alpine plant community type in the Uinta Mtns. (Brown 2006).

Carex interior Bailey — inland sedge — Uinta Mtns.; The few specimens seen are from Bear River, Rock Creek, Uinta, and Whiterocks drainages; boggy meadows and along streams; 7,000-7,400 ft; July-Aug.

Carex lasiocarpa Mackenzie (*C. lanuginosa* Michaux) — woolfruit sedge, slender sedge — Uinta Mtns.; common to abundant, and forming closed stands in alpine moraine potholes of Ashley Creek, Sheep Creek, and probably Carter Creek drainages, also at Meadow Park, Eagle Creek drainage; margins of and in small ponds (potholes), forming floating mats in peatlands; July-Sept. Often associated with *C. utriculata,* but not extending into as deep of water.

Carex lenticularis Michaux (*C. kelloggii* W. Boott) — Kellogg sedge — Uinta Mtns.; occasional in wet areas at 7,000-11,000 ft. Sometimes mistaken for *C. aquatilis.*

Carex leporinella Mackenzie — Sierra hare sedge — w. of Lake Fork (s. slope) and w. of Henrys Fork (n. slope) in the Uinta Mtns.; locally common in wet and drying meadows, often in drying mud of ephemeral pools or at the edge of ponds and lakes; 9,000-10,200 ft; July-Sept.

Carex leptalea Wahlenberg — bristlestalked sedge — Uinta Mtns.; known from small populations in Whiterocks and Uinta Canyons and in seeps of Eagle Creek and Lake Creek drainages in Daggett Co. where it is locally abundant in and around sphagnum bogs and other wet places at 7,200-7,800 ft; July-Aug.

Carex limosa L. — mud sedge, quaking bog sedge — Uinta Mtns.; locally abundant in quaking-bogs and wet meadows at 9,300-11,000 ft; July-Aug. Similar to *C. magellanica.*

Carex livida (Wahlenberg) Willdenow — livid sedge — Uinta Mountains; known from a few collections from Big Park, Ashley Creek Drainage, near Twin Lakes in the Dry Fork drainage and near Chepeta Lake in the Whiterocks drainage; wet meadows where soil is saturated to the surface for much of the growing season.

Carex magellanica Lamarck in J. Lamarck et al. (*C. paupercula* Michaux) — poor sedge — Uinta Mtns.: The few specimens seen are from sphagnum bogs and streambanks at 9,000-10,400 ft; June-Aug. Similar to and easily confused with *C. limosa.*

Carex microglochin Wahlenberg — subulate sedge — Uinta Mtns.; specimens seen are from a calcareous bog in South Fork Rock Creek, Sheep Creek Park and other meadows in the Hickerson Park area. One specimen is from a subalpine meadow near Dime Lake, Uinta drainage; 9,000-10, 600 ft; June-Aug.

Carex microptera Mackenzie (*C. festivella* Mackenzie; *C. limophila* Herm.) — smallwing sedge — Widespread; common to abundant along ephemeral or permanent water courses, moist but usually not wet meadows, and aspen and coniferous woods; 7,200-11,200 ft; May-Aug. Cronquist (IMF 6: 165) recognized the similarities of *C. microptera, C. ebenea,* and *C. haydeniana* to be such that the 3 might be regarded as varieties of a single species. Such a combination could also include *C. subfusca* W. Boott., *C. pachystachya,* and other species. They are all similar. Intergradation between *C. ebenea* and *C. haydeniana* has been found in plants of the Big Park area of N. Fork Ashley Creek. However, such intergradation is not common.

Carex multicostata Mackenzie — many-rib sedge — The 3 specimens seen are from the Uinta Mtns., (Hickerson and Sheep Creek Parks, Daggett Co. and Pat Carrol Park, Uintah Co.); wet meadows and edge of wet meadows; 8,600-9,300 ft; July-Aug. Perhaps specimens assigned here belong to other members of the Ovales Section.

Carex nardina Fries (*C. hepburnii* F. Boott) — spikenard sedge — The few specimens seen are from Rock Creek and westward in the Uinta Mtns.; above timberline on dry slopes and ridges; June-Aug. Might easily be confused with *C. elynoides.*

Carex nebrascensis Dewey — Nebraska sedge — (orthographic variant: *C. nebraskensis*) Widespread; common to abundant in wet meadows, marshy places, in seeps and springs, and along ditches, washes, and other water courses; 4,800-9,000 ft; May-Aug.

Carex nelsonii Mackenzie — Nelson sedge — Uinta Mtns.; occasional or locally common in meadows and rocky slopes; mostly above timberline; June-Aug. Similar to small specimens of *C. nova,* but distinguished by inflated and rounded rather than flattened perigynia.

Carex neurophora Mackenzie — alpine nerved sedge — Known from the Currant Creek and West Fork Duchesne River drainages, and one specimen seen (Lewis 731) is from 1-2 mi n. of Duchesne Ridge in Lake Creek, Provo River drainage. The closely related *C. jonesii* Bailey might be expected.

Carex nigricans C. A. Meyer — black alpine sedge — Uinta Mtns.; occasional in meadows, along streams, in coniferous forests where perhaps most common along ecotones between meadows and forest communities and forming nearly closed stands in alpine snowbeds where the soil is kept moist or wet through most or all of the growing season (Brown 2006); (8,000) 9,000-12,100 ft; June-Aug.

Carex norvegica Retz. (*C. media* R. Brown; *C. vahlii* Schkuhr) — Scandinavian sedge — Across the Uinta Mtns.; most common in the lodgepole pine-Engelmann spruce belt, occasionally at lower elevations in the major canyons, quite specific for moist sites, often along small streams; 7,400-11,050 ft; June-Aug.

Carex nova Bailey (*C. pelocarpa* F. Hermann) — new sedge — Uinta Mtns. at 9,160 -12,100 ft; June-Aug. With 2 vars.:

1   Perigynia not granular-roughened, the margins smooth; achenes long-stipitate; scales acuminate, the midrib prominent; culms flexuous, the inflorescence often nodding; plants most common on well drained, alpine slopes .... **var. pelocarpa** (F. Hermann) R. Dorn

1 Perigynia granular-roughened, the margins ciliate-scabrous; achenes short-stipitate; scales acute to short-cuspidate; with the midrib almost obsolete; culms stiff, the inflorescence erect; plants more common in riparian areas, meadows, and mesic forests than alpine . . . . **var. nova**

Carex obtusata Liljeblad — obtuse sedge — Widespread; occasional and locally common in sagebrush, aspen, and dry meadow communities; 6,500-9,000 ft; June-Aug.

Carex occidentalis Bailey — western sedge — Uinta Mtns. and Blue Mt.; occasional in open coniferous forests or where these forests have been opened by logging or fire, perhaps in other communities; 7,200-10,600 ft; June-Aug.

Carex oederi Retz. Green s. (*C. viridula* Michaux) — green sedge — Known from the Whiterocks and Uinta drainages; rare or infrequent along streams-and in wet places; 5,800-7,300 ft; June-Aug.

Carex pachystachya Chamisso — chamisso sedge — Strawberry Valley and Uinta Mtns.; infrequent in wet meadows and wet places in woods; 7, 000-9,000 ft; June-Aug. Similar to *C. microptera* but with smaller perigynia that are more readily deciduous.

Carex paysonis Clokey (*C. podocarpa* R. Brown misapplied; *C. tolmiei* F. Boott misapplied) — Payson sedge — Occasional and locally common in the central Uinta Mtns. at (10,500) 11,000-12,400 ft. This is the dominant in a minor alpine plant community type in the Uinta Mtns. (Brown 2006). June-Aug.

Carex pellita Willdenow (*C. lanuginosa* Michaux misapplied – the type specimen belongs to *C. lasiocarpa*) — woolly sedge — Widespread; common to abundant along streams, ditches, in marshes, and other wet places; about 4,800-9,500 ft; June-Aug.

Carex petasata Dewey — Liddon sedge — Widespread; common in sagebrush and dry aspen communities, and open grassy slopes; 7,500-9,000 ft; June-Aug. Similar to and passing into *C. praticola*.

Carex phaeocephala Piper — dunhead sedge — Uinta Mtns.; common on exposed, rather dry, often rocky places at (9,300) 10,000-11,900 ft; July-Aug. Difficult identification of this species is demonstrated by the number of times (5) this occurs in keys to species in FNA (23: 341-346).

Carex praeceptorum Mackenzie — slope sedge — widespread in the Uinta Mtns.; infrequent in wet, boggy places, often associated with sphagnum moss; (10, 000) 10,600-11,200 ft; July-Aug. Similar to and *C. canescens* with which it is easily confused.

Carex praegracilis W. Boott — blackcreeper sedge — Widespread; common to abundant in marshy places, irrigated pastures, along ditches, washes, and other water courses, around seeps and springs, occasionally in rather dry places, tolerant of alkali; 4,700-8,000 ft; May-Aug.

Carex praticola Rydberg — meadow sedge — Uinta Mtns. and W. Tavaputs Plateau; occasional in moist openings or open coniferous or aspen woods; 7,300-10,000 ft; June-Aug. Easily confused with and possibly passing into *C. petasata.*

Carex pyrenaica Wahlenberg — Pyrenean sedge, snowbed sedge — Known from isolated sites across the alpine of the Uinta Mountains where restricted to late-persisting snowbeds. This plant has capability to persist in snowbeds where in some years there is no release from snow cover. This is the dominant in an alpine plant community type in the Uinta Mtns. (Brown 2006).

Carex raynoldsii Dewey — Raynolds sedge — Locally common from w. of Rock Creek and s. of Grandaddy Basin in the Uinta Mtns. and w. of the Avintaquin drainage of the W. Tavaputs Plateau, infrequent at the e. end of the Uinta Mtns.; open slopes and in grass-forb communities, and openings in woods; July-Aug.

Carex rossii Boott (*C. brevipes* W. Boott misapplied) — Ross sedge — Widespread; occasional to common in sagebrush-grass, pinyon-juniper, mt. brush, and ponderosa pine communities, also in lodgepole pine-Engelmann spruce forests where it often becomes abundant following logging or after fire, rarely above timberline; 7,000-11,300 ft; May-Aug. Perhaps *C. geophila* Mackenzie enters the Uinta Basin at the se. corner in the Piceance Basin. The feature used to separate this from *C. rossii* (bract of lowest non-basal pistillate spike not leaf-like and shorter than the inflorescence) hardly warrants separation at the species level.

Carex rupestris Allioni (*C. drummondiana* Dewey) — curly sedge, rock sedge — common to dominant across the alpine of the Uinta Mtns. and less common at lower elevations including the windy summit of Cold Spring Mtn. at 8,550 ft; in dry exposed places where snow cover is blown away or greatly reduced during winter; June-Aug. In an alpine plant community classification study in the Uinta Mtns., curly sedge ranked 5th in amount of canopy cover (Brown 2006).

Carex saxatilis L. (*C. physocarpa* Presl) — russet sedge — Uinta Mtns.; occasional and locally abundant in shallow water and wet meadows; 9,000-12,500 ft; July-Sept.

Carex scirpoidea Michaux (*C. pseudoscirpoidea* Rydberg) — northern single-spike sedge — Uinta Mtns.; common to abundant on quartz-rich sandstones of the Uinta Mtn. Group and rare in limestone areas at the head of Blind Stream and S. Fork Rock Creek; various communities above timberline, and occasionally in ecotones between coniferous forest and meadows; 10,300-13,000 ft; June-Aug. Our plants belong to var. *pseudoscirpoidea* (Rydberg) Cronquist — western single-spike sedge. In an alpine plant community classification study in the Uinta Mtns., this was the 2nd most frequently encountered species with the 4th highest percent canopy cover (Brown 2006).

Carex scopulorum T. H. Holm — Rocky Mt. sedge — Locally common in the Green Draw-Pipe Creek area at the e. end of the Uinta Mtns.; moist and wet meadows and along streams; 8,040-8,200 ft; July-Aug. Specimens assigned here need additional study.

Carex siccata Dewey (*C. foenea* Svenson) — dry sedge, silvertop sedge — The several specimens seen are from lodgepole pine-Engelmann spruce and aspen communities toward the e. end of the Uinta Mtns. and from open slopes at the head of Hill Creek, E. Tavaputs Plateau; June-Aug.

Carex simulata Mackenzie — analogue sedge, short-beaked sedge — Widespread; wet, marshy, and boggy meadows, perhaps more-common in calcareous than noncalcareous places; 6,000-9,400 ft; June-Aug.

Carex stipata Muhlenberg in Willdenow — prickly sedge — Reported for area now covered with water of Flaming Gorge Reservoir at Sheep Creek, Daggett Co. by Flowers et al. (1960). No specimen was found in Utah Herbaria to verify this listing (Albee et al 1988). The closest known collection is from Weber Co. Utah. Not listed for southern Wyoming by Dorn (2001). It seems highly likely the report for Daggett Co. was based on a misidentification.

Carex straminiformis Bailey — Shasta sedge — Known from the head of Blind Stream w. of Rock Creek, Uinta Mtns. and Red Ledge Ridge above Strawberry Valley and from Blacks Fork drainage; grassy openings in coniferous forests and sagebrush-grass communities; 9,000-10,300 ft; June-Aug.

Carex subnigricans Stacey — dark alpine sedge — Widespread in the Uinta Mtns. but apparently limited to isolated populations in moist and wet meadows where forms dense stands in relatively small areas; 9,750-10,400 ft; June-Aug.

Carex tahoensis Smiley (*C. xerantica* Bailey misapplied) — dryland sedge — Two specimens seen (Vickery & Wiens 1683 UT, Hill Cr., E. Tavaputs Plateau, aspen-fir; Goodrich 6291, Strawberry Peak, W. Tavaputs Plateau open ridge top, 10,300 ft).

Carex utahensis Reznicek & Murray (*C. parryana* Dewey misapplied) — Utah sedge — The 2 specimens seen (Atwood 24445, 25765) are from along the Green River in Desolation Canyon; 7,600-9,720 ft; June-Aug.

Carex utriculata F. Boott (*C. rostrata* Stokes misapplied) — beaked sedge — Common across the Uinta Mtns., occasional on the Tavaputs Plateau; wetlands of canyon bottoms, subalpine wet meadows, in and around ponds, abundant to dominant in "pot holes" of alpine moraine where it is a pioneer in hydrach succession and grows in water up to 1 or 2 ft deep; 7,100-10,500 ft; June-Aug.

Carex vallicola Dewey — valley sedge — widespread; occasional to common in sagebrush-grass, mt. brush, and dry aspen communities, also in grass-forb communities of open slopes; 7,000-9,500 (10,000) ft; June-Aug.

Carex vesicaria L. — blister sedge, inflated sedge — Uinta Mtns., rarely collected, likely more common than number of collections indicate, in water and wet meadows, locally common along and just below the high-water line of East Park Reservoir; 9,850-10,000 ft; July-Aug.

Cyperus L. — Flatsedge

Plants annual (ours) or perennial with mostly solid, triangular stems; leaves basal or on the lower ¼ of the stem; inflorescence subtended by a whorl of leaf-like involucral bracts that usually exceed the inflorescence; spikelets several or many in capitate or spicate clusters, the terminal cluster sessile or nearly so, the lateral ones on short or long rays that originate in the axils of the sheathless involucral bracts; scales of spikelets 2-ranked in vertical rows, the lowest one empty and different from the fertile ones; flowers perfect, without perianth; stamens 1-2 (3); styles 2-3; achenes lenticular or trigonous.

1 Spikelets 1-4 cm long; at least some leaf-like bracts of inflorescence over 10 cm long . . . . ***C. erythrorhizos***

1 Spikelets to 1 cm long; leaf-like bracts of inflorescence less than 10 cm long 2

2 Awns of scales straight; inflorescence capitate-contracted, ovoid or nearly globose . . . . ***C. michelianus***

2 Awns spreading to slightly recurved; inflorescence various . ***C. squarrosus***

Cyperus erythrorhizos Muhlenberg — redroot flatsedge — Plants annual, 10-70 cm tall; leaf blades 2-10 cm wide; involucral bracts as wide or wider than the leaves, to 40 cm long; rays of inflorescence 2-6 and 1-10 (28) cm long cm long; spikelets 1-4 cm long, borne in elongate clusters, scales 1-1.5 mm long, blunt apically or the midrib slightly excurrent as a mucro; stamens 3; styles 3-branched; achenes trigonous. Specimens seen (Goodrich 27707 & 27732) are from a sandy banks of the Green River near Ouray.

Cyperus michelianus (L.) Link — Michel flatsedge — Plants annual, 2-15 cm tall; involucral bracts as long or longer than the leaf blades, widely spreading; inflorescence ovoid or nearly globose; rays short and hidden in the dense spikelets; scales with a short, erect awn; stamens (1) 2; stigmas 2 (3); achenes plano-convex or trigonous, surface with white hyaline cells. The one specimen seen (Goodrich 27731) is from sandy bars and banks of the Green River near Ouray. Introduced from Eurasia likely by waterfowl.

Cyperus squarrosus L. (*C. aristatus* Rottboell) — awned flatsedge — Plants annual, 1-15 cm tall; leaf blades 0.5-2.5 mm wide; involucral bracts mostly 1-6 cm long; rays of inflorescence lacking or 1-3 (6) and 0.4-4 cm long; spikelets 4-10 mm long, borne in dense head-like clusters, the clusters on ray-like peduncles or sometimes solitary and sessile, scales 1-3 mm long, awn-tipped, deciduous; stamens 1; style trifid; achene trigonous. Most of the specimens seen are from floodplains of the Green and Yampa Rivers with one specimen from a heavily grazed meadow at Dry Gulch; 4,650-5,100 (6,600) ft.

Eleocharis R. — Brown Spikerush

Plants perennial sometimes appearing annual, from rhizomes or fibrous roots, stoloniferous in 1 species; stems simple, slender, angled to terete or compressed; leaves reduced to bladeless sheaths, these at the base of the stem; inflorescence a solitary, terminal spikelet, lacking true involucre bracts, the lowest scale of the spike sometimes slightly modified; flowers bisexual, each one subtended by a scale; perianth of 0-6 (9) retrorsely scabrous bristles; style thickened at the base into a tubercle, this sometimes appearing like a cap on the achene; stigmas 2 or 3; achenes lenticular or 3-angular. *Eleocharis ovata* (Roth) Roemer & Schultes (*E. engelmannii* Steudel) has been reported, but we have not seen specimens supporting such reports. With its annual habit and 2 stigmas it is quite distinct from the taxa listed below.

---

1   Stigmas 2; achenes lenticular; our most common spikerush . . . **E. palustris**

1   Stigmas 3; achenes trigonous. 2

2   Base of style enlarged, forming a cap on the achene; culms often tufted . 3

2   Base of style not much enlarged, not forming a cap on the achene 4

3   Achenes longitudinally many ribbed, with cross ridges between the ribs; scales green or straw colored, the lowest one bearing a flower, not much if any broader than the upper ones; plants 3-12 cm tall, sometimes appearing annual . . . . **E. acicularis**

3   Achenes 3-ribbed, the surface cellular roughened; upper scales mostly dark except for the hyaline tip, the lower sterile ones mostly much wider than the upper ones; plants 8-30 cm tall . . . . **E. bolanderi**

4   Achenes 0.9-1.3 mm long; scales 1.5-2.5 mm long; spikelets 2.5-4 (6) mm long, with 2-9 (20) flowers; plants 2-6 (10) cm tall, sometimes appearing annual . . . . **E. parvula**

4 Achenes 1.9-2.8 mm long; scales 2.5-5.5 mm long; spikelets various; plants (5) 10 cm tall or taller 5

5 Some culms arching, taking root at the tip, (20) 40-100 cm long, more or less flattened distally, 1-2 mm wide; spikelets (5) 8-13 mm long, with (5) 10-25 flowers; lowest scale empty . . . . *E. rostellata*

5 Culms erect, not taking root at the tip, 5-15 (20) cm tall, seldom over 1 mm thick: not flattened; spikelets 4-8 mm long, with 3-9 flowers, the lowest scale not empty . . . . *E. quinqueflora*

Eleocharis acicularis (L.) Roemer & Schultes — needle spikerush — widespread but other than from Desolation Canyon no specimens seen from the Tavaputs Plateau; mostly in wet places, often on mud flats of drying ponds and reservoirs; (4,300) 7,185-10,500 ft; June-Sept.

Eleocharis bolanderi A. Gray — Bolander spikerush — Specimens seen are from scattered locations across the Uinta Mtns. including Cold Spring Mtn.; locally common in wet and ephemerally wet meadows; 8,100-8,500 ft; July-Aug.

Eleocharis palustris (L.) Roemer & Schultes (*E. calva* Torrey; *E. macrostachya* Britton in Small) — common spikerush — Our most widespread and common spikerush; wetlands, tolerant of alkali; from the bottom of the Uinta Basin to about 10,500 ft; May-Aug. The specimen referred to as *E. montana* H.B.K. (*E. parishii* Britt) by Flowers and others (1960) belongs here as does Graham 7768 reported as *E. arenicola* Torrey by Graham (1937).

Eleocharis parvula (Roemer & Schultes) Link (*E. coloradoensis* Britton) — dwarf spikerush — The few specimens seen are from the edge of ponds, lakes, and wet lowlands of the Basin in Uintah Co.; inconspicuous and likely more common than collections indicate; June-Sept. With reduced perianth bristles and dull, minutely roughened achenes, our plants belong to var. *anachaeta* (Torrey) Svenson.

Eleocharis quinqueflora (Hartman) O. Swartz [*E. pauciflora* (Lightfoot) Link] — few-flowered spikerush — Uinta Mtns.; locally common to abundant in boggy meadows and other wet places at 6,840-11,100 ft; July-Sept. See comment under *Scirpus cespitosus*.

Eleocharis rostellata (Torrey) Torrey — beaked spikerush — The 3 specimens seen are from irrigated pastures at Neola, Mt. Home, and wetlands along the Duchesne River near Tabiona; May-Aug.

Eriophorum L. — Cotton-grass; Cotton-sedge

Plants perennial, from long, scaly rhizomes; stems solid, simple; leaves with closed sheaths and grass-like blades or upper blades reduced; spikelets 1-several in a terminal inflorescence, subtended by 1-several leafy or scale-like involucral bracts; flowers bisexual, each subtended by a scale; the perianth of numerous whitish, much elongate (in fruit) bristles, these giving the inflorescence the appearance of a tuft of cotton; stamens 3; stigmas 3; achenes triangular in cross-section.

---

1 Spikelets (2) 3-6 (8), borne on slender pedicels; lowest bract more or less leaf-like, 2.5-5.5 cm long; anthers 2-4 mm long; upper leaves with well-developed blades mainly 2-6 mm wide; scales acute . . . . . *E. angustifolium*

1   Spikelet 1, sessile at the apex or the culm; bracts scale-like, less than 1 cm long; anthers less than 2 mm long; leaf blades obsolete or to about 1 mm wide; scales acuminate .... *E. scheuchzeri*

Eriophorum angustifolium Honckeny (*E. polystachion* L.) — manyhead cottongrass — The 19 specimens seen are from the Uinta drainage east to the Brush Creek drainage on the s. slope of the Uinta Mtns. and Daggett Co. on the n. slope; locally common in wet and boggy meadows and along streams at 8,990-11,800 ft; July-Aug.

Eriophorum scheuchzeri Hoppe (*E. chamissonis* C. A. Meyer) — onehead cottongrass — The 24 specimens seen are from Rock Creek east to the Whiterocks drainage on the s. slope of the Uinta Mtns. and Blacks Fork to Henrys Fork on the n. slope; wet and boggy meadows, along streams, and in wet rocky places at 10,900-12,500 ft; July-Aug.

Kobresia Willdenow

Plants perennial, caespitose; culms obtusely triangular, solid; leaves narrow or filiform, basal or borne near the base of the culm (in ours); inflorescence a spike or of spikes, each spike with several spikelets; spikelets subtended by a small scale and consisting of a single pistillate flower and one or more staminate flowers; flowers without a perianth, the pistillate consisting of a pistil only, the staminate reduced to stamens, these subtended by and loosely folded in a scale, the scale (corresponding to the perigynia of *Carex*) with open margins and not closed around the achene.

1   Plants 25-35 cm tall, of wet places, below timberline; spikes (1) 3-12 .... *K. simpliciuscula*

1   Plants 5-20 cm tall, from well-drained, exposed sites near or above timberline; spike solitary 2

2   Spike to 3 mm wide .... *K. myosuroides*

2   Spike greater than 3 mm wide .... *K. sibirica*

Kobresia myosuroides (Villars) Fiori & Paoli [*K. bellardii* (Allioni) Degland] — Bellardi kobresia — Occasional to locally common along the crest of the Uinta Mtns. on flat or slightly convex alpine slopes and ridges that are mostly free of snow cover by June at 10,500-12,350 ft; June-Aug. Rather easily confused with *Carex elynoides* (q.v.).

Kobresia sibirica Turczaninow — Siberian kobresia — The 2 specimens seen (Goodrich 22759 and 25932) are from within 1 mile of Gilbert peak, Uinta Mtns. on the edge of leeward sides of alpine ridges at 12,100-12,700 ft.

Kobresia simpliciuscula (Wahlenberg) Mackenzie — wetland kobresia — Uinta Mtns.; known from a calcareous fen in the S. Fork of Rock Creek (Duchesne Co.), and boggy meadows at and near Sheep Creek Park and Hickerson Park (Daggett Co.), Blacks Fork, Smiths Fork, and Henrys Fork drainages (Summit Co.) were usually found on or near limestone at 8,500-9.860 ft; June-Aug.

Scirpus L. — Bulrush

Annual or perennial plants; leaves reduced to bladeless sheaths or with well-developed blades; involucre of 1-many scale-like or leaf-like bracts; spikelets solitary to numerous, in a spicate, capitate, umbellate, or paniculate inflorescence; scales of spikelets spirally arranged, with or without an excurrent awn; flowers bisexual; perianth of 1-6 bristles, these sometimes obsolete; stamens 3 (rarely fewer); stigmas 2 or 3; achenes lenticular or trigonous, with or without a stylar apiculus, but without a tubercle.

---

1 Spikelets solitary; involucre of 2-3 modified empty scales less than 1 cm long; plants 4-20 cm tall, perennial, densely tufted, of wet and boggy subalpine meadows in the Uinta Mtns. . . . . *S. cespitosus*

1 Spikelets 2-many; involucre of 1-several more or less leaf-like bracts; plants various but often taller, of lower elevations 2

2 Plants annual, tufted, from fibrous roots, 3-25 cm tall; achenes dark gray-brown to blackish, conspicuously cross-ridged . . . . *S. supinus*

2 Plants perennial, from rhizomes, often over 25 cm tall; achenes mostly lighter, not cross-ridged. 3

3 Involucre with 2 or more leaf-like, spreading bracts. 4

3 Involucre with only 1 leaf-like bract, this erect or nearly so 7

4 Spikelets small and numerous; mostly 3-6 mm long and over 100 in an open umbellate inflorescence 5

4 Spikelets larger and fewer, mostly 10-35 mm long and 3-40 (rarely more) in a compact or somewhat open umbellate inflorescence. 6

5 Stigmas 2; midrib of scales abruptly contracted into a short mucro; leaf sheaths mostly anthrocyanic; plants mostly of canyons . . . *S. microcarpus*

5 Stigmas 3; midrib of scales tapered into a short awn; plants known from low elevations of the Uinta Basin . . . . *S. pallidus*

6 Stigmas 2; spikelets 3-25, all in a sessile cluster or in 1-4 additional clusters borne on peduncles to 6 cm long; plants common, widespread . . . . *S. maritimus*

6 Stigmas 3; spikelets commonly 10-40, at least a few usually borne singly, the inflorescence usually umbellate with few to several ray-like peduncles to 7 cm long, plants rare, known from Daggett Co. . . . . *S. fluviatilis*

7 Inflorescence open, with conspicuous branches; spikelets numerous, more than 20; culms terete, commonly 1 cm or more in diameter, 80-300 cm tall; leaves reduced to bladeless sheaths or the blades short and erect or nearly so 8

7 Inflorescence head-like, with the sessile spikelets in a sessile cluster; spikelets 3-15; culms triangular or if subterete then usually with well-developed ascending-spreading leaf blades, commonly less than 1 cm in diameter, 10-116 cm tall. 9

8 Spikelets appearing dull orange or reddish-brown, the scales with striolae about the same color as the rest of the scale; scales mostly 2-3 (3.5) mm long; plants rather uncommon . . . . *S. validus*

8 Spikelets appearing dull gray-brown, the scales with prominent red-brown striolae that contrast with the gray-white color of the rest of the scale;

larger scales mostly 3.5-4 mm long; plants common . . . . **S. acutus**
9 Culms sharply triangular, conspicuously concave in cross-section . . . .
. . . . **S. americanus**
9 Culms rounded to triangular but not strongly concave in cross-section . .
. . . . **S. pungens**

Scirpus acutus Muhlenberg ex Bigelow [*S. occidentalis* (S. Watson) Chase; *Schoenoplectus acutus* (Muhlenberg ex Bigelow) A. & D. Love] — hardstem bulrush, tule bulrush — Widespread; common to dominant and often forming closed clones in marshes at low elevations; not expected above 6,500 ft; June-Aug.

Scirpus americanus Persoon [*S. olneyi* A. Gray; *Schoenoplectus americanus* (Persoon) Volkart ex Schinz & B Keller] — American threesquare — Known from along the Green River at Little Hole and near Steamboat Rock (Wolf 5277 as listed by Bradley 1950). Reported to have been frequent along the Green River in the area now covered by water of the Flaming Gorge Reservoir by Flowers et al. (1960). However, no specimen was found in Utah Herbaria by Albee et al. (1988) to verify this listing.

Scirpus cespitosus L. [*Trichophorum cespitosum* (L.) Hartman] — deerhair bulrush — Uinta Mtns.; locally abundant in subalpine, boggy meadows; 9,500-11,200 ft; July-Aug. With a single, terminal spike and nearly bladeless leaves, this is easily mistaken for an *Eleocharis*. However, the small bract that simulates a continuation of the culm can usually be seen. This is also distinguished from *Eleocharis* by the small blade (to 6 mm long) often on the uppermost sheath. It is separated from most species of *Eleocharis* by its high elevation habitat, but it is often associated with or growing near *Eleocharis quinqueflora* from which it is distinguished by caespitose rather than a spreading-rhizomatous habit and paler color of the inflorescence. It often forms hummocks in meadow areas that dry late in the season, while *Eleocharis quinqueflora* does not form hummocks and is most common where meadows are permanently wet.

Scirpus fluviatilis (Torrey) A. Gray [*Bulboschoenus fluviatilis* (Torrey) Sojak] — river bulrush — known from along the Green River at Browns Park, bulrush cattail community at 5,410 ft (Neese 14802 at BRY); July-Aug.

Scirpus maritimus L. [*S. paludosus* A. Nelson; *S. robustus* Pursh misapplied; *Bulboschoenus maritimus* (L.) Palla in Koch] — alkali bullrush — Widespread; common in wetlands and along ditches and other waterways; up to about 7,000 ft; June-Aug. Our plants belong to the rather ill-defined var. *paludosus* (A. Nelson) Kukenthal.

Scirpus microcarpus Presl (*S. rubrotinctus* Fernald) — panicled bulrush — Occasional in canyons of the Uinta Mtns. (one specimen seen from low in the Uinta Basin); wet places; up to about 8,000 ft; June-Aug.

Scirpus pallidus (Britton) Fernald (*S. atrovirens* Willdenow misapplied) — pale bulrush — The several specimens seen are from Uintah Co.; along ditches, washes, and the Green River, tolerant of alkali; not expected above 7,000 ft; June-Aug.

Scirpus pungens Vahl [*Schoenoplectus pungens* (Vahl) Palla] — common threesquare — Widespread; common in wet places at low elevations and up to about 7,000 ft, tolerant of alkali; May-Aug. Reports of *S. nevadensis* S. Watson are probably based on misidentified specimens of *S. pungens.* Our plants belong to var. *longispicatus* (Britton) Cronquist.

Scirpus supinus L. [*Schoenoplectus saximontanus* (Fernald) J. Raynal] — sharpscale bulrush — Known from ephemeral ponds on the floodplain of the Green River near Ouray; July-Sept.

Scirpus validus Vahl [*S. lacustris* L. ssp. *validus* (Vahl) Koyama; *Schoenoplectus tabernaemontani* (C. C. Gmelin) Palla] — softstem bullrush — Rather uncommon but apparently widespread; wet places; 5,380-7,120 ft; June-Aug.

DIPSACEAE Teasel Family

Dipsacus L. — Teasel

Dipsacus sylvestris Hudson [*D. fullonum* L. ssp. *sylvestris* (Clapham) Clapham] — teasel — Robust biennial herbs 50-200 cm tall; stems angled, with prickles (these often recurved) on the angles; leaves of the basal rosette oblanceolate, crenate, to 60 cm long, usually withered in the second season; stem leaves usually lanceolate, crenate-serrate or the upper entire, sessile, connate, 10-30 cm long, prickly on the midrib below; inflorescence a dense (brush-like) head subtended by few to several linear, elongate, prickly involucral bracts shorter than or exceeding the head; each flower subtended by a chaffy awned bract that exceeds the flower; calyx a tuft of silky hairs about 1 mm long; corolla tubular, 8-12 mm long, white to light purple, 4-lobed; stamens 4, slightly exserted; ovary inferior, 1-chambered; style elongated; stigma entire; fruit an achene. One specimen seen (L. C. Higgins & Welsh 14763) is from Grouse Canyon, Daggett Co. Other specimens are from the floodplain of the Green River at Little Hole; riparian communities at 5,600-6,600 ft; July-Aug.

Usage of *Dipsacus sylvestris* rather than *D. fullonum* is based on a concept of wild (*D. sylvestris*) and cultivated (*D. fullonum*) populations being different species. However, usage based on this concept has been reversed in different treatments, and usage is likely to remain divided (IMF 4: 550-551).

ELAEAGNACEAE Oleaster Family

Shrubs or trees; herbage conspicuously scurfy with silver or reddish-brown to golden stellate or scale-like hairs; leaves alternate or opposite, simple, entire; flowers bisexual or unisexual, radially symmetrical, mostly axillary; perianth of a single whorl; tepals 4, arising with the stamens at the apex of a floral tube; pistil 1; ovary superior, surrounded by the floral tube; fruit drupe-like or berry-like, consisting of a nutlet or achene closely invested by the floral tube.

---

1   Leaves alternate; flowers bisexual; stamens 4; fruit dry and mealy, covered with scurfy scales ....*Elaeagnus*

1   Leaves opposite; flowers unisexual; stamens usually 8; fruit succulent, not covered with scales ....**Shepherdia**

Elaeagnus L. — Oleaster; Silverberry

---

1   Widest leaves over 15 mm wide; leaves and twigs with brown dots as well as silvery scurfy; plants of Daggett Co., 1-2 (5) m tall ....**E. commutata**

1   Leaves less than 15 mm wide, except on vigorous young shoots; leaves and
     twigs without brown-dots; plants widespread, common, to 10 m tall . . . . .
     . . . . *E. angustifolia*

Elaeagnus angustifolia L. — Russian olive — Introduced from Europe, cultivated,
escaping, naturalized and weedy; along water courses, roadsides, fence lines,
and neglected pastures and fields; up to about 7,000 ft; May-June.

Elaeagnus commutata Bernhardi — silverberry, silverbush — The few specimens
seen are from Henrys Fork Park, Burnt Fork, and Sheep Creek on the n. flank
of the Uinta Mtns. and one specimen is from Lake Fork Canyon of the s. slope;
along streams and reported for dry hillsides; 6,500-8,000 ft; May-June.

Shepherdia Nuttall — Buffaloberry

---

1   Leaves silver-gray; shrubs or trees 2-7 m tall; mostly growing below 7,500
     ft . . . . *S. argentea*
1   Leaves green and nearly glabrous above, the lower surface lighter and
     densely dotted with copper-colored or brownish stellate, scurfy scales;
     shrubs 0.5-1.5 m tall; mostly growing above 7,500 ft . . . . *S. canadensis*

Shepherdia argentea (Pursh) Nuttall [*Lepargyraea argentea* (Pursh) Greene] —
silver buffaloberry — Widespread; common to abundant along ditches, washes,
fencelines, powerlines, and in abandoned fields; mostly below 7,500 ft; May-
June.

Shepherdia canadensis (L.) Nuttall [*Lepargyraea canadensis* (L.) Greene] — russet
buffaloberry — Uinta Mtns. and Tavaputs Plateau; occasional to common in
aspen, fir, lodgepole pine, and spruce woods, apparently more common on
basic substrates but also on quartz-rich sandstones of the Uinta Mt. Group;
8,200-10,600 ft; July-Aug.

ELATINACEAE Waterwort Family

Annual herbs; leaves opposite; flowers axillary, inconspicuous, bisexual; sepals
and petals 2-5, free or united at the base; stamens as many or twice as many as
the petals; ovary superior; fruit a septicidal capsule with several or many
seeds.

---

1   Plants glandular-puberulent; leaves denticulate, elliptic to oblanceolate;
     flowers 5-merous; sepals coarsely glandular-denticulate, pubescent, pointed
     . . . . ***Bergia***
1   Plants glabrous; leaves entire, sessile or nearly so, linear to narrowly
     spatulate; flowers 2- or 3-merous; sepals not as above . . . . ***Elatine***

Bergia L. — Bergia

Bergia texana (Hooker) Seubert — Texas bergia — Erect to decumbent annual
with simple or freely branched stems 2-30 cm long; leaves 1-3 (4) cm long,
gradually tapered or rather abruptly tapered to a petiole; stipules conspicuous,
glandular-denticulate to glandular-pectinate; sepals about 3 mm long, strongly

carinate; petals white, shorter than the sepals. The 2 specimens seen are from mud flats on the floodplain of the Green River at Leota Bottom near Ouray (Thorne & Goodrich 3578), and below the high waterline of Steinaker Reservoir (Goodrich 22007). Several repeat visits to Steinaker Reservoir have failed to locate more specimens. The one specimen appears to a waif introduced by waterfowl.

Elatine L. — Waterwort; Mud-purslane

Elatine rubella Rydberg — three lobed waterwort — Annual plants 2-5 (15) cm tall, with ascending to erect branches; leaves 5-12 mm long, entire, linear to narrowly spatulate, the tips truncate to emarginate; flowers bisexual, 2- or 3-merous, usually solitary in the axils of leaves, minute; sepals and petals not united or the sepals united at the base; fruit a septicidal capsule with axil placentation; seed numerous, reticulate in longitudinal rows. The one specimen seen (Neese 15104) is from the Uinta Mtns., Big Park on the road to Leidy Peak; drying mud and shallow water of pot holes; 9,880 ft; Aug. Our plants have commonly been referred to as *E. triandra* Schkuhr.

EPHEDRACEAE Ephedra Family

Ephedra L. — Mormon Tea; Ephedra; Jointfir; Brigham Tea

Freely branched dioecious (rarely monoecious) shrubs; stems jointed, longitudinally furrowed, green; leaves minute, scale-like, united at the base, whorled or opposite, seldom long-persisting; cones paired or in whorls, borne at the nodes of young branches, small, of several whorls of more or less membranous bracts, each bract of the staminate cones subtending a 2-lipped fleshy perianth with 5-8 anthers that are free or united into a column, the ovulate cones with 2 terminal ovules, each ovule enclosed in 2 integuments; seeds paired, enclosed in a hardened envelope.

1   Leaves and bracts principally in 3's; seeds slender, acute apically, about twice as long as wide; stems divaricately branched, bluish-green, glaucous
.... *E. torreyana*

1   Leaves and bracts in 2's; seeds plump, rather blunt apically 2

2   Base of mature leaves light brown or gray, not persistent; branches divergent, glaucous-green .... *E. nevadensis*

2   Base of mature leaves usually dark brown, persistent; branches erect, fastigiate, broom-like, bright yellowish or green..................... 3

3   Ovulate cones usually evidently pedicellate; plants with spreading rhizomes, forming hummocky clumps or colonies in deep sand; known from near the Colorado border e. of Raven Ridge .... *E. cutleri*

3   Ovulate cones subsessile, the peduncles if present mostly less than 5 mm long; plants mostly solitary, not forming colonies not rhizomatous, widespread .... *E. viridis*

Ephedra cutleri Peebles [*E. viridis* var. *viscida* (Cutler) L. D. Benson] — Cutler Mormon tea, Cutler ephedra, Navajo ephedra — Locally common in sand dunes near the Colorado border e. of Raven Ridge at about 5,700 ft; May-June.

This is not always clearly distinct from *E. viridis;* also reported to hybridize with *E. torreyana.*

Ephedra nevadensis S. Watson — Nevada Mormon tea — Occasionally reported for the Uinta Basin, but most of the material is better assigned to E. viridis. Neese & Fullmer 11354 (9 mi e. of Jensen; juniper mixed shrub community), Harrison & Larsen 7748, (town of Mount Emmons, Duchesne Co.), Goodrich 4070 (Tridell) and Lambert 78 (Dry Gulch, 7,500 ft) (both specimens at USUUB) are the only specimens retained here. The distribution map for this species in FNA (2: 433) excludes northern Utah. Perhaps the above listed specimens belong to *E. viridis* or *E. torreyana.*

Ephedra torreyana S. Watson — Torrey Mormon tea, Torrey ephedra — Widespread: frequent in shadscale, greasewood, mixed desert shrub, sagebrush, and pinyon-juniper communities below 6,000 ft; May-June.

Ephedra viridis Coville — green Mormon tea, green ephedra — Widespread; common in shadscale, sagebrush, pinyon-juniper, pinyon-mt brush-sagebrush communities, often on rocky, gravelly slopes, or sandy places; to 8,000 ft; May-July. See note under *E. cutleri.*

EQUISETACEAE Horsetail Family

Equisetum L. — Horsetail; Scouringrush

Annuals or rhizomatous perennials; stems green (or, in *E. arvense,* the fertile ones pale pinkish or brown, without chlorophyll and ephemeral), mostly erect, simple, consisting of a series of jointed, sheathing, grooved segments; leaves reduced to a ring of confluent scales forming sheaths at the nodes; reproduction by spores; spores borne in 5-10 sporangia clustered around each of numerous spore-bearing stalks, these covered by polygonal caps and closely spaced around a central axis to form a terminal cone-like structure (strobilus).

---

1 Sterile stems with whorls of branches, green; fertile stems precocious and preceding the sterile ones, unbranched, pinkish or brownish . . *E. arvense*

1 Stems all alike, unbranched. 2

2 Sheaths black-banded above and below an ash-gray center . . . . *E. hyemale*

2 Sheaths mostly green, black-banded only at the tip, or sometimes the lower sheaths banded at the tip and base. 3

3 Leaf sheaths mostly 7-15 mm long, the teeth usually deciduous; stems mostly 20-100 cm tall, 3-9 mm thick, with 15-34 vertical ridges; strobilus 1.5-2.5 cm long, usually blunt at the tip . . . . *E. laevigatum*

3 Leaf sheaths mostly 2-4 mm long, the teeth persistent; stems mostly 10-30 cm tall, 1-3 mm thick, with 5-12 vertical ridges; strobilus to 1 cm long, with a rigid pointed tip . . . . *E. variegatum*

Equisetum arvense L. — field horsetail — Widespread; common to abundant along water courses; up to about 8,000 ft; April-June. The short-lived fertile stems simulate the other species in form but they lack chlorophyll.

Equisetum hyemale L. (*E. prealtum* Rafinesque) — common scouringrush — Widespread; common along water courses and other wet places; up to 7,500 ft; May-Oct. Our plants belong to var. *affine* (Engelmann) A. A. Eaton.

Equisetum laevigatum A. Braun (*E. kansanum* Schaffner) — smooth scouringrush — Common, widespread, along water courses and in other wet places; up to about 8,050 ft; May-June.

Equisetum variegatum Schleicher ex Weber & Mohr — variegated scouring rush — To be expected in similar habitat as the above species. Listed for Daggett, Duchesne, and Summit Counties (AUF 3). However, only 7 specimens are listed in AUF for all of Utah. This indicates the species can be expected to be relatively uncommon. Plotted for Moffet County but not other counties within our area in TPD. The distribution map in FNA shows the Uinta Basin at the southern end of the range of this species. More collections of this species from our area are needed to better determine distribution.

ERICACEAE Heath Family

Shrubs or subshrubs; leaves usually evergreen, alternate, opposite, or whorled, simple; inflorescence various; petals free or united; stamens equal to or twice as many as the corolla lobes, rarely fewer, arising from the outer margin of a disk or pouch of the corolla, the anthers inverted or erect, usually opening by terminal pores; pistil 1; ovary superior or inferior, (1) 4-12 chambered; styles erect; fruit various.

1 Leaves mostly opposite; corolla rose-purple, 10-15 mm wide . . . . . . *Kalmia*

1 Leaves alternate; corolla various 2

2 Twigs and lower side of leaves with few to numerous yellow resin-dots; leaves slightly to strongly revolute, dark green above, pale beneath; flowers in terminal umbel-like racemes; petals not united . . . . *Ledum*

2 Twigs and leaves not as above; corolla united at least about ½ the length 3

3 Leaves serrate with small teeth. 4

3 Leaves entire or nearly so 5

4 Corolla about ½ united . . . . *Gaultheria*

4 Corolla united to near the tip . . . . *Vaccinium*

5 Stems creeping, mostly less than 15 cm tall 6

5 Stems ascending, not creeping 7

6 Stems scarcely woody, not over 1.5 mm thick, mostly less than 20 cm long . . . . *Gaultheria*

6 Stems woody, mostly over 1.5 mm thick, mostly over 20 cm long . . . . *Arctostaphylos*

7 Leaves leathery, persistent through winter; fruit not succulent, not blue; plants of well drained soil . . . . *Arctostaphylos*

7 Leaves thin, deciduous, fruit blue and succulent; plants of wet places . . . . *Vaccinium*

Arctostaphylos Adanson — Manzanita

Shrubs usually with smooth, reddish or brown bark; leaves alternate, evergreen, entire, thick, leathery; flowers borne in terminal racemes or panicles, each subtended by a bract; sepals nearly distinct, corolla united, urn-shaped, the lobes spreading; stamens (8) 10, included in the corolla; anthers with 2, curved to reflexed, horn-like appendages; ovary superior, 5 (4-10) chambered; fruit berry-like, with about 5 stony, 1-seeded nutlets.

---

1   Leaves widest at the middle, mostly (0.6) 1.5-3 cm wide; stems ascending; shrubs 30-150 cm tall; fruit white or brown . . . . *A. patula*

1   Leaves widest toward the apex, 0.3-1.2 cm wide; stems creeping, the ascending branches or tips seldom over 15 cm tall; fruit red   . . *A. uva-ursi*

Arctostaphylos patula Greene [*A. parryana* Lemmon misapplied; *A. platyphylla* (A. Gray) Kuntze] — greenleaf manzanita — West end and south slope of the Uinta Mtns. from the Provo River drainage east to the Uinta drainage and on Blue Mt., no specimens seen from the north slope of the Uinta Mtns.; locally common to abundant in ponderosa pine and mt. brush communities; 7,000-9,000 ft; June-July. Plants with features of both *A. patula* and *A. uva-ursi* are occasional to locally common in the canyons of the south slope of the Uinta Mountains. The intermediate features of these plants strongly suggest hybridization of *A. patula* and *A. uva-ursi.* These intermediate plants may be the basis for the report of *A. nevadensis* A. Gray in Graham (1937).

Arctostaphylos uva-ursi (L.) Sprengel — bearberry, kinnikinnick — Common across the Uinta Mtns. and Cold Springs Mt.; aspen and dry, coniferous forests, and particularly common with lodgepole pine; 7,000-10,000 ft; June-July. See *A. patula.* The foliage of bearberry is rarely used by ungulates, but the fruits are used birds, rodents, bears, and wild ungulates.

Gaultheria L.

Gaultheria humifusa (Grab.) Rydberg — alpine wintergreen — Stems creeping, rooting at the nodes, scarcely woody, not over 1.5 mm thick; leaves evergreen, shiny, orbicular or broadly elliptic, 7-15 (20) mm long, about as wide, rounded at both ends, entire to obscurely serrulate, sometimes setose-serrulate toward the apex, the teeth hardly if at all visible without magnification; flowers solitary in the axils of leaves, 3-4 mm long, pink or white; fruit red, subglobose, 5-7 mm in diameter; Uinta Mtns.; infrequent in moist coniferous woods and meadows; 9,000-11,000 ft; July-Aug.

Kalmia L. — Laurel

Kalmia microphylla (Hooker) A. Heller [*K. polifolia* Wang. var. *microphylla* (Hooker) Rehder in Bailey] — bog laurel — subshrubs, not over 15 cm tall; older branches often appearing alternate, but young twigs and leaves opposite; leaves evergreen, green above, pale beneath, with revolute margins; flowers stalks slender, to 25 mm long, sometimes reddish; corolla united, 8-10 mm long, 10-15 mm wide, rose to light purple; fruit a 5-valved capsule, less than 1 cm long. Uinta Mtns.; occasional in moist and wet meadows, margins of pond and lakes and wet places in coniferous forests; 9,650-11,4500 ft; June-Aug.

Ledum L. — Labrador Tea

Ledum glandulosum Nuttall — western Labrador tea — Upright shrubs to 1 m tall; bark of older twigs marked with leaf scars; young twigs often with yellow resin-dots; leaves 1-4 cm long, 4-18 mm wide, dark green above, pale and often dotted with yellow resin beneath; petals separate, about 1 cm long, white; fruit a 5-valved capsule, the valves separating at maturity. Occasional and locally common in glaciated canyons of the Uinta Mtns.; wet ground in woods and sphagnum bogs; 7,400-11,000 ft; June-Aug.

Vaccinium L. — Blueberry; Bilberry; Huckleberry; Whortleberry

Shrubs or subshrubs, often associated with acidic soils derived from quartzite; leaves alternated, simple, thin, deciduous (ours), serrulate or entire; flowers axillary or terminal, solitary or racemose; calyx more or less 4-6 lobed; corolla united, globose or urn-shaped, the 4-5 lobes reflexed; anthers with dorsal awns; ovary inferior; fruit a reddish or bluish berry.

---

1   Leaves entire; plants sometimes over 30 cm tall, growing in wet meadows and along streams; fruit blue .... ***V. uliginosum***

1   Leaves serrulate, the teeth usually conspicuous without magnification; plants various, often of somewhat drier places; fruit red or blue 2

2   Leaves (2) 3-7 cm long, (1) 2-3.5 cm wide; plants 10-60 cm tall, rather uncommon .... ***V. membranaceum***

2   Leaves mostly less than 3 (3.5) cm long, seldom over 1.5 cm wide, plants less than 30 cm tall, occasional to common 3

3   Young branches bright green and strongly angled longitudinally; leaves often widest below the middle; fruit red, blue, or black 4

3   Young branches not green or dull brownish-green, not strongly angled; leaves often widest above the middle; fruit blue .... ***V. caespitosum***

4   Leaves 5-18 mm long, 3-9 mm wide; fruit bright red when ripe; corolla 2-4 mm long; branches usually numerous, slender, narrow and rigidly branched to produce a broom-like aspect, glabrous or sparsely, finely puberulent .. .. ***V. scoparium***

4   Larger leaves 15-40 mm long, 7-16 mm wide; fruit dark red, blue, or black when ripe; corolla about 4-5 mm long; branches usually puberulent, less numerous than in the following species .... ***V. myrtillus***

Vaccinium caespitosum Michaux — dwarf bilberry, dwarf huckleberry — Uinta Mtns.; common to abundant in mixed coniferous and Engelmann spruce forests, and lower alpine; June-July.

Vaccinium membranaceum Douglas ex Torrey (*V. globulare* Rydberg) — blue huckleberry — Known from scattered populations in the Uinta Mountains.; aspen and coniferous woods; June-July. Within our populations are small plants with small leaves that approach those of *V. myrtillus*. Plants from Lightning Ridge, N. Fork Duchesne drainage and in Wolf Creek are commonly larger than those from the eastern Uinta Mountains.

Vaccinium myrtillus L. — low bilberry — Uinta Mtns.; most common along narrow streamside meadows in the lodgepole pine-Engelmann spruce belt, and at the ecotone between meadows and woods; June-July.

Vaccinium scoparium Leiberg ex Coville — grouse whortleberry — Uinta Mtns.; often forming a dense ground cover under lodgepole pine and Engelmann spruce, occasionally lower alpine at 8,900-11,200 ft, closely associated with acidic soils derived from Precambrian sandstones of the Uinta Mtn. group and rare or absent on limestone; June-July.

Vaccinium uliginosum L (*V. occidentale* A. Gray) — western bog blueberry — Locally common to abundant toward the west in the Uinta Mountains, and increasingly less common toward the east, no specimens seen east of the Dry Fork drainage on the south slope; wet and boggy meadows and along streams; 9,000-11,500 ft; June-July.

EUPHORBIACEAE Spurge Family

Annual or perennial forbs with milky juice; flowers unisexual without calyx or corolla; pistillate and staminate flowers borne together in cup-like structures (involucres), the pistillate flower solitary and central, the staminate flowers 1-several, borne on the edge of the involucre, and consisting of a single stamen, the pistillate and staminate flowers and involucre collectively referred to as a cyathium, at casual observation appearing as a single flower; edge of the involucre often with glands, and with or without petaloid appendages; styles 3, usually 3-cleft; fruit a 3-celled, pedicellate, usually nodding capsule.

---

1   Plants upright, perennial; inflorescence an umbel at least terminally; leaves below the inflorescence alternate except for a whorl just below the umbel, 0.5-6 cm long, more or less symmetrical at the base; bracts of the umbel opposite .... ***Euphorbia***

1   Plants prostrate or loosely ascending; flowers axillary; leaves opposite, 3-15 mm long, more or less asymmetrical at the base .... ***Chamaesyce***

Chamaesyce A. Gray

Stipules present and eglandular; glands of the cyathium 4, usually with petaloid appendages.

---

1   Plants perennial, from few to much branched caudices; stems to 15 cm long; leaves entire, some broadly ovate to orbicular .... ***C. fendleri***

1   Plants annual, from simple, slender taproots; stems 5-35 cm long; leaves entire or often serrate at the apex, ovate or narrower 2

2   Seeds coarsely trans-corrugated, the corrugations extending across the longitudinal ridges; leaves thick-margined, entire to denticulate, linear-oblong, mostly widest at the middle or below, strongly asymmetrical at the more or less truncate base; involucres more or less turbinate ....
    .... ***C. glyptosperma***

2   Seeds smooth on the faces to wrinkled or pitted but not coarsely transcorrugated; leaves not thick-margined, serrate at least at the apex, more nearly obovate-oblong or ovate-oblong and some broadest above the middle and tapering to a more or less wedge-shaped, weakly asymmetrical base; involucres nearly campanulate .... ***C. serpyllifolia***

Chamaesyce fendleri (Torrey & A. Gray) Small (*Euphorbia fendleri* Torrey & A. Gray) — Fendler euphorb — Widespread and common in desert shrub, sagebrush, and pinyon-juniper communities; 4,700-6,700 ft; May-Aug.

Chamaesyce glyptosperma (Engelmann) Small (*Euphorbia glyptosperma* Engelmann in Torrey) — ridge-seeded spurge — Widespread; roadsides and other disturbed places, and sagebrush, juniper, and ponderosa pine communities; up to 6,850 ft; July-Oct. See *E. serpyllifolia.*

Chamaesyce serpyllifolia (Persoon) Small [*Euphorbia serpyllifolia* Persoon] — thyme-leaved spurge — To be expected across the lower elevations of our area in gardens, flower beds, roadsides, riparian sandbars, and desert shrub, black sagebrush, and juniper communities; July-Oct. Similar to *E. glyptosperma* but generally less compact in habit as well as differing by the features of the key.

Euphorbia L. — Spurge; Euphorbia

Stipules wanting, or small inconspicuous glands; glands of the cyathium 1-5 (4 in ours), without petaloid appendages.

1 Leaves (0.5) 1-2 cm long, ovate to oblong; plants from deep taproots and branched crowns, native, of indigenous plant communities, 10-30 cm tall; umbel with 3-5 primary rays .... *E. brachycera*

1 Leaves 2-6 cm long, linear; plants from robust rootstocks, introduced, weedy, mostly of roadsides and other disturbed areas and spreading into native plant communities; umbel with 4-17 primary rays .... *E. esula*

Euphorbia brachycera Engelmann [*E. robusta* (Engelmann) Small ex Britton & Brown; Tithymalus montanus (Engelmann) Small; T. robustus (Engel.) Small] — shorthorn spurge, Rocky Mt. spurge, robust euphorbia — Widespread; occasional in desert shrub, sagebrush, pinyon-juniper, ponderosa pine and dry aspen communities; 5,000-7,800 ft; May-June (July).

Euphorbia esula L. [*Tithymalus esula* (L.) Scopoli] — leafy spurge — Introduced from Eurasia; weedy with disturbance and highly invasive in native plant communities; known in our area from Ashley Valley, Dry Fork Canyon, Greendale, East of Duchesne, and spreading down the Yampa River, and to be expected elsewhere; ditch banks, gardens, fields, and native plant communities at 5,200-8,000 ft. One of the worst invasive species of our area and in much of the world. The rather extensive infestations along the Yampa River will be extremely difficult to control, and they pose a major threat to riparian communities of Dinosaur National Monument.

FABACEAE (LEGUMINOSAE) Pea Family

Plants herbs (our native species), shrubs or trees; leaves alternate, simple (rarely in ours) or pinnately or palmately compound, stipulate; flowers bisexual, irregular, usually in racemes; calyx 5-lobed; corolla papilionaceous (sweetpea type), with 5 petals but these appearing 4, the uppermost one (banner) the longest, the 2 lateral ones (wings) turned about 90 degrees to the banner, the lower 2 (keel) fused and appearing as a single keel-shaped petal, this parallel with and folded between the wings; stamens 10, the filaments free, or all united (monadelphous) into a tube that surrounds the pistil, or 9 united with the 10th one free (diadelphous); pistil 1; ovary superior; style and stigma 1; fruit (pod) dry, several-seeded, with 1 carpel, 1-or 2-loculed, and dehiscent along both sutures, or sometimes indehiscent, not constricted between the seeds (legume) or constricted between the seeds and braking into 1-seeded segments at maturity (loment).

1 Plants shrubs mostly over 1 m tall; flowers yellow .... *Caragana*

1 Plants herbaceous, woody only at the base if at all; flowers yellow various
.... 2

2 Leaves simple .... *Astragalus*

2 Leaves with 3 or more leaflets 3

3 Leaves pinnately compound with more than 3 leaflets 4

3 Leaves trifoliate or palmately compound 14

4 Leaves even-pinnate, the rachis ending in a tendril and not a leaflet 5

4 Leaves odd-pinnate, the rachis ending in a terminal leaflet, this sometimes nearly filiform and a mere continuation of the rachis but not a tendril .. 6

5 Style bearded down one side as in a toothbrush; wings of corolla essentially free from the keel .... *Lathyrus*

5 Style bearded at the apex with a tuft or ring of hairs as in a bottlebrush; wings of the corolla adherent to the keel .... *Vicia*

6 Pods covered with hooked, bristle-like appendages, more or less cocklebur-like; flowers dingy-white or cream; plants 40-120 cm tall ..... *Glycyrrhiza*

6 Pods without hooked, bristle-like appendages, sometimes with bristle-like spines but not at all cocklebur-like; plant height various. 7

7 Keel petals much longer than the wings; fruit strongly constricted between the seeds or with only 1 seed 8

7 Keel petals sub-equal to or shorter than the wings; fruit not strongly constricted between the seeds, with more than 1 seed 9

8 Stamens diadelphous; pods over 1 cm long with (2) 4-several seeds and strongly constricted between the seeds; plants native .... *Hedysarum*

8 Stamens monadelphous but the upper filament partly free and appearing diadelphous; pods less than 1 cm long, with 1 seed; plants introduced ....
.... *Onobrychis*

9 Filaments of stamens all free to the base; flowers 15-27 mm long, blue-purple to blue; pods terete and constricted between the seeds; herbage silvery silky sericeous .... *Sophora*

9 Stamens diadelphous; flowers various; pods not constricted between the seeds; herbage various 10

10 Flowers and fruit in head-like umbels 11

10 Flowers and fruit in racemes 12

11 Flowers yellow, leaflets mostly 5 .... *Lotus*

11 Flowers pink-purple; leaflets mostly 9-27 .... *Coronilla*

12 Keel with a forward-pointed beak; ventral suture of pods forming a partial or complete partition; leaves mostly all basal .... *Oxytropis*

12 Keel beakless; ventral suture of pods usually not produced internally; some leaves born on stems or occasionally all basal 13

13 Flowers red-orange when fresh .... *Sphaerophysa*

13 Flowers pinkish, purplish or whitish, not red-orange .... *Astragalus*

14 Flowers in heads; corollas persistent, enveloping the pods ..... *Trifolium*

14 Flowers usually in racemes; corollas not persistent, not enveloping the pods
....15

15 Leaves palmately compound with 5 or more leaflets; flowers mostly in racemes..16

15 Leaves trifoliate 18

16 Leaves all basal or nearly so..17

16 Some leaves usually on stems ....*Lupinus*

17 Plants perennial from tuberous roots, glandular-dotted, strongly malodorous; pod 1-seeded, included in the calyx ....*Pediomelum*

17 Plants annual from taproots, not glandular-dotted, not malodorous; pod with more than 1 seed, exserted from the calyx ....*Lupinus*

18 Leaflets toothed (sometimes only minutely so); flowers 2-7 (10) mm long .
....19

18 Leaflets entire; flowers various.....................................20

19 Leaflets toothed only on the upper third; racemes compact, seldom more than twice as long as wide ....*Medicago*

19 Leaflets toothed on the upper 1/2 or more; mature racemes much elongate, several times longer than wide ....*Melilotus*

20 Flowers yellow; pods 25-70 mm long ....*Thermopsis*

20 Flowers not yellow; pods various 21

21 Plants 15-68 cm tall; flowers born on leafy stems; leaflets 14-50 mm long; pods with 1 seed ....*Psoralidium*

21 Plants less than 15 cm tall, otherwise not as above in all features ....
....*Astragalus*

Astragalus L. — Locoweed; Milkvetch; Vetch

Annual or perennial, caulescent or acaulescent herbs, mostly with alternate, odd-pinnate compound leaves, or rarely with simple or trifoliate leaves; stipules adnate to the petiole base, free or connate-sheathing around the stem; inflorescence a raceme; flowers papilionaceous, each subtended by a single bract and sometimes with 1 or 2 bracteoles attached at the base of the calyx or on the pedicels; calyx 5-toothed; petals 5, pink, lavender, pink-purple, ochroleucous, or white, the keel shorter than the wings; stamens diadelphous; ovary enclosed in the staminal sheath, the style glabrous; pods unilocular or bilocular, sessile or stipitate, extremely variable in shape, size, and pubescence. Major portions of the keys are adapted from those of Welsh (1978).

1 Leaflets stiffly spinulose tipped, 3-9 per leaf; plants mat forming or erect; flowers 4.5-10 mm long, pink-purple, white, or ochroleucous; pods sessile, 4-10 mm long, declined or spreading at maturity; plants caulescent .....
....*A. kentrophyta*

1 Leaflets not spinulose tipped or weakly so in *A. detritalis,* but then plants different from above in other features 2

2 Leaves mostly simple; pods erect 3

2 Leaves mostly with 3 or more leaflets; pods various 5

3   Leaves not all basal, ovate to orbicular, 1.5-6.5 cm long, 1.5-5 cm wide, glabrous; flowers 17-27 mm long, purplish or ochroleucous; pods 25-35 mm long, 11-16 mm thick, inflated, stipitate, the stipe 10-21 mm long . . . . *A. asclepiadoides*

3   Leaves all basal, grass-like, linear or narrowly oblanceolate or spatulate, usually strigose with malpighian hairs; pods 4-13 mm long, 1.5-3 mm thick, sessile. 4

4   Leaves 0.8-6 cm long; racemes usually with less than 8 flowers, 0.2-3 cm long in fruit; plants 1.5-9 cm tall, widespread . . . . *A. spatulatus*

4   Leaves 1-13 (17) cm long; racemes usually with more than 8 flowers, 4.5-24 cm long in fruit; plants 5-24 cm tall, apparently restricted to Uintah Co. below 6,500 ft and n. of Hwy. 40, confined to sandstone . . . . *A. chloodes*

5   Terminal leaflet confluent with the rachis, filiform or linear and appearing like a continuation of the rachis; lateral leaflets usually filiform or linear or leaves sometimes reduced to a simple filiform or linear rachis; plants rush-like, caulescent, mostly restricted to Uintah Co. except *A. convallarius* . . . . **KEY 1**

5   Terminal leaflet jointed to the rachis, expanded into a blade 6

6   Plants without conspicuous leaf-bearing stems. . . . . **KEY 2**

6   Plants with well-developed leaf-bearing stems 7

7   Keys based mainly on features of the flowers. 8

7   Keys based mainly on features of the fruit 9

8   Flowers 12.5-25 mm long . . . . **KEY 3**

8   Flowers 5-12.5 mm long . . . . **KEY 4**

9   Pods inflated, the valves often papery and easily crushed or "popped" when ripe . . . . **KEY 5**

9   Pods not inflated 10

10 Pods with stipe over 2 mm long, strongly spreading or pendulous . . **KEY 6**

10 Pods sessile or the stipe less than 2 mm long . . . . **KEY 7**

— KEY 1 —
Terminal leaflet confluent with the rachis, filiform or linear and appearing like a continuation of the rachis

1   Pods bladdery-inflated, 12-30 mm long, half to nearly as wide as long, pendulous, stipitate, the stipe 1-3.3 mm long, the valves papery, usually bright red-mottled; flowers 6.3-9.5 mm long, purplish to pink, rarely white; pubescence of malpighian hairs; stems weak, sprawling to erect, from an elongate rhizome-like caudex . . . . *A. ceramicus*

1   Pods not bladdery-inflated, 13-50 mm long and linear or nearly so except in *A. nelsonianus,* the valves not papery; flowers sometimes larger than above, or if as small (*A. convallarius*) then usually ochroleucous; pubescence of basifixed hairs 2

2   Flowers 13-30 mm long; stems usually few-many together from a branched subterranean caudex 3

2   Flowers 6.5-12.5 mm long; stems usually solitary or few together from a slender or rhizome-like caudex. 6

3 Flowers 24-30 mm long, white; pods 13-33 mm long, 6-12 mm thick, sessile, rounded in cross-section; leaflets 3-13; plants of Daggett Co **A. nelsonianus**

3 Flowers 18-24 mm long, whitish or purple; pods 4-7.5 mm thick, if over 6 mm thick then long stipitate; leaflets (1) 3-9; plants of Uintah Co........ 4

4 Pods with a stipe 9-12 mm long, the body 25-35 mm long, dorsiventrally compressed; flowers ochroleucous, sometimes purple tinged; leaflets 2-7 mm wide; calyx gibbous 5

4 Pods sessile, 15-35 mm long, laterally compressed; flowers pink-purple; leaflets 5-2 mm wide; plants odoriferous selenophytes .... **A. saurinus**

5 Flowers 20-24 mm long; broadest leaflets 4-8 (10) mm wide . **A. hamiltonii**

5 Flowers 13-20 mm long; broadest leaflets1.5-2.5 (3.5) mm wide ....
   .... **A. lonchocarpus**

6 Flowers pink-purple with white wing-tips; pods 20-35 mm long, dorsiventrally compressed in the lower 1/2, becoming laterally compressed toward the apex; plants known from 4,700-5,400 ft ...... **A. duchesnensis**

6 Flowers ochroleucous, sometimes tinged with purple; pods 13-50 mm long, laterally compressed; plants from 4,900-8,500 ft .... **A. convallarius**

— KEY 2 —
Flowers born among basal leaves or on scapes or peduncles; plants without conspicuous leafy stems

1 Flowers 6-8 mm long, 2 or 3 per raceme; pods 4-5 mm long. leaves 0.6-2 cm long .... **A. aretioides**

1 Flowers and pods longer than above; at least some leaves usually over 2 cm long 2

2 Leaflets 3-7, narrowly oblanceolate to linear, 3-30 mm long, 0.5-2.5 mm wide; flowers 13-20 mm long; pods linear-oblong, 15-38 mm long, 2-3.5 mm wide, erect .... **A. detritalis**

2 Leaflets mostly more than 7, mostly elliptic or ovate; pods elliptic or ovoid, mostly over 3.5 mm wide 3

3 Banner whitish; wings and keel often whitish with bluish or purple tips . 4

3 Banner pink or pink-purple; wings and keel also usually pinkish........ 7

4 Plants with malpighian, strigose hairs; pods glabrate or with hairs less than 1 mm long, brightly mottled with reddish blotches some of which are over 1 mm across .... **A. chamaeleuce**

4 Plants with basifixed hairs; pods permanently hairy with hairs over 1 mm long, not mottled or the mottles less than 1 mm across 5

5 Leaflets strigose; axis of fruiting racemes 1-12 mm long; plants known from Strawberry Valley .... **A. eurekensis**

5 Leaflets villose or pilose; axis of fruiting racemes 2-25 mm long; plants widespread and common 6

6 Hair of leaves and calyx bent or curved; valves of pods hidden by dense, white hair .... **A. purshii**

6 Hair of leaves and calyx straight; valves of pods clearly visible through the hairs .... **A. argophyllus**

7   Pods bladdery-inflated, 35-65 (70) mm long, 15-31 mm thick, stipitate, the stipe 2-4 mm long, the valves thin and papery, often mottled; flowers 15-23 mm long .... *A. megacarpus*

7   Pods not bladdery-inflated, sessile or with a stipe to 2.5 mm long, the values not thin and papery. 8

8   Pods 10-14 mm long, 3.6-6 mm wide; flowers 12-16 mm long; upper surface of leaflets often glabrous along all or at least some the midrib; peduncles or scapes more-or-less erect; plants known only from near Horseshoe Bend of the Green River .... *A. equisolensis*

8   Pods 13-50 (70) mm long; flowers 15-31 mm long; upper surface of leaflets pubescent along the midrib. 9

9   Pods descending; peduncles or scapes erect; racemes 7-20 flowered, the axis 1.5-18 cm long in fruit; leaves 2-28 cm long with 15-35 leaflets; plants 6-45 cm tall .... *A. mollissimus*

9   Pods, ascending; peduncles or scapes spreading; racemes 2-8 flowered, the axis 0.3-2.6 mm long in fruit; leaves with 5-21 leaflets, 1.5-12 cm long; plants 1.5-15 cm tall 10

10 Valves of pods clearly visible through the hairs of the surface 11

10 Valves of pods mostly hidden in dense hair 13

11 Plants with basifixed hairs; pods permanently hairy with hairs over 1 mm long .... *A. argophyllus*

11 Plants with malpighian, strigose hairs; pods glabrate or with hairs less than 1 mm long. 12

12 Flowers usually bicolored, ochroleucous or purple tinged to pink-purple, 20-40 mm long; pods spongy in texture, with a thick mesocarp separating the endocarp and papery exocarp, the wall 1-3 mm thick, with 37-46 ovules, with reddish blotches over 1 mm across; axis of raceme 1-2 cm long in fruit .... *A. chamaeleuce*

12 Flowers usually bright pink-purple, 16-31 mm long; pods leathery or woody in texture, sometimes succulent or fleshy when young, readily deciduous at maturity, the walls less than 1 mm thick, with 26-70 ovules, reddish blotches (if present) mostly less than 1 mm across; axis of racemes, 1-6.5 cm long in fruit .... *A. amphioxys*

13 Hairs of pods 4-8 mm long and about equal to the diameter of the pod; flowers 23-31 mm long; pods 17-30 mm long, 5.5-7.5 mm wide, sessile or with a stipe to 2.5 mm long; plants of Duchesne and Wasatch Cos. .... *A. utahensis*

13 Hairs of pods 1.5-5 mm long and shorter than the diameter of the pod; flowers 19-26 mm long; pods 13-26 mm long, 5-11 mm wide, sessile; plants of Daggett and Summit Cos. .... *A. purshii* var. *glareosus*

— KEY 3 —
Flowers 12.5-24 mm long

1   Flowers pink or pink-purple at least toward the base. 2

1   Flowers white or ochroleucous, sometimes with faint purplish markings 11

2   Pubescence malpighian 3

2   Pubescence basifixed. 4

3   Racemes with 16-50 flowers . . . . *A. laxmannii*

3   Racemes with 2-13 flowers . . . . *A. amphioxys*

4   Racemes 0.5-2.7 cm long, with 1-15 flowers 5

4   Racemes mostly over 2.7 cm long, often with more than 15 flowers 7

5   Pedicels 3.5-8 mm long; ovaries stipitate. . . . . *A. megacarpus*

5   Pedicels 1-2.5 mm long; ovaries subsessile 6

6   Leaflets glabrous above except at the margins; calyx strigose . . *A. cibarius*

6   Leaflets strigose above; calyx villous . . . . *A. agrestis*

7   Pubescence malpighian; peduncle (4) 10-18 cm long; flowers 13-16 mm long
    . . . . *A. laxmannii*

7   Pubescence basifixed. 8

8   Calyx gibbous; racemes commonly with 25-80 flowers . . . . . . *A. bisulcatus*

8   Calyx not gibbous; racemes with 5-30 flowers 9

9   Ovaries sessile . . . . *A. lentiginosus*

9   Ovaries stipitate. 10

10 Flowers 7.2-11.5 mm long; leaflets glabrous above . . . . *A. robbinsii*

10 Flowers 12-21 mm long; leaflets strigose on both sides . . . . *A. coltonii*

11 Plants 2-10 cm tall, of marly ridges and slopes of Green River shale;
    peduncles 0.5-2 cm long; leaves with 15-27 crowded to imbricate, strongly
    folded leaflets; flowers 9-17 mm long . . . . *A. lutosus*

11 Plants mostly over 10 cm tall; peduncles 1.5-22 cm long 12

12 Flowers 12-17.5 mm long 13

12 Flowers 17-30 mm long. 16

13 Leaflets glabrous or glabrate on both sides, sometimes ciliate; ovaries
    stipitate . . . . *A. racemosus*

13 Leaflets strigose at least below; ovaries sessile or subsessile 14

14 Pubescence malpighian; flowers nodding at anthesis 15

14 Pubescence basifixed; flowers declined to ascending at anthesis; axis of
    racemes elongating to 7 cm in fruit; plants introduced, widespread, seeded
    in roadside and rangeland seedings . . . . *A. cicer*

15 Leaflets 6-16 mm wide; racemes often with more than 30 nodding flowers,
    the axis elongating to 22 cm long in fruit; plants 15-120 cm tall, of Daggett
    Co. . . . . *A. canadensis*

15 Leaflets 0.5-6 mm wide; racemes with 6-30 ascending flowers, the axis 2-12
    cm long in fruit; plants 5-35 cm tall, widespread . . . . *A. flavus*

16 Flowers 24-30 mm long; plants of Daggett Co.; see also lead 3 of KEY 1 . . . .
    . . . . *A. nelsonianus*

16 Flowers 17-25 mm long. 17

17 Ovaries sessile or stipe less than 1 mm long 18

17 Ovaries stipitate, the stipe over 1 mm long. 19

18 Flowers erect or ascending; calyx villous . . . . *A. agrestis*

18 Flowers declined nodding; calyx strigulose . . . . *A. pattersonii*

19 Leaflets glabrous on both sides, often ciliate; plants of the E. Tavaputs

Plateau (see also lead 2 in KEY 6 to compare with *A. megacarpus*) . . . .
. . . . *A. oophorus*

19 Leaflets villose-pilose below; plants known from Strawberry Valley . . . . .
. . . . *A. drummondii*

— KEY 4 —
Flowers 5-12.5 mm long

1   Plants annual, usually of sandy places; calyx 2.7-3.8 mm long; flowers 5-7.6 mm long . . . . *A. geyeri*

1   Plants perennial, or if biennial then the calyx and flowers longer than above 2

2   Leaves with 5-27, crowded to imbricate, strongly folded leaflets; peduncle 0.5-2 cm long; flowers 2-10, ochroleucous with purple-tipped keel; plants 2-10 cm tall, of shaly ridges and talus slopes of Green River shale*A. lutosus*

2   Plants not as above in all features. 3

3   Calyx and usually leaflets sub-villous to villous, not strigose; leaflets 5-15; flowers pink-purple or ochroleucous. 4

3   Calyx and leaflets strigose or if calyx pilose than flowers white or yellow 5

4   Ovaries soon pilose, sessile; plants common to abundant at 5,200-7,500 ft; in desert shrub and pinyon-juniper communities . . . . *A. pubentissimus*

4   Ovaries glabrous, usually stipitate; plants uncommon, known from 9,350-11,900 ft . . . . *A. australis*

5   Flowers white, ochroleucous or yellow, sometimes with purplish markings

5   Flowers pink-purple or bluish. 10

6   Calyx gibbous; plants ill-scented selenophytes; plants 15-75 cm tall; calyx 3.5-9.6 mm long . . . . *A. bisulcatus*

6   Calyx not gibbous; plants not scented as above except   sometimes in   *A. flavus* 7

7   Stipules all free; calyx 4.6-6.2 mm long; racemes with 2-9 flowers; plants rare, known from the Yampa River area, 4-26 cm tall . . . . *A. wetherillii*

7   At least the lower stipules connate-sheathing; racemes with (1) 3-30 flowers; plants widespread and common, 3-52 cm tall. 8

8   Calyx 5.5-9.5 mm long; pubescence malpighian . . . . *A. flavus*

8   Calyx 2.1-5.2 mm long; pubescence basifixed 9

9   Stipules turning black in age; peduncles 0.2-4 cm long; ovaries sometimes stipitate . . . . *A. tenellus*

9   Stipules not turning black; peduncles 2-14 cm long; ovaries sessile . . . . *A. miser*

10   Stipules all free; stems arising singly or few together from a slender, rhizome-like caudex; plants of desert shrub and juniper communities, not above 5,500 ft; flowers 8.5-12.5 mm long; leaflets 5-15 . . . . *A. duchesnensis*

10   At least the lower 1-3 pairs of stipules connate-sheathing; plants of higher elevations; flowers various; leaflets sometimes more than 15. 11

11   Flowers 9-12 mm long, pink-purple; plants rare, known from 8,800 ft on the E. Tavaputs Plateau; leaves with 15-26 leaflets . . . . *A. alpinus*

11 Flowers 5.5-8 mm long or if to 10 mm (A. miser) then bluish or blue-purple
.... 12

12 Flowers bluish or bluish-purple or whitish; spreading-declined; plants widespread and common, but the blue-flowered phase seldom encountered except at higher elevations of the Uinta Mtns. .... *A. miser*

12 Flowers bright pink-purple, ascending; plants rare, just entering the Uinta Basin on the Tavaputs Plateau...................................... 13

13 Flowers spreading at anthesis, 7-11 mm long; calyx 3.3-5.8 mm long, the tube 2.7-4.3 mm long; ovaries often pubescent (see also leads 9 and 10 in KEY 8) .... *A. flexuosus*

13 Flowers ascending at anthesis, 5.5-8 mm long; calyx 2.5-3.7 mm long, the tube 1.5-2.6 mm long; ovaries glabrous .... *A. wingatanus*

— KEY 5 —
Pods inflated

1 Pods with stipe 2-8.5 mm long...................................... 2

1 Pods sessile or the stipe less than 2 mm long 3

2 Stems 1-5 cm long, the internodes mostly concealed by stipules; racemes with 3-5 (8) flowers, shorter than the basal leaves; ovaries and pods strigose; pods 35-65 mm long, 15-31 mm thick, with a stipe 2-4 mm long; plants known from the W. Tavaputs Plateau in Duchesne Co., Lookout Mtn., Moffat Co., and Hickey Mtn., Uinta Co. .... *A. megacarpus*

2 Stems 5-20 cm long, rarely shorter, the internodes mostly longer than and not concealed by the stipules; racemes with 4-13 flowers, mostly exceeding the lowermost leaves; ovaries and pods glabrous; pods 25-55 mm long, 10-30 mm thick, with a 3.5-8.5 mm long stipe; plants known from the E. Tavaputs Plateau in Rio Blanco Co. .... *A. oophorus*

3 Plants annual; pods strigose .... *A. geyeri*

3 Plants perennial, sometimes appearing annual or biennial in *A. pubentissimus* but then the pods shaggy-villous 4

4 Plants 2-10 cm tall, of shaly ridges and talus slopes of Green River shale; leaflets 15-27, crowded to imbricate, strongly folded; pods 15-37 mm long
.... *A. lutosus*

4 Plants (5) 10-70 cm tall, not restricted as above; leaflets 7-17 or more in *A. cicer* but then plants taller and pods smaller 5

5 Pods pilose or villose with spreading hairs 6

5 Pods glabrous or strigose with appressed hairs 7

6 Pods strongly curved becoming U-shaped; leaves with 5-17 leaflets ......
.... *A. pubentissimus*

6 Pods not strongly curved; leaves with 17-31 leaflets .... *A. cicer*

7 Pods unilocular; flowers white, sometimes tinged with lavender, 7.5-11 mm long; plants of the Yampa River drainage................. *A. wetherillii*

7 Pods bilocular; flowers mostly blue-purple, (12.6) 14-21 mm long; plants known from Daggett Co. .... *A. lentiginosus*

— KEY 6 —
Pods not inflated, strongly spreading or pendulous, with stipe 2-15 mm long

1   Plants ill-scented selenophytes; pods triangular or obcordate in cross-section 2

1   Plants not with the odor of selenium, pods rounded or flattened in cross-section, the stipe-2-6 mm long 3

2   Pods triangular in median cross-section, glabrous, 9-45 per raceme, the stipe 4-6 mm long; calyx teeth 3.3-6 mm long . . . . *A. racemosus*

2   Pods obcordate (bisulcate) in median cross-section, often strigose, 25-80 per raceme, the-stipe 3.5-9.6 mm long; calyx teeth 1-3 mm long . . . .
. . . . *A. bisulcatus*

3   Stipules turning black in drying; pods unilocular, strongly flattened . . . . .
. . . . *A. tenellus*

3   Stipules not turning black; pods semi-bilocular 4

4   Calyx deep purple throughout . . . . *A. coltonii*

4   Calyx not as above 5

5   Body of pods 22-35 mm long; flowers (15) 20-40 mm long 6

5   Body of pods 10-27 mm long; flowers 7.5-12.6 mm long 8

6   Calyx and pedicels with black hairs; plants montane . . . . . . *A. scopulorum*

6   Calyx and pedicels with white hairs, these sometimes mixed with blackish hairs 7

7   Broadest leaflets 4-8 (10) mm wide; fruit round or nearly so in cross-section when young and at maturity, over 5 mm wide at maturity; raceme 2-12 cm long in fruit, with 7-30 pods . . . . *A. hamiltonii*

7   Broadest leaflets1.5-2.5 (3.5) mm wide; fruit tending to be flat when young, round in cross-section when mature but mostly less than 5 mm wide (in Uinta Basin plants); axis of some racemes often over 12 cm and up to 45 cm long in fruit with 7-40 (50) pods . . . . *A. lonchocarpus*

8   Leaves with 7-15 leaflets; calyx and usually leaflets villous; pods glabrous; plants from the W. Tavaputs Plateau and Uinta Mtns. . . . . *A. australis*

8   Leaves with 15-26 leaflets; calyx and leaflets (at least below) strigulose; pods strigulose 9

9   Plants known from the E. Tavaputs Plateau . . . . *A. alpinus*

9   Plants known from the W. Fork of the Bear River . . . . *A. robbinsii*

— KEY 7 —
Pods not inflated, sessile or with stipe less than 2 mm long

1   Pubescence malpighian 2

1   Pubescence basifixed 5

2   Pods 7-13 mm long, erect 3

2   Pods 15-50 cm long 4

3   Flowers yellow or white; plants sometimes ill-scented selenophytes . . . . .
. . . . *A. flavus*

3   Flowers pink-purple; plants without odor of selenium . . . . . . *A. laxmannii*

4 Stems arising singly or few together from rhizomes, mostly 15-100 cm tall; racemes often with more than 13 flowers; pods erect, straight . . . . ***A. canadensis***

4 Stems usually several from a taproot or caudex, 2-35 cm tall; racemes with 2-13 flowers; pods spreading, curved . . . . ***A. amphioxys***

5 Pods ascending or erect 6

5 Pods spreading or pendulous. 7

6 Axis of the raceme 0.5-2.7 cm long in fruit; flowers pink-purple ***A. cibarius***

6 Axis of the raceme 2-15 cm long in fruit; flowers white, the keel sometimes purplish . . . . ***A. pattersonii***

7 Pods silky-villous, 7-10 mm long, with 14-26 ovules, with a stipe 0.3-1 mm long; calyx 7-12.5 mm long, villous; raceme head-like, the axis 0.5-2.5 cm long in fruit . . . . ***A. agrestis***

7 Pods strigose or glabrous, longer than 10 mm, or if shorter then with only 3-9 ovules; calyx 2.1-5.2 mm long, strigose; racemes sometimes over 2.5 cm long in fruit. 8

8 Pods 6-12 mm wide, nearly terete in cross-section; plants of Daggett Co. (see lead 3 in KEY 1) . . . . ***A. nelsonianus***

8 Pods 2-4.5 mm wide, usually flattened except in *A. flexuosus* 9

9 Pods rounded in cross-section; plants known from Rio Blanco Co. . . . . ***A. flexuosus***

9 Pods somewhat to strongly flattened; plants of various distribution. . . . 10

10 Pods glabrous, 9-15 mm long with 4-8 ovules, subsessile or with stipe to 1.7 mm long; plants just entering the Uinta Basin on the W. Tavaputs Plateau; flowers mostly pink-purple . . . . ***A. wingatanus***

10 Pods strigose; plants widespread; flowers white, ochroleucous, or occasionally pink purple in A. miser; plants widespread 11

11 Stipules turning black in drying; pods 7-16 mm long, with only 3-9 ovules; peduncles 0.2-4 cm long . . . . ***A. tenellus***

11 Stipules not turning black; pods 12-37 mm long, with 6-31 ovules; peduncles 2-19 cm long. 12

12 Plants from within and below the pinyon-juniper belt . . . ***A. duchesnensis***

12 Plants known from above the pinyon-juniper belt . . . .

13 Calyx tube 3.2-4.5 mm long; plants rare, known from W. Fork Bear River . . . . ***A. robbinsii***

13 Calyx tube 1.9-2.9 mm long; plants common, widespread . . . . ***A. miser***

Astragalus agrestis Douglas ex G. Don (*A. hypoglottis* L.) — field milkvetch — Widespread; occasional to common in sagebrush, aspen, and riparian communities; 6,000 -9,000 ft; June-July (Sept.).

Astragalus alpinus L. — alpine milkvetch — Known in the Uinta Basin by a single collection (N. Holmgren et al. 2321) from Florence Creek, E. Tavaputs Plateau, this from an aspen community at 8,800 ft, to be expected elsewhere; June-July.

Astragalus amphioxys A. Gray — crescent milkvetch — The one specimen seen (Goodrich 28713) is from sandy soil near Pelican Lake. This is common to the s.

of the Uinta Basin (Barneby 1964; AUF 416). This is sometimes mistaken for *A. chamaeleuce*. Previously reported for the Uinta Basin, based on Harrington 3881 (CS!) but this specimen belongs to *A. chamaeleuce*. Key 2 provides several characters that help differentiate the 2 taxa, but mature fruit provides the best means of separation. Plants of the population near Pelican Lake commonly have leafy stems. Our plants belong to var. *vespertinus* (Sheldon) M. E. Jones.

Astragalus aretioides (M. E. Jones) Barneby (*A. sericoleucus* A. Gray var. *aretioides* M. E. Jones) — cushion orophaca, cushion milkvetch — Specimen seen are from volcanic ash barrens at 5,800 ft in Browns Park, Daggett Co; June. Pulvinate caespitose plants; leaves with 3 leaflets 3-7.5 mm long.

Astragalus argophyllus Nuttall ex Torrey & A. Gray — silverleaf milkvetch — With 2 vars.:

1 Flowers 22-26 mm long; petals bright pink-purple; plants of meadow and riparian communities .... **var. *argophyllus***
1 Flowers 18-22.5 mm long, at least the banner pale purple or white; plants of sagebrush; pinyon-juniper, and mtn. brush communities **var. *martinii***

Var. *argophyllus* — meadow milkvetch — Listed for Daggett, Summit, and Uintah Cos. (AUF 417) for meadows, streambanks, and lake shores up to 6,460 ft. Also along the Blacks Fork River in Sweetwater Co. Wyoming.

Var. *martinii* M. E. Jones — silverleaf milkvetch — Widespread; common; sagebrush, pinyon-juniper, mt. brush, and occasionally aspen communities; 6,000-8,000 ft; May-June (Oct.).

Astragalus asclepiadoides M. E. Jones — milkweed milkvetch — Widespread, infrequent or occasional; desert shrub communities, often on clay, or formations that weather to badlands; 4,800-6,200 ft; May-June.

Astragalus australis (L.) Lamarck (*A. aboriginorum* Richardson) — subarctic milkvetch, Indian milkvetch — Known from the Strawberry Peak area, W. Tavaputs Plateau in Wasatch Co. and Farm Creek Peak, of the south slope of the Uinta Mtns. and Blacks Fork drainage of the north slope; sagebrush-grass, spruce-fir, and alpine communities at 9,350-11,900 ft; June-July. Weight 84 referred here by Graham (1937) belongs to *A. miser*.

Astragalus bisulcatus (Hooker) A. Gray — two-grooved milkvetch — With 2 vars.:

1 Corolla bright pink-purple, rarely white, 12-15 mm long; calyx purplish . ... **var. *bisulcatus***
1 Corolla ochroleucous or whitish, the keel often purple-tipped, 6-8 mm long; calyx not purplish .... **var. *haydenianus***

Var. *bisulcatus* — pink two-grooved milkvetch — Occasional to rather common in Daggett and e. Uintah Cos. and eastward; streamside, pinyon-juniper, sagebrush, and desert shrub communities; 5,000-7,400 ft; May-July.

Var. *haydenianus* (A. Gray) Barneby (*A. bisulcatus* var. *major* Welsh, *A. haydenianus* A. Gray) — white two-grooved milkvetch — Locally common at isolated populations across our area in desert shrub, sagebrush and mt. brush communities; 6,500-8,550 ft; June-Aug.

Astragalus canadensis L. — Canada milkvetch — Two vars. as follows:

1 Pods and ovaries glabrous, terete at maturity, not sulcate on the back; calyx teeth 2.5-4.1 mm long; plants mostly 40-120 cm tall, perhaps present in Daggett Co. .... **var. *canadensis***

1 Pods and ovaries pubescent, sulcate on the back at maturity; calyx teeth
   mostly 1-2.5 mm long; plants 10-50 cm tall .... **var. brevidens**

Var. brevidens (Gandoger) Barneby — pasture milkvetch — Occasional; all
specimens seen are from Birch Creek, Sheep Creek, Dutch John Canyon, Goslin
Mtn., and near Manila in Daggett Co. and Beaver Creek in Summit Co.; along
ditch banks and stream sides; 6,000-7,000 ft; May-June.

Astragalus ceramicus Sheldon — painted milkvetch — Occasional from near
Maeser and e., mostly n. of Hwy. 40; mostly on sandy ground, often with
scattered junipers; 5,000-6,000 ft; May-early June. Our plants generally belong
to var. ceramicus of Utah, sw. Colorado, New Mexico, and Arizona, but some
are somewhat intermediate to var. *filifolius* (A. Gray) Hermann (*A. c.* var.
imperfectus Sheldon) that is from Idaho to S. Dakota and s. to w. Nebraska and
New Mexico.

Astragalus chamaeleuce A. Gray in Ives [*A. cymboides* M. E. Jones misapplied?
(Graham 1937)] — cicada milkvetch — Widespread; common in desert shrub,
sagebrush, and pinyon-juniper communities; 4,700-7,500 (8,000) ft; April-July
(November with fall precipitation). Leaflets 5-21, rarely less, 4-15 mm long, 2-
10 mm wide, obovate to oblanceolate, not spinulose-tipped; flowers 20-40 mm
long; pods 20-40 mm long, 5-15 mm thick, ascending but hardly erect; stipules
mostly all free.

Astragalus chloodes Barneby — grass milkvetch — Occasional or locally common;
endemic to Uintah Co., from near Dry Fork to the Colorado-Utah line (possibly
in Colorado) n. of Hwy. 40; crevices of sandstone outcrops; 5,000-6,000 ft; May-
June (July). The plant indicated as an undescribed taxon and listed as *A. moen-
coppensis* M. E. Jones by Graham (1937) belongs here.

Astragalus cibarius Sheldon — browse milkvetch, silky milkvetch — Strawberry
Valley (Brotherson 1881), Birch Cr. at 7.1 mi s. of McKinnon, (Atwood 13586), 9
mi s. of Robertson (Ostler 111), 9 mi nw. of Rangely (Neese & Welsh 7490).
Other specimens seen are from adjacent to our area in Sweetwater and Uinta
Cos., Wyoming; sagebrush, pinyon-juniper and mtn. brush communities at
5,700-8,200 ft; May-June.

Astragalus cicer L. — chickpea milkvetch — Introduced from Eurasia; used on
roadsides, pipelines and powerline corridors, and in rangeland seedings in
sagebrush, mt. brush, pinyon-juniper, aspen, and coniferous forest communities,
to be expected throughout our area; June-July. Plants at 9410 ft in E. Fork
Blacks Fork were short but vigorous. Perhaps this is near the upper elevational
limit for the species.

Astragalus coltonii M. E. Jones — Colton milkvetch — Adjacent to our area near
Lonetree and Cedar Mtn. in Uinta and Sweetwater Cos. Wyoming; Wyoming
big sagebrush and juniper communities.

Astragalus convallarius Greene (*A. diversifolius* A. Gray misapplied) — lesser
rushy milkvetch, timber poisonvetch — Widespread; many plant communities
including: desert shrub, sagebrush, pinyon-juniper, riparian, and ponderosa
pine; 4,900-8,500 ft; May-July. Our plants belong to var. *convallarius*. Increasing
and becoming abundant following fire.

Astragalus detritalis M. E. Jones — debris milkvetch — Endemic, occasional to
locally common across the Tavaputs Plateau and adjacent in the Basin from
near Starvation Reservoir to the White River drainage and Vermillion Bluffs in

Colorado; desert shrub, sagebrush, and pinyon-juniper communities; 5,600-9,000 ft; April-May.

Astragalus duchesnensis M. E. Jones — Duchesne milkvetch — Endemic to and locally common across the bottom of the Uinta Basin from near Ioka and Myton to the Colorado-Utah line and Raven Ridge of ne. Rio Blanco Co., also in Browns Park area of Daggett Co.; salt desert shrub and juniper communities in sandy areas and gravelly pediments; 4,700-5,400 (6,000 on Raven Ridge) ft; late April-early June.

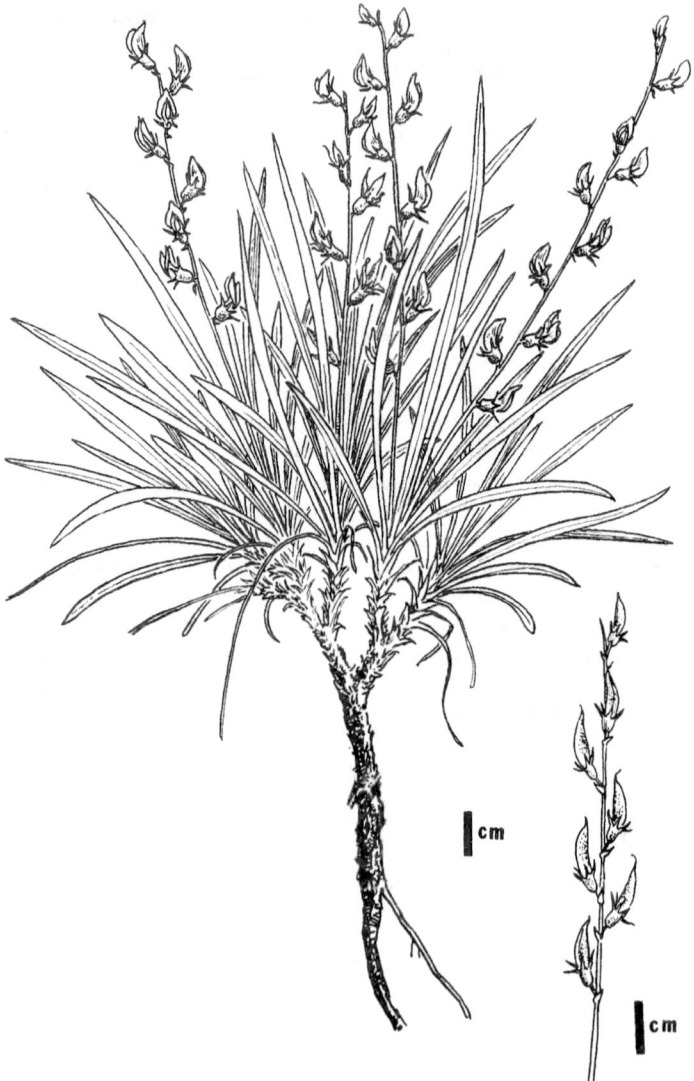

Astragalus chloodes Barneby — grass milkvetch

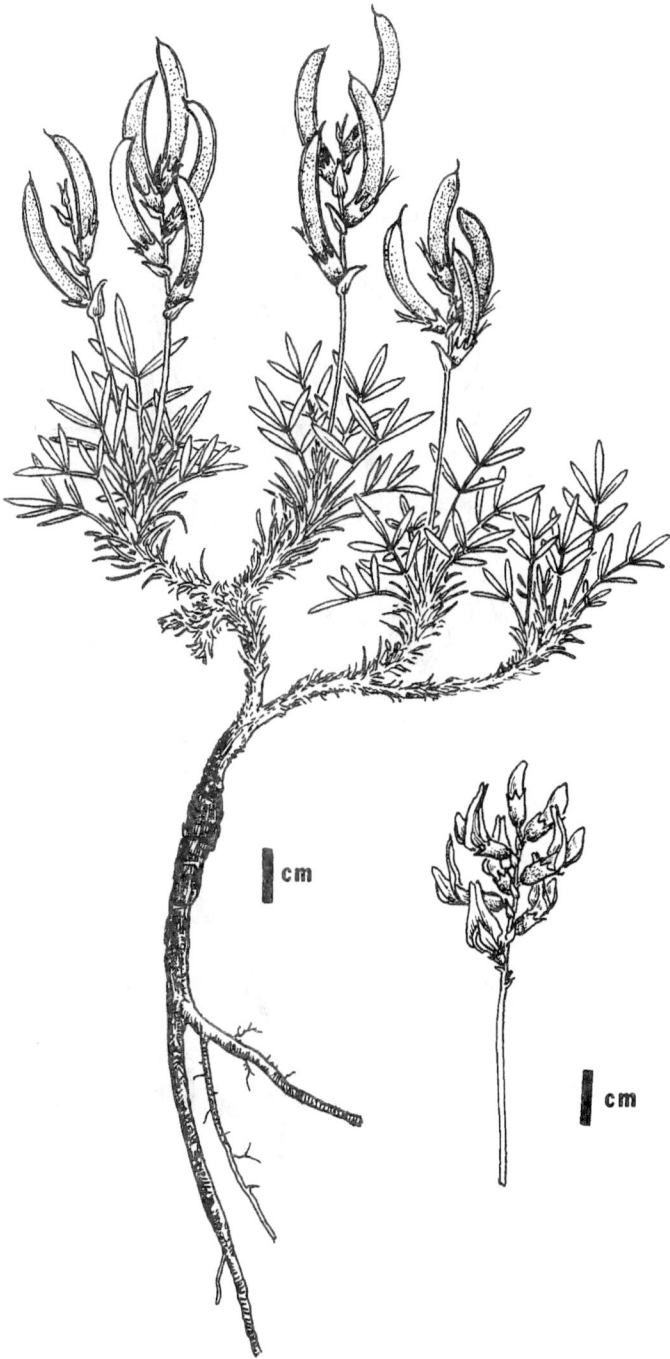

Astragalus detritalis M. E. Jones — debris milkvetch

Astragalus duchesnensis M. E. Jones — Duchesne milkvetch

Astragalus equisolensis Neese & S. L. Welsh (*A. desperatus* var. *neeseae* Barneby) — Horseshoe Bend milkvetch — Endemic to the vicinity of Horseshoe Bend, Uintah Co.; on river terrace sand and gravels and Duchesne River Formation, desert shrub communities; 4,700-5,200 ft; April-May and capable of considerable vegetative growth into November in years with abundant fall precipitation.

Astragalus eurekensis M. E. Jones — Eureka milkvetch — The one specimen seen (Brotherson 1882) is from the w. side of Strawberry Valley on foothill knolls; April-June.

Astragalus flavus Nuttall in Torrey & A. Gray (*A. confertiflorus* A. Gray) — yellow milkvetch — Widespread; common to locally abundant; desert shrub and Utah juniper communities, usually on clays of Mancos Shale, Duchesne River, and other formations that weather to bad lands; 4,800-5,700 ft; May-June. Our plants belong to var. *flavus*.

Astragalus flexuosus (Hooker) Don. — flexile milkvetch — The one specimen seen (Kelley & Sigstedt 82-86) is from Cow Creek, T4 S, R94S, S17) Rio Blanco Co.; shale ridge top with sagebrush; 8,100 ft; July. Reported by Graham (1937) for Middle Fork Stuart Creek. Reported for Uintah Co. (Barneby 1964) based on erroneous label data of Harrison 5605 - the specimen actually from the San Rafael Swell, Emery Co. (AUF 428) Except for shape and pubescence of the pods, this is much like *A. wingatanus*.

Astragalus geyeri A. Gray — Geyer milkvetch — Widespread; occasional and locally common; desert shrub, sagebrush, and juniper communities; 4,500 - 6,000 ft, usually, but not always, on sandy soils; May.

Astragalus hamiltonii C. L. Porter — Hamilton milkvetch — Endemic, the 49 specimens seen are from Whiterocks Bench to Steinaker Reservoir and 7 mi sw. of Vernal and from a disjunct population 18 miles e. of Vernal; most often on raw, eroding slopes of the Duchesne River Formation; desert shrub and juniper communities and advantageous on roadcuts; 5,500-6,740 ft; May-June. Closely related to *A. lonchocarpus* but apparently isolated from the nearest populations of *A. lonchocarpus* by 35 miles or more.

Astragalus kentrophyta A. Gray Kentrophyta. With 3 vars.:

1 Pubescence entirely of basifixed hairs; plants prostrate, of high elevations, forming mats . . . . **var. *tegetarius***
1 Pubescence mostly of malpighian hairs; plants prostrate or erect, of low elevations 2
2 Pods beakless or nearly so . . . . **var. *jessiae***
2 Pods incurved into a definite beak . . . . **var. *elatus***

Var. *elatus* S. Watson [*A. tegetarius* S. Watson var. *elatus* (S. Watson) Barneby] — tall kentrophyta, barbed-wire milkvetch — The 2 specimens seen are from Evacuation Creek, and Argyle Canyon; desert shrub and pinyon-juniper communities; 5,000-6,000 ft; July-Sept. The specimen from Evacuation Creek is annotated with a + by S. L. Welsh 2004. Most specimens identified as var. *elatus* have been annotated as var. *jessiae*. Mature fruit is needed for positive identification which most herbarium specimens lack.

Var. *jessiae* (Peck) Barneby — Jessie kentrophyta — Six of the specimens seen are from Sheep Creek and Boars Tusk and 2 from Browns Park, Daggett Co., others are from, near Bonanza, Pariette Bench, Wells Draw, and Nine Mile Canyon in southern Uintah and Duchesne Cos., and Little Snake River drainage in Moffat Co.; sagebrush and juniper communities; 5,800-6,700 ft; July-Sept.

Astragalus hamiltonii C. L. Porter — Hamilton milkvetch

Specimens from Uintah and Duchesne Cos. are intermediate to var. *elatus*. Future collections with mature fruit could demonstrate most of the Uintah and Duchesne Co. plants belong to var. *elatus*. Habit of stems (erect or diffuse) appears to be of little value for separating plants of var. *jessiae* and var. *elatus*.

Var. *tegetarius* (S. Watson) Dorn [*A. e.* var. *implexus* (Canby) Barneby; (*A. tegetarius* S. Watson var. *implexus* Canby] — mtn. kentrophyta — specimens seen are from exposed ridges and rocky slopes of the Uinta Mtns. and high points of Current Creek-W. Fork Duchesne area; 9,000-10,700 ft; June-Aug.

Astragalus laxmannii Jacquin [*A. adsurgens* Pallas; *A. striatus* Nuttall ex Torrey & A. Gray] — standing milkvetch — Specimens seen are all from within 2-8 mi of Manila and from Dry Canyon, on the south slope of the Uinta Mtns., also known from Carbon Co. outside our area; juniper and sagebrush communities, roadsides, and gravel pits; 6,500-8,500 ft; May-July. Our plants belong to ssp. *robustior* (Hooker) Barneby & S. L. Welsh.

Astragalus lentiginosus Douglas ex Hooker — freckled milkvetch — Three specimens with pink-purple flowers and fruits over 2 times longer than wide are from Kingfisher Island, Boars Tusk, and Red Canyon in Daggett Co. These are referable to var. *diphysus* (A. Gray) M. E. Jones [var. *araneosus* (Sheldon) Barneby], disjunct from southern Utah and the Great Basin. Variety *chartaceous* M. E. Jones [var. *platyphyllidius* (Rydberg) Peck] (broadleaf milkvetch) with white flowers and fruits about as wide as long is listed for Daggett Co. in AUF (434) and IMF (3B 160). Variety *salinus* (Howell) Barneby with flowers 9-13 mm long and thin, papery pods is listed for Summit Co. (AUF 434) and Uinta Co., Wyoming (Dorn 2001).

Astragalus lonchocarpus Torrey — great rushy milkvetch — Specimens seen are from Coyote Basin, Uintah Co., White River drainage in Rio Blanco Co., and near Deerlodge, Moffet Co.; desert shrub and Utah juniper communities; 5,550-5,800 ft.

Astragalus lutosus M. E. Jones — Dragon milkvetch — White River drainage in Rio Blanco and Uintah Cos., most common in the Piceance Basin, also known from isolated populations on Reservation Ridge and in Willow Creek drainage of the W. Tavaputs Plateau; occasional and locally common on the Green River Formation on ridges and talus slopes in salt desert shrub, pinyon-juniper-Salina wildrye, limber pine, and Douglas-fir communities; 5,400-8,600 ft; May-Aug. depending on elevation.

Astragalus megacarpus (Nuttall) A. Gray — great bladdery milkvetch — Specimens seen are from the W. Tavaputs Plateau from Avintaquin Canyon to Chokecherry Canyon and Argyle Canyon in s. Duchesne Co., Lookout Mtn., Moffat Co., and Bridger Formation at Hickey Mtn. Uinta Co. Wyoming; sagebrush, pinyon-juniper, ponderosa pine and Douglas-fir communities; 7,500-9,000 ft; May-June.

Astragalus miser Douglas ex Hooker [*A. decumbens* (Nuttall) A. Gray] — weedy milkvetch — Common to abundant across our area in aspen and coniferous woods, also in meadow, riparian, sagebrush, and alpine communities; 7,800-10,700 ft; June-Aug. Our plants belong to var. oblongifolius (Rydberg) Cronquist High elevation plants often differ in being smaller and having smaller leaflets, and often have blue rather than ochroleucous flowers.

Astragalus mollissimus Torrey — woolly locoweed, woolly milkvetch — Widespread; occasional or locally common in desert shrub, sagebrush and pinyon-juniper communities; 4,800-6,000 ft; April-early June. Our plants belong to var. thompsonae (S. Watson) Barneby. Woolly locoweed differs from our other species with wooly pods in the erect peduncle or scape. Capable of considerable green-up in late fall with abundant precipitation.

Astragalus nelsonianus Barneby — Nelson milkvetch — The 5 Utah specimens seen are all from within 5 mi of Manila in Daggett Co. and Peterson & Wilken 83-283 (CS) from 2 mi se. of community of Powder Wash, Moffat Co., several specimens seen from adjacent to our area from Sweetwater Co., Wyoming; sagebrush and juniper communities; 6,200-6,900 ft; May-early June.

Astragalus oophorus S. Watson — egg milkvetch — Apparently on the E. Tavaputs Plateau in Colorado. Bradley (1950) listed Graham 9118 from the head of Deep Channel Creek at the eastern edge of our area and Porter 3593 from 30 mi w. of Craig to the east of our area in Moffet Co.

Astragalus pattersonii A. Gray in Brandegee — Patterson milkvetch — Six specimens seen are from 1 mile north of Brush Creek near Highway 191, and 2 specimens from 3 miles s. of Jensen and 1 each from Evacuation Creek and base of Blue Mtn. all from Uintah Co.; desert shrub, greasewood, sagebrush, and, juniper communities; 4,800-6,900 ft; May.

Astragalus pubentissimus Torrey & A. Gray — Green River milkvetch — Common across the lower and middle elevations of the Tavaputs Plateau from Duchesne to Rangely, also adjacent to our area in Sweetwater Co. Wyoming; desert shrub and pinyon-juniper communities, also colonizing in the drawdown basin of Flaming Gorge Reservoir; 5,200-7,500 ft; May-June or to Sept. with late season precipitation. Our plants belong to var. *pubentissimus*.

Astragalus purshii Douglas ex Hooker — woollypod milkvetch — With 2 vars.:

1 Flowers pink-purple; plants of Daggett Co. . . . . **var. *glareosus***
1 Flowers whitish or ochroleucous with the keel purple-tipped **var. *purshii***

Var. *glareosus* (Douglas) Barneby — woollypod milkvetch — Known in our area from Daggett Co. in sagebrush communities.

Var. *purshii* — woollypod milkvetch — Widespread; occasional to common in desert shrub, sagebrush, and pinyon-juniper communities; 4,900-7,200 ft; April-May (early June). Less likely to green-up with fall rains than *A. chamaeleuce* and *A. mollissimus*.

Astragalus racemosus Pursh — alkali milkvetch — Occasional across the Uinta Basin from Duchesne to near the Colorado-Utah line, apparently lacking from our part of Colorado; desert shrub, sagebrush, and pinyon-juniper communities; 5,400-7,200 ft; May-June. Our plants belong to var. *treleasei* C. L. Porter.

Astragalus robbinsii (Oakes) A. Gray — Robbins milkvetch — Known from 3 collections taken in 1914, 1923 (C. H. McDonald sn USUUB) and 1984 in W. Fork Bear River near Whitney Reservoir; with willows at 8,960-9,100 ft. Our plants belong to var. *minor* (Hooker) Barneby.

Astragalus saurinus Barneby — dinosaur milkvetch — Endemic to Uintah Co. from Twelve Mile Wash near Lapoint to Dinosaur National Monument and s. to Red Wash and Kennedy Wash; on Duchesne River, Morrison, Chinle, Moenkopi, and other formations that weather to badlands; 4,700-5,600 ft; May-June. This was referred to as *A. coltonii* M. E. Jones by Graham (1937). Wolf & Dever 5005 and Weber 5318 DINO!, referred to as *A. rafaelensis* M. E. Jones (Holmgren 1962) belong here.

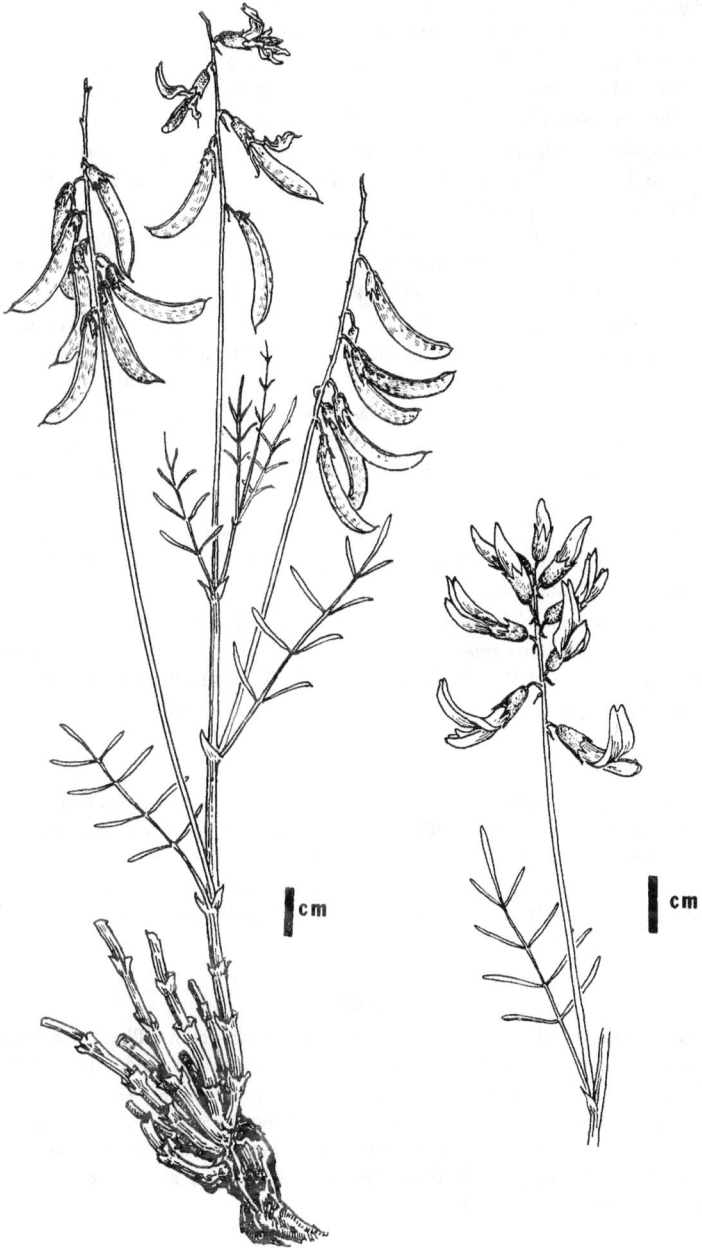

Astragalus saurinus Barneby — dinosaur milkvetch

Astragalus scopulorum T. C. Porter — Rocky Mtn. milkvetch — Locally common along the paved road west of Strawberry Reservoir about 1 mile s. of Highway 40; mtn. big sagebrush community at 7,700 ft elevation. Also on Strawberry Ridge at the edge of the Uinta Basin. Previous report of *A. drummondii* Douglas ex Hooker is based on specimens of *A. scopulorum*.

Astragalus spatulatus Sheldon [*A. simplex* Tidestrom; *A. simplicifolius* (Nuttall) A. Gray] — draba milkvetch, tufted milkvetch — Widespread; common in desert shrub, black sagebrush, big sagebrush, pinyon-juniper, and mt. brush communities, on numerous substrates but often on rocky or shallow soils; 4,800-8,000 (8,600) ft; April-June (Oct.). White-flowered plants are common on Blue Mtn. Elsewhere in our area they are mostly pink-purple.

Astragalus tenellus Pursh — looseflower milkvetch, pulse milkvetch — Widespread; common; various plant communities including: pinyon-juniper, sagebrush, Douglas-fir, and aspen; (5,000) 6,000-9,500 ft; June-July.

Astragalus utahensis (Torrey) Torrey & A. Gray — Utah milkvetch — Occasional from Strawberry Valley and east to Indian Canyon and Tabiona; sagebrush, pinyon-juniper, mt. brush, and aspen-fir communities; 6,100-7,500 ft; May-June.

Astragalus wetherillii M. E. Jones — Wetherill milkvetch — Known from the Yampa River area in Moffat Co. and Range Creek of the E. Tavaputs Plateau (Welsh 14932); May-June.

Astragalus wingatanus S. Watson — Fort Wingate milkvetch — The 4 collections seen are from Indian Canyon and Lake Canyon drainages, W. Tavaputs Plateau, Duchesne Co., and Hill Creek-Willow Creek, Tavaputs Plateau, Grand Co.; pinyon pine and Douglas-fir communities; 7,000 ft; May-June.

Caragana Fabricus — Pea shrub

Caragana arborescens Lamarck — Siberian pea shrub — Shrubs to 4 m tall or more; leaves 4-10 cm long; leaflets 8-12, 12-25 mm long, 5-15 mm wide, villous, becoming glabrate in age; stipules slender, occasionally persisting as spines; flowers 17-23 mm long, yellow; pods sessile, straight, 35-55 mm long, 4-7 mm wide, glabrous. Introduced from central Siberia, cultivated as an ornamental and for erosion-control.

Coronilla L.

Coronilla varia L. [*Securigera varia* (L.) Lassen] — crown-vetch — Plants perennial from a taproot or caudex, 30-150 cm tall; leaves pinnate compound with 9-27 leaflets; flowers in head-like umbels, pink or pink-purple, 9-12 mm long; stamens 10, diadelphous; fruit erect or spreading, 35-50 cm long. Introduced from Europe, planted as an ornamental and for erosion control, escaping and persisting. Sparingly known from the Uinta Basin. However, additional introduction might be expected.

Glycyrrhiza L. — Wild Licorice

Glycyrrhiza lepidota Pursh — American wild licorice — Perennial caulescent herbs from stout, deep-seated roots, 40-12 cm tall; leaves 8-19 cm long; leaflets 13-19, 8-53 mm long, 3-15 mm wide, mucronate, glabrous above, gland-dotted and puberulent beneath; racemes 20-50 flowered; flowers 9-13 mm long, dingy-white to cream; pods 13-20 mm long, the body 5-7 mm wide, densely covered with hooked prickles, cocklebur-like. Widespread; occasionally and locally abundant; along water courses, moist low lands, ditch banks, and about seeps; 4,700-7,900 ft.

Hedysarum L. — Sweetvetch

Perennial caulescent herbs from taproots and caudices; leaves odd pinnate compound; flowers in axillary racemes, red-purple to pink or pink-purple; stamens 10, diadelphous; fruit a loment with 2-8 segments.

---

1   Veins of leaflets readily apparent; fruit segments winged; calyx lobes unequal, shorter than the tube; plants known from the N. slope of the Uinta Mtns., Range Creek drainage, W. Tavaputs Plateau and from near Tabiona .... *H. occidentale*

1   Veins of leaflets not apparent; fruit segments not winged; calyx lobes subequal, longer than the tube; plants widespread .... *H. boreale*

Hedysarum boreale Nuttall (*H. cinerascens* Rydberg; *H. utahense* Rydberg) — northern sweetvetch — Desert shrub, sagebrush, pinyon-juniper, and mt. brush communities, sometimes on barrens. Plants referable to var. *boreale* have glabrous loments and are widespread from 4,700-8,400 ft; May-July. Plants referable to var. *gremiale* (Rollins) Northstrom & S. L. Welsh have small spines on the surface of the loments, and are endemic from Tridell to Dinosaur National Monument from 5,500-5,800 ft; June.

Hedysarum occidentale Greene — western sweetvetch — Specimens seen are from the north slope of the Uinta Mountains, Range Creek drainage of the W. Tavaputs Plateau and near Tabiona; pinyon-juniper and mt. brush communities; 6,800-7,500 ft; June-July. The plants from the north slope of the Uinta Mtns. and Tavaputs Plateau with leaflets mostly over 3 times longer than wide and dark pink-purple flowers mostly less than 20 mm long belong to var. *occidentale*. Those from near Tabiona with leaflets 2-3 times longer than wide and white and pinkish flowers mostly 20-24 mm long belong to var. *canone* S. L. Welsh.

Lathyrus L. — Peavine; Sweetpea

Perennial caulescent herbs, clambering, trailing, or climbing; leaves pinnately compound, the rachis ending in a tendril or bristle; flowers in axillary racemes; stamens 10, diadelphous; style bearded down one side as in a toothbrush; fruit a legume, oblong, several seeded.

---

1   Leaflets 2; stems winged; plants introduced, cultivated and escaping .... *L. latifolius*

1   Leaflets 4 or more; stems not winged; plants native 2

2   Keel shorter than the wings; calyx glabrous or the teeth merely ciliate, the lower tooth usually longer than the tube; stipules foliaceous; petals pink-purple, rarely white .... *L. pauciflorus*

2   Keel usually subequal to the wings; calyx often hairy, the lower tooth shorter than the tube; stipules not foliaceous; flowers white or pink-purple .... *L. lanszwertii*

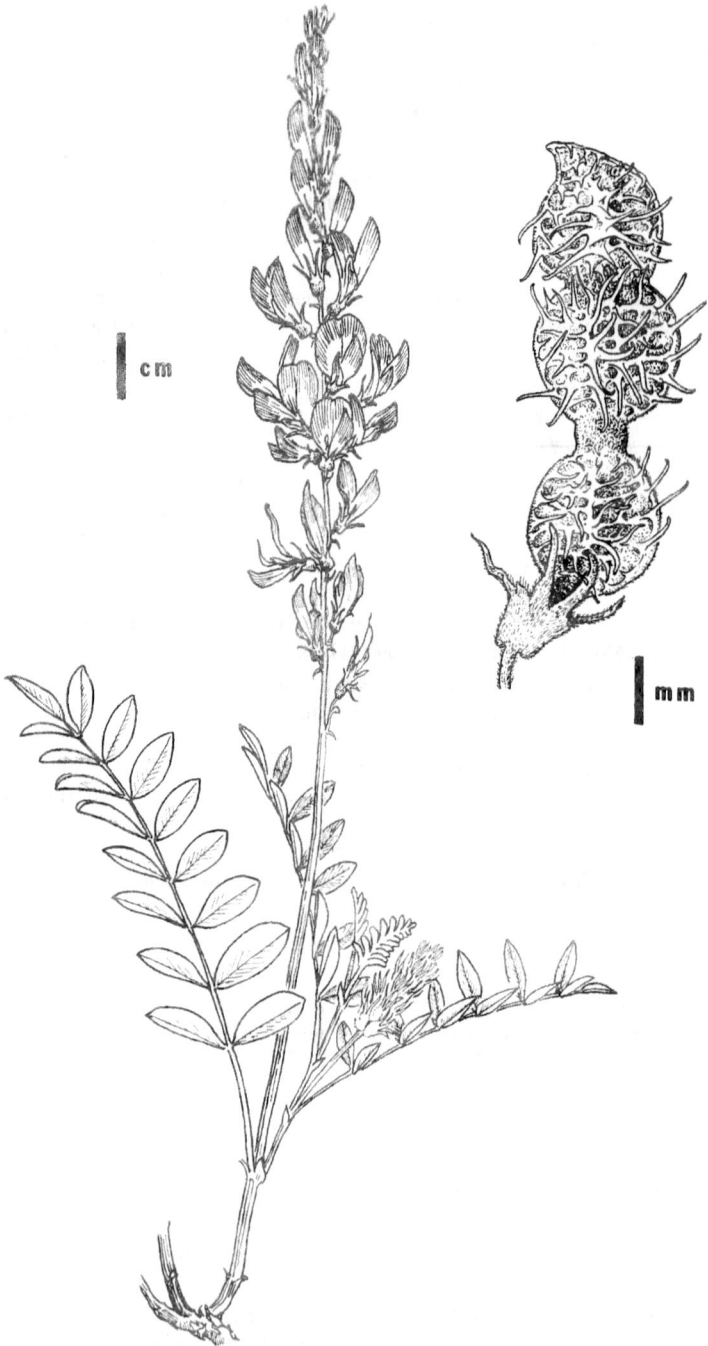

cm

mm

Hedysarum boreale Nuttall — northern sweetvetch

Lathyrus lanszwertii Kellogg — mtn. peavine — With 4 intergrading vars.

1 Tendrils reduced to a straight (rarely coiled) stock . . . . . . **var. *leucanthus***
1 Tendrils coiled 2
2 Leaflets mostly less than 3 times as long as wide . . . . **var. *laetivirens***
2 Leaflets mostly over 3 times as long as wide 3
3 Wings and keel white and fading a yellow-tan . . . . **var. *pallescens***
3 Wings and keel pink-purple or purple-tinged . . . . **var. *lanszwertii***

Var. *laetivirens* (Greene) S. L. Welsh [*L. leucanthus* var. *laetivirens* (Greene) C. L. Hitchcock] — aspen peavine, plateau peavine — The one specimens seen (S. L. Welsh & K. Taylor 15068) is from Range Creek, E. Tavaputs Plateau; aspen-Douglas-fir community at 7,000; June-Aug.

Var. *lanszwertii* (*L. coriaceus* White) — mtn. peavine — The 23 specimens seen are from West Tavaputs Plateau, Strawberry Valley and w. end of the Uinta Mtns. east to the N. Fork Duchesne drainage on the south slope and to east to the Bear River drainage on the n. slope; 7,800-10,150 ft; June-Aug.

Var. *leucanthus* (Rydberg) Dorn — Nevada peavine — The 2 specimens seen (Atwood 32369 and 32375) are from Atchee Ridge, East Tavaputs Plateau, mtn. brush communities: 8,100-8,300 ft; June.

Var. *pallescens* Barneby — palled peavine — Specimens seen are from the West Tavaputs Plateau; aspen, aspen-conifer parklands.

Lathyrus latifolius L. — perennial sweetpea — Introduced from Europe, cultivated ornamental, persisting and escaping; the 1 record seen (Neese & Moore 7953) is from cultivation at Vernal.

Lathyrus pauciflorus Fernald (*L. utahensis* M. E. Jones) — fewflower peavine, Utah sweetpea — The 10 specimens seen are from Strawberry Valley to Wolf Creek in the Uinta Mtns and to the Avintaquin drainage of the E. Tavaputs Plateau.; aspen, yellowbrush, mtn. big sagebrush, tall forb, and Douglas-fir communities; 8,000-9,265 ft; July-Aug.

Lotus L. — Trefoil

Lotus corniculatus L. — birds-foot trefoil — Plants herbaceous, perennial, glabrous, 20-60 cm tall; stipules reduced to glands; leaves sessile, palmate; leaflets 3.5-15 mm long1.5-7 mm wide; inflorescence a raceme with 2-5 (8) flowers; corolla 7-12 mm long, yellow; fruit 25-30 mm long. Introduced from Europe and cultivated as a forage crop and somewhat persistent; the few specimens seen are from the Lapoint and Ft. Duchesne areas. *Lotus tenuis* Waldstein & Kitabel ex Willdenow might also be planted in our area. Plants of these 2 taxa are difficult to separate.

Lupinus L. — Lupine

Annual or perennial herbs; leaves palmately compound; flowers borne in terminal racemes, blue, blue-purple, or rarely white (albino specimens); stamens 10, monadelphous, with 5 long filaments alternating with 5 short ones; fruit a laterally compressed legume, 2-several seeded. This is a difficult genus in which numerous morphologically intergrading and hybridizing taxa have been described. Wide-ranging taxa tend to intergrade with other taxa they contact, and identification of some specimens is arbitrary.

1   Plants annual, 3-24 cm tall; leaflets glabrous above except sometimes along the margins. 2

1   Plants perennial, usually over 24 cm tall or else leaflets pubescent above 3

2   Racemes 1-2.5 cm long in flower, the axis 1.5-3 cm long in fruit; flowers 5.2-7 mm long; leaves mainly in a basal tuft; leaflets 5-18 mm long . . . .
. . . . **L. brevicaulis**

2   Racemes 1-17 cm long in flower, the axis 4-21 cm long in fruit; flowers 8.5-12 mm long; leaves mainly cauline; leaflets 11-48 mm long   . . . . **L. pusillus**

3   Plants 3-12 cm tall, caespitose; racemes 1-4 cm long in flower, the axis 2-6 cm long in fruit; leaflets 3-25 mm long, 1-6 mm wide, pilose on both sides; peduncles 0-2 cm long; flowers 7-8.5 mm long . . . . **L. lepidus**

3   Plants 13-90 cm tall; racemes 9-30 cm long in flower, the axis 10-35 cm long in fruit; leaflets 7-95 mm long, glabrous or pilose; peduncles 1.5-15 cm long; flowers 8-16 mm long, or if only 5-7 mm long then leaflets glabrous above
. . . . 4

4   Banner recurved at or below the midpoint and more than 3 mm from the apex 5

5   Banner glabrous dorsally; basal leaves tufted and with long petioles, persistent at anthesis . . . . **L. polyphyllus**

4   Banner recurved above the midpoint and less than 3 mm from the apex . 6

5   Banner pubescent dorsally; basal leaves not tufted, usually deciduous by anthesis . . . . **L. sericeus**

6   Pedicels 1-2.5 mm long at anthesis; racemes densely flowered with internodes 1-3 (5) mm long after anthesis; stems with widely spreading hairs and dense, appressed hairs . . . . **L. leucophyllus**

6   Some pedicels usually over 2.5 mm long at anthesis; racemes loosely flowered with at least some internodes commonly over 5 mm long after anthesis; with rare exceptions stems glabrate or with appressed, upward pointing hairs . . . . **L. argenteus**

Lupinus argenteus Pursh (*L. laxiflorus* Douglas in part) — silver lupine — The most common and widespread lupine of our area; highly variable; with 3 vars. Most specimens can be assigned to variety by use of the following key:

1   Calyx with a gibbous-saccate spur at the base; wings or keel or both ciliate below the claws (rarely glabrous) . . . . **var. utahensis**

1   Calyx at most gibbous at base; wings and lower edge of keel glabrous . . . 2

2   Stems with solitary, terminal inflorescence, rarely with 1 or 2 depauperate lateral racemes . . . . **var. rubricaulis**

2   Stems commonly branched with the terminal raceme and those of the branches maturing about the same time . . . . **var. argenteus**

Var. argenteus [*L. tenellus* Douglas; *L. a.* var. *tenellus* (Douglas) Dunn] — silver lupine — Widespread; occasional or locally common in desert shrub, sagebrush, pinyon-juniper, mt. brush, cottonwood, grass, and Douglas-fir and other coniferous forest communities; 4,900-7,500 (10,100) ft; June-July.

Var. utahensis (S. Watson) Barneby (*L. aduncus* Greene, *L. caudatus* Kellogg.) — tailcup lupine, spurred lupine — Widespread; occasional or locally common in desert shrub, sagebrush, and pinyon-juniper communities; 5,000-8,000 ft; June-Sept.

Var. rubricaulis (Greene) S. L. Welsh [*L. alpestris* A. Nelson; *L. a.* var. *boreus* (C. P. Smith) S. L. Welsh] — subalpine lupine — Uinta Mtns. and Tavaputs Plateau where locally common; streamside, aspen, coniferous forest, forb-grass, and alpine communities; 7,000-11,100 ft; July-Sept.

Lupinus brevicaulis S. Watson — shortstem lupine — Widespread but no specimens seen from Duchesne Co.; occasional or locally common in desert shrub, sagebrush, and juniper communities; 4,700-5,800 ft; May-June.

Lupinus lepidus Douglas ex Lindley — Pacific lupine — (*L. caespitosus* Nuttall in Torrey & A. Gray). Occasional in Strawberry Valley and on the W. Tavaputs Plateau w. of the Avintaquin drainage; and eastern Uinta Mtns., grass, sagebrush, and aspen-fir-spruce communities and on barren shaly slopes; 7,600-9,000 ft; mid June-July. Our plants belong to var. *utahensis* (S. Watson) C. L. Hitchcock

Lupinus leucophyllus Douglas in Lindley — velvet lupine, white-leaved lupine — The 8 specimens seen are from Strawberry Valley and s. flank of the Uinta Mtns. from Lake Fork to Whiterocks on flats of glacial outwash; sagebrush communities; 6,000-8,200 ft; June-Sept.

Lupinus polyphyllus Lindley (*L. wyethii* S. Watson) — showy lupine — Two native vars. are found in the Uinta Basin as follows. In addition, var. *polyphyllus* (native to the Pacific Northwest) is planted as an ornamental.

1 Upper surface of leaflets glabrous or pubescent only at the margin . . . . .
. . . . **var. ammophilus**
1 Leaflets silky-strigose on both sides, gray or silvery when fresh, reddish brown when dry . . . . **var. humicola**

Var. ammophilus (Greene) Barneby (*L. ammophilus* Greene) — sand lupine — Specimens seen are from extreme se. Duchesne Co (Wrinkle Road-Sand Wash), and Big Pack Mtn. area, Willow Creek, Red Wash, Raven Ridge, and Snake John Reef in Uintah Co., and from Rio Blanco Co.; desert shrub, sagebrush, grease-wood-rabbitbrush, and juniper communities, sometimes abundant on wash bottoms; 4,670-5,800 ft. May-June. Most specimens from Uintah Co. seem better placed with var. *ammophilus* (AUF 467). However, Red Wash specimens have leaves with upper side pubescent.

Var. humicola (A. Nelson) Barneby (*L. wyethii* S. Watson) — wash bottom lupine — Listed for Uintah Co. (AUF 467, IMF 3B: 244) The only specimen seen from Uintah Co. (Neese 14369) is from 2 mi w. of Dry Fork Settlement on the north side of Little Mtn., from a juniper-mtn. brush-sagebrush community at 7,050 ft.

Lupinus pusillus Pursh — dwarf lupine, rusty lupine — Widespread, occasional in desert shrub, sagebrush, and juniper communities; 4,700-6,000 ft; May-June. Most of the our plants belong to var. pusillus with peduncles commonly 1-3.5 cm long, inflorescence usually longer than the leaves, and banner 6-10 mm wide. Plants with peduncles seldom more than 1 cm long, inflorescence shorter than the leaves, and banner 5 mm wide or less belong to var. *intermontanus* (A. Heller) C. P. Smith which is known from Browns Park, Daggett Co. and 5 mi s. of Myton, Duchesne Co (Neese & B. Welsh 7227). References to var. *rubens* (Rydberg) S. L. Welsh [*L. rubens* Rydberg; *L. p.* ssp. *rubens* (Rydberg) Dunn], a plant from s. of our area, are not supported by specimens.

Lupinus sericeus Pursh (*L. barbiger* S. Watson) — silky lupine — Occasional or infrequent on the E. and W. Tavaputs Plateaus, Strawberry Valley, and s. slope of the Uinta Mtns. in sagebrush communities at 7,300-9,050 ft; June-Aug. Our plants belong to var. *sericeus*.

Medicago L. — Alfalfa; Lucerne; Medic

Annual or perennial caulescent herbs; leaves pinnately trifoliate, the leaflets serrate toward the apex; flowers in axillary, pedunculate racemes; fruit a legume, curved to spirally coiled, 1-several seed, indehiscent, reticulate or spiny.

1   Flowers 2-3 mm long, yellow; inflorescence less than 10 mm long in flower, but to 25 mm long in fruit; pods coiled through a single spiral, 1-seeded, unarmed; plants annual, prostrate to decumbent, rarely erect   **M. lupulina**

1   Flowers 4-10 mm long, variously colored; inflorescence usually over 1 cm long in flower or pods armed with prickles 2

2   Flowers 4-5 mm long, yellow, 2-5 per raceme; racemes less than 1 cm long; pods armed with prickles, several-seeded; plants annual, introduced from Europe, weedy on cultivated lands, to be expected in our area (bur clover) (*M. hispida* Gaertner) .... **M. polymorpha** L.

2   Flowers 6-10 mm long, yellow, blue, lavender, pink, purple, purple-black or white, 6-many on at least some racemes; racemes longer than 1 cm; pods unarmed, several-seeded; plants perennial 3

3   Pods curved or hooked but not coiled; flowers yellow or mostly yellow ... . **M. falcata**

3   Pods coiled; flowers blue, lavender, purple, purple-black, or infrequently white .... **M. sativa**

Medicago falcata L. — yellow alfalfa — Introduced from Europe, sparingly cultivated, known from Uintah Co. Yellow alfalfa forms hybrids with M. sativa, and it sometimes considered a phase of that species.

Medicago lupulina L. — black medic, hop clover — Introduced from Europe, widespread; weedy in lawns, fields, roadsides, and riparian and other indigenous communities where usually associated with disturbance; 5,100-7,500 ft and probably higher; June-Sept.

Medicago sativa L. — alfalfa, lucerne — Introduced from Eurasia; widely planted as a forage crop, and on rangelands and roadsides, persisting and escaping, over a wide elevational range, but seldom above 9,500 ft; June-Sept. Any of a number of cultivars can be expected as this plant has undergone intensive selection in development of plants suitable for heavy production on irrigated and fertilized fields and for use on rangelands. Among the cultivars used on rangelands are Ladak, Nomad, and Rambler. These have been highly successful in seedings following fire in pinyon-juniper communities.

Melilotus Miller — Sweetclover

Annual or biennial caulescent herbs, from taproots; leaves pinnately trifoliate, the leaflets dentate on the upper 1/2-3/4; flowers in racemes; pods straight, with 1 or 2 seeds, indehiscent. Our species contain coumarin that can cause hemorrhaging and death in livestock. This substance or one of its isomers is used in rat and mouse poison.

---

1   Petals white, the banner distinctly longer than the wings . . . . *M. alba*
1   Petals yellow, the banner and wings about equal . . . . *M. officinalis*

Melilotus alba Medicus — white sweetclover — Introduced from Europe, widely established, but not so common as M. officinalis; often along roadsides; June-Sept. The range of flowers per raceme (38-115) is greater than that of *M. officinalis* (20-65). This plant is an excellent source for honey for domestic bees.

Melilotus officinalis (L.) Pallas — yellow sweetclover — Introduced from Europe or the Mediterranean; widespread but mostly below 7,000 ft; often along roadsides; May-Sept.

Onobrychis Adanson — Sainfoin

Onobrychis viciifolia Scopoli — common sainfoin — Perennial caulescent herbs from caudices and taproots, 20-45 cm tall; leaves 3-12 cm long; leaflets 11-21 (27), 8-25 mm long, 2-7 mm wide, pilose mainly along the veins beneath, glabrous above; peduncles 8-19 cm long; racemes with 14-39 (50) flowers, the axis 4-15 cm long in fruit; flowers 10-13 mm long, red-purple, lavender or pink; stamens 10, essentially diadelphous; fruit a loment but this reduced to 1 segment with only 1 or 2 seeds, armed with prickles. Introduced from Europe; planted in rangeland and roadside seedings, sometimes escaping and persisting.

Oxytropis de Candolle — Crazyweed; Loco; Locoweed; Oxytrope; Pointvetch; Stemless Loco

Perennial caulescent or more often acaulescent (ours) herbs from taproots and caudices; leaves alternate or often all basal; flowers in racemes, pink, pink-purple or white, the keel tip produced into a porrect beak; fruit a legume.

---

1   Flowers 5-12 mm long 2
1   Flowers 11-26 mm long; pods ascending to erect. 3
2   leaflets 23-41; pods spreading-declined . . . . *O. deflexa*
2   Leaflets 7-17; pods erect . . . . *O. parryi*
3   Racemes with 1-5 flowers, subcapitate; plants 1-5 (10) cm tall 4
3   Racemes with 6-40 flowers, elongate, only rarely subcapitate; plants 8-50 cm tall 6
4   Calyx swollen at full anthesis, becoming inflated and finally enclosing the pod; pods 6-10 mm long; plants of Daggett and Duchesne Cos., below 7,500 ft . . . . *O. multiceps*
4   Calyx campanulate, not inflated nor enclosing the fruit; pods 13-25 mm long; plants not known from Daggett Co. 5
5   Pods oblong-ellipsoid, not inflated, leathery in texture; flowers 7.5-12 mm long; plants of known from above 8500 ft in Wasatch and Duchesne Cos. . . . . *O. parryi*

5 Pods ovoid, inflated, papery in texture; flowers (11) 14-16.5 mm long; plants of the E. Tavaputs Plateau .... *O. oreophila*

6 Petals bright pink-purple... 7

6 Petals white or tinge with purple 8

7 Plants pubescent with malpighian hairs; calyx not swollen; plants of the Tavaputs Plateau .... *O. lambertii*

7 Plants with basifixed hairs; calyx somewhat swollen; plants of Daggett and Moffat Cos. .... *O. besseyi*

8 Banner 14-18 mm long; plants from the alpine ridge between N. Fork Duchesne and Blind Stream, and Henrys Fork River on the n. slope of the Uinta Mtns... *O. campestris*

8 Banner 18-26 mm long; plants of broad distribution but not as above ....
.... *O. sericea*

Oxytropis besseyi (Rydberg) Blank. (*O. obnapiformis* C. L. Porter) — red loco — Occasional in Daggett and n. Moffat Cos., and L. Boyd 66 is from Pariette Draw, Uintah Co; desert shrub, sagebrush, and pinyon-juniper communities, mostly in sandstone or in sandy or sandy-clay soils; 5,800-6,900 ft; June. Plants with leaves 10-20 cm long are referable to var. *obnapiformis* (C. L. Porter) S. L. Welsh. Those with leaves 2-8 cm long belong to var. *ventosa* (Greene) Barneby.

Oxytropis campestris (L.) de Candolle — yellow locoweed, field locoweed — Alpine on the limestone plateau between North Fork Duchesne and Rock Creek, and Henrys Fork River at 8,920 feet in the Uinta Mtns. The Henrys Fork specimen indicates a population higher up in the Henrys Fork drainage.

Oxytropis deflexa (Pallas) de Candolle — drop-pod locoweed — With 2 vars.:

1 Flowers 9-10.5 mm long, blue-purple; inflorescence 15-22 mm wide at anthesis .... **var.** *pulcherrima*

1 Flowers 5-9 mm long, whitish, lilac or blue-purple; inflorescence less than 17 mm wide at anthesis .... **var.** *sericea*

Var. *pulcherrima* S. L. Welsh & A. Huber. — pretty oxytrope — Specimens seen are from Log Hollow-Rock Creek divide, Rock Creek-Brown Duck divide, Uinta Mtns.; some specimens from the north flank of the Uinta Mtns. in Daggett and Summit Cos. appear to belong here and seem transitional to the following var.; open slopes with grasses and forbs, krummholz spruce, and talus; 9,240-11,300 ft; July-Aug.

Var. *sericea* Torrey & A. Gray — shortstem oxytrope — Specimens seen are from the E. Tavaputs Plateau, Grand Co. and e. end and north slope of the Uinta Mtns.; meadows, aspen woods, and drying streams; 8,500-9,000 ft; June-Aug.

Oxytropis lambertii Pursh — purple locoweed — The few specimens seen are from the Avintaquin drainage to Range Creek on the W. Tavaputs Plateau and Piceance Basin of the E. Tavaputs Plateau; sagebrush, Douglas-fir-snowberry, and aspen communities, sometimes on open slopes of Green River shale; 7,100-10,050 ft; June-Aug. Our plants belong to var. *bigelovii* A. Gray.

Oxytropis multiceps Torrey & A. Gray — Rocky Mt. oxytrope — Specimens seen are from Lodgepole Creek, Sheep Creek, Kingfisher Island, and Green Lakes, Daggett Co.; Vermillion Gap and Little Snake River, Moffat Co.; and Red Creek,

Duchesne Co.; semi-barrens and sagebrush, pinyon-juniper, and ponderosa pine communities; 6,050-7,450 ft; May-June. Graham 8989 cited by Graham (1937) belongs to *O. jonesii.*

Oxytropis oreophila A. Gray — mtn. oxytrope — The few specimens seen are from the E. Tavaputs Plateau, Grand and Uintah Cos.; shale and marl barrens, desert shrub and pinyon-juniper communities; 6,400-8,100 ft; June-Aug. Our plants are referable to var. *jonesii* (Barneby) Barneby (*O. jonesii* Barneby).

Oxytropis parryi A. Gray — Parry oxytrope — Specimens seen are from W. Tavaputs Plateau (Strawberry Peak), Red Creek Mt. and Uinta Mtns. (as far e. as the Lake Fork drainage on the s. slope and the Blacks Fork drainage on the n. slope); windswept ridges and talus; 10,200-11,300 ft; July-Aug.

Oxytropis sericea Nuttall in Torrey & A. Gray [*O. albiflora* (A. Nelson) K. Schumann] — silky locoweed, white locoweed — The most common oxytrope our area, Vermillion Bluffs, Moffat Co., Tavaputs Plateau and its flank, and n. slope of the Uinta Mtns. but no specimens seen from the s. slope of the Uinta Mtns.; desert shrub, pinyon-juniper, Salina wildrye, sagebrush, and lodgepole pine communities at 4,750-8,900 ft; late April-June. Our plants belong to var. *sericea.*

Pediomelum Rydberg — Breadroot

Pediomelum megalanthum (Wooton & Standley) Rydberg (*Psoralea megalantha* Wooton & Standley.) — large-flowered breadroot — Subacaulescent to caulescent malodorous herbs with slender caudex branches arising from deep-seated tuberous roots, 4-25 cm tall; herbage strigose; leaves palmate compound; leaflets 5 (8), 9-34 mm long, 4-23 mm wide, punctate; inflorescence a raceme, with 6-24 flowers, 2-5 cm long; flowers 12-21 mm long, commonly purplish; pods included in the calyx. Occasional across Uintah Co. and into Colorado, only one specimen seen from Duchesne Co., and none from Daggett or Summit Cos.; desert shrub, sagebrush, but most often juniper communities; 4,800-6,100 ft; May-June. Plants referred to as *Psoralea mephitica* S. Watson by Graham (1937) belong here.

Psoralidium Rydberg — Scurfpea

Psoralidium lanceolatum (Pursh) Rydberg (*Psoralea lanceolata* Pursh) — dune scurfpea — Caulescent herbs 15-68 cm tall, from rhizomes, glabrous or strigose; leaves commonly 3-foliate; leaflets 14-50 mm long, 1-9 mm wide, entire; inflorescence a raceme with 5-24 flowers; flowers 5-6 mm long, blue, white, or both; pods 1-seeded, glandular. Specimens seen are from near Lapoint to Dinosaur National Monument and s. to near Bonanza Canyon, and from Daggett and Moffat Cos.; along washes and in desert shrub and juniper communities, usually on sandy soil; 4,900-6,300 (8,500?) ft; May-June. Our plants belong to var. *lanceolatum.*

Sophora L. — Sophora

Sophora stenophylla A. Gray in Ives. — silvery or fringeleaf sophora — Perennial, caulescent, silvery pubescent herbs from deep-seated rhizomes, 13-41 cm tall; leaflets 9-15, linear or linear-oblong, 7-27 mm long, 0.5-4 mm wide; racemes 10-40-flowered; flowers 15-27 mm long, blue-purple or blue; pods spreading-declined, with a stipe 8-16 mm long, the body 15-60 mm long, 6-8 mm wide; terete but strongly constricted between the seeds when the seeds are more than 1; seeds 1-5. The 4 specimens seen are from Jensen to Ouray, Uintah Co.; desert shrub communities, sandy soil; 4,800-5,200 ft; May-June. A beautiful plant.

Sphaerophysa de Candolle — Sphaerophysa

Sphaerophysa salsula (Pallas) de Candolle [*Swainsonia salsula* (Pallas) Thubert in Engler & Prantl] — alkali globepea — Perennial caulescent herbs from rhizomes, 40-70 cm tall; leaves 3-10 cm long; leaflets 15-25, 3-18 mm long, 1-7 mm wide, retuse to obtuse and apiculate, strigose beneath, glabrous above; flowers 5-17 per raceme, 12-14 mm long, dull-red, fading lavender to brown; pods ascending to declined, on a stipe 4-7 mm long, bladdery-inflated, 13-24 mm long, 9-20 mm wide. Introduced from Asia. Graham (1937) reported Doutt 4312 from Red Creek, 2 mi n. of Fruitland where it appears to have not persisted, but this plant is well established on the floodplain of the Green River near Ouray; June-Aug.

Thermopsis R. Brown — Thermopsis; False Lupine; Yellow Lupine; Golden Pea; Yellow Pea

---

1   Pods straight, erect or ascending; larger leaflets 3-5.5 times longer than wide,; plants mostly 20-70 cm tall or more, widespread throughout our area, mostly of moist and wet places . . . . **T. montana**
1   Pods strongly curved, somewhat constricted between the seeds, spreading or recurved; larger leaflets mostly less than 3 times as long as wide; plants mostly 10-40 cm tall, somewhat restricted, mostly of dry places or of seeps that dry in summer . . . . **T. rhombifolia**

Thermopsis montana Nuttall in Torrey & A. Gray [*T. pinetorum* Greene; *T. rhombifolia* var. *montana* (Nuttall) Isely] — golden pea, yellow pea — Widespread; occasional or locally abundant in meadow, pasture, ditch bank, willow-streamside, and aspen communities; 5,300-8,400 ft; May-July.

Our plants belong to var. *montana*. *Thermopsis ovata* (Robinson) Rydberg [*T. montana* ssp. *ovata* Robinson ex Piper; *T. m.* var. *ovata* (Robinson ex Piper) St. John; *T. rhombifolia* var. *ovata* (Robins:-ex-Piper) Isely] has been listed for our area, but this is a plant from Washington to California and east to Montana and Nevada. *Thermopsis divaricarpa* A. Nelson [*T. rhombifolia* var. *divaricarpa* (A. Nelson) Isely] has been indicated for our area, but this is a plant from east of the Uinta Basin where it is distributionally and morphologically intermediate between *T. montana* and *T. rhombifolia*, and it is the basis for Isely's inclusion of the *T. montana* complex into *T. rhombifolia* (Isely 1981). Apparently the inter-mediates are not found in our area. *Thermopsis montana* is also included in *T. rhombifolia* in IMF (3B: 236).

Thermopsis rhombifolia Nuttall in Richards — circle-pod pea — Daggett Co. and Pine Ridge and Asphalt Ridge and east to Dinosaur National Monument and s. to Cliff Ridge and Hell's Hole Canyon in Uintah Co. and e. into Moffat and probably Rio Blanco Cos.; isolated populations in desert shrub, sagebrush, and pinyon-juniper communities, especially common on Mowry Shale, but also on sandstone and other materials; 5,000-6,600 ft; May-June. This is readily distinguished from *T. montana* when in fruit by the strongly curved pods.

Trifolium L. — Clover

Annual or perennial herbs from taproots, caudices or rhizomes; leaves 3-foliate or rarely palmate with 4-7 leaflets, commonly serrate, rarely entire; flowers borne in terminal or axillary pedunculate or sessile heads or subcapitate racemes; pods usually shorter than the calyx, indehiscent, 1-several seeded.

1   Leaves all basal; flowers born among the leaves or on scapes 2

1   Stems with leaves, and with 1 or more elongate internodes, introduced in all but *T. longipes* 6

2   Flowers solitary or 2-4 per head, 15-23 mm long; heads essentially sessile; calyx glabrous; plants above timberline on the Uinta Mtns. . . . . ***T. nanum***

2   Flowers 5-20 per head, either shorter than above or heads pedunculate; distribution various including above timberline on the Uinta Mtns. 3

3   Plants densely pulvinate-caespitose and mat forming, 0.3-5 cm tall; leaflets toothed only on the upper 1/3, or subentire . . . . ***T. andinum***

3   Plants not pulvinate-caespitose, not or hardly mat forming, 2-25 cm tall; leaflets various 4

4   Leaflets entire; calyx strigose . . . . ***T. dasyphyllum***

4   Leaflets toothed 5

5   Calyx pubescent with soft hairs; flowers 8.5-11 mm long; heads not subtended by involucral bracts . . . . ***T. gymnocarpon***

5   Calyx glabrous, flowers 12-17 mm long; heads subtended by involucral bracts . . . . ***T. parryi***

6   Heads immediately subtended by a trifoliate sessile leaf-like bract; flowers 13-20 mm long, reddish; plants 18-60 cm tall . . . . ***T. pratense***

6   Heads without a trifoliate leaf-like bract; flowers 4-13 mm long, white, pink (or if reddish only 5-9 mm long); plants 5-37 cm tall. 7

7   Leaflets of main leaves commonly over 3 times as long as wide; flowers 11-13 mm long; plants native . . . . ***T. longipes***

7   Leaflets mostly less than twice as long as wide; flowers 5-10 mm long; plants introduced 8

8   Heads subtended by an involucre; flowers 4-6 mm long, soon enclosed in the bladdery-inflated calyx . . . . ***T. fragiferum***

8   Heads without an involucre; flowers 5-13 mm long, not enclosed by the ·calyx; the calyx not inflated. 9

9   Plants creeping, rooting at the nodes; calyx tube glabrous or hairy only at the base, often with a purple spot below the sinus between the teeth, the teeth generally equal to or shorter than the tube; petals white . . ***T. repens***

9   Plants erect or ascending, rarely decumbent; calyx tube with a few hairs below the sinus between the teeth, but without a purple spot, the teeth generally longer than the tube; petals white or pink to red . . ***T. hybridum***

Trifolium andinum Nuttall in Torrey & A. Gray — Andean clover — Specimens seen are from the north flank of the Uinta Mountains and lower elevations of Daggett Co., Blue Mtn. in Uintah Co. and Peterson & Deardorff 83-91 (CS) reported for Vermillion Creek Gap, Moffat Co.; sagebrush, mt. brush, pinyon-juniper, and ponderosa pine communities, usually in rocky places and locally dominant on crests of hogbacks; 6100-8,850 ft; May-June. Our plants belong to var. *andinum.*

Trifolium dasyphyllum Torrey & A. Gray — whiproot clover, Uinta clover — Occasional, and locally abundant in the Uinta Mtns. where wind limits snow cover to less than 1.5 ft for most of the winter (6,850) 11,000-12,000 ft; July-Aug. Our plants belong to var. *uintense* (Rydberg) S. L. Welsh. Found on Bridger Formation with a veneer of pediment gravels at Mtn. View, Wyoming at 6,800 ft.

Trifolium fragiferum L. — strawberry clover — Introduced from Europe, probably widespread but seldom collected in our area; pastures, meadows, roadsides, ditch banks, seeps, and riparian communities; 5,100-5,600 ft and probably higher; June-Sept.

Trifolium gymnocarpon Nuttall in Torrey & A. Gray — hollyleaf clover, Nuttall clover — Widespread; occasional and locally common in desert shrub-juniper, sagebrush, pinyon-juniper, mt. brush, ponderosa pine, Douglas-fir, and lodgepole pine communities; 5,350-9,000 ft; May-June.

Trifolium hybridum L. — alsike clover — Introduced from Europe, cultivated and escaping and to be expected throughout our area; pastures, meadows, along roadside ditches, and riparian and other plants communities; up to about 9,000 ft; June-Sept.

Trifolium longipes Nuttall in Torrey & A. Gray (*T. rydbergii* Greene) — longstalk clover — Specimens seen are all from the Uinta Mtns. where occasional or locally common in meadow and aspen communities; 6,500-9,500 ft; June-Aug. Our plants belong to var. *reflexum* A. Nelson. Calyx pubescent or rarely glabrous. Rare specimens with glabrous calyx could be confused for *T. beckwithii* Brewer ex. S. Watson that is not known from our area.

Trifolium nanum Torrey — dwarf clover — Uinta Mtns.; infrequent or occasional in alpine curly sedge-cushion plant and curly sedge-alpine avens communities; July-Aug.

Trifolium parryi A. Gray — Parry clover — Across the Uinta Mtns; occasional or locally common in aspen, lodgepole pine, Engelmann spruce, meadow, and alpine communities including snowbeds; 9,000-12,000 ft; June-Aug. Our plants belong to var. *montanense* (Rydberg) S. L. Welsh.

Trifolium pratense L. — red clover — Introduced from Europe, widespread; pastures, meadows, roadsides, ditches, and riparian communities; up to about 8,000 ft; May-Sept.

Trifolium repens L. — white clover, Dutch clover — Introduced from Europe, widespread; lawns, pastures, roadside ditches, meadows, around seeps and springs, and in riparian communities, included in seed mixtures for lawns; up to about 9,200 ft; June-Aug.

Vicia L. — Vetch

Plants clambering, trailing, or climbing; leaves alternate, even-pinnate compound, the rachis ending as a tendril; wings adnate to the keel; style with a ring of hairs below the stigma; pods stipitate.

---

1   Flowers 15 or more in secund racemes, pink-purple or reddish violet; plants
    annual or biennial . . . . **V. villosa**
1   Flowers 3-10 per raceme; pink to pink-purple; plants perennial . . . .
    . . . . **V. americana**

Vicia americana Muhlenberg ex Willdenow — American vetch. With 2 vars:

1   Leaflets 2-5 mm wide, nearly linear or narrowly elliptic, the lateral veins
    prominent and diverging from the midrib at a narrow (about 10°-20°)
    angle; plants mostly less than 40 cm tall, known from 5,200-6,500 ft . . . . .
    . . . . **var. minor**
1   Leaflets 1-19 mm wide, linear to broadly elliptic, the lateral veins
    prominent or not, diverging from the midrib at a wide (about 30°-40°)
    angle; plants mostly over 40 cm tall, known from 7,000-9,800 ft . . . .
    . . . . **var. americana**

Var. americana (*V. oregana* Nuttall; *V. truncata* Nuttall in Torrey & A. Gray) — American vetch — Widespread; occasional in mt. brush, aspen, Douglas-fir, meadow, and Engelmann spruce-sub alpine fir communities; 7,000-9,800 ft; June-Aug.

Var. minor Hooker (*V. linearis* (Nuttall) Greene; *V. trifida* Dietrich) — plains vetch — Tridell to Island Park and Red Wash in Uintah Co., Daggett Co. and into Colorado; occasional in desert shrub and juniper communities, often on Duchesne River, Morrison, Mowery, and other formations that weather to badlands; 5,200-6,500 ft; May-June.

Vicia villosa Roth — hairy vetch — Specimens seen are from a roadcut along Highway 40 about 20 miles west of Maybell and a roadside at Farm Creek north of Neola. Introduced from Europe, cultivated as a forage crop and escaping along roadsides and fence rows.

FAGACEAE Beech Family

Quercus L. — Oak

Quercus gambelii Nuttall [*Q. gunnisonii* (Torrey) Rydberg; *Q. novomexicana* (A. de Candolle) Rydberg; *Q. utahensis* (A. de Candolle) Rydberg] — Gambel oak, scrub oak — Shrubs or small trees to about 5 m tall; leaves alternate, simple, (4) 6-10 cm long, pinnately lobed over 1/2 way to the midrib, the lobes rounded to acute, darker green above; flowers small, without petals, unisexual, the staminate in pendulous catkins and with 5-10 stamens and a perianth of 4-7 lobes, the pistillate solitary or clustered in an involucre of flat scales and with a 6-lobed perianth; ovary 1-celled with 3 styles, superior; fruit a nut (acorn) about 12-20 mm long, partly enveloped in a cup (involucre of scales) 10-15 mm long. Current Creek east to Tabiona and Blind Stream on the s. flank of the Uinta Mtns., e. to the Avintaquin drainage of the W. Tavaputs Plateau, and common on the E. Tavaputs Plateau; occasional to abundant and forming clones in the mt. brush belt; May-June.

    The absence of Gamble oak in the central and eastern Uinta Mtns. is puzzling. It is common to the east of these mtns. in Moffat County at equal latitude and in the Sierra Madre Mtns. of Wyoming of higher latitude. Other members of the genus are planted as shade trees or ornamentals but rarely in our area.

FUMARIACEAE Fumitory Family

Annual to perennial herbs; leaves compound, glabrous and glaucous; flowers bisexual, irregular; sepals 2; petals 4, in 2 series, the outer 2 erect to spreading and one or both pouched or spurred at the base the inner 2 smaller, usually fused and crested at the apex, not spurred; stamens 6, in 2 sets of 3 each; style 1, entire or 2-lobed; fruit a capsule with several seeds.

1   Leaves basal; flowers solitary and terminal; outer petals widely spreading to recurved at the apex, and pouched at the base . . . . *Dicentra*

1   Leaves alternate; flowers in racemes; outer petals erect or nearly so, one of them spurred at the base . . . . *Corydalis*

Corydalis Medicus— Scrambled Eggs

Corydalis aurea Willdenow [*C. montanum* (Engelmann) Britton; *Capnoides aureum* Willdenow] — golden smoke, golden corydalis — Plants annual or biennial from a taproot; stems prostrate to ascending, 10-50 cm long, generally branched; herbage glabrous, glaucous, succulent; leaves 2-several times pinnate with blades 2-8 cm long, the ultimate segments linear-oblong to elliptic, 3-8 mm long, 0.5-2 mm wide; flowers in axillary 1-6 cm long racemes; sepals 1-3 mm long, often deciduous by anthesis; corolla yellow, fading to white, 8-18 mm long, petals 4, the outer 2 more or less fused below with the upper petal, these broader than the lower and spurred, expanded to winged at the apex, the inner 2 petals erect, expanded and fused at the apex, clawed at the base; style elongate; fruit a capsule; seeds many, dark, shiny. Widespread; occasional to locally common in sagebrush, mt. brush, ponderosa pine, aspen, and fir communities, somewhat adventive with disturbance including fire and roadcuts; 7,000-9,500 ft; May-Aug. Apparently our plants belong to var. *occidentalis* Engelmann in A. Gray. This species is found in abundance 1-3 years after fire in some burned areas. This indicates the hard, black seeds persist as seed-banks in the soil that are activated by fire.

Dicentra Bernhardi — Bleeding heart

Dicentra uniflora Kellogg — steer's head — Scapose perennial from fascicled tuberous roots, 2-10 cm tall; base of petioles and scapes etiolated, subterranean, easily detached from the tuberous roots; leaves basal, the blades 1-5 cm long, divided into 3-5 leaflets, the leaflets with 3-9 deep lobes; scapes with 1 or 2 small bracts below the solitary terminal flower; sepals 2, green, deciduous in age; petals 4, white to pinkish, the outer 2 narrowly elongate with the upper 1/2 slender and widely spreading to recurved, the inner 2 erect, arrowhead-shaped and clawed, usually purple at the fused tips; capsules 1-1.3 cm long; seeds black and shiny. Known from Strawberry Ridge at the w. margin of our area; aspen, fir, and forb-grass communities; 8,000-9,000 ft; blooms with snowmelt and shortly after.

GENTIANACEAE Gentian Family

Annual or perennial, mostly glabrous herbs; leaves simple, entire, opposite or whorled, sessile and often fused at the base, stipules lacking; flowers bisexual, radially symmetrical, 4-5 merous; calyx more or less deeply lobed; corolla united or of nearly separate petals, the lobes often bearing conspicuous nectar glands or fringed appendages near the base or in the throat; stamens 4-5, arising from the corolla tube, alternate with the corolla lobes; pistal 1; ovary superior; style elongate or obsolete; stigmas 2; fruit a capsule, usually 2-valved; seeds numerous.

1   Corolla lobes free to near the base 2
1   Corolla united with the lobes shorter than the tube 4
2   Leaves 5-25 mm long .... *Lomatogonium*
2   At least some leaves over 25 mm long 3
3   Corolla with 5 lobes .... *Swertia*
3   Corolla with 4 lobes .... *Frasera*
4   Corolla with 4 lobes... 5
4   Corolla with 5 lobes... 6
5   Corolla 7-16 mm long, white or tinged with blue ...... *Gentianella tenella*
5   Corolla over 16 mm long, deep blue .... *Gentianopsis*
6   Corolla 23-48 mm long .... *Gentiana*
6   Corolla mostly less than 23 mm long 7
7   Corolla with numerous hair-like appendages in the throat .... *Gentianella*
7   Corolla without hair-like appendages in the throat 8
8   Corolla pink to rose; leaves 10-30 (40) mm long; plants annual, known from
    below 7,000 ft .... *Centaurium*
8   Corolla blue (rarely white); leaves 4-12 mm long; plants perennial, known
    from above 7,500 ft .... *Gentiana prostrata*

Centaurium Hill — Centaury

Centaurium exaltatum (Grisebach) Wight — exalted centaury, Great Basin centaury — Glabrous annual, stems 1-several, erect, simple or sparingly branched, 5-30 cm tall; basal leaves few or lacking, soon withering; stem leaves elliptic, lanceolate or oblanceolate, 5-30 mm long, entire; flowers solitary or in terminal cymes, the pedicels slender; calyx 6-10 mm long, cleft to near the base, the lobes narrow; corolla salverform, 7-20 mm long, the tube white to yellowish, about twice as long as the calyx when mature, the lobes rose-pink, rarely white, to 5 mm long; stamens 4 (5) arising at about mid length in the corolla. Scattered across the lower elevations of our area, usually in alkaline, moist areas, not expected much over 6,500 ft; July-Sept.

Frasera Walt. — Green Gentian

Frasera has been tossed back and forth in and out of the genus *Swertia*. The treatment in IMF (4:16-23) is followed here in separating *Frasera* from *Swertia*.

1   Plants 50-150 cm tall; leaves in whorls of 3-5 (6), over 7 cm wide ....
....*F. speciosa*
1   Plants 5-25 cm tall; leaves opposite, 2-7 mm wide ....*F. ackermanae*

Frasera ackermanae Newberry and Goodrich — Ackerman gentian — This narrow endemic is known from south of Brush Creek and west of Highway 191 on semi-barrens of Chinle Formation and less commonly on alluvium washed from the Chinle Formation at about 5800 ft.

Frasera speciosa Douglas [*Swertia radiata* (Kellogg) Kuntze] — elkweed — Robust perennial from a simple caudex and stout taproot; stem solitary, erect, 40-100 (200) cm tall, to 3 cm thick at the base; basal leaves mostly oblanceolate, forming a rosette, generally withered by anthesis, stem leaves in whorls, sometimes opposite, the lower ones 10-30 cm long; flowers few to several, in axillary pedunculate cymes; calyx lobes 1-2 cm long, lobed to near the base; corolla lobes about as long as the calyx lobes, greenish-white to yellow-green, usually spotted with purple, each with a fringed nectary gland; stamens 4; style 3-5 mm long. Widespread in sagebrush, aspen, riparian, and coniferous forest communities, also on open rocky slopes and on talus; 7,500-10,500 ft; June-Aug.

Gentiana L. — Gentian

Herbs from taproots or rhizomes; leaves entire; flowers solitary to several in cymose clusters; calyx and corolla each united, the lobes (4) 5, the corolla often with pleated folds that bear teeth at the sinuses; stamens inserted on the corolla-tube; style short or none; stigmas 2: capsules 1-celled.

1   Corolla 8-22 mm long; leaves 3-8 mm long ....*G. prostrata*
1   Corolla mostly over 25 mm long; leaves over 15 mm long 2
2   Basal leaves 4-9 (12) cm long, basal and stem leaves over 3 times as long as wide; corolla white or yellowish, spotted and striped with purple *G. algida*
2   Leaves all cauline, 1.2-3 (4) cm long and less than 3 times as long as wide; corolla blue or purplish, occasionally with yellowish margins or white in albino phases 3
3   Plants glabrous ....*G. calycosa*
3   Leaves and calyx lobes finely ciliolate; stems finely puberulent in lines below the leaves 4
4   Calyx tube 0.4-1 cm long; anthers 1.6-3 mm long ....*G. affinis*
4   Calyx tube 1-1.8 mm long; anthers 3.5-5 mm long ....*G. parryi*

Gentiana affinis Grisebach [*G. forwoodii* A. Gray; *Pneumonanthe affinis* (Grisebach) Greene] — Rocky Mt. pleated gentian — In and adjacent to the Uinta Mtns. including Blue Mt.; occasional in moist or wet places often at the ecotone of dry and wet meadows; from about 6,000 ft to above timberline; July-Sept.

Gentiana algida Pallas [*G. romanzovii* Ledebour; *Gentianodes algida* (Pallas) Love & Love] — whitish gentian, arctic gentian — Locally common across the alpine of the Uinta Mtns. in tufted hairgrass, timber oatgrass, alpine avens, and kobresia communities, and less common in subalpine meadows (10,000) 11,000-13,000 ft; July-Sept.

Gentiana calycosa Grisebach [*G. parryi* Engelmann misapplied (Graham 1937)] — explorer's gentian, mt. bog gentian, Rainier pleated gentian — Uinta Mtns.; common in dry and wet meadows and alpine tundra; 10,000-11,500 ft; July-Sept.

Gentiana parryi Engelmann [*Pneumonanthe parryi* (Engelmann) Greene] — Parry gentian — The distribution map in Albee et al. (1988) shows a widely disjunct dot in Wasatch Co. that appears to be in the Strawberry drainage. The main body of the species is in southern Utah with the nearest location to the Wasatch Co. specimen about 150 miles to the sw. in the Tushar Mtns. in Beaver Co. Utah.

Gentiana prostrata Haenke in Jacquin [*G. fremontii* Torrey; *Ciminalis fremontii* (Torrey) W. A. Weber; *C. prostrata* (Haenke) A. & D. Love; *Chondrophylla prostrata* (Haenke) in J. P. Anderson] — moss gentian — Uinta Mtns.; occasional in wet meadows and moist places in coniferous woods; 9,000-11,600 ft; July-Sept.

Gentianella Moench — Dwarf Gentian

Similar to *Gentiana* except the corollas not pleated, mostly smaller, and the lobes without appendages or teeth in the sinuses, but with a row of hair-like fimbriae across the face of the lobes, and plants annual.

1 Flower solitary on a naked peduncle 2-10 cm long . . . . ***G. tenella***

1 Flowers 3 or more and borne on short peduncles or pedicels from axils of leaves. 2

2 Corolla 5.5-8.5 mm long; plants 2-5 (10) cm tall . . . . ***G. tortuosa***

2 Corolla 9-24 mm long; plants 5-60 cm tall or taller 3

3 Calyx lobes about equal; appendages inside the corolla tube at the base of the lobes with wholly separate fimbriae . . . . ***G. amarella***

3 Calyx lobes unequal, some leaf-like and more or less enveloping the smaller ones; appendages inside the corolla tube at the base of the lobes with fimbriae united at the base . . . . ***G. heterosepala***

Gentianella amarella (L.) Borner [*Gentiana amarella* L.; *G. plebeia* Chamisso ex Bunge; *G. strictiflora* (Rydberg) A. Nelson; *Gentianella amarella* ssp. *acuta* (Michaux) Gillett] — northern gentian, felwort — Widespread; occasional or locally common along streams in aspen and coniferous woods and alpine tundra; 7,320-11,500 ft; July-Sept.

Gentianella heterosepala (Engelmann) Holub (*Gentiana amarella* ssp. *heterosepala* J. S. Gillett; *G. heterosepala* Engelmann) — unequal sepal gentian — Widespread; occasional; sagebrush, aspen, and coniferous forest communities; 7,400-10,650 ft; July-Sept.

Gentianella tenella (Rottboell) Borner [*Gentiana monantha* A. Nelson; *G. tenella* Rottboell; *Comastoma tenellum* (Rottboell) Toyokuni] — Lapland gentian — Apparently uncommon; the few specimens seen are from subalpine dry and wet meadows, and turf of alpine tundra on the Uinta Mtns.; July-Aug. Our plants belong to ssp. *tenella*.

Gentianella tortuosa M. E. Jones — limestone gentian — Known in our area on sparsely vegetated hills and slopes of Green River shale, Willow Creek and Argyle drainages of W. Tavaputs Plateau and E. Tavaputs Plateau in Garfield Co. July-Aug.

Gentianopsis Ma — Fringed Gentian

Similar to *Gentiana* but the corollas not pleated.

---

1   Plants perennial; flowers closely subtended by a pair or bract-like leaves; taproot yellowish . . . . *G. barbellata*

1   Plants annual; flowers on naked peduncles 2-16 cm long; taproot not particularly yellowish . . . . *G. detonsa*

Gentianopsis barbellata (Engelmann) Gillett (*Gentiana barbellata* Engelmann) — barbellate gentian — Uinta Mtns.; uncommon, mostly in small isolated populations, Rock Creek and Lake Fork drainages of s. slope and Blacks Fork and Mill Creek drainages on the n. slope; often associated with limestone;10,500-11,300 ft.

Gentianopsis detonsa (Rottboell) G. Don (*Gentiana detonsa* Rottboell; *G. thermalis* Kuntze) — meadow gentian — Apparently uncommon and restricted along the south slope of the Uinta Mtns. and around Strawberry Valley, and locally common in drainages of the north slope of the Uinta Mountains; wet places at 6,300-10,300 ft. Our plants belong to var. *elegans* (A. Nelson) N. Holmgren.

Lomatogonium A. Br.

Lomatogonium rotatum (L.) Fries ex Nyman — marsh felwort — Known from Hickerson Park, Sheep Creek Park, and Beaver Creek near Brownie Lake, n. slope, Uinta Mtns. at 8,300-9,000 ft; wet meadows including areas under water for a time in spring and early summer.

Swertia L. — Swertia

Swertia perennis L. (*S. palustris* A. Nelson) — star swertia, alpine bog swertia — Perennial from short rhizomes; stem usually solitary, erect, 10-50 cm tall; usually unbranched; basal leaves prominent with distinct petioles, (2) 4-20 cm long, obovate or spatulate; stem leaves few, opposite or a few alternate, sessile, much reduced upward; flowers 1-3; calyx 3-10 mm long, the lobes lanceolate; corolla dull blue-purple, occasionally nearly white, generally streaked with green or purple, the lobes 6-14 mm long, widely spreading, each bearing small, fringed nectary-glands about 1 mm long; capsule about 1 cm long. Uinta Mtns. and extreme w. of W. Tavaputs Plateau; occasional in wet meadows, seeps and springs, and along streams; 8,500-11,500 ft; July-Aug.

GERANIACEAE Geranium Family

Annual or perennial herbs; leaves alternate or opposite; flowers bisexual, radially symmetrical, in cymes or umbels; sepals 5, separate, often awn-tipped; petals 5, separate; stamens usually 10, the filaments often dilated and sometimes fused at the base; pistil 1; ovary superior, with 3-5 (8) weakly fused carpels, the styles 5 and fused around an extension of the receptacle and forming a much elongate stylar column with 5 free stigmas at the apex; fruit a schizocarp, the 5 mericarps separating at maturity, each splitting upward from the base and with an elongate, coiled or twisted persistent style.

1   Leaves pinnately dissected .... *Erodium*
1   Leaves palmately lobed to divided .... *Geranium*

Erodium L'Heritier — Storksbill; Filaree

Erodium cicutarium (L.) L'Heritier — redstem storksbill, heronsbill, alfileria, filaree — Winter annual forming a prominent rosette; stems erect to prostrate 3-60 cm long; herbage often turning red in age, often with gland-tipped hairs; stem leaves opposite, few, much reduced, pinnately divided or compound, but somewhat less dissected than the basal ones; inflorescence an umbel with (1) 2-12 flowers; sepals 3-7 mm long, persistent; petals 4-11 mm long, pink to rose-violet, readily deciduous; stamens 5, alternating with 5 scale-like staminodes; stylar column 1-5 (7) cm long; mericarps 4-7 mm long. Widespread in many plant communities, often rather weedy, usually on disturbed ground, lower elevations and up to about 8,000 ft; native of the Mediterranean region and appearing to be less adapted to our area with cold winters as in warmer areas of the Intermountain Region. However, this annual appears to have increased considerably in distribution and abundance since 1986 when the first edition of this flora was published, and in places it is displacing native plants; April-Oct.

Geranium L. — Wild Geranium

Plants perennial from a caudex; stems frequently dichotomously branched; leaves long-petiolate and usually opposite near the base of the stem, sessile and alternate in the inflorescence, mostly palmately lobed or divided, the blades more or less orbicular in outline; flowers 1 or 2 or more in cymes, borne on axillary peduncles; stamens 10, all fertile or rarely 3-5 reduced to staminodes; flowers and fruit as described in the family description.

1   Plants perennial; petals 12-22 mm long; stylar column 2-3.5 cm long 2
1   Plants annual; petals 6-8 mm long; stylar column about 1-1.8 mm long ...
    .... **G. bicknellii**
2   Petals pink to purple, rarely white; stigmas (4) 5-7 mm long in fruit; mericarps 5-6 mm long; taproot rope-like with loose fibers, the pithy center brownish .... **G. viscosissimum**
2   Petals white with pinkish or purplish veins; stigmas 4-5 mm long in fruit; mericarps 3-5 mm long; taproot without loose fibers, the pithy center pinkish... **G. richardsonii**

Geranium bicknellii Britton (*G. carolinianum* L. var. *longipes* S. Watson) — annual geranium — This plant is locally common for 1 or 2 years following fire in lodgepole pine and other coniferous forest communities in the eastern Uinta Mountains. After 3 or 4 years following fire, it is essentially not seen. This appears to be a seedbank species with long-lived seeds that are activated by fire.

Geranium richardsonii Fischer & Trautvetter — white geranium, Richardson geranium — Widespread; occasional or common in meadows, aspen and coniferous woods, and rarely talus, generally of more moist or shaded areas than *G. viscosissimum;* (6,000) 7,200-10,600 ft; June-Sept.

Geranium viscosissimum F. & M. (*G. caespitosum* James misapplied; *G. fremontii* Torrey misapplied) — sticky geranium — Widespread; most common in open grass-forb parklands of the aspen and coniferous forest belts, also in sagebrush, and mt. brush communities, occasionally in aspen and coniferous woods, but not nearly as common in shade of trees as is *G. richardsonii;* 7,200-10,400 ft; June-Sept. Our plants belong to var. *incisum* (Torrey & Gray) N. Holmgren [*G. nervosum* Rydberg; *G. v.* var. *nervosum* (Rydberg) C. L. Hitchcock].

GROSSULARIACEAE Currant or Gooseberry Family

Ribes L. — Currant; Gooseberry

Shrubs with or without bristles and spines; leaves alternate, simple, palmately lobed, toothed; flowers bisexual (1) 2-many in axillary clusters or short to elongate racemes; sepals (4) 5 separate, arising from the apex of a saucer-shaped or cylindrical floral tube, the conspicuous, showy part of the flower; petals usually shorter than the sepals; stamens arising alternately with and at or just below the base of the petals; ovary inferior or partly so; fruit a berry, often crowned with the persistent floral parts.

Short Key

1 Plants armed with spines and/or prickles...........................2
1 Plants without spines or prickles 5
2 Berries and base of flowers with bristle-like, gland-tipped hairs 3
2 Berries and base of flowers without gland-tipped hairs...............4
3 Racemes with 3-8 flowers; berries red ....**R. montigenum**
3 Racemes (5) 7-15 flowered; berries black or purple-black ......**R. lacustre**
4 Superior part of the floral tube 4-5 mm long; nodal spines usually 3 .....
   ....**R. oxyacanthoides**
4 Superior part of the floral tube 2-3.5 mm long; nodal spines 1 (rarely 3) ..
   ....**R. inerme**
5 Flowers bright yellow, often reddish in part in age, glabrous ...**R. aureum**
5 Flowers not yellow, often pubescent or glandular.....................6
6 Leaf blades, ovaries, and fruits stipitate-glandular 7
6 Leaf blades not stipitate-glandular; ovaries and fruits stipitate glandular in
   R. wolfii only 8
7 Flowers pinkish; berries red ....**R. cereum**
7 Flowers greenish white to cream; berries blackish ......**R. viscosissimum**
8 Flowers 1-3 per raceme; plants common and widespread, often with spines
   and also keyed above in lead 4 ....**R. inerme**
8 Flowers 8-30 per raceme; plants uncommon in our area 9
9 Racemes 4-10 (17) cm long with about 20-30 flowers .....**R. hudsonianum**

9   Racemes about 1-4 cm long, with about 8-16 flowers . . . . **R. wolfii**

Expanded Key

1   Plants armed with spines and/or prickles; styles divided to near the base; flowers various; pedicels various 2

1   Plants without spines or prickles; styles mostly united to above the middle and often to near the apex; flowers (1) 3-many per raceme; pedicels jointed just below the ovary, often disarticulating in fruit (currants). 5

2   Berries and base of flowers with bristle-like, gland-tipped hairs; floral cup saucer shaped; racemes with 3-15 flowers; styles not pilose; pedicels jointed; disarticulating in fruit (gooseberry currants). 3

2   Berries and base of flowers without gland-tipped hairs; floral tube narrow; racemes with 1-3 flowers; styles pilose; pedicels not jointed, not disarticulating (gooseberries) 4

3   Racemes with 3-8 flowers; berries red; leaves with glandular and non glandular hairs, the blades cleft 3/4 the way to nearly all the way to the base; plants common, widespread . . . . **R. montigenum**

3   Racemes (5) 7-15 flowered; berries black or purple-black; leaves glandular only or with a few hairs along the veins and margin, the blades lobed 2/3 the way to the base or less; plants uncommon, known from Duchesne Co.
. . . . **R. lacustre**

4   Superior part of the floral tube 4-5 mm long; nodal spines usually 3; branchlets finely puberulent and sometimes with internodal bristles; berries purple-black, plants from the s. slope of the Uinta Mtns. . . . .
. . . . **R. oxyacanthoides**

4   Superior part of the floral tube 2-3.5 mm long; nodal spines 1 (rarely 3) or lacking; branchlets usually glabrous; rarely with internodal bristles; berries reddish-purple; plants common, widespread . . . . **R. inerme**

5   Flowers bright yellow, often reddish in part in age, glabrous; ovaries and berries glabrous; leaves not cordate, with 3 (rarely 5) primary lobes, the lobes seldom more than 3-lobed or 3-toothed; berries black or golden . . .
. . . . **R. aureum**

5   Flowers not yellow, often pubescent or glandular; ovaries and berries with sessile or stipitate glands (except R. inerme); leaves often pubescent, mostly cordate basally, with (3) 5-7 primary lobes, these with usually more than 3 teeth or lobes; berries various. 6

6   Leaf blades, flowers, and fruit stipitate-glandular, if sparingly so then flowers pinkish and berries red; floral tube 4-11 mm long, campanulate to cylindrical; anthers glandular apically 7

6   Leaf blades not stipitate-glandular, glabrous or with hairs on the veins, or with sessile glands; flowers not pinkish; berries red only in R. inerme; floral tube less than 4 mm long, variously shaped; anthers not glandular. 8

7   Flowers (1) 2 or 3 per raceme, pinkish, the floral tube less than 3 mm wide; berries red; leaf blades 7-30 (44) mm wide, not pilose-hirsute . . **R. cereum**

7   Flowers 4-12 per raceme, greenish white to cream, the floral tube 3-6 mm wide; berries blackish; larger leaf blades 30-100 mm wide, pilose-hirsute and stipitate-glandular . . . . **R. viscosissimum**

8   Flowers 1-3 per raceme; styles pilose; divided to below the middle; floral
     tube pilose within; berries reddish or purplish; leaves without sessile,
     crystalline glands; ovaries and berries glabrous or at least not glandular as
     in the following; plants common and widespread, often with spines and also
     keyed above in lead 4 .... *R. inerme*

8   Flowers 8-30 per raceme; styles not pilose, mostly united to above the
     middle; floral tube not pilose within; berries blackish; leaves often with
     sessile crystalline glandular dots; ovaries and berries either with stipitate
     or with sessile glands; plants uncommon 9

9   Ovaries, berries, floral tube, petioles, lower surface of leaves, and young
     twigs with scattered, crystalline, yellowish, sessile, glandular dots; racemes
     4-10 (17) cm long with about 20-30 flowers; plants strongly aromatic; larger
     leaf blades 5-12 cm wide .... *R. hudsonianum*

9   Ovaries and berries with stipitate glands; floral tube not with glandular
     dots; petioles and young twigs puberulent, without or with inconspicuous
     glandular dots; lower surface of leaves with inconspicuous clear, crystalline,
     sessile, glandular dots; racemes about 1-4 cm long, with about 8-16 flowers;
     plants not strongly aromatic; larger leaves 4-8 cm wide .... *R. wolfii*

Ribes aureum Pursh — golden currant — Widespread and locally common
along washes and streams and other moist or wet places, sometimes cultivated
for the fruit from which jelly and syrup are made; 4,800-8,500 ft; May-June.

Ribes cereum Douglas (*R. inebrians* Lindley) — wax currant — Widespread;
locally common in riparian, mtn. big sagebrush, mtn. brush, Rocky Mtn.
juniper, ponderosa pine, Douglas-fir, limber pine, and uncommonly in alpine
communities; 6,500-11,000 ft;-May-July.

Ribes hudsonianum Richardson (*R. oxyacanthoides* L.; *R. petiolare* Douglas) —
wild black currant — Specimens seen are from Rock Creek, Lake Fork (Moon
Lake), and Mill Hollow of the Provo River drainage, Uinta Mtns.; riparian communities; 7,500-8,620 ft; May-July. No specimen was found at UT in 1984 to
support the listing by Flowers and others (1960) for this plant at Carter Creek,
Daggett Co. Plotted in Strawberry drainage by Albee et al. (1988).

Ribes inerme Rydberg — whitestem gooseberry — Common in mountains and
canyons across our area; often along water courses and at seeps and springs;
7,000-11,000 ft; May-July.

Ribes lacustre (Persoon) Poiret — prickly swamp currant — The few specimens
seen are from N. Fork Duchesne (Hades Creek), Rock Creek, Lake Fork, and
Yellowstone Canyons of the s. slope of the Uinta Mtns., Mill Hollow of the
western Uinta Mtns., and Smith and Morehouse drainage of the n. slope of the
Uinta Mtns.; moist coniferous and aspen woods and along streams; 7,680-9,275
ft; June-July.

Ribes montigenum McClatchie — gooseberry currant — Occasional to abundant
across much of the mountainous part of our area, more common in the
western half of the Uinta Mtns. than toward the east, no specimens seen from
the E. Tavaputs Plateau; coniferous woods, and on rocky slopes; 8,500 ft to
above timberline; June-Aug.

Ribes oxyacanthoides L. (*R. setosum* Lindley) — Missouri gooseberry — The few
specimens seen are from Lower Stillwater of Rock Creek, Lake Fork, Uinta

Canyon, and Brush Creek; aspen and riparian communities; 5,600-7,550 ft; May-July. Our plants belong to var. *setosum* (Lindley) Dorn

Ribes viscosissimum Pursh — sticky currant — Occasional to common across much of our area, no specimens seen from the E. Tavaputs Plateau; aspen and coniferous woods, often increasing on burned areas; 7,500-10,000 ft; June-July.

Ribes wolfii Rothrock (*R. mogollonicum* Greene) — Rothrock currant — Reported for Baxter Pass, E. Tavaputs Plateau at the se. edge of our area (Graham 1934), also in the Wasatch Mtns. and Wasatch Plateau to the s. and w. of our area.

HALORAGACEAE Water Milfoil Family
Myriophyllum L. — Water Milfoil

Aquatic perennial herbs; leaves 3 or 4 per node of the slender stems, 1-3 cm long, pinnately dissected into 10-34 filiform segments; flowers verticillate, axillary or in an interrupted, usually emergent spike, unisexual or unisexual and bisexual ones mixed in the same spike, the upper ones usually staminate; sepals 2-4; petals 2-4, quickly deciduous, about 2.5 mm long; stamens 8; fruit an achene, about 2-3 mm long. Occasional; seldom collected; to be expected throughout our area in quiet streams and still waters of ponds and lakes, often brackish water, mostly at lower or mid-elevations; July-Sept. *Myriophyllum spicatum* L. has been reported. This European species is not known from our area. However, it has been collected from Utah and might be expected to be introduced to the our area.

---

1   Floral bracts strongly dissected, even those subtending the staminate flowers evidently cleft or laciniate-toothed, mostly longer than the flowers and fruit . . . . *M. verticillatum*
1   Pistillate floral bracts entire or toothed, equal to or shorter than the fruit, staminate bracts entire, mostly shorter than the anthers . . . . *M. sibiricum*

Myriophyllum sibiricum Komarov [*M. spicatum* var. *exalbescens* (Fernald) Jepson] — common water-milfoil — Specimens seen are from Lake Fork Canyon and Dahlgren Cr. drainage; 7,550-9,150 ft. To be expected to be widespread. Aquatic in still and slow-moving (often alkaline) water.

Myriophyllum verticillatum L. — whorl-leaf water-milfoil — Specimens seen are from the Uinta Mtns. in Daggett and Uintah Cos. at 7,200-9.680 ft. To be expected to be widespread. Aquatic in slow moving streams and ponds.

HIPPURIDACEAE Marestail Family
Hippuris L. — Marestail

Hippuris vulgaris L. — common marestail — Perennial, submerged or emersed, usually aquatic herbs from creeping rootstocks, 10-49 (60) cm tall; leaves whorled, (4) 6-12 per node, 8-35 (50) mm long, 1-2 mm wide, linear, entire, glabrous; flowers mostly bisexual, without a conspicuous perianth, axillary in the whorls of leaves; fruit nut-like, 1-seeded, about 2 mm long. Widespread; locally common in the Uinta Mtns., rarely collected from the Tavaputs Plateau; ponds and slow-moving streams, sometimes in mud; 5,000-10,000 ft; June-Aug.

HYDROCHARITACEAE Tape-grass Family

Elodea Michaux — Waterweed

Submersed aquatic perennial herbs with fibrous roots; stems slender, branching, often rooting at the nodes; leaves sessile, entire, or finely serrulate; flowers small, borne on thread-like perianth tubes (these appearing as pedicels), the tubes often elongate and usually bringing the flower to the water surface, the base of the tube enclosed in a tubular bilobed spathe, regular, mostly unisexual, axillary, solitary or the staminate ones rarely 2-3; sepals 3, greenish; petals 3, whitish; stamens 9; stigmas 3, usually bilobed; fruit dry, indehiscent, ripening under water.

---

1   Upper and middle leaves opposite, the larger ones mostly (17) 20-26 mm long; spathes of pistillate flowers 3-7 cm long; petals of pistillate flowers 3.5-4 mm long; spathes of staminate flowers 2-5 cm long ..... *E. bifoliata*

1   Upper and middle leaves in whorls of 3, 5-17 mm long; spathes of pistillate flowers 1-2 cm long; petals of pistillate flowers 2.3-2.6 mm long; spathes of staminate flowers to 1.5 cm long .... *E. canadensis*

Elodea bifoliata S. John (*E. longivaginata* St. John) — long-sheath waterweed — The one specimen seen (B. F. Harrison 7980) is from Strawberry Reservoir, to be expected across our area.

Elodea canadensis Michaux — Canada waterweed — Specimens seen are from Stuntz Reservoir on Blue Mtn., Sheep Creek Bay of Flaming Gorge Reservoir, Brownie Lake, Red Fleet Reservoir, and the Green River near Jensen at 4,735-8,500 ft; to be expected across our area.

HYDRANGEACEAE Hydranga Family

---

1   Petals 4, 9-17 mm long; sepals 4, rarely 5; stamens 20 or more; petioles 1-3 mm long .... *Philadelphus*

1   Petals 5, about 3-4 mm long; sepals 5; stamens 10; leaves sessile *Fendlerella*

Fendlerella A. Heller

Fendlerella utahensis (S. Watson) A. Heller — Utah fendlerella, yerba desierto — Diffusely branched shrubs to 1 m tall; twigs strigose; leaves numerous, opposite, fascicled on older twigs, oblong to spatulate, strigose, 5-15 mm long, with 3 faint nerves; sepals 1-1.5 mm long; petals 5, 2-4 mm long, whitish; stamens 10; styles 3; fruit a capsule, about 3-4 mm long. Four specimens seen are from Weber Sandstone in or near Dinosaur National Monument, an additional specimen is from limestone in Irish Canyon, Moffat Co. 4,800-6,430 ft; May-June.

Philadelphus L. — Mockorange

Philadelphus microphyllus A. Gray [*P. occidentalis* A. Nelson; *P. microphyllus* var. *occidentalis* (A. Nelson) C. L. Hitchcock] — littleleaf mockorange — Shrubs 80-110 cm tall; bark reddish brown to light brown, shredding from older stems; leaves 8-35 mm long, mostly entire; flowers borne singly or few together at the ends of branches, rather showy, pleasantly scented; petals 4, rarely 5, 9-17 mm long, white; stamens 20-60; styles 3-5, fruit a leathery or woody capsule. Whiterocks Canyon and Sheep Creek and eastward in the Uinta Mtns., probably to Blue Mtn.-Douglass Mt., also on Tavaputs Plateau as far w. as Mt. Bartles; mostly in rocky canyons; 6,000-8,500 ft; June-July. The report of *Fendlera rupicola* A. Gray for along the Yampa River (Potter and others 1983) is based on a specimen belonging here.

HYDROPHYLLACEAE Waterleaf Family

Annual or perennial herbs, often rough-hairy or stipitate-glandular; leaves alternate or lowermost opposite, simple, entire, toothed, or pinnately compound; stipules lacking; flowers bisexual, solitary or in modified cymes, these often coiled at first and elongating with age; calyx cleft to the middle or more; corolla united, 5-lobed, regular or nearly so; stamens 5, alternating with the corolla-lobes, attached to the tube; ovary superior, the style usually cleft to divided, with 2 branches; fruit a capsule with 1-many seeds.

1   Leaves all basal; flowers solitary at the end of naked peduncles ....
    .... *Hesperochiron*
1   Leaves not all basal; flowers not solitary, or if so then in the axils of leaves
   2
2   Stamens exserted from the corolla....................................3
2   Stamens included in the corolla 4
3   Flowers in capitate clusters, these well exceeded by the leaves; plants perennial from fleshy fibrous roots, without a taproot or caudex, known from w. of Duchesne .... *Hydrophyllum*
3   Plants not as above in all features .... *Phacelia*
4   Leaves toothed to pinnately divided 5
4   Leaves entire 7
5   Some or all of the flowers solitary or in pairs in the axils of the leaves, born on pedicles mostly longer than 2 mm especially in fruit 6
5   Flowers in terminal helicoid cymes, the pedicles less than 2 mm long ....
    .... *Phacelia ivesiana*
6   Calyx with reflexed appendages between the lobes .... *Nemophila*
6   Calyx without reflexed appendages .... *Ellisia*
7   Leaves linear or nearly so, sessile or at least without well marked petioles
    ....*Nama*
7   Leaves not linear, with blades well set apart from petioles ...... *Phacelia*

Ellisia L. — Aunt Lucy

Ellisia nyctelea (L.) L. — Aunt Lucy — Annual herb 5-40 cm tall; leaves pinnatifid with mostly 3-6 pairs of entire or few-toothed lateral segments; flowers solitary or in pairs in axils of leaves or less commonly the stem ending in a few-flowered cyme; calyx cleft to below the middle, the lobes ciliate; corolla white or lavender, equal to or slightly longer than the calyx; capsule with 1 chamber; seeds mostly 4. Included on a list of plants of the Ouray National Wildlife Refuge and plotted for Moffat Co. in TPD.

Hesperochiron S. Watson

Hesperochiron pumilus (Grisebach) T. C. Porter [*Capnorea pumila* (Douglas) Greene] — low hesperochiron — Acaulescent, perennial herb from a taproot; leaves all basal, the blades oblanceolate to elliptic or ovate, 2.5-5 cm long; peduncles few, 2-5 (8) cm long; calyx lobes 4-9 mm long, glabrous or nearly so except for the ciliate margin; corolla 6-15 mm long, rotate, white or tinged with lavender; the limb 1.5-3 cm wide, the tube densely hairy inside; style shortly 2-cleft; capsule, unilocular, many-seeded. The single specimen seen (C. DeMoisy Jr. 12 RM) is from Whiterocks; 6,000 ft, dry level ground. Low hesperochiron is well distributed along the Wasatch Front with isolated populations in w. and S. Utah and West to California and North to British Columbia. It appears the DeMoisy collection is from the margin of the range where it might not have persisted.

Hydrophyllum L. — Waterleaf

Hydrophyllum capitatum Douglas ex Bentham — ballhead waterleaf — Perennial herbs, 10-40 cm tall; stems delicately attached to a short, rather deep rhizome to which a cluster of fleshy-fibrous roots is attached; leaves few, the blades 5-15 cm long, to 10 cm wide, pinnately divided into 5-11 segments, the upper segment confluent; flowers in 1-several globose clusters, in the axils of leaves; corolla 5-9 mm long, lavender-blue to white, bell-shaped; stamens well exserted. Strawberry Valley to Currant Creek, and in the Uinta Mtns. w. of Rock Creek on the south slope and w. of Blacks Fork on the north slope; occasional in aspen and aspen-conifer communities, but also in sagebrush, oak, and snowbed communities; 7,200-10,730 ft; May-June. Our plants belong to var. *capitatum*.

Nama L.

Annual plants with branched stems; leaves alternate, entire; inflorescence terminal nonscorpioid cymes; sepals subequal; corolla purple or lavender, deciduous; stamens borne unequally on the corolla tube; styles included in the corolla, 2-lobed or divided.

---

1   Styles solitary, shallowly lobed at apex; stems prostrate and commonly dichotomously branched . . . . *N. densum*
1   Styles 2, distinct to the base; stems erect or ascending . . . . . . *N. retrorsum*

Nama densum Lemmon — compact nama, leafy nama — Duchesne, Moffat and Uintah Cos.; occasional in desert shrub, pinyon-juniper, and sagebrush communities; 4,800-5,700 ft; May-Aug. Our plants belong to var. *parviflorum* (Greenman) C. L. Hitchcock. Corolla white to lavender, mostly 2.5-7 mm long, the limb 1-4 mm wide.

Nama retrorsum J. T. Howell. — retrorse nama — Known from a single collection (S. Goodrich & T. Nauman 27251 BRY!) from aeolian sand north of Orchard Draw, Uintah Co. at 5,400 ft. Corolla purple, funnelform, 4-7 mm long.

Nemophila Nuttall

Nemophila breviflora A. Gray — woodlove — Weak-stemmed annual, 3-30 cm tall, decumbent to erect; stems with minute, reflexed prickles; herbage otherwise glabrate; leaves alternate or the lower most opposite, 1.5-6 cm long including the petiole, 1.5-4 cm wide, pinnately divided into 3-7 segments, these usually entire; calyx lobed nearly to the base, the lobes 3-5 mm long, ciliate; flowers solitary in the axils of leaves, on ascending curved or declined pedicels 5-15 mm long in fruit; corolla white or lavender, about 2 mm long, usually surpassed by the calyx; stamens included; style shallowly bifid; capsule 3-5 mm wide, 1-seeded, the seed 2-4 mm in diameter, reddish-brown, smooth to faintly pitted. The few specimens seen are from Strawberry Valley to Currant Creek, and 1 from Goslin Mt., Daggett Co., and 1 from E. McKee Draw of the Uinta Mtns., and reported (Bradley 1950) for Cold Springs Mt.; mostly with aspen; April-June.

Phacelia Jussieu — Phacelia

Annual or perennial herbs, usually pubescent and sometimes glandular; flowers 5-merous, often borne in coiled, secund cymes that unfold and straighten as the flowers mature, the inflorescence then appearing racemose or paniculate; calyx lobed to near the base; corolla tubular to rotate, sometimes quickly deciduous; style shortly to deeply bifid; capsules with 1-many seeds.

1   Stamens and styles included in the corolla; plants annual, 2-25 cm tall . . 2

1   Stamens and styles exserted from the corolla; plants biennial or perennial, sometimes over 25 cm tall 6

2   Corollas yellow at least when fresh, persistent, 2.5-4 (5) mm long, about equaling or shorter than the calyx; plants approaching our area in Sweetwater Co., Wyoming . . . .
      P. lutea (Hooker & Arnot) J.T. Howell **var.** *scopulina* (A. Nelson) Cronquist

2   Corolla not yellow, except sometimes for the tube, the limb white, blue, or lavender, deciduous 3

3   Leaves toothed to pinnately divided 4

3   Leaves entire or undulate. 5

4   Corolla 6-10 mm long; plants known from Wasatch Co. . . . . . . *P. franklinii*

4   Corolla 2-4 mm long; plants not known from Wasatch Co. . . . . . *P. ivesiana*

5   Limb of corolla 2-3 mm wide, white or bluish, the tube white or yellowish; leaves not clustered toward the ends of the branches . . . . *P. incana*

5   Limb of corolla 5-11mm wide, lavender-violet to purple, the tube often vivid yellow; leaves clustered toward the ends of the branches . . . . . *P. demissa*

6   Leaves entire, or with 1-2 (4) pairs of lobes or leaflets near the base 7

6   Leaves toothed or pinnately lobed to pinnatifid the entire length 8

7   Plants biennial or short-lived perennial from a taproot, usually with a single, erect stem over 50 cm tall or this surrounded by several ascending shorter stems; some of the middle and lower leaves usually with 1-2 (4) pairs of lateral lobes or leaflets at the base; herbage bristly hirsute or hispid, hardly densely silvery-pubescent; flowers white to yellow-white . . . .
      . . . . *P. heterophylla*

7   Plants perennial from a taproot, surmounted by a branching caudex, usually with more or less equal, ascending to prostrate stems seldom over 50 cm tall; leaves all entire, rarely some of them with a pair of small lateral lobes

near the base; herbage often densely silvery-pubescent with short hairs; flowers white to lavender .... *P. hastata*

8   Plants with stipitate-glandular hairs, often ill-scented 9

8   Plants not stipitate glandular 11

9   Plants to 16 cm tall; known from Argyle Canyon .... *P. argylensis*

9   Plants often over 16 cm tall; distribution various 10

10 Leaves simple, pinnately lobed but not cut to the midrib; plants widespread .... *P. crenulata*

10 At least some of the leaves cut to the midrib or compound with 3 or more leaflets; plants of e. Uintah, Rio Blanco, and perhaps Moffat Cos. .... *P. glandulosa*

11 Filaments (1.5) 2-3 times longer than the corolla; corolla pubescent within; plants widespread .... *P. sericea*

11 Filaments less than 1.5 times longer than the corolla; corolla glabrous within; plants known from Wasatch Co. .... *P. franklinii*

Phacelia argylensis N. D. Atwood & S. L. Welsh — Argyle Canyon phacelia — Known only from Argyle Canyon in wash bottoms in Green River Shale in pinyon-juniper-serviceberry-Douglas-fir communities at 7595 ft elevation.

Phacelia crenulata Torrey in S. Watson (*P. corrugata* A. Nelson) — crenulate phacelia — Widespread south of the Uinta Mtns.; occasional to locally common in some years in desert shrub, sagebrush, and pinyon-juniper communities; 4,700-6,500 (7,200) ft; May-June (July). Our plants with stipitate-glandular hairs belong to var. *corrugata* (A. Nelson) Brand. Apparently no collections from Rio Blanco Co.

Phacelia demissa A. Gray — brittle phacelia — Widespread in Duchesne, Uintah, and Rio Blanco cos., Vermillion Bluffs in Moffat Co., and reported for Daggett Co. (Albee et al. 1988); locally common to abundant in desert shrub communities and often on clay barrens; 4,900-6,200 ft; May-June. Plants of our area belong to var. *minor* Atwood.

Phacelia franklinii (R. Brown) A. Gray — Franklin phacelia — Known from Strawberry Valley and Currant Creek and Willow Creek drainages of the Strawberry River Where disjunct from the main body of the species from Idaho n. to Alaska and Yukon; silver sagebrush communities, often on recent alluvium; 7,700-8,500 ft; July-Aug.

Phacelia glandulosa Nuttall — glandular phacelia — Extreme e. Uintah Co. se. of Bonanza, and Rio Blanco Co.; apparently infrequent; desert shrub, sagebrush, and mt. brush communities, and roadcuts, usually on clay soils and gravelly slopes of Green River Formation; 5,000-8,400 ft; May-June. We have been unable to find Nielson 85, which was reported by Bradley (1950) for Lookout Mt., Moffat Co. Our plants belong to var. *glandulosa*. Variety *deserta* (A. Nelson) Brand is found adjacent to our area in sw. Wyoming.

Phacelia hastata Douglas ex Lehmann (*P. leucophylla* Torrey) — lanceleaf phaceila — Common from Strawberry Valley across the Uinta Mtns. and flanks of these mountains and to Zenobia Peak and to the Avintaquin drainage on the W. Tavaputs Plateau and disjunct in the Mt. Bartles area; sagebrush, riparian, meadow, aspen, and open coniferous forest communities and a little above timberline, often on rocky ground; 7,000-10,600 ft; late June-early Oct.

Phacelia heterophylla Pursh — varileaf phacelia — Occasional in Strawberry Valley to Tabiona, and on the E. Tavaputs Plateau in Hill and Stuart Creek drainages, and Harrington 1426 CS reported by Bradley (1950) for 10 mi w. of Maybell; mostly in aspen, aspen parklands, and mt. brush communities; 7,000-9,800 ft; June-July. Our plants belong to var. *heterophylla*. Apparently hybridizing with *P. hastata* (IMF 4: 182).

Phacelia incana Brand. — hoary phacelia — The 6 specimens seen are from Weaver Ridge, Weaver Canyon, Cowboy Canyon and Willow Creek of the E. Tavaputs Plateau, Raven Ridge, and Jones Hole Fish Hatchery; barren shale ridges and desert shrub communities; 5,200-5,830 ft; late May-June.

Phacelia ivesiana Torrey in Ives. — Ives phacelia — Widespread; occasional in desert shrub, sagebrush, and juniper-mt. brush communities, mostly in sandy soil; 4,800-6,000 ft; May-June.

Phacelia sericea (Graham) A. Gray (*P. idahoensis* Henderson misapplied) — silky phacelia — Widespread; occasional in sagebrush, aspen, and open coniferous forest communities; 7,000-10,200 ft; mid May-July. Our plants belong to var. ciliosa Rydberg. *Phacelia idahoensis* (sometimes reported for our area) is a plant of n. and c. Idaho with stamens included or barely exserted from the corolla. Our plants have the stamens well-exserted.

HYPERICACEAE (GUTTIFERAE) St. Johnswort Family

Herbaceous plants, sometimes woody at the base; leaves opposite, simple, entire, punctate with translucent or dark colored glandular dots; flowers bisexual, regular, cymose; sepals 5 (rarely 4); petals separate, 5 (rarely 4), yellow; stamens few or many, often in 3-5 clusters with the filaments united below; ovary superior, 1-celled; fruit a capsule.

Hypericum L. — St. Johnswort

---

1 Stems mostly prostrate, 3-15 (20) cm long; petals mostly less than 4 cm long, about equal to the sepals . . . . *H. anagalloides*

1 Stems erect, 20-60 cm tall; petals 7-14 mm long, longer than the sepals . . . . *H. formosum*

Hypericum anagalloides C. & S. — bog St. Johnswort — Uinta Mtns.; infrequent; wet meadows, along streams, and in moist woods; 8,500-10,500 ft; July-Aug.

Hypericum formosum H. B. K. — western St. Johnswort — Uinta Mtns. and Strawberry Valley; occasional; meadows and along streams; 7,000-9,000 ft; June-Aug.

IRIDACEAE Iris Family

Perennial herbs (ours); leaves mostly basal, linear or nearly so, usually in 2 ranks, sheathing at the base, the lower ones enfolding the upper, equitant; flowers bisexual, subtended by spathe-like bracts; perianth of 6 segments in 2 series of 3 each; stamens 3; ovary inferior, 3-chambered; style 3-cleft, distinct; fruit a many-seeded capsule.

1   Flowers over 5 cm wide, irregular; capsules 3-5 cm long . . . . *Iris*
1   Flowers not over 2 cm wide, regular; capsules less than 1 cm long . . . .
     . . . . *Sisyrinchium*

Iris L. — Flag; Iris

Iris missouriensis Nuttall — Rocky Mtn. iris, western blue flag — Stems from thick rootstocks, 20-50 cm tall; leaves light green, glaucous, somewhat shorter to about as long as the flowering stem, up to 1 cm wide; sepals about 6 cm long, 2 cm wide, the claw yellowish-white, blade lilac or purple-veined on a paler background; petals somewhat shorter than the sepals, not veined; capsule 3-5 cm long. Widespread; common to locally abundant in meadows, along small streams, around seeps and springs, in sunny places but also locally common under aspen; up to about 8,000 ft; May-July.

Sisyrinchium L. — Blue-eyed Grass

Caespitose perennial herbs with clusters of fibrous roots; stems slender, compressed, flattened, often conspicuously winged; leaves grass-like, sheathing at the base, the blades flat, linear; inflorescence enclosed (when young) or subtended (at maturity) by 2 bracts (spathes); flowers solitary or few; perianth bluish to violet; stamens with filaments united into a tube surrounding the style.

1   Stems mainly unbranched, the spathe bracts unequal, the inner bract 13-35 mm long, the outer bract (15) 20-75 mm long . . . . *S. idahoense*
1   Stems mainly branched with a leaf-like bract subtending 2 or more pedunculate spathes, the peduncles up to 14 cm long; spathe bracts subequal, the inner one 12-24 mm long, the outer one 13-26 mm long . . . .
     . . . . *S. demissum*

Sisyrinchium demissum Greene (*S. radicatum* Bicknell) — blue-eyed grass — Occasional along the s. flank of the Uinta Mtns. and adjacent in the Uinta Basin from Rock Creek to Lapoint and in Range Creek, W. Tavaputs Plateau; meadow and streamside communities and in seeps; 5,400-7,500 ft; June-Aug.

Sisyrinchium idahoense Bicknell (*S. occidentale* Bicknell) — Idaho blue-eyed grass — widespread; occasional; wet meadow and streamside communities; 7,500-8,000 ft; June-Aug. Reports of *S. angustifolium* Miller, *S. montanum* Green, and *S. halophilum* Greene are likely based on plants of *S. idahoense.*

ISOETACEAE Quillwort Family

Isoetes L. — Quillwort; Merlin Grass

Submerged or emergent perennial herbs; stems reduced to subterranean, corm-like structures with slender fibrous roots; leaves (sporophylls) onion-like, clustered at the summit of the corm-like stems, soft, linear, the main portion of the blade with a central vascular bundle (seen in cross-section) flanked by 4 longitudinal air cavities and transversed by numerous partitions; reproduction by spores; sporangia borne on the expanded basal portion of the leaves, more or less covered by a velum; spores of 2 types [microspores (typically borne on inner leaves) and megaspores (typically borne on the outer leaves)]. Feature of the megaspores used in the following key are not detectable by means ordinarily available in the field. Power of 40x or greater is helpful and perhaps necessary.

1 Megaspores with separated tubercles, about 0.3-0.45 mm in diameter; leaves 6-25, 5-10 (15) cm long, stomata present but few; ligules cordate; velum narrow, covering about 1/4-1/3 of the sporangia; sporangia orbicular or oval, 3-4 (6) mm long . . . . ***I. bolanderi***

1 Megaspores with spines or with confluent ridges, about 0.5-0.65 mm in diameter; leaves 5-20 cm long; ligules deltoid or short-triangular 2

2 Megaspores with spines; leaves 10-35, erect or recurved, not especially rigid; stomata present but few; velum covering 1/2 to all of the sporangia; sporangia oblong, spotted, 4-7 mm long . . . . ***I. echinospora***

2 Megaspores with confluent ridges becoming somewhat reticulate; leaves 9-30, rarely to 60, ridged; stomata lacking; velum narrow, covering about 1/3 of the sporangia, sporangia orbicular, 5-6 mm long . . . . ***I. occidentalis***

Isoetes bolanderi Engelmann — Bolander quillwort — Widespread across the Uinta Mtns.; permanent and ephemeral ponds, and shallow lakes; 8,100-10,820 ft.

Isoetes echinospora Durieu (*I. muricata* Durieu) — spiny quillwort — Listed for the Uinta Mtns. (IMF 1:183).

Isoetes occidentalis L. F. Henderson (*I. lacustris* L. apparently misapplied) — western quillwort — Listed or mapped for the Uinta Mtns. (IMF 1: 184 and FNA 2: 74).

JUNCACEAE Rush Family

Perennial or annual grass-like herbs; stems terete or flattened, not jointed as in grasses, caespitose or arising singly or few together from rhizomes; leaves alternate and/or all basal, mostly 2-ranked, the blades linear, sometimes much reduced or lacking; inflorescence head-like to open-paniculate, subtended by an involucral bract; branches of the inflorescence, and the pedicles often subtended by scarious or hyaline bractlets; flowers bisexual, sometimes subtended by 2 bracteoles that are borne atop the pedicels and directly beneath the flowers; perianth much reduced, the petals and sepals hardly if at all different and referred to here as tepals, the tepals membranous, rather scale-like, greenish or brownish, 6, in an outer and an inner series of 3 each; stamens (3) 6; pistil 1; ovary superior, with 1 or 3 chambers; style 1 with 3 stigmas; fruit with 1 or 3 chambers.

1 Bracteoles subtending the flowers entire . . . . ***Juncus***

1 Bracteoles subtending the flowers lacerate or dentate . . . . ***Luzula***

Juncus L. — Rush; Wiregrass

Perennial or annual herbs; stems terete or flattened; leaf blades flat, strongly folded, or terete and hollow with cross membranes at intervals (septate), or reduced to a bristle, or lacking; flowers as in the family.

1 Plants annual 2

1 Plants perennial.. 3

2 Stems scapose with one flower or if with 2 or 3 flowers, the lowest flower subtended by scarious bracts shorter than the flower; perianth segments 1.5-2 (3) mm long; plants 0.5-3 (5) cm tall ....*J. bryoides*

2 Stems usually with one or more leaves, and with (1) 3-20 flowers, the lowest flower usually subtended by a leaf or leaf-like bract longer than the flower; perianth segments 3-6 (8) mm long; plants (1) 5-30 cm tall ....*J. bufonius*

3 Leaves all reduced to bladeless sheaths, the upper ones-sometimes tipped with a bristle, this not over 5 mm long 4

3 At least the uppermost leaf of most stems with a well-developed leaf blade well over 1 cm long 6

4 Inflorescence with (1) 2-3 flowers ....*J. drummondii*

4 Inflorescence usually with more than 3 flowers........................ 5

5 Involucral bract about as long or longer than the stem; inflorescence appearing about or below the middle of the plant; stems seldom over 1 mm thick; plants 5-40 cm tall, of the Uinta Mtns., above 9,500 ft ..*J. filiformis*

5 Involucral bract mostly shorter than the stem; inflorescence appearing well above the middle of the plant; stems often over 1 mm thick, mostly arising singly or few together from robust, dark rhizomes; plants 20-100 cm tall, widespread, mostly below 9,000 ft ....*J. arcticus*

6 Flowers 1 and terminal or 2-3 (5) in a solitary terminal head; involucral bracts 2, about equaling or slightly exceeding the head, spathe-like; plants 3-15 (25) cm tall, uncommon............................... **J. triglumis**

6 Flowers either more numerous or not in a terminal head; plants various but generally differing from the above in 1 or more other features 7

7 Flowers born singly, not in heads; leaves borne on the lower 1/5 of the plant, usually 1-3 per culm, not hollow, not septate; rhizomes lacking or short and plants caespitose 8

7 Flowers borne in 1 or more heads; at least the uppermost blade borne well above the lower 1/3 of the plant or else hollow and septate; rhizomes usually well developed 12

8 Seeds tailed at each end, the tails 1/2 as long to longer than the body; uppermost leaf with a well-developed blade, the lower ones often ending in a bristle; the blades terete or channeled on the upper surface; inflorescence with (1) 2-7 flowers; plants from 9,000-11,000 ft or higher . 9

8 Seeds not tailed; stem leaves usually all with well-developed blades, the blades involute or strongly channeled; inflorescence often with more than 7 flowers; plants from 4,800-10,000 ft 10

9 Capsules retuse at the apex; tepals 4-5 (5.5) mm long; only the uppermost sheath with a blade (blades entirely lacking on a few stems) ......*J. hallii*

9 Capsules acute; tepals 5-7 mm long; upper and occasionally a lower sheath with a blade ....*J. parryi*

10 Outer tepals obtuse, sometimes incurved or with hooded tips, 1.5-2.2 mm long .... *J. compressus*

10 Outer tepals acute, straight, 3-5 mm long 11

11 Capsules retuse at the apex, tepals with hyaline margins extending to the apex of the acute tip; panicles mostly less than 2 cm long; plants commonly from 7,000-10,000 ft .... *J. confusus*

11 Capsules blunt but not retuse; outer tepals with hyaline-margins not extending onto the acuminate or acuminate-attenuate tip; panicles often over 2 cm long; plants commonly from 5,400-8,100 ft .... *J. tenuis*

12 Leaves flat or strongly folded, not terete nor hollow, the sheaths with hyaline margins; capsules not exserted beyond the tepals 13

12 Leaves terete, hollow, often with internal cross-membranes at intervals (septate), if terete and hollow only toward the tip then the sheaths without hyaline margins and capsules exserted well beyond the tepals......... 16

13 Leaves strongly folded with the narrow edge oriented toward the flattened stem; scarious margins of the sheaths extending well beyond the juncture with the stem and gradually tapering to inconspicuous auricles, or the auricles lacking, the blade more or less united beyond these scarious margins; tepals 2.5-4 mm long .... *J. ensifolius*

13 Leaves flat, the flat surface oriented toward the terete stem, the scarious margins of the sheaths not extending beyond the juncture with the stem; tepals various.. 14

14 Seeds with tail-like appendages at both ends, the tails equaling the length of the body; heads sometimes more than 10-flowered; tepals granular-papillate on the back.................................... *J. regelii*

14 Seeds without tails; heads seldom more than 10-flowered; tepals smooth on the back.. 15

15 Stamens 6; tepals 5-6 mm long; heads 1-8; plants widespread . *J. longistylis*

15 Stamens 3; tepals 2-3.5 mm long; heads 2-40; plants known from the Piceance Basin....................................... *J. marginatus*

16 Capsule tapering almost from the base into a mostly non dehiscent stylar beak, spreading in all directions in the mature heads; tepals greenish, brownish or tawny, acuminate or acuminate-subulate; rhizomes sometimes swollen at the nodes 17

16 Capsules rather abruptly narrowed above, ascending to slightly spreading in the heads, or if spreading in all directions then the tepals deep brown or black-purple, rounded or acute; rhizomes not swollen at the nodes 18

17 Auricles 1.5-5 mm long; tepals mostly 4-5 mm long, rigid at the tip, acuminate-subulate; heads 10-15 mm in diameter, nearly globose; capsules about equaling the tepals; plants (20) 40-100 cm tall; stems to 6 mm thick; rhizomes with fusiform-tuberous nodes .... *J. torreyi*

17 Auricles not over 1 mm long; tepals mostly 2.5-4 mm long, acuminate but hardly subulate; heads 6-12 mm in diameter; capsules distinctly longer than the tepals; plants 10-40 (58) cm tall; stems 1-2 mm thick; rhizomes with slightly swollen nodes .... *J. nodosus*

18 Seeds tailed on both ends, the tails longer than the body; leaves folded or rolled, not truly terete at the base, becoming terete and hollow distally; the blades poorly differentiated from the sheath, the sheaths without hyaline

margins; auricles lacking; capsules distinctly longer than the tepals . . . . .
. . . . *J. castaneus*

18 Seeds not tailed, or tails shorter than the body; leaves terete their whole length, blade distinct from the sheath, the sheaths with hyaline margins; auricles well developed 19

19 Tepals 3-5 mm long, or if shorter than the inflorescence with only 1 or rarely 2 heads, equal to or longer than the capsules; anthers various . . . 20

19 Tepals 1.5-2.8 mm long, shorter than the capsules; inflorescence with (1) 3-25 heads; anthers to about 0.7 mm long 21

20 Heads 1 or rarely 2, globose or nearly so, with 5-40 or more flowers; anthers 0.5-1 mm long, shorter than the filaments; tepals purple-black . . . .
. . . . *J. mertensianus*

20 Heads (1) 2-13, not or hardly globose, with about 3-15 flowers; anthers 1-2 mm long, longer than the filaments; tepals light brown to purplish-black
. . . . *J. nevadensis*

21 Outer tepals more or less acute, equal or shorter than the inner; capsules acute; branches of the inflorescence mostly spreading . . . . . . *J. articulatus*

21 Outer tepals rounded, longer than the inner; capsule rounded at the apex; branches of the inflorescence mostly strongly ascending to erect *J. alpinus*

Juncus alpinus Villars — alpine rush — Rarely collected, the few specimens seen are from the Uinta Mtns. and flanks of these Mtns.; wet places; (5,350) 6,100-9,500 ft; July-Aug.

Juncus arcticus Willdenow (*J. balticus* Willdenow) — wiregrass — Common to abundant throughout the wetlands of our area from the lowest elevations up to about 9,000 ft and rarely above 9,000 ft, tolerant of alkali, persistent under heavy grazing; May-July. *Juncus arcticus* forms a variable complex. In FNA (22: 216-217) 2 vars. are treated that supposedly occur in our area. One without leaf bearing blades [var. *balticus* (Willdenow) Trautvetter] and the other with 1 or 2 upper leaves with blades [var. *mexicanus* (Willdenow ex Roemer & Schultes) Balslev]. In TPD our plants are included in ssp. *littoralis* (Engelmann) Hulten with *J. mexicanus* Willdenow ex Roemer & Schultes treated at species level and excluded from Utah but included for New Mexico. San Juan Basin plants of New Mexico and Utah are all included in var. *balticus* where this var. is keyed with leaves reduced to bladeless sheaths (Jameson 2013). Specimens we have seen from our area have leaves reduced to bladeless sheaths. It appears our plants all belong to var. *balticus* (Baltic rush).

Juncus articulatus L. — jointleaf rush — Specimens from the floodplain of the Green River at Little Hole appear to belong here.

Juncus bryoides F. J. Hermann (*J. kelloggii* Engelmann misapplied) — minute rush, moss rush — The 7 specimens seen are from Diamond Mt., Eagle Creek, and Green Lakes, Daggett Co.; Pine Ridge, Uintah Co.; and Cold Spring Mtn, Douglas Mtn., and Zenobia Peak, Moffat Co.; ephemeral pools and on spring-fed sandy soil or sandstone; 7,500-8,225 ft; June-July. This small plant is easily overlooked. It is likely more common than the number of collections indicate.

Juncus bufonius L. (*J. sphaerocarpus* Nees misapplied) — toad rush — Widespread; occasional and locally common on moist or ephemerally wet ground at low and moderate elevations; May-Sept.

Juncus castaneus Smith — chestnut rush — The 5 specimens seen are from Garfield Basin, head of Yellowstone, Little East Fork Blacks Fork, n. side of Bald Mtn. of the Blacks Fork-Smiths Fork drainage, and Lakeshore Basin, reported by Graham (1937) for head of W. Fork of Whiterocks; 11,200-11,800 ft; moist and wet meadows and seeps of talus slopes.

Juncus compressus Jacquin — round fruit rush — Specimens seen are from the floodplain of the Green River and draw-down basin of Flaming Gorge Reservoir at 4,800-6040 ft where this species appears to have increased since the first collection from our area was made in 1988; June-Aug. Apparently introduced from Eastern United States. A specimen taken in 2011 from along Highway 191 at 8,465 ft in the Uinta Mountains indicates this species can be expected to expand its range within our area.

Juncus confusus Coville — Colorado rush — Occasional across our area; wet places; 7,000-9,800 ft; June-Aug. Similar to and closely related to *J. tenuis,* and might be considered a var. of that species (IMF 6:50).

Juncus drummondii E. Meyer in Ledebour — Drummond rush — Common to abundant in the Uinta Mtns.; open woods, meadows, and open slopes of the spruce-pine belt, and on rocky ground, talus and meadows above timberline where it is most abundant where snow cover likely exceeds 3 ft most of the winter; 9,000-11,400 ft; June-Aug.

Juncus ensifolius Wikstrom (*J. brunnescens* Rydberg; *J. saximontanus* A. Nelson; *J. tracyi* Rydberg; *J. xiphioides* E. Meyer var. *montanus* Engelmann) — swordleaf rush — Widespread and common; wet places across our area; 7,000-10,500 ft; May-Aug. A variable plant that has been divided into several taxa based on inconsistent features. Three vars. in our area:

1 Stamens usually 3, sometimes 6; plants rare; the few specimens from isolated stations may represent odd plants of var. *montanus* . . . . **var. *ensifolius***
1 Stamens 6; plants common. 2
2 Heads seldom more than 10, each with 10-25 flowers, mostly 8-15 mm thick . . . . **var. *montanus*** (Engelmann) C. L. Hitchcock
2 Heads seldom less than 10, each with 4-12 flowers, usually 3-8 mm thick; plants wholly intergrading into the preceding . . . . **var. *brunnescens*** (Rydberg) Cronquist

Juncus filiformis L. — thread rush — Infrequently collected but widespread in the Uinta Mtns.; wet meadows, streambanks, and other moist places; 9,000-10,500 ft; June-Aug. The report by Potter and others (1983) for along the Green River is based on a specimen (Potter G 268!) of *J. bufonius.* Thread rush is highly selected by cattle for forage in our area.

Juncus hallii Engelmann — Hall rush — Occasional in the Uinta Mtns.; often at the edge of moist meadows and open woods in the lodgepole pine-Engelmann spruce belt; 9,700-11,000 ft; June-Aug. Similar to *J. parryi* but distinct by the retuse capsules.

Juncus longistylis Torrey — longstyle rush — Widespread and common; moist or wet places; 5,300-11,000 ft; June-Aug. *J. orthophyllus* Coville was listed for Sheep Creek, Daggett Co. in an area now covered by water of the Flaming Gorge Reservoir (Flowers et al. 1960). However, this species is not listed for Utah or Wyoming in IMF (6:34), AUF, TPD, or Dorn (2001). This report was likely based on *J. longistylis.*

Juncus marginatus Rostkov — grassleaf rush — A single specimen (WRE 799 CS) from the upper Piceance Basin has been identified as *J. marginatus*. However, this species is shown only for Boulder Co., Colorado in TPD. This specimen might belong to *J. longistylis.*

Juncus mertensianus Bongard — one-head rush — Occasional across the Uinta Mtns.; open moist woods, streamside and wet meadow communities in the pine-spruce belt and alpine; (7,200) 8,000-11,400 ft; June-Aug.

Juncus nevadensis S. Watson (*J. badius* Suksdorf) — Nevada rush — Strawberry Valley and Uinta Mtns.; occasional in meadows and along streams; 7,600-10,000 ft; June-Aug.

Juncus nodosus L. — knotted rush — Widespread but no specimens seen from Daggett Co.; occasional in wet areas at 5,500-7,600 ft; July-Sept.

Juncus parryi Engelmann — Parry rush — Uinta Mtns.; moist, often rocky, coniferous woods, and on rocky slopes above timberline, common to dominant in relatively early-melting, well-drained snowbeds; 9,800-11,000 ft; June-Aug.

Juncus regelii Buchenau — Regel rush — The few specimens seen are from near Mirror Lake, and Governor Dern Lake, Uinta Mtns.; July-Aug.

Juncus tenuis Willdenow (*J. dudleyi* Wiegand; J. interior Wiegand) — poverty rush — Widespread; occasional in moist and wet places; 5,400-8,100 ft; June-Aug.

Juncus torreyi Coville — Torrey rush — Widespread but no specimens seen from Summit Co.; locally common along ditches and washes, and in wet lowlands, tolerant of alkali; 4,700-6,000 (7,000?) ft; May-July.

Juncus triglumis L. [*J. albescens* (Lange) Fernald] — three-flowered rush — Occasional across the Uinta Mtns.; wet and boggy meadows; 9,200-12,500 ft; July-Aug. Our plants belong to var. *albescens* Lange.

Luzula de Candolle — Woodrush

Like *Juncus,* but different by the features in the key.

---

1 Inflorescence on open panicle; flowers not congested; leaves glabrous at maturity, the blades flat, 3-13 mm wide; plants 25-70 cm tall . **L. parviflora**

1 Inflorescence of congested head-like spike(s); leaves pubescent with long hairs along the margins near the collar; plants 5-20 (30) cm tall 2

2 Flowers borne in a terminal compound spike-like inflorescence; leaves mostly 1-3 mm wide; plants below and above timberline . . . . **L. spicata**

2 Inflorescence with 1 or more lateral spikes as well as the terminal one, some of the lateral ones often borne on elongated peduncles; leaves 2-6 mm wide, plants mostly below timberline . . . . **L. campestris**

Luzula campestris (L.) de Candolle [*L. intermedia* (Thuillier) A. Nelson; *L. multiflora* (Retz.) Lejeune] — hairy woodrush — Uinta Mtns.; occasional in woods and along streams at 8,000-11,000 ft; June-Aug.

Luzula parviflora (Ehrhart) Desvaux [*L. piperi* (Coville) Henry misapplied; *L. wahlenbergii* Ruprecht misapplied; *Juncoides parviflorum* (Ehrhart) Coville] — millet woodrush — Uinta Mtns. and Strawberry Valley; occasional in moist and wet places, mostly along streams; 7,500-11,400 ft; June-Aug.

Luzula spicata (L.) de Candolle — spike woodrush — Uinta Mtns.; occasional in subalpine coniferous forests and more common in alpine communities where found in at least 34 of the 50 alpine plant communities described by Brown (2006) ranging from deep snowbeds to fell fields and moist meadows to talus but always averaging low percent crown cover; 10,400-12,500 ft; June-Aug.

JUNCAGINACEAE Arrowgrass Family
Triglochin L. — Arrowgrass

Plants perennial, herbaceous, glabrous, scapose, usually from rhizomes; leaves with a membranous ligule at the junction of the sheathing base and subterete to flattened blade; flowers bisexual, in bractless, spike-like racemes; tepals 6, green or purple-tinged, rather inconspicuous; stamens 6; carpels 3 or 6; style lacking or short; fruit of 1-seeded follicles, these united at first but later separating from the base upward.

1 Fruit linear to clavate, tapered to the base, 5-9 mm long; stigmas and fertile carpels 3; leaves less than 2 mm wide; ligules bilobed to the base; plants 15-30 (60) cm tall, mostly montane (above 7,500 ft) . . . . *T. palustris*
1 Fruit oblong, rounded at the base, 3-6 mm long; stigmas and fertile carpels 6; plants of valleys, up to 7,500 (8,600) ft 2
2 Ligules entire or only slightly bilobed, (1) 1.5-5 mm long; leaf blades obcompressed, 1.5-2.5 (4) mm wide; stems closely tufted on a proliferating rhizome (10) 30-100 (120) cm tall, common . . . . *T. maritima*
2 Ligules bilobed to the base or emarginate, 0.5-1 mm long; leaf blades nearly terete, 0.5-1-(1.5) mm wide; stems well spaced on the rhizomes, 15-30 (40) cm tall, rarely collected . . . . *T. concinna*

Triglochin concinna Davy — low arrowgrass — Specimens seen are from below 7,000 ft from Spring Creek inlet of Flaming Gorge Reservoir, Daggett Co. near the Utah-Wyoming line (Neese & Peterson 5437 & Goodrich 26961), Yellow Creek, Rio Blanco Co. (Goodrich 28424), reported for Hells Half Mile along the Green River (Holmgren 1962); alkaline and saline, riparian communities; June-Aug.

Triglochin maritima L. — shore arrowgrass — widespread and locally common in wet, alkaline meadows and along streams at 4,800-7,000 (8,600) ft; May-Sept.

Triglochin palustris L. — marsh arrowgrass — Specimens seen are from across the n. slope of the Uinta Mtns. and from S. Fork Rock Creek of the s. slope and E. Tavaputs Plateau; wet meadows and along streams; 8,000-9,800 ft, perhaps to be expected down to 7,000 ft; July-Aug.

LAMIACEAE (LABIATAE) Mint Family

Annual to perennial herbs; stems generally 4-angled; leaves opposite, simple; stipules lacking; flowers bisexual, irregular (often 2-lipped) or occasionally nearly regular, the corolla united, the upper lip entire or 2-lobed and the lower lip often 3-lobed; stamens 2 or 4; pistil 1; ovary superior, 4-lobed; style 2-4 cleft; fruit of 4 nutlets, these laterally or basally attached and 1-seeded.

1   Inflorescence terminal, the flower whorls subtended by bracts much smaller than the stem leaves, the flowers usually so crowded as to conceal most of the internodes of the inflorescence 2

1   Inflorescence axillary, the flower whorls in the axils of well developed stem leaves, usually not concealing the internodes 10

2   Leaves entire 3

2   Leaves toothed 4

3   Plants strongly odorous; stamens exserted; corolla white to pink-purple; leaves sessile or nearly so .... *Monardella*

3   Plants not strongly odorous; stamens included; corolla blue or violet, rarely white or pink; lower leaves petiolate .... *Prunella*

4   Corollas 25-35 mm long; inflorescence a head .... *Monarda*

4   Corollas 7-14 mm long; inflorescence usually spike-like. 5

5   Stamens conspicuously exserted beyond the corolla .... *Agastache*

5   Stamens included or scarcely exserted 6

6   Leaves with fewer than 10 teeth .... *Prunella*

6   Leaves with more than 10 teeth 7

7   Leaves sessile or some of the lower ones with petioles less than 1 cm long
    .... *Stachys*

7   Leaves with distinct petioles 8

8   Bracts of the inflorescence spinulose toothed; calyx 9-12 mm long ....
    .... *Dracocephalum*

8   Bracts of the inflorescence not spinulose toothed; calyx 5-6.5 mm long .. 9

9   Plants canescent tomentose .... *Nepeta*

9   Plants glabrous or glandular .... *Mentha*

10 Leaves entire 11

10 Leaves toothed or lobed or toothed and lobed 12

11 Corollas 7-11 mm long .... *Hedeoma*

11 Corollas 14-21 mm long .... *Scutellaria*

12 Leaves lobed, the sinuses cut ¼ or more the way to the midrib......... 13

12 Leaves toothed with sinuses of teeth cut less than ¼ the way to the midrib
    14

13 Leaves with 3 or 5 palmate lobes, with conspicuous petioles .... *Leonurus*

13 Leaves pinnately lobed, sessile .... *Lycopus*

14 Bracts of inflorescence spinulose; calyx 9-12 mm long .... *Dracocephalum*

14 Bracts not spinulose; calyx mostly shorter than 9 mm 15

15 Corollas 11-20 mm long 16

15 Corollas less than 11 mm long.................................... 17

16 Flowers usually 1 or 2 per node of the rachis .... *Scutellaria*

16 Flowers usually more than 2 per node of the rachis ....

17 Stamens well exerted beyond the corolla; plants strongly aromatic *Mentha*

17 Stamens included in the corolla; plants not strongly aromatic......... 18

18 Leaves sessile .... *Lycopus*

18 Leaves with conspicuous, but sometimes short, petioles . . . . . . *Marrubium*

Agastache Clayton ex Gronovius — Giant Hyssop; Horsemint

Agastache urticifolia (Bentham) Kuntze — nettle-leaf horsemint — Strongly aromatic herbs; stems 1-many, 40-100 (150) cm tall; herbage short-hairy to subglabrous, glandular in the inflorescence and on lower surfaces of leaves; leaves petiolate, 2-10 cm long, 2-9 cm wide, cordate to lanceolate, coarsely crenate or serrate; inflorescence a terminal spike, 3-15 cm long; calyx 15 nerved, 5 toothed; corolla white or pale pink. Strawberry Valley to Blind Stream in the Uinta Mtns. and to Avintaquin Canyon, W. Tavaputs Plateau; occasional and locally abundant in aspen, tall forb, and mesic sagebrush communities, sometimes in snowbed areas; 7,000-8,500 ft; late June-early Oct. Our plants belong to var. *urticifolia.*

Dracocephalum L. — Dragonhead

Dracocephalum parviflorum Nuttall [*Moldavica parviflora* (Nuttall) Britton in Britton & Brown] — American dragonhead — Non-aromatic annual or short-lived perennial herb, 15-60 (80) cm tall; herbage glabrate to pubescent; leaf blades 1-6 (8) cm long, 1-2.5 cm wide, coarsely serrate-spinulose; inflorescence 2-10 cm long, spike-like, bracteate, the bracts leaf-like, spinulose toothed, secondary spikes sometimes arising from axils of upper leaves; calyx 15-nerved, gland-dotted and villous-hirsute, the lobes spinulose; corolla blue to purplish. Widespread; infrequent or occasional in sagebrush, aspen, Douglas-fir, meadow, and streamside communities, often on recent alluvium, roadsides, and burned areas; 7,600-9,800 ft; July-Aug. Apparently a seedbank species released by fire.

Hedeoma Persoon — Pennyroyal

Hedeoma drummondii Bentham — Drummond pennyroyal — Plants perennial, from taproots, somewhat woody at the base, 10-25 cm tall, puberulent throughout, with several stems, the stems branching; leaves linear to narrowly elliptic; calyx strongly ribbed, the ribs about 13, the lower teeth surpassing the upper; corolla about 1 cm long. Specimens seen are from Stewart Gulch (J. Walker & M. Waters 82-277) and South Canyon (Thorne 15122) of the E. Tavaputs Plateau; juniper-rabbitbrush and aspen-sagebrush-Douglas-fir community; 6,900-8,200 ft; also reported for Duchesne Co. (IMF: 314); June.

Leonurus L. — Motherwort

Leonurus cardiaca L. — common motherwort — Perennial from a branched caudex and fibrous roots, 40-150 cm tall; stems retrorsely strigose-puberulent on the angles; leaf blades 5-10 cm long, about as wide, the petiole about equaling the blade, the upper ones less palmately cleft than the middle ones and sometimes entire; calyx tube 5-angled with 5 spinulose lobes, the lower 2 lobes strongly spreading to deflexed; corolla about 1 cm long, pale pink, the upper lip white-villous. Introduced from Asia, formerly cultivated for a home remedy, only 4 specimens seen but expected to be widespread; ditch banks, riparian communities, and persisting where cultivated; up to 8,000 ft; June-Aug.

Lycopus L. — Bugleweed; Water-Horehound

Plants perennial from rhizomes, not aromatic, usually gland-dotted; calyx 5-13 nerved, inconspicuously 2-lipped, with 5 subequal teeth; corolla regular, the limb 4-lobed, the upper lobe slightly broader than the others and often notched at the apex; nutlets 3-angled.

---

1   Leaves shallowly to deeply pinnately lobed, the blades mostly tapering to a
    short petiole .... *L. americanus*
1   Leaves serrate, sessile or nearly so .... *L. asper*

Lycopus americanus Muhlenberg — American bugleweed — Widespread but no specimens seen from Daggett or Summit Cos; occasional in riparian, meadow, and marsh communities, and along ditch banks; up to about 7,600 ft; July-Aug.

Lycopus asper Greene (*L. lucidus* Turczaninow in Bentham misapplied) — rough bugleweed — Widespread but no specimens seen from Daggett or Summit Cos; occasional in riparian, meadow, and marsh communities, along ditch banks, and edges of ponds and lakes, tolerant of alkali; up to 6,000 or perhaps 7,000 ft; July-Aug.

Marrubium L. — Horehound

Marrubium vulgare L. — common horehound — Non-aromatic perennial herb from a taproot, 2-5 (8) dm tall; stems 1-several, erect to nearly prostrate; herbage gray or white with woolly hairs except sometimes on the upper leaf surfaces; leaves petiolate, ovate to suborbicular, 1.5-4 (6) cm long, about as wide, crenate; calyx 10-nerved with recurved teeth and with a ring of exserted hairs arising from within the throat; corolla 5-6 mm long, white or nearly so, the lips about equal. Introduced from Europe, somewhat weedy, seldom collected but expected to be widespread; roadsides, dry waste places, often associated with disturbances; up to about 8,000 ft; May-Sept.

Mentha L. — Mint

Aromatic perennial herbs from rhizomes; leaves petiolate, serrate; inflorescence spike-like; calyx regular or nearly so (in ours); corolla nearly regular; stamens 4; style exserted.

---

1   Flowers in several axillary clusters .... *M. arvensis*
1   Flowers in few terminal clusters and occasional few axillary clusters ....
    .... *M. citrata*

Mentha arvensis L. [*M. penardi* (Briquet) Rydberg] — Field mint — Widespread; locally common in riparian, meadow, marsh, and streamside communities; 4,800-8,200 ft; mid June-Sept. Our plants belong to var. *canadensis* (L.) Kuntze.

Mentha citrata Ehrhart — lemon mint, bergamot mint — Cultivated spice plant introduced from Europe. Known from along ditches at Jose Morris Cabin at Cub Creek, Dinosaur National Monument where locally abundant and apparently long-persistent.

Monarda L.

Monarda fistulosa L. (*M. menthaefolia* Bentham) — wild bergamot — Perennial from creeping rhizomes, 3-7 dm tall, the herbage finely puberulent; leaf blades 2.5-8 cm long, 1-3 cm wide, lance-triangular to narrowly ovule; petioles mostly less than 1 cm long, the upper ones sometimes nearly obsolete; calyx 7-11 mm long, 13-15 nerved, white-hairy within and sometimes without; corolla 25-35 mm long, purple, puberulent, the upper lip exceeding the lower. Known from the Piceance Basin and Larson 14 (RM!) from Kimberly-Williams Fork road 4 mi s. of Yampa River, Moffat Co.; moist places; July-Aug. Our plants belong to var. *menthaefolia* (Graham) Fernald.

Monardella Bentham

Monardella odoratissima Bentham — ultra-odor-mint, stinking horsemint — Strongly aromatic perennial herb from a branching woody caudex and taproot, 10-35 cm tall; stems usually several to many, somewhat woody at the base; herbage glandular-punctate, puberulent; leaves 1-3.5 mm long, 2-12 mm wide, lanceolate to narrowly elliptic or ovate; inflorescence a head or head-like, 1.5-3 cm wide, subtended by an involucre of membranous, imbricate, broadly ovate bracts; bracts 7-15 mm long, greenish or whitish at the base, purplish at the apex, ciliate; calyx 10-15 nerved, the teeth ciliate to densely hairy within and without; corolla rose-lavender, violet, or rarely whitish. Known from the W. Tavaputs Plateau and Yellow Creek drainage of the East Tavaputs Plateau; steep slopes, exposed ridges, snowbed areas in various plant communities at (5,720) 9,000-10,000 ft; July-Sept. Our plants belong to var. *glauca* (Greene) St. John.

Nepeta L.

Nepeta cataria L. — catnip — Aromatic, perennial herb from a taproot, 3-10 dm tall; herbage gray or white, velvety-pubescent, and gland-dotted at least above; leaves 2-8 cm long, 1.5-5 cm wide, the blades triangular-ovate or triangular-lanceolate, the base more or less cordate, coarsely crenate or serrate; flowers crowded or sometimes remote in spike-like terminal clusters, the clusters sometimes compound and paniculate; calyx 4-7 mm long; 15-nerved, with 5 subequal teeth; corolla white or pinkish, often purple spotted, 7-12 mm long, strongly 2-lipped, the lower lip with a dilated central lobe. Introduced from Eurasia, more or less weedy, seldom collected, to be expected to be widespread; roadsides, ditches, and waste places; up to 7,450 ft; June-Oct. Catnip contains a feline attractant (nepetalactone) that has strong behavioral effects on some domestic and wild cats.

Prunella L. — Self-heal

Prunella vulgaris L. — common self-heal, heal-all — Perennial herb from a short, slender rhizome, 10-50 cm tall; stems generally unbranched, with crinkled, multicellular hairs at least above; leaves 1-7 cm long, 5-30 mm wide; spike solitary, 2-5 cm long, 1-2 cm wide; bracts subtending the dense verticils of flowers about 1 cm long, orbicular or fan-shaped, ciliate, the midrib sometimes projected as a tooth, membranous and reticulate-veined in age; calyx usually 10-nerved, 2-lipped, the upper lip broad and shallowly 3-toothed, the lower lip deeply 2-toothed, the lower teeth narrower than the upper teeth;

corolla mostly blue or violet, conspicuously 2-lipped, the upper lip hooded and entire, the lower lip spreading, 3-lobed, the central lobe the largest and minutely toothed. Specimens seen are from the s. slope of the Uinta Mtns. and adjacent in the Uinta Basin, to be expected across our area; riparian and meadow communities, and along streams in woods; 5,600-8,100 ft; June-Aug.

Scutellaria L. — Skullcap

Plants perennial, commonly rhizomatous, not aromatic; calyx campanulate in flower, splitting to the base at maturity, strongly 2-lipped, corolla well exerted from calyx, 2-lipped, the upper lip arched, the lower lip spreading or deflexed; stamens 4 in 2 pairs, ciliate or bearded.

1   Plants mostly 5-10 cm tall; leaves 1-2 cm long, entire . . . . **S. nana**
1   Plants mostly 20-80 cm tall; leaf blades 2-5 cm long, crenate-serrate . . . . .
    . . . . **S. galericulata**

Scutellaria galericulata L. — little skullcap — The 4 specimens seen are from Strawberry Valley, Spilt Mtn. Campground, and s. of the town of Whiterocks; riparian and meadow communities; 4,800-7,600 ft; June-Sept

Scutellaria nana A. Gray — dwarf skullcap — The 3 specimens seen (Huber 4009, 4268, and 5100) are all from along the Cottonwood Wash road ne. of Altamont. This plant is perhaps a waif at this location. It is otherwise known in Utah from Iron and Washington Cos. The Huber collections span 12 years. This seems to indicate the species can be expected to persist here.

Stachys L. — Hedge Nettle

Stachys palustris L. (*S. scopulorum* Greene) — marsh hedge nettle, woundwort — Perennial herbs from rhizomes, 2-7 dm tall, with rank odor, hairy throughout; stems with spreading or retrorse hairs on the angles and with viscid or gland-tipped hairs on the sides; leaves 3.5-9 cm long, 1-4 cm wide, elliptic to lanceolate, not or hardly cordate at the base, finely crenate with teeth less than 2 mm long; calyx 6-9 mm long, pubescent with both slender, gland-tipped hairs and long, stout, glandless ones; corollas 11-16 mm long, purplish, with an internal ring of hairs. Widespread but seldom collected; riparian and meadow communities up to about 6,500 ft; June-Aug. Our plants belong to var. *pilosa* (Nuttall) Fernald

LEMNACEAE Duckweed Family

Lemna L. — Duckweed

Small floating plants without stems, consisting of a leaf-like thallus and usually a single, tiny, dangling root; flowers rarely developed, and plants usually propagating by buds; flowers (when developed) unisexual, inconspicuous, without perianth, 3 borne together in pouches on the upper surface or margins of the thallus, 2 of them staminate (each reduced to a single stamen), and 1naked pistillate; fruit an utricle. The following taxa seem to be separated on trivial overlapping features. More work is indicated and more collections are needed to understand this group in our area.

1   Fronds veinless or 1-veined on the upper surface, flat, smooth, 1-2.5 mm
     long, 0.7-1.5 mm wide, solitary or in pairs . . . . **L. minuta**
1   Fronds with 3 or more veins on the upper surface with evident (with
     magnification) protuberances (except in L. turionifera), sometimes longer
     or wider than in L. minuscula, solitary or 2-5 cohering in colonies 2
2   Fronds not reddish on the lower surface (or at least much less so than on
     the upper surface), with greatest distance between lateral veins near the
     middle 3-6 mm long, 1.5-4 mm wide, in groups of 2-5 . . . . **L. minor**
2   Fronds often reddish on the lower surface (more intensely so than on the
     upper), with greatest distance between lateral veins distal to the middle,1.5-
     3.5 mm long, 1-2.5 mm wide, solitary or in groups of 2. 3
3   Fronds flat, mostly with distinct papillae on midline of upper surface; seeds
     with 30-60 indistinct ribs . . . . **L. turionifera**
3   Fronds often gibbous, with distinct papillae above node and near apex on
     the upper surface but not between the node and apex; seeds with 10-16
     distinct ribs . . . . **L. obscura**

Lemna minor L. — lesser duckweed — Widespread, seldom collected; floating
on ponds, lakes and slow moving streams; up to 8,580 ft.

Lemna minuta Kunth in H.B.K. (*L. minuscula* Herter) — tiny duckweed —
Reported for Daggett and Duchesne Cos. (AUF 917).

Lemna obscura (Austin) Daubs — obscure duckweed — The 2 specimens seen
are from ponds on the floodplain of the Green River at Browns Park and
Stewart Lake; 4,900-5,410 ft.

Lemna turionifera Landolt — turion duckweed — Plotted for Duchesne and
Wasatch Cos. (TPD). Plants sometimes form small, olive to brown, rootless
turions that sink to the bottom.

LENTIBULARIACEAE Bladderwort Family
Utricularia L. — Bladderwort

Aquatic herbs with stems mostly submerged; leaves alternate or some whorled,
2-3 or more times pinnately dissected, bearing small urn-shaped bladders
with a valve-like opening that trap insects and minute crustacea; flowers
bisexual, in racemes; peduncles emergent, with auricled scales; corolla united,
2-lipped, yellow; stamens 2, twisted; fruit a capsule with 2-4 valves; seeds several.

1   Lower lip of flowers 10-20 mm long; ultimate segments of leaves filiform;
     bladders numerous; plants free floating, often in deep water . . **U. vulgaris**
1   Lower lip of flowers commonly 4-8 mm long; ultimate segments of leaves
     slender but flat, not filiform; bladders few; plants commonly of shallow
     water . . . . **U. minor**

Utricularia minor L. — lesser bladderwort — Known from Pelican Lake (Neese
15319a) and near Sims Peak, Uinta Mtns. (Tuhy 1210 UTC); aquatic in standing
or slow-moving water; 4,800-10,100 ft.

Utricularia vulgaris L. — common bladderwort — Apparently widespread, seldom collected; aquatic in standing or slow-moving water, up to 9,935 ft. June-Aug.

LILIACEAE   Lily Family

Perennial herbs from bulbs, corms, rootstocks, or woody caudices; leaves simple, mostly entire with parallel veins; flowers regular or nearly so, mostly bisexual, the perianth of 6 united or separate tepals (the sepals and petals often alike and referred to collectively as tepals); stamens mostly 6; ovary superior or partly inferior in *Zigadenus elegans;* styles 3 or at least stigmas 3-lobed; fruit a capsule or a berry. As treated here Liliaceae includes Asparagaceae (*Asparagus*), Melanthiaceae (*Veratrum* and *Zigadenus*), and Uvulariaceae (*Disporum* and *Streptopus*). The family could be broken into several families. However, separation at tribal level seems much more user friendly.

---

1   Leaves dispersed upward on the stem, not basal or crowed toward the base 2

1   Leaves basal or crowded toward the base of the plant 9

2   Leaves linear, 0.5-8 mm wide...3

2   Leaves lanceolate or wider, some well over 8 mm wide 6

3   Leaves 3-20 mm long; plants mostly over 70 cm .... *Asparagus*

3   Leaves over 20 mm long; plants less than 70 cm tall 4

4   Perianth segments 2.5-6.3 cm long, white, pink, or bluish .... *Calochortus*

4   Perianth segments 0.8-2.5 cm long...................................5

5   Flowers white; plants alpine .... *Lloydia*

5   Flowers purplish or greenish-brown to chocolate-brown with yellow or pale spots; plants not alpine .... *Fritillaria atropurpurea*

6   Flowers in terminal racemes or panicles.............................7

6   Flowers 1-3 in axils of leaves or at ends of branches...................8

7   Leaves 10-20 cm wide; flowers 8-17 mm long .... *Veratrum*

7   Leaves 1.5-8 (10) cm wide; flowers 2-7 mm long .... *Smilacina*

8   Flowers 1 (2) in axils of leaves and on slender, bent, glabrous pedicels; style 4-5 mm long; fruit 10-12 mm long, yellow or red; stamens not exserted, unequal, the inner set longer than the outer set .... *Streptopus*

8   Flowers 1-2 (3), on pendulous, pubescent pedicels; style 9-12 mm long; fruit 7-10 mm long, orange or red; stamens slightly exserted, about equal .... *Disporum*

9   Flowers yellow or orange 10

9   Flowers white, pink, or bluish.....................................12

10   Perianth segments 5-9 cm long; plants 80-120 cm tall .... *Hemerocallis*

10   Perianth segments 1-3.5 cm long; plants 6-50 cm tall 11

11   Perianth segments reflexed, 20-40 mm long; anthers exserted; leaves strictly basal or appearing opposite, 1-5 cm wide; capsules 3-6 cm long... .... *Erythronium*

11 Perianth segments not reflexed, 12-22 mm long; anthers included; leaves
    alternate to semi whorled, 0.2-1.2 cm wide; capsules 1.7-3 cm long . . . . . .
    . . . . ***Fritillaria pudica***

12 Inflorescence with 1 or 2 flowers; plants 5-20 cm tall, alpine . . . . . . ***Lloydia***

12 Inflorescence generally with more than 2 flowers 13

13 Flowers in racemes or panicles 14

13 Flowers in umbels 15

14 Perianth segments 15-30 mm long, blue or less commonly white ***Camassia***

14 Perianth segments 3-11 mm long, white or greenish . . . . ***Zigadenus***

15 Perianth segments 4-13 mm long; plants with odor of onion . . . . . . ***Allium***

15 Perianth segments 12-27 mm long; plants without odor of onion. 16

16 Leaves 1.5-2.5 mm wide; plants from east of Strawberry Valley . . . .
    . . . . ***Androstephium***

16 Leaves 3-10 mm wide; plants known from Strawberry Valley . . . . ***Triteleia***

Allium L. — Onion

Onion-scented herbs from bulbs; leaves sheathing, linear or terete, basal or
nearly so; inflorescence capitate or a solitary terminal involucrate, false umbel;
tepals persistent even when dry, white to rose-purple, separate or nearly so;
fruit a capsule.

---

1 Leaves about twice as long as the scape; plants 2-10 (13) cm tall 2

1 Leaves shorter than to slightly exceeding the scape; plants mostly 10-70 cm
    tall 2

2 Leaves 2, concave-convex or nearly flat, not curled at the tip; scape slightly
    flattened; plants of mountains . . . . ***A. brandegei***

2 Leaf 1, terete, often curled at the tip, but the curled portion often broken
    off; plants mostly of deserts . . . . ***A. nevadense***

3 Umbel nodding; tepals 4-6 mm long, shorter than the exserted stamens;
    leaves usually more than 3 per stem, 1-7 mm wide; involucral bracts
    deciduous by anthesis . . . . ***A. cernuum***

3 Umbel not nodding; tepals 5-16 mm long, longer than the stamens; leaves
    various; involucral bracts persistent 3

4 Scapes flattened and narrowly winged; leaves flat, 2-8 mm wide; tepals 10-
    13 mm long, all about equal; plants from slender bulbs that are attached to
    a short, stout rhizome, in moist woods in mtns. . . . . ***A. brevistylum***

4 Scapes terete; leaves not over 4 mm wide; plants from bulbs without
    rhizomes 4

5 Bulb not covered with coarse, brown fibers; tepals 7-12 (16) mm long, the
    outer 3 longer than the inner 3; involucral bracts 2, usually with 3 or more
    nerves . . . . ***A. acuminatum***

5 Bulbs covered with a dense, net-like coat of course, brown fibers; tepals 5-
    8 (10) mm long, the outer ones not longer than the inner ones; involucral
    bracts 2 or 3, mostly 1-nerved 5

6   Tepals mostly pink, seldom white; scapes 15-60 cm tall; leaves 3 or
      occasionally 2 per stem; plants montane, mostly in moist meadows or along
      streams .... *A. geyeri*
6   Tepals white, seldom pink; scapes 6-20 (30) cm tall; leaves 2 or occasionally
      3 per scape; plants of deserts and dry hills .... *A. textile*

Allium acuminatum Hooker — tapertip onion — Strawberry Valley and the
flanks and lower slopes of the Uinta Mtns. to the Yampa Plateau, and probably
elsewhere; common in sagebrush, mt. brush, and aspen communities; 7,200-
8,200 ft; May-June. Flowers et al. (1960) reported *A. biceptrum* for an area now
covered by water of Flaming Gorge Reservoir. However, this species is listed
from west of our area (AUF 919). The Flowers (1960) report is likely based on *A.
acuminatum*.

Allium brandegei S. Watson — Brandegee onion — Strawberry Valley, W.
Tavaputs Plateau (Avintaquin drainage and w.), and Uinta Mtns. to Blue and
Douglas Mtns.; locally common in sagebrush, aspen, and open coniferous
forest communities; 7,000-10,600 ft; May-June depending on elevation. In
addition to features of the key, the following features contrast with those of *A.
brandegei* and *A. nevadense*.

  1   Tepals 5-8 mm long at anthesis, white; inner bulb coats white, dark red, or
        purplish; outer bulb coats not contorted reticulate; ovary crestless; bulbs
        mostly 1-4 cm below the collar of the leaves .... *A. brandegei*
  1   Tepals 8-12 mm long at anthesis, pink or white; inner bulb coats white;
        outer bulb coats contorted reticulate; ovary strongly crested; bulbs mostly
        4-8 cm below the collar of the solitary leaf .... *A. nevadense*

Allium brevistylum S. Watson — shortstyle onion — Across the Uinta Mtns.;
occasional along streams and other moist places in aspen and coniferous
woods, occasionally alpine; 7,500-11,000 ft; June-Aug. Sometimes confused
with *A. geyeri*, but different by features of the key and with tepals 10-14 mm
long and leaves 2-8 mm wide compared to tepals 7-10 mm long and leaves 1-4
mm wide in *A. geyeri*.

Allium cernuum Roth. — nodding onion — Uinta Mtns. and Tavaputs Plateau;
occasional and locally common in sagebrush, dry aspen, ponderosa pine,
lodgepole pine, and meadow communities; 7,400-8,800 ft; July-Aug.

Allium geyeri S. Watson (*A. rubrum* Osterhout) — Geyer onion — Uinta Mtns.
including Cold Spring Mtn., no specimens seen from Rio Blanco Co.; occasional
in ephemerally and permanently wet meadows and riparian, sagebrush, aspen,
and ponderosa pine communities; 7,500-9,000 ft; May-June. Plants with all
normal flowers belong to var. *geyeri*. Plants with some flowers replaced by
bulblets belong to var. *tenerum* M. E. Jones for which our one specimen is from
a wet meadow in the Bear River drainage at 8,720 ft.

Allium nevadense S. Watson — Nevada onion — Widespread south of the Uinta
Mtns. but no specimens seen from Rio Blanco Co.; infrequent in desert shrub,
sagebrush, and juniper communities; 5,000-6,000 (7,400) ft; May-June.

Allium textile Nelson & Macbride — textile onion — Widespread; common in
desert shrub, sagebrush, and juniper communities, on heavy clay to sandy
soils; 4,700-6,500 (7,500) ft; May-June.

Androstephium Torrey — Funnel-lily

Androstephium breviflorum S. Watson (*Brodiaea paysonii* A. Nelson) — Purple
funnel-lily — Scapose herbs from ovoid, fibrous-coated corms; scapes 10-30
cm tall; leaves several, as long or longer than the scape; flowers 3-12 in
umbels; tepals united below, 15-20 mm long, light purple or whitish; stamens
6, adnate to the tepals, the filaments united. Widespread but no specimens
seen from Rio Blanco Co.; occasional in desert shrub communities and the
lower fringe of the juniper belt, often on clayey soils; up to about 5,700 ft;
April-early June.

Asparagus L. — Asparagus

Asparagus officinalis L. — asparagus — Herbs from branching rootstocks;
young stems simple, succulent, thick, edible; older stems much branched,
commonly 1-2 m tall; leaves scale-like, 0.5-2 cm long, with clusters of axillary,
needle-like branches 1-2 cm long; flowers solitary or in pairs in axils of leaves
and needle-like branches, bell shaped, borne on spreading or drooping, filiform
pedicels; tepals 3-7 mm long; fruit a red berry, 6-8 mm in diameter. Introduced
from Eurasia, cultivated and escaping; often along ditches, fences, roadsides,
and in pastures and riparian communities from 4,300-6,200 ft; July-Aug.

Calochortus Pursh — Mariposa Lily; Sego Lily

Glabrous herbs from deep-seated bulbs; leaves few, linear, reduced upward on
the stem; flowers large and showy; tepals in 2 distinct sets, the inner set
(petals) white, pink, lavender, or bluish, larger than the outer greenish set
(sepals), with densely bearded conspicuous glands on the lower 1/2; fruit a 3-
angled, elongate capsule.

---

1 Glands of petals surrounded by forked hairs; petals white or pale blue, with
   a purplish band above the gland and purple blotch on the claw; anthers
   acute; plants known from the rim of the Tavaputs Plateau .. **C. gunnisonii**
1 Gland on petals surrounded by simple hairs; petals white to pinkish-
   lavender, with a reddish brown band above the gland; anthers rounded;
   plants widespread and common .... **C. nuttallii**

Calochortus gunnisonii S. Watson — Gunnison mariposa lily or sego lily —
Known from the Tavaputs Plateau from Argyle Canyon and Anthro Mtn.
eastward and mapped for Moffet Co. in TPD; sagebrush, mt. brush, and aspen
communities; 7,800-9,000 ft; (May) July-Aug. Our plants belong to var. *gunnisonii*.

Calochortus nuttallii Torrey & A. Gray in Beckwith — sego lily — Widespread;
occasional to locally abundant in desert shrub, sagebrush, and juniper communities, and infrequent in mt. brush, dry aspen, and ponderosa pine communities; up to 6,000 (7,500) ft; May-June. Plants with lavender or purple petals
have been referable to *C. ciscoensis* S. L. Welsh & N. D. Atwood — Cisco sego lily
— However, the wide distribution of plants with white and with lavender
petals along with mixed populations indicate a single species for the complex.
Sego lily is the state flower of Utah.

Camassia Lindley — Camas

Camassia quamash (Pursh) Greene — blue camas — Perennial herbs from deep-seated ovoid bulbs, 20-70 cm tall; leaves basal, 10-50 cm long, 1-2.5 cm wide; flowers usually numerous, in bracteate racemes 5-30 cm long; tepals deep blue or bluish-purple, rarely white, 1.5-4 cm long; fruit a ovoid capsule 1-2.5 cm long. Known from Strawberry Valley and Deep Creek, Wasatch Co.; sagebrush-grass and meadow communities; 7,000-7,880 ft; June-July. Our plants belong to var. *utahensis* (Gould) C. L. Hitchcock. The pasty texture and rather bland flavor of the root hardy seems worth going to war. However, due to the size of the bulb and abundance of the plant in some places this was a major food source for native Americans in the Pacific Northwest, and it was the root of conflict of at least one war.

Disporum Salisbury ex D. Don — Fairybells

Disporum trachycarpum (S. Watson) Bentham & Hooker (*Prosartes trachycarpa* S. Watson) — roughfruit fairybells — Herbs 30-60 cm tall; stems flexuous; leaves alternate, 3-10 cm long, 2-9 cm wide, ovate to ovate-oblong; flowers white, (8) 10-15 mm long; anthers white; fruit a berry. Uinta Mtns.; infrequent along streams and in woods; 7,500 ft to near timberline; June-Aug.

Erythronium L.— Dogtooth-violet

Erythronium grandiflorum Pursh [*E. parviflorum* (S. Watson) Gooding] — dog-tooth-violet — Perennial herbs from deep-seated, elongate corms; stems 10-30 (35) cm tall, with a pair of seemingly opposite, nearly basal, subsessile to petiolate leaves, the blades 10-20 cm long; flowers 2-4 (5), on nodding, naked peduncles; fruit a capsule. Strawberry Ridge e. to the Rock Creek drainage in the Uinta Mtns., and reported for Douglas Mt. (Bradley 1950); occasional to common in mtn. big sagebrush, mtn. brush, aspen and coniferous forests, and parklands within aspen and coniferous forests; 7,800-10,600 ft; May-June.

Fritillaria L. — Fritillary

Erect, glabrous, perennial herbs with simple stems from small bulbs; leaves linear, alternate or appearing opposite or whorled; inflorescence of 1-4 spreading or nodding flowers, 10-20 (25) mm long; fruit a membranous, 6-angled or winged capsule.

---

1   Flowers yellow or orange, fading red or purple; style 1; stigmas discoid or
      obscurely 3-lobed .... **F. pudica**

1   Flowers purplish or greenish-brown to chocolate-brown with yellow or pale
      spots; styles 3, united only at the base, the stigmas elongate ....
      .... **F. atropurpurea**

Fritillaria atropurpurea Nuttall — leopard or chocolate lily — Widespread; infrequent; seldom collected; many plant communities; 7,000-9,500 ft; May-June.

Fritillaria pudica (Pursh) Sprengel — Yellow fritillary, yellow bell — Parks & Tauzer 318 (CS) is from O-Wi-Yu-Kuts Flats 8,760 ft, Beetle & Porter 5546 (RM!) and T. Naumann 421 (DINO) are from the summit of Blue Mt., and Refsdal 330 is from Christmas Meadows Road on the n. slope of Uinta Mtns.; to be expected elsewhere including Strawberry Valley; May-June, usually flowering immediately after snow melt.

Hemerocallis L.

Hemerocallis fulva (L.) L. — daylily — This plant (introduced from Europe) is cultivated, persists, and escapes mainly along ditches. This is clearly distinct from our native lilies by the 5-15 large (5-15 cm long), dull orange or reddish funnel-shaped flowers with yellowish centers and long-exserted stamens and by the stout scape 0.5-1.2 m tall with several linear leaves clustered at the base that are about as long as the scape and about 1-3 cm wide.

Lloydia Salisbury ex Reichenbach

Lloydia serotina (L.) Salisbury ex Reichenbach — alpine lily — Slender, erect, perennial herbs from short rhizomes, these sometimes simulating bulbs; leaves 2-few, the basal blades 2.4-10 (20) cm long, 0.8-1 mm wide; tepals white with greenish or purplish veins; fruit a capsule, 6-8 mm long. Infrequent across the alpine of the Uinta Mtns. at 11,000-12,500 ft; June-Aug. Recorded in only 7 of the 308 Uinta Mtn. alpine plots included in Brown (2006) where this plant was found with reticulate willow, alpine avens, curly sedge, and *Bellardi kobresia*. Inconspicuous except when in flower, and perhaps more common than indicated above.

Smilacina Desfontaines — Solomon-plume; False Solomon Seal

Perennial herbs from creeping rhizomes, with simple stems; leaves alternate; flowers white; fruit a globose berry, greenish to reddish or purplish to black in age, 5-10 mm in diameter.

---

1 Flowers rather loosely arranged in a few-flowered raceme; tepals (3) 5-7 mm long; stamens about equal to the tepals; leaves 4-17 cm long, 1.5-5 cm wide; plants widespread across our area .... *S. stellata*

1 Flowers densely arranged in many-flowered panicles; tepals 1-2 mm long; stamens longer than the tepals; leaves 7-20 cm long, (3) 4-10 cm wide; plants apparently uncommon and restricted .... *S. racemosa*

Smilacina racemosa (L.) Desfontaines [*S. amplexicaulis* Nuttall; *Maianthemum racemosum* (L.) Link] — false Solomon-seal, western Solomon-seal — The 7 specimens seen are from the Strawberry and Currant Creek drainage and east in the Uinta Mtns. to Whiterocks Canyon; mt. brush-tall forb, aspen, Douglas-fir, and Engelmann spruce communities; 8,500-10,500 ft; June-July.

Smilacina stellata (L.) Desfontaines [*Maianthemum stellatum* (L.) Link] — starry or stellate smilacina — Widespread and common; many plant communities, but mostly along streams and at the edge of lakes and ponds, often in shade of woods or rocks; 4,700-10,500 ft; May-June.

Streptopus Michaux — Twisted Stalk

Streptopus amplexifolius (L.) de Candolle — twisted stalk — Perennial herbs,

(30) 50-80 (120) cm tall; leaves lanceolate to ovate, 5-12 (15) cm long, 2.5-5 cm wide, the base cordate clasping; flowers white with a greenish or yellowish tinge, 9-15 mm long; fruit a berry. Uinta Mtns. and Strawberry Valley; infrequent or occasional along streams, usually in aspen and coniferous woods; 7,300-10,600 ft; June-Aug. Our plants belong to var. *chalazatus* Fassett.

Triteleia Douglas ex Lindley

Triteleia grandiflora Lindley (*Brodiaea douglasii* Ways.; *B. grandiflora* Macbride) — wild hyacinth — Perennial scapose glabrous herbs, 20-70 cm tall, from globose corms 1-2.5 cm long and 1-3 cm thick; leaves basal, linear, 25-50 cm long, 3-10 mm wide, persistent; flowers in simple umbels; pedicels 1-5 cm long; corolla united about 1/2 the length, about 2-3 cm long; capsules 6-12 mm long. Strawberry Valley and one specimen without collector (UT 93874) from n. of Roosevelt; infrequent in sagebrush-grass and meadow communities; June-July. Our plants belong to var. grandiflora. Apparently mistaken as *Dichelostemma pulchellum* (Salisbury) A. A. Heller (*Brodiaea capitata* Bentham) by Graham (1937).

Veratrum L. — False-Hellebore; Skunk Cabbage; Cow Lily

Veratrum californicum Durand — California false-hellebore — Perennial herbs from thick rhizomes; stems often over 2 cm thick, (50) 100-200 cm tall, mostly glabrous and simple below, becoming tomentose and branched above; leaves several, strongly nerved, 25-40 cm long, 10-20 cm wide, the upper ones smaller; inflorescence of many-flowered panicles; tepals 8-17 mm long, whitish or greenish, with greenish Y-shaped glands; fruit a 3-lobed ovoid capsule Locally abundant around Strawberry Valley and western Uinta Mountains but rare or lacking east of the Yellowstone drainage; in seeps, around springs and ponds, along streams, tall forb, aspen, and spruce-fir communities; 7,500-11,000 ft; July-Aug.

Zigadenus Michaux — Death Camas

Perennial, simple-stemmed, glabrous herbs from deep-seated bulbs; leaves grass-like, sheathing, mostly basal, the cauline ones much reduced; flowers small, greenish to yellow-white, or white, unisexual or bisexual, the tepals with a greenish or yellowish gland near the base; stamens with flattened filaments; fruit a 3-lobed capsule.

1 Tepals (6) 7-11 mm long with obcordate glands 2
1 Tepals 3-7 mm long, the glands entire 3
2 Flowers cream to greenish, racemose or less commonly paniculate; plants of mountains . . . . *Z. elegans*
2 Flowers white, paniculate; plants of canyons of the Green and Yampa Rivers . . . . *Z. vaginatus*
3 Inflorescence mostly racemose; plants of moist or wet places *Z. venenosus*
3 Inflorescence paniculate at least in part; plants mostly of well drained soil . . . . *Z. paniculatus*

Zigadenus elegans Pursh [*Anticlea elegans* (Pursh) Rydberg] — mtn. death camas — Common in the Uinta Mtns. and infrequent on the W. Tavaputs Plateau; wet and dry meadows, along streams, aspen and coniferous forest communities, rocky slopes and alpine tundra at 5,600-8,500 ft; July-Aug.

Zigadenus paniculatus S. Watson [*Toxicoscordion paniculatum* (S. Watson) Rydberg] — foothill death camas — Widespread; mostly in and pinyon-juniper, mtn. brush, and mtn. big sagebrush communities; about 5,000-8,500 ft; May-June.

Zigadenus vaginatus (Rydberg) Macbride (*Anticlea vaginata* Rydberg) — alcove death camas — Hanging gardens and other shady places in the canyons of the Green and Yampa Rivers (Palmquist 2011).

Zigadenus venenosus S. Watson [*Toxicoscordion venenosum* (S. Watson) Rydberg] — meadow death camas — Widespread; locally common in ephemerally wet or moist meadows and edges of permanently wet meadows at 7,560-8,000 ft; June-July. Inflorescence usually dense and short (mostly less than 10 cm long) compared to *Z. paniculatus* with panicles commonly10-30 cm long.

LIMNANTHACEAE False Mermaid Family

Floerkea Willdenow — False Mermaid

Floerkea proserpinacoides Willdenow — false mermaid — Annual, glabrous herbs, 4-30 cm tall; leaves 1-6 cm long, alternate, pinnately compound, with 3-5 leaflets, the leaflets 5-12 mm long and entire or 2-cleft; flowers regular, bisexual, solitary, axillary, on peduncles 1-3 cm long; sepals 2-3 mm long or longer in fruit; petals separate, only about 1/2 as long as the sepals, white; stamens 6; ovary superior; fruit of 2 or 3 obovoid-globose, drupe-like nutlets, about 2.5 mm long, tuberculate. The 3 specimens seen are from Currant Creek and Strawberry Valley; wet meadows and aspen communities; 7,500-8,000 ft; June.

LINACEAE Flax Family

Linum L. — Flax

Plants herbaceous, glabrous or puberulent, often glaucous, annual or perennial; leaves alternate, simple, entire, sessile; flowers bisexual, regular, in racemes or cymose-panicles; sepals 5, imbricate, persistent; petals 5, separate, quickly deciduous; stamens 5, alternate with the petals, the filaments united at the base; ovary superior, of 5 united carpels; styles 5, distinct or united below; capsule 10-chambered (each of the 5 carpels with a false septum), each of the chambers with 1 seed.

---

1 Petals blue; stigmas elongate or at least longer than wide 2
1 Petals yellow or orange; stigmas capitate 3
2 Flowers homostylic (anthers and styles of about equal length) .. **L. lewisii**
2 Flowers heterostyilc (anthers longer than the styles) .... **L. perenne**
3 Styles free throughout; sepals not exceeding the capsule; petals 6-10 mm long; plants mostly above 8,000 ft .... **L. kingii**
3 Styles mostly united; sepals usually longer than the capsule; petals various but often over 10 mm long; plants of lower elevations 4

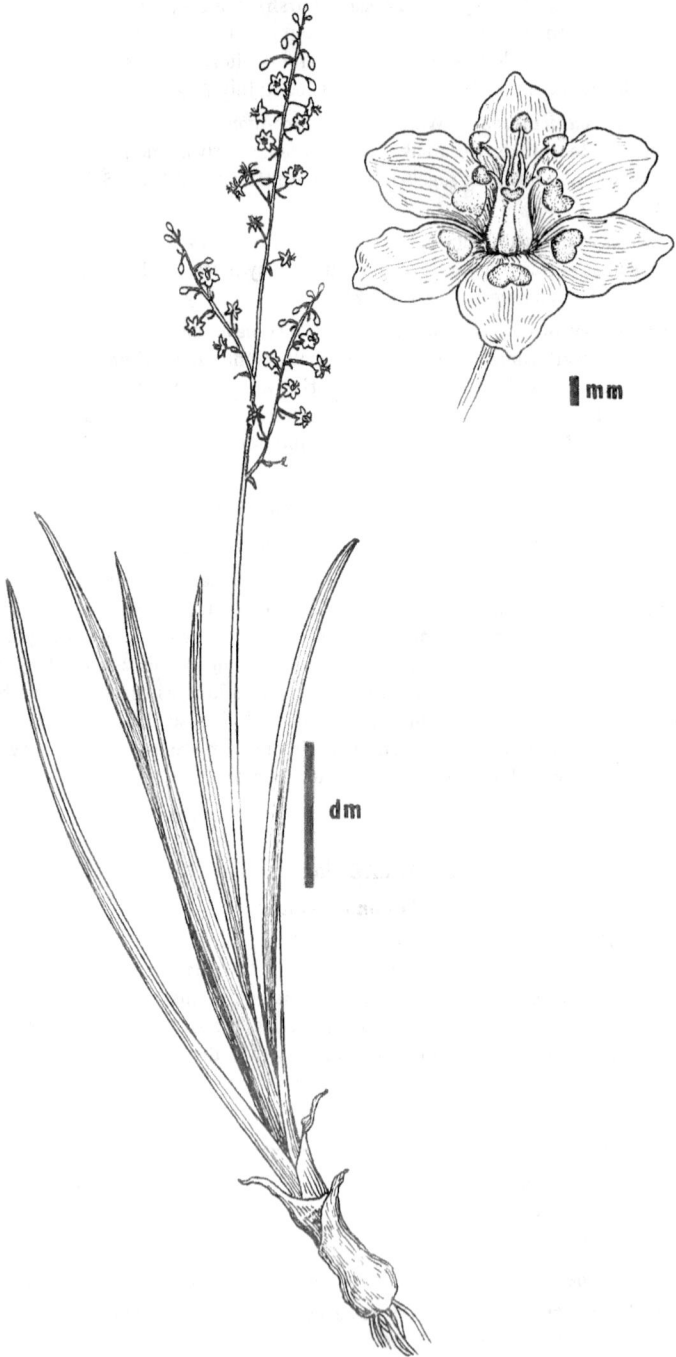

Zigadenus vaginatus (Rydberg) Macbride — alcove death camas

4   Stems and pedicels densely puberulent; plants annual . . . . . *L. puberulum*
4   Stems and pedicels glabrous or nearly so; plants short-lived perennial . . .
  . . . . *L. subteres*

Linum kingii S. Watson — King yellow flax — Specimens seen are from the Uinta Mtns. and Tavaputs Plateau; on Mississippian limestone, Green River Formation, and other basic substrates (Huber 1995) in mtn. big sagebrush, mt. brush, Salina wildrye, Douglas-fir, and spruce communities, and open gravelly slopes; 7,900-10,800 ft; June-Aug.

Linum lewisii Pursh — wild blue flax — Widespread and common; many plant communities, including black sagebrush, mt. brush, Salina wildrye, meadow, aspen, and Douglas-fir; 5,700-9,000 ft; May-July. Our plants belong to var. *lewisii* (Pursh) Eaton & Wright.

Linum perenne L. — blue flax — Introduced from Eurasia, and widely used in seedings of power and pipeline corridors, roadsides, and in burned area reclamation. Compared to wild blue flax the flowers are consistently darker blue.

Linum puberulum (Engelmann) A. Heller — puberulent yellow flax — The 1 record seen (Neese & Chatterley 9921) is from near Starvation Reservoir, Duchesne Co.; roadside at 5,800 ft. This species is more common to the s. of the Uinta Basin.

Linum subteres (Trelease) Winkler (*L. aristatum* Engelmann misapplied; *L. rigidum* Pursh misapplied) — Utah yellow flax — The 8 specimens seen are from the bottom of the Uinta Basin and lower elevations of the Tavaputs Plateau from Gate Canyon e. to Hill Creek; desert shrub, and pinyon-juniper communities; 4,980-6,800 ft; June-July.

LOASACEAE Stickleaf Family

Mentzelia L. — Blazingstar; Mentzelia; Stickleaf

Annual or perennial herbs; stems often glossy-white, the epidermis exfoliating; leaves alternate, simple, but sometimes deeply pinnatifid, covered with minute many-barbed hairs; flowers terminal or axillary, bracteate, bisexual, regular; calyx tube adnate to the ovary, the 5 narrow lobes free; petals 5 or appearing 10 with 5 of the outer filaments expanded and petaloid, free, cream or pale to bright yellow; stamens 10-many, the filaments free or fused into groups; ovary inferior; style 1; fruit a capsule.

---

1   Petals 2-10 mm long, 5 in number; filaments of stamens all filiform; capsules about 2-3 mm thick, 5-30 mm long; seeds angular, not flattened, not winged; plants annuals, mostly with slender stems 2
1   Petals 7-80 mm long, 5 or 10; capsules 5-10 mm thick or thicker; seeds flattened, often winged; plants annual, biennial, or short-lived perennial .
  4
2   At least the lower leaves pinnatifid, basal rosette present . . . **M. albicaulis**
2   Leaves entire or sometimes toothed; basal rosette lacking or poorly developed 3
3   Seeds grooved on the 3 angles, the faces smooth or minutely muricate, not obviously tuberculate, arranged in a single row the full length of the

capsules; plants entering our area on the Tavaputs Plateau ... *M. dispersa*

3 Seeds not especially grooved on the angles, tuberculate on the faces, irregularly arranged and not in a single row in the upper part of the capsule; plants of semi-barrens of Mancos Shale in Uintah Co. ..... *M. thompsonii*

4 Petals (2.5) 3-8 cm long; capsules 3-5 cm long, about 7-13 mm thick; plants robust, 30-100 cm tall, with thick stems; leaves to 15 cm long ....
....*M. laevicaulis*

4 Petals 0.7-2 cm long; capsules, plants and leaves various but often smaller than above 5

5 Plants perennial, with few or many stems, from a branching caudex; stems usually much branched; leaves mostly all pinnatifid into linear segments or the upper ones reduced to a simple, linear rachis................... 6

5 Plants biennial or short-lived perennial, from a simple taproot; stems often solitary, erect, nearly simple below or somewhat branched; petals mostly over 2 times as long as wide; seeds usually winged 7

6 Calyx segments (7) 9-13 mm long; petals 5, 12-20 mm long, with 5 petaloid stamens about equal to the petals, 12-20 mm long .... *M. goodrichii*

6 Calyx segments 5-9 mm long; petals 5, 7-16 mm long; petaloid stamens about ½ as long as the petals .... *M. multicaulis*

7 Leaves all pinnatifid; petals 7-12 mm long, about 2-3 mm wide 8

7 Leaves entire to lobed not pinnatifid; petals mostly either longer or wider than above 9

8 Petals acute apically, seeds narrowly winged, plants of Daggett Co. ......
....*M. pumila*

8 Petals rounded or obtuse apically, seeds evidently winged; plants of the Tavaputs Plateau .... *M. multiflora*

9 Capsules 15-30 mm long; calyx tube 10-18 mm long; petals 16-20 mm long, pale straw-yellow; leaves 5-8 cm long; plants 30-80 cm tall ..... *M. rusbyi*

9 Capsules 8-15 mm long; calyx tube 6-10 mm long; petals 7-20 mm long, golden-yellow; leaves 2-5 6 cm long; plants 10-40 cm tall............. 10

10 Capsules turbinate, 5-13 mm wide when pressed, less than 3 times as long as wide; petals 10-20 mm long; plants not known from Daggett Co. ......
....*M. pterosperma*

10 Capsules cylindrical, about 5 mm wide, 3-4 times longer than wide; petals to 10 mm long; plants known only from Daggett Co. .... *M. pumila*

Mentzelia albicaulis Douglas in Hooker [*M. montana* (Davidson) in Davidson & Moxley; *Acrolasia albicaulis* (Douglas) Rydberg]. — white-stem blazingstar, small-flowered blazingstar — Widespread; occasional and locally common to abundant; desert shrub, sagebrush, and pinyon-juniper communities; mostly below 6,500 ft; May-early June.

Mentzelia dispersa S. Watson [*Acrolasia dispersa* (S. Watson) Davidson] — Nevada blazingstar, entire-leaved mentzelia — The 2 specimens seen are from the W. Tavaputs Plateau, Carbon Co., listing for near the Quarry, Dinosaur National Monument (Welsh 1957) is likely based on specimens of *M. thompsonii*. Also listed for nw. Colorado (Harrington 1954).

Mentzelia goodrichii K. Thorne & S.L. Welsh — Tavaputs blazingstar — Known from eroding slopes of Green River and Uinta Formations of the West Tavaputs Plateau from the Avintaquin drainage to the Badland Cliffs at 6,440-8,800 ft. July-Aug.

Mentzelia laevicaulis (Douglas) Torrey & A. Gray [*Nuttallia laevicaulis* (Douglas) Greene] — big blazingstar, beautiful blazingstar — Specimens seen are from Strawberry Pinnacles, Duchesne Co. and Jesse Ewing Canyon, Daggett Co. where the plant is locally common on shaly slopes and roadcuts 6,000-6,800 ft; June-Sept.

Mentzelia multicaulis (Osterhout) Darlington [*M. humilis* (A. Gray) Darlington misapplied; *Nuttallia multicaulis* (Osterhout) Osterhout] — many-stem blazingstar — Scattered across the s. 3/4 of Uintah Co., extreme se. Duchesne Co. and w. Rio Blanco Co.; desert shrub, and pinyon-juniper communities, on nearly barren, eroding, calcareous substrates; 5,000-7,000 (9,000) ft; June-Sept. Plants of the Uinta Basin belong to var. *uintahensis* N. H. Holmgren & P. K. Holmgren.

Mentzelia multiflora (Nuttall) A. Gray. (*M. pumila* var. *lagarosa*; *Nuttallia multiflora* (Nuttall) Greene] — desert blazingstar — The 9 specimens seen are from the W. and E. Tavaputs Plateaus from the Avintaquin drainage east to Douglas Pass; sagebrush, juniper, pinyon-juniper, and oak communities, usually on light colored, marly or gravelly soils; 6,560-7,600 ft. June-July (Sept.).

Mentzelia pterosperma Eastwood [*M. integra* (M. E. Jones) Tidestrom; *Nuttallia pterosperma* (Eastwood) Greene] — wing-seed blazingstar, wing-seed stickleaf — The 5 specimens seen are from the E. Tavaputs Plateau (Willow Creek), near Jensen, near Ouray, and Diamond Mt. rim, no specimens seen from the Colorado part of our area; 4,900-7,300 ft; May-June.

Mentzelia pumila (Nuttall) Torrey & A. Gray — golden blazingstar, Wyoming stickleaf — The few specimens seen are from the vicinity of lower Sheep Creek, Daggett Co.; silty clay of Woodside Shale and Park City Formations; 6,300-6,400 ft; June-July.

Mentzelia rusbyi Wooten [*M. nuda* var. *rusbyi* (Wooten) Harrington; *Nuttallia rusbyi* (Wooton) Rydberg] — creamy blazingstar — Scattered on the Tavaputs Plateau, 3 specimens seen from near Hanna, and one from a roadcut in Dry Fork Canyon, Uinta Mtns.; sagebrush and mt. brush communities, usually on raw shaly slopes or areas of disturbance; 6,000-8,100 ft; July-Aug. Graham 9775 (CM!) listed by Graham (1937) as *M. chrysantha* Engelmann belongs here.

Mentzelia thompsonii Glad [*Acrolasia thompsonii* (Glad) W. A. Weber] — Mancos Shale stickleaf — Specimens seen are from Red Fleet Dam s. to Kennedy Wash and e. to Dinosaur National Monument and 8 mi nw. of Rangely; desert shrub communities, Mancos Shale, Duchesne River Formation, and perhaps other formations that weather to badlands; 4,880-6,200 ft; May.

LYTHRACEAE Loosestrife Family

Annual glabrous herbs; leaves simple, opposite, entire; flowers bisexual, regular, sessile or subsessile in the axils of leaves or bracts; calyx united, with 4-7 lobes; petals free, 4, small, often quickly deciduous; stamens 4; ovary superior; style 1; fruit a capsule; placentation axile.

1 Petals 7-12 mm long, pink-purple . . . . *Lythrum*
1 Petals 1-3 mm long, pink or white 2
2 Style usually exserted from the calyx, at least 1 mm long; leaves sessile, cordate-clasping; flowers 1-5 per axil; capsules about 4 mm long; petals purplish, about 2-3 mm long . . . . *Ammannia*
2 Style not exserted from the calyx, less than 1 mm long; leaves tapered at the base not cordate-clasping; flowers usually only 1 per axil; capsules about 3 mm long; petals white, about 1 mm long . . . . *Rotala*

Ammannia L. — Ammannia

Ammannia robusta Heer & Regel (*A. coccinea* Rottboell misapplied) — purple ammannia — Stems 4-angled, 4-50 cm tall, erect; leaves linear or linear-lanceolate, 2-4 (7) cm long, 2-8 mm wide, sessile, cordate-clasping at base; flowers 1-5 per axil; sepals 4, 3-4 mm long; petals about 1-3 mm long, readily deciduous; capsule nearly globose, about 4 mm long; style persistent; stigma capitate. The few specimens seen are from the margins of Pelican Lake and the floodplain of the Green River near Ouray; drying mud; 4,650 ft.

Lythrum L. — Loosestrife

Lythrum salicaria L. — purple loosestrife — Plants herbaceous, colonial, 50-150 (200) cm tall from long rhizomes; leaves opposite or a few lower nodes with 3 leaves, sessile, entire, 3-10 cm long, lanceolate or nearly linear or some elliptic-oblong; inflorescence showy, spike-like with numerous flowers; petals rose-purple; stamens mostly 12, attached well down in the floral tube. Cultivated ornamental with showy inflorescences, introduced from Eurasia, escaping and persisting in moist and wet sites, forming dense stands and capable of displacing native vegetation, listed as a noxious weed in Utah and in many other states of the U. S., found in most Canadian provinces and nearly all states of the U. S. except Arizona, Louisiana, Florida, Georgia, and S. Carolina (TPD). The one specimen seen from our area (J. Spencer 1643) is from 5.7 miles s. of Roosevelt from along an irrigation canal where purple loosestrife is well established. Some small infestations at other locations are being treated or have been controlled. This attractive plant is well established along the Wasatch Front and in Emery Co, Utah, and additional introductions to our area can be expected.

Rotala L.

Rotala ramosior (L.) Koehne in Mart. — lowland rotala — Stems angled, 4-15 cm long; leaves 1.5-3 cm long, the blades lanceolate to oblanceolate, usually attenuate or short-petiolate; flowers 1 (2-3) per axil, with an accrescent, campanulate hypanthium 3-4 mm long in fruit; stamens and style included in the hypanthium; capsule globose. No specimens seen, but to be expected on the floodplain of the Green River.

MALVACEAE Mallow Family

Plants herbaceous, usually pubescent with branched or stellate hairs; leaves alternate, simple, palmately veined, often palmately lobed; inflorescence various; flowers bisexual, rarely unisexual, regular, sometimes with an involucel of sepal-like bractlets; sepals 5, separate; petals 5, separate, united to a staminal sheath; stamens numerous, united by their filaments; ovary superior; fruit a schizocarp, with several carpels. The introduced hollyhock (*Althaea rosea* (L.) Cavanilles) is cultivated but apparently does not persist. This is different from native members of the family in our area with flowers 6-10 cm broad and plant height of up to 2 m.

1 Petals orange .... *Sphaeralcea*
1 Petals not orange..2
2 Petals white or cream with dark purple base; calyx inflated ..... *Hibiscus*
2 Petals not with dark purple base...................................3
3 Petals yellow; plants 50-150 cm tall; leaf blades 3-25 cm long .... *Abutilon*
3 Petals not yellow or plants less than 50 cm tall 4
4 At least the upper leaf blades cleft over 1/2 the way to the midrib; sepals not subtended by sepal-like bractlets .... *Sidalcea*
4 Leaf blades not lobed more than 1/2 the way to the midrib; sepals subtended by an involucel or sepal-like bractlets 5
5 Petals 20-37 mm long; plants montane, perennial, erect .... *Iliamna*
5 Petals not over 15 mm long; plants seldom montane; stems decumbent to prostrate 6
6 Plants perennial from elongate rhizomes; fruit with 5-10 carpels *Malvella*
6 Plants annual from a taproot; fruit with 10-15 carpels .... *Malva*

Abutilon Miller

Abutilon theophrasti Medikus — velvet-leaf — Annual with stems 50-150 cm tall, velvety with short, soft hairs; leaves long-petiolate, the blades 3-25 cm long and nearly as wide, cordate at the base; calyx lobes broadly ovate-acuminate; petals yellow, 6-12 mm long; carpels of fruit 10 or more, each with a bent awn. This weedy annual was recently collected in Vernal. The plant was eradicated, but new infestations can be expected. Velvet-leaf is mapped for all states of the U. S. and lower provinces of Canada in TPD. This indicates possibility for additional introductions from many potential sources.

Hibiscus L.

Hibiscus trionum L. — flower-of-an-hour — Annual herb with slender taproot, 15-50 cm tall; stems coarsely hispid to glabrate; leaves 3-lobed or more commonly 3-5 parted; flowers axillary; bractlets usually 10, linear; calyx inflated at maturity, strongly ribbed; corolla 3-6 cm wide, the petals white or cream with dark purple base; fruit a capsule with 5 carpels. Known from a ditch and oat field in Jensen at 4,745 ft elevation (Goodrich 27685). This weedy species was introduced from central Africa.

Iliamna Greene

Iliamna rivularis (Douglas) Greene [*Sphaeralcea rivularis* (Douglas) Torrey] — wild hollyhock — Perennial herbs, sparingly and minutely stellate-hairy; leaf blades cordate to truncate at the base, 3-7 lobed, 2-16 cm wide, the lobes crenate-serrate; carpels 6-10 mm long at maturity, hispid and stellate. widespread; infrequent or occasional following fire in sagebrush, mt. brush, ponderosa pine, and Douglas-fir communities and rarely in meadows; 7,000-9,000 ft; June-Oct.

Malva L

Malva neglecta Wallroth — cheese, cheese-weed, dwarf mallow — Annual or biennial herbs, more or less sparsely stellate; stems prostrate-spreading, 10-60 cm long; leaf blades 2-6 cm wide, rounded-reniform, crenate, often with 5-9 shallow lobes; calyx 3-6 mm long; petals about 6-12 mm long; schizocarp with 10-15 rounded carpels. Introduced from Europe, somewhat weedy on disturbed ground, to be expected across the lower elevations of our area; May-Sept.

Malvella Jaubert & Spach

Malvella leprosa (Ortega) Krapovickas [*Disella hederacea* (Douglas) Greene; Sida hederacea (Douglas) Torrey] — alkali-mallow — Perennial herbs from rhizomes, grayish stellate; stems 10-40 cm long; leaf blades reniform to orbicular, dentate, crenate or sinuate, 3-25 (30) mm long; calyx 5-7 mm long; petals 10-12 mm long; schizocarp with 5-10, 1-seeded, reticulate carpels. The few specimens seen are from the floodplain of the Green River at Stewart Lake and near Ouray and around Pelican Lake where the plant is locally common to abundant, and reported by Flowers and others (1960) for Lucerne Valley, Daggett Co. 5,900 ft (area now inundated by Flaming Gorge Reservoir); July-Oct.

Sidalcea A. Gray — Checker Mallow

Erect, perennial herbs from taproots or short rhizomes, usually stellate; leaves alternate; lowermost leaves sometimes merely lobed, the upper ones commonly cut over 1/2 the way to the midrib; flowers in spicate racemes; calyx 5-cleft; carpels 5-10, 1-seeded.

---

1 Petals white or merely pinkish tinged, often drying yellow; anthers bluish-pink or purple .... *S. candida*
1 Petals pink or lavender; anthers usually yellow or white.............. 2
2 Calyx, pedicels and stems with simple hairs .... *S. neomexicana*
2 Calyx, pedicels, and stems with stellate hairs .... *S. oregana*

Sidalcea candida A. Gray — white checker mallow — The 5 specimens seen are from Strawberry Valley e. to Rock Creek and disjunct in Greens Draw in the Uinta Mtns. and the E. Tavaputs Plateau; streamside and meadow communities; 6,900-8,050 ft; July-Sept.

Sidalcea neomexicana A. Gray — New Mexico checker mallow — Flood plain head of Piceance Cr (Goodrich 23154). Harrington 1566 (CS) was reported by Bradley (1950) for edge of trees along a creek, Browns Park, 5,000 ft. To be expected in marshy, alkaline meadows (Weber 1987).

Sidalcea oregana (Nuttall) A. Gray — Oregon checker mallow. Specimens seen are all from Strawberry Valley; forb-grass and aspen-tall forb communities;

8,000-9,000 ft; July-Aug. This and *S. neomexicana* are noted to intergrade (AUF 505).

Sphaeralcea J. St. Hilaire — Globemallow

Plants perennial, glabrescent to canescent with stellate hairs; leaves toothed to palmately lobed or dissected; flowers in racemose to thyrsoid cymes, orange; schizocarp with 8-20 carpels each with 1 or 2 seeds, the carpels united at the reticulate, indehiscent base and free at the dehiscent apical portion. This is a confusing genus in which boundaries between species are often vague.

1   Sepals not subtended by bractlets or rarely so; leaves dissected, the major divisions cut to the midrib; inflorescence racemose, rarely with more than 1 flower per node . . . . **S. coccinea**

1   Sepals subtended by bractlets; leaves divided, lobed or merely crenate; inflorescence often with more than 1 flower per node 2

2   Leaves divided to or nearly to the base . . . . **S. grossulariifolia**

2   Leaves rarely divided more than 1/2 the way to the base 3

3   Foliage green; plants sparingly pubescent; carpels nearly orbicular, obtuse or slightly mucronate at apex . . . . **S. munroana**

3   Foliage gray-white; plants densely pubescent; carpels broadly ovate, usually acute or cuspidate at the apex . . . . **S. parvifolia**

Sphaeralcea coccinea (Nuttall) Rydberg — scarlet globemallow — Widespread and common; desert shrub, sagebrush, pinyon-juniper, and mt. brush communities; 4,700-8,300 ft; mid May-June and through Sept. with summer and early fall precipitation. Sprouting and flowering in abundance shortly after fire. Most of our material is referable to ssp. *dissecta* (Nuttall) Kearney, but a few specimens from 7,000-7,200 (8,200) ft in Strawberry Valley have broad leaf divisions that suggest introgression with *S. munroana*. These plants might suggest ssp. *elata* (Baker) Kearney. However, plants of that taxon are not known from the Intermountain area (IMF 2B: 36).

Sphaeralcea grossulariifolia (Hooker & Arnot) Rydberg — gooseberry-leaf globemallow — Three specimens seen from desert shrub communities at 4,700-5400 ft; May-June. Of the 3 specimens seen, 1 from Pariette Draw (Neese 4449) has but 1 flower at most nodes and some are ebracteate and this seems allied to *S. coccinea*. Another specimen from the junction of Ouray Road and Highway 40 (England 1937) is more typical. The third specimen (Goodrich 27843) from near Lapoint was taken where typical plants of *S. coccinea* and *S. parviflora* were growing together on disturbed ground. Only a few plants with features of *S. grossulariifolia* were found among many of *S. coccinea,* and the specimen from this location is highly suspect of being a hybrid of *S. coccinea* and *S. parviflora*. *S. grossulariifolia* is abundant in the Great Basin and to the north and west. The rarity of random plants that appear to be this species in the Uinta Basin seems to indicate that plants of the Uinta Basin with features of *S. grossulariifolia* are hybrids of *S. coccinea* and *S. parviflora.*

Sphaeralcea munroana (Douglas) Spach — Munro globemallow — Occasional to locally abundant; Strawberry Valley and Blind Stream-Log Hollow area of the Uinta Mtns.; mtn. big sagebrush communities and along roads; 7,300-8,000;

July-Aug. Specimens from the Blind Stream and Log Hollow drainages have lobed leaves and grade toward *S. coccinea*. Separation in this case seems to depend upon the presence or absence of bractlets subtending the sepals. Specimens from Strawberry drainage have simple crenate or undulate margined leaves as in *S. parvifolia*. This latter phase is much like *S. parvifolia* except for the less pubescent leaves. However, the 2 are considered to have different geographic ranges with *S. munroana* found from British Columbia to w. Montana and s. to California and Utah, and *S. parvifolia* found from Colorado to California and s. to New Mexico and Arizona. In our area, *S. munroana* is found at higher, more mesic areas than *S. parvifolia*.

Sphaeralcea parvifolia A. Nelson — small-leaf globemallow — Widespread; common to locally abundant in desert shrub and sagebrush communities and along roads; 4,800-5, 400 ft; May-Aug. See *S. munroana*. The report of *S. fendleri* A. Gray for Island Park along the Green River (Potter and others 1983) is based on a specimen belonging here. This species makes extensive flower displays in the deserts of the Uinta Basin in years of abundant spring precipitation.

MARSILEACEAE   Pepperwort Family
Marsilea L. — Pepperwort; Water-clover

Marsilea vestita Hooker & Greville — pepperwort, water clover — Aquatic, emergent (or stranded in drying pools), plants consisting of long-petiolate, 4-foliate leaves scattered along a slender, creeping, superficial rhizome (stem) and axillary sporocarps; petioles 2-15 cm long; leaflets approximate, simulating a 4-leaf clover, 5-17 long, 3-17 mm wide, obdeltoid or fan-shaped, floating or emergent, more or less pubescent or the emergent ones sometimes glabrous; sporocarp solitary on short peduncles arising from the rhizomes, elliptic to ovate, 4-8 mm long, densely pubescent, commonly with 2 caudate basal teeth. The 3 specimens seen are from mud flats and drying pools of the floodplain of the Green River near Ouray.

MENYANTHACEAE   Buckbean Family
Menyanthes L. — Buckbean; Bogbean

Menyanthes trifoliata L. — buckbean — Plants perennial, aquatic or growing in mud, from creeping rootstock; leaves alternate but mostly basal, the petioles 5-25 cm long, sheathing at base, the blades 3-foliate; leaflets 5-10 cm long, oval to elliptic, thick, glabrous, entire; flowers bisexual, racemose or paniculate; calyx deeply 5-parted, short; corolla united, short-funnel form, about 10-15 mm long, white or tinged with rose or purple, 5-cleft, the lobes covered on inner surface with long white hairs; stamens 5, alternate with the corolla lobes; ovary superior, 1-celled; style 1; stigma 2-lobed; fruit capsule-like but indehiscent or finally rupturing irregularly. Uinta Mtns.; infrequent but locally abundant in shallow pools, ponds, bogs, and lakes in the lodgepole pine-spruce belt; July-Aug. Often included in Gentianaceae.

MONOTROPACEAE   Indian-pipe Family
Pterospora Nuttall — Pinedrops

Pterospora andromedea Nuttall — woodland pinedrops — Saprophytic, glandular herbs, without chlorophyll and not green; stems simple, 30-100 cm tall, brownish red, fleshy at anthesis, fibrous in age and remaining standing for 1 or more years after dry; leaves reduced to narrow bracts; inflorescence a raceme, usually equal in length to the rest of the stem, loosely flowered; flowers 5-8 mm long, on pendulous or recurved pedicels 5-15 mm long, from the axils of linear bracts; sepals about 1/2 as long as the glabrous, pale yellow, globose-urceolate united corolla; fruit a depressed-globose capsule, 8-12 mm wide. Uinta Mtns.; occasional in coniferous woods, mostly in humus; June-Aug.

MONTIACEAE Montia Family

More or less succulent, usually glabrous annual or perennial herbs; leaves alternate, opposite, or basal, simple, entire; flowers bisexual, regular or nearly so; sepals mostly 2 (6-8 in Lewisia); petals separate, mostly 4-5, imbricate; stamens usually as many as the petals and opposite them; ovary superior or partly inferior, 1-celled, the styles 2-5 or sometimes more; fruit a loculicidal or circumscissile capsule; seeds 1-many, often smooth and shining.

1   Flowers in a head-like cluster, numerous; leaves basal or nearly so, forming rosettes; plants rare, known from 11,900 ft .... *Calyptridium*

1   Flowers not in a head-like cluster.................................. 2

2   Leaves all basal or all crowded at the base of the stem; stem scape-like, sometimes with opposite or whorled bracts less than 11 mm long....... 3

2   Leaves not all basal 4

3   Scape-like stems without bracts .... *Phemeranthus*

3   Scapes with a pair or a whorl of bracts .... *Lewisia*

4   Stem with one orbicular, perfoliate united leaf ...... *Claytonia perfoliata*

4   Stem leaves not as above 5

5   Stem with 1 pair of opposite or subopposite bracts born near or above the middle of the stem .... *Claytonia*

5   Stem not as above 6

6   Stem leaves 2-3 (5), semi whorled, near ground level, 1-2.5 mm wide .....
    .... *Lewisia triphylla*

6   Stem with 2 or more pairs of distinctly opposite leaves with some borne well above ground level, 2-18 mm wide .... *Montia*

Calyptridium Nuttall

Calyptridium umbellatum (Torrey) Greene [*Cistanthe umbellata* (Torrey) Hershkovitz; Spraguea umbellata Torrey] — Mt. Hood pussypaws — Plants annual or short lived perennial, 2-10 cm tall, glabrous; leaves 1-3 cm long; heads 2-4 cm in diameter; sepals 2, white to pink, enlarging and often purplish in age, 4-10 mm long; petals 4, much smaller than the sepals, quickly withered, stamens 3; styles 2-lobed; capsules fully dehiscent by 2 valves; seeds several. Occasional in talus-creep and pond margins above treeline in the Uinta Mtns. in Duchesne and Summit Cos. Our specimens are referable to var. *caudicifera* A. Gray.

Claytonia L. — Spring-beauty

Perennial glabrous plants; stems usually with 1 pair of opposite leaves; inflorescence of terminal racemes, 1-2 bracts at the base of the racemes; sepals 2, persistent; petals 5; stamens 5; styles 3; ovary superior; capsules dehiscent downward by 3 valves; seeds 1-6.

---

1   Stem with a perfoliate, orbicular leaf just below the inflorescence . . . .
    . . . . *C. perfoliata*
1   Stem without conspicuous leaves or with a pair of leaves that are conspicuously longer than wide 2
2   Plants from fusiform, fleshy taproots; basal leaves several, spatulate, or orbicular; inflorescence not exceeding the leaves . . . . *C. megarrhiza*
2   Plants from globose corms; basal leaves few or none, oblanceolate to ovate; inflorescence exceeding the leaves . . . . *C. lanceolata*

Claytonia lanceolata Pursh — lanceleaf spring-beauty, western spring beauty — Occasional to locally abundant in Strawberry Valley, Blue Mtn., Diamond Mtn. and Uinta Mountains w. of Rock Creek and e. of Dry Fork, no specimens seen from between these drainages, no specimens seen from the Tavaputs Plateau; sagebrush aspen and various coniferous forest communities, often at the edge of melting snow at 7,600-10,200 ft; May-July. Our plants belong to var. *lanceolata*.

Claytonia megarrhiza Pursh — alpine spring beauty — Isolated sites across the alpine of the Uinta Mtns. in late melting snowbeds. This plant appears to be able to persist in snowbeds where there is no release of snow cover in some years.

Claytonia perfoliata Donn ex Willdenow [*Montia perfoliata* (Donn) Howell] — miners lettuce — Three specimens seen from Strawberry Valley, one from the Whitney drainage on the north slope of the Uinta Mtns., and T. Naumann 414 (DINO) from Rippling Brook Campground on the Green River below Hells Half-mile.

Lewisia Pursh — Bitterroot, Lewisia

Perennial, somewhat fleshy, succulent, scapose plants from corms or thick fleshy roots and a short caudex; inflorescence scapose; scapes often with a pair or whorl of bracts; sepals, petals, stamens, and styles various in number; ovary superior; capsule circumscissile at the base then splitting upward.

---

1   Flowers (1) 2-10 (15) in a subumbellate inflorescence; stems delicate, tapering to and easily detached from a bulb-like, globose corm; petals 4-7 mm long; stamens (3) 5; styles 3-5 . . . . *L. triphylla*
1   Flowers solitary; stems from caudices or fleshy roots; petals 6-35 mm long; stamens 4-50; styles 3-8 2
2   Sepals 4-9, 10-25 mm long, petal-like; petals 12-18, 10-35 mm long; flowers usually exceeding the leaves; stems with a whorl of bracts; stamens 20-50; styles 4-8 . . . . *L. rediviva*

2 Sepals 2, 2-12 mm long; petals 5-9, 6-17 mm long; leaves usually exceeding the flowers; stems with a pair of bracts; stamens 4-14; styles 3-6 ....
.... *L. pygmaea*

Lewisia pygmaea (A. Gray) Robinson [*L. nevadensis* (A. Gray) B. L. Robinson; *Oreobroma nevadensis* (A. Gray) Howell; *O. pygmaea* (A. Gray) Howell] — least lewisii — Uinta Mtns., Blue Mt., and Strawberry Valley, no specimens seen from the Tavaputs Plateau except where adjacent to Strawberry Valley; occasional or locally abundant; many plant communities; 7,500 ft to above timberline; May-Aug, flowering with snow-melt. Plants of the Uinta Basin area belong to var. *pygmaea*.

Lewisia rediviva Pursh — bitterroot — Infrequent or occasional; specimens seen are all from Daggett and Moffat Cos., common on Cold Spring Mtn., Dinosaur National Monument at Zenobia Basin (B. Neely 4340 DINO) and Hackings Springs (W. Kelley 88-76 DINO); sagebrush and pinyon-juniper communities, rocky hills and ridges; 7,200-8,400 ft; June-July. Our plants belong to var. *rediviva*.

Lewisia triphylla (S. Watson) Rob. in A. Gray [*Erocallis triphylla* (S. Watson) Rydberg] — threeleaf lewisia — Known the head of the Provo and Duchesne drainages near Mirror Lake, and Silver Meadows, Uinta Mtns., at 9,000-10,600 ft; meadows, open coniferous forests and open rocky slopes, often flowering near melting snow; June-Aug.

Montia L.

Montia chamissoi (Ledebour. ex Sprengel) Greene [*Crunocallis chamissoi* (Ledebour) Cockerell] — water spring-beauty — Glabrous, somewhat succulent herbs; sepals 2, persistent; petals usually t (2-6), stamens 2-5, opposite the petals; styles deeply 3-cleft; capsules 3-valved; seeds 1-3 per capsule. The 3 specimens seen are from; Strawberry Valley and Whitney Reservoir area, silver sagebrush-grass, meadow, and aspen-fir parkland communities; 7,600-9,280 ft; June-July.

Phemeranthus Rafinesque

Phemeranthus confertiflorus (Greene) Hershkovitz [*Phemeranthus parviflorus* (Nuttall) Kiger; *Talinum parviflorum* Nuttall misapplied] — small-flowered flameflower — Perennial herbs mostly 5-19 cm tall with fleshy roots; leaves succulent, crowded at the base of the plant, linear or terete,15-50 mm long, 0.8-2.5 mm wide; inflorescence cymose, borne on slender scape-like peduncles that are much longer than the stems; pedicels 1-4 mm long; sepals 2.7-4 mm long; petals 5.5-7 mm long, pink or pink-purple; stamens 4-8; style longer than the stamens, stigma subcapitate; fruit a capsule 3.5-4.5 mm long. Known from Docs Valley to the north of Daniels Canyon, Blue Mtn. (Franklin 7413 BRY) and Dry Woman Plateau (D. J. Tilley sn DINO); sandstone slick rock; 6,690-7,380 ft.

NAJADACEAE Naiad Family

Najas L. — Naiad

Najas guadelupensis (Sprengel) Magnus — southern waternymph — Aquatic, submerged herbs with slender, branched stems; leaves simple, opposite or whorled, l2-25 mm long, 0.5-1 mm wide, usually inconspicuously serrate and tipped with 1 or 2 weak short spines; flowers unisexual, inconspicuous, the staminate ones consisting of l stamen and a spathe-like involucre, the pistillate ones consisting of a pistil with 2-4 stigmas; fruit a 1-seeded nutlet. Two specimens seen are from Big Park of the Ashley Creek drainage at about 9,800 ft and Tungsten Lake at 11,350 ft. To be expected in quiet water at other locations.

NYCTAGINACEAE Four-o'clock Family

Annual or perennial, mostly glabrous or glandular, often glaucous herbs; stems usually swollen at the nodes; leaves simple, usually opposite, entire; flowers bisexual, regular, subtended by bracts, these sometimes united into a calyx-like involucre; perianth of a single whorl, corolla-like, united, 3-5 lobed, the lower part of the tube closely investing the fruit; ovary superior but sometimes appearing inferior, 1-celled, with 1 ovule; fruit an achene, indehiscent, usually angled, ribbed, or winged.

1 Flowers solitary or 2-8 in a cluster; perianth rotate campanulate to campanulate-funnelform .... *Mirabilis*
1 Flowers borne in heads, many per head; perianth salverform........... 2
2 Plants annual; involucral bracts pointed; wings of the fruit 5-10 mm wide .... *Tripterocalyx*
2 Plants perennial; involucral bracts rounded; wings of the fruit (if present) 1-2 mm wide .... *Abronia*

Abronia Jussieu — Sand-verbena

Perennial herbs; leaves opposite, sometimes all basal or nearly so, 2-9 cm long, variable in shape but mostly ovate-oblong to orbicular, viscid-puberulent to glabrous; flowers in heads, the heads long-pedunculate; perianth funnelform to salverform, white to pinkish, the tube slender and elongate, the limb 5-lobed; stamens usually 5.

1 Stems absent or nearly so, the internodes very short and hidden by marcescent leaf bases, or internodes none; plants often caespitose; heads appearing scapose .... *A. nana*
1 Stems and internodes evident; plants tufted or open and spreading 2
2 Body of fruit glabrous, without wings; herbage glabrous; leaves orbicular to elliptic .... *A. glabrifolia*
2 Body of fruit villous-glandular, more or less winged; herbage glabrous or glandular; leaves mostly elliptic .... *A. fragrans*

Abronia glabrifolia Standley (*A. argillosa* S. L. Welsh & Goodrich) — clay-verbena — Specimens seen are from Grand Co. from south of our area. Perhaps to be excepted in the Uinta Basin.

Abronia elliptica A. Nelson (*A. fragrans* Nuttall ex Hooker misapplied) — sand-verbena — Widespread; occasional to common in desert shrub, sagebrush, pinyon-juniper, Salina wildrye, and ponderosa pine communities, on a great variety of sites including shaly slopes, clay flats, and sand dunes; 4,700-8,150 ft; May-mid June. Plants of *A. elliptica* are often (not always) rhizomatous.

Abronia nana S. Watson — dwarf sand-verbena — Reported for Duchesne Co. (IMF 2A: 597).

Mirabilis L. — Four-o'clock

Plants perennial; leaves opposite, often glaucous; flowers axillary or terminal, subtended by 5 separate involucral bracts or these united; perianth white, rose, or reddish-purple; stamens 3-5, the filaments united below.

---

1 Leaves linear to narrowly lanceolate, rarely to 1 cm wide, 3-10 cm long; involucral bracts united into a calyx-like cup, scarious, subtending 1-3 flowers; perianth about 1 cm long, the limb to about 1 cm wide **M. linearis**

1 Leaves ovate-oblong to suborbicular, mostly well over 1 cm wide; involucral bracts free or united at the base, greenish, leaf-like, subtending 4-8 flowers; perianth 2-5 cm long, the limb over 1 cm wide 2

2 Perianth 2-2.5 cm long; involucral bracts free or united at the base; plants 20-40 cm tall . . . . **M. alipes**

2 Perianth commonly 3-5 cm long; the involucre united for 1/2 or more of its length; plants 30-100 cm tall . . . . **M. multiflora**

Mirabilis alipes (S. Watson) Pilz (*Hermidium alipes* S. Watson; *M. alipes* var. *pallidum* C. L. Porter) — winged four-o'clock — Common in Uintah and Rio Blanco Cos., rare w. to Wells Draw in Duchesne Co., mostly s. of Hwy. 40 and s. to the flank of the Tavaputs Plateau; desert shrub and pinyon-juniper communities; 4,700-6,350 ft; April-May.

Mirabilis linearis (Pursh) Heimerl [*Allionia linearis* Pursh; *Oxybaphus diffusa* A. Heller; *O. linearis* (Pursh) Robinson] — narrowleaf umbrellawort — Widespread; infrequent or occasional in desert shrub, pinyon-juniper, and mt. brush communities; 4,800-7,400 ft; June-Sept. Plants of our area belong to var. linearis. The report of *Oxybaphus lanceolatus* (Rydberg) Standley (Potter and others 1983) is based on a fragmentary specimen (Y 186!) apparently belonging here.

Mirabilis multiflora (Torrey) A. Gray in Torrey — Colorado four-o'clock — The one specimen seen (England 2017) is from the head of Taylor Canyon, Bookcliff Divide, Grand Co.; road cut in Green River Formation; 8,200 ft; July. Apparently our plants belong to var. *glandulosa* (Standley) Macbride.

Tripterocalyx Hooker ex Standley

Annual, succulent, glabrous to viscid-pubescent herbs; leaves opposite; flowers in axillary, pedunculate, capitate clusters; perianth narrow-tubular with flaring limb; fruit usually with 3 showy wings.

---

1 Surface of the fruit body indurate (not easily scraped and broken), the 1-3 ribs prominent; perianth 10-29 mm long, the limb 6-19 mm wide; stems and peduncles glabrous, glabrate, or pubescent . . . . **T. carneus**

1   Surface of the fruit body papery-spongy (easily scraped or broken), the
    spongy tissue more or less obscuring the ribs; perianth 7-15 mm long, the
    limb 3-4 mm wide; stems and peduncles pubescent . . . . *T. micranthus*

Tripterocalyx carneus (Greene) L. A. Galloway [*T. pedunculatus* (M. E. Jones)
Standley] — winged sandpuff — Listed for Uintah Co. (IMF 2A: 602). Our plants
belong to var. *pedunculatus* (M. E. Jones) Spellenberg.

Tripterocalyx micranthus (Torrey) Hooker [*Abronia micrantha* Torrey] — small
flower sandpuff, annual sand-verbena — Widespread; infrequent in desert
shrub, sagebrush, and juniper communities, on various soils from sand dunes
to silty clay; 4,600-5,800 ft; May-June.

NYHPHAEACEAE Water-lily Family

Nuphar J. E. Smith — Cow lily; Yellow Pond-lily; Yellow Water-lily

Nuphar polysepala Engelmann [*N. lutea* ssp. *polysepala* (Engelmann) Beal;
Nymphaea polysepala (Engelmann) Greene; spelling of the specific name includes
polysepala and *polysepalum*] — yellow pondlily — Perennial aquatic herb from
thick creeping rhizomes; leaves all basal with long submerged terete petioles
and large cordate usually floating blades, these 10-45 cm long and about 3/4 as
broad; flowers solitary, emersed, borne on long submerged terete peduncles,
bisexual, regular; sepals usually 9 (7-12), 3-6 cm long, rounded to slightly
retuse, the outer ones leathery, greenish, and shorter than the bright yellow
to reddish tinged inner ones; petals lanceolate, thick, inconspicuous, scarcely
equaling the numerous reddish or purplish stamens; stigmas 2-2.5 cm wide;
fruit a many-seeded, several-loculed, leathery, ovoid to nearly cylindrical
capsule, 5-9 cm long, prominently ribbed. Uinta Mtns.; aquatic in ponds and
shallow lakes, mostly in water 1 to 5 ft deep; 9,000-11,000 ft; July-Aug.

OLEACEAE Olive Family

Fraxinus L. — Ash

Fraxinus anomala Torrey in S. Watson — singleleaf ash — Shrubs or small
trees, 2-8 m tall; twigs 4-angled; leaves 2-5 cm long, simple or sometimes 3-
foliate, ovate to rhombic-elliptic, cuneate to subcordate at the base, entire or
crenate; flowers polygamous, in panicles; calyx minute, 4-lobed or 4-toothed;
corolla none; stamens 1-3; fruit a 1-seeded, single samara, 12-25 mm long,
about 8-10 mm wide, winged all around. Rocky, sandy canyons along the
Green River in the Split Mtn.-Dinosaur National Monument area and in
Desolation Canyon; May-June. Other species of ash are cultivated in our area
for shade or ornament, but these all have pinnately compound leaves.

ONAGRACEAE Evening Primrose Family

Annual or perennial herbs (ours); leaves alternate, opposite or all basal, simple
or pinnately compound, entire or toothed to pinnatifid; flowers bisexual,
mostly regular; sepals and petals (2) 4, arising with the (2) 8 stamens from a
floral tube or receptacle, free or the sepals united at the base; pistil 1; ovary
inferior; style 1, the stigma discoid or 4-lobed; fruit a capsule (ours).

1 Flowers 2-merous, 1-2 mm long; fruit with hooked hairs; leaves opposite, more or less cordate, long-petiolate; plants rare .... *Circaea*

1 Sepals and petals 4; stamens 8; fruit without hooked hairs 2

2 Petals conspicuously bilobed at the apex; at least the lower leaves opposite; sepals not reflexed .... *Epilobium*

2 Petals entire or somewhat obcordate but not bilobed; leaves alternate, basal, or sometimes opposite but then sepals reflexed at anthesis 3

3 Petals 6-50 mm long; plants mostly perennial 4

3 Petals 1-4 (6) mm long; plants annual or perennial in *Gaura coccinea* 8

4 Stigmas divided into 4 linear lobes; petals white, pink or yellow 5

4 Stigmas capitate or shallowly lobed; petals yellow 7

5 Plants with leafy stems.. 6

5 Plants with leaves all basal .... *Oenothera*

6 Floral tube lacking; seeds with tufts of hair at the apex; petals pink-purple or rarely white; leaves entire or nearly so .... *Chamerion*

6 Floral tube 1-5 cm long; seeds without tufts of hair; petals yellow or white and fading to pink; leaves entire, toothed, or lobed .... *Oenothera*

7 Plants with stems 5-20 cm long with well developed leaves 0.5-1.5 cm long and entire; petals 13-22 mm long .... *Calylophus*

7 Plants without stems; leaves all basal, (3) 5-20 cm long and entire to deeply sinuate-pinnatifid; petals to 15 mm long .... *Camissonia*

8 Petals yellow, sometimes reddish in age .... *Camissonia*

8 Petals white or pink 9

9 Filaments of stamens and style equal to or longer than the petals .. *Gaura*

9 Filaments and style shorter than the petals 10

10 Lower leaves petiolate, lanceolate to elliptic; capsules often strongly bent .... *Camissonia minor*

10 Leaves sessile or nearly so, linear; capsules straight .... *Gayophytum*

Calylophus Spach.

Calylophus lavandulifolius (Torrey & A. Gray) Raven (*Oenothera lavandulifolia* Torrey & A. Gray) — lavenderleaf evening-primrose — Plants suffrutescent, caespitose, grayish strigose; flowers solitary in upper axils; floral tube 2.5-5 cm long; sepals 8-12 mm long; petals yellow, reddish in age; stamens subequal; capsules 8-20 mm long. Widespread, mostly occasional but locally abundant on semi-barrens of Uinta Formation on the Tavaputs Plateau; desert shrub, sagebrush, and pinyon-juniper communities; 5,400-8,300 ft, often but not always on substrates that weather to semi-barrens; mid May-June.

Camissonia Link — Camissonia

Annual or perennial herbs (ours); leaves basal, alternate or occasionally a few opposite; flowers usually opening near sunrise, sometimes in afternoon, withering in less than a day, reddish or purplish as they wither; stamens and style yellowish; capsules straight to coiled.

1   Petals 6-13 mm long; floral tube 1-9 cm long; plants perennial..........2

1   Petals 1-6 mm long; floral tube about 0.1-0.3 cm long; plants annual.....3

2   Floral tube 1-2.5 cm long; petals 6-8 (10) mm long; leaves pinnatifid; capsules hairy .... *C. breviflora*

2   Floral tube 3-9 cm long; petals 8-13 mm long; leaves entire to dentate or lobed; capsules glabrous .... *C. subacaulis*

3   Capsules borne on pedicels 4-20 mm long; leaves often abruptly petiolate.
   4

3   Capsules sessile or on pedicels to 2 mm long; leaves sessile or gradually tapered to a petiole 6

4   Leaves 0.1-0.6 cm wide, entire; petals white distally, yellowish toward the base, 1.5-2.5 mm long; capsules 1.2-1.8 cm long; pedicels 4-8 mm long ....
    .... *C. pterosperma*

4   Leaves 0.5-3 cm wide, entire, toothed, or pinnatifid; petals yellow, 1.7-5 mm long; capsules (1) 1.5-3 cm long; pedicles 5-20 mm long................5

5   Plants more or less copiously beset with spreading white hairs about 0.5-1.5 mm long; pods mostly 1.25-2 mm thick; plants apparently rare and restricted .... *C. walkeri*

5   Plants glabrous or puberulent but not pubescent as above; pods mostly 2-3 mm thick; plants common, widespread .... *C. scapoidea*

6   Plants with naked capillary stems, each bearing a crowded leafy inflorescence at its apex; capsules strongly flattened, 5-10 mm long; plants apparently rare in Moffat Co. .... *C. andina*

6   Plants not as above; capsules not flattened, 12-40 mm long; plants of various distribution..7

7   Leaves (2) 4-15 mm wide, not linear; plants strigose, not glandular; petals about 2 mm long, white or cream; capsules 12-25 mm long, slightly enlarged at the base, often contorted or even coiled .... *C. minor*

7   Leaves 1-2 mm wide, linear; plants glabrous or with scattered usually glandular hairs; petals 2.5-4 mm long, yellow; capsules 20-40 mm long, mostly linear and not enlarged at the base, often straight ..... *C. parvula*

Camissonia andina (Nuttall) Raven [*Oenothera andina* (Nuttall) ex Torrey & A. Gray] — Andean camissonia — Approaching our area in Wyoming and entering it on Blue Mtn., Moffat Co (Weber 1984), and reported for Wasatch Co (AUF 526).

Camissonia breviflora (Torrey & A. Gray) Raven (*Oenothera breviflora* Torrey & A. Gray) — short-flowered camissonia — The few specimens seen are from Strawberry Valley; along streams and in meadows; 7,500-8,000 ft; June-July.

Camissonia minor (A. Nelson) Raven [*Oenothera minor* (A. Nelson) Munz] — small-flowered camissonia — Widespread, seldom collected; desert shrub, and sagebrush communities; 4,700-6,700 ft; May-June.

Camissonia parvula (Nuttall) Raven [*C. contorta* (Douglas) Raven misapplied; *Oenothera contorta* (Douglas) Kearney var. *flexuosa* (A. Nelson) Munz; *Sphaerostigma contortum* (Douglas) Walpers var. *flexuosum* A. Nelson] — tiny camissonia — Duchesne, Uintah, and Moffat Cos.; infrequent in desert shrub, sagebrush, and pinyon-juniper communities, mostly in sandy areas; 4,750-5,700 (8,100) ft; May-July.

Camissonia pterosperma (S. Watson) Raven (*Oenothera pterosperma* S. Watson) — wingseed camissonia — The 3 specimens seen are from Uintah Co., near Brush Creek and Doc's Beach n. of Vernal and from Bitter Creek; desert shrub, sagebrush, and juniper communities; 5,300-6,000 ft; May-June. The fruit is similar to that of *C. scapoidea*, but the leaves are strikingly different.

Camissonia scapoidea (Torrey & A. Gray) Raven (*Chylismia scapoidea* (Torrey & Gray) Small; *Oenothera scapoidea* Torrey & A. Gray) — barestem camissonia — Widespread and locally common in desert shrub, sagebrush, and pinyon-juniper communities; up to 6,500 ft; May-June.

Camissonia subacaulis (Pursh) Raven [*Oenothera heterantha* Nuttall; *O. subacaulis* (Pursh) Garrett] — long-leafed camissonia — Specimens seen are from Currant Creek, Wolf Creek, Strawberry Valley and Moffat Co; aspen, sagebrush-aspen, silver sagebrush-meadow, and willow-streamside communities and along roads; 7,500-8,000 ft; June.

Camissonia walkeri (A. Nelson) Raven (*Chylismia walkeri* A. Nelson) — Walker camissonia — The one specimen seen (Brotherson 807) is from Split Mt. Gorge Campground.

Chamerion Rafinesque Ex Holub

| 1 | Plants commonly over 30 cm tall, widespread: style 1-2 cm long . . . . *C. angustifolium* |
| 1 | Plants commonly less than 30 cm tall, most common on alpine talus slopes; style not over 1 cm long . . . . *C. latifolium* |

Chamerion angustifolium (L.) Holub (*Chamerion danielsii* D. Love; *Chamaenerion angustifolium* (L.) Scopoli; *Epilobium angustifolium* L.) — fireweed — Strawberry Valley and Uinta Mtns., occasional and sometimes becoming abundant after fire or logging; lodgepole pine, spruce, and fir communities; 8,000-11,500 ft; July-Sept. Our plants likely all belong to var. *canescens* (A. W. Wood) N. H. Holmgren & P. K. Holmgren.

Chamerion latifolium (L.) Holub (*Chamaenerion latifolium* (L.) Sweet; *Epilobium latifolium* L.) — broadleaf willowherb — Mostly known from alpine, talus slopes of the Uinta Mountains, rarely along streams at lower elevations; July-Aug. This plant is adapted to talus-creep and is usually found in comparatively small-diameter talus subject to creep.

Circaea L. — Enchanter's Nightshade

Circaea alpina L. (*C. pacifica* Ascherson & Magnus) — enchanter's nightshade — Plants perennial from slender rootstocks, 5-25 cm tall; stems glabrous below, sparsely strigose to short-pilose above and in the inflorescence; leaf blades cordate-ovate to ovate, 2-6 cm long, longer than the petioles; flowers in racemes, 2-merous, 1-2 mm long; sepals spreading to reflexed, white; petals notched, white; fruit 1 (2) seeded, more or less pear-shaped, about 2 mm long, covered with short, hooked hairs, about 2 mm long, borne on spreading to reflexed pedicels. The 3 specimens seen are from Whiterocks Canyon, Uinta

Mtns.; very wet places in shade of willows and alder; 7,650-7,680 ft; no specimen was found in Utah herbaria by Albee et al. (1988) to support the Flowers and others (1960) report for Eagle Creek and Carter Creek, 5,760 ft (area now inundated by Flaming Gorge Reservoir); July-Aug.

Epilobium L. — Willowherb

Herbs, annual from taproots or perennial from rhizomes, the rhizomes sometimes with turions (small bulb-like shoots); leaves opposite (at least the lower ones) or less commonly all alternate, sessile or short-petiolate, entire or toothed; floral tube well developed; petals white, pink, purple, or red; stamens 8, those alternate the petals shorter than the opposite ones; stigmas oblong or 4-lobed; fruit an elongate, usually slender capsule, 4-sided, 4-chambered, dehiscent; seeds wind-dispersed, with tufts of hair at the apex. The *E. alpinum* and *E. ciliatum* groups are complex. Recognition of the several so-called taxa within them requires the use of overlapping features. Identification is further vexed by hybrids. The treatment presented here is conservative.

---

1 Sepals, petals, and floral tube scarlet-red flowers bilaterally symmetrical and somewhat penstemon-like a with the floral tube expanded and simulating tubular flower, 2-3 cm long including the floral tube; stamens exserted; plants from the extreme w. part of our area . . . . *E. canum*

1 Flowers not as above; plants widespread. 2

2 Plants annual; some or most leaves alternate. 3

2 Plants perennial; leaves mostly all opposite 4

3 Capsules 18-25 mm long; leaves not crowded . . . . *E. brachycarpum*

3 Capsules 5-9 mm long; leaves crowded . . . . *E. pygmaeum*

4 Inflorescence and usually much of the stem canescent with minute hairs; leaves entire or nearly so, often more or less revolute, sessile or nearly so
. . . . 5

4 Inflorescence and stem glabrous to glandular or pubescent but not canescent; leaves entire or serrate, not revolute, sessile or petiolate 6

5 Upper leaf surface finely pubescent . . . . *E. leptophyllum*

5 Upper leaf surface glabrous . . . . *E. palustre*

6 Leaf veins conspicuous; seeds, as seen at 20X, longitudinally finely ridged; plants 20-120 cm tall, growing from 5,000-10,000 ft elevation . . *E. ciliatum*

6 Leaf veins inconspicuous; seeds, as seen at 20X, merely finely papillate or finely cellular reticulate; plants mostly less than 50 cm tall 7

7 Plants with underground turions 8

7 Plants without turions 9

8 Fruits sessile or subsessile, the pedicels not over 5 mm long; leaves mostly sessile and clasping (*E. brevistylum* Barb.; *E. rubescens* Rydberg) . . . .
. . . . *E. saximontanum*

8 Fruits with pedicels mostly 5-15 mm long or longer; leaves sessile or short-petiolate but not clasping [*E. glandulosum* var. *tenue* (Trelease) C. L. Hitchcock] . . . . *E. halleanum*

9 Plants 1-20 (25) cm tall; stems often S-shaped; leaves mostly 1-2.8 cm long
. . . . *E. alpinum*

9   Plants 10-40 (50) cm tall; stems usually not S-shaped; leaves mostly 1.5-4
    cm long . . . . *E. hornemannii*

Epilobium alpinum L. Strawberry Valley and Uinta Mtns.; occasional to common
in wet places, snowbeds and talus; 7,250 -11,340 ft; June-Sept. With 2 intergrading
and sympatric vars. as follows:

1   Inflorescence nodding at least when young; capsules linear, about 1 mm
    thick; seeds smooth, 0.7-1.4 mm long: ***E. anagallidifolium*** Lamarck **var.**
    alpinum — alpine willowherb
1   Inflorescence more or less erect; capsules subclavate, about 1.5-2 mm thick;
    seeds papillate 1.4-2 mm long (*E. clavatum* Trelease): **var. *clavatum***
    (Trelease) C. L. Hitchcock — talus willowherb, clavate willowherb

Epilobium brachycarpum Presl (*E. paniculatum* Nuttall ex Torrey & A. Gray) —
tall annual willowherb, autumn willowherb — Apparently widespread but no
specimens seen from the Tavaputs Plateau; occasional in sagebrush, mt. brush,
forb-grass, ponderosa pine, and aspen communities, disturbed ground and
talus slopes; 7,300-9,000 ft; June-Sept.

Epilobium canum (Greene) Raven [*Zauschneria garrettii* A. Nelson; *Z. latifolia*
(Hooker) Greene var. *garrettii* (A. Nelson) Hilend] — Garrett firechalice, hum-
mingbird flower, hummingbird trumpet, wild fuschia — The 2 specimens seen
(Goodrich 16107 and 27483) are from the Currant Creek drainage, Wasatch Co.;
crevices of sandstone outcrops; also observed near Soldier Creek Dam in the
Strawberry drainage; July-Sept. Our plants belong to var. *garrettii* (A. Nelson)
N. H. Holmgren & P. K. Holmgren.

Epilobium ciliatum Rafinesque — common willowherb, fringed willowherb —
This appears to be the most widespread and common willowherb of our area;
aspen, coniferous forest, riparian, and meadow communities; 4,900-9,500
(10,000) ft; June-Sept. Common willowherb likely hybridizes with *E. halleanum,*
E. hornemannii, E. leptophyllum, and *E. saximontanum.* The following 2 taxa are
sometimes treated at species level:

1   Plants with underground turions, less often of disturbed places; petals 4.5-
    12 (14) mm long, rose purple to rarely white; inflorescence mostly leafy
    and not branched (*E. glandulosum* Lehmann) . . . .
    . . . . **ssp. *glandulosum*** Hoch & Raven
1   Plants without underground turions, more often in disturbed places than
    above; petals 2-6 (9) mm long, white to pink; inflorescence not leafy, open
    and well branched (*E. adenocaulon* Haussknecht; *E. watsonii* Barb.) . . . .
    . . . . **ssp. *ciliatum***

Epilobium halleanum Haussknecht — glandular willowherb — Specimens seen
are from Twelve Hundred Dollar Ridge, Strawberry Valley, and Uinta Mtns.,
none seen from the E. Tavaputs Plateau; aspen, lodgepole pine, spruce-fir, and
sedge-grass communities at 8,300-11,000 ft.

Epilobium hornemannii Reichenbach — willowherb — With 2 varieties:

1   Fruiting pedicels 0.5-2 cm long; petals pink, purplish, or rarely white, 5-8
    mm long: **var. *hornemannii*** — Hornemann willowherb — Sagebrush, aspen,
    lodgepole pine, spruce-fir communities, mainly in wet places at 8,200-
    11,250 ft.
1   Fruiting pedicels 2-4.5 cm long; petals white or rarely pink, 2-4 (5) mm
    long: **var. *lactiflorum*** (Haussknecht) D. Love — milkflower willowherb —

Aspen, Douglas-fir, spruce-fir, and sedge-forb communities, mainly in wet places at 6,200-11,000 ft.

Epilobium leptophyllum Rafinesque [*E. palustre* var. *gracile* (Farwell) Dorn] — bog willowherb, swamp willowherb — The 3 specimens seen are from near Ouray, Roosevelt, and along the Duchesne River between Duchesne and Tabiona; reported for Daggett Co. (AUF 524; IMF 3A:232) wet places, tolerant of alkali; 4,670-6,000 ft; Aug.-Sept.

Epilobium palustre L. — marsh willowherb — Reported for wet low ground in Duchesne and Uintah Counties (IMF 3A:232). Perhaps some specimens we have assigned to E. leptophyllum belong here. Reported by Bradley (1950) based on Boyd 74 (CSI), but this specimen belongs to *E. halleanum*.

Epilobium pygmaeum (Spegazzini) Hoch & P. H. Raven. [*Boisduvalia glabella* (Nuttall) Walpers] — pygmy willowherb — Annual tap-rooted herbs, 5-30 cm tall; stems simple or branched below, the branches sometimes opposite; leaves mostly alternate, crowded, 1-1.5 cm long, serrulate; flowers axillary; sepals 2 mm long; petals 2-4 mm long, purplish; capsules 6-8 mm long, straight; seeds without a coma. The one specimen seen (Neese 14837) from Diamond Mt. Plateau; mud at margin of Crouse Reservoir; 7,220 ft is likely the basis for the report for Uintah Co. (IMF 3A:242). No specimens found in Utah herbaria by Albee et al. (1988) to support the Flowers and others (1960) report for Sheep Creek at 5,900 ft (area now inundated by Flaming Gorge Reservoir); Aug. Plants of this taxon differ from the rest of *Epilobium* in our area by annual habit, leaves mostly alternate, axillary flowers, and most importantly in seeds without a coma. This is reasonably included in *Boisduvalia* in AUF (524).

Epilobium saximontanum Haussknecht — Rocky Mtn. willowherb — Strawberry Valley; Uinta Mtns. uppermost elevations of Tavaputs Plateau; occasional in sagebrush, mtn. brush, willow-cottonwood, lodgepole pine, and spruce fir communities at 6,800-11,220 ft.

Gaura L. — Gaura

Leaves alternate; flowers in terminal spikes or racemes, subtended by small bracts, ephemeral; sepals reflexed at anthesis; petals whitish at first, becoming pink or reddish; stamens and styles exserted beyond the petals; fruit small, hard, indehiscent, with 1-few seeds.

---

1   Plants perennial, known from Daggett Co., with clustered stems, rarely over 0.5 m tall; leaves 1-4 cm long; sepals 5-9 mm long; petals 3-7 mm long; floral bracts usually persistent .... **G. coccinea**

1   Plants annual or biennial, widespread; stems solitary, often branched from a taproot, to 2 m tall; larger leaves 1.5-3 mm long; floral bracts caducous . ... **G. parviflora**

Gaura coccinea Pursh — scarlet gaura — Known from Daggett Co. in pastures at and near Manila.

Gaura parviflora Douglas in Hooker — small-flowered gaura — widespread; infrequent but locally common along roadsides and other disturbed places; up to about 7,000 ft; June-Aug.

Gayophytum Jussieu — Groundsmoke

Annual herbs; stems slender, simple to much branched, the ultimate branches capillary; leaves alternate or occasionally the lower ones opposite, mostly sessile or nearly so, linear, entire; flowers borne in leaf axils, small; capsules terete or flattened, 2-celled, 4-valved. *Gayophytum* forms a polyploidy complex in which most of the specific boundaries are obscure (IMF 3A: 222). This is a difficult genus with poorly marked species (AUF 535, IMF 3A: 222-223), and the following key is tentative at best. Arnow et al. (1980) recognized only 2 of the 5 taxa listed below.

---

1   Mature capsules sessile or on pedicels to 3 mm long or these sometimes longer in the lower axils. 2

1   Mature capsules mostly on pedicels (1) 4-15 mm long 3

2   Well developed plants with a few long simple branches below the middle
. . . . ***G. racemosum***

2   Well developed plants with several long branches from above as well as below the middle . . . . ***G. decipiens***

3   Capsules 3-6 (9) mm long; pedicels often spreading or deflexed . . . .
. . . . ***G. ramosissimum***

3   Capsules 6-12 (15) mm long; pedicels various. 4

4   Well developed plants with a few long, ascending, well spaced, simple or nearly simple branches separated by 2 or more nodes; first flower mostly 1-3 (5) nodes above the base . . . . ***G. decipiens***

4   Well developed plants more freely branched, some of the branches at adjacent nodes or separated by only one node; first flower often 5 or more nodes above the base . . . . ***G. diffusum***

Gayophytum decipiens Lewis & Szweykowski — deceptive groundsmoke — Known from Duchesne, Uintah, and Wasatch Counties in sagebrush, mtn. brush, pinyon-juniper and coniferous forest communities.

Gayophytum diffusum Torrey & A. Gray — diffuse groundsmoke — Widespread in sagebrush, mtn. brush, ponderosa pine, aspen, and coniferous forest communities.

Gayophytum racemosum Torrey & A. Gray (*G. helleri* Rydberg) — blackfoot groundsmoke — Widespread; Occasional to locally abundant in sagebrush, pinyon-juniper, mt. brush, ponderosa pine, aspen, fir, and spruce communities; 6,000-10,300 ft; May-Aug.

Gayophytum ramosissimum Nuttall ex Torrey & A. Gray (*G. lasiospermum* Greene; *G. nuttallii* Torrey & A. Gray misapplied) — branched groundsmoke — Widespread; occasional to locally abundant in sagebrush, pinyon-juniper, mt. brush, forb-grass, aspen, Douglas-fir, and ponderosa pine communities; 6,000-9,100 ft; June-Aug. Plants often branched from the base such that there is no well defined central axis.

Oenothera L. — Evening-primrose

Annual to perennial herbs; leaves alternate or basal; flowers regular, mostly opening in the evening; petals mostly somewhat obcordate; capsules membranous to woody, straight or coiled, 4-celled.

1   Plants with leaves all basal or nearly so; floral tube 4-15 cm long 2
1   Plants with leafy stems; floral tube 1.5-4.5 cm long. 5
2   Petals white or pink when fresh, drying reddish or pinkish-lavender, 2.5-4
    cm long; capsules not winged . . . . ***O. caespitosa***
2   Petals yellow when fresh, often drying orange-red or purple; capsules more
    or less winged. Note: our 2 perennial taxa of *Camissonia* will key here except
    for their capitate stigmas. They also differ from the following 3 taxa in
    having mostly shorter petals (6-13 mm long) that remain yellow in age, and
    in having a persistent floral tube that is not deciduous from the capsule . 3
3   Leaves entire to undulate-toothed or lobed, glabrous; sepals 3-5 cm long;
    petals 3-5 cm long, drying reddish-orange . . . . ***O. howardii***
3   Leaves rather sharply toothed to runcinate-pinnatifid. 4
4   Sepals 1-2.5 cm long; petals 1-3 cm long, drying to purple; at least the upper
    1/2 of leaf blades 5-12 mm wide excluding the teeth; capsules (10) 20-40 mm
    long, the wings 2-5 mm wide . . . . ***O. flava***
4   Sepals 2.5-5 cm long; petals 2.8-5 cm long, drying purplish-brown; leaf
    blades mostly narrow-linear, mostly 2-6 mm wide excluding the teeth or
    lobes; capsules 14-22 mm long, the wings 1-2 mm wide . . . . . ***O. acutissima***
5   Petals yellow when fresh; stems 30-150 cm tall, often reddish; floral tube
    2.5-4.5 cm long; leaves entire or dentate, 4-20 cm long 6
5   Petals white or pink when fresh, sometimes rose-pink upon drying; stems
    5-50 cm long; floral tube 1.5-3.5 cm long; leaves pinnatifid or less commonly
    entire or nearly so. 7
6   Petals 1-2 cm long; sepals 1-1.5 cm long . . . . ***O. villosa***
6   Petals 2.5-4 cm long; sepals 2.5-4 cm long . . . . ***O. elata***
7   Floral tube with a conspicuous, dense ring of hairs at the throat; capsules
    fusiform, not curved, not contorted, 10-15 mm long, about 5 mm wide,
    membranous; leaves except the lower ones pinnatifid to the midrib, 1.5-2.5
    cm long; petals 7-15 mm long . . . . ***O. coronopifolia***
7   Floral tube without a ring of hairs; capsules nearly linear, broadest at the
    base and gradually tapering to the apex, straight, curved, or contorted,
    about 20-40 mm long, to 3 mm wide, leathery or nearly woody; leaves entire
    to deeply pinnatifid, 2-6 cm long; petals about 10-20 mm long . . ***O. pallida***

Oenothera acutissima Wagner [*O. flava* var. *acutissima* (W. L. Wagner) S. L.
Welsh] — acute-leaf evening-primrose — Endemic, e. Uinta Mtns., Blue Mt.,
Cold Spring Mt. and Yampa Plateau; sandy, summer-dry stream beds, dry
rocky meadows, and rocky areas adjacent to streams; ponderosa pine and
mtn. big sagebrush communities; 7,000-8,355 ft; June-July.

Oenothera caespitosa Nuttall — Tufted evening-primrose, sand-lily — With 2
intergrading ssp.:
  1   Floral tube (6) 7-14+ cm long; petals (2.5) 3-5 (6) cm long; leaf blades more
      or less lobed; capsules cylindrical or lance cylindrical . . . . ***ssp. marginata***
  1   Floral tube 3-7 (8.5) cm long; petals 2-4 (4.5) cm long; leaf blades nearly
      entire, undulate or shallowly lobed; fruit oblong cylindrical . . . .
      . . . . ***ssp. navajoensis***

Ssp. maginata (Nuttall) Munz (*O. maginata* Nuttall) — evening-primrose — Widespread; occasional in desert shrub, sagebrush and mt. brush communities and roadcuts up to 7,800 ft; May-June (July).

Ssp. navajoensis Wagner, Stockhouse & W. Klein — evening-primrose — Widespread; occasional in desert shrub, sagebrush, and pinyon-juniper communities; 4,700-6,500 ft; May-June. Perhaps more common at lower elevations than ssp. *marginata*.

Oenothera coronopifolia Torrey & A. Gray — hairy-ring evening-primrose — Widespread; locally common; dry meadows, sagebrush, mt. brush, rabbit-brush-Basin wildrye, juniper, and ponderosa pine communities, and roadsides; 7,100-8,500 ft; June-Aug.

Oenothera elata H.B.K. [*O. hookeri* Torrey & A. Gray var. *angustifolia* Gates] — elated evening-primrose — Widespread; occasional in riparian, sagebrush-rabbit brush,,ponderosa pine, Douglas-fir,,and aspen communities, usually on floodplains, road-cuts, roadside ditches, cut-banks of streams, gullies, and rocky drainages; 4,700-8,650 ft; July-Sept. Our plants belong to var. *hirsutissima* (A. Gray ex S. Watson) Cronquist.

Oenothera flava (A. Nelson) Garrett — long-tube evening-primrose — Widespread; open ground about ephemeral pools and streams and rocky snowbed areas, roadbeds, ponderosa pine communities, and dry and wet meadows; 7,600-9,800 ft; July-Aug.

Oenothera howardii (A. Nelson) Wagner [*O. brachycarpa* (A. Gray) Britton misapplied] — bronze evening-primrose — Widespread; infrequent in riparian, pinyon-juniper, and sagebrush communities, often on substrates that weather to semi-barrens; 5,800-6,800 ft; June.

Oenothera pallida Lindley (*O. trichocalyx* Nuttall in Torrey & Gray) — pale evening primrose — Widespread, common and locally abundant; desert shrub, sagebrush, pinyon-juniper, and mtn. brush communities; 4,815-8,200 ft; May-Aug. Features of plants of our area include glabrous, sparsely villose, densely villose, and stipitate glandular calyces; glabrous and strigose leaves with both cases sometimes found on the same plant; with habit of annual, tap-rooted perennial, and strongly rhizomatous. Some plants of Daggett and Moffat Cos. are rhizomatous and others are annual. Most plants from the Uinta Basin seem to be annual.

Based on pubescence of leaves and calyces separation at var. and ssp. levels is available in various floras. Plants with glabrous calyces are found growing side by side with those with thinly to densely pubescent calyces. In some cases glabrous and pubescent calyces are found on the same plant. Separation of vars. does work well with plants of our area, and no attempt to do so is offered here. Difficulty in recognizing taxa within the complex is also discussed in AUF 540-541. In IMF 3A: 214-218, varieties recognized are admittedly confluent. A specimen from Moffat Co. with stipitate glandular calyces (Goodrich & Huber 28071) is an anomaly.

Oenothera villosa Thunberg [*O. biennis* L. misapplied; *O. rydbergii* House] — common evening-primrose — Specimens seen are from canyons of the Yellowstone, Uinta, and Whiterocks drainages and adjacent in the Uinta Basin east to Vernal, and one is from the E. Tavaputs Plateau; riparian, sagebrush, and ponderosa pine communities, roadsides, fencelines, and ditch banks; 5,600-7,700 ft; July-Sept. Our plants belong to var. *strigosa* (Rydberg) Dorn [*O. strigosa* (Rydberg) Mackenzie & Bush].

OPHIOGLOSSACEAE Adder's-tongue Family

Botrychium Sw. — Grape-fern

Perennial, fern-like herbs from clustered fleshy roots; stems simple, fleshy, surrounded by a sheath of brown scaly leaf-bases; leaf solitary with a common stalk or petiole bearing one sterile blade (trophophore) and one fertile one (sporophore) bearing a spicate or branched cluster of globose sporangia in 2 rows.

1 Sterile blade 2-4 times compound, broadly deltoid, commonly broader than long . . . . *B. multifidium*

1 Sterile blade pinnatifid to pinnate pinnatifid, often longer than broad. . . 2

2 Division of trophophore and sporophore near ground level, the common stalk short . . . . *B. simplex*

2 Division of trophophore and sporophore above ground level, the common stalk longer. 3

3 Lower pair of leaflets of sterile blade simple or bifid, shorter than or about equal to the rachis segment. 4

3 Lower pair of leaflets of sterile blade pinnatifid, mostly longer than the rachis segment. 6

4 Leaflets longer than wide, bifid at apex or simple . . . . *B. lineare*

4 Leaflets as wide or wider than long 5

5 Sterile blade over 5 cm long and over 2 cm wide, with 4-9 pairs of leaflets, the lower 2 or 3 pairs overlapping or nearly so; plants know from Daggett, Duchesne, and Summit Cos. . . . . *B. lunaria*

5 Sterile blade less than 4 cm long and less than 1.5 mm wide, with 2-5 pairs of leaflets, the lower 2 or 3 pairs remote, not overlapping; plants known from Wasatch Co. . . . . *B. crenulatum*

6 Sterile blade dull or glaucous; sporangia dull green before spore release . . . . *B. echo*

6 Sterile blade lustrous; sporangia bright yellow-green prior to spore release 7

7 Upper leaflet bases obtuse to cordate; leaflet apices rounded, sporophore stalk equal to trophophore length . . . . *B. pinnatum*

7 Upper leaflet bases acute; leaflet apices angular; sporophore stalk shorter than trophophore length . . . . *B. lanceolatum*

Botrychium crenulatum W. H. Wagner — scalloped moonwort — Known from Silver Meadow, Wasatch Co., wet meadow at 9,440 ft.

Botrychium echo W. H. Wagner. — reflected grapefern — Plotted for Summit and Daggett Cos. in TPD, but mapped for sw. Utah not including Summit or Daggett Cos. in FNA 2: 96).

Botrychium lanceolatum (S. G. Gmelin) Angstrom — lanceleaf grapefern — Reported for Summit Co. (TPD).

Botrychium lineare W. H. Wagner — narrowleaf grapefern — One specimen from the head of Indian Canyon on the West Tavaputs Plateau is the only specimen for Utah. The location is over 200 miles from the nearest known location in Colorado. Perhaps the location is a record book or label error. However, sporadic distribution for this species is demonstrated in TPD where the plant is plotted for only one county in each of Alaska, California, Idaho, Oregon, and Washington.

Botrychium lunaria (L.) Sw. — common moonwort — Uinta Mtns.; the 1 record seen (Harrison 7666) is from near Moon Lake; wet meadows and stream sides; 8,100 ft. Also reported for Summit and Daggett Counties in TPD.

Botrychium multifidium (S. G. Gmelin) Ruprecht — leathery grape-fern — Known from Yellowstone Canyon, Uinta Mtns., riparian communities on the canyon bottom at 8,100-8,300 ft.

Botrychium pinnatum H. St. John — northern moonwort — Reported for Summit Co. (TPD).

Botrychium simplex E. Hitchcock — little grapefern — Reported for Summit Co. (TPD).

ORCHIDACEAE Orchid Family

Perennial herbs; leaves mostly sheathing, linear to orbicular, often coarsely parallel-ribbed; flowers usually bisexual, minute to showy and zygomophic; sepals and petals 3 each, sometimes fused in part, similar or more often the petals more showy, the lower petal (lip) larger and unlike the other perianth segments, often cylindrical, sometimes bladder-like, sometimes prolonged at the base into a spur; fertile stamens 1 (2), an additional staminode present in some genera; ovary inferior, with 1-3 chambers; stigmas 3; fruit a 3-valved, dry capsule; seeds minute, numerous.

---

1 Leaves reduced to bladeless sheaths; plants not green, lacking chlorophyll except in *Corallorhiza trifida* .... ***Corallorhiza***

1 Leaves with at least 1 blade well developed, green 2

2 Flowers mostly solitary, showy; stem from a globose (or nearly so) corm, with only 1 well-developed blade-bearing leaf, this also arising from the corm .... ***Calypso***

2 Flowers 2-many, spicate to racemose; plants not from corms; blade-bearing leaves arising from the stem, often more than 1 3

3 Blade-bearing leaves usually 2, these opposite or subopposite, borne about midway on the stem; additional reduced or bract-like leaves sometimes present, these alternate 4

3 Blade-bearing leaves not as above 5

4 Leaves 4-9 cm long; sepals 12-20 mm long .... ***Cypripedium***

4 Leaves 1.5-3.5 cm long; sepals about 2-5 mm long .... ***Listera***

5 Sepals 12-15 mm long, greenish with brownish veins; petals brownish-purple, the lip 15-18 mm long, not prolonged into a spur; stems leafy throughout; plants known from below 7,000 ft .... ***Epipactis***

5 Sepals 2-12 mm long, greenish or white; petals greenish or white, the lip 3-12 mm long, spurred or not; stems sometimes with reduced or bract-like

leaves above; plants mostly from above 7,000 ft 6

6   Scapes and flowers glandular pubescent; plants rhizomatous ... *Goodyera*

6   Plant glabrous, from fibrous or fleshy, fascicled roots 7

7   Lip prolonged at the base into a spur; leaves basal or cauline, reduced leaves and bracts below the inflorescence (if present) usually not sheathing, or if so then solitary .... *Habenaria*

7   Lip not prolonged into a spur; leaves with well-developed blades basal or on the lower 1/3 of the stem, usually sheathing, the sheath sometimes rather loosely enveloping the stem; upper leaves bract-like, these, at least the lower, usually sheathing .... *Spiranthes*

Calypso Salisbury

Calypso bulbosa (L.) Oakes — fairy slipper, Venus slipper — Plants from a perennial, fleshy, globose or ellipsoid corm, 5-15 (20) cm tall; a solitary blade-bearing leaf arising from the corm, the blade 3-6 cm long, broadly elliptic, ovate or cordate, the petiole about as long as the blade; stem scapose, enveloped by long-sheathing bladeless or bract-tipped sheaths; sepals and petals similar, purplish, with deeper colored veins (rarely whitish), lanceolate or linear, 15-23 mm long; lip ovate-oblong, slightly longer than the other perianth segments, about 10 mm wide, whitish, yellowish or pale reddish-brown in age, spotted and bearded on the upper side, the basal opening hooded by the petaloid, sub-orbicular, 7-12 mm long column. Widespread in coniferous forests in the Uinta Mtns, seldom collected, but locally common in some years; May-June.

Corallorhiza Chat. — Coralroot

Plants perennial, saprophytic and without chlorophyll, herbaceous, fleshy when young, scapose, the scapes yellow, reddish, or reddish-brown; leaves reduced to sheaths, these enveloping the stem; inflorescence a bracteate raceme; pedicels erect or spreading in flower, reflexed in fruit, 1-2.5 mm long; sepals 3, the lateral pair sometimes extended under the base of the lip and forming a small pouch or spur; petals 3, the lateral pair sepal-like, the lip short-clawed; anther 1; capsule rounded, pendulous.

1   Plants yellow to greenish-yellow, 5-30 cm tall; scapes slender, seldom over 3 mm wide; sepals whitish, 5-6 mm long; corolla whitish .... *C. trifida*

1   Plants purplish or reddish-brown (yellow in albino forms), 15-70 cm tall; scapes mostly over 3-mm wide at least toward the base; sepals 6-16 mm long; corolla marked with purple 2

2   Corolla lip entire, sometimes over 6 mm wide, white with prominent purple stripes, not spotted; sepals also striped with purplish lines ..... *C. striata*

2   Corolla lip undulate or erose along the margins or lobed toward the base, mostly less than 6 mm wide, often purple spotted, not strongly striped with purple lines; sepals lightly if at all striped 3

3   Lip with an evident lobe on either side at the base; plants common ......
.... *C. maculata*

3   Lip without lateral lobes; plants rare .... *C. wisteriana*

Corallorhiza maculata Rafinesque [*C. mertensiana* Bongard misapplied by Graham (1937)] — spotted coralroot — Most specimens seen are from the Uinta Mtns., but a few are from the Tavaputs Plateau; most common in dry lodgepole pine and ponderosa pine woods, occasionally in aspen woods, June-Aug.

Corallorhiza striata Lindley — hooded coralroot — The 5 specimens seen are from near Bruin Point of the West Tavaputs Plateau, Strawberry Ridge, Rock Creek, Hole In The Wall Canyon, Ashley Gorge, and Blue Mt.; riparian, aspen and coniferous forest communities; 6,400-10,000 ft; June-July.

Corallorhiza trifida Chatelain — early coralroot — The 13 specimens seen are from across the Uinta Mtns.; along streams and in bogs in coniferous woods and in duff of lodgepole pine communities at 8,600-9,560 ft; July-Aug.

Corallorhiza wisteriana Conrad — spring coralroot — Well scattered across the Uinta Mtns at 8,000-10,000 ft; of the 18 specimens seen, 17 were from lodgepole pine communities and one specimen was from a spruce community, commonly growing in conifer needle-litter; June-July (Aug.).

Cypripedium L. — Lady's Slipper

Cypripedium fasciculatum Kellogg — clustered lady's slipper — Stems arising singly from short rhizomes, 4-12 cm tall, densely viscid-villous, with a pair of opposite leaves borne at about mid-length, the blades broadly elliptic or orbicular, 4-9 cm long; 1-2 bladeless sheaths enveloping the base of the stem; 1 or more bract-like leaves below the inflorescence; sepals lanceolate, 1-2 cm long, greenish-brown or greenish purple; petals similar to sepals but usually wider; lip ovoid, appearing inflated, shorter than the sepals and petals, greenish-yellow, with deep purple margins. Known from Mosby Mt. and eastward to the East Draw drainage on the s. slope of the Uinta Mtns. and from near Elk Park east to Gorge Creek on the n. slope of the Uinta Mtns. with a few small outlier populations toward the west end of the Uinta Mtns.; lodgepole pine and much less commonly spruce-fir communities; 8,000-10,000 ft; June. This plant is most common in lodgepole pine stands that burned with stand-replacing fire within the last 120 years.

Epipactis Sw. — Helleborine

Epipactis gigantea Douglas [*Amnesia gigantea* (Douglas) A. Nelson] — stream orchid — Stems arising singly or several together from creeping rhizomes, 30-70 (100) cm tall, glabrous below, pubescent in the inflorescence; leaves sessile, 5-20 cm long, the sheaths enveloping the stem, the lower most reduced to sheaths, the lower blades ovate, the upper ones lanceolate or linear, greenish with brownish veins; petals sepal-like, 15-18 mm long, brownish-purple; lip 15-18 mm long, the sac with purplish lines, 3-lobed at the base, the blade or central lobe curved downward, greenish-yellow; capsules 2-2.5 cm long, ovoid, reflexed, yellowish with dark brown ridges. Specimens seen are from Big Sand Wash Reservoir (E. Neese 7655 & 14844), Desolation Canyon (A. Ross sn UTC), Firewater Canyon (D. Atwood 25767 BRY), reported by Graham (1937) from near Headquarters, Dinosaur National Monument and several specimens (at DINO) seen from Dinosaur National Monument; moist to wet often shady areas including hanging gardens; 4,800-6,000 ft; June-Aug.

Goodyera R. Brown — Rattlesnake Plantain

Goodyera oblongifolia Rafinesque — rattlesnake-plantain — Plants 20-40 cm tall; from short rhizomes; leaves basal, the blades 3-7 cm long, ovate-lanceolate to elliptic-lanceolate, thick, dark green and usually striped with white, especially along the midrib; scape with 2-4 small, membranous, non-green, sheathing bracts; flowers in a spike-like raceme, mostly secund and often spiraled, usually pale greenish white, pubescent, about 6-10 mm long. The few specimens seen are from the eastern Uinta Mountains in Daggett and Uintah Cos.; dense lodgepole pine woods at 7,800-8,470 ft.

Habenaria Willdenow — Bog or Rein Orchid

Glabrous perennial herbs from tuberous or fleshy roots or occasionally rhizomes; leaves alternate, entire, solitary to several, gradually reduced upward, mostly with a sheathing base; inflorescence a spike or spike-like raceme, loosely to densely flowered; upper sepals often united with the petals; lateral sepals spreading to reflexed; lip curved upward to strictly descending; spur usually conspicuous, borne behind and longer or shorter than the lip. The last 4 species in the following key form a complex that has been separated by minor (if not trivial), inconsistent features. The variability recognized in the 3 vars. of *H. dilatata* is as great or greater than is found in the entire complex. Indeed the variations found in *H. dilatata* nearly encompass the entire complex as do the variations found in H. hyperborea. The 4 so-called species apparently hybridize freely (IMF 6: 555). This creates an array of hybrid variants that further complicates identification. Separation of troublesome specimens may serve more to waste time than for any useful purpose.

1 Stems with a solitary basal leaf; scapes without bracts, or with a solitary sheathing bract near the middle; inflorescence 3-7 cm long . . . **H. obtusata**

1 Stems with 3 or more leaves; inflorescence often over 7 cm long 2

2 Leaves clustered at or near the base of the stem, usually withering at or before flowering; stems with scale-like bracts; bracts subtending the flowers inconspicuous . . . . **H. unalascensis**

2 Leaves scattered on the stem (sometimes mostly basal), usually green at flowering; stems with leaves or leaf-like bracts; bracts subtending the flowers often conspicuous, at least the lower ones usually longer than the flowers. 3

3 Lip tridentate at the apex, about 3 mm wide, pendulous; bracts subtending each flower leaf-like, green, mostly 2-4 cm long except the upper ones; spur to 1/2 as long as the lip . . . . **H. viridis**

3 Lip entire at the apex; bracts subtending the flowers less than 2 cm long (see discussion at the end of the generic description above). 4

4 Spur saccate to clavate; lip not at all widened toward the base; racemes usually open . . . . **H. saccata**

4 Spur not saccate or clavate; lip usually widened toward the base, sometimes only slightly so; raceme mostly congested 5

5 Flowers mostly white, rarely pale green; lip rhombic-lanceolate, prominent and rather abruptly dilated at the base . . . . **H. dilatata**

5   Flowers greenish; lip linear to broadly lanceolate, not prominently dilated
     at the base

6   Spur 1.5-2 times longer than the lip . . . . **H. zothecina**

6   Spur shorter than or slightly exceeding the lip 7

7   Raceme usually elongate and loosely flowered; lip linear to linear-elliptic,
     often with a fleshy ridge in the center below the middle   . . . **H. sparsiflora**

7   Raceme usually short and densely flowered; lip linear-lanceolate to
     lanceolate or slightly broader, without a fleshy ridge . . . . . . **H. hyperborea**

Habenaria dilatata (Pursh) Hooker [*Limnorchis dilatata* (Pursh) Rydberg;
Platanthera dilatata (Pursh) Lindley ex Beck] — white bog orchid — Uinta Mtns.;
wet places, about springs, in meadows and bogs, and along streams; about
7,300-10,000 ft; June-Aug. Three vars. have been recognized for the Intermountain
Region (IMF 6: 558). Separation is largely made on length of the spur in
relation to length of the lip.

Habenaria hyperborea (L.) R. Brown [*Platanthera hyperborea* (L.) Lindley] —
northern bog orchid — Specimens seen are from the Uinta Mtns., plotted for
upper Hill Cr. or Willow Cr, E. Tavaputs Plateau (Albee et al. 1988); wet places;
6,500-10,000 ft and rarely lower elevations in the Uinta Basin; June-Aug.

Habenaria obtusata (Banks) Richards [*Lysiella obtusata* (Banks) Britton &
Rydberg; *Platanthera obtusata* (Banks ex Pursh) Lindley] — northern small bog
orchid — The few specimens seen are from S. Fork Rock Creek, Uinta Mtns.;
along streams and in boggy places in Engelmann spruce forests; 8,600-9,560 ft;
June-July.

Habenaria saccata Greene [*Limnorchis saccata* (Greene) Love & Simon; *Platanthera
stricta* Lindley] — slender bog orchid — Like *H. hyperborea,* rather tentatively
separated by the criteria given in the key, suggested as possibly only a variety
of *H. hyperborea* (IMF 6: 562). The one specimen seen (E. Neese & B. Welsh 7492)
is from a canal bank 6 miles n. of Duchesne.

Habenaria sparsiflora S. Watson [*Limnorchis sparsiflora* S. Watson; *Platanthera
sparsiflora* (S. Watson) Schlechter] — canyon habenaria — Like *H. hyperborea,*
separation is arbitrary at best (Arnow and others 1980). Specimens seen are
from Rock Creek to Whiterocks Canyon in the Uinta Mtns.; riparian and bog
communities usually in aspen and coniferous woods; 7,250-9,560 ft; July-Aug.

Habenaria unalascensis (Sprengel) S. Watson — Alaska rein *orchid* — [*Piperia
unalascensis* (Sprengel) Rydberg] The few specimens seen are from the Uinta
Mtns where it is widespread and occasional to common in dry or moist
coniferous woods; 7,800-10,000 ft; June-July.

Habenaria viridis (L.) R. Brown. [*H. bracteata* (Willdenow) R. Brown; *Coeloglossum
viride* (L.) Hartman; *Dactylorhiza viridis* (L.) R. M. Bateman, A. M. Pridgeon & M.
W. Chase] — long-bracteate bog orchid — The few specimens seen are from
Little Elk Creek, Daggett Co., Uinta Canyon, Duchesne Co., and Provo River
drainage; 7,400-7,900 ft; June-July.

Habenaria zothecina L. C. Higgins & S. L. Welsh [*Platanthera zothecina* (L.C.
Higgins & S. L. Welsh) Kartesz & Gandhi] — alcove bog orchid — Streambanks,
seeps, and hanging gardens, Uintah and Moffat Cos. in canyons associated
with the Green and Yampa Rivers, not expected over 6,500 ft.

Listera R. Brown — Listera; Twayblade

Slender, inconspicuous herbs from small rhizomes; leaves 2, opposite or sub-opposite, sessile, borne near the middle of the stem; flowers small, greenish or purplish in age, racemose; sepals and petals similar, spreading to reflexed; lip somewhat longer then the sepals, pointing forward, usually with a pocket-like nectary under the stigma.

1   Lip cleft for about 1/2 its length into linear-lanceolate, more or less divergent lobes, glabrous, 3-6 mm long; leaves subcordate, opposite . . . . .
. . . . *L. cordata*

1   Lip retuse or shallowly notched at the apex, with rounded lobes, puberulent at least along the margins, 8-13 mm long; leaves elliptic to suborbiculate.
. . . . 2

2   Lip noticeably tapered to the base; petals strongly recurved; leaves ovate to suborbicular, opposite . . . . *L. convallarioides*

2   Lip not much if at all tapered to the base; petals spreading-reflexed; leaves lanceolate to ovate elliptic, subopposite . . . . *L. borealis*

Listera borealis Morong — northern twayblade — The few specimens seen are from S. Fork Rock Creek, Uinta Mtns.; mossy, boggy places in shade of coniferous trees; 9,320-9,560 ft; July-Aug.

Listera convallarioides (SW.) Torrey — broadleaved twayblade — Known from wet meadows and woods, Hades Canyon, Uinta Mtns., Duchesne Co. (Flowers 1457 UT).

Listera cordata L. R. Brown — heartleaf twayblade — specimens seen are from the S. Fork Rock Creek, Uinta Canyon and Broadhead Meadows, Provo River drainage; along streams, in seeps and wetlands usually in shade of coniferous trees; 7,500-9.560 ft; July-Aug.

Spiranthes Rich — Ladies-tresses

Perennial herbs from fascicled fleshy roots; stems 1-several, simple, erect, glabrous below, sometimes glandular-pubescent above; leaves mostly on the lower portion of the stem, alternate, sheathing, the blades mostly linear, occasionally oblong, rarely obovate, those of the stem reduced upward to bladeless sheaths; flowers cream, white, or greenish-white, spirally arranged in a dense spike; sepals glandular-hairy, united with the lateral petals into a tubular hood.

1   Lip constricted toward the middle (violin-shaped), recurved, sharply differentiated from the other perianth segments; plants known from above 7,500 ft . . . . *S. romanzoffiana*

1   Lip not constricted toward the middle, ovate, not recurved, not so sharply differentiated from the other perianth segments; plants known from below 7,500 ft . . . . *S. diluvialis*

Spiranthes romanzoffiana Chamisso. — hooded ladies-tresses — Uinta Mtns.; occasional in woods and meadows, and clearings in woods made from logging or fire at about 7,200-11,000 ft; July-early Sept.

Spiranthes diluvialis Sheviak [*S. romanzoffiana* var. *diluvialis* (Sheviak) S. L. Welsh] — Ute ladies-tresses — Specimens seen are from riparian communities of the Green River from Little Hole to Browns Park and outwash plains and stream pediments of drainages of the south slope of the Uinta Mountains, and from along irrigation canals and gravel pits at 5,185-7,200 ft; July-Sept.

OROBANCHACEAE Broomrape Family

Orobanche L. — Broomrape; Cancer-root

Herbs perennial (ours), parasitic on the roots of other plants, lacking chlorophyll, whitish, yellowish to brown, often more or less fleshy; herbage glandular-hairy; leaves alternate or occasionally subopposite, bract-like; flowers solitary or in usually dense bracteate spikes or racemes; calyx often closely subtended by bractlets, with cylindrical or bell-shaped tube and 4-5 lobed limb or divided to the base; corolla united, 2-lipped, the upper lip 2-lobed, the lower 3-lobed and at least as long as the upper; stamens 4, mostly included in the corolla; ovary superior, 1-chambered, the stigma 2-4 lobed or disk shaped; fruit a capsule.

---

1 Flowers sessile or on pedicels to about 2 cm long, closely subtended by 2 bractlets; calyx deeply cleft, the lobes much exceeding the tube 2

1 Flowers on scape-like pedicels 2-15 cm long, not closely subtended by 2 bractlets; calyx lobes shorter or a little longer than the tube 3

2 Inflorescence a short corymb, 2.5-5 cm long; calyx lobes 9-16 mm long; plants 3-10 cm tall; anthers woolly . . . . *O. corymbosa*

2 Inflorescence an elongate spike or raceme, (3) 4-10 cm long; calyx lobes 5-10 (12) mm long; anthers glabrous or slightly pubescent . . . *O. ludoviciana*

3 Flowers (3) 4-10, pedicels about as long or shorter than the true stem; calyx lobes mostly equal to or shorter than the tube . . . . *O. fasciculata*

3 Flowers 1-3, pedicels much longer than the stem; calyx lobes longer than the tube . . . . *O. uniflora*

Orobanche corymbosa (Rydberg) Ferris [*O. californica* C. & S. var. *corymbosa* (Rydberg) Munz.] — flat-topped broomrape — The few specimens seen are from Island Park and the lower edge of the Tavaputs Plateau and Uinta Basin from areas of white, calcareous sandstone and marly mudstone and Dinosaur National Monument at Iron Springs Bench and w. of Island Park, and B. Corbin 1156 is from Slate Creek of the Provo River drainage;. The listing of *O. multiflora* Nuttall for w. of Island Park (Holmgren 1962) apparently is based on Bradley 5339 DINO! that belongs to *O. corymbosa;* desert shrub and pinyon-juniper communities; June-Aug. The anthers are reported to be wooly pubescent. In some of our specimens the anthers appear to go glabrous with age.

Orobanche fasciculata Nuttall [*Thalesia fasciculata* (Nuttall) Britton] — clustered broomrape — Widespread; infrequent or occasional in desert shrub, sagebrush, pinyon-juniper, and Douglas-fir-snowberry communities, often parasitic on sagebrush roots, but also on several other plants; 5,000-9,200 (10,000) ft; May-July.

Orobanche ludoviciana Nuttall — Louisiana broomrape — The 2 specimens seen are from Douglas Canyon, w. of Dinosaur Quarry (Graham 7556 CM!) and Browns Park (Refsdal 6416); plotted for the E. Tavaputs Plateau in Albee et al. (1988) 5,000-5,480 ft. July-Aug.

Orobanche uniflora L. — ghost-pipe, naked broomrape — Uinta Mtns. (no specimens seen from Daggett Co.); infrequent in many plant communities including desert shrub, sagebrush, pinyon-juniper, aspen, and spruce-fir; 6,160-10,800 ft; may-July. Our plants are referable to var. *occidentalis* (Greene) Taylor & Macbride [*O. u.* var. *minuta* (Suksdorf) Beck].

PAPAVERACEAE Poppy Family

Papaver L. — Poppy

Papaver coloradense (Fedde) Fedde ex Wooton & Standley [*P. uintaense* S. L. Welsh. The following names of arctic and northern Rocky Mtn. plants have been misapplied to the Uinta Mtn. plants: *P. nudicaule* L. var. *radicatum* de Candolle; *P. radicatum* ssp. *kluanense* (D. Love) D. F. Murray; *P. pygmaeum* Rydberg; *P. radicatum* Rottboell var. *pygmaeum* (Rydberg) S. L. Welsh] — Rocky Mtn. poppy, Uinta poppy — Plants caespitose; scapes 5-15 cm long, blackish hirsute; leaves basal, 2-10 cm long, deeply lobed or parted, the divisions sometimes incised-toothed to cleft, rarely entire; calyx densely black-hirsute; flowers solitary; sepals 2; petals 4, 1-3 cm long, yellow; stamens numerous; stigmas disk-like; fruit a capsule dehiscent by pores just under the stigma, about 1 cm long. Known from several, isolated, alpine locations, above 11,000 ft along or near the crest of the Uinta Mtns. where most common in semibarrens of creeping talus; July-Aug.

PINACEAE Pine Family

Evergreen, monoecious, resinous trees or shrub-like in krummholz condition at treeline; leaves (needles) needle-like; each ovule borne on the face of a scale or bract, not enclosed in an ovary; the scales several, firm to woody, borne along a central axis and forming a cone, the cones unisexual, the staminate cones rather soft, the scales not woody.

---

1   Leaves borne in fascicles of 2-5 together; scales of pistillate cones woody, rigid, often with a prickly point; twigs alternate . . . . *Pinus*

1   Leaves borne singly on the branches; pistillate scales softer more flexible than above, without a prickly point; twigs often opposite 2

2   Leaves more or less 4-angled in cross-section (the angles detectable to the touch when leaves are rolled between thumb and fore finger), soon deciduous in dried specimens, borne on woody, peg-like bases, these pegs persistent on the twigs after the leaves have fallen; pistillate cones pendulous, the scales persistent, not subtended by a bract . . . . *Picea*

2   Leaves flat in cross-section, not soon deciduous in dried specimens, not borne on peg-like bases, leaving rounded, rather sunken scars when fallen; scales various 3

3   Pistillate cones erect, yellow-green to purplish-black; cone-scales deciduous, the subtending bracts concealed by the scales; leaves not

twisting nor narrowed at the base, often curving outward and upward from the twigs; winter buds not reddish, blunt; twigs, glabrous, nearly all opposite .... *Abies*

3   Pistillate cones pendulous, greenish when young, brown in age, the scales not deciduous, subtended and exceeded by a conspicuous 3-pronged bract; leaves slightly twisted and narrowed at the petiolate base, not strongly curving upward from the twigs; leaf scars slightly raised; winter buds reddish, pointed; twigs usually pubescent at least for some years, alternate and opposite .... *Pseudotsuga*

Abies Mill — Fir

Trees; bark smooth when young, with resin blisters in age; leaves flat, linear, inserted singly, the basal scars circular; pistillate cones erect, with thin, deciduous scales, and with bracts shorter than and concealed by the scales; seeds winged.

---

1   Young branches glabrous; resin ducts of leaves next to the epidermis of the lateral margins; some leaves of the lower branches usually over 3 cm long; pistillate cones gray or yellow-green .... *A. concolor*

1   Young branches pubescent; resin ducts of leaves midway between the lateral margin and the midvein; leaves not over 3 cm-long; pistillate cones blackish-purple .... *A. lasiocarpa*

Abies concolor (Gordon & Glendinning) Lindley ex Hildebrand — white fir — The few specimens seen are from Rock Creek and w. in the Uinta Mtns. and from the Avintaquin drainage and w. on the W. Tavaputs Plateau, also reported for Argyle Canyon drainage; 7,500-9,000 ft.

Abies lasiocarpa (Hooker) Nuttall — subalpine fir — Forming solid stands and even thickets to the exclusion of most other vegetation in concave drainage-heads at upper elevations of the W. Tavaputs Plateau, and found as a dominant in small, isolated, usually wet places in the Uinta Mtns. but mostly mixed with lodgepole pine and Engelmann spruce on the vast areas of quartz-rich sandstones of the Uinta Mtn. Group; 7,500-11,000 ft. This tree is sometimes referred to as balsam or balsam fir, but this name is more properly applied to a closely related tree [*A. balsamea* (L.) Miller] that grows in ne. United States and across Canada.

Picea A. Dietrich — Spruce

With features of the family and as listed in the key. In addition to the 2 native species listed below Norway spruce [*P. abies* (L.) Karsten] with ovulate cones 10-18 cm long is commonly cultivated as an ornamental and shade tree.

---

1   Pistillate cones 4-6 (7.5) cm long; young twigs usually puberulent; needles moderately sharp to rather blunt to the touch; plants mostly above 8,500 ft .... *P. engelmannii*

1   Pistillate cones 6-10 cm long; young twigs glabrous; needles sharp to the touch; plants mostly below 8,500 ft. .... *P. pungens*

Picea engelmannii Parry in Engelmann — Engelmann spruce — Uinta Mtns., no specimens seen from the Tavaputs Plateau east of Willow Creek of the Strawberry drainage; codominant with lodgepole pine at 9,000-10,500 ft, dominant from 10,400 ft up to timberline (about 11,000 ft) where the trees are reduced to windswept shrubs (krummholz).

Picea pungens Engelmann — Colorado blue spruce., blue spruce — Widespread in the Uinta Mtns. along drainages and on limestone hills of the Brush Creek drainage and on the Tavaputs Plateau along drainages and forming upland stands in the headwaters of Slab Canyon and Beaver Canyon and perhaps elsewhere; 7,000-8,500 ft; a handsome tree, often cultivated as an ornamental.

Pinus L. — Pine

Trees; conspicuous leaves borne in fascicles of 2-5, bound together at the base by a membranous sheath, needle-like or narrowly linear; pistillate cones with woody thickened scales, these often with prickles at the apex; seeds winged.

1 Leaves 4 or 5 per fascicle 2
1 Leaves 2 or 3 per fascicle 3
2 Cones scales ending in small but sharp prickles; leaves to 4 cm long, persisting for many years and providing a dense covering on the branches; trees known from the Tavaputs Plateau .... *P. longaeva*
2 Cone scales with smooth margins; leaves 3-7 cm long, usually borne near the tip of the branches; trees widely distributed in our mountains .... *P. flexilis*
3 Leaves mostly greater than 7 cm long, 2-3 per fascicle; cones mostly greater than 5 cm long .... *P. ponderosa*
3 Leaves mostly less than 7 cm long, 2 per fascicle; cones less than 5 cm long 4
4 Cone scales ending in short prickles; trees often over 8,000 ft except in canyons, mostly over 15 m tall at maturity; seeds less than 2 mm thick ... . *P. contorta*
4 Cone scales smooth; trees mostly below 8,000 ft, seldom exceeding 6 m tall; seeds over 2 mm thick .... *P. edulis*

Pinus contorta Douglas ex Loudon (*P. murrayana* Balfour) — lodgepole pine — The dominant tree of much of the montane and lower subalpine elevations of the Uinta Mtns.; closely associated with soils derived from quartz-rich sandstones of the Uinta Mtn. Group, forming thickets of stunted trees by prolific seed germination following fire or logging; common to dominant and forming solid stands at (7,500 ft in canyons) 8,000-9,500 ft, mostly in association with Engelmann spruce at 9,500-10,500 ft, occasionally above 10,500 ft and rarely above timberline. Lodgepole pine is the basis of the local pole and mine prop industry. It is somewhat small for a saw timber tree, but because of its abundance, it has been an important lumber tree of our area. Our trees belong to var. *latifolia* Engelmann. ex S. Watson. Leaves (2) 3-7 cm long and averaging longer than those of *P. edulis*.

Pinus edulis Engelmann in Wislizenius — pinyon, pinyon pine, Colorado pinyon pine — Widespread: codominant with Utah juniper and forming solid stands at higher elevations of the pinyon-juniper thermal belt, apparently more sensitive to cold temperatures of long duration than Utah juniper and not extending down as far into areas of cold air drainage and more common on the Tavaputs Plateau and on the n. slope of the Uinta Mtns. than on the s. slope of the Uinta Mtns. at (6,000) 6,500-8,000 ft. The seeds (pinyon nuts) are edible. The wood is pitchy, and that of long-dead trees is prized as firewood. Leaves are 1.5-4 (5) cm long and averaging shorter than those of *P. contorta*.

Pinus flexilis James — limber pine — Widespread in the Uinta Mtns. and Tavaputs Plateau on windswept dry ridges, and canyons of the s. slope of the Uinta Mtns., usually with Douglas-fir and other coniferous species; 7,000-10,000 ft.

Pinus longaeva D. K. Bailey (*P. aristata* Engelmann misapplied) — Intermountain bristlecone pine — Confined to marly slopes and ridges of the West Tavaputs Plateau in the Avintaquin, Indian Canyon, Lake Canyon, Sowers Canyon, and Argyle Canyon drainages; 7,500-9,000 ft.

Pinus ponderosa Douglass ex Lawson and Lawson [*P. scopulorum* (Engelmann) Lemmon] — ponderosa pine, yellow pine — Widespread; mostly 7,000-8,000 ft on the Uinta Mtns. and the Yampa Plateau, to 9,000 ft on the Tavaputs Plateau, down to 6,000 ft along drainages such as the Whiterocks and Uinta Rivers, and at a few isolated stations well into the Uinta Basin on Mowry Shale. Most abundant and forming the largest stands on the n. slope of the Uinta Mtns. in Daggett Co. and with a sizable area of dominance from Lake Fork Canyon to Uinta Canyon on the south slope.

Pseudotsuga Carr. — Douglas-fir; Red pine

Pseudotsuga menziesii (Mirbel) Franco D. [*P. mucronata* (Rafinesque) Sudworth; *P. taxifolia* (Poiret) Britton] — Douglas-fir, red pine — Trees to 50 m tall; bark of young trees gray, smooth and thin, but becoming reddish-brown or brown and deeply fissured on older trees; limbs generally lacking on the lower portion of mature trees, more irregular in size in comparison to the uniform branches of fir and spruce; leaves 1.5-3 cm long, flat in cross-section, slightly twisted in the minute petiolate base; cones 4-7 cm long, brown, each scale subtended by a prominent 3-pronged bract, the middle prong longer and narrower than the lateral 2. Widely distributed and abundant on the Tavaputs Plateau where it often forms closed stands on n. exposures below elevations favored by subalpine fir, and sometimes on all exposures at higher elevations; widespread in the Uinta Mtns. from 7,000-9,000 ft where associated with other conifers and aspen or forming solid stands on soils derived from limestone, shale, and Weber Sandstone, comparatively rare on soils derived from sandstones of the Uinta Mtn. Group except in canyons associated with Red Canyon of the Green River;. Our trees belong to var. *glauca* (Beissner) Franco (inland Douglas-fir), and they are slower growing and much smaller than the huge trees of var. *menziesii* of the Pacific Northwest.

PLANTAGINACEAE Plantain Family

Plantago L. — Plantain

Annual or perennial herbs; leaves basal (ours), often in rosettes, usually strongly nerved, simple, entire, or occasionally toothed or lobed; flowers small, 4-merous, in congested bracteate spikes of heads; corolla united, scarious, not at all showy, persistent, often remaining on top of the mature fruit, the lobes spreading or deflexed; stamens (2) 4; ovary superior, appearing inferior, 1-4 chambered; style 1; fruit mostly a circumscissile capsule, with 1-4 seeds in each chamber.

---

1 Plants annual; leaves filiform to linear-oblanceolate, 1-5 (7) mm wide . . . 2

1 Plants perennial; leaves lanceolate or wider, some over 7 mm wide 3

2 Inflorescence glabrous; leaves threadlike . . . . *P. elongata*

2 Inflorescence sparsely to densely hairy; leaves linear to linear lanceolate
. . . . *P. patagonica*

3 Leaf blades broadly ovate, 1-3 times as long as wide, abruptly contracted to the petiole; seeds 6-30 per capsule; spikes 5-30 cm long; plants not woolly at the base . . . . *P. major*

3 Leaf blades lanceolate to elliptic, 2-10 times as long as wide, gradually contracted to the petiole; seeds 2-4 per capsule; spikes either less than 8 cm long or else root crown bearing dense tufts of long hairs 4

4 Root crown not bearing dense tufts of long hairs, sometimes sparsely brown-hairy . . . . *P. tweedyi*

4 Root crown bearing dense tufts of long hairs 5

5 Root crown with tufts of rust-colored hairs; capsule circumscissile below the middle; outer 2 sepals like the others, free; bracts acute . . . *P. eriopoda*

5 Root crown with tufts of white or tan hairs; capsule circumscissile at the middle; outer 2 sepals of at least the upper flowers fused, the resulting segment usually 2-notched at the apex; bracts acuminate or caudate-acuminate . . . . *P. lanceolata*

Plantago elongata Pursh — slender plantain — The one specimen seen (B. Welsh & G. Moore 17) is from near Red Wash ne. of Rangely. Listed for Uintah Co. in AUF (548) apparently based on specimens belonging to *P. patagonica*. Weber (1987) lists this as locally abundant in alkali flats in Mesa Co., Colorado outside our area. (IMF 4: 336) lists habitat as "alkaline marshes and somewhat saline lowlands"; May-June. Our plants belong to var. *elongata*. Capsules 4 (6) seeded; stamens generally 2.

Plantago eriopoda Torrey — alkali plantain, redwool plantain — Most specimens seen are from Manila to Browns Park in Daggett Co., one specimen is from Tabiona in Duchesne Co.; floodplains, riparian communities; 5,910-6,400 ft; June-Aug.

Plantago lanceolata L. — buckhorn plantain, English plantain — Introduced from Europe; gardens, lawns, pastures, seeps, meadows; 5200-7,900 (9,500) ft; May-Sept.

Plantago major L. — broadleaf plantain, common plantain — Native of Eurasia and perhaps parts of North America, widespread in our area; a weed of lawns and in moist and wet, often disturbed places; up to 8,800 ft; June-Sept.

Plantago patagonica Jacque. (*P. purshii* Roemer & Schultes) — Indian-wheat, woolly plantain — Widespread; locally abundant in salt desert shrub, sagebrush, and pinyon-juniper communities; May-June. The name *P. patagonica* var. *spinulosa* (Decaisne) A. Gray is available for plants with long bracts that may be a result of introgression with *P. aristata* Michaux. Capsules 2-seeded; stamens 4.

Plantago tweedyi A. Gray — Tweedy plantain — Most of the specimens seen are from Strawberry Valley, Wolf Creek Pass and e. to the Rock Creek on the s. slope of the Uinta Mtns., and e. to the Blacks Fork drainage on the n. slope; occasional along roads, about ponds, dry to wet sites in parklands of aspen and coniferous forests, and open woods, 9,055-10,400 ft; June-Sept.

PLUMBAGINACEAE Leadwort Family

Armeria Willdenow

Armeria maritima (Miller) Willdenow — sea-pink, thrift — Plants scapose herbs; leaves all basal, linear to narrowly lanceolate, 1.5-10 cm long, 1-2 mm wide, glabrous or ciliate; scapes mostly 10-15 cm tall with the apex enveloped by a sheathing bract; inflorescence a head-like cluster of flowers 1.5-3 cm wide, subtended by scarious bracts; flowers borne in clusters of 3 subtended by paired transparent bracts, short pedicellate; calyx 4-7 mm long, pubescent at the base and along the ribs; corolla included in the calyx, pinkish. Known from alpine tundra nw. of Gilbert Peak and on Kabell Ridge on the north slope of the Uinta Mtns. Our plants belong to var. *sibirica* (Turczaninow ex Boissier) A. Blytt.

POACEAE (GRAMINEAE) Grass Family

Plant annual or perennial, herbaceous, arising from fibrous roots or rhizomes, sometimes stoloniferous; stems mostly terete, rarely flattened, the internodes hollow or rarely solid, the nodes often swollen and prominent; leaves alternate, simple, the blades mostly linear or filiform, parallel-veined, strongly sheathing at the base, the sheath open or closed; auricles sometimes developed at the juncture of the sheath and blade (collar); ventral surface of the collar projected into a ligule, this mostly a membranous, translucent scale or occasionally a fringe of hairs or rarely lacking; inflorescence a spike, raceme, panicle, or panicle of racemose or spike-like branches; flowers bisexual or unisexual in a few species, some sterile or rudimentary in just a few of our species, solitary or 2-many in a spikelet; spikelets consisting of a rachilla (axis), (1) 2 glumes (empty scales that subtend or enclose 1 or more florets), and 1 or more florets; florets consisting of 2 scales (a lemma and palea, the lemma usually enclosing the palea), usually 1-3 stamens, and 1 pistil with 3 stigmas; fruit a caryopsis, dry and indehiscent.

The separation of *Elymus, Hordeum,* and *Sitanion* is laden with exceptions and undermined by hybridization. The synonymy listed herein reflects several alternative treatments. The inclusion of *Sitanion* and most species previously treated as *Agropyron* into *Elymus* follows Gould (1947) and Arnow (1987) where section level rather than generic level treatment seems more user-friendly.

1   Inflorescence an open or contracted panicle, sometimes with 2 or more spicate branches, if contracted the branches readily evident upon teasing the inflorescence apart...2

1   Inflorescence a single spike or spike-like raceme, or dense, spike-like panicle in which the branches are not readily evident (species with spike-like panicles are keyed both ways)...................................7

2   Spikelets borne in burs or cup-like structures........................3

2   Spikelets not borne in burs or cup-like structures...................4

3   Burs spiny, not closely subtended by leaf-like bracts; plants sometimes over 20 cm tall ....*Cenchrus*

3   Bur-like or cup-like structures awned but not spiny, closely subtended and sometimes partly enveloped by leaf-like bracts; plants not over 20 cm tall ....*Buchloe*

4   Inflorescence of 2-10 digitate or spicate branches, these often strongly one-sided with spikelets born on one side ....**KEY 6**

4   Inflorescence not as above...5

5   Spikelets with 2 or more well-developed fertile florets (1 or more florets sometimes staminate)..6

5   Spikelets with 1 well-developed fertile floret ....**KEY 7**

6   Glumes shorter than the first floret ....**KEY 1**

6   One or both glumes longer than the lowest floret to as long as the spikelet ....**KEY 5**

7   Ligules of hairs..8

7   Ligules membranous...12

8   Spikelets borne in the axils of leaves, more or less hidden in the leaves ....*Munroa*

8   Spikelets usually elevated above the leaves 9

9   Glumes 9-23 mm long ....*Danthonia unispicata*

9   Glumes less than 9 mm long 10

10 Inflorescence not one sided or comb-like ....*Erioneuron*

10 Branches of the inflorescence one-sided and comb-like...............11

11 Inflorescence staminate ....*Buchloe*

11 Inflorescence with staminate and pistillate ....*Bouteloua*

12 Spikelets borne in burs or cup-like structures ....(see lead 3 above)

12 Spikelets not borne in burs or cup-like structures.................13

13 Inflorescence densely long villous, the hairs 3-5 mm long ....*Hilaria*

13 Inflorescence not villous as above 14

14 Spikelets with1floret, not over 6 mm long ....**KEY 9**

14 Spikelets with more than1floret and often over 6 mm long ....**KEY 10**

— KEY 1 —

1   Tall stout reeds generally over 1.5 m tall; rachilla with silky hairs as long as the spikelets ....*Phragmites*

1   Plants not as above; hairs of rachilla lacking or shorter than above 2

2   Florets converted into bulblets; plants with bulbous bases . . . *Poa bulbosa*
2   Florets not converted into bulblets; plants without bulbous bases except in
    Melica and *Dactylis* 3
3   Lemmas (at least some) awned or awn pointed, the awn over 2 mm long .
    . . . . **KEY 2**
3   Lemmas awnless, or if awn pointed then the point less than 2 mm long . . 4
4   Ligule a fringe of hairs; sheaths often with a tuft of pilose hairs at the
    summit and sometimes on the collar . . . . **KEY 3**
4   Ligule not a fringe of hairs; sheaths various . . . . **KEY 4**

— KEY 2 —

1   Upper lemmas with dorsal twisted awns and bristle-like teeth . *Ventenata*
1   Lemmas awned from near the tip 2
2   Spikelets crowded and arranged 1-sided on the panicle branches; glumes
    and lemmas stiff-ciliate (use 10× magnification) . . . . *Dactylis*
2   Spikelets not as above; glumes not ciliate; lemmas sometimes soft-ciliate .
    3
3   Callus of lowest floret bearded with hairs 1-3 mm long . . . . *Schizachne*
3   Callus of florets not bearded as above 4
4   Lemmas (excluding awns) averaging over 7 mm long . . . . *Bromus*
4   Lemmas (excluding awns) averaging less than 7 mm 5
5   Lemmas awned from a notched apex; ligule 2-7 mm long . . . . . *Leptochloa*
5   Lemmas not notched, ligule mostly less than 2 mm long 6
6   Plants perennial; spikelets at maturity with 2-4 (8) florets . . . . *Festuca*
6   Plants annual; spikelets at maturity with 5-17 florets . . . . *Vulpia*

— KEY 3 —

1   Inflorescence of 2-4 spikelets, these exceeded by a tuft of leaves . . *Munroa*
1   Inflorescence with more than 4 spikelets or spikelets well exerted above
    the leaves 2
2   Lemmas densely hairy at least at the base. 3
2   Lemmas without hairs 4
3   Inflorescence 20-50 cm long . . . . *Redfieldia*
3   Inflorescence 1.5-4.5 cm long . . . . *Erioneuron*
4   Plants with long, stout scaly rhizomes . . . . *Distichlis*
4   Plants from fibrous roots, rhizomes lacking . . . . *Eragrostis*

— KEY 4 —

1   Spikelets with (1) 2 (3) florets 2
1   Spikelets with 3-25 florets 3
2   Glumes blunt and erose; leaves 3-8 (12) mm wide . . . . *Catabrosa*
2   Glumes acute or awn-pointed, not erose; leaves 1-4 mm wide . . . .
    . . . . *Muhlenbergia*

3 Ligules ciliate, the hairs longer than the membranous base . . . . . **Distichlis**

3 Ligules membranous 4

4 Glumes 0.5-3.5 mm long; lemmas 1.5-3 (4) mm long, glabrous 5

4 Glumes and/or lemmas mostly longer than above except sometimes in *Poa*
   and then the lemmas mostly pubescent 7

5 Plants tufted, without rhizomes, often growing in alkaline soil or salted
   roadsides; leaf sheaths open with membranous margins; leaf blades 0.5-4.5
   mm wide, often becoming involute . . . . **Puccinellia**

5 Plants strongly rhizomatous with culms arising singly or few together, of
   non alkaline places and not on salted roadsides; leaf sheaths open or closed,
   the margins not membranous; leaf blades 2-15 mm wide, usually flat or
   folded 6

6 Leaf sheaths open for nearly their entire length; lemmas with 5 nerves . .
   . . . . **Torreyochloa**

6 Sheaths closed from just below the collar; lemmas with (5) 7-9 nerves . . .
   . . . . **Glyceria**

7 Spikelets crowded and arranged 1-sided on the panicle branches; glumes
   and lemmas stiff-ciliate . . . . **Dactylis**

7 Spikelets not crowded nor arranged 1-sided on the panicle branches;
   glumes and lemmas not ciliate 8

8 Plants with bulbous bases; glumes papery; upper florets empty . . . **Melica**

8 Plants without bulbous bases. 9

9 Spikelets 7-15 mm wide, 12-32 mm long; plants annual . . . .
   . . . . **Bromus brizaeformis**

9 Spikelets less than 7 mm wide or else plants perennial 10

10 Spikelets over 19 mm long . . . . **Bromus**

10 Spikelets less than 17 mm long 11

11 Spikelets mostly 3-10 mm long 12

11 Spikelets 10-14 mm long (unusual specimens of *Poa* will also key here) . 13

12 Lemmas rounded to acute at the tip . . . . **Poa**

12 Lemmas narrowed to a pointed tip or short awn . . . . **Leucopoa**

13 Panicle narrow, the branches short and bearing spikelets to near their base;
   spikelets with 3 (5) florets . . . . **Leucopoa**

13 Panicle spreading at the base, some of the branches elongate and not
   bearing spikelets at the base; spikelets with (4) 5-13 florets . . . .
   . . . . **Schedonnardus**

— KEY 5 —

1 Glumes 18-29 mm long. 2

1 Glumes less than 18 mm long 3

2 Panicles mostly with less than 6 spikelets . . . . Danthonia

2 panicles mostly with over 6 spikelets . . . . Avena

3 Spikelets 7-20 mm long; awns 5-20 mm long 4

3 Spikelets 3-7 mm long (excluding awns); awns lacking or to 6 mm long . . 6

4   Spikelets with 5-7 or more flowers; awns flattened, 5-11 mm long . . . . ***Danthonia***

4   Spikelets mostly with 2 florets, the second floret smaller, staminate or perhaps sterile; awns not flattened, 8-18 mm long 5

5   Plants less than 30 cm tall, from above timberline . . . . ***Helictotrichon***

5   Plants over 30 cm tall, from well below timberline . . . . ***Arrhenatherum***

6   Leaf blades of fertile culms mostly less than 2 cm long; plants from long slender rhizomes; lemmas pubescent, awnless . . . . ***Hierochloe***

6   At least some leaf blades of fertile culms over 2 cm long; rhizomes lacking or short 7

7   Lemmas awned and slightly bearded at the base 8

7   Lemmas awnless or seldom with a minute awn, glabrous 9

8   Lemma awned from above the middle or awnless; ligules blunt to rounded, not over 4 mm long; panicles narrow, the branches ascending . . ***Trisetum***

8   Lemmas awned from below the middle; ligules pointed, often over 4 mm long; panicles narrow or with widely spreading branches . . . ***Deschampsia***

9   Rachilla pubescent . . . . ***Trisetum wolfii***

9   Rachilla glabrous 10

10 Second glume widest toward the apex, contrasting in shape and size with the first glume; disarticulation below the glumes; plants annuals or short-lived perennials . . . . ***Sphenopholis***

10 Second glume not wider toward apex, similar to first glume; disarticulation above the glumes; plants caespitose perennials . . . . ***Koeleria***

— KEY 6 —

1   Spikelet bearing branches not over 1 cm long; plants stoloniferous, not over 20 cm tall . . . . Buchloe

1   Some of the spikelet bearing branches over 1 cm long, or plants not stoloniferous and usually over 20 cm tall 2

2   Pedicles and rachis densely long villous, the hairs 3-5 mm long; plants sparingly introduced. 3

2   Pedicles and rachis not long villous as above 4

3   Pedicels and rachis segments flat to rounded, not grooved down the center; first glume of sessile spikelet glabrous or scaberulous below mid-length; inflorescence with 2-7 more or less erect branches . . . . ***Andropogon***

3   Pedicels and rachis segments (at least toward the apex of branches) vertically grooved; first glume of sessile spikelet usually pubescent below mid-length; inflorescence of 2-10 ascending or spreading branches . . . . ***Bothriochloa***

4   Inflorescence more or less digitate with 2-16 branches 3-20 cm long . . . . ***Digitaria***

4   Inflorescence not at all digitate. 5

5   Ligule membranous or lacking 6

5   Ligule a fringe of hairs 8

6   Leaf blades 0.6-2 (3) mm wide, 2-12 cm long . . . . ***Schedonnardus***

6   Leaf blades 3-30 mm wide, some often over 12 cm long 7
7   Spikelets with rigid hairs; ligule lacking .... *Echinochloa*
7   Spikelets glabrous or at most scabrous; ligule 4-11 mm long .. *Beckmannia*
8   Branches of the inflorescence strongly divergent to reflexed; culms solid; plants of dry places .... *Bouteloua*
8   Branches of the inflorescence erect or ascending; culms hollow; plants of moist or wet places .... *Spartina*

— KEY 7 —

1   Glumes 1.4-2.5 mm long with awns about 2 or more times longer than the body .... *Polypogon*
1   Glumes longer or awnless or with awns not longer than the body 2
2   Plants annual or short-lived perennials, mostly in cultivated fields, or weedy in gardens and waste places; fertile florets rather plump, partly enveloped by a sterile floret that resembles the second glume 3
2   Plants perennial or annual, usually not weedy as above; sterile florets none or much reduced 5
3   Ligules lacking; spikelets nearly sessile; sterile floret awned .. *Echinochloa*
3   Ligules present, often of a fringe of hairs; spikelets often on long pedicels; florets awnless 4
4   Plants mostly 1-2 m tall; leaf blades 1.2-5 cm wide; panicle with ascending branches .... *Sorghum*
4   Plants mostly less than 1 m tall; leaf blades 0.4-1.5 (2) cm wide; panicle with slender, widely spreading branches .... *Panicum*
5   Glumes fan-like, widest at the middle or toward the apex (at least the second one); disarticulation below the glumes 6
5   Glumes not fan-like, widest at base or toward the middle; disarticulation above the glumes except in *Cinna* .... **KEY 8**
6   Glumes about equal, widest near the middle; panicle branches bearing spikelets on 2 sides; ligules 5-8 mm long; spikelets 1-flowered *Beckmannia*
6   Second glume widest at the apex; first glume smaller and linear; panicle branches not as above; ligules 1-4 mm long; spikelets 2-flowered ....
      .... *Sphenopholis*

— KEY 8 —

1   Lemmas awned from the back, the awn sometimes hidden in callus hairs .
      .... *Calamagrostis*
1   Lemmas awnless or awned from near the tip 2
2   Lemmas with awn over 2 mm long 3
2   Lemmas awnless or awn less than 2 mm 6
3   Lemmas with 3 awns .... *Aristida*
3   Lemmas with 1 awn.. 4
4   Glumes 2.5-3.6 mm long; plants rhizomatous ....
      .... *Muhlenbergia* (M. andina, M. thurberi in part)
4   Glumes 4 mm long or longer; plants tufted, not rhizomatous 5

5 Lemma pubescent with an awn 3-240 mm long .... *Stipa*

5 Lemma scabrous with an awn 1-5 mm long .... *Festuca dasyclada*

6 Lemmas pubescent 7

6 Lemmas glabrous or at most scabrous 9

7 Leaves 6-25 mm wide, some usually well over 15 cm long .... *Phalaris*

7 Leaves 1-6 mm wide, mostly 1-15 cm long. 8

8 Plants tufted, without rhizomes; ligules not ciliate; panicle often open ...
   .... *Blepharoneuron*

8 Plants strongly rhizomatous; ligules ciliate; panicle narrow *Muhlenbergia*

9 Lemmas (4) 5-6 mm long .... *Festuca dasyclada*

9 Lemmas 1.5-5 mm. long 10

10 Glumes longer than the lemma .... *Agrostis*

10 One or both glumes shorter than the lemma 11

11 Ligule a ring of hairs .... *Sporobolus*

11 Ligule membranous, sometimes ciliate 12

12 Leaf blades (5) 7-20 mm wide; plants of riparian communities in glacial
    canyons of Uinta Mtns. .... *Cinna*

12 Leaf blades mostly less than 6 mm wide .... *Muhlenbergia*

— KEY 8 —

1 Ligules a fringe of hairs 2

1 Ligules membranous. 3

2 Spikelets subtended by 1-several awn-like bristles .... *Setaria*

2 Spikelets without subtending bristles .... *Crypsis*

3 Glumes with awns longer than the body .... *Polypogon*

3 Glumes awnless or with awns shorter than the body 4

4 Lemmas awned from the back; glumes awnless; disarticulation below the
   glumes .... *Alopecurus*

4 Lemmas awnless; glumes awn-pointed; disarticulation above the glumes .
   .... *Phleum*

KEY 10

1 Spikelets mostly 1 per node of the rachis, scattered to dense in the spike 2

1 Spikes with 2 or more spikelets per node of the rachis; some spikelets
   reduced to awns in *Hordeum* 9

2 Spikelets turned edgewise to the rachis, with only 1 glume (the side turned
   into the rachis without a glume) .... *Lolium*

2 Spikelets not turned as above; glumes 2 3

3 Plants annual, introduced, weedy or cultivated and occasionally escaping
   4

3 Plants perennial, introduced or native 7

4 Spike not over 2 cm long .... *Eremopyrum*

4 Spike over 2 cm long. 5

5   At least the upper glumes with awns over 1 cm long: margins of leaf sheaths
    ciliate; auricles inconspicuous .... *Aegilops*

5   Awns of glumes less than 1 cm long; leaf sheaths not ciliate; auricles usually
    well developed 6

6   Florets 2-5 per spikelet; glumes mostly 3-nerved .... *Triticum*

6   Florets 2 per spikelet; glumes 1-nerved, stiff, subulate .... *Secale*

7   Rachis tardily disarticulating; culms not prostrate; plants sterile hybrids;
    glumes as well as lemmas with awns over 1 cm long .... *Elymus* (hybrids)

7   Rachis not disarticulating, or if so culms prostrate; plants growing at high
    elevations, not sterile hybrids...................................... 8

8   Spikelets densely crowded and turned edgewise to the rachis; rachis
    internodes less than 2 mm long .... *Agropyron*

8   Spikelets not as above; rachis internodes over 2 mm long .... *Elymus*

9   Spikelets 3 per node of the rachis, the 2 lateral spikelets slightly
    pedunculate, empty, sometimes reduced to awns; central spikelet sessile
    and fertile with 1 floret, if florets all sessile (*Hordeum vulgare*) then plants
    annual and cultivated .... *Hordeum*

9   Spikelets 2 or more per node of the rachis, all alike, each with 2 or more
    florets (only 1 floret often present as they are sometimes readily deciduous);
    plants perennial, native, not cultivated............................ 10

10 Awns of lemmas not over 6 mm long .... *Elymus*

10 Awns of lemmas over 6 mm long 11

11 Spikes about 5 mm wide; body of the lemmas 6-8 mm long; rachis
    disarticulating at the nodes .... *x Elyhordeum*

11 Spikes wider and body of lemmas usually longer than above ..... *Elymus*

Aegilops L. — Goatgrass

Aegilops cylindrica Host [*Cylindropyrum cylindrica* (Host) Love] — jointed
goatgrass — Plants annual, 20-50 cm tall; ligules less than 1 mm long; leaf
blades flat or sometimes rolled, 2-5 mm wide; inflorescence a spike 4-12 cm
long, more or less cylindrical; spikelets solitary at each node of the rachis,
cylindrical, somewhat sunken in the concave internodes of the rachis, 2-5
flowered; glumes (6) 10-14 mm long, about as long as the spikelets, awned and
toothed at the apex, the awn 1-20 mm long; lemmas 8-11 mm long, awned
from a notched apex, the awn 1-5 mm long, awns of the terminal spikelet up to
6 (8) cm long. Introduced from Eurasia; usually on disturbed ground at lower
elevations; June-Aug.

Agropyron Gaertner — Wheatgrass

Agropyron cristatum (L.) Gaertner [*A. desertorum* (Fischer) Schultes; *A. sibiricum*
(Willdenow) Beauvios] — crested wheatgrass — Plants perennial, tufted; auricles
usually present; blades flat or loosely involute upon drying; Inflorescence a
more-or less flattened spike with densely crowded spikelets; spikelets disarticulating above the glumes and between the florets. Introduced from Eurasia,
used extensively in rangeland and roadside seedings in the sagebrush and
pinyon-juniper belts; May-June. Intergrading taxa of this complex are complicated
by selective breeding designed to develop new cultivars. While many taxonomic

treatments include *A. desertorum* (standard wheatgrass) and *A. sibiricum* (Siberian wheatgrass) as synonymous with *A. cristatum,* there are important agronomic differences including drought tolerance and growth form.

Agrostis L. — Bentgrass; Redtop

Plants perennial with or without rhizomes; culms arising singly or in tufts; leaf sheaths open, ligules membranous; auricles lacking; inflorescence an open or compact panicle; spikelets small, with a single floret; glumes awnless or awn-tipped, longer than the lemma.

Species of this genus presents are difficult to separate in the field. The absent or presence and length of the pale and other minute characteristics are critical criteria. The palea is not only small but it is also often concealed by the lemma and glumes, and a dissecting scope is helpful if not essential for positive separation of species. Plants of the various taxa show little variation in the spikelets, and it seems highly likely that the genus has been rather overworked by separating taxa on variations in the panicle (i.e. compact to open and small to large) and by the presence or length of a minute rachilla (i.e. lacking or 0.1-0.5 mm long). The length of the palea does seem to provide a consistent means for separation, but the features of the panicle and rachilla seem quite dubious in some cases.

---

1   Palea over 1/2 as long as the lemmas; plants sometimes with rhizomes . . 2

1   Palea less than 1/2 as long as the lemmas (plants tufted and without rhizomes or these short and thus unlike *A. stolonifera;* see discussion under *A. humilis* for further contrast) 4

2   Plants, with long creeping rhizomes 30-80 (130) cm tall, seldom growing above 9,500 ft; anthers 1 mm long or longer; panicle often over 5 cm long
. . . . *A. stolonifera*

2   Rhizomes lacking or short; plants 5-25 (35) cm tall, often found above 9,500 ft; anthers less than 1 mm long; panicle often less than 5 cm long 3

3   Plants often 15-35 cm tall; leaf blades 1-4 mm wide, flat; panicle 1.5-15 cm long, the branches 0.5-5 cm long, typically angled and scabrous; glumes narrowly elliptic, barely exceeding the lemma; lemmas whitish . . . .
. . . . *A. thurberiana*

3   Plants mostly less than 15 cm tall; leaf blades to 2 mm wide, rolled or flat; panicle to 6 cm long, the branches to 2 cm long, terete and mostly smooth; glumes lanceolate to ovate, averaging 0.2 mm longer than the lemmas; lemmas dark purple . . . . *A. humilis*

4   Panicles more or less open to diffuse, often over 2 cm wide, the mostly ascending to widely spreading branches not bearing spikelets to the base and not concealing the main axis of the panicle . . . . *A. scabra*

4   Panicles contracted, often less than 2 cm wide, the erect branches often bearing spikelets to near the base and often dense enough to conceal parts or much of the main axis of the panicle. 5

5   Panicles green, sometimes with pale purple markings; plants mostly growing below 9,500 ft; some of the wider leaves 2-5 mm wide . *A. exarata*

5   Panicles purplish; plants mostly growing above 9,500 ft; leaves 0.5-2 mm wide . . . . *A. variabilis*

Agrostis exarata Trinius — spike bentgrass — Apparently widespread; infrequent along streams and in other wet places; 7,000-9,500 ft; June-Sept.

Agrostis humilis Vasey — alpine bentgrass — Uinta Mtns.; occasional to common in wet meadows, bogs, and moist slopes; 9,700-11,500 ft; July-Sept. This has been separated form A. thurberiana by the length of the rachilla, slightly smaller stature, and slightly smaller panicles. Our specimens of these taxa show a continuum in rachilla length from lacking or through 0.1-0.5 (0.7) mm long. Plants with somewhat open panicles are sometimes mistaken for the smaller phase of A. scabra (A. idahoensis). The palea remains the positive means of separation, but rarely does the panicle or the stature approach even the small phase of A. scabra. Sympatric with and easily confused with A. variabilis (q.v.).

Agrostis scabra Willdenow [A. hyemalis (Walt.) B.S.P. var. tenuis (Tuckerman) Gleason] — ticklegrass — Occasional to common in Strawberry Valley and across the Uinta Mtns.; one specimen seen from the E. Tavaputs Plateau; wet to rather dry places in many plant communities; 6,100-10,800 ft; July-Aug. A. idahoensis Nash has been separated from A. scabra as follows:

1 Plants 5-30 (40) cm tall; spikelets 1.3-2.3 (2.6) mm long; panicles not diffused, the branches usually forked below the middle . . . . **A. idahoensis**

1 Plants (20) 30-90 cm tall; spikelets (2.2) 2.5-3.2 mm long; panicles diffuse, the branches forked beyond the middle . . . . **A. scabra**

In the examination of numerous specimens in the field as well as in the herbarium no real morphological or ecological boundary is seen between the 2 forms. However, the trends in the overlapping features of the above key are apparent, and the extremes are different. Hitchcock and others (1969) reported numerous intermediate specimens. Some intermediate plants may be the result of hybridization between A. scabra and narrow-panicled species.

Agrostis stolonifera L. (A. alba L. misapplied; A. palustris Hudson) — redtop, creeping bentgrass — Native to Eurasia and north Africa, introduced and widespread in our area; common to abundant along ditch banks and roadsides and in pastures, wet meadow and riparian communities up to about 9,000 ft; June-Sept. This species has been divided into varieties (var. stolonifera and var. palustris Hudson). The width of panicle is often used in separation of these taxa, but this is highly subject to phonological stage. The panicle is often narrow when young and spreading at anthesis. Some specimens have been collected in which separate panicles of the same plant would key the plant to the 2 different vars. Some of the variation might be due to selective breeding as this is a cultivated pasture grass.

Agrostis thurberiana A. S. Hitchcock — Thurber bentgrass — Uinta Mtns. Infrequent or locally common in moist or wet meadows; about 10,000-11,000 ft; July-Sept. Probably much less common than but sympatric with A. humilis q.v.

Agrostis variabilis Rydberg (A. rossiae Vasey misapplied) — mtn. bentgrass — Uinta Mtns.; occasional or common along streams, in meadows, dry coniferous woods, clear cuts, and open slopes above timberline; 9,750-11,500 ft; July-Sept. Small specimens of A. variabilis are easily mistaken for those of the sympatric A. humilis. Apparently the palea provides the only reliable means of separation, but a few other features of use are listed below:

1 Plants mostly 5-15 cm tall, often with slender although short rhizomes, apparently nearly always growing in wet places; panicles seldom over 4

cm long, densely to loosely flowered: rachilla 0.3-0.7 mm long; lemmas not
awned .... ***A. humilis***
1 Plants 5-40 cm tall, tufted, without rhizomes, growing in wet places or on
well-drained soil; panicles 2-12 cm long, densely flowered; rachilla lacking;
lemmas sometimes awned .... ***A. variabilis***

Alopecurus L. — Foxtail

Perennial plants from fibrous roots or rhizomes; leaf sheaths open, blades
usually flat; ligules membranous, auricles lacking; inflorescence a cylindrical
or ovoid spike-like panicle, the branches short, erect, hidden in the dense
spikelets and not evident without teasing the inflorescence apart and then
hardly visible without magnification; spikelets 1-flowered, flattened, disarticulating
below the glumes; glumes as long or longer than the lemma; lemmas awned
from the back at or below the middle, the awn exserted from or hidden in the
glumes.

---

1 Inflorescence less than 7 mm wide; spikelets 2-4 mm long 2
1 Inflorescence mostly over 7 mm wide; spikelets 3-6 mm long........... 3
2 Awn not or scarcely exceeding the glumes, 0.7-2.5 mm long, nearly straight
.... ***A. aequalis***
2 Awn exserted beyond the glumes, (3) 5-8 mm long, abruptly bent ....
.... ***A. geniculatus***
3 Glumes pubescent throughout; inflorescence 1-4 (5.5) cm long . ***A. alpinus***
3 Glumes ciliate on keel and sometime nerves, otherwise glabrous;
inflorescence 3-10 cm long... 4
4 Glumes divergent at apex, suffused with dark purple; inflorescence turning
dark brown or blackish in age; leaf blades 8-12 mm wide; ligule 1-5 mm long
.... ***A. arundinaceus***
4 Glumes parallel or converging at the apex, typically not suffused with dark
purple; inflorescence not turning dark in age; leaf blades mainly 4-8 mm
wide; ligule less than 2.5 mm long .... ***A. pratensis***

Alopecurus aequalis Sobolevski — short-awn foxtail — Widespread in the Uinta
Mtns. rare or lacking on much of the Tavaputs Plateau; locally common in wet
places, sometimes partly submerged for some of the season, often in recently
drained beaver ponds, also in places that are wet in spring but dry in summer;
(5,800) 7,000-10,500 ft; June-Sept.

Alopecurus alpinus J. E. Smith — alpine foxtail — Uinta Mtns. from the
Whiterocks drainage eastward on the south slope and from Bear River drainage
and eastward across the north slope.; occasional in wet meadows and along
streams; 7,600-10,200 ft; June-Aug.

Alopecurus arundinaceus (Poiret.) in Lamarck (*A. ventricosus* Persoon misapplied)
— creeping foxtail — Rather recently introduced from Eurasia, widely planted
in pastures and as hay and spreading along roadsides and ditches below 7,000
ft.

Alopecurus geniculatus L. — water foxtail, marsh foxtail — Introduced from
Eurasia and known from the n. slope of the Uinta Mtns. in Summit Co. Perhaps
to be expected elsewhere in wet meadows and seeps.

Alopecurus pratensis L. — meadow foxtail — Introduced from Eurasia; meadows, pastures, and edge of woods; to about 9,000 ft; June-Aug. Meadow foxtail appears to be increasing in subalpine meadows of the eastern Uinta Mtns. No specimens seen from the Tavaputs Plateau, but to be expected there.

Andropogon L. — Bluestem

Andropogon gerardii Vitesay (*A. hallii* Hackel) — big bluestem, sand bluestem, turkey-foot — Plants perennial, glaucous, 5) 100-200 cm tall, from robust rhizomes; ligules 2-4mm long, decurrent, ciliate at apex; leaf blades flat or loosely rolled, 3-9 mm wide; inflorescence of 2-5 digitate spicate, branches 4-8 (10) cm long; spikelets 7-12 mm long, 2-flowered (1 fertile and 1 sterile), in pairs, 1 sessile and 1 pedicellate, the pedicel 3.5-6 mm long with yellow hairs 3-5 mm long; fertile lemma 5-8 mm long, slightly shorter than the sterile lemma, awned, the awn about 3-5 mm long. Known from near Starvation Reservoir (Neese 8445) and planted as an ornamental, common e. of the Rocky Mtns. in the Plains; July-Sept.

Aristida L. — Three-awn

Aristida purpurea Nuttall (*A. fendleriana* Steudel; *A. longiseta* Steudel) — purple three-awn — Perennial plants growing in tufts, 15-40 cm tall; leaf blades mostly basal, not over 2 mm wide, mostly involute, pilose at the collar; ligules a ring of hairs, less than 1 mm long; panicle 2-6 cm long, narrow; glumes unequal, the first about 8 mm long, the second about twice as long, both awnless; lemmas 9-12 mm long, each with 3 awns, the awns 2-8 cm long, divergent. Widespread; occasional and locally common in desert shrub, sagebrush, and pinyon-juniper communities; up to about 7,000 ft; May-June. This is a complex taxon [IMF (6:456), Arnow et al. (1980)].

Arrhenatherum Beauvios — Tall oatgrass

Arrhenatherum elatius (L.) Presl — tall oatgrass — Culms loosely tufted, 60-150 cm tall, sometimes bulbous at the base; leaf blades 3-10 mm wide, flat; ligules 1-3 mm long; panicle 15-25 cm long, narrow with ascending branches, or branches spreading at anthesis; spikelets 2-flowered, the first floret staminate and larger than and often enclosing the second, which is pistillate or bisexual; glumes unequal, the first 4-6 mm long, the second 7-9 mm long and as long as the spikelet; lemma of first floret about 6-8 mm long, awned from below the middle on the back with a bent awn 8-18 mm long; lemma of the second floret a little shorter or about equal to the first lemma, awned from above the middle on the back with a awn 3-4 mm long. Introduced from Europe, used extensively in the n. and e. United States as a pasture and hay crop, rather infrequent in our area, seeded in Log Hollow, Uinta Mountains in1955 where it did not persist and apparently recently seeded along Utah Highway 150 in the Bear River drainage where it was fairly common in 2008.; May-Aug.

Avena L. — Oats

Robust annuals; leaf sheaths open, the blades flat; ligule membranous; auricles lacking; inflorescence an open panicle; spikelets 2-3 flowered, large, disarticulating above the glumes; florets bisexual, or the uppermost rudimentary; glumes subequal, longer than the first floret; lemmas awned from about mid length on the back. The 2 species listed below are treated at varietal level in AUF (826).

---

1   Two florets per spikelet with stout, strongly bent awns; lemmas with brown
    hairs .... *A. fatua*
1   Awns lacking or present only on 1 floret and then straight or curved but
    not strongly bent; lemmas glabrous or scabrous .... *A. sativa*

Avena fatua L. — wild oats — Introduced from Eurasia or North Africa; weedy in fields, along roadsides and waste places, sometimes sprouting in montane places where introduced as a contaminant in feed for livestock, but seldom maturing at the higher elevations; May-Sept.

Avena sativa L. (*A. fatua* L. var. *sativa* Haussknecht) — cultivated oats — Introduced from Europe, cultivated, rarely escaping and apparently not persisting; June-Aug. Similar to *A. fatua*, but distinguished by the features given in the key.

Beckmannia Host — Sloughgrass

Beckmannia syzigachne (Steudel) Fernald — American sloughgrass — Plants robust annuals, often stoloniferous, 40-120 cm tall; ligules 5-9 mm long; blades flat, 5-12 mm wide; panicles about 10-30 cm long, narrow, congested; spikelets 1-(2)-flowered; glumes about equal, 2-3 mm long, fan shaped and widest near the middle, strongly keeled; lemma 2-4 mm long, lanceolate, usually projected just beyond the glumes. Widespread; occasional along water courses, marshes, and shorelines of ponds and reservoirs; 5,000-7,240 ft; June-Sept.

Blepharoneuron Nash — Hairy Dropseed

Blepharoneuron tricholepis (Torrey) Nash — hairy dropseed, pine dropseed — Tufted perennial plants; culms 20-60 cm tall; basal leaves numerous, short, rolled and often curled when dry, culm leaves few; ligules about 1 mm long; panicles 5-10 cm long, somewhat open but with ascending branches; spikelets grayish-green or lead colored; glumes 2-3 mm long; lemmas slightly longer than the glumes, densely pubescent on the nerves. Strawberry Valley and across the s. slope of the Uinta Mtns. and Mill City Creek and West Fork Blacks Fork of the n. slope; infrequent or occasional in mtn. big sagebrush, silver sagebrush, ponderosa pine, lodgepole pine, and Engelmann spruce communities; 7,400-9,600 ft; July-Oct.

Bothriochloa Kuntz

Bothriochloa ischaemum (L.) H. Keng — yellow bluestem — Introduced from Eurasia and known from roadsides near Midview Reservoir (northeast of Bridgeland) in Duchesne Co. A limited number of plants seen at the site indicates a recent introduction. On subsequent visits to the area, yellow bluestem was not seen at the site. It appears the plant was not persistent. Additional introductions to our area might be expected.

Bouteloua Lag. — Gramma Grass

Plants annual or perennial; leaf sheaths open; ligule a fringe of hairs; inflorescence of 1-many spikes; spikelets in 2 rows on 2 sides of a 3-angled rachis, with 1 bisexual floret and 1-3 staminate or rudimentary florets above.

1 Plants annual . . . . *B. simplex*

1 Plants perennial. 2

2 Spikelets densely crowded in 1-3 comb-like spikes, more than 10 per spike; plants native . . . . *B. gracilis*

2 Spikelets borne in 20-80 spikes, less than 10 per spike; plants introduced . . . . *B. curtipendula*

Bouteloua curtipendula (Michaux) A. Gray — side-oats gramma — Known from near Starvation Reservoir where it was introduced but apparently did not persisted, occasionally planted in lawns and as an ornamental, native and common in the Plains and southwestern United States; July-Sept.

Bouteloua gracilis (H. B. K.) Lag. — blue gramma, curlygrass — Occasional to common on the Tavaputs Plateau, infrequent but scattered across the Uinta Mtns.; sagebrush and pinyon-juniper communities; 5,500-8,000 ft and extending up to 9,500 ft on exposed ridges; July-Oct.

Bouteloua simplex Lag. — mat gramma — Known from a single collection from Indian Canyon along Highway 191 where the plant was a waif. Subsequent visits to the site verified the population did not persist. However, the introduction indicates additional populations might be found along roads within the Uinta Basin.

Bromus L. — Brome; Chess

Annual or perennial plants; leaf sheaths closed for most their length, the blades flat; ligules membranous; auricles rarely present; inflorescence an open or compact panicle; spikelets many flowered, rather larger, disarticulating above the glumes; glumes unequal, shorter than the lower most lemma; lemmas awned from the tip between 2 small scarious teeth, or awnless in 2 of our species.

1 Plants annual 2

1 Plants biennial or perennial. 5

2 Lemmas awnless or awns less than 2 mm long . . . . *B. brizaeformis*

2 Awns of lemmas over 2 mm long 3

3 Second glume 8-11(13) mm long; lemmas 8-18 mm long . . . . . . *B. tectorum*

3 Second glume 5-9 mm long; lemmas not over 10 mm long. 4

4 Pedicels mostly longer than the spikelets, awns 7-14 mm long; panicle open, the branches spreading . . . . *B. japonicus*

4 Pedicels mostly shorter than the spikelets; awns 4-9 mm long; panicle rather compact, the branches erect or ascending . . . . *B. hordeaceus*

5 Lemmas awnless (awns to 4 mm in one cultivar); plants with rhizomes . . . . *B. inermis*

5   Lemmas awned; plants without rhizomes 6

6   Spikelets flattened; lemmas keeled, 11-17 mm long; second glume 9-13 mm
     long; upper panicle branches not nodding; ligules 1-4 mm long B.
     marginatus

6   Spikelets rounded; lemmas not keeled, 9-13 mm long; second glume 6-10
     (12) mm-long; upper panicle branches usually nodding; ligules 0.3-1 (1.5)
     mm long 7

7   Awns of lemmas (2.5) 3-5 mm long; glumes glabrous or scabrous on the
     midnerve, the lowest one mostly 1-nerved or occasionally 3-nerved at the
     base, gradually tapered from near the base; culms glabrous on and near the
     nodes; leaves 3-13 mm wide; lemmas usually more densely hairy along the
     margins than over the back . . . . **B. ciliatus**

7   Awns of lemmas 1.5-3 mm long; glumes mostly pubescent, the lowest one
     often 3-nerved, lateral nerves as well as the central nerve extending to near
     the tip, usually widest near the middle; culms pubescent on and near the
     nodes; leaves to 2.5 mm wide; lemmas not or only slightly more pubescent
     along the margins than over the back . . . . **B. anomalus**

Bromus anomalus Ruprecht ex Fournier [*Bromopsis anomala* (Fournier) Holub.;
B. porteri (Coulter) Holub — nodding brome or chess — Widespread, common in
mtn. big sagebrush, mt. brush, and ponderosa pine, communities and less
common in Douglas-fir, aspen, and spruce-fir parkland communities, more
often in openings in woods than in shade of trees; 7,400-10,500 ft; June-Aug.

Bromus brizaeformis Fisher & Meyer — rattlesnake chess — Introduced from
Eurasia; Apparently uncommon; The 3 specimens seen are from near Massadona
in Moffat Co. and se. of Vernal, and below Steinaker Dam. Plotted by Albee et
al. (1998) for Daggett Co.; desert shrub and mtn. brush communities; 5,000-
5700 ft; May-July. This plant has been present in our area for many years
without being highly invasive.

Bromus ciliatus L. [*Bromopsis canadensis* (Michaux) Holub] — fringed brome —
Widespread in the Uinta Mtns and in Strawberry Valley and w. end of the W.
Tavaputs Plateau., only one specimen seen from the E. Tavaputs Plateau;
common to abundant in shade of willow, aspen, cottonwood, lodgepole pine,
and Engelmann spruce communities, much less common without shade of
trees; 7,400-11,000 ft; July-Aug. Plants of *B. ciliatus* and *B. anomalus* are rather
easily confused, but are different in a number of features. The narrower-
leaves in even robust specimens of *B. anomalus* are quite different than those
of *B. ciliatus*. About 5 percent of the specimens examined seem to have
intermediate features.

Bromus hordeaceus L. (*B. mollis* L.; *B. racemosus* L.) — soft chess — Introduced
from Eurasia; more or less weedy, various plant communities, often in places
where the ground has been disturbed; Daggett and Uintah Cos. and eastward
in Colorado, no specimens seen from Duchesne Co.; 5,570-7,240 ft; May-July.

Bromus inermis Leysser [*Bromopsis inermis* (Leysser) Holub] — smooth brome —
Introduced from Eurasia, used extensively to seed roadsides, borrow areas,
and rangelands in the aspen-mtn. sagebrush belt, and as a pasture and hay
crop in cultivated areas where it has escaped along ditches, roadsides, and
washes. Occasionally a cultivar with awns to 4 mm long has been planted; July-

Sept. More persistent and even displacing intermediate wheatgrass in and above the aspen belt.

Bromus japonicus Thunberg (*B. commutatus* Schrader) — Japanese chess — Introduced from Eurasia; weedy in gardens, fields and along ditch banks, roadsides, and other disturbed areas, seldom in indigenous plant communities; June-Oct. See Arnow and others (1980) for a discussion of *B. japonicus* and *B. commutatus.* Found in burns in the pinyon-juniper belt in Daggett Co. that had been seeded. Seed was likely a contaminant in seed purchased for these burns. However, the species apparently did not persist in this setting. This indicates Japanese chess to be much less aggressive than cheatgrass in our area. With 5-18 florets/spikelet compared to 4-12 for *B. hordeaceus* and *B. secalinus.*

Bromus marginatus Nees ex Steudel [*B. carinatus* Hooker & Arnot misapplied; *B. polyanthus* Scribner; *Ceratochloa marginata* (Nees) Jackson] — mtn. brome — Uinta Mtns. and Tavaputs Plateau, most common w. of Rock Creek in the Uinta Mtns., common to abundant in aspen and coniferous woods and parklands; 8,000-10,000 ft; June-Sept. Our plants are part of a large complex from which several species have been separated. The part of the complex involving plants of our area have been treated as a single species under *B. carinatus* (Allred 2013; Arnow and others 1980; IMF 6: 187;). Some of our specimens have glabrous lemmas and glabrous leaf sheaths, some have pubescent lemmas and pubescent leaf sheaths, and others have pubescent leaf sheaths and glabrous lemmas. There is no clear geography pattern associated with these features.

Bromus tectorum L. [*Anisantha tectorum* (L.) Nevski] — cheatgrass, downy brome or chess — Introduced from Eurasia, widespread, weedy and abundant in many plant communities, often on disturbed ground and especially on warm aspects and after fire; to about 9,000 ft; April-June. Cheatgrass is a major driver of plant community dynamics in our area and other parts of the West.

Buchloe Engelmann — Buffalo-grass

Buchloe dactyloides (Nuttall) Engelmann — buffalograss — Plants unisexual, perennial, matted, with creeping stolon-like stems, not over 20 cm tall; ligules a fringe of hairs; leaves pilose-hirsute, the blades curled, not over 5 cm long; staminate spikes 1-5, about 1 cm long, strongly 1-sided, with 8-20 spikelets; spikelets 4-6 mm long, with 2 florets; glumes 1-nerved, about 3 mm long; lemmas 3-nerved, about 5 mm long; pistillate spikelets partly enclosed in bur-like, urn- or cup-shaped, thickened involucre-like structures, these 5-7 mm long with 3-5 awn-like lobes. Introduced from the Great Plains and seeded in lawns and other areas. Buffalo-grass has persisted at Antelope Campground and Lucerne Campground in Daggett Co. after being seeded in the early 1960's.

Calamagrostis Adanson — Reedgrass

Plants perennial, from rhizomes or fibrous roots; leaf sheaths open, the blades flat or rolled; ligules membranous; auricles lacking; inflorescence an open or spike-like panicle; spikelets 1-flowered, disarticulating above the glumes; glumes longer than the floret, lemmas awned from the back, with a hardened callus base with tufts of long hairs.

1 Awns of lemmas bent, often protruding from the tips or sides of the glumes; callus hairs less than 1/2 as long as the lemmas 2
1 Awns of lemmas mostly straight, included within the glumes; callus hairs 1/2 to as long as the lemma 3
2 Awns over 4 mm long, mostly exserted .... *C. purpurascens*
2 Awns to 4 mm, slightly exserted or included in the glumes .. *C. rubescens*
3 Panicle open or occasionally contracted, the branches to 5 cm long .... *C. canadensis*
3 Panicle narrow, the branches erect, seldom over 3 cm long 4
4 Rhizomes short or absent; glumes 4-6.5 mm long; awns lacking or attached above the middle of the lemmas; plants of well-drained soil *C. scopulorum*
4 Rhizomes long creeping; awns attached below the middle of the lemmas 5
5 Callus hairs about 2/3 as long to as long as the lemma; plants of wet soils .... *C. stricta*
5 Callus hairs about 1/2 as long as the lemma; plants of well-drained soil .. .. *C. montanensis*

Calamagrostis canadensis (Michaux) Beauvios — bluejoint — Common to abundant across the Uinta Mtns.; one specimen seen from the E. Tavaputs Plateau; streambanks, wet meadows, bogs, and in rather dry to moist woods; 7,400-11,000 ft; July-Aug. Of wet places, with rhizomes, and with small spikelets as in plants of *C. stricta,* but with spreading to open panicles and more often in shady places.

Calamagrostis montanensis Scribner & Vasey. — plains reedgrass — Known from Bare Top Mtn. and Goslin Mtn., Daggett Co. in pinyon-juniper and mtn. big sagebrush communities; 7,400-7,550 ft; June-Aug.

Calamagrostis purpurascens R. Brown — purple reedgrass — Uinta Mtns.; occasional to common on rocky, timbered slopes and alpine; (7,500) 9,000-13,000 ft; July-Aug.

Calamagrostis rubescens Buckley — pinegrass — East end of the Uinta Mtns. and particularly in the Cart Cr. and Pipe Cr. drainages; occasional to dominant in the understory of aspen and lodgepole pine communities; 8,100-8,500 ft; July-Sept.

Calamagrostis scopulorum M. E. Jones — tussock reedgrass — Rather common from limestone hills in the Blind Stream and Rock Creek drainages of the Uinta Mtns., specimens also seen from Chain Lakes, Whiterocks Lake Dam, and Ashley Gorge, Uinta Mtns., Weber sandstone at Split Mt., Red Creek Mt. in Wasatch Co., and Book Cliffs of the E. Tavaputs Plateau, and Badlands Cliffs of the West Tavaputs Plateau; mt. brush, ponderosa pine, Douglas-fir, and open spruce communities; 7,500-10,800 ft; July-Sept.

Calamagrostis stricta (Timm) Koeler [*C. inexpansa* A. Gray; *C. neglecta* (Ehrhart) Beauvios misapplied] — northern reedgrass — Widespread but no specimens seen from the E. Tavaputs Plateau; occasional to common; wet meadows, along streams, and other wet places, rarely in rather dry, rocky places; 5,500-10,500 ft; July-Aug.

Catabrosa Beauvios

Catabrosa aquatica (L.) Beauvios — brookgrass — Perennial plants rooting at the nodes of decumbent bases; upright portion of culms 15-40 cm tall; leaf sheaths closed about 1/2 their length, the blade flat, 2-13 mm wide, prow-shaped at the tip; ligules membranous, 2-8 mm long; auricles lacking; panicles open with both long and short branches at the lower nodes; florets falling at maturity and leaving tiny glumes (about 1 mm long) at the ends of the panicle branches; lemmas awnless, blunt, 3-nerved, the nerves parallel and not converging at the apex. Widespread; aquatic in fresh water springs and slow-moving streams, and sometimes in mud; from low elevations up to 10,600 ft; a reliable indicator of perennial springs; June-Sept.

Cenchrus L. — Sandbur

Cenchrus longispinus (Hackel) Fernald — longspine sandbur — Plants annual, native, 10-50 cm tall, occasionally taller; culms often prostrate and branched at the base, bent at the nodes; ligules a dense fringe of hairs, about 1 mm long or the marginal hairs up to 3 mm long; leaf blades flat or folded, 3-8 mm wide; inflorescence a spike-like panicle, 3-10 cm long, the axis geniculate, flattened and angled, bearing 4-15 (25) burs; the burs urceolate, 3-8 mm long, with numerous spines, the longer spines 3-7 mm long; spikelets borne inside the burs, sessile, about 5-8 mm long; first glume 1-4 mm long; second glume 4-6 mm long; sterile lemmas 5-7 mm long; fertile lemma 5-7 (8) mm long. Weedy in sandy waste places at lower elevations, apparently uncommon in the Uinta Basin, the 2 specimens seen are from Duchesne and Uintah Cos.; July-Sept.

Cinna L. Woodreed

Cinna latifolia (Treviranus) Grisebach in Ledebour — drooping woodreed — Plants perennial from rhizomes, 0.5-1(2) m tall; leaf sheaths open; leaf blades lax and flat, 5-17 (20) mm wide; ligule membranous, brownish,2-8 mm long; inflorescence an open, nodding panicle, 10-30 cm long; spikelets 2-4.5 mm long, disarticulating below and sometimes above the glumes, the rachilla prolonged behind the palea and bristle-like; glumes about equal to or exceeding the floret, usually strongly nerved along the midrib, strongly scabrous or short-ciliate on the keel, scaberulous over the back; lemma glabrous or often scaberulous, awnless or with a sub terminal awn to 1.2 mm long. Specimens seen are from the Provo River, Lake Fork, Uinta, and Whiterocks drainages; riparian communities on bottoms of glacial canyons and in seeps and springs on canyon sides; 6,800-9,700 ft; July-Sept.

Crypsis Aiton

Annuals with prostrate to erect culms; sheaths open, leaf blades mostly short; ligule of hairs; inflorescence a spike-like panicle similar to those of *Phleum* and *Alopecurus;* spikelets 1-flowered, disarticulating below (sometimes above) the glumes.

---

1   Inflorescence at least 5 times longer than wide, free of the upper expanded sheath; spikelets often black tinged; lemma averaging less than 3 mm long
. . . . ***C. alopecuroides***

1 Inflorescence averaging less than 5 times longer than wide, typically remaining partially enclosed within the expanded upper sheath; spikelets pale to purple-tinged; lemmas often averaging 3 mm or longer ....
.... *C. schoenoides*

Crypsis alopecuroides (Piller & Mitterpacher von Mitterburg) Schrader — foxtail picklegrass — Introduced from Eurasia, no specimens seen from our area but known from the Wasatch Front and to be expected to be introduced to similar habitats as the following species.

Crypsis schoenoides (L.) Lamarck — swamp picklegrass — Introduced from Eurasia, specimens seen are from the floodplain of the Green River and drawdown basins of Pelican Lake and Steinaker Reservoir. Apparently only recently introduced to our area, and to be expected to spread.

Dactylis L. — Orchard grass

Dactylis glomerata L. — orchard grass — Perennial plants growing in small to large bunches; culms 30-90 cm tall; cauline leaf blades flat and lax, those of the innovations strongly folded; ligules of mature leaves about 5 mm long, lacerate especially in age; panicle 5-15 cm long, narrow, erect at first with lower branches spreading at maturity; lower branches somewhat distant, upper branches short and more congested; spikelets congested and arranged one-sided on the panicle branches, about 7 mm long, 2-5 flowered; glumes rather unequal, the second one longer and broader and more prominently ciliate on the keel; lemmas stiff ciliate on margin and keel. Introduced from Europe, widely used as pasture and hay on cultivated lands, and in rangeland seedings, and for stabilizing roadsides in the aspen and coniferous forest belts, naturalized along ditch banks and other moist areas; not expected much over 10,000 ft; June-Aug.

Danthonia de Candolle in Lamarck & de Candolle — Oatgrass

Tufted perennial plants; leaf sheaths open; ligule a fringe of short hairs; auricles lacking; inflorescence an open or contracted panicle, or reduced to a single spikelet; spikelets several flowered; disarticulation above the glumes; glumes unequal, generally longer than and mostly enclosing the florets; lemmas awned from just below the teeth of a bifid apex, the awn flat, twisted, and more or less bent.

1 Sheaths long-pilose throughout with hairs up to 5 mm long; spikelets 1 or 2. .... *D. unispicata*
1 Sheaths not as above... 2
2 Inflorescence with 4-12 spikelets with erect pedicels .... *D. intermedia*
2 Inflorescence with (1) 2-5 spikelets with spreading pedicels 5-30 mm long
.... *D. californica*

Danthonia californica Bolander — California oatgrass — Strawberry Valley and eastern Uinta Mtns. including Blue Mtn. and Cold Spring Mtn.; occasional or locally abundant in montane meadows and rocky drainages within the elevations of mtn. big sagebrush, ponderosa pine, aspen, and lodgepole pine communities, 6,900-9,700 ft; June-Aug.

Danthonia intermedia Vasey — timber oatgrass, dry meadow oatgrass — Uinta Mtns.; common to abundant in subalpine meadows, along streams, open coniferous forests, and alpine where locally abundant and a plant community dominant (Brown 2006) up to 11,200 ft and infrequent up to 11,600 ft. where often associated with tufted hairgrass and western single spike sedge and lacking in curly sedge-cushion plant and most wetland and snowbed communities.

Danthonia unispicata (Thurber) Munro — onespike oatgrass — Infrequent or locally common Cold Spring Mtn. and Blue Mtn. and e. of Dry Fork and Sheep creek drainages and rarely west of these drainages in the Uinta Mountains; sagebrush and ponderosa pine communities at 6,900-8,400 ft; June-Sept.

Deschampsia Beauvios — Hairgrass

Plants perennial, tufted; leaf sheaths open; ligules membranous; auricles lacking; inflorescence an open or contracted panicle; spikelets small (1) 2 (3) flowered, disarticulating above the glumes, the rachilla prolonged beyond the upper floret as a hairy bristle; glumes exceeding at least the lower floret; lemmas awned from about mid length, the callus hairy.

1	Plants annual, known from eastern Daggett Co. . . . . ***D. danthonioides***	
1	Plants perennial, distribution various	
2	Panicle branches widely spreading . . . . ***D. caespitosa***	
2	Panicle branches ascending, somewhat appressed to the rachis and producing a narrow inflorescence. . . . . ***D. elongata***	

Deschampsia caespitosa (L.) Beauvios (*Aira caespitosa* L.) — tufted hairgrass — Strawberry Valley and across the Uinta Mtns.; one specimen seen from the Tavaputs Plateau, and 1 from along the Green River at Browns Park; most common in moist to wet meadows and long streams, also in coniferous woods and increasing on areas cleared of timber, and alpine in turf and meadows; (5,500) 7,000-12,500 ft; July-Sept. In an alpine plant community classification study in the Uinta Mtns., this was the most frequently encountered species where it was found in 41 of the 50 communities described by Brown (2006).

Deschampsia danthonioides (Trinius) Munro ex Bentham — annual hairgrass — Locally common to abundant in the ponderosa pine belt and infrequently in the lodgepole pine belt from Green Lakes east to Gorge Creek in Daggett Co. in ephemeral pools, meadows, and ponderosa pine communities at 7,200-8,050 ft. Sometimes confused with *D. elongata*. At maturity awns of *D. danthonioides* are bent and the panicle branches are ascending to spreading. At maturity awns of *D. elongata* are straight and the panicle branches remain more-or-less erect.

Deschampsia elongata (Hooker) Munro — slender hairgrass — Occasional from Strawberry Valley to Whitney Reservoir and the canyon bottom of the N. Fork Duchesne drainage in the western Uinta Mtns.; dry meadows, parklands, aspen-tall forb, and spruce-fir communities; 7,400-9,950 ft; June-Sept.

Digitaria Heiser — Crabgrass

Digitaria sanguinalis (L.) Scopoli — hairy crabgrass — Plants annual, prostrate
to ascending, rooting at the nodes and occasionally forming mats, 10-100 cm
tall; leaf sheaths and blades pilose with 2-3 mm long hairs, the blades 2.5-8 (10)
mm wide; ligules 0.5-2.5 mm long, membranous; inflorescence of few to several
more or less digitately arranged spicate racemes (2) 6-16 cm long; spikelets in
pairs, on pedicels about 1-3 mm long, nearly appressed to the raceme branches,
each with 1 sterile and 1 fertile floret; glumes 0.1-2 mm long; lemmas 2.5-3.5
mm long, the sterile one 5-nerved, the fertile one 1-3 nerved. A weed of lawns,
gardens, sidewalks, vacant lots, and other places; Aug.-Oct.

Distichlis Rafinesque — Saltgrass; Alkaligrass

Distichlis spicata (L.) Greene [*D. stricta* (Torrey) Rydberg] — saltgrass — Plants
unisexual, arising from robust, scaly rhizomes; culms usually low, the staminate
somewhat taller than the pistillate, usually not over 30 cm tall; leaves distributed
about equally up the entire length of the culm, conspicuously 2 ranked, and
rather harsh, the sheaths often hairy at the summit; ligules short, less than 1
mm long, membranous but ciliate and appearing to be of hairs; panicles
narrow, the branches short erect; staminate spikelets straw colored and 10-25
mm long; pistillate spikelets light green, 8-15 mm long, several flowered.
Widespread; common to locally abundant; moist, low alkaline areas, occasional
on upland sites where soil pH is high; May-Sept.

Echinochloa Beauvios — Barnyard Grass; Cockspur

Echinochloa crus-galli (L.) Beauvios — barnyard grass — Robust annual plants,
15-100 cm tall; culms usually bent at the lower nodes; leaf blades flat, 4-12 (16)
mm wide, sometimes papillate pubescent on the margins near the collar,
otherwise mostly glabrous; ligules lacking; panicles 5-25 cm long, of 5-15
spicate or racemose, somewhat 1-sided branches; spikelets 3-3.5 mm long,
consisting of 2 glumes, a sterile glume-like lemma, and a fertile floret; glumes
about 1-3 mm long, awnless or awn-pointed; sterile lemma with an awn 1-30
mm long; fertile lemma 2-3.5 mm long, awnless or awn-pointed. Weedy in
gardens, fields, ditch banks, roadsides, and other disturbed places; mostly
below 7,000 ft; July-Oct.

X *Elyhordeum* (Vasey) Barkworth & D. R. Dewey
(X *Agrohordeum* Camus)

X *Elyhordeum macounii* (Vasey) Barkworth & D. R. Dewey [X *Agrohordeum
macounii* (Vasey) Lepage; *Elymus macounii* Vasey; ×*Elytesion macounii* (Vasey
Barkworth & D. R. Dewey] — Macoun wildrye — Tufted perennial plants; culms
0-80 cm tall; leaves erect, flat to rolled, 2-6 mm wide; ligules not over 1 mm
long; spike 4-12 cm long; spikelets (1) 2 per node of the rachis; glumes narrow
with awns 5-10 mm long; lemmas 6-9 mm long, awned, awns 10-20 mm long.
Occasional on drainage bottoms of the Tavaputs Plateau and along ditches,
fence lines, and in waste places in the Uinta Basin; June-Sept. This is considered
to be a sterile hybrid of *Agropyron trachycaulum* and *Hordeum jubatum.* The
chief *Agropyron* features are the tendency for 1 spikelet per node of the rachis
and the much reduced awns. *Hordeum* features are 2 spikelets per node of the

rachis, disarticulating rachis, and narrow glumes with awns. Some populations in the Uinta Basin seem far removed from populations of *A. trachycaulum,* and the *Elymus* parent of these populations could be some other member of the genus, such as *E. repens* or *E. lanceolatus.*

Elymus L. — Wildrye

Plants perennial, from rhizomes or fibrous roots, often densely tufted; leaf sheaths open; ligules membranous; auricles usually present; inflorescence typically a solitary terminal spike with persistent or rarely disarticulating rachis; spikelets sessile or short-pedicellate or occasionally rather long pedicellate, 1-4 (6) per node of the rachis, with 2-6 (10) florets, disarticulating above the glumes; glumes mostly shorter than the lemmas; lemmas awnless or awned from the tip. Elymus is treated herein a broad sense. Species of this complex often hybridize with each other with the hybrids usually sterile.

1 Spikelets mostly 1 per node of the rachis, sometimes 2 at a few of the nodes .... **KEY 1**

1 Spikelets 2 or more per node of the rachis .... **KEY 2**

— KEY 1 —

1 Lemmas with awns over 6 mm long 2

1 Lemmas awnless or with awns less than 6 mm long 7

2 Inflorescence densely flowered, glumes and lemmas with long, divergent awns, and rachis breaking up at maturity; culms ascending to prostrate, rarely more than 40 cm tall, known from high elevations ..... *E. scribneri*

2 Inflorescence not as above; plants mostly erect, often over 40 cm tall, not prostrate 3

3 Glumes 0.6-1.5 mm wide, the 1st widest at the base and tapering from base 4

3 Glumes mostly wider than 1.5 mm, usually widest near the middle...... 5

4 Culms several or many arising from a caespitose base ... *E. wawawaiensis*

4 Culms arising singly or few together from creeping rhizomes .. *E. simplex*

5 Anthers not over 2 mm long; glumes usually 2/3-3/4 as long as the mature spikelets (excluding awns) .... *E. trachycaulus*

5 Anthers over 3 mm long; glumes about 1/2 as long as the mature spikelets (excluding-awns) 6

6 Leaves flat, mostly over 5 mm wide, awns straight .... *E. repens*

6 Leaves rolled (sometimes flat when young), less than 5 mm wide; awns divergent when dry .... *E. spicatus*

7 Glumes awn-like the full length or subulate (tapering from near the base and awn-pointed) 8

7 Glumes broader and tapering from the middle or above; culms arising singly or in bunches 10

8 At least the lower glume with 3-7 nerves, usually over 1 mm wide at the base .... *E. smithii*

8 Glumes with 0-1 (3) obscure nerves, usually not over 1 mm wide........ 9

9 Awns of lemmas 2.5-8 mm long; culms arising singly or few together from long creeping rhizomes; plants known from along the Green River and perhaps along the Yampa River, on alluvium of floodplains .... *E. simplex*

9 Awns of lemmas obsolete or to 3 mm long; culms few to several and tufted; plants widespread, common; often on residuum and colluvium of canyon sides and fans .... *E. salinus*

10 Lemmas pubescent over the back................................... 11

10 Lemmas glabrous or ciliate 12

11 Glumes and lemmas pointed; leaf sheaths not ciliate .... *E. lanceolatus*

11 Glumes and lemmas blunt, sometimes mucronate; leaf sheaths often ciliate (use 10 times magnification) .... *E. hispidus*

12 Glumes blunt or mucronate, lemmas often blunt; at least 1 side of the leaf sheath usually ciliate .... *E. hispidus*

12 Glumes and sometimes lemmas pointed; sheaths not conspicuously ciliate or occasionally so in E. lanceolatus................................. 13

13 Leaf blade flat, 5-15 mm wide 14

13 Leaf blades involute (sometimes flat when young), mostly less than 5 mm wide 15

14 Glumes about 1/2 as long as the spikelet; spike not secund ..... *E. repens*

14 Glumes 2/3 to as long as the spikelet; spike often secund *E. × pseudorepens*

15 Spikelets 3-4 times as long as the internodes of the rachis .... *Agropyron cristatum*

15 Spikelets not over 3 times as long as the internodes of the rachis 16

16 Some lower spikelets often equaling or shorter than the internodes; upper spikelets mostly less than 1.5 times as long as the internodes (excluding awns if present) .... *E. spicatus*

16 Spikelets toward the center of the spike often 2 times as long as the internodes or longer 17

17 Spikes (15) 20-40 cm long; plants 70-150 (200) cm tall; culms mostly over 4 mm thick near the base .... *E. elongatus*

17 Plants different from above in 1 or more features..................... 18

18 Glumes about 1/2 as long as the spikelets; anthers (2.5) 3-5 mm long .... *E. lanceolatus*

18 Glumes usually 2/3 to 3/4 as long as the spikelets; anthers not over 2 mm long .... *E. trachycaulus*

— KEY 2 —

1 Awns of lemmas over 9 mm long 2

1 Awns of lemmas less than 9 mm long............................... 5

2 Glumes awn-like the entire length, 2-12 cm long including awns; rachis readily breaking up; awns of lemmas 2-8.5 cm long .... *E. elymoides*

2 Glumes expanded at the base; rachis not readily breaking apart 3

3 Awns of glumes averaging over 5 mm (4-20) mm long; awns of lemmas 10-40 mm long, spreading in age; spike flexuous to nodding ... *E. canadensis*

3 Awns of glumes 1-5 mm long; awns of lemmas 1-30 mm long; spikes stiffly

erect.. 4

4   Plants of woods, often associated with aspen, leaf blades flat or slightly involute, 4-12 mm wide; rachis usually not breaking up at maturity .....
.... *E. glaucus*

4   Plants mostly of saline or alkaline open places; leaf blades mostly firm and sub-involute, 2-5 mm wide; rachis usually breaking up at maturity ......
.... *x Elyhordeum*

5   Spikes with 2-8 spikelets at most nodes of the rachis, densely flowered or interrupted below 6

5   Spikelets 1-2 per node of the rachis, spike usually slender and often interrupted with spikelets rather distant; plants often with long creeping rhizomes 8

6   Plants strongly rhizomatous .... *E. giganteus*

6   Plants strongly caespitose 7

7   Spikelets mostly 2 per node of the rachis, often less than 12 mm long, mostly 3-flowered; ligules about 1 mm long; plants mostly less than 80 cm tall ..
.... *E. junceus*

7   Spikelets 3-5 per node of the rachis, mostly over 12 mm long, commonly 3-4 but up to 7-flowered; ligules over 2 mm long; plants often over 80 cm tall
.... *E. cinereus*

8   At least the lower glume with 3-7 nerves, usually over 1 mm wide at the base .... *E. smithii*

8   Glumes with 1 nerve, usually not over 1 mm wide..................... 9

9   Awns of lemmas 2.5-8 mm long; culms arising singly or few together from long creeping rhizomes; plants known from along the Green River and perhaps along the Yampa River, on sandy alluvium of floodplains ....
.... *E. simplex*

9   Awns of lemmas obsolete or to 3 mm long 10

10 Plants with few or many culms arising from a tufted base, widespread, common .... *E. salinus*

10 Culms arising singly or few together from slender rhizomes; plants rarely collected in our area .... *E. triticoides*

Elymus canadensis L. — Canada wildrye — Occasional or common; widespread; mostly along ditch banks, fence rows, and water courses; up to about 7,000 ft; June-Sept.

Elymus cinereus Scribner & Merrill [*E. condensatus* Presl misapplied; *Leymus cinereus* (Scribner & Merrill) A. Love] — basin wildrye — Widespread; canyon bottoms (especially of the Tavaputs Plateau), upland mima mounds in Strawberry Valley, along water courses, most common on alluvial soils but included in rangeland and burn-area seedings in uplands; up to 8,000 ft; June-Aug. Specimens with compound spikes have occasionally been observed. This large grass forms hybrids with *E. elymoides* and *E. triticoides* (Arnow 1987) See *E. salinus.* It produces voluminous forage of low to moderate palatability.

Elymus elongatus (Host) Runemark [*Agropyron elongatum* (Host) Beauvious; *Elytrigia pontica* (Podpera) Holub; *Lophopyrum elongatum* (Host) Love; *Thinopyrum ponticum* (Podpera) Barkworth & D. R. Dewey] — tall wheatgrass — Introduced from Eurasia, used in pasture and roadside seedings, most suited to soil that is wet for some part of the growing season and particularly useful in seeding wet or moist saline or alkaline lowlands, but livestock have low preference for this plant, and they are not likely to use much of it unless forced by fencing by or lack of other forage; June-Aug. Similar to *E. hispidus* but different as in the following key:

1 Spikes (15) 20-40 cm long; plants 70-150 (200) cm tall, forming large bunches, without rhizomes; culms mostly over 4 mm thick near the base; leaf sheaths not ciliate . . . . **E. elongatus**
1 Spikes 8-20 cm long; plants 40-100 cm tall, culms seldom over 4 mm thick, arising singly or in small bunches; rhizomes long-creeping; one side of the leaf sheath commonly ciliate . . . . **E. hispidus**

Elymus elymoides (Rafinesque) Swezey [*Sitanion hystrix* (Nuttall) J. G. Smith; *S. longifolium* J. G. Smith] — squirreltail — Tufted perennial plants without rhizomes,-10-50 cm tall; leaf sheaths open, the blades flat or rolled, 1-3 (5) mm wide, often strongly nerved; ligules membranous, less than 1 mm long; auricles lacking to well developed; inflorescence a densely flowered spike, 4-10 cm long, bristly from the long divergent awns of glumes and lemmas; spikelets 2 per node of the rachis, 1-6 flowered, those at the base of the spike often reduced to glume-like structures, disarticulating at each node of the rachis; glumes awn-like their whole length, 5-9 cm long; lemmas 7-10 mm long, with divergent awns about as long as those of the glumes. Widespread; common in a variety plant communities over a wide elevational range from salt desert shrub communities to above timberline; May-Oct.

Hybridizing with other species of *Elymus*. See *Elymus salinus*. *Elymus multisetus* M. E. Jones (*Sitanion jubatum* J. G. Smith) has been reported for Utah based on occasional plants with some glumes 3-cleft. Wilson (1963) reported *E. multisetus* to be generally restricted to states west of Utah. Arnow (1987) suggests the sporadic occurrence of plants with some glumes 3-cleft could be the result of introgressive hybridization between *E. glaucus* and *E. elymoides*.

Elymus giganteus Vahl [*Leymus racemosus* (Lamarck) Tzvelev — mammoth wildrye, Volga wildrye — Known from Danforth Hills at 6,320 ft where planted for stabilization of blowing sand. Also seen along Highway 40 at Craig, Colorado.

Elymus glaucus Buckley — blue wildrye — Widely distributed in the Uinta Mtns. and Strawberry Valley where it is most common under aspen and lodgepole pine; to be expected on the Tavaputs Plateau but no specimens seen from there; June-Aug.

Elymus hispidus (Opiz) Melderis [*Agropyron intermedium* (Host) Beauvios; *A. trichophorum* (Link) Richter; *Elytrigia intermedia* (Host) Nevski; *Thinopyrum intermedium* (Host) Barkworth & D. R. Dewey] — intermediate wheatgrass — Introduced from Eurasia, used extensively in rangeland and burned area seedings and for cover on roadsides and other disturbed places where more persistent and even displacing smooth brome below the aspen belt; most common in sagebrush-grass, pinyon-juniper, and ponderosa pine communities; June-Aug. Plants with glabrous lemmas [var. *intermedium* (intermediate wheatgrass)] and plants with pubescent lemmas [var. *trichophorum* (Link) Halacsy (pubescent wheatgrass)] have been planted in our area. Plantings seeded exclusively with certified pubescent wheatgrass seed show a mixture of pubescent and glabrous phases. In general the pubescent phase is more drought-tolerant.

Elymus junceus Fischer [*Psathyrostachys juncea* (Fischer) Nevski] — Russian wildrye — Introduced from Siberia, widely used for seeding rangelands in pinyon-juniper, sagebrush, and mt. brush communities, well adapted to greasewood and other communities of canyon bottoms of the Tavaputs Plateau where it has increased as crested wheatgrass has decreased; May-July; highly selected by cattle and elk, but avoided by horses.

Elymus lanceolatus (Scribner & Smith) Gould [*Agropyron dasystachyum* (Hooker) Scribner; *A. albicans* Scribner & Smith; *A. griffithsii* Scribner & Smith ex Piper; *Elymus albicans* (Scribner and Smith) A. Love] — thickspike wheatgrass — Widespread; common to abundant in sagebrush-grass and pinyon-juniper communities; up to about 9,500 ft, becoming abundant after fire; June-Aug. Plants with glabrous or scabrous lemmas have been separated as *E. riparius* Gould [*Agropyron riparium* Scribner & Smith, *A. dasystachyum* var. *riparium* (Scribner & Smith) Bowden]. Hybrids between this and *E. spicatus* have been designated as *E. albicans*. Frequently misidentified as *E. smithii*, but generally of higher elevations than *E. smithii*. The Critana cultivar with poorly or mildly developed rhizomes commonly forms tufts rather than having stems arising singly along elongated rhizomes as is typical for rhizomatous species. Current and past floras treat thickspike wheatgrass as strongly rhizomatous. It appears that an expanded concept of *E. lanceolatus* is due in floras. Critana thickspike wheatgrass has been used for roadside, oil and gas pad, and other reclamation seedings. It establishes rapidly and is well adapted to areas with annual precipitation of 10-20 inches, and it has been used with some success at 7 inches annual precipitation.

Elymus × pseudorepens (Scribner & Smith) Barkworth & D. R. Dewey (*Agropyron × pseudorepens* Scribner & Smith) — slender wheatgrass × thickspike wheatgrass — Like *E. repens*, but with glumes 2/3 to as long as the spikelet. A hybrid of *E. trachycaulus* and *E. lanceolatus* (IMF 6: 320). The specimens seen are robust plants to as much as 1-1.5 m tall with rather strongly secund spikes. This form has been treated as *Agropyron pseudorepens* var. *magum* Scribner & Smith . The few specimens seen are from the Tavaputs Plateau; woods and slopes of the aspen-spruce-fir belt; June-Aug.

Elymus repens (L.) Gould [*Agropyron repens* (L.) Beauvios; *Elytrigia repens* (L.) Nevski] — quackgrass — Introduced from Eurasia; widespread; often weedy along ditch banks, in fields and lawns, floodplains, and drawdown basins, mostly below 8,000 ft, occasionally along roads and around logging, livestock, and hunters camps in mountains where the seed has been introduced as a contaminant in hay to about 9,000 ft; May-June. Quackgrass is less weedy on upland rangelands than in areas of cultivation.

Elymus salinus M. E. Jones [*E. ambiguus* Vasey & Scribner var. *salina* (M. E. Jones) C. L. Hitchcock; *Leymus salinus* M. E. Jones) A. Love] — Salina wildrye — Abundant on the Tavaputs Plateau where often the dominant species on dry, ridges, steep canyon sides and debris fans of canyon bottoms where it is associated with sagebrush, pinyon-juniper and mountain brush, most abundant on soils derived from marly mudstones, but also on soils derived from calcareous sandstone, also locally common on Morrison Formation at Sheep Creek Gap and perhaps other points in Daggett Co. and occasionally in desert shrub communities on Duchesne River, Uinta, and other formations in the Uinta Basin, and adjacent to our area on the Green River Formation in Sweetwater Co., Wyoming; 5,600-9,200 ft; May-July.

Salina wildrye, hard grass, bullgrass, and that old bunch grass are all common names that have been applied to the species. The first 3 names depict the toughness of the foliage, which cures very well, and culms with spikes remain standing through 2 or more growing seasons. Some florets also persist, but most fall leaving the narrow glumes. Traditionally this has been treated as a species without rhizomes (Hitchcock and Chase 1950; Harrington 1954). It has also been described recently this way (Hitchcock and Cronquist 1973), but Holmgren and Holmgren (IMF 6: 303) noted the type specimen as well as most of the topotype specimens have rhizomes. Even though the plants often produce robust and occasionally long rhizomes, the rhizomes are often excluded in collecting. The caespitose nature of the species has probably also helped to obscure its rhizomatous nature as caespitose species are often considered non-rhizomatous.

Failure to recognize the rhizomatous nature of this species has helped to make this one of the most frequently confused taxa in our area. Rhizomatous plants key to *E. simplex* or *E. triticoides* Buckley in some manuals. The treatment of Holmgren and Holmgren in IMF (6: 302-303) does much to clarify the problem. Specimens that appear much like a cross between this and *Elymus elymoides* and between this and *E. cinereus* have been collected from the W. Tavaputs Plateau. Such plants can be referred to the hybrid genus *Elysitanion* Bowden.

Elymus scribneri (Vasey) M. E. Jones (*Agropyron scribneri* Vasey) — spreading wheatgrass, Scribner wheatgrass — Infrequent on windswept ridges and talus at or above timberline on the Uinta Mtns. and Cathedral Bluffs of the E. Tavaputs Plateau; June-Aug.

Elymus simplex Scribner & Williams [*E. triticoides* Buckley var. *simplex* Scribner & Will.; *Leymus simplex* (Scribner & Williams) D. R. Dewey] — alkali wildrye — Reported for sandy areas along the Green River in Daggett Co. (IMF 6: 302 & AUF 850). Much of the Daggett Co. habitat of this species is under water of the Flaming Gorge Reservoir. It is locally abundant on the floodplain of the Blacks Fork River above the Flaming Gorge Reservoir in Sweetwater Co., Wyoming. To be expected along the Yampa River in Moffat Co.

Elymus smithii (Rydberg) Gould [*Agropyron smithii* Rydberg; *Elytrigia smithii* (Rydberg) A. Love; *Pascopyrum smithii* (Rydberg) A. Love] — western wheatgrass, bluestem wheatgrass — Widespread; occasional to common and locally abundant in desert shrub, sagebrush, and riparian communities, and along roads, tolerant of alkaline and saline conditions; mostly below 6,000 ft but along roadsides to 8,000 ft. June-Aug. This and *E. lanceolatus* are similar and are distinguished by the shape of the glumes. Also, the anthers of *E. lanceolatus* (3-5 mm long) average longer than those of *E. smithii* (2.5-3.5 mm long). Specimens with 2 spikelets at some of the nodes of the rachis are relatively common. Plants of this species are sometimes confused with *E. lanceolatus.*

Elymus spicatus (Pursh) Gould [*Agropyron spicatum* (Pursh) Scribner & Smith; *A. inerme* (Scribner & Smith) Rydberg; *A. spicatum* var. *inerme* (Scribner & Smith) A. Heller; *Elytrigia spicata* (Pursh) D. R. Dewey; *Pseudoroegneria spicata* (Pursh) A. Love] — bluebunch wheatgrass — Widespread; common to abundant in sagebrush-grass, pinyon-juniper, mt. brush, and ponderosa pine communities, increasing rather rapidly after fire; June-Aug. Plants with awnless or short-awned lemmas have been separated at species and varietal levels. These plants are like long-awned plant in all other respects, and a continuum is found In awn length. Awnless and long-awned forms are found side by side. However, the awned form may be more common on mesic sites. Both forms are described without rhizomes in older manuals, but in recent years the rhizomatous nature of the species has been recognized (Hitchcock and Cronquist 1973). Plants of our area sometimes have rhizomes and occasionally have robust, long-creeping rhizomes. This might be a result of crossing with rhizomatous species of the genus. These plants might be referred to as *E. albicans.* See *E. lanceolatus.*

Elymus trachycaulus (Link) Gould ex Shinners [*Agropyron trachycaulum* (Link) Malte; *A. caninum* L. ssp. *majis* (Vasey) C. L. Hitchcock; *A. latiglume* (Scribner & Smith) Rydberg; *A. pauciflorum* (Schweinitz) A. S. Hitchcock; *A. subsecundum* (Link) A. S. Hitchcock; *A. subsecundum* var. *andinum* (Scribner & Smith) A. S. Hitchcock; *Elymus subsecundus* (Link) A. & D. Love] — slender wheatgrass — Widespread, common in montane communities; typically on mesic sites, occasional under aspen, abundant on open slopes, occasional in coniferous woods where often becoming abundant following clear cutting of timber, also common above timberline, extending to rather low elevations in the dry drainages of the Tavaputs Plateau where moisture collects in depressions and from road run-off; June-Sept.

Forming a diverse complex in which several species have been recognized, but similarities (Hitchcock and Chase 1950) and intergradation (Harrington 1954) have long been recognized; included in the *A. caninum* complex by Hitchcock and Cronquist (1973), but the name *A. trachycaulum* is retained for our plants in IMF (6: 329-331) and by Arnow and others (1980). The whole complex is distinguished from most of our native species of the genus by short anthers and the glumes being nearly as long as the spikelets. Specimens with 3-4 spikes per culm are rarely found. The following key may help with identification of variety for most specimens. In TPD var. *glaucum* and var. *unilaterale* are lumped under ssp. *subsecundus* (Link) A. Love & D. Love.

1 Lemmas awnless, or awns less than 6 mm long........................2
1 Lemmas with awns (5) 10-30 mm long 3
2 Spikelets scarcely imbricate, the tips rarely reaching the base of those above on the same side of the spike; spikes mostly more than 10 cm long; valleys to mid montane habitats .... **var. *trachycaulus***
2 Spikelets mostly closely imbricate; spikes mostly less than 10 cm long; mid montane to well above timberline .... **var. *latiglume*** (Scribner & Smith) .... Beetle [*E. alaskanus* (Scribner & Merrill) A. Love]
3 Glumes 6-10 mm long; plants 20-30 cm tall, decumbent at the base; awn of lemma (5) 7-16 (20) mm long; spikes slender, 3-10 mm thick; high elevations .... **var. *glaucum*** (Pease & Moore) Malte
3 Glumes 10-18 mm long; plants usually over 30 cm tall, erect or ascending; awn of lemma 17-40 mm long; spikes 6-13 mm thick; valleys and mid montane, often under aspen .... **var. *unilaterale*** (Cassidy) Malte

Elymus triticoides Buckley — Beardless, wildrye, creeping wildrye — Reported for Daggett and Uintah Cos. (AUF 851); saline meadows, desert shrub, and pinyon-juniper communities.

Elymus wawawaiensis J. Carlson & Barkworth — Snake River wheatgrass, Secar wheatgrass — Native to the Snake River drainage in Washington, Oregon, and northern Idaho, cultivated for seed production, and widely used in rangeland seedings. This was a major part of the seed mix in both the Mustang Fire in Daggett Co. in 2002 and the Neola North Fire in Uintah and Duchesne Counties in 2007. Plants of this species closely resemble those of *Elymus spicatus,* but the glumes are narrower and sharper and the spike is more imbricate. Also the seedlings have pubescent leaf-sheaths that become glabrous in age. Seedlings of *E. spicatus* have glabrous sheaths.

Elymus hybrids

1 Awns of the lemmas (14) 18-37 mm long, spreading, often recurved; internodes of the rachis mostly 7-10 mm long .. *E. elymoides* × *E. spicatus*

1 Awns of the lemmas 4-17 mm long, straight; internodes of the rachis 2.5-6 (7) mm long .... *E. elymoides* × *E. trachycaulus*

Elymus elymoides × **Elymus spicatus** [*Agropyron saxicola* (Scribner & Smith) Piper, *Agrositanion saxicola* (Scribner & Smith) Bowden, *Elymus saxicolus* Scribner & Smith in Scribner; *Pseudelymus saxicola* (Scribner & Smith) Barkworth & D. R. Dewey] — bluebunch squirreltail — A hybrid between *Elymus elymoides* and *Elymus spicatus* and probably other species of *Elymus,* with the disarticulating rachis and longer awns of *Elymus elymoides* and with 1 spikelet per node of the rachis as in *Elymus spicatus.* Most abundant in burned areas. Specimens with long rhizomes that belong to this taxon have been collected from Reservation Ridge of the Tavaputs Plateau. The rhizomatous *Elymus lanceolatus* is present in that area, and it a possible parent in this case.

Elymus elymoides × **E. trachycaulus** [*Agrositanion saundersii* (Vasey) Bowden; *Elymus saundersii* Vasey] — slender squirreltail — A hybrid between *Elymus trachycaulus* and *Elymus elymoides* with more or less disarticulating rachis and lowest floret sometimes reduced to a glume-like structure as in *Sitanion* and with 1 spikelet per node of the rachis as in *Elymus trachycaulus.*

In addition to the above, hybrids of *Elymus cinereus* and *E. lanceolatus, E. cinereus* and *E. salinus, E. trachycaulus* and *Hordeum jubatum* [× *Elymus macounii* Vasey, × *Elyhordeum macounii* (Vasey) Barkworth & D. R. Dewey (Macoun wildrye)], and likely other crosses are found in our area.

Eragrostis Beauvios — Lovegrass

Caespitose annuals (ours) or perennials; leaf sheaths open, tufted at the throat with hairs; ligules of short dense hairs; inflorescence an open or contracted panicle; spikelets with 6-14 florets, disarticulating above the glumes; glumes shorter than the first floret; lemmas 3-nerved, awnless; palea as long as the lemma.

1 Plants decumbent to prostrate and often rooting at the nodes; panicles often less than 4 cm long .... *E. hypnoides*

1 Plants ascending to erect, not rooting at the nodes; panicles often over 4 cm long 2

2 Plants with minute glandular depressions on the panicle branches and

sometimes on the keels of lemmas, 10-25 cm tall rarely taller; panicles 4-15 (23) cm long 3

2   Plants not glandular on the panicle branches nor the lemmas, 15-80 cm tall or taller; panicles various. 4

3   Spikelets 2.5-3 mm wide; panicles usually dense; anthers 0.5 mm long; glands on lemmas prominent . . . . *E. cilianensis*

3   Spikelets about 1.5 mm wide; panicles open; anthers 0.2 mm long; glands on lemmas sometimes obscure . . . . *E. minor*

4   Caryopsis ventrally flattened or more commonly grooved throughout the length; plants commonly 50-80 cm tall or taller; panicles 15-30 (36) cm long; spikelets 0.6-1.5 mm wide . . . . *E. mexicana*

4   Caryopsis rounded ventrally; plants 15-60 (75) cm tall; panicles 5-20 (35) cm long; spikelets 1.2-2 mm wide . . . . *E. pectinacea*

Eragrostis cilianensis (Allioni) Mosher — stink grass — Introduced from Europe; weedy in gardens, ditch banks, fields, and roadsides; not expected over 7,000 ft; July-Oct.

Eragrostis hypnoides (Lamarck) Britton, Sterns, & Poggenburg — teal lovegrass, creeping lovegrass — Introduced from Africa. Known from along the floodplain of the Green River.

Eragrostis minor Host (*E. poaeoides* Beauvios) — minor lovegrass — The one specimen seen (Neese et al. 11022) is from a sidewalk in Roosevelt.

Eragrostis mexicana (Hornemann) Link (*E. orcuttiana* Vasey) — Mexican lovegrass — Weedy in gardens, roadsides, and other disturbed areas; July-Oct.

Eragrostis pectinacea (Michaux) Nees (*E. diffusa* Buckley) — tufted lovegrass — The few specimens seen are from the floodplain of the Green River, along roads and ditches, gardens, and other areas of disturbance; July-Oct.

Eremopyrum (Ledebour) Juab. & Spach

Eremopyrum triticeum (Gaertner) Nevski (*Agropyron triticeum* Gaertner) — annual wheatgrass — Plants annual; culms arising singly or 2-3 together, glabrous except retrorse hirsute below the spike; auricles usually conspicuous, ligules to 1 mm long; leaf blades flat to involute, 1-4 (6) mm wide; spikes 0.8-2 cm long, more or less densely flowered, the rachis sometimes disarticulating in age; spikelets with 3-6 flowers, ascending to spreading; glumes 4-7.5 mm long, 1-nerved, awn-tipped; lemmas 5-7.5 mm long, awn-tipped. Introduced from Eurasia. In preparation of the first edition of this flora in 1986, only one specimen (Neese & Trent 11804) was seen. Since then this weedy annual has been found in many locations usually along roads and other sites of disturbance from 4,700-5,300 ft; May-June. Rapid expansion in range is indicated for this species.

Erioneuron Nash

Erioneuron pilosum (Buckley) Nash [*Dasyochloa pulchella* (H.B.K.) Rydberg; *Tridens pilosus* (Buckley) A. S. Hitchcock; *Triodia pilosa* Merrill] — hairy tridens — Plants perennial but weakly rooted; culms 10-20 (30) cm tall, usually with a single node above the tuft of basal leaves; ligule a fringe of hairs, about 0.5 mm long; leaves basal or on the lower 1/3 of the culm, sparingly pilose on the collar, the blades flat or folded, less than 2 mm wide, 1-6 cm long; panicles 1.5-3 (5) cm long, ovoid, head-like or racemose with 3-9 nearly sessile spikelets; spikelets 8-14 (20) mm long, 3-6 mm wide, with 6-12 (18) flowers; glumes about 4-8 mm long, pilose on the 3-nerves, the mid nerve extending into a 1-2 mm long awn; the one specimen seen (Neese et al. 4267) is from Sand Wash, also observed at Hog Canyon east of Vernal; May-July.

Festuca L. — Fescue

Densely tufted plants; leaf sheaths open or partly closed; ligules membranous; auricles lacking or well developed; inflorescence a raceme or narrow to open panicle; spikelets 2-12 (20) flowered, disarticulating above the glumes; glumes shorter than the lowermost lemma; lemmas awnless, awn-pointed, or awned from the tip. In addition to taxa listed below, *Festuca rubra* L., noted for its shade tolerance, might also be cultivated in lawns and escaped.

1   Leaf blades flat, averaging over 3 mm wide; lemmas awnless or with awns to 2(3) mm long; plants 50-100 cm tall 2

1   Leaf blades mostly rolled, up to 2 mm wide 3

2   Lemmas averaging less than 7 (4-7.5 mm long); leaf blades rarely more than 7 (4-7.5) mm wide; auricles typically glabrous; panicle branches 2 at the lower most node, together rarely bearing more than 9 spikelets . . . .
. . . . *F. pratensis*

2   Lemmas averaging at least 7 (6-10) mm long; leaf blades 2-12 mm wide; auricles typically ciliate; panicles branches 2 or 3 at the lower most node, together usually bearing 10-30 spikelets . . . . *F. arundinaceae*

3   Lemmas awnless; plants 40-90 cm tall or taller; ligule 2.5-8 mm long . . . . .
. . . . *F. thurberi*

3   Lemmas awned, or if only awn-tipped then plants usually less than 25 cm tall; ligules essentially lacking or up to 2 mm long 4

4   Plants annual, mostly in and below the pinyon-juniper belt; spikelets with 5-13 flowers . . . . *Vulpia octoflora*

4   Plants perennial. 5

5   Florets 1-3 per spikelet; panicles spreading, the branches strongly angled, and densely ciliate on the angles; plants 20-50 cm tall . . . . *F. dasyclada*

5   Florets mostly more than 3 per spikelet or else the plants less than 20 cm tall; panicle open or closed, the branches not as above 6

6   Lemma awns, at least some, 3-5 mm long . . . . *F. ovina* var. *idahoensis*

6   Lemma awns 1-3 mm long 7

7   Anthers when dry 2-3.5 mm long; plants introduced . . . . *F. brevipila*

7   Anthers when dry 0.4-1.2 mm long; plants native . . . . *F. ovina*

Festuca arundinaceae Schreber, [*Schedonorus phoenix* (Scopoli) Holub; *S. arundinaceae* (Schreber) Dumortier; *Lolium arundinaceum* (Schreber) S. J. Darbyshire)] — tall fescue — Introduced from Europe, cultivated for pasture and hay, and used in wildland plantings mostly in areas of greater than 35 cm annual precipitation. In the examination of Utah specimens of *F. arundinaceae* and *F. pratensis* the features used to separate the 2 taxa were found to occur at random in more than ½ of the specimens (Arnow 1987). This plant is an invader of lawns. Tall fescue is host to a fungus that can cause reproductive difficulties in horses and lameness and ultimately gangrene of hind feet of susceptible cattle. Fungus-free materials can be grown that do not cause these difficulties.

Festuca brevipila Tracey [*F. trachyphylla* (Hack.) Krajina; *F. duriuscula* L. misapplied; *F. ovina* var. *duriuscula* (L.) W. D. J. Koch misapplied] — hard fescue — Introduced and widely planted in pinyon-juniper communities and along roads and other disturbed sites. Plants of this species remain green longer than members of the sheep fescue complex and the leaves are more lax.

Festuca dasyclada Hackel ex Beal [*Argillochloa dasyclada* (Hackel) Weber] — Utah fescue, Green River Shale fescue — Occasional to locally common in the Piceance Basin and Roan Cliffs of the E. Tavaputs Plateau and Willow Creek drainage of the W. Tavaputs Plateau; most common on shaly slopes of the Green River Formation; 7,120-8,600 ft; June-Aug.

Festuca ovina L. — sheep fescue — Forming a variable complex in which extreme forms are different, but a continuum is found through the entire complex. Several schemes have been devised to separate taxa at the species level and varietal level, but they contain overlapping criteria. Sometimes separation of taxa in the complex becomes an exercise in frustration.

Harrington (1954) noted frequent intergradation between the small and intermediate forms. *Festuca idahoensis* (the large form) was placed as a var. of *F. ovina* when first published in 1896 (*F. ovina* var. *ingrata* Hackel ex Beal). Gould (1968) indicated that it was no more than a var. of *F. ovina* as did Arnow (1987). The intergrading vars. seem to be expressions of ecology. Separation may be attempted with the following key, and many specimens can be reasonably assigned to a var., but this distinction is arbitrary with many other specimens especially in those of intermediate habitat.

In an alpine plant community classification study in the Uinta Mtns., sheep fescue (represented by var. *breviflora*) was the 7th most frequently encountered species (Brown 2006).

1   Panicles open (5) 7-15 cm long; spikelets (4) 5-7 flowered; first glume 2.5-5 mm long; second glume 4-6.5 mm long; lemmas 4.5-7.5 mm long; awns 2-5 mm long; anthers 2.5-4 mm long; basal leaves often over 10 cm long; plants (30) 40-100 cm tall, mostly in mesic sagebrush (sometimes silver sagebrush) communities, and meadows; primarily of moderate elevations in Strawberry Valley and Uinta Mtns.: **var. *ingrata*** Hackel ex Beal (*F. idahoensis* Elmer) — Idaho fescue

1   Panicles narrow, 1.5-7 (10) cm long; spikelets 2-4 (5) flowered; first glume 2-4 mm long, second glume 2.8-5 mm long; lemmas 3-5.5 mm long; awns 1-3 mm long or lacking; anthers 0.7-1.7 mm long; basal leaves mostly less than 10 cm long; plants 4-35 (40) cm tall 2

2   Culms mostly over 25 cm tall, 2-3 times the height of the basal leaves; anthers over 1 mm long; plants of sagebrush communities and open woods and dry meadows to near timberline, common to abundant on the Uinta Mtns., infrequent on the Tavaputs Plateau: ..... **var. *saximontana*** (Rydberg) Gleason [*F. ovina* var. *rydbergii* St-Yves; *F. saximontana* Rydberg] — mountain fescue

Festuca dasyclada Hackel ex Beal
Utah fescue, Green River Shale fescue

2 Culms mostly 4-25 cm tall, usually less than 2 times the height of the basal
   leaves; anthers mostly not over 1 mm long; plants of woods and rocky
   slopes in the upper pine-spruce belt to well above timberline on the Uinta
   Mtns.: **var. brevifolia** (R. Brown) S. Watson [*F. brachyphylla* Schultes &
   Schultes; *F. ovina* ssp. *brachyphylla* (Schultes) Piper] — shortleaf fescue

Festuca pratensis Hudson [*F. elatior* L.; *Lolium pratense* (Hudson) S. J. Darbyshire;
Schedonorus pratensis (Hudson) Beauvios] — meadow fescue — Introduced from
Europe, planted in pastures and for hay, escaping and persisting along ditches,
roadsides, and other places; up to about 9,000 ft; June-Aug.

Festuca thurberi Vasey — Thurber fescue — Known from upper elevations of
the Tavaputs Plateau from s. and e. of Minnie Maud (Nine Mile) Creek; oak,
aspen, and open coniferous communities; July-Aug.

Glyceria R. Brown — Mannagrass

Plants perennial, rhizomatous, of wetlands; stems erect or decumbent and
rooting at the nodes; leaf sheaths closed at least in upper leaves; ligules membranous; auricles lacking; inflorescence an open, often drooping panicle;
spikelets 3-14 flowered, often purple tinged, disarticulating above the glumes;
glumes unequal, shorter than the first lemma; lemmas awnless, with 5-9
rather prominent nerves, the nerves sometimes slightly ridged, parallel and
not converging at the apex of the lemma. In general appearance *Torreyochloa
pallida* looks like a *Glyceria* and might be mistaken for *G. grandis* or *G. striata*.
However it has open leaf sheaths and generally fewer nerves on the lemmas.

---

1 Spikelets linear, 4-5 times longer than wide; panicle narrow, the branches
   erect .... **G. borealis**
1 Spikelets less than 3 times longer than wide; panicle open 2
2 Ligules of upper stem leaves mostly 4-9 mm long; glumes acute, the first
   1.2-2 mm long; lemmas 2-2.7 mm long, mostly elliptic, usually purple at
   maturity, not distinctly membranous at the tips; stamens (2) 3; leaf blades
   4.5-12 (15) mm wide .... **G. grandis**
2 Ligules of upper stem leaves 1-4 (6) mm long; glumes mostly blunt, the first
   0.4-1.3 mm long; lemmas 1.4-2.3 mm long, mostly obovate, green or purple
   tinged, distinctly membranous margined at the tips; stamens 2; leaf blades
   2-12 mm wide .... **G. striata**

Glyceria borealis (Nash) Batchelder — northern mannagrass — Widespread in
the Uinta Mtns., 20 of the 23 specimens seen are from the south slope with
only 3 from the north slope; wet places and in standing water up to 1 ft deep;
7,360-10,760 ft; July-Aug.

Glyceria grandis S. Watson — American mannagrass — Widespread but highly
localized in the Uinta Mtns and along the Green River as far downstream as
Browns Park; streams, seeps and springs; 5,500-8,020 ft; June-Aug. Similar to *G.
striata,* but averaging somewhat taller, otherwise different by the features
given in the key.

Glyceria striata (Lamarck) A. S. Hitchcock [*G. elata* (Nash) M. E. Jones] — fowl
mannagrass — Locally common; Uinta Mtns., Cold Springs Mtn., Strawberry

Valley, and perhaps on the Tavaputs Plateau; mostly along streambanks in woods and other wet places; 7,000-10,500 ft; May-Sept.

Helictotrichon Besser — Perennial oatgrass

Helictotrichon mortonianum (Scribner) Henrard — alpine oat — Tufted perennial plants, 5-20 cm tall; leaf sheaths open, the blades rolled, about 1-2 mm wide; ligules about 1 mm long, membranous, ciliolate; panicle 2-7 cm long, narrow, the short branches erect usually with a single spikelet; spikelets mostly 2-flowered; the second floret smaller than the first, with a shorter awn, apparently staminate or sterile, borne on a densely long-hairy rachilla about as long as the floret; glumes 8-12 mm long, equal or exceeding the lemmas; lemmas 6-9 mm long, the callus bearded with hairs 1-2 mm long, the awns 10-15 mm long, attached near the middle of the lemma. Occasional across the Uinta Mtns.; alpine communities; 11,000-12,600 ft; July-Aug.

Hierochloe — Sweetgrass

Hierochloe hirta (Schrank) Borbas [*H. odorata* (L.) Beauvios misapplied] — sweetgrass, vanilla grass — Culms arising singly or few together from slender rhizomes, 20-50 (70) cm tall; leaf sheaths open, blades of the culm often reduced, the upper one acute-triangular, not over 1 cm long, the lower ones to 3 cm long, blades of sterile shoots more elongate; ligules about 1.5-6 (8) mm long; panicle 3-10 (12) cm long, more or less open; spikelets 3-6 mm long, 2-3 (4) mm wide; glumes about as long as the spikelets, translucent or greenish toward the base; lemmas about as long or longer than the glumes, greenish at first but maturing to yellow-brown, 3 per spikelet, the central one bisexual and glabrous except at the tip, the 2 lateral ones staminate and pubescent over the back, all 3 lemmas awnless. Specimens seen are all from the Uinta Mtns.; occasional on canyon bottoms along water courses and in wet meadows of the pine-spruce belt to perhaps somewhat above timberline, 7,400-11,500 ft, but to be expected along the floodplains of the Green River and other water courses at low elevations; June-Aug. This grass is often fragrant at maturity.

Hilaria H. B. K. — Galleta

Hilaria jamesii (Torrey) Bentham (*Pleuraphis jamesii* Torrey) — galleta — Perennial plants from large, scaly rhizomes, 20-40 cm tall; culms solid, the nodes often pubescent; leaves mostly basal, the sheaths open, the blades soon involute, reduced in size upward on the culm, basal leaves often curled when dry; collars usually pilose; ligules 1-4 mm long, ciliate or lacerate; inflorescence a spike, often purplish but pale upon drying; spikelets long pilose at the base, arranged in groups of 3, each group 6-11 mm long; 2 lateral spikelets staminate and more than 1-flowered, central spikelet fertile and 1- (2)-flowered, the spikelets falling as a unit after maturity leaving a bare, zigzag rachis; glumes of lateral spikelets single-awned, 1 awned from near the tip, the others awned from below the middle; glumes of central spikelet with several awns; lemmas 4-7 (9) mm long; awns of lemmas mostly inconspicuous without magnification, and hidden in the groups of spikelets, but some to about 5 mm long and visible without magnification. Widespread; dry hills and flats, desert shrub, greasewood, sagebrush, and pinyon-juniper communities; particularly abundant in the Kennedy Wash-Red Wash area and south to the White River where it is often a community dominant with greasewood and sagebrush and in desert grasslands; 4,700-6,800 ft; May-Aug.

Hordeum L. — Barley

Plants annual or perennial, tufted; leaf sheaths open; ligules membranous; auricles lacking or well developed; inflorescence a dense terminal, solitary spike, the rachis readily disarticulating at the nodes at maturity; spikelets 3 at each node of the rachis, 1-flowered, the central spikelet generally sessile with a well-developed, fertile floret, the lateral spikelets short-pedicellate with sta-minate or rudimentary florets; glumes awn-like their whole length or in some species expanded toward the base; lemmas usually awned from the tip. *Hordeum vulgare* (cultivated barley) differs from the above description in having a persistent rachis and sessile lateral spikelets with fertile florets. Sometimes hybridizing with species of *Agropyron* and *Elymus*.

1 Plants perennial. 2

1 Plants annual 3

2 Awns of lemmas to 15 mm long . . . . **H. brachyantherum**

2 Awns of lemmas over 20 mm long . . . . **H. jubatum**

3 Awns of lemmas 5-16 cm long or lacking, rachis not disarticulating; leaf blades 5-15 mm wide; plants cultivated, occasionally escaping, 60-130 cm tall . . . . **H. vulgare**

3 Awns of central lemma 0.2-2 cm long; rachis disarticulating; blades 1-5 mm wide; plants 10-40(60) cm tall, not cultivated. 4

4 Awn of central lemma 8-20 mm long . . . . **H. marinum**

4 Awn of central lemma 2-8 mm long . . . . **H. pusillum**

Hordeum brachyantherum Nevski [*H. nodosum* L. misapplied; *Critesion brachyan-therum* (Nevski) Barkworth. & D. R. Dewey] — meadow barley — Widespread; occasional to common; typically in meadows and other places where water other than direct precipitation is available to keep the ground moist for much of the growing season; (6,000) 6,500-10,500 ft; June-Aug. A cross between this species and *Elymus trachycaulus* has been collected from the E. Slope in the Blind Stream Drainage of the Uinta Mtns. *Hordeum* characteristics of the cross are narrow, awned glumes, disarticulating rachis, and some tendency for more than 1 spikelet per node of the rachis. *Elymus* characteristics are spikelets all sessile, predominantly 1 spikelet per node of the rachis, and more than 1 floret per spikelet.

Hordeum jubatum L. [*Critesion jubatum* (L.) Nevski] — foxtail, foxtail barley — Widespread; occasional to abundant; weedy along roadsides and in waste places, abundant in heavily grazed, degraded pastures and in drawdown basins of reservoirs; most often in alkaline places at lower elevations, occasionally in montane places where sometimes introduced in hay at hunting, livestock, and logging camps; May-Sept.

Hordeum marinum Hudson (*H. geniculatum* Allioni; *H. gussonianum* Parl.) — rabbit barley — Introduced from Europe; the 2 specimens seen (Goodrich 27558 & 28064) are from the Leota Bottom near the Green River at 4,700 ft elevation and a parking lot in Vernal. Also reported for the Flaming Gorge Reservoir Basin (Flowers et al. 1960).

Hordeum pusillum Nuttall [*Critesion pusillum* (Nuttall) A. Love] — little barley — The few specimens seen are from desert shrub communities on heavy clay soils in the Island Park area and Kennedy Wash at 5120-5,530 ft, to be expected in other areas at low elevations; April-June.

Hordeum vulgare L. — barley — Cultivated and occasionally growing along roadsides and waste places for a season and used for reclamation on oil and gas pads, but apparently not persisting without cultivation; June-Aug.

Koeleria Persoon — Junegrass

Koeleria macrantha (Ledebour) Schultes [*K. cristata* (L.) Persoon misapplied; *K. gracilis* Persoon; *K. nitida* Nuttall] — Junegrass — Tufted perennial plants 20-60 cm tall; leaves mostly basal, the sheaths open, the blades flat or folded, 1-3 mm wide, seldom over 10 cm long; ligules about 0.2-2 mm long, ciliate along the margins (use 10X magnification); panicle dense and spike-like, opening at anthesis then closing again, the branches appressed, 1-30 mm long; glumes 3-6 mm long, the second glume about as long as the spikelet, with translucent margins and green center; lemmas about 5 mm long, similar to glumes in color. Widespread; occasional to abundant; most common in sagebrush, piny-on-juniper, mt. brush, dry meadow, moist meadow, aspen, and Douglas-fir communities; 5,600-9,000 ft. Listed for dry meadows above timberline on the Uinta Mtns. (Lewis 1970); June-Aug.

The translucent nature of the spikelet parts gives a shining appearance to the panicle. This is particularly noticeable at anthesis when the translucent paleas are exposed and again after the plants cure and the green centers of the glumes and lemmas have faded. A few specimens have short rhizomes. This is uncommon for the species, but has been also reported by Harrington (1954). See Arnow and others (1980) for discussion of synonymy. North American plants have been referred to as *K. pyramidata* (Lamarck Beauvios (Correll and Johnston 1970), but the relationship of our plants to this European species is yet to be clarified. The minutely soft-hairy pedicels and panicle branches (the hairs visible at 10X) distinguish this from the sometimes similar *Poa fendleriana* and *Trisetum wolfii.*

Leptochloa Beauvios — Sprangletop

Leptochloa fascicularis (Lamarck) A. Gray [*L. fusca* (L.) Kunth ssp. *fascicularis* (Lamarck) N. W. Snow] — bearded sprangletop — Tufted annual, 10-70 cm tall; ligules 2-7 mm long, membranous; leaf sheaths open the blades 1-6 mm wide, flat or rolled; panicles 7-40 cm long, with a few or many spike-like or racemose branches arranged along the main axis; spikelets 5-12 flowered, (5) 7-12 mm long, linear; glumes 2-6 mm long; lemmas 3-6 mm long or uppermost smaller, appressed-pubescent at least at the base of each of the 3 nerves, minutely bifid at the apex, awn-pointed or with awns up to about 4 mm long. The few specimens seen are from near Myton, Pelican Lake, and from the floodplain of the Green River near Jensen and Ouray; margins of ponds and lakes 4,725-5,100 ft; July-Oct.

Leucopoa Grisebach

Leucopoa kingii (S. Watson) W. A. Weber [*Festuca kingii* Cassidy; *Hesperochloa kingii* (S. Watson) Rydberg] — spike fescue — Plants perennial, usually unisexual, densely tufted, 0-65 (80) cm tall, sometimes with short rhizomes; leaf sheaths open, the bases persistent, the blades flat or rolled in age, 2-9 mm wide, firm and ascending to erect; ligules membranous, about 1-4 mm long; auricles lacking; inflorescence a narrow panicle with short erect branches; spikelets pale, with 3-6 florets, about 6-12 mm long, disarticulating above the glumes, the staminate ones usually maturing before the pistillate; glumes 3-7 mm long, awnless, scarious, only the midrib greenish; lemmas 4-8 mm long, scabrous, awnless or awn-tipped. Locally common to abundant, especially on Reservation Ridge and associated ridges of the W. Tavaputs Plateau, occasional from Strawberry Valley to Blind Stream in the Uinta Mtns., apparently rare e. of Blind Stream but specimens are from Uinta Canyon and Ashley Gorge, also found in the ponderosa pine belt on the north slope of the Uinta Mtns. and Blue Mtn.; sagebrush, aspen, oak, ponderosa pine, Douglas-fir, and spruce communities; 7,400-10,000 ft; May-Aug. The more or less scarious glumes help to distinguish spike fescue from species of *Poa* and *Festuca*.

Lolium L. — Ryegrass

Lolium perenne L. (*L. multiflorum* Lamarck) — ryegrass — Plants annual, biennial or short-lived perennial, sometimes appearing annual; culms tufted, mostly 30-80 cm tall, rarely taller; leaf sheaths open, the blades flat or slightly involute, 2-8 (10) mm wide; ligules membranous, to 1.5 mm long; auricles generally well developed; inflorescence a spike, 7-25 cm long; spikelets with (3) 4-15 flowers, 8-15 (20) mm long, laterally compressed, solitary at each node, turned edgewise to the rachis, the side next to the rachis without a glume, both glumes present on the terminal spikelet; lemmas 5-8 mm long, awnless or awned from the tip, the awn up to 8 mm long. Introduced (native to Eurasia and Africa), sometimes used in lawn mixtures as a nurse crop with emergence of seedlings in about 7 days compared to 21 day for Kentucky bluegrass, sometimes a contaminant in commercial seed used for other plantings; May-July. I have followed Arnow and others (1980) in placing *L. multiflorum* in synonymy here. Apparently the differences are largely a function of selective breeding for development of agronomic cultivars.

Melica L. — Oniongrass

Perennial plants from bulbous bases and/or short, thick rhizomes; leaf sheaths closed; ligules membranous; auricles lacking; inflorescence a contracted or spreading panicle; spikelets with 2-several fertile florets and 1-4 progressively reduced florets above (the terminal floret often reduced to a mostly club-shaped rudiment), disarticulating above or below the glumes; glumes mostly shorter than the lowermost lemmas; lemmas firmer than the glumes.

---

1 The first glume mostly less than 1/2 as long as lower lemma, 3.5-5 (6) mm long; ligules 1-3.2 mm long; bulbs spaced along a slender rhizome at intervals of about 1-3 cm, the rhizome generally forming a "tail" on the bulb, but this easily broken and often lacking in herbarium specimens ...
.... *M. spectabilis*

1   The first glume mostly over 1/2 as long as the lower lemma, 5-10 mm long; ligules 2-5 (7) mm long; bulbs generally tightly clustered on a short, thick rhizome, or rhizome lacking .... *M. bulbosa*

Melica bulbosa Geyer ex Porter & Coulter [*Bromelica bulbosa* (Geyer) Weber] — oniongrass — Occasional; widespread; sagebrush, mt. brush, ponderosa pine, and aspen communities; 7,700-10,500 ft; June-Aug.

Melica spectabilis Scribner [*Bromelica spectabilis* (Scribner) Weber] — purple oniongrass — Most of the 18 specimens seen are from Strawberry Ridge and east to the North Fork Duchesne drainage on the s. slope. and to the Whitney drainage on the n. slope of the Uinta Mtns. and east to Horse Ridge of the West Tavaputs Plateau, and one is from Greens Draw, in the eastern Uinta Mtns., but apparently rare or lacking in much of the central Uinta Mountains [Wolf & Dever 5171 DINO!, reported for Moffat Co. (Bradley 1950) belongs with *M. bulbosa*]; grassy slopes, open woods, and aspen and spruce-fir parklands; 8,280-9,800 ft; June-Aug.

Muhlenbergia Schreber — Muhly

Annual or perennial plants from fibrous roots or rhizomes; culms simple or sometimes branched; leaf sheaths open; ligules membranous; auricles lacking; inflorescence an open or contracted to spike-like panicle; spikelets almost always 1-flowered, disarticulating above the glumes; glumes shorter or occasionally slightly longer than the lemma; lemma awnless or awned, glabrous or with bearded callus.

1   Plants annual, appear perennial when forming mats, 5-20 (40) cm tall; growing in wet places mostly in mountains; rhizomes lacking **M. filiformis**

1   Plants perennial, 5-100 cm tall, with robust scaly rhizomes except in M. wrightii 2

2   Lemmas glabrous 3

2   Lemmas pubescent at least at the base; panicles narrow 6

3   Panicle narrow, 1-15 cm long, the branches short, erect 4

3   Panicle open and diffuse at maturity; plants 10-50 (70) cm tall; leaves flat or folded 5

4   Glumes awn-tipped; plants without rhizomes .... **M. wrightii**

4   Glumes awnless; plants with extensive, scaly rhizomes ... **M. richardsonis**

5   Lemmas awnless; spikelets 1.5-2 mm long; plants common, widespread of moist to wet, alkaline places .... **M. asperifolia**

5   Lemmas awned, the awn about 1 mm long; spikelets 2.5-4 mm long; plants rare, known from Moffat Co., dry sandy places .... **M. pungens**

6   Leaves 1-2 mm wide .... **M. thurberi**

6   At least some leaves over 2 mm wide 7

7   Callus hairs as long or longer than the lemma; awn of lemma 2-8 mm long .... **M. andina**

7   Callus hairs much shorter than the lemmas; lemmas awnless or awn-tipped 8

8   Internodes dull, puberulent; ligules to 0.6 mm long; anthers 0.8-1.5 mm
     long; lemma pilose at the base and margins .... *M. glomerata*
8   Internodes polished, except near apex; ligules 0.6-1.5 mm long; anthers 0.4-
     0.8 mm long; lemmas pilose at the base only .... *M. racemosa*

Muhlenbergia andina (Nuttall) A. S. Hitchcock — foxtail muhly — Widespread, apparently uncommon, the 6 specimens seen are from widely scattered locations in the Uinta Mtns., East Tavaputs Plateau (Moon Ridge), along the Green River, and base of Blue Mtn.; riparian communities and seeps; 4,900-10,080 ft; July-Sept.

Muhlenbergia asperifolia (Nees & Meyer) Parody — scratchgrass, alkali muhly — Widespread; most often in moist or ephemerally moist alkaline soil of low elevations, heavily grazed pastures, margins of ponds, along ditches and streams, roadsides, weedy in lawns, extending up the drainages of the Tavaputs Plateau to about 7,000 ft; June-Sept.

Muhlenbergia filiformis (Thurber) Rydberg — pull-up muhly — Strawberry Valley and the Uinta Mtns.; wet meadows, bogs, stream sides, and other wet places, occasionally in rather dry coniferous woods or sagebrush-grass communities; (5,300) 7,500-10,500 ft; June-Aug.

Muhlenbergia glomerata (Willldenow) Trinius — spiked muhly — The 2 specimens seen are from a sphagnum bog in Whiterocks Canyon, Uinta Mtns.; 7,360 ft; July-Sept. Similar to and perhaps not distinct from *M. racemosa,* but our plants differ as listed in the key and are found in wet, indigenous communities rather-than on dry often disturbed ground as is typical for *M. racemosa.*

Muhlenbergia pungens Thurber in A. Gray — sandhill muhly — Two specimens at RM (Dorn 3865, T9N R102W S15; Nielson 96, Cold Springs Mt., near Browns Park, 7,000 ft) are from the nw. corner of Moffat Co.; sandy places.

Muhlenbergia racemosa (Michaux) B.S.P. — creeping muhly, green muhly — Two specimens seen: Goodrich sn USUUB is from a dry rocky slopes in the N. Fork Duchesne and N. Holmgren et al. 467 DINO! is from Echo Park, Dinosaur National Monument; 5,000-7,500 ft; July-Sept.

Muhlenbergia richardsonis (Trinius) Rydberg [*M. squarrosa* (Trinius) Rydberg] — mat muhly — Widespread; occasional or locally common in sagebrush, cottonwood, and moist to dry meadow communities, often along roads and in other disturbed places where the soil has been compacted; 5,800-9,500 ft; July-Sept.

Muhlenbergia thurberi Rydberg (*M. curtifolia* Scribner) — Thurber muhly — One specimen seen (V. Swain sn UTC) is from top of Willam Creek (Willow Creek?) Uintah Co.; dry soil; another specimen (Holmgren et al. 441 DINO!) is from Echo Park, Dinosaur National Monument; July-Sept.

Muhlenbergia wrightii Vasey ex. Coulter — spike muhly — Two specimens seen from Daggett Co. One is from Bare Top Mtn. at 7,380 ft, sagebrush-bitterbrush community. The other is from bedrock at the south rim of Red Canyon. Also listed for Duchesne, and Uintah Cos. in AUF (873-874).

Munroa Torrey

Munroa squarrosa (Nuttall) Torrey — false buffalograss — Plants low, sprawling annuals; culms seldom over 20 cm long, leafless and often stolon-like except with tufts of leaves toward the apex, the tufts of leaves subtending or partly concealing the inflorescence; leaf sheaths ciliate near the collar; ligules a fringe of hairs, not over 1 mm long; leaf blades flat or rolled, 1-3 (4) cm long, not over 3 mm wide; panicles much reduced and often partly concealed in tufts of leaves, with 2-4 spikelets; spikelets 6-8 mm long, with 3-5 florets; glumes about 3-4 mm long; lemmas 3-5 mm long, 3-nerved, the central nerve prolonged into a short awn 0.5-2 mm long. The 4 specimens seen are from near Duchesne, Myton, and Asphalt Ridge; desert shrub and pinyon-juniper communities, and roadsides; 5,100-5,800 ft. Sept.

Oryzopsis Michaux — Ricegrass

Caespitose perennials without rhizomes; leaf sheaths open; ligules membranous; auricles lacking; inflorescence an open or contracted panicle; spikelets with 1 floret, disarticulating above the glumes; glumes equal to or a little shorter than the lemma; lemmas often plump, rather firm, the callus sometimes bearded, awned from the tip, the awn straight or bent, readily deciduous.

---

1   Lemmas 2.7-5.5 mm long; awns 7-18 mm long; rare sterile hybrids involving *Stipa hymenoides* and other species of *Stipa* . . . . see **Stipa hymenoides**

1   Lemmas either less than 2.7 mm long or else the awns less than 7 mm long
. . . . 2

2   Panicle narrow, erect, racemose, the branches seldom over 2 cm long . . . 3

2   Panicles open at maturity with some branches usually over 2 cm long . . . 4

3   Some leaves flat, mostly (3) 5-9 mm wide; lemmas 5-7 mm long; glumes 5-8 mm long . . . . *O. asperifolia*

3   Leaves rolled, less than 3 mm wide; lemmas 3-5 mm long; glumes 3-6 mm long . . . . *O. exigua*

4   Spikelets 2-4 mm long excluding the awn; lemmas glabrous; ligules 0.2-2 mm long . . . . *O. micrantha*

4   Spikelets 6-8 mm long excluding the awn; hairs of lemmas about equal or longer than the body; ligules 3-8 mm long . . . . *Stipa hymenoides*

Oryzopsis asperifolia Michaux — roughleaf ricegrass — The several specimens seen are from moist woods of the s. slope and e. end of the Uinta Mtns., and from Eagle Creek of the n. slope; 7,400-9,000 ft; July-Sept.

Oryzopsis exigua Thurber — little ricegrass — Sporadic across the Uinta Mtns.; mostly with lodgepole pine and ponderosa pine, rather common in clearcuts and recent burns, apparently a seedbank species that is released by fire; 8,300-8,960 ft; July-Aug.

Oryzopsis micrantha (Trinius & Ruprecht) Thurber — littleseed ricegrass — Most common on the Tavaputs Plateau where occasional to locally frequent on alluvial soils of the canyon bottoms in partial shade of tall shrubs or in rocky places, a few specimens also seen from widely scattered locations across the Uinta Mtns. to Blue Mt. and near Browns Park and Irish Canyon where mostly found in rocky places; 6,500-9,100 ft; June-Aug.

Panicum L. — Panic grass

Annuals or perennials; leaf sheaths open; ligules a ring of hairs; auricles lacking; inflorescence a compact to open panicle; spikelets dorsiventrally compressed to terete, disarticulating below the glumes, with 2 florets, the lower floret staminate or much reduced and sterile, much like the second glume; fertile lemma usually hardened, often shiny and smooth, the margins rolled around the edges of the hardened palea, dorsiventrally compressed, awnless.

1 Plants perennial . . . . **P. virgatum**
1 Plants annual 2
2 Second glume and sterile lemma short pilose throughout; basal leaves forming a winter rosette . . . . **P. acuminatum**
2 Second glume and sterile lemma glabrous except sometimes on the nerves apically; basal leaves similar to the culm leaves, not forming a winter rosette 3
3 Spikelets 4-6 mm long; panicle branches ascending, the panicle usually over twice as long as broad . . . . **P. miliaceum**
3 Spikelets less than 4 mm long; panicle branches widely spreading; the panicle mostly less-than twice as long as wide . . . . **P. capillare**

Panicum acuminatum Swartz [*P. huachuacae* Ashe; *P. lanuginosum* Ell.; *P. tennesseense* Ashe; *Dichanthelium accuminatum* (Swartz) Gould & Clark; *D. lanuginosum* (Elliott) Gould] — bundle panic — The 2 specimens seen are from near Whiterocks (Neese & Goodrich 8188) and Echo Park on the Green River (N. Holmgren et al. 473 DINO!); riparian communities; 5,100-5,770 ft; July-Sept. Holmgren (1962) referred the latter specimen to *P. scribnerianum* Nash (*P. oligosanthes* Schultes).

Panicum capillare L. — old witchgrass — Widespread and weedy, gardens, fields, ditch banks, and roadsides; not expected over 7,000 ft; July-Sept.

Panicum miliaceum L. — broom corn millet, hog millet — Infrequently cultivated and sometimes spontaneous for a season or two, but apparently not persisting; July-Sept. Broom corn millet is a common component of bird seed mixtures. It might be expected to sprout where birds are fed.

Panicum virgatum L. — switchgrass — Known from Duchesne Co. at Coyote Draw (L. Rasmusssen 580 NRCS) and Uintah Co. at Bennett (J. Brown 670 NRCS). Sparingly planted as an ornamental.

Phalaris L. — Canarygrass

Phalaris arundinacea L. [*Phalaroides arundinacea* (L.) Rauschert] — reed canarygrass — Perennial plants with robust rhizomes, 50-130 (200) cm tall; leaf sheaths open or closed near the base; ligules membranous, 2-5 (10) mm long; leaf blades flat, 5-15 (25) mm wide; panicles 5-25 (30) cm long, narrow, densely flowered, rather spike-like, the short branches appressed; spikelets appearing as if with a single floret, but with 2 inconspicuous florets (about 1 mm long) that are reduced to sterile lemmas, which are often appressed to the base of the fertile floret where they appear as pubescent lines; glumes 3.5-7.5

mm long; lemmas of fertile floret 3-4 mm long, hard and with a shining, varnish-like surface, glabrous or scattered pubescent on the back to rather densely ciliate along the margins and especially toward the tip; paleas mostly enclosed in the lemma, but sometimes partly exposed, similar to the lemmas in pubescence. Widespread; abundant along ditches, streams and other wet places; from low elevations up to about 7,000 ft and infrequent to about 8,000 ft; June-Aug.

Phleum L. — Timothy

Perennial plants; leaf sheaths open, the blades flat; ligules membranous; auricles minute or lacking; inflorescence a spike-like, cylindrical or ellipsoid panicle in which the branches are not visible without both teasing the panicle apart and magnification; spikelets with 1 floret, disarticulating above the glumes; glumes equal or nearly so, longer than the lemmas, prominently keeled, the keel prolonged into a short awn-point and beset on the back with stiff ciliate hairs, the hairs about equal to the width of the glumes.

---

1   Panicles seldom over 4 cm long, often 1 cm wide or wider; awns of the glumes (1.2) 1.5-3.2 mm long, the terminal ones often noticeably exserted from the panicle; culms mostly less than 50 cm tall, not bulbous at the base
. . . . *P. alpinum*

1   Panicles mostly over 4 cm long, usually less than 1 cm wide; awns of the glumes (0.6) 1-1.6 mm long, the terminal ones not noticeably exserted from the panicle; culms often over 50 cm tall, usually bulbous at the base . . . . .
. . . . *P. pratense*

Phleum alpinum L. (*P. commutatum* Gaudin) — alpine timothy — Occasional to common in the Uinta Mtns. and Strawberry Valley, apparently rare or absent over much of the Tavaputs Plateau; meadows, woods, along streams, and above timberline; 7,500-11,500 ft; June-Aug.

Phleum pratense L. — timothy — Introduced from Europe, widely seeded in pastures and as a hay crop and in wildland seedings, escaping and persisting in moist places; 4,700-9,000 ft; June-Sept.

Phragmites Trinius— Reed

Phragmites australis (Cavanilles) Trinius ex Steudel (*P. communis* Trinius) — common reed — Stout perennial plants 2-3 m tall with large creeping rhizomes to 5 m long and to over 1 cm thick; leaf sheaths open; ligules membranous, 1-3 mm long, short ciliate at apex; leaf blades 1-4 cm wide, flat; panicle 15-40 cm long, plume-like from the long silky hairs of the rachillas; spikelets 10-15 mm long, disarticulating. above the glumes, usually 3-6 flowered; first glume 3-7 mm long, the second glume 6-10 mm long; lemmas 8-15 mm long, tapering to an awn-like apex. Widespread; locally common to dominant; water courses, marsh lands, margins of ponds and lakes, and other wet places; not expected over 7,000 ft; July-Oct. Both native and introduced populations are found in our area. The introduced material is highly aggressive and displaces native vegetation. The native material appears to be less aggressive and apparently does not form the very dense, aggressive stands common in the introduced material.

Poa L. — Bluegrass

Annual or perennial plants; culms arising singly or few together from rhizomes or densely tufted from fibrous roots; leaf sheaths open for at least half their length, the blades usually folded and keeled with a prow-like tip, or flat to tightly rolled; ligules membranous; auricles lacking; inflorescence an open to contracted panicle; spikelets compressed to terete, small, rarely over 10 mm long, (1) 2-15 flowered, disarticulating above the glumes; glumes shorter than the lowest floret; lemmas keeled or rounded, awnless.

This is a perplexing if not an exasperating genus. It is comprised of numerous species that are distinguished by minor features. Compressed and keeled as opposed to rounded lemmas and degree of pubescence of lemmas are used as critical features. Use of these features require experience and a usually a magnifying lens. Immediate results should not be expected with the first few attempts at species separation. See IMF (6: 224-225) and Arnow and others (1980) for further discussions of problems in this genus. *Poa arida* Vasey (*P. glaucifolia* Scribner & Williams) has been indicated for our area, but plants of this taxon are from far to the e. of here.

---

1 Plants bulbous at the base; florets often converted into bulblets with awned tips; bulblets subtended by awned bracts . . . . **P. bulbosa**

1 Plants not bulbous at the base; florets not converted into bulblets, awnless
. . . . 2

2 Plants low spreading annuals, may appear to be perennials when growing in tufts; panicle open, the branches 1 or 2 at a node, 1-2 (3) cm long, spreading at maturity or even reflexed; lemmas pubescent on keel and marginal nerves; habitat often roadsides, around watering places for livestock or other moist places where the soil has been disturbed **P. annua**

2 Plants perennial, not as above in all other features. 3

3 Spikelets rounded or little (rarely strongly) compressed, narrow, 7-10.5 mm long; glumes not keeled; lemmas rounded on the back, not keeled or scarcely so, glabrous or puberulent; rachilla internodes 0.6-1.9 mm long; anthers 1-4.2 mm long; bunchgrasses without rhizomes (an extremely variable complex) . . . . **P. secunda**

3 Spikelets compressed or sometimes subterete in P. arctica (a rhizomatous species), commonly less than 6 mm long or else plants commonly rhizomatous; glumes keeled; lemmas keeled, glabrous or often pilose or even with a tuft of cobwebby hairs at the base; rachilla internodes mostly less than 0.6 mm long; anthers mostly less than 1 mm long; rhizomes various or lacking 4

4 Rhizomes present 5

4 Plants without rhizomes. 12

5 Panicle 2-3 cm long, the branches less than 1 cm long; lemmas glabrous . .
. . . . **P. cusickii**

5 Panicle or branches usually longer than above; lemmas usually pubescent
. . . . 6

6 Panicle narrow, the branches appressed, erect or somewhat spreading at anthesis. 7

6 Panicle open at maturity, most of the branches spreading 9

7   Plants unisexual, bunchgrass with long-lived bases; some spikelets usually 6-8 (10) mm long; panicles more or less shiny . . . . ***P. fendleriana***

7   Plants bisexual, culms arising singly or few together; spikelets mostly less than 6 mm long; panicles not especially shiny 8

8   Culms strongly flattened, 2-edged, sometimes slightly winged on the edges; lemmas usually sparsely if at all webbed, usually bronze-tipped, 2-3 mm long, and nearly coriaceous; plants rather uncommon; panicle narrow, the branches remaining erect or nearly so at maturity . . . . ***P. compressa***

8   Culms terete, not winged; lemmas copiously webbed at the base with long tangled hairs, 2.5-4 mm long, not firm nor coriaceous; plants common; panicle spreading at maturity . . . . ***P. pratensis***

9   Lemmas glabrous to pilose on marginal nerves and keel, lacking a tuft of cobwebby hairs at the base, or if web weakly developed then plants likely hybrids; florets pistillate or bisexual; spikelets 4-10 (11) mm long; lemmas 4-6 mm long 10

9   Lemmas either with a tuft of cobwebby hairs at base or pubescent over the back as well as on the keel and marginal nerves; florets bisexual; spikelets 3-7 (8) mm long; lemmas 2.5-5 (6) mm long 11

10 Panicle internodes elongate, the lower most generally averaging at least 3 cm long, at flowering, most of the branches horizontally spreading to deflexed; sheaths glabrous or scabrous; florets pistillate or bisexual . . . . . .
. . . . ***P. arnowiae***

10 Panicle internodes variable, the lowermost usually averaging less than 3 cm long, the branches ascending to spreading but not reflexed; sheaths (sometimes only the upper) retrorsely hairy; florets mostly pistillate . . . .
. . . . ***P. nervosa***

11 Lemmas copiously cobwebby at the base, glabrous above or nearly so except on keel and marginal nerves; lower panicle branches (2) 4-5 (9) per node; first glume 2-3 mm long; ligules 1-2 mm long or to 3 mm on upper culm leaves; plants mostly growing below 9,500 ft . . . . ***P. pratensis***

11 Lemmas scantly if at all cobwebby at base, pubescent on midnerves and inter-nerves as well as on the keel and marginal nerves; lower branches of the panicle (1) 2 (4) per node; first glume 2.5-5 mm long; ligules 1-4 mm long; plants mostly growing above 9,500 ft . . . . ***P. arctica***

12 Panicles 2-6 (9) cm high, about as broad as high; spikelets about 4-7 mm long, about 1/2 as broad as long; lemmas sparsely to densely white-hairy on the keel and marginal nerves and usually sparsely hairy between the nerves; blades of leaves 2-6 cm long, flat for nearly the entire length, those of the culm 2-6 mm wide; plants mostly 5-20 (30) cm tall, in upper coniferous forest belt and above timberline . . . . ***P. alpina***

12 Plants not as above in all respects 13

13 Panicle open, the branches spreading or reflexed; lemmas mostly with a tuft of cobwebby hairs at the base 14

13 Panicle contracted, the branches short, not spreading, not reflexed; lemmas without a tuft of cobwebby hairs at the base 17

14 Lower panicle branches 1-2 per node, often reflexed at maturity, bearing spikelets on the outer 1/2; leaf sheaths closed about 1/2 their length; anthers mostly less than 1 mm long. 15

14 Lower panicle branches mostly more than 2 per node, seldom reflexed, bearing spikelets for over 1/2 their length; leaf sheaths closed about 1/4 their length; anthers usually 1 mm long or longer 16

15 First glume tapered to a finely acute, nearly awn-like tip; lemmas 3-5 mm long, glabrous between keel and marginal nerves; palea glabrous or scabrous to ciliate on the keels with minute, stiff hairs; spikelets 4-7.5 mm long .... *P. leptocoma*

15 First glume usually abruptly acute at the tip; lemmas 2.2-3 (3.5) mm long, at least some of them sparsely hairy between keel and marginal nerves; palea ciliate with soft, crinkled hairs on the keels, sometimes minutely so; spikelets 3-4 (5) mm long .... *P. reflexa*

16 Ligules of upper stem leaves acute to obtuse, 2-6 mm long; panicles (7) 10-25 cm long usually, drooping; plants mostly 15-130 cm tall, growing in wet or mesic sites up to about 9,000 ft .... *P. palustris*

16 Ligules of upper stem leaves obtuse to truncate, 0.5-2 mm long; panicles 4-10 (17)-cm long; plants usually less than 40 cm tall, growing on dry or mesic sites, often above 8,500 ft .... *P. glauca*

17 Spikelets all pistillate, rarely all staminate, 4-12 mm long; plants (10) 15-50 (70) cm tall; panicle often pale 18

17 Spikelets bisexual, 2-7 mm long; plants 2-25 (50) cm tall; panicle often purplish from the purple-tipped lemmas 19

18 Lemmas glabrous on the back; plants alpine and near alpine ... *P. cusickii*

18 Lemmas pubescent on the keel and marginal nerves at least below; plants mostly from sagebrush and other lower elevation communities, but from a wide range of habitats and elevations including above timberline .... *P. fendleriana*

19 Lemmas glabrous; plants 2-7 (12) cm tall, growing well above timberline; panicles 0.7-2 (3) cm long, densely flowered .... *P. lettermanii*

19 Lemmas pubescent, sometimes scarcely so; plants 5-25 cm tall or taller, of moderate to high elevations; panicles 1-9 cm long, loosely to densely flowered 20

20 Lemmas 2-3.5 mm long; lower leaf blades 1-5 cm long; second glumes averaging less than 3.5 mm long .... *P. glauca*

20 Some lemmas usually over 3.5 mm long; at least some lower leaf blades usually over 5 cm long; second glumes averaging more than 3.5 mm long .... *P. pattersonii*

Poa alpina L. — alpine bluegrass — Widespread in the Uinta Mtns.; occasional to common in meadows, open woods and alpine; (9,000) 10,000-13,000 ft; July-Aug.

Poa annua L. — annual bluegrass — Widespread; occasional in moist places, along streams, around livestock watering places, roadsides and other areas where the soil has been disturbed; up to about 9,000 ft; June-Sept.

Poa arctica R. Brown (*P. grayana* Vasey) — arctic bluegrass — Uinta Mtns. and W. Tavaputs Plateau; occasional on exposed ridges or slopes and in snowbeds; (8,400) 10,000-13,000 ft; July-Aug.

Poa arnowiae Soreng (*P. curta* Rydberg misapplied) — Wasatch bluegrass — Occasionally from Strawberry Valley east to the Rock Creek drainage in the Uinta Mtns. and less common eastward in the Uinta Mtns; forb-grass-sedge meadow, tall forb, aspen, and spruce communities; 7,500-10,500 ft; June-Aug. See discussion under *P. nervosa.*

Poa bulbosa L. — bulbous bluegrass — Introduced from Europe, widespread; occasional to locally abundant on disturbed soil, sagebrush, pinyon-juniper, ponderosa pine, and aspen-Douglas-fir communities; 7,000-8,100 ft; May-July. Both normal and modified bulblet type florets may occur in the same panicle, or entire panicles often consist of just 1 type of floret.

Poa compressa L. — Canada bluegrass — Apparently widespread; occasional in moist places in meadows or along water courses, but occasionally in mtn. big sagebrush and ponderosa pine communities; 7,200-8,820 ft; June-Aug.

Poa cusickii Vasey — Cusick bluegrass — Infrequent or locally common in the Uinta Mountains with 3 vars. as follows:

1 Plants known from hummocks of peatlands in Sheep Creek Park, with very slender rhizomes that are difficult to collect .. **var. *pallida*** (Soreng) Dorn
1 Plants not in peatlands 2
2 Leaves tightly rolled; specimens seen are from subalpine and alpine meadows and alpine ridges of the Uinta Mtns. . . . . **var. *cusickii***
2 Leaves folded or flat; specimens seen are generally from more protected habitats and slightly lower elevations but sometimes growing side by side with plants of the above taxon . . . . . **var. *epilis*** (Scribner) C. L. Hitchcock

Poa fendleriana (Steudel) Vasey (*P. longiligula* Scribner & Williams) — muttongrass — Widespread; common; sagebrush-grass, pinyon-juniper, mt. brush, and ponderosa pine communities, and above timberline; May-July depending upon elevation. Of the many specimens examined, all are pistillate but 1. The staminate specimen is from a population that has rhizomes up to 10 cm long. The presence of rhizomes and anthers may reflect influence of a rhizomatous member of the genus (IMF 6: 240). Arnow and others (1980) have combined *P. epilis* Scribner and *P. cusickii* Vasey with *P. fendleriana.* Not only are plants of these taxa morphologically similar, but they are sexually alike in having mostly pistillate florets. However, strong ecological differences correlate with the morphological differences, and separation at some level seems appropriate. They have been separated by criteria given the key. Most of the specimens seen can be separated by the key, but in the examination of numerous specimens a continuum was found.

Poa glauca Vahl [*P. interior* Rydberg; *P. nemoralis* L. var. *interior* (Rydberg) Butt. & Abbe; *P. rupicola* Nash] — Inland bluegrass — Occasional to common; Uinta Mtns. and W. Tavaputs Plateau; Vasey sagebrush, aspen, subalpine meadow, and alpine tundra communities, and open coniferous woods and talus slopes; about 8,000 ft to well above timberline; June-Sept. With 2 intergrading phases that have been separated as follows:

1 Lemmas scantly webbed at the base; plants 20-50 (70) cm tall, mostly below timberline; panicles 4-15 cm long, with spreading-ascending branches . .
. . . . ***P. interior***
1 Lemmas not webbed but villous at the base; plants 5-25 (35) cm tall, mostly near or above timberline; panicles 1-5 (7) cm long, with ascending or appressed branches . . . . ***P. rupicola***

At what point the lemmas cease to be webbed and begin to be villous is highly interpretive, and the plants of the complex intergrade in all other features. I have followed Arnow and others (1980) in reducing these taxa to synonymy under the circumboreal *P. glauca,* which shows the same range of variation in Eurasia as in America. The complex is further complicated by large specimens approaching those of *P. palustris.*

Poa leptocoma Trinius — bog bluegrass, nodding bluegrass — Thirty specimens seen from across the Uinta Mtns. and one from Twelve Hundred Dollar Ridge of the W. Tavaputs Plateau; occasional in aspen and coniferous forests and parklands, moist and wet meadows and along streams, mostly below timberline but occasional in krummholz and alpine; (8,040) 9,000-11,000 (11,700) ft; June-Aug. In habitat similar to *P. reflexa* but much less common in alpine snowbeds.

Poa lettermanii Vasey — Letterman bluegrass — Specimens seen are from the crest or near the crest of the Uinta Mtns. from the Uinta, Yellowstone, W. Fork Beaver Creek, and Lake Fork drainages; rocky places above timberline at12,000-13,498 ft. Komarkova (1979) found Letterman bluegrass above 11,500 ft (3,500 m) in the Indian Peaks area of Colorado.

Poa nervosa (Hooker) Vasey — Wheeler bluegrass — Widespread and common from Strawberry Valley and across the Uinta Mtns. (no specimens seen from the Tavaputs Plateau east of the Avintaquin drainage); aspen, lodgepole pine, Engelmann spruce, and less often tall forb communities; 7,500-11,200 ft; June-Sept. Plants of our area belong to var. *wheeleri* (Vasey) C. L. Hitchcock. Some specimens do not yield to traditional criteria given for separation of *P. nervosa* and *P. arnowiae,* and considerable disagreement among various authors on descriptions of these taxa has been found while preparing this treatment. Arnow (1987) kept the 2 separate. The key provided for the genus is adapted from her work. By using the features listed by Arnow, our plants are quite reasonably separated. However, the distinction is not always clear as occasional specimens with short internodes have glabrous sheaths, and some populations (apparently rarely) have sheaths that vary from glabrous to pubescent. Plants (especially those of *P. arnowiae*) might be confused with those of *P. leptocoma* and *P. reflexa,* but they have pistillate florets, stamens lacking or much reduced and translucent or functional and over 1 mm long; lemmas glabrous or pubescent but not webbed at the base; rhizomes often well developed. In the latter 2 taxa, plants have bisexual florets, with anthers less than 1 mm long; lemmas webbed (sometimes scantly) at the base; and rhizomes lacking.

Poa palustris L. — fowl bluegrass — Occasional in Strawberry Valley and across the Uinta Mtns. apparently rare or infrequent on the Tavaputs Plateau; riparian communities; 6,400-8,000 ft; July-Aug.

Poa pattersonii Vasey [*P. abbreviata* R. Brown var. *pattersonii* (Vasey) Love] — Red Pine Shale bluegrass — Uinta Mtns.; apparently rather uncommon; upper coniferous forest belt and above timberline, often found on semi-barrens of Red Pine Shale; July-Aug. With rather large and strongly flattened spikelets and long-lived bases, this could pass for a small phase of *P. fendleriana,* but the flowers are bisexual.

Poa pratensis L. — Kentucky bluegrass — Native and introduced, the principal species used in lawn mixtures in our temperate climate, aggressive and well established in many indigenous communities at low and moderate elevations (up to about 9,000 ft), often the dominant understory plant in stands of rubber rabbitbrush on canyon bottoms of the Tavaputs Plateau; May-Oct. England 786 (UI!), the basis for listing *P. trivialis* L. in Goodrich and others (1981), belongs here.

Poa reflexa Vasey & Scribner — nodding bluegrass — Forty one specimens seen from Strawberry Valley and across the Uinta Mtns. (no specimens seen from the Tavaputs Plateau), along streams in coniferous forests, subalpine parklands and meadows, and alpine where most common in late persisting snowbeds; (9280) 10,000-12,400 ft. Holmgren and Holmgren (IMF 6: 238) suggested that it may be more realistic to recognize *P. leptocoma* and *P. reflexa* at the varietal level, and the combination *P. leptocoma* var. *reflexa* (Vasey & Scribner) M. E. Jones is available.

Poa secunda Presl [*P. sandbergii* Vasey; *P. scabrella* (Thurber) Bentham; *P. confusa* Rydberg] — Sandberg bluegrass — Common to abundant; widespread; salt desert shrub, sagebrush, pinyon-juniper, and ponderosa pine communities, and exposed, dry sites to above timberline; April-July depending on elevation.

We have followed Arnow (1981) in placing *P. sandbergii* in synonymy. *Poa secunda* seems to be the basic element of a rather large apomictic complex in which there are no clear boundaries, but in which the extremes are so different that this is a perplexing group to work with. As many as 16 or more taxa have been described. Many of these are generally accepted in synonymy, but several have been listed in recent manuals. Intergradation in the complex is frequent, and such a continuum is formed that separation at the species level is impractical and often serves more to waste time than for any useful purpose. Separation of taxa has largely been based on pubescence or lack of pubescence of the lemmas. Based on this character alone, 2 groups (Nevadenses & Scabrellae) with 4 species each have been separated. However, this feature is not consistent enough within the complex to provide confident separation of species let alone groups of species. Although there appears a rather strong trend for plants of dry habitats to have pubescent lemmas and those from moist or wet habitats to have glabrous lemmas, a complete range from glabrous to pubescent can be found in single populations and to a lesser degree in different spikelets of the some plant and even in different lemmas of the same spikelet.

Stature of plants, length of leaves and length of ligule have also been used as means of separation, but these features also form a continuum. As noted by Hitchcock (in Hitchcock and others 1969) one of the most consistent combinations of features is that plants with open panicles have glabrous lemmas and short ligules. However, in one form or another, nearly all of the segregates intergrade into each other, and the variability in the whole complex is not any greater than found in many sexual species, and he suggested that the whole complex could appropriately be regarded as a single species. Kellogg (1985) has treated nearly all the complex as a single species.

Although we prefer a single species approach to this complex, some discussion of the so called taxa within the complex seems helpful to describing the variation. Thus, the following key, patterned after a traditional approach, is provided for some of the phases:

1 Lemmas scabrous over the back 2
1 Lemmas glabrous over the back. 4
  2 Panicle open, the branches spreading, divergent at anthesis; plants 2-6 dm tall, mostly of moderate to high elevations . . . . ***P. gracillima***

2 Panicle contracted, the branches appressed or ascending or somewhat spreading at anthesis. 3

3 Culms seldom over 30 cm tall, greatly exceeding the basal leaves; plants maintaining a distinct identity over large areas in desert shrub and black sagebrush communities, but intergrading into other taxa of the complex in more mesic communities; leaves usually rolled; inflorescence often marked with purple . . . . *P. secunda*

3 Culms mostly over 30 cm tall; plants of rather mesic communities; basal leaves folded; inflorescence greenish or slightly marked with purple . . . .
. . . . *P. canbyi*

4 Ligules long decurrent, often over 3 mm long . . . . *P. nevadensis*

4 Ligules short, 0.5-3 mm long, not long decurrent. 5

5 Leaf blades involute, mostly less than 1.5 mm wide, greenish; plant usually on alkaline soils . . . . *P. juncifolia*

5 Leaf blades flat or folded, 1.5-3.5 mm wide, often glaucous; plants growing on alkaline-free soils . . . . *P. ampla*

Poa ampla Merrill — big bluegrass — Mostly taller (40-120 cm) than *P. secunda* with more open inflorescence, larger spikelets, and shorter ligules; similar in stature to *P. canbyi* and intergrading into that taxon in pubescence of lemmas, but usually with more open panicles; confluent with *P. gracillima* in stature, open panicle, and pubescence of lemmas, and clearly allied to *P. secunda* and *P. canbyi* through this phase. Sagebrush and aspen communities in mountains; June-Aug. Materials selected from this tall phase are sold under the name of Sherman big bluegrass. This phase was seeded in the Mustang Fire of 2002 in Daggett Co. where it has been highly competitive with cheatgrass and with time has increased by continued seedling establishment. Similar trend is apparent in the Neola North Fire of 2007.

Poa canbyi (Scribner) Howell — Canby bluegrass — More robust than *P. secunda*, 30-100 cm tall, often with wider and denser panicle, and with less purple color. Intergrading in stature and other features into *P. scabrella* and then in turn into *P. secunda* through the *P. scabrella* phase. Sagebrush and aspen communities in mountains; June-Aug.

Poa gracillima Vasey — slender bluegrass, Pacific bluegrass — Slightly to considerably larger than *P. secunda*, and with more open panicle and less rolled leaves, similar to *P ampla* (q.v.) in the open inflorescence and intergrading into that taxon in pubescence of lemmas. The spreading of the panicle is dependent to some degree on phenology, and at anthesis the panicle of even the desert form of *P. secunda* is spreading. Open woods and slopes at rather high elevations, often near timberline; June-Aug.

Poa juncifolia Scribner — alkali bluegrass — Similar to *P. ampla* but with involute leaves, greenish not glaucous foliage, and shorter on the average, 2-7 (9) dm tall. Usually in alkaline lowlands; May-July.

Poa nevadensis Vasey — Nevada bluegrass — Separated from *P. secunda* by glabrous lemmas and taller stature, but completely intergrading into *P. secunda* in these and all other features including the long decurrent ligule. Sagebrush, aspen, and moist to wet meadow communities; June-Aug.

Polypogon Desfontaines — Beardgrass

Polypogon monspeliensis (L.) Desfontaines — rabbitfoot grass — Plants annual,

3-50 (80) cm tall; ligules 2-6 (10) mm long; leaf blades usually flat, 2-10 mm wide; panicle 1-10 cm long, dense and somewhat spike-like; spikelets 1-flowered; glumes about equal, 2-3 mm long, scabrous on the keel, slightly lobed and awned at the apex, the awns 4-10 mm long; lemmas about 1/2 as long as the glumes, smooth and shiny, minutely toothed at the apex, awnless or with a short awn. Widespread and locally common; moist places, often in but not confined to saline or alkaline areas; not expected much over 7,000 ft; May-Aug.

Puccinellia Parl. — Alkaligrass

Plants tufted, without rhizomes, perennial; leaf sheaths open; ligules membranous; auricles lacking; inflorescence an open to contracted panicle; spikelets 3-9 flowered, subterete, disarticulating above the glumes; glumes usually shorter than the lowermost lemma; lemmas awnless, with 5-7 indistinct to conspicuous parallel nerves, the lateral nerves not extending through the membranous margins of the lemmas.

1   Lower panicle branches to 15 cm long, spreading but rarely reflexed at maturity; lemmas 1.8-3.5 mm long, glabrous or minutely hairy; ligules 1-3 mm long . . . . ***P. nuttalliana***

1   Lower panicle branches rarely over 6 cm long; ascending when young, often reflexed at maturity; lemmas to 1.4-3 mm long, with minute hairs at the base, these hairs visible only with magnification; ligules 0.9-2.2 mm long . . . . ***P. distans***

Puccinellia distans (L.) Parl. — weeping alkaligrass — Widespread; occasional; wet or moist soil along streams or near springs or seeps, roadsides, alkaline and rarely non alkaline meadows; 5,400-7,600 ft or up to about 8,000 ft in canyons of the Tavaputs Plateau, and following salted highways up to 8,500 ft in Uinta Mtns., usually associated with saline or alkaline conditions; May-Sept.

Puccinellia distans and *P. nuttalliana* are reported to intergrade in Colorado (Harrington 1954). Arnow and others (1980) found it is often impossible to separate individual specimens of the two taxa by means of ligules or by components of the spikelets. The shorter deflexed panicle branches of *P. distans* provides a more reliable means of separation, but the panicle branches are not deflexed in young plants. Also herbage of *P. distans* is reported to be blue-green while that of *P. nuttalliana* is yellow-green (Allred 2013).

Puccinellia nuttalliana (Schultes) A. S. Hitchcock [*P. airoides* (Nuttall) S. Watson& Coulter in A. Gray] — Nuttall alkaligrass — Widespread; Infrequent in moist and wet alkaline or saline lowlands, along streams, and desert shrub communities; 5,000-6,200 ft; June-Aug. See P. distans.

Redfieldia Vasey

Redfieldia flexuosa (Thurber) Vasey — blowout grass — Plants perennial from long creeping rhizomes, 50-120 cm tall; leaf sheaths imbricate, open; ligule a fringe of hairs, about 1 mm long; leaf blades loosely rolled, 2-5 mm wide; panicle 20-50 cm long, diffuse; spikelets 5-8 mm long, with (1) 2-6 florets;

glumes 2-5 mm long; lemmas 4-6 mm long, strongly keeled, 3-nerved, nerves sometimes projecting slightly beyond the lemma, the callus with silky hairs. Infrequent or locally common; Devils Playground, Snake John Wash, Kennedy Wash, and Coyote Wash drainages in e. Uintah Co. and perhaps into Colorado; desert shrub, Indian ricegrass, and juniper communities, sandy soils, and in dunes; 4,900-5,700 ft; late June-Sept.

Schedonnardus Steudel

Schedonnardus paniculatus (Nuttall) Trelease. — tumblegrass — Plants tufted perennials, 10-40 cm tall; leaf sheaths open; spikelets membranous, 1-4 mm long; leaf blades usually folded, 1-2 mm wide; inflorescence 10-20 cm long, consisting of a few spreading or divaricate branches with sessile spikelets that are borne on one side of and slightly depressed in the branches; spikelets 3-5 mm long, 1-flowered; glumes 1.5-5 mm long; lemmas 3-5 mm long, awnless This is not indicated for our area by Harrington (1954), and the one specimen seen (Goodrich sn UT) from a roadside at the town of Tridell could be a waif. It has not been found at this location in recent years.

Schizachne Hackel

Schizachne purpurascens (Torrey) Swallen — false melica — Perennial plants with loosely tufted bases; culms 30-70 cm tall; leaf sheaths closed; ligules membranous, about 1-2 mm long; leaf blades flat, 2-5 mm wide; panicle 6-17 cm long, the branches spreading, sometimes drooping, each bearing 1-3 spikelets; spikelets with 3-7 florets, about 2-3 cm long including the awns, often purplish, disarticulating above the glumes; glumes shorter than the first floret, 4-9 mm long; lemmas 8-10 (12) mm long, the callus pilose with hairs 1-2 mm long, awned from between a bifid apex as in Bromus; awns 8-15 mm long, more or less divergent. Glacial Canyons of the s. slope of the Uinta Mtns. from Rock Creek to Whiterocks Canyon and scattered locations on the n. slope including Elk Creek and Eagle Creek; occasional in riparian and aspen-coniferous communities at 7,400-8,300 ft; June-Aug.

Secale L. — Rye

Secale cereale L. — winter rye — Tall annual plants, 60 to over 100 cm tall; leaf sheaths open, the blades flat, 3-10 mm wide; ligules membranous, 1-2 mm long; auricles to about 1 mm long; inflorescence a spike, 5-15 cm long, dense; spikelets mostly 2-flowered; glumes narrow in comparison to the broad lemmas, 5-12 mm long; lemmas to about 15 mm long with an awn 1-5 cm long.

A cultivated cereal crop, apparently not often used in our area and apparently not so commonly escaping and persisting as along the Wasatch Front, rather weedy in fields and on roadsides, spontaneous in some years especially on disturbed sites but not persisting for many years, occasionally used in rangeland seedings following fire or other disturbance for temporary cover mostly within the sagebrush and pinyon-juniper belts where it has not persisted in our area, perhaps not reaching maturity in montane places of greater than about 8,000 ft; June. This plant can be highly successful in establishing in difficult areas including blow-out sandy areas where can function as a nurse crop for establishing perennial plants. Numerous cultivars have been developed *Secale montanum* Gussone has been included in wildland plantings where it might persist. This

Asiatic species is reported to be perennial with a disarticulating rachis, and the awns (1-2.5 cm long) shorter than in *S. cereale.*

Setaria Beauvios — Bristlegrass

Plants annual; leaf sheaths open, the blades flat or folded to involute; ligules membranous-based but with a distal fringe of hairs; auricles lacking; inflorescence a spike-like panicle in which the branches are hardly visible without teasing the inflorescence apart and then hardly visible without magnification, densely flowered, cylindrical, bristly; spikelets 2-flowered, the lower floret sterile or staminate, the upper fertile, disarticulating below the glumes, but above the subtending bristles; glumes unequal, the first less than 1/2 as long as the spikelet, the second about as long as the spikelet and similar to the lower lemma. Both the glumes and the lemmas are awnless. The bristly appearance of the spikelets is a function of the bristles that subtend the spikelets or groups of spikelets.

1   Sheaths glabrous; each spikelet subtended by 4-20 antrorsely scabrous bristles 3.5-9 mm long; second glume about 1/2 as long as the spikelet; fertile lemma strongly cross wrinkled at maturity . . . . *S. glauca*

1   Sheaths ciliate toward the summit; each spikelet subtended by 1-4 bristles; second glume as long or nearly so as the spikelet, appearing like an empty lemma; fertile lemmas smooth to obscurely wrinkled 2

2   Bristles retrorsely scabrous, 2-6 mm long, 1 (2) subtending each spikelet; panicle branches usually conspicuously whorled . . . . *S. verticillata*

2   Bristles antrorsely scabrous, (4) 5-11 mm long, 1-3 (4) subtending each spikelet; panicle branches not conspicuously whorled . . . . *S. viridis*

Setaria glauca (L.) Beauvios — yellow bristlegrass — Introduced from Europe, more or less weedy; the one specimen seen (Neese & Nelson 15031a) is from a garden at Vernal; Aug.

Setaria verticillata (L.) Beauvios — bur bristlegrass — Weedy in gardens, fields, ditch banks, roadsides; not expected over 7,000 ft; July-Sept.

Setaria viridis (L.) Beauvios — green bristlegrass — Weedy in gardens, fields, ditch banks, and roadsides; not expected over 7,000 ft; July-Sept. Setaria italica (L.) L. Beauv.) — foxtail millet — also with antrorse bristles is cultivated and escapes. This has larger and more lobate panicles than *S.viridis.*

Sorghum Moench

Sorghum bicolor (L.) Moench (*S. vulgare* Persoon) — sorghum — Robust annuals, 1-2 (3) m tall; leaves glabrous, the blades 1-4 (5) mm wide; ligules 1.5-5.5 mm long; membranous below, fringed above; Introduce from Eurasia, cultivated primarily for silage, rarely escaping, probably not persisting without cultivation in the Uinta Basin.

Spartina Schreber — Cordgrass

Perennials from robust scaly rhizomes; leaf blades flat but rolled in age, firm, tough; sheaths open; ligule a ring of hairs; inflorescence a racemose panicle with 2-many spicate branches; spikelets densely crowded on one side of the comb-like branches, sessile, in 2 rows, 1-flowered, disarticulating below the glumes.

---

1   Second glume awned, the awn 1-7 mm long, conspicuously exceeding the lemmas; spicate branches 4-15 cm long; plants (0.7) 1-1.3 m tall; leaf blades 6-11 mm wide; ligules 1-3 mm long . . . . **S. pectinata**
1   Second glume awnless or mucronate, the mucro to 0.5 mm long, not conspicuously exceeding the lemmas; spicate branches 2-5 (7) cm long; plants 0.3-0.7 (1) m tall; leaf blades 2.5-5 (8) mm wide; ligules 0.5-1.5 mm long . . . . **S. gracilis**

Spartina gracilis Trinius — alkali cordgrass — The few specimens seen are from 10 mi sw. of Bonanza, along the White River, from Browns Park along the Green River and Red Creek, and Strawberry River at Duchesne, and near Robertson on the floodplain of the Blacks Fork River, also along the Blacks Fork in Sweetwater Co. Wyoming; riparian communities at 5,250-7,400 ft; June-Aug.

Spartina pectinata Link — prairie cordgrass — Occasional to locally common at isolated places on floodplains of the Yampa River, Green River, and White River; below 5,600 ft; June-Sept.

Sphenopholis Scribner — Wedgegrass; Wedgescale

Sphenopholis obtusata (Michaux) Scribner — prairie wedgescale — Plants annual or short-lived perennial, 20-90 cm tall, rarely taller; leaf sheaths open, the blades flat, 1.5-8 (12) mm wide; panicle 3-20 cm long, dense and spike-like or open; spikelets 1.5-5 mm long, 2-flowered, disarticulating below the glumes and falling as a unit; glumes 1-3 (4) mm long, the first linear, the second widest near the tip, much wider than the first; lemmas 2-3 (4.4) mm long, awnless. Widespread but infrequent; the 5 specimens seen are from Chandler Canyon-Desolation Canyon, Browns Park, Tridell, Upalco, and the seeps n. of Vernal; riparian and wet meadow communities and in pastures and along ditches; 4,600-7,170 ft; July-Sept.

Sporobolus R. Brown — Dropseed

Plants tufted perennials; leaf sheaths open, often ciliate on margins near summit; ligule a ring of hairs or short ciliate membrane; auricles lacking; inflorescence an open or congested panicle, sometimes enveloped in the up-permost leaf sheaths; spikelets 1-flowered, disarticulating above or below the glumes; glumes thin, translucent, mostly shorter than the lemma; lemma awnless; fruit hard, rounded, small, falling from the floret at maturity.

---

1   Spikelets 4-6 mm long; anthers 1.5-2.5 mm long; plants rare . . . . . **S. asper**
1   Spikelets 1.5-2.5 mm long; anthers 0.4-1.5 mm long; plants various 2
2   Panicle open and diffuse, the longer branches to over 10 cm long; leaf sheaths ciliate pilose only on the margin near the collar, the collar glabrous; ventral surface of leaf blades often with a row of pilose hairs to 5 mm long just above the ligule . . . . **S. airoides**

2 Panicles contracted or often included in the upper leaf sheath for the entire length, the branches if spreading mostly less than 5 cm long; leaf sheaths ciliate pilose well below the collar, the collar long pilose well around to the opposite side of the culm from the sheath margins; ventral surface without a row of pilose hairs as above 3

3 Panicle branches spreading if not confined within the upper sheath .....
.... *S. cryptandrus*

3 Panicle strongly contracted, densely flowered, nearly cylindrical, the branches appressed even when exserted from the upper leaf sheath .....
.... *S. contractus*

Sporobolus airoides (Torrey) Torrey — alkali saccaton — Widespread, locally abundant; alkaline or saline bottom lands including floodplains of the major drainages, occasional at alkaline seeps in canyons of the Tavaputs Plateau, and along roads and highways, seeded as a reclamation plant at oil and gas wells and along pipelines in desert locations; 4,700-6,000 (7,000) ft; June-Oct.

Sporobolus asper (Michaux) Kunth — tall dropseed — The 2 specimens seen are from near the Green River about 4 mi se. of Dinosaur Quarry; juniper-snakeweed community; 4,900 ft; Sept. (Neese & Trent 12319) and near Davis Spring at the base of Blue Mtn (Goodrich 25641). The panicle is contracted even when exserted from the upper sheath as in *S. contractus* from which it differs by the features of the key and by glabrous sheaths.

Sporobolus contractus A. S. Hitchcock — spike dropseed — Four specimens seen as follows: S. L. Welsh 380 from the e. side of Split Mt. Gorge, Dinosaur National Monument, L. England 1054 from Browns Park, B. Welsh & G. Moore 237 from 2 mi e. of Ouray, and J. McDonald & W. Smith sn. from vicinity of Uintah Basin; saltbush-ricegrass community, alluvial and sandy soils including sand dunes; 5,000-5,800 ft; June-Sept. This is occasional on wind-sand in the Kennedy Wash-Red Wash area of the White River drainage where it grows to a height of 1 m or more.

Sporobolus cryptandrus (Torrey) A. Gray — sand dropseed — Widespread; occasional to abundant; usually in sandy soil, desert shrub, Wyoming big sagebrush, and pinyon-juniper communities and following salted highways up to 8,500 ft.

Stipa L. — Needlegrass

Tufted perennial plants without rhizomes; leaf sheaths open; ligules membranous; auricles lacking; inflorescence an open or contracted panicle; spikelets 1-flowered, disarticulating above the glumes; glumes exceeding the body of the lemma; lemma awned from the tip, usually with overlapping margins, usually pubescent at least at the base, the callus well developed and pubescent with hairs usually longer than those of the body and glabrous at the tip; awns often bent, the lowest segment often twisted.

---

1 Awns of lemmas deciduous 2

1 Awns of lemmas persistent 4

2 Awns of lemmas 7-20 mm long .... *S. × bloomeri*

2 Awns of lemmas 3-6 mm long 3

3   Panicle open, the branches divaricate at maturity . . . . *S. hymenoides*
3   Panicle contracted, often enclosed at the base by the upper leaf sheath . .
    . . . . *S. arnowiae*
4   Awns of lemmas 6-18 (24) cm long; glumes 15-20 (40) mm long . *S. comata*
4   Awns of lemmas to 5 cm long 5
5   Lower segment of awn pubescent; glumes 14-25 mm long . . . . . *S. speciosa*
5   Awns glabrous or nearly so; glumes mostly less than 14 mm long 6
6   Palea glabrous, less than 1/2 as long as the lemma; florets plump, 5-6 times
    longer than wide; leaf sheaths usually villous at the throat; plants sparingly
    introduced . . . . *S. viridula*
6   Palea pubescent; florets nearly linear, usually over 6 times longer than wide;
    leaf sheaths glabrous or sparsely villous at the throat; plants native. 7
7   Lemma densely villous, the whitish hairs 1-4 mm long . . . . . . *S. pinetorum*
7   Lemma not densely villous, the hairs not over 2 mm long 8
8   Summit of sheaths pubescent on margins and on the back of the collar;
    lower panicle nodes villous; plants known from along Wyoming Highway
    414 near Lonetree . . . . *S. robusta*
8   Summit of sheaths glabrous or pubescent on margins but not on the back
    of the collar; panicle nodes not villous; plants widespread 9
9   Palea at least 2/3 as long as the lemma; leaf blades involute, 1-1.5 mm wide,
    the sheaths glabrous; awns of lemmas (10) 15-22 (25) mm long; callus blunt,
    pubescent except at the base . . . . *S. lettermanii*
9   Palea about 1/2 as long as the lemma; leaf blades rolled or flat, some usually
    over 2 mm wide, the sheaths commonly pubescent on at least 1 of the
    margins; awns of lemmas 20-30 (50) mm long; callus sharp pointed with a
    curved glabrous base . . . . *S. nelsonii*

Stipa arnowiae S. L. Welsh & N. D. Atwood [*Achnatherum arnowiae* (S. L. Welsh & N. D. Atwood) Barkworth] — Arnow ricegrass — Listed for Uintah Co. (AUF 900). Perhaps this is not different from S. contracta B. L. Johnson) W. A. Weber.

Stipa × bloomeri Bolander [*Achnatherum × bloomeri* (Bolander) Barkworth; × *Stiporyzopsis × bloomeri* (Bolander) B. L. Johnson] — Bloomer ricegrass — Listed for Duchesne and Uintah Cos. (AUF 900). Indian ricegrass rarely crosses with needlegrasses. As used here, this name applies to crosses of Indian ricegrass with all species of needlegrass. The one hybrid specimen seen (Goodrich 6641) is from Reservation Ridge, Tavaputs Plateau, from a cutover Douglas-fir stand.

Stipa comata Trinius & Ruprecht [*Hesperostipa comata* (Trinius & Ruprecht) Barkworth] — needle-and-thread — With 2 intergrading varieties as follows:

1   Terminal segments of the awn flexuous, usually over 5 cm long; lower
    branches of the panicle usually partly included in the sheath **var. *comata***
1   Terminal segments of the awn nearly straight, rather firm, usually less
    than 5 cm long; panicle usually exserted from the sheath **var. *intermedia***

Var. *comata* Widespread; common to abundant; desert shrub, black sagebrush, Wyoming sagebrush, pinyon juniper, and occasionally mtn. big sagebrush communities; 4,800-7,200 ft; May-June.

Var. *intermedia* Scribner & Tweedy Widespread; common to abundant; mtn. big sagebrush, aspen-sagebrush, and ponderosa pine communities; 7,000-10,000 ft; June-Aug.

Stipa hymenoides Roemer & Schultes [*Oryzopsis hymenoides* (Roemer & Schultes) Ricker; *Achnatherum hymenoides* (Roemer and Schultes) Barkworth] — Indian ricegrass — Widespread; common to abundant; desert shrub, mt. brush, sagebrush, pinyon-juniper, and ponderosa pine communities, increasing rapidly and becoming abundant following burning and chaining in pinyon-juniper communities; 4,700-10,000 ft; June-Sept.

Stipa lettermanii Vasey [*Achnatherum lettermanii* (Vasey) Barkworth] — Letterman needlegrass — Widespread; occasional to common; sagebrush, dry meadow, and alpine communities, openings in aspen and coniferous woods, dry ridges and rim-rock; 8,000-11,500 ft; June-Oct. Usually conspicuously distinct from *S. nelsonii*, but much like and possibly intergrading into that species. The length of the awns is the most obvious floral difference. However, this feature is only useful at the extremes. The shape of the callus provides a consistent difference, but this feature requires considerable experience and frequent review. The most reliable feature, offering the greatest contrast, is the length of the palea. Unfortunately, observation of this feature requires teasing the hardened lemma apart while viewing the parts under a dissecting scope. Rolled leaves (*S. lettermanii*) contrasted with some leaves flat (*S. nelsonii*) provides a useful means of separation in the field.

Stipa nelsonii Scribner [*S. williamsii* Scribner; *S. columbiana* Macoun misapplied; *Achnatherum nelsonii* (Scribner) Barkworth] — Nelson needlegrass, Columbia needlegrass — Plants of this taxon have long been known as *S. columbiana* but the Holotype of *S. columbiana* belongs with *S. lemmonii* (Vasey) Scribner, a plant not known from our area. Two vars. as follows:

1 Awns 20-25(30) mm long; callous somewhat rounded, usually glabrous only at tip . . . . **var. *dorei***
1 Awns mostly 25-30 (50) mm long; callous sharp-pointed, glabrous at the tip and upward on the ventral side . . . . **var. *nelsonii***

Var. *dorei* (Barkworth & Maze) Dorn — Dore needlegrass — The one specimen seen (Goodrich 28467) is from Gilbert Meadow, north slope, Uinta Mtns. at 9,670 ft. Likely more common than indicated by the one specimen.

Var. *nelsonii* — Nelson needlegrass — Widespread; common to abundant in a variety of montane plant communities, possibly attaining the greatest abundance in aspen communities and on open slopes, occasionally to near timberline; 6,500-10,500 ft; June-Aug.

Stipa pinetorum M. E. Jones [*Achnatherum pinetorum* (M. E. Jones) Barkworth] — pine needlegrass — The few specimens seen are from the W. Tavaputs Plateau and from Bear Top Mt., Daggett Co.; sagebrush and pinyon-juniper-mt. brush communities and shale barrens; 7,600-8,820 ft; July-Aug.

Stipa robusta (Vasey) Scribner [*Achnatherum robustum* (Vasey) Barkworth] — sleepy grass — Known at the margin of our area along Wyoming Highway 414 about 9 miles n. of Lonetree (Goodrich (25774). Likely introduced from the southwestern U.S. The species appeared to be successfully spreading in 1997. However, persistence is yet to be determined. Very much like a robust specimens of *S. nelsonii*, but with palea over 2/3 as long as lemma and lower nodes of panicle villous.

Stipa speciosa Trinius & Ruprecht [*Achnatherum speciosum* (Trinius & Ruprecht) Barkworth] — desert needlegrass — Listed for Uintah Co. (AUF 903, Albee et al. 1988).

Stipa viridula Trinius [*Nassella viridula* (Trinius) Barkworth] — green needlegrass — Used in experimental seedings on lands disturbed in oil shale development and more recently in reclamation of burned areas and found along roadsides; June-Aug. Bradley (1950) reported Harrington 176 for Blue Mt., 8,000 ft, and indicated that specimen to be at FC (CS). This specimen was not found there in 1985, but Boyd 176, from the same location and elevation, was found at CS. This specimen belongs to *S. nelsonii*.

Torreyochloa Church

Torreyochloa pallida (Torrey) Church [*Torreyochloa pauciflora* (Presl) Church; *Glyceria pauciflora* Presl; *Puccinellia pauciflora* (Presl) Munz] — weak mannagrass, pale false mannagrass — Plants rhizomatous; ligules 3-9 mm long; leaf sheaths open, the blades flat, 4-15 mm wide; lemmas prominently nerved. Found across the Uinta Mtns. where rather uncommon along streams and other wet places; not expected over 10,500 ft; June-Aug. Our plants belong to var. *pauciflora* (J. Presl) J. I. Davis.

This species has been included in *Glyceria* and in *Puccinellia*. This species looks like a *Glyceria*. It is rhizomatous and the lemmas have prominent nerves as in *Glyceria*. It grows in *Glyceria* habitat rather than habitat typical of *Puccinellia*. It has been included in *Puccinellia* by the open leaf sheaths, which is probably no more significant than the features by which it is aligned with *Glyceria*. The draft treatment in FNA is followed in placing this in *Torreyochloa*.

Trisetum Persoon — Trisetum

Plants tufted perennials; leaf sheaths open; ligule membranous; auricles lacking; inflorescence an open or spike-like panicle; spikelets 2 (3-5) flowered, disarticulating above (occasionally below) the glumes; rachillas long-hairy; glumes shorter or longer than the lowest floret; lemmas awnless or awned from the back above mid-length, the callus glabrous to short-hairy.

---

1   Lemmas with awns 3-7 mm long .... *T. spicatum*
1   Lemmas awnless or awn not over 2 mm long .... *T. wolfii*

Trisetum spicatum (L.) Richter [*T. montanum* Vasey; *T. spicatum* var. *montanum* (Vasey) W. A. Weber] — spike trisetum — Common to abundant in the Uinta Mtns. and Strawberry Valley, rare or infrequent on the Tavaputs Plateau; many plant communities including coniferous and aspen forests, openings in woods, stream sides, and alpine; 7,200-13,400 ft; June-Aug. Two forms have been recognized. They have been separated by features of the following key:

   1 Panicle dense and spike-like; leaves usually basal; culms 5-50 cm tall; plants densely tufted .... *T. spicatum*
   1 Panicle loose and open, culms usually leafy (30) 40-80 cm tall; plants loosely tufted .... *T. montanum*

Plants clearly with features of *T. spicatum* are common in the upper coniferous forest belt in open places and above timberline. Plants with features of *T. montanum* are common in aspen and coniferous forests. However, there seems to be no consistent morphological difference. The features of the above key are all overlapping as well as features given in some other manuals treating the 2 taxa. In coniferous forests of the Uinta Mtns., plants of this complex have the growth form of *T. montanum* (culms single or a few arising together in small tufts, and panicles interrupted). A few years after clear cutting, plants of the other phase with tufted culms and dense panicles are the common type. The differences seem to be a function of shading, and our plants appear to belong to a single taxon. *Trisetum montanum* is treated as a synonym of *T. spicatum* in TPD and indicated to be such in AUF (905).

Trisetum wolfii Vasey — meadow trisetum — Occasional across the Uinta Mtns. in meadows, along streams, and in moist open woods; 9,000-11,100 ft; June-Aug.

Triticum L. — Wheat

Triticum aestivum L. — wheat — Plants annual or winter annual, 50-100 cm tall, occasionally taller; culms often branched near the base; ligules about 1 mm long; auricles prominent; leaf blades flat, 2-20 mm wide; spikes 3-18 cm long; spikelets 9-12 mm long, with 2-5 (9) florets; glumes 6-11 mm long, with antrorsely scabrous awns 1-6 mm long; lemmas 7-12 mm long, with awns 2-20 (the beardless form) or 45-75 cm long (the bearded form). Cultivated and occasionally growing along roadsides and waste places for a season, but not persisting without cultivation; June-Aug. Cultivars of wheat are used in reclamation seedings in oil and gas fields of the Uinta Basin.

Ventenata Koel. — Ventenata

Ventenata dubia (Leers) Cosson in Durieu — North African grass — Plants winter annuals 10-70 cm tall; culms wiry with few leaves and with reddish black nodes; leaf blades slender; ligules 1-8 mm long; inflorescence an open panicle; spikelets with 2-3 florets; glumes shorter than the 1st floret; lemmas with bearded callus, dorsally awned, the awns of lower lemmas 1-3 mm long, those of the upper lemma 10-16 mm long, twisted and bent. Introduced from southern Europe and Africa. First U. S. record apparently from Washington in 1952. This invasive species appears to be on a track similar to that of cheatgrass and halogeton with current distribution from British Columbia and Alberta south to California and Utah and eastern provinces of Canada and some north eastern states of the U. S. Known in Utah from Cache Co. (AUF 905). Not yet known from our area but to be expected here in the future.

Vulpia C. C. Gmelin — Annual Fescue

Vulpia octoflora (Walter) Rydberg [*Festuca octoflora* Walter] — sixweeks fescue — Daggett Co. and from Vernal and e. in Uintah Co. and in Colorado, no specimens seen from Duchesne Co.; occasional or common in desert shrub, sagebrush, pinyon-juniper, mt. brush, ponderosa pine, and dry meadow communities; 4,900-7,000 ft; May-June.

POLEMONIACEAE Phlox Family

Annual or perennial herbs or low shrubs; leaves simple or compound, alternate, opposite, or appearing whorled; flowers bisexual; calyx united, (4) 5-lobed; corolla united, (4) 5-lobed; stamens (4) 5, attached to the corolla tube; ovary superior; style simple, usually with a 3-lobed stigma; fruit a (2) 3-chambered capsule.

1 Leaves simple, entire or rarely lobed 2
1 Leaves palmatifid, pinnatifid, or pinnately compound 7
2 Stems leafless except for a pair of persistent, 2-3 mm long cotyledons at the base and a whorl of sessile or connate bracts just below the inflorescence; plants 1-5 cm tall . . . . *Gymnosteris*
2 Stems leafy 3
3 Plants perennial; leaves opposite, not trilobate, linear to awl-shaped; corolla limb 8 mm wide or wider . . . . *Phlox*
3 Plants annual, or if perennial then with alternate or trilobate leaves; corolla limb mostly less than 8 mm wide 4
4 Plants tomentose or woolly . . . . *Ipomopsis*
4 Plants not tomentose or woolly. 5
5 Flowers in head-like clusters, rarely solitary in depauperate plants . . . . . .
   . . . . *Collomia*
5 Flowers (1) 2 at the tips of stems and branches or in axils of leaves and branches 6
6 Calyx 3-4 mm long; style longer than the corolla tube; plants stipitate glandular, the glands often blackish; plants known from Cold Spring Mtn. . . . . *Collomia tenella*
6 Calyx 4.5-8 mm long; style shorter than the corolla tube; lower parts of plants usually not stipitate glandular; plants widespread and common . . . . *Microsteris*
7 Some or all of the leaves opposite or appearing whorled, sessile, palmatifid into linear or needle-like segments, the segments sometimes appearing as whorls of filiform leaves. 8
7 Leaves basal and alternate, pinnatifid or pinnately compound. 10
8 Plants annual, from taproots, herbaceous throughout; corollas not over 6 mm long . . . . *Linanthus*
8 Plants perennial, woody at the base; corollas 8-25 mm long 9
9 Plants not woody above the base, not matted; leaves soft, not pungent . . .
   . . . . *Linanthastrum*
9 Plants woody above the base or else matted; leaves ridged and pungent . .
   . . . . *Leptodactylon*
10 Corolla lobes yellow; leaves pinnatifid into firm, more or less prickly segments . . . . *Navarretia*
10 Corolla lobes not yellow; leaves not firm or prickly 11
11 Corolla red . . . . *Ipomopsis aggregata*
11 Corolla lobes white, pink, or blue, the tube sometimes with some yellow .12

12 Corolla lobes over 5 mm long .... *Polemonium*

12 Corolla lobes not over 5 mm long 13

13 Stamens and styles well exserted from the corolla tube .... *Gilia*

13 Stamens and styles shorter than to slightly exceeding the corolla tube . 14

14 Inflorescence open with flowers borne singly or 2-3 together .... *Gilia*

14 Inflorescence congested with few to several flowers borne in head-like clusters .... *Ipomopsis*

Collomia Nuttall — Collomia

Annual or perennial herbs; leaves alternate or the lower ones opposite, entire (rarely toothed in C. debilis); flowers mostly in head-like clusters; corolla funnelform or nearly so, lavender to pink, rarely white; stamens arising at the same or different levels within the corolla tube; capsule with 1 (2) seeds per chamber.

1 Plants perennial, from branched caudices and slender taproots, more or less mat-forming; rare, known from Wasatch Co.; corolla 15-25 (35) mm long, broadly funnelform .... *C. debilis*

1 Plants annual, from simple taproots, widespread; corolla narrowly funnelform 2

2 Flowers solitary or paired in axils of leaves and branches ...... *C. tenella*

2 Flowers in capitate terminal clusters 3

3 Corollas 15-30 mm long, the lobes 5-10 mm long .... *C. grandiflora*

3 Corollas 7-15 mm long, the lobes mostly less than 5 mm long .. *C. linearis*

Collomia debilis (S. Watson) Greene [*Collomiastrum debile* (S. Watson) S. L. Welsh] — alpine collomia — Specimen seen are from Willow Creek, Buffalo Canyon and Racetrack Hollow, W. Tavaputs Plateau; from crusted, calcareous slopes of Green River shale; 8,400 ft; July-Aug. Occasional in the Wasatch Range to the w. of our area. Our plants belong to var. *debilis.*

Collomia grandiflora Douglas ex Lindley — large collomia — Known from Baxter Pass area of Rio Blanco and Garfield Cos and at the western edge of the Uinta Basin on Strawberry Ridge; June-Aug.

Collomia linearis Nuttall — narrowleaf collomia, small collomia — Widespread; occasional to common in sagebrush, pinyon-juniper, mt. brush, dry meadow, ponderosa pine, and aspen communities, often on disturbed soil; 7,000-7,800 ft; June-Aug.

Collomia tenella A. Gray — slender collomia — The one specimen seen (Huber & Goodrich 5135) is from a depauperate chokecherry community in a snowbed at 8,200 ft on the leeward side of Limestone Mtn. of Cold Spring Mtn., Moffat Co.

Gilia R. & P. — Gilia

Annual, biennial, or perennial herbs from taproots; leaves alternate, entire or pinnatifid to palmatifid, often in a basal rosette and more or less well developed along the stem; flowers solitary and axillary, paniculate, thyrsoid, or capitate;

calyx usually membranous with green ribs, 5-lobed; corolla funnelform to salverform, regular or nearly so, 5-lobed; stamens attached at the same, or nearly the same, level in the corolla tube; capsules with 1-many seeds per chamber.

1 Stamens exserted beyond the corolla. 2

1 Stamens about equaling the corolla throat 3

2 Stems usually much branched, sometimes from near the base; inflorescence open with branches commonly over 5 cm long, not at all spike-like or raceme-like; corolla lobes blue; capsules 2.5-4 mm long . . . . *G. pinnatifida*

2 Stems mostly simple (in ungrazed specimens); inflorescence often spike-like or raceme-like, or with branches to about 5 cm long; corolla white or pale blue-white or blue; capsules 4-6 mm long . . . . *G. stenothyrsa*

3 Plants, especially in the lower part, with some cobwebby tomentum on the leaves or at least in the axils of leaves; leaves pinnatifid, appearing compound, the primary divisions often toothed or lobed; corollas (5) 7-12 mm long, the lobes not tridentate; inflorescence with gland-tipped hairs, the glands often purplish or blackish . . . . *G. inconspicua*

3 Plants without tomentum; leaves pinnately lobed, or merely toothed, the lobes mostly entire; corollas 4-6 mm long, the lobes often tridentate; gland-tipped hairs of the inflorescence sometimes not purplish or blackish . . . .
   . . . . *G. leptomeria*

Gilia inconspicua (Smith) Sweet (*G. ophthalmoides* A. Brand; *G. sinuata* Douglas) — shy gilia, sinuate gilia — Widespread, occasional or locally common in desert shrub, sagebrush and pinyon-juniper communities; 4,750-7,100 ft; May-June. We have followed IMF (4: 116) in using the name of *G. inconspicua* in a broad sense.

Gilia leptomeria A. Gray [*G. micromeria* A. Gray; *Alciella leptomeria* (A. Gray) J. M. Porter] — sand gilia — Occasional in Daggett and Uintah Cos. and e. into Colorado, seldom collected from Duchesne Co.; desert shrub, sagebrush, and pinyon-juniper communities; 4,700-6,200 ft; May-June.

Gilia pinnatifida Nuttall in A. Gray [*G. calcarea* M. E. Jones; *G. mcvickerae* M. E. Jones; *Alciella pinnatifida* (Nuttall ex A. Gray) J. M. Porter] — sticky gilia — Occasional in the e. 1/4 of Uintah Co., Browns Park in Daggett Co., and e. into Colorado; Uinta and Green River Formations, Madison Limestone, and outcrops of the Uinta Mtn. Group in desert shrub, sagebrush, and pinyon-juniper communities; 5,200-8,400 ft; June-July. Rodeck 4425 DINO!, apparently the basis for the Holmgren (1962) report of *G. haydenii* A. Gray, belongs here.

Gilia stenothyrsa A. Gray [*Alciella stenothyrsa* (A. Gray) J. M. Porter; *Ipomopsis stenothyrsa* (A. Gray) W. A. Weber] — Uinta Basin gilia — Rio Blanco, Duchesne and Uintah Cos.; occasional and locally common in desert shrub, sagebrush, and pinyon-juniper communities; 5,600-7,900 ft; May-Sept. Most specimens have white or pale blue-white flowers, but those seen from the W. Tavaputs Plateau in s. Duchesne Co. have blue flowers. Uinta Basin gilia is also known from Carbon and Emery Counties, Utah from outside our area.

Gymnosteris Greene — Gymnosteris

Gymnosteris parvula (Rydberg) A. Heller — small-flowered gilia — Annual herbs, 1-4 cm tall; stems leafless; bract-like leaves 3-13 mm long, narrowly lanceolate to ovate; flowers sessile in a compact terminal cluster or solitary in small plants; calyx 2.5-4 mm long; corolla white or pinkish, the tube 2-5 mm long, the lobes 0.7-1.5 mm long. The few specimens seen are from Taylor Mtn., Diamond Mtn, Blue Mtn. and Yampa Plateau; sagebrush, ponderosa pine, and lodgepole pine communities, often in sandy places; 7,400-9,000 ft; June.

Ipomopsis Michaux

Annual, biennial, or perennial herbs from taproots; leaves alternate, entire or pinnatifid to palmatifid, often in a basal rosette and more or less well developed along the stem; flowers paniculate, thyrsoid, or capitate, rarely solitary and axillary, calyx tube campanulate, usually membranous throughout, 5-lobed; corolla 5-lobed, campanulate to salverform; stamens more or less equally inserted on the corolla tube, the filaments subequal to unequal in length; capsules descent apically; seeds 1 to many per locule.

---

1 Corolla red, 1.5-5 cm long . . . . *I. aggregata*
1 Corolla not red, less than 1.5 cm long 2
2 Calyx less than 3/4 as long as the corolla tube; plants annual . . . *I. pumila*
2 Calyx 3/4 or more the length of the corolla tube 3
3 Inflorescence closely subtended by a cluster of leaves about as large as those in the cluster at the base of the plant; plants annual . . . . *I. polycladon*
3 Stems usually without a terminal cluster of larger leaves; plants perennial 4
4 Plants with a single unbranched stem (in ungrazed specimens), known from Daggett and Moffat Cos.; inflorescence conspicuously longer than wide; flowers not in head-like clusters (in ungrazed or unbroken specimens) . . . . *I. spicata*
4 Plants often branched, widespread; flowers in head-like clusters and the ends of branches 5
5 Styles including stigmas equaling or exceeding the corolla tube; calyx villous or villous-tomentose; occasionally only sparsely so; corolla bright white even in dried specimens (the lobes sometimes flecked with blue-purple), the tube slightly if at all surpassing the calyx; anthers shorter than their filaments . . . . *I. congesta*
5 Styles including the stigmas shorter than and well included in the corolla tube; calyx with short glandular hairs only, or rarely with a few short-villous hairs; corolla cream white at least in dried specimens; the tube surpassing the calyx; anthers about as long as their filaments . . *I. roseata*

Ipomopsis aggregata (Pursh) V. Grant [*Gilia aggregata* (Pursh) Sprengel; *G. pulchella* Douglas] — scarlet gilia — Widespread and common in sagebrush, pinyon-juniper, rabbitbrush-wildrye, mt. brush, ponderosa pine, and Douglas-fir communities; 5,500-9,500 (10,400) ft; mid May-Sept. Our plants belong to var. *aggregata*.

Ipomopsis congesta (Hooker) V. Grant (*Gilia congesta* Hooker) — ballhead gilia, many-flowered gilia — With 2 vars. in our area:

1 Leaves mainly trifid or pinnatifid, less commonly some of them simple; plants widespread .... **var. *congesta***
1 Leaves mainly simple, less commonly some of them trifid or pinnatifid; plants of Green River and Uinta Formations .... **var. *goodrichii***

Var. *congesta* Widespread and common in desert shrub, sagebrush, and pinyon-juniper communities, on a wide range of soil types and substrates including: sand dunes, rock crevices, and clay barrens; 5,000-7,000 ft; May-July. See *I. roseata.*

Var. *goodrichii* S. L. Welsh — Tavaputs ballhead gilia — Several specimens seen are from Indian Canyon to Wells Draw on the West Tavaputs Plateau and two from the East Tavaputs Plateau (Weaver Ridge and File Mile Creek); occasional or locally common on semi-barren slopes and ridge tops of Green River and Uinta Formations with desert shrubs, pinyon-juniper, Douglas-fir, and limber pine; 5,800-8,700 ft; May-July.

Ipomopsis polycladon (Torrey) V. Grant (*Gilia polycladon* Torrey in Emory) — spreading gilia — Stems spreading or nearly prostrate, with elongate internodes; leaves with 5-9 teeth or lobes. Widespread and occasional in desert shrub, sagebrush, and juniper communities, often on sandy soil; 4,800-5,500 ft; May-June.

Ipomopsis pumila (Nuttall) V. Grant (*Gilia pumila* Nuttall) — dwarf gilia — Plants tomentose throughout and especially in the inflorescence; flowers rather congested in head-like clusters; stems leafy; leaves entire or more often 3-parted or occasionally with up to 5 segments. Widespread and occasional or locally common in desert shrub, sagebrush, and juniper communities, often on sandy ground; 4,700-6,000 ft; May-June. Sometimes mistaken for *Eriastrum diffusum* (A. Gray) Mason, a plant known from far s. of our area.

Ipomopsis roseata (Rydberg) V. Grant (*Gilia roseata* Rydberg) — roseate gilia, San Rafael gilia — Occasional n. of Hwy. 40 and s. of the Uinta Mtns. from Rock Creek and Duchesne and e. Tridell; sagebrush, pinyon-juniper, and rarely desert shrub communities, associated with but not restricted to glacial outwash from the major canyons of the Uinta Mtns.; 5,500-7.600 ft; June-July. Specimens from Dinosaur National Monument and northward and eastward referred here (Bradley 1950; Welsh 1957) belong to the widespread and similar *I. congesta* which is apparently rare or lacking from the range of *G. roseata* in the Uinta Basin.

Ipomopsis spicata (Nuttall) V. Grant (*Gilia spicata* Nuttall) — spike gilia — The few specimens seen are from Daggett Co. from Birch Creek to Antelope Flat and Moffat Co. (also adjacent Sweetwater Co., WY); desert shrub and juniper communities; 6,200-8,900 ft; June. Our plants belong to var. *spicata.*

Leptodactylon Hooker & Arnot — Prickly phlox

Plants more or less woody at the base; leaves palmately parted into needle-like divisions, lower ones opposite, the upper ones opposite, subopposite, or alternate, with smaller leaves fascicled in the axils of the primary ones; flowers sessile, solitary or a few in a cluster, showy; calyx ruptured by the maturing fruit; corolla white, cream-colored, or purplish in the throat, salverform with funnelform throat, the tube longer than the calyx, the lobes often loosely spirally closed during the day.

---

1 Flowers 4-merous; plants depressed-pulvinate, more or less mound-forming; not over 5 cm tall; corollas 12-16 mm long; leaf division 1-7 mm long . . . . *L. caespitosum*

1 Plants not pulvinate; stems over 5 cm long; corollas 15-25 cm long; leaf divisions often over 7 mm long 2

2 Flowers mostly 6-merous; leaves all opposite, the divisions 7-15 mm long; corollas 18-25 mm long, the lobes 9-12 mm long; plants woody only at the base . . . . *L. watsonii*

2 Flowers mostly 5-merous; upper leaves subopposite or alternate, the divisions 3-9 mm long; corollas 15-20 mm long, the lobes 6-8 mm long; plants often woody well above the base . . . . *L. pungens*

Leptodactylon caespitosum Nuttall [*Linanthus caespitosus* (Nuttall) J. M. Porter & L. A. Johnson] — mat prickly phlox — The few specimens seen and reports are from Lodgepole Creek (s. of Manila), Browns Park, Vermillion Creek Gap and 2 mi e. of Shell Creek Ranch, also plotted for Rio Blanco Co. in TPD; white tuffaceous semi-barrens, Moenkopi, and other formations that weather to badlands; 5,700-7,100 ft; June.

Leptodactylon pungens (Torrey) Nuttall [*Linanthus pungens* (Torrey) J. M. Porter & L. A. Johnson] — common prickly phlox — Widespread and occasional to common in desert shrub, sagebrush, pinyon-juniper, and ponderosa pine communities; 4,700-8,150 ft; May-July. Our plants belong to var. *pungens.*

Leptodactylon watsonii (A. Gray) Rydberg [*Linanthus watsonii* (A. Gray) J. M Porter & L. A. Johnson] — rock prickly phlox — Three specimens seen are from Cross Mt. Canyon, other specimens from Yampa River (Parks and others 12052 BRY), 2 mi w. of Neola (A. Collotzi & M. Collotzi 456 UTC), Canyon of Lodore (A. Holmgren & N. Holmgren 14236 UTC), vegetative plants also seen at the west end of Bare Top Mtn., Goodrich & Huber 28576 from near Dudly Bluffs, Piceance Creek is from a population of glaucous plants on semi-barrens of Green River Shale, also adjacent to our area in Sweetwater Co., Wyoming; crevices of cliffs and rocky canyons and hills; 5,500-5,800 ft; June-July.

Linanthastrum Ewan — Linanthastrum

Linanthastrum nuttallii (A. Gray) Ewan [*Linanthus nuttallii* (A. Gray) Greene ex Milliken; *Leptosiphon nuttallii* (A. Gray) M. porter & L. A. Johnson] — flaxflower — Perennial herbs with woody bases, 12-25 (30) cm tall, much branched; leaves opposite, (3) 5-9 parted, into spinulose-tipped linear segments, the segments 1-1.5 (2) cm long; flowers in small head-like or loose clusters, scarcely exserted from a tuft of leaves; calyx 6-9 mm long, ruptured by the maturing capsules; corolla 7-14 mm long, funnelform or salverform, white or cream-colored, or yellow, the tube woolly pubescent externally; the limb about 1 cm wide; stamens about equal, inserted equally at the base of the short corolla throat; capsules with 2-4 seeds per chamber but sometimes only 1 developing. Known from n. of Maeser, juniper-shrub-grass community, sandy soil (Atwood 7228 BRY, Goodrich 26389 BRY) and from near Granddaddy Lake in rock crevices, 10,500 ft (Garrett 133 UT); June-Aug. Our plants belong to ssp. *nuttallii* var. *nuttallii.*

Linanthus Bentham — Linanthus

Annual herbs; leaves opposite, palmately divided into 3-5 filiform or linear segments; flowers mostly solitary on filiform pedicels; calyx ruptured by the maturing fruit; corolla with a funnelform throat and short tube; stamens equally inserted and included in the corolla throat; capsules with 2-5 seeds in each chamber.

---

1   Corolla 1.5-3.5 mm long, less than 1.5 times longer than the calyx . . . .
. . . . ***L. harknessii***
1   Corolla 2.5-5 mm long, about 1.5 times longer than the calyx . . . .
. . . . ***L. septentrionalis***

Linanthus harknessii (Curran) Greene [*Leptosiphon harknessii* (Curran) J. M. Porter & L. A. Johnson] — Harkness flaxflower — Apparently widespread but no specimens seen from the Tavaputs Plateau, infrequent in meadow, sagebrush, mtn. brush, aspen, and coniferous forest parkland communities; 7,150-9,400 ft; June-Aug.

Linanthus septentrionalis Mason [*L. harknessii* (Curran) Greene var. *septentrionalis* (Mason) Jepson & Bailey; *Leptosiphon septentrionalis* (Mason) J. M. Porter & L. A. Johnson] — northern linanthus — Widespread but no specimens seen from the W. Tavaputs Plateau, infrequent or occasional in sagebrush, juniper, and mt. brush communities, often on sandy soils; 6,200-8,000 ft; late May-July.

Microsteris Greene — Microsteris

Microsteris gracilis (Hooker) Greene [*M. humilis* (Douglas) Greene; *M. micrantha* (Kellogg) Greene] — pink microsteris, little polecat — Annual herbs 3-15 (25) cm tall, puberulent or glandular-puberulent at least above; leaves opposite or alternate, 6-50 mm long to 8 mm wide, linear to lanceolate or oblanceolate, entire; flowers mostly in pairs at the ends of stems and branches, one subsessile the other conspicuously pedicellate, or sometimes solitary; calyx ruptured by the developing capsule; corolla salverform, 5-15 mm long, the tube white or yellowish, the limb pink or lavender; stamens unequally inserted on the corolla; capsules with 1 seed per chamber. Widespread but no specimens seen from the Tavaputs Plateau, infrequent or occasional in sagebrush, mt. brush, and ponderosa pine communities, often on disturbed ground; 7,000-8,600 ft; May-July. Our plants belong to var. *humilior* (Hooker) Cronquist.

Navarretia R. & P. — Navarretia

Navarretia breweri (A. Gray) Greene — Brewer navarretia, yellow-flower navarretia, pincushion plant — Annual herbs, 1-10 cm tall, glandular-puberulent; leaves alternate, 1-3 cm long, pinnatifid into pungent-tipped, linear segments 3-20 mm long; flowers sessile or nearly so in clusters at the ends of branches and stems, the clusters closely subtended by bracts that are similar to the leaves; calyx sometimes ruptured by the maturing capsules, the lobes longer than the tube, pungent-tipped; corolla 5-8 mm long, funnelform to salverform, yellow; stamens unequally inserted on the corolla; capsules with (1) 2-3 seeds per chamber. Widespread but no specimens seen from the central Uinta Mtns.; infrequent and locally common in sagebrush, pinyon-juniper, ponderosa pine, forb-grass, and aspen communities, often on open or disturbed ground; 7,000-9,000 ft; June-July.

Phlox L. — Phlox; Sweet William; Sweet William moss

Perennial herbs from a taproot and sometimes branching caudex; stems sometimes woody at the base; leaves opposite or the upper ones sometimes alternate, sessile, entire, linear or needle-like, the primary ones often with fascicles of axillary leaves; flowers solitary or variously clustered, usually terminal; calyx tube ruptured by the developing capsule; corolla salverform, white, pink, or blue; stamens unequally inserted on the corolla tube; capsules with 1-several seeds per chamber.

---

1   Plants not tufted or mat-forming, sometimes over 10 cm tall; principal leaves 1.5-8 cm long, more or less well spaced on the stems and not concealing the internodes .... *P. longifolia*

1   Plants forming mats or cushions, mostly less than 10 cm tall; leaves mostly less than 1.5 cm long, usually crowded 2

2   Calyx with gland-tipped hairs 3

2   Calyx glabrous or pubescent but not with gland-tipped hairs 4

3   Style 5-8 mm long; plants known from the pinyon-juniper belt .... *P. albomarginata*

3   Style 2-6 mm long; plants subalpine and alpine .... *P. pulvinata*

4   Calyx 9-15 mm long, glabrous or nearly so, the teeth longer than the tube; plants of eastern Uinta Mountains .... *P. multiflora*

4   Calyx 4-10 mm long, the teeth mostly shorter than the tube, glabrous or villose or tomentose with crinkly or cobwebby hairs 5

5   Limb of corolla mostly less than 12 mm in diameter.................. 6

5   Limb of corolla 13-23 mm in diameter 7

6   Leaves (2.5) 4-10 (13) mm long, loosely overlapping, linear lanceolate, the larger ones over 5 mm long; plants loosely caespitose, not mound-forming, not as arachnoid as in the following; widespread, common; corolla lobes 4-8 mm long .... *P. hoodii*

6   Leaves 2-5 mm long, strongly overlapping, triangular; plants densely caespitose, more or less mound-forming, strongly arachnoid; known from Moffat Co.; corolla lobes 3-5 mm long .... *P. muscoides*

7   Plants glabrous or nearly so, subalpine and alpine in the Uinta Mountains .... *P. pulvinata*

7   Plants pubescent at least in part 8

8   Leaves 2-7 mm long .... *P. opalensis*

8   Leaves (5) 8-15 (20) mm long .... *P. austromontana*

Phlox albomarginata M. E. Jones. — whitemargin phlox — Known from the north side of Bare Top Mtn., Jarvies Canyon, and Dutch John Draw, Daggett Co. at 6,250-6,520 ft in burned and unburned pinyon-juniper communities.

Phlox austromontana Coville (*P. densa* Brand; *P. diffusa* Bentham ssp. *subcarinata* Wherry) — desert phlox — Common on the E. Tavaputs Plateau and-flank of the plateau to the Piceance Basin, and locally abundant on the W. Tavaputs Plateau as far w. as the Timber Canyon drainage; one specimen seen from Jones Hole; sagebrush, pinyon juniper, mt. brush, and Douglas-fir communities; 6,000-8,000 ft; May-June (Oct. in recent burns). *Phlox diffusa* in the strict sense is well removed to the n. and w. of our area. Reports of *P. diffusa* from our area are based on ssp. *subcarinata*, which belongs with *P. austromontana.*

Phlox hoodii Richardson (*P. canescens* Torrey & A. Gray) — Hood phlox — Widespread; common in Wyoming big sagebrush communities, also in desert shrub, pinyon-juniper, mt. brush, and ponderosa pine communities; 4,800-8,400 ft.; March-May (Oct). Occasionally small specimens of *P. hoodii* approach those of *P. muscoides* (q.v.). Apparently these taxa hybridize.

Phlox longifolia Nuttall [*P. longifolia* ssp. *calva* Wherry; *P. longifolia* ssp. *cortezana* (A. Nelson) Wherry; *P. grahamii* Wherry] — wild sweet william, longleaf phlox — Widespread and common in desert shrub, sagebrush-grass, pinyon-juniper, rabbit brush, mt. brush, and rarely in aspen-coniferous forests communities; 4,700-8,000 ft; April-July(Oct). *Phlox grahamii* Wherry was described from rust-infested specimens (Graham 7884 CM!) of *P. longifolia* from the mouth of Sand Wash on the w. side of the Green River. The specimens were referred to by Graham (1937) as *P. speciosa* Pursh.

Phlox multiflora A. Nelson — many-flowered phlox, flowery phlox — Common from the Whiterocks drainage and e. on the s. slope of the Uinta Mtns and from eastern Summit, Daggett and Moffat Cos.; mtn. big sagebrush-grass, mt. brush, and aspen, and coniferous forest communities and rocky ridges; 7,100-9,500 ft; June-July (Oct). Our material is referable to ssp. *depressa* (A. Nelson) Wherry.

Phlox muscoides Nuttall (*P. bryoides* Nuttall) — moss phlox — Specimens seen (Peterson & Wilken 83-311, Goodrich & Huber 28114) are from Lookout Mt. at the edge of Vermillion Bluffs on a ridge top at 7,100 ft where intermediate plants (*P. muscoides* Nuttall) were found just below the ridge tops and plants of *P. hoodii* (q.v.) were found below among shrubs on deeper soils (Wilken, personal communication); also seen on Limestone Ridge of Cold Spring Mtn.; May-June. This taxon also approaches the northern edge of our area in Sweetwater Co., Wyoming where it is quite common.

Phlox opalensis Dorn — opal phlox — Widespread and common in Daggett, Uintah, and Moffat Cos. and in the Bridger Formation near Lonetree and Mtn. View, Wyoming in many plant communities including greasewood, saltbush, black sagebrush, big sagebrush, pinyon-juniper, and mtn. brush at 5,200-8,000 ft. In AUF (567) this is treated as a large flowered species that is sympatric with *P. hoodii* into which it could be placed as a synonym. The large flowered plants share pubescent features (arachnoid pubescent) with *P. hoodii*. If maintained as a separate species based on width of the corolla limb *P. opalensis* is a widespread and common plant in Uintah and Daggett Cos, Utah, and Uinta Co, Wyoming where, in many areas, it is much more common than small-flowered plants of *P. hoodii*. The breaking point of 12 mm diameter of the corolla limb seems to provide a means to separate the two taxa. However, within populations and in individuals of *P. opalensis* there are flowers with corolla limbs only 10 mm wide mixed with larger flowers. Plants of *P. opalensis* are generally not as compact as those of *P. hoodii*.

Phlox pulvinata (Wherry) Cronquist (*P. caespitosa* Nuttall misapplied) — alpine tufted phlox — Strawberry Valley and Uinta Mtns. as far e. as the Whiterocks drainage and in the uplands of Moffat Co.; lodgepole pine-spruce communities, open slopes, ridges, and fell fields, often on sandy or rocky ground; (7,500) 9,000-11,200 ft; June-Aug. Most specimens have glandular hairs, but some specimens are glabrous or nearly so.

Polemonium L. — Jacobsladder; Polemonium

Perennial, usually glandular, often malodorous herbs; leaves alternate, pinnatifid or more often pinnately compound; flowers in terminal or axillary clusters or rarely solitary; calyx not ruptured by the maturing capsule; corolla broadly funnelform, blue, white, or purple; stamens about equally inserted on the corolla tube, the filaments usually hairy near the base; capsules with 1-10 seeds per chamber.

1   Leaflets 1-6 mm long, divided into 3-5 segments, these appearing as whorls of tiny leaves; corolla tube exceeding the calyx and longer than the corolla lobes; plants 5-20 (30) cm tall, from above 10,000 ft on the Uinta Mtns. . . . .
. . . . ***P. viscosum***

1   Leaflets mostly 10-50 mm long, entire; corolla tube not or scarcely exceeding the calyx, about equal to or to 1/2 as long as the corolla lobes; plants of various elevations 2

2   Styles about 10 mm long, the branches to 1.5 mm long; plants from 9, 000-11,000 ft . . . . ***P. pulcherrimum***

2   Styles about 15 mm long, the branches various; plants from 7,400-9,500 ft
. . . . 3

3   Corolla purple; style branches to about 2 mm long; plants of wet places; upper few leaves greatly reduced . . . . ***P. caeruleum***

3   Corolla white or rarely blue; style branches 2-3.5 mm long; plants of well-drained soil; upper leaves not greatly reduced . . . . ***P. foliosissimum***

Polemonium caeruleum L. (*P. occidentale* Greene) — western Jacobsladder, western polemonium, skunkweed — Locally common in Strawberry Valley and across the Uinta Mtns. but no specimens seen from Uintah Co.; wet and boggy meadows, around seeps and springs, and in streamside-willow communities; 7,000-10,050 ft; June-Aug. Our plants belong to var. *pterosperma* Bentham.

Polemonium foliosissimum Gray — leafy Jacobsladder — With 2 vars. in our area:

1 Corolla white . . . . **var.** *alpinum*
1 Corolla blue . . . . **var.** *foliosissimum*

   Var. *alpinum* A. Brand (*P. albiflorum* Eastwood) — white leafy Jacobsladder — Strawberry Valley and e. on the W. Tavaputs Plateau to the Avintaquin drainage, and to the Rock Creek drainage in the Uinta Mtns.; occasional to common in tall forb, aspen, mtn. big sagebrush-snowberry, meadow, and streamside willow communities; 7,500-9,600 ft; June-Aug.

Var. *foliosissimum* — blue leafy Jacobsladder — Occasional; E. Tavaputs Plateau including Cathedral Bluffs, and Mosby Canyon of Uinta Mtns.; Vasey sagebrush, rabbitbrush-basin wildrye communities, and streambanks and shaly barrens; 7,400-8,400 ft; June-July.

Polemonium pulcherrimum Hooker (*P. delicatum* Rydberg) — pretty Jacobsladder, skunk-leaf polemonium — Occasional across the Uinta Mtns.; one specimen seen is from Bruin Point, W. Tavaputs Plateau; meadows, along streams, and among rocks, aspen, lodgepole pine, fir, and spruce woods; 9,000-11,100 ft; June-Aug. Our plants belong to var. *delicatum* (Rydberg) Cronquist.

Polemonium viscosum Nuttall — sticky sky-pilot, skunk polemonium — Occasional across the Uinta Mtns.; talus, fell fields, rock stripes, and Engelmann spruce and alpine tundra communities; 10,200-12,700 ft; July-Aug.

POLYGALACEAE Milkwort Family

Polygala L. — Milkwort

Annual herbs or subshrubs with simple, entire, alternate or whorled leaves; flowers bisexual, legume-like with wings and keel; stamens commonly 8 in 2 series, the filaments united; fruit a capsule.

---

1   Plants perennial subshrubs; leaves alternate . . . . *P. subspinosa*
1   Plants annual; leaves whorled at some of the nodes . . . . *P. verticillata*

Polygala subspinosa S. Watson [*Rhinotropis subspinosa* (S.watson) J. R. Abbott] — cushion milkwort, showy milkwort — Two specimens seen are from 5 mi se. of Florence Creek Lodge, Grand Co. (J. Brown 772 specimen at Natural Resources Conservation Service Herbarium, Roosevelt, UT). And Little Desert s. of Pariette Draw and w. of the Green River (M. Lewis 7387. Also reported for Duchesne County (AUF 570), but we have seen no specimens from Duchesne Co. Flowers 8-12 mm long, pink-purple with yellow keel.

Polygala verticillata L. — whorled milkwort — Annual herbs, 8-40 cm tall; stems erect, divergently branched; leaves 1-2.3 cm long, 1-3 mm wide, simple, entire, linear, at least the lower and sometimes the upper in whorls; lower branches of the inflorescence usually opposite or whorled; inflorescence of dense conic or cylindrical spike-like racemes 6-15 mm long; flowers inconspicuous, whitish or greenish; tepals 5, the outer 3 herbaceous and minute, the inner 2 slightly larger (about 1 mm long), free or united at the base; fruit a capsule, about 2 mm long. The one specimen seen (Hutchings 285) is from 2 mi s. of Whiterocks; meadow with *Juncus, Carex,* and *Agrostis*; 6,800 ft; July.

POLYGONACEAE Knotweed Family

Herbs or shrubs; leaves alternate, basal, rarely opposite or whorled, simple, usually entire; with or without stipules (the scarious, united stipules of some genera are also referred to as an "ocreae"; flowers bisexual or unisexual, the perianth small, of 3-6 tepals in 1-2 series; tepals free or united below the middle, scarious or petaloid; stamens (3) 6-9; pistil 1, the ovary superior, 1-chambered; styles 2-3 (4) or lacking; stigmas sessile; fruit a 3-angled or lenticular achene.

1   Stems without stipules; flowers borne few-several together in involucres
.... ***Eriogonum***

1   Stems with scarious stipules or ocreae; flowers not borne in involucres. . 2

2   Leaves all basal or the stem with 1 leaf just below the inflorescence, the blades cordate to reniform, about as wide or wider than long, with the petiole about as long or longer than the blade; plants montane to alpine at (9,600) 10,500-13,000 ft .... ***Oxyria***

2   Leaves not all basal, the blades linear, elliptic to oblong, sessile to petiolate; distribution various. 3

3   Tepals 3; plants of alpine, mossy sites in the Uinta Mtns. .... ***Koenigia***

3   Tepals 5 or 6; habitat not as above 4

4   Tepals 5, remaining about equal in size in fruit; flowers axillary or in spikes or spicate panicles; stigmas sub-globose .... ***Polygonum***

4   Tepals 6, the inner set enlarging in fruit and forming small to large scarious wings on the fruit; flowers borne in crowded to remote whorls or panicles; stigmas with filiform branches .... ***Rumex***

Eriogonum Michaux — Wild buckwheat; Eriogonum

Annual or perennial herbs or shrubs; leaves basal or cauline and mostly alternate, opposite or whorled in a few taxa, simple, entire, lacking stipules; inflorescence capitate, racemose, umbellate, or cymose; involucres subtending the flowers, campanulate, cylindrical or turbinate, 4-8 toothed or lobed, sessile or pedunculate; flowers bisexual, each with a thread-like pedicel, the pedicel included or exserted from the involucre; perianth parted or deeply cleft into 6 small, more or less petaloid segments; stamens 6-9; styles 3-parted; achene 3-angled, or 3-winged. A complex genus with several intergrading and hybridizing taxa. Flowers are reduced and quite uniform. Separation of taxa is largely based on flower color and vegetative features.

1   Plants annual; leaves all basal except in *E. salsuginosum* and *E. divaricatum*
.... **KEY 1**

1   Plants perennial; leaves basal and/or cauline. ....

2   Plants tall forbs with mostly simple, strigose stems from a more or less chambered taproot; achenes winged, exserted beyond the flowers, usually yellowish; flowers yellow .... ***E. alatum***

2   Plants not as above; achenes not winged, usually blackish or brownish; flowers white or yellow ....

3   Inflorescence sessile or nearly so in a basal rosette, the flowers not elevated above the leaves; plants densely pulvinate caespitose .... **KEY 3**

3   Inflorescence on a conspicuous stem or peduncle, the flowers elevated above the leaves; plants various ....

4   Flowers with a 1-2 mm long stipe-like base about as slender as the pedicel
.... **KEY 2**

4   Flowers sessile, expanded immediately from the joint with the pedicel . . 5

5   Flowers borne in solitary heads 6

5   Flowers not borne in heads 7

6   Leaf blades not over 2 cm long .... **KEY 3**

6   Leaf blades over 2 cm long.................................... **KEY 5**

7   Woody stems not dying back to the ground level each year; leaves not at all basal and not basally disposed; plants shrubs.................... **KEY 4**

7   Aerial stems dying back to or near ground level each year, mostly not woody; leaves basal or basally disposed; plants herbs or subshrubs . **KEY 5**

— KEY 1 —

1   Leaves tomentose to lanate at least beneath; flowers white or cream, glabrous outside; stems glabrous; peduncles strongly spreading to deflexed 2

1   Leaves glabrous to pilose but not tomentose or lanate; flowers yellow and pubescent outside except in E. gordonii and then the stems pilose at least in the lower part; peduncles erect to strongly spreading 3

2   At least some of the involucres pedunculate, the peduncles to 2 cm long; tepals not cordate at the base; white or pinkish; leaf blades 4-22 mm wide .... *E. cernuum*

2   Most of the involucres sessile or nearly so, the peduncles rarely over 1 mm long except in an occasional one; tepals cordate at the base, yellowish to reddish; leaf blades mainly 8-40 mm wide .... *E. hookeri*

3   Flowers glabrous outside, white but turning pinkish or reddish in age ... . *E. gordonii*

3   Flowers hirsute to pilose outside, yellowish or reddish 4

4   Leaves basal and cauline, the basal ones ephemeral 5

4   Leaves all basal; plants glandular or pilose at least below 6

5   Plants glabrous or nearly so; stem leaves sessile .... *E. salsuginosum*

5   Plants pubescent; some stem leaves with petioles .... *E. divaricatum*

6   Stems usually inflated, glabrous, glaucous; branches of the inflorescence glabrous; peduncles straight or slightly curved; leaves long-pilose beneath, often glabrous above, not glandular .... *E. inflatum*

6   Stems not inflated, often glandular; branches of the inflorescence stipitate-glandular; peduncles often strongly bent; leaves sometimes glandular, pilose or not .... *E. flexum*

— KEY 2 —

1   Flowers pubescent on the outside; plants from a thickened caudex with short branches .... *E. jamesii*

1   Flowers glabrous on the outside; plants from a much-branched stoloniferous or rhizomatous caudex, the branches elongate and spreading 2

2   Stems with a whorl of leaf-like bracts near the middle; flowers white to cream .... *E. heracleoides*

2   Stems scapose, without bracts as above or rarely with a solitary bract; flowers yellow or white to cream .... *E. umbellatum*

— KEY 3 —

1 Plants 5-30 cm tall; scapes exserted well above the leaves; leaves ovate to orbicular, 4-20 mm wide .... *E. ovalifolium*

1 Plants mostly less than 5cm tall 2

2 Leaves (1.5) 2-6 mm wide, 4-12 mm long, oblanceolate to spatulate; heads more or less elevated above the leaves on a scape .... *E. shockleyi*

2 Leaves 0.7-2 mm wide, 3-6 mm long; heads sessile slightly elevated on a short peduncle 3

3 Perianth segments pubescent internally as well as externally, whitish ... . *E. tumulosum*

3 Perianth segments glabrous internally, pubescent externally, pale yellow when fresh .... *E. acaule*

— KEY 4 —

1 Flowers bright yellow; a series of crosses involving *E. corymbosum*, *E. brevicaule*, and *E.* × *duchesnense* .... (see *E. corymbosum*)

1 Flowers white, pink or pale cream 2

2 Leaves 4-35 mm long, 1-8 mm wide, linear or narrowly elliptical .... *E. microthecum*

2 Leaves longer and/or wider than above 3

3 Scapose portion of stem and branches of the inflorescence tomentose; leaves not basally disposed .... *E. corymbosum*

3 Scapose portion of stem and branches of the inflorescence glabrous; leaves more or less basally disposed .... *E. lonchophyllum*

— KEY 5 —

1 Flowers white or pink, rarely cream 2

1 Flowers yellow 7

2 Flowers in a solitary head .... *E. ovalifolium*

2 Flowers not as above 3

3 Inflorescence a raceme .... *E. racemosum*

3 Inflorescence not a raceme 4

4 Scapose portion of stem and branches of the inflorescence tomentose; involucres 3.5-4 mm long; flowers 3.4-4.5 mm long, white; plants mostly of the Bad Land Cliffs of the Tavaputs Plateau *E. corymbosum* var. *hylophilum*

4 Scapose portion of stem and branches of the inflorescence mostly glabrous 5

5 Leaf blades about 1.5 times longer than wide, (1.5) 2-4 cm long .... *E. batemanii*

5 Leaf blades mostly over 2 times longer than wide 6

6 Plants of the E. Tavaputs Plateau; lowest node of the inflorescence often with 1 or more involucres of flowers with a peduncles less than 1 cm long in addition to longer branches; flowers bright-white ... *E. lonchophyllum*

6 Plants known from the West Tavaputs Plateau, near Altamont, and n. of the Uinta Mtns.; lowest node of the inflorescence often without short

pedunculate involucres of flowers; flowers often cream-white ....
.... *E. brevicaule*

7 Leaves ovate to orbicular; flowers in a solitary head .... *E. ovalifolium*

7 Leaves linear or lanceolate; flowers various 8

8 Scapes and branches densely tomentose; inflorescence capitate to umbellate; plants of Duchesne and Wasatch Cos. .... *E. brevicaule*

8 Scapes and branches of the inflorescence glabrous and greenish; inflorescence cymose or umbellate 9

9 Leaves 1-9 mm wide, oblanceolate or spatulate or less commonly linear, slightly or not at all revolute, rather densely pubescent on the upper as well as the lower surface; plants of Summit, Daggett, Moffat, and Rio Blanco Cos. and Uinta Co., Wyoming .... *E. brevicaule*

9 Leaves 1-2 (3) mm wide, linear or narrowly oblanceolate, loosely to tightly revolute, glabrous to rather sparsely pubescent on the upper surface; plants of Duchesne, Uintah, and extreme w. Rio Blanco Cos................. 10

10 Leaves 1-7 cm long, tightly revolute; woody branches of the caudex and leaf-bearing portions of stems elongate, often over 1.5 cm long and up to 4 (6) cm long; inflorescence 0.5-1.5 (2.2) times as long as the green scapose part of the stem, the branches about equal; flowers bright yellow; plants common across the bottom of the Basin from Duchesne to the Green River, uncommon e. of the Green River .... *E. brevicaule* **var.** *viridulum*

10 Leaves 1-2.5 cm long, sometimes not tightly revolute; woody branches of the caudex and leaf-bearing portions of stems short, seldom over 1.5 cm long; inflorescence 1.5-3.6 (5) times as long as the greenish scapose portion of the stem; forks of the inflorescence sometimes with a much-reduced branch; flowers white, cream or pale yellow; plants apparently confined to e. of the Green River .... *E. ephedroides*

Eriogonum acaule Nuttall — stemless buckwheat — One specimen seen (Peterson & Deardorft 83-131) is from 2 mi ne. of confluence of Shell Creek and Hells Canyon, Moffat Co.; barren hillsides with saltbush and wheatgrass; 6,820 ft; June. Also common at the edge of our area on and near Hickey Mtn., Uinta Co. Wyoming on semi-barrens of Bridger Formation.

Eriogonum alatum Torrey [*Pterogonum alatum* (Torrey) Gross] — winged buckwheat — Widespread; occasional in sagebrush, pinyon-juniper, and mt. brush communities; 7,000-8,500 ft; late June-July (Aug.). Our plants belong to var. *alatum*.

Eriogonum batemanii M. E. Jones — Bateman buckwheat — Widespread in Duchesne, Uintah, and Rio Blanco Cos. and apparently rather restricted in Moffat Co. in desert shrub, sagebrush, pinyon-juniper, and mt. brush communities at 4,650-6,500 (8,250) ft; mid June-Oct. Based on Welsh 377 and Bradley 5304 DINO! Welsh (1957) listed *E. spatulatum* A. Gray and *E. bicolor* M. E. Jones for Dinosaur National Monument. These specimens appear to be the basis for listing of these species by Holmgren (1962). Both these specimens belong to *E. batemanii*.

Eriogonum brevicaule Nuttall — shortstem buckwheat — With 3 vars.:

1 Leaves strongly revolute such that the lower surface is nearly all hidden by the rolled margins .... **var.** *viridulum*

1 Leaves not or somewhat revolute, the lower surface readily visible 2
2 Scapes and branches densely tomentose; inflorescence capitate to
  umbellate; plants of Duchesne and Wasatch Cos. . . . . **var. laxiflorum**
2 Scapes and branches glabrous, green or blue-gray glaucous; inflorescence
  cymose; plants of Daggett, Moffat, and Rio Blanco Cos. . . . **var. brevicaule**

Var. brevicaule [*E. campanulatum* (Nuttall) Stokes] — shortstem buckwheat
— White-flowered populations are scattered across the n. flank of the Uinta
Mtns in Summit Co. Utah and Uinta Co, Wyoming. Yellow-flowered plants are
common from Sheep Creek and east across Daggett Co. through Moffat Co.,
and from near Rangely to Massadona, Rio Blanco Co. in desert shrub, sagebrush,
and pinyon-juniper communities; 5,600-6,600 ft; (late June) July-Aug. Some
specimens from Rio Blanco Co. with linear leaves (i.e. B. Welsh & Moore 301)
are much like those of var. *viridulum*.

Var. laxiflorum (Torrey & A. Gray) Reveal (*E. brevicaule* var. *huberi* S. L.
Welsh; *E. brevicaule* var. *promiscuum* S. L. Welsh; *E. medium* Rydberg) — varying
buckwheat — Represented by 2 phases that are occasionally seen in the same
population. The phase with a capitate inflorescence (*E. chrysocephalum* A. Gray)
is common across the W. Tavaputs Plateau from Argyle Canyon w. to Strawberry
Valley and n. to Tabiona; sagebrush, pinyon-juniper, Salina wildrye, and open
Douglas-fir communities, usually on shaly or marly limestone slopes and
ridges; 6,800-8,800 ft; late May-Sept. The phase with open inflorescence (var.
pumilum Stokes ex M. E. Jones) is occasional across Duchesne Co., n. and e. of
the above var. from Duchesne to the foothills of the Uinta Mtns. near the town
of Whiterocks where populations include capitate as well as open inflorescences;
desert shrub, pinyon-juniper, and open aspen and coniferous forest communities;
5,800-7,400 ft; June-Sept.

White-flowered plants on Mt. Bartles and on the divide between Soldier
Creek and Nine Mile Creek have been referred to as *E. brevicaule* var. *promiscuum*.
These plants likely intergrade into *E. lonchophyllum* as well as *E. corymbosum*
var. *hylophilum*. White flowered plants of *E. brevicaule* var. *brevicaule* are found
near Altamont and on the north slope of the Uinta Mtns. The widely scattered
distribution of white flowered plants suggests independent origin of different
populations.

Var. viridulum (Reveal) S. L. Welsh (*E. viridulum* Reveal) — Uinta basin
buckwheat — Endemic, locally common to abundant from Starvation Reservoir
e. to Island Park and Raven Ridge near Bonanza in Duchesne and Uintah Cos.;
salt desert shrub and juniper communities, most common on raw, eroding
slopes of the Duchesne River and Uinta Formations, but also on other formations
that typically weather to badlands including Dakota, Frontier, Morrison,
Mowery, and Mancos Shale, commonly colonizing on roadcuts; 4,800-6,500
(7,000?) ft; mid Aug.-Oct.

In addition to features of the key the cymes are more densely flowered than
in plants of the other vars. Plants of this taxon are closely related to and
perhaps intergrade into *E. ephedroides*. Uinta basin buckwheat occasional
crosses with *E. corymbosum* and possibly with *E. microthecum*. Crosses of *E.
corymbosum* and *E. brevicaule* var. *viridulum* (i.e. Neese 8496-8500; Goodrich
14988-14993) will key here. Plants described as *Eriogonum* × *duchesnense* Reveal
are found within a 20 mi radius n. and w. of Duchesne; sagebrush and pinyon-
juniper communities; 6,000-7,000 ft; Aug.-Sept. The name applies to plants of

hybrid origin involving *E. corymbosum* and the yellow flowered *E. brevicaule* including var. *viridulum*.

Eriogonum cernuum Nuttall — nodding eriogonum, nodding buckwheat — Widespread; common in many plant communities including desert shrub, sagebrush, and pinyon-juniper, often on areas of disturbances; 4,700-7,600 ft; July-Sept. Our plants belong to var. *cernuum.*

Eriogonum corymbosum Bentham in de Candolle — big buckwheat — With 2 vars. as follows. A third var. [var. *glutinosum* (M. E. Jones) M. E. Jones; *E. aureum* M. E. Jones] has been reported for the Uinta Basin based on a vegetative specimen (Graham 1937). This yellow-flowered taxon is known from well s. of our area.

1 Leaf blades not over 3 cm long or if so then over 1 cm wide, usually crenate-revolute . . . . **var. corymbosum**
1 Leaf blades 3-9 cm long, less than 1 cm wide, usually revolute but not crenate . . . . **var. hylophilum**

Var. corymbosum (*E. divergens* Small; *E. c.* var. *erectum* Reveal & Brotherson; *E. effusum* Nuttall misapplied) — big buckwheat — Widespread and occasional or locally common in desert shrub, sagebrush, Salina wildrye, and pinyon-juniper communities; 5,150-8,700 ft; Aug.-Sept. Goodrich 14991 comes from a single yellow flowered plant in a large population of white flowered plants. Plants described as *Eriogonum* × *duchesnense* Reveal are found within a 20 mi radius n. and w. of Duchesne; sagebrush and pinyon-juniper communities; 6,000-7,000 ft; Aug.-Sept. The name applies to hybrids involving *E. corymbosum* and the yellow flowered *E. brevicaule.*

Var. hylophilum (Reveal & Brotherson) S. L. Welsh (*E. hylophilum* Reveal & Brotherson) — Gate Canyon buckwheat — Endemic, locally common in the Bad Land Cliffs of the W. Tavaputs Plateau near the summit of Gate Canyon in se. Duchesne Co. near the Carbon Co. line; sagebrush and pinyon-juniper communities; 6,260-8,250 ft; late July-Aug. Specimens from lower McCook Ridge (Neese 6576) and head of Buck Canyon (Goodrich 27917) of the E. Tavaputs Plateau appear to be this or a cross of *E. corymbosum* and *E. lonchophyllum* (q.v.), which might be the origin of var. *hylophilum*. These specimens are likely the basis for the listing of *E. corymbosum* var. *revealianum* (S. L. Welsh) Reveal for Uintah Co. in IMF (2A: 272). We include these specimens within the concept of *E. lonchophyllum.*

Eriogonum divaricatum Hooker — spreading buckwheat — Listed for Uintah Co. and adjacent to our area in Sweetwater co. Wyoming (AUF 580-581).

Eriogonum ephedroides Reveal — ephedra buckwheat — Endemic, locally common e. of the Green River from Nutters Hole to McCook Ridge, also along lower Evacuation Creek in se. Uintah Co. and on Raven Ridge in adjacent Rio Blanco Co.; desert shrub, pinyon-juniper, and sagebrush communities, mostly on semi-barren hillsides on Green River Formation; 5,000-6,400 ft; mid July-Aug. Much like *E. brevicaule* var. *viridulum* from which it is distinguished by a series of minor and mostly overlapping features.

Eriogonum flexum M. E. Jones [*Stenogonum flexum* (M. E. Jones) Reveal] — bent buckwheat — Infrequent or locally common in Uintah Co. from Tridell e. to Dinosaur National Monument and to Rangely in Rio Blanco Co.; desert shrub and scattered juniper communities, on raw eroding clay or sandy-clay soil of the Duchesne River and Mancos Shale Formations; 4,800-5,900 ft; May-June.

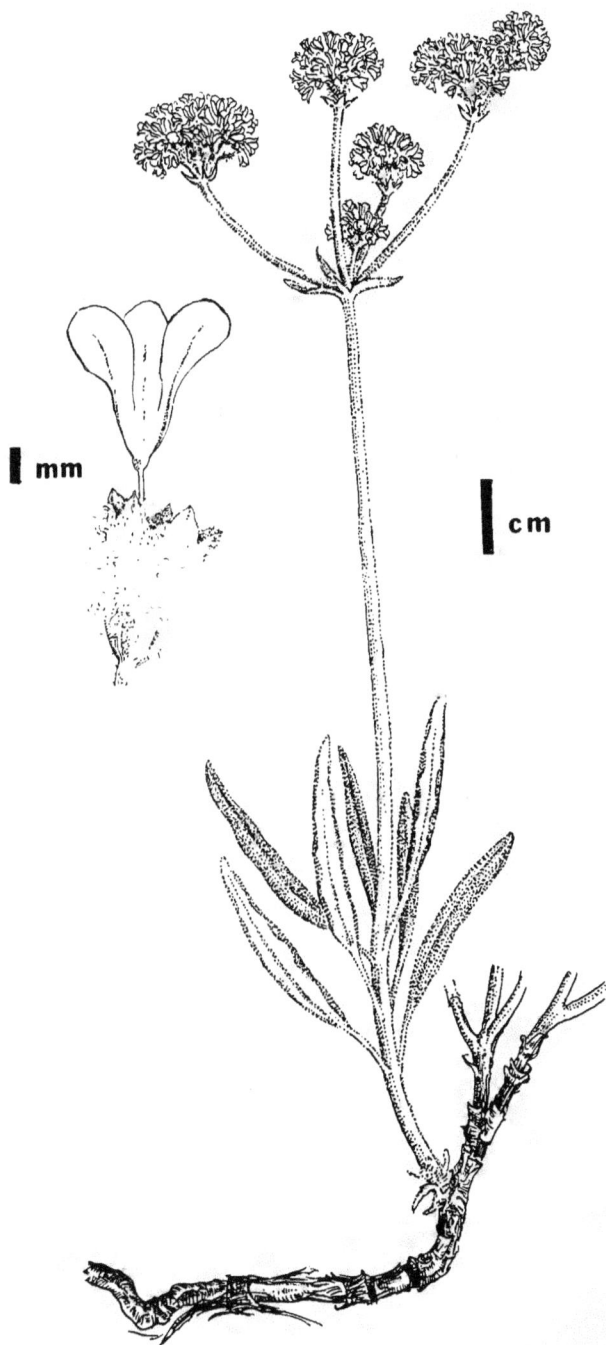

Eriogonum corymbosum var. *hylophilum* (Reveal & Brotherson) S. L. Welsh
Gate Canyon buckwheat

Eri-

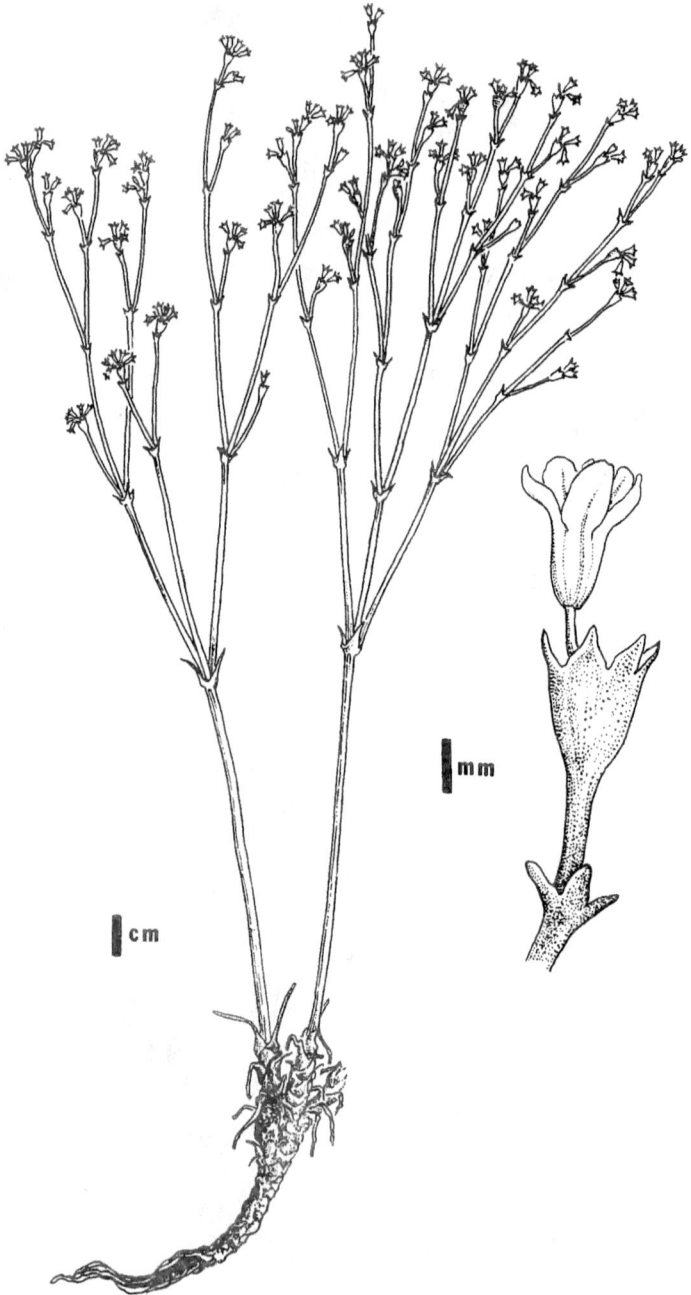

Eriogonum ephedroides Reveal — ephedra buckwheat

ogonum gordonii Bentham in de Candolle — Gordon buckwheat — Uintah and Daggett Cos. and east into Moffat and Rio Blanco Cos.; desert shrub communities, mostly on Duchesne River, Mancos Shale, and other formations that weather to badlands; 4,800-6,300 ft; mid June-July (Oct.).

Eriogonum heracleoides Nuttall — whorled buckwheat — Common from Strawberry Valley across the Uinta Mtns. and to Cold Spring Mt., Moffat Co.; sagebrush, mt. brush, dry aspen, ponderosa pine, and silver sagebrush-grass-meadow communities; 7,200-8,300 ft; mid June-July. Our plants belong to var. *heracleoides*.

Eriogonum hookeri S. Watson — Watson buckwheat, Hooker buckwheat — Infrequent from Duchesne to Rangely and from Dinosaur National Monument s. to Hill Creek of E. Tavaputs Plateau; desert shrub, sagebrush, and pinyon-juniper communities; 4,800-6,000 ft; mid July-early Sept.

Eriogonum inflatum Torrey & Fremont in Fremont (*E. fusiforme* Small) — desert trumpet, bottlestopper — Occasional or locally common across the bottom of the Uinta Basin from Arcadia e. to Rangely in desert shrub communities on formations that weather to badlands; 4,700-5,400 ft; June-mid July. Our plants belong to var. *fusiforme* (Small) Reveal.

Eriogonum jamesii Bentham (*E. arcuatum* Greene) — James buckwheat — Occasional or locally common in Wells Draw and Minnie Maud Creek to Nutters Ridge on the Tavaputs Plateau and on Blue Mt.; sagebrush and pinyon-juniper communities; 6,000-7,500 ft; late June-early Sept. Our plants with leaves 1-3 cm long and 5-15 mm wide, and inflorescence divided 1-3 times or more belong to var. *flavescens* S. Watson

Eriogonum lonchophyllum Torrey & A. Gray — longleaf buckwheat — With 2 vars. as follows:

1 Leaves (2) 4-10 mm wide; flowers white or pale greenish-cream, seldom at all pinkish, with pale to dark green midrib; plants 30-50 cm tall, sometimes woody above ground level, from e. Uintah Co . . . . **var. *saurinum***
1 Leaves 1-4 (6) mm wide; tepals white, often pinkish tinged especially toward the base, the midrib often dark green, pinkish or reddish-brown in age; plants to 30 cm tall, herbaceous from near the base or somewhat woody above ground level, of the E. Tavaputs Plateau and n. into Moffat Co. . . . . **var. *lonchophyllum***

Var. lonchophyllum (*E. intermontanum* Reveal; *E. sarothriforme* Gandoger; *E. scoparium* Small; *E. tristichum* Small) — longleaf buckwheat — Occasional to locally common across the Roan Cliffs from Range Creek, Emery Co. e. into the Piceance Basin where locally abundant and n. into Moffat Co., and with a disjunct anomalous phase near the old mining town of Rainbow; sagebrush and mt. brush communities, usually on shaly ground; (6,000) 7,000-9,000 ft; mid July-Sept. One specimen (Neese 6576) from lower McCook Ridge is intermediate between *E. lonchophyllum* and *E. corymbosum* var. *hylophilum* with involucres the size of the former and flowering stems and branches of the inflorescence tomentose as in the latter.

Except for the more-or-less geographic isolation, the 2 vars. of *E. lonchophyllum* and *E. corymbosum* var. *hylophilum* form a complex that is hardly separable. Plants from near the old mining town of Rainbow are as much like the plants of var. *saurinum* from near Vernal as they are like those of the Roan Cliffs, and their range of distribution is intermediate. This Rainbow phase possibly

Eriogonum lonchophyllum var. lonchophyllum — longleaf buckwheat

represents hybridization of *E. corymbosum* and *E. lonchophyllum* (Welsh 1984c). This is the likely origin of var. *saurinum* and *E. corymbosum* var. *hylophilum* with the features of *E. lonchophyllum* dominant in *E. l.* var. *saurinum* and the features of *E. corymbosum* dominant in *E. c.* var. *hylophilum*. The plants from the Roan Cliffs (*E. intermontanum*) are similar to those of the Piceance Basin. Through this basin these plants are more or less continuous through short-leaved phases (*E. scoparium* and *E. tristichum*) with *E. lonchophyllum*. The only apparent differences are in the degree by which leaves extend up the stems and in the tendency toward clustered involucres.

In plants of the Tavaputs Plateau, the leaves are basal or nearly so, with the tomentose, leaf-bearing portion of the stem mostly less than 1.5 cm long. In plants from other parts of Colorado and northern New Mexico, the tomentose, sheath-covered, leafy portion of the stem is often over 1.5 cm long, but this feature is not consistent, and plants with nearly all basal leaves occasionally occur as far away as New Mexico, and *E. intermontanum* does not seem to be distinct from *E. lonchophyllum*. Small plants without woody stems above ground level are common in places on the E. Tavaputs Plateau. Larger plants with woody stems well above ground level area also common.

Var. *saurinum* (Reveal) S. L. Welsh (*E. saurinum* Reveal) — dinosaur or Mowey buckwheat — Endemic, Maeser and Asphalt Ridge east to Island Park in desert shrub, sagebrush, juniper, and juniper-serviceberry communities, forming nearly pure stands on acidic Mowery Shale, but also on Carmel, Curtis, Entrada, Mesa Verde, and other formations that are not acidic; 5,200-6,200 ft; late July-early Oct. Plants of var. *saurinum* have the glabrous features of var. *lonchophyllum* (q.v.) and some of *E. brevicaule* and the more leafy and woody features of *E. corymbosum*. A population on Raven Ridge (Goodrich 21997-22001) with yellow as well as whitish flowers strongly suggests a close relationship to *E. brevicaule*. Specimens referred to as *E. effusum* Nuttall by Welsh (1957) belong here.

Eriogonum microthecum Nuttall — slender buckwheat — Widespread and occasional in desert shrub, sagebrush, pinyon-juniper, and mt. brush communities; 4,800-8,000 ft; July-Sept. Plants with flat leaves (1) 2.5-6 (8) mm wide belong to var. *laxiflorum* Hooker, and those with revolute leaves 1-2 (2.5) mm wide belong to var. *simpsonii* (Bentham) Reveal. One collection (Neese 14531) appears to be a hybrid between *E. microthecum* and *E. brevicaule*.

Eriogonum ovalifolium Nuttall — cushion buckwheat — Widespread; infrequent or common in desert shrub, sagebrush, and pinyon-juniper communities on many kinds of substrates; 4,800-7,500 ft, perhaps higher; late April-June (July). This taxon has white flowered and yellow flowered (var. *multiscapum* Gandoger) phases. The 2 phases are sympatric over a wide range in w. North America and have been found in the same populations. The white phase is most common in the Uinta Basin. Both color phases of our area belong to var. *ovalifolium*.

Eriogonum racemosum Nuttall — redroot buckwheat — Infrequent along the s. flank of the Uinta Mtns. and apparently most common w. of Mosby Canyon; sagebrush communities; 6,300-7,000 ft; mid Aug.-Oct.

Eriogonum salsuginosum (Nuttall) Hooker (*Stenogonum salsuginosum* Nuttall) — smooth buckwheat — Common from Tridell e. to Dinosaur National Monument and to Bonanza (possibly to Rangely) and s. to Ouray, mostly in Uintah Co.; one specimen from extreme e. edge of Duchesne Co. in Pleasant Valley, also at Sheep Creek Gap in Daggett Co.; desert shrub communities and often on clayey

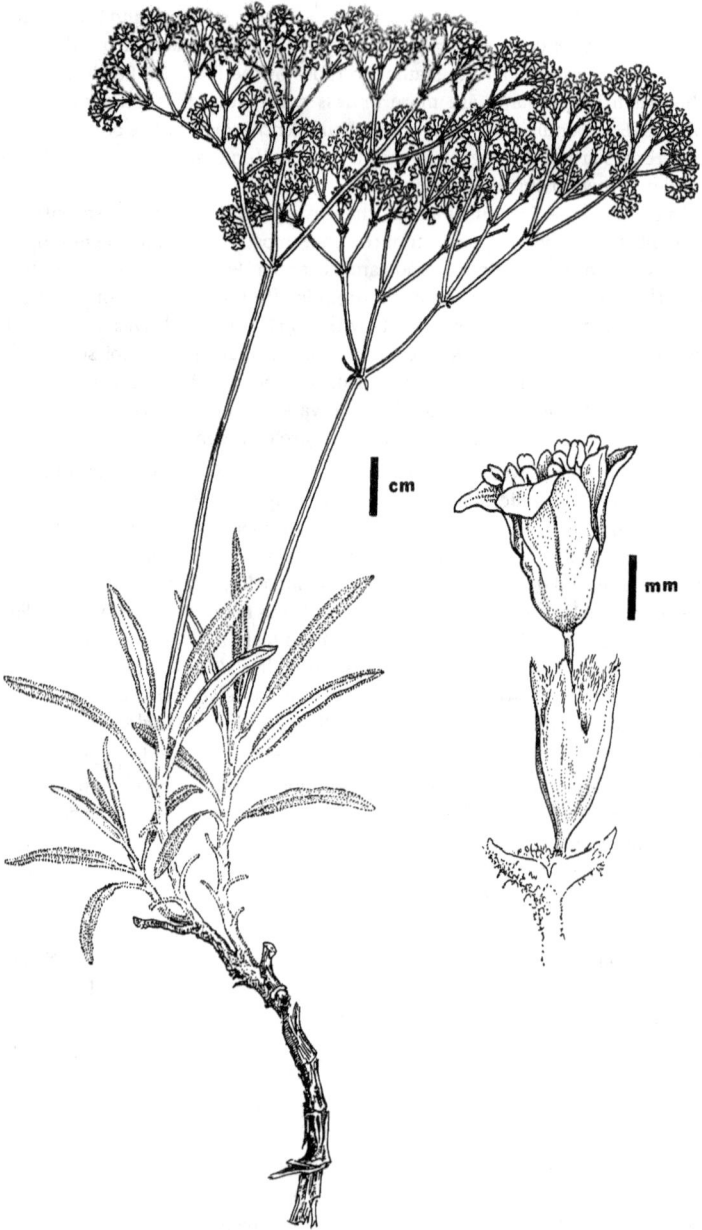

Eriogonum lonchophyllum* var. *saurinum (Reveal) S. L. Welsh
dinosaur or Mowey buckwheat

soils of formations that weather to badlands; 4,700-5,600 (6,300) ft; mid May-
June (early July).

Eriogonum shockleyi S. Watson — Shockley buckwheat — Occasional; widespread
in Duchesne, Uintah, and Moffat Cos., 2 specimens seen from e. Daggett Co.;
desert shrub, sagebrush, and pinyon-juniper communities on a variety of substrates; 4,800-6,200 ft; June. With whitish rather than yellowish flowers, our
plants belong to var. *longilobum* (M. E. Jones) Reveal.

Eriogonum tumulosum (Barneby) Reveal — tumulose buckwheat — Endemic,
the following specimens seen: 10 from within a 10 mi radius nw. of Duchesne;
1 from near Hanna, Duchesne Co.; 1 from near Steinaker Reservoir, Uintah Co.;
and 6 from the Gates of Lodore to Vermillion Gap, Moffat Co.; desert shrub,
sagebrush-juniper, and pinyon-juniper communities; 5,625-7,520 ft; May-June.

Eriogonum umbellatum Torrey — sulfur buckwheat — Comprised of about 40
vars. in the western U. S. and Canada with 16 of these in the Intermountain
Region (IMF 2A: 321-322) with 4 vars. in our area as follows:

1 Flowers cream or white . . . . **var. *majus***
1 Flowers yellow 2
2 Leaves tomentose below, floccose to glabrous above . . . **var. *umbellatum***
2 Leaves glabrous on both sides or loosely floccose toward the margins . . 3
3 Inflorescence a head or head-like, not much over 1 cm long; stems usually
   not much over 5 cm long; plants of the Uinta Mtns.; 9,200-11,500 ft . . . . .
   . . . . **var. *porteri***
3 Inflorescence mostly umbellate, rarely head-like, over 1 cm long; stems to
   50 cm long; plants rare in the Uinta Mtns. . . . . **var. *aureum***

Var. *aureum* (Gandoger) Reveal — sulfur buckwheat — The few specimens
seen are from the Tavaputs Plateau, upper Strawberry drainage and from
Moffat Co.; sagebrush, Salina wildrye, forb-grass, aspen, and coniferous forest
communities; mostly 8,300-9,500 ft; (late June) July-Aug. Included in var. *umbellatum* in AUF (591).

Var. *majus* Hooker (*E. umbellatum* var. *deserticum* Reveal; *E. umbellatum* var.
dichrocephalum Gandoger; *E. subalpinum* Greene) — cream buckwheat — Common
in Strawberry Valley, apparently rare (or seldom collected) across the Uinta
Mtns. to Daggett Co. (no specimens seen from Uintah Co.) and e. across the W.
Tavaputs Plateau to Indian Canyon, to be expected elsewhere; sagebrush, mt.
brush, dry meadow, and open lodgepole pine communities, and on open rocky
slopes; 7,000-10,000 ft; July-Aug. In IMF (2A: 326) vars. *deserticum* and *dichrocephalum* are recognized as separate taxa from var. *majus*.

Var. *porteri* (Small) Stokes (*E. porteri* Small) — ballhead sulfur buckwheat —
Common in the Uinta Mtns. from the Uinta River drainage and westward on
the s. slope and W. Fork Blacks Fork and west on the n. slope; on open, rocky
ground, dry meadows, and open coniferous forests; 8,200-11,500 ft; July-Aug.
This intergrades into var. *aureum* and is listed in synonymy under var. *aureum*
in TPD. However, many populations show little intergradation, and the 2 vars.
are maintained in IMF (2A: 321).

Var. *umbellatum* (*E. neglectum* Greene) — sulfur buckwheat — Common
across the Uinta Mtns. And their flanks to Blue and Douglas Mtns., occasional
on the Tavaputs Plateau from Argyle Canyon e. to the Piceance Basin; sagebrush,
pinyon-juniper, mtn. brush, ponderosa pine, Douglas-fir, and Salina wildrye
communities; 5,700-8,600 ft; June-July.

Koenigia L.

Koenigia islandica L. — island purslane, koenigia — Diminutive annual herb, 1-5 cm tall; stems slender, simple or branched; leaves alternate and with distinctive tubular ochreae, or opposite and with perfoliate ochreae, the blades 2-9 mm long, 1-3 mm wide; inflorescence an umbel with few to several flowers; perianth about 1 mm long, greenish, whitish, or reddish; stamens usually 3, alternating with the perianth segments; achenes glabrous, enclosed by the perianth. The one specimen seen (Goodrich et al. 26308) is from an alpine, mossy seep on Gilbert Bench, Uinta Mtns. at 12,100 ft. This small plant is easily overlooked, and it could be more common than the one specimen seen would indicated. However, the Gilbert Peak area is home to other species of very limited distribution within our area.

Oxyria Hill — Mountain Sorrel

Oxyria digyna (L.) Hill — mtn. sorrel — Perennial herbs with sour juice, from a taproot and branched crown, 6-20 cm tall rarely taller, scapose or stems with 1 leaf; leaves 1-5 cm wide; inflorescence 3-10 cm long, rarely longer; flowers bisexual; tepals 4, 1-2.5 mm long, green to bright red in age, the outer 2 tepals strongly keeled, reflexed by the mature fruit, the other 2 tepals flat, erect, deciduous; stamens 6; styles 2, with expanded, fringed stigmas; achenes flattened, winged, 4-6 mm wide at maturity. Occasional and locally abundant on the Uinta Mtns. and apparently rare on Strawberry Ridge; lodgepole pine, Engelmann spruce, and most common in rocky, late melting, alpine-snowbed communities, (rarely down to 9,600 ft on shady cliffs) 10,500-13,000 ft; June-Aug.

Polygonum L. — Knotweed; Smartweed; Ladysthumb

Annual or perennial herbs, often cosmopolitan weeds; stems often with swollen joints; leaves alternate or alternate and basal, simple, entire; stipules (ocreae) thinly membranous, fused and sheathing the stem, usually fringed, split, or lacerate distally; flowers perfect (functionally dioecious in *P. cuspidatum* and *P. viviparum*), the perianth segments usually more or less petaloid, 5-parted, the outer often largest; achenes enclosed in the persistent perianth, lenticular or trigonous, often smooth and shining.

---

1 Robust cultivated ornamentals to 1.5 m or more tall, occasionally adventive or persisting; leaves ovate, broadly rounded or truncate at base, the larger ones 8 cm or more wide 2

1 Adventive weeds or indigenous plants, not or rarely cultivated, generally less than 1 m tall; leaves various but usually much narrower than 8 cm. . 3

2 Plants annual; flowers perfect, rose to crimson, borne in terminal densely flowered, showy, drooping racemes; achene lenticular . . . . ***P. orientale***

2 Plants perennial from rhizomes, forming clumps, dioecious; flowers white or greenish-white, borne in showy paniculate racemes from the upper axils; achene trigonous . . . . ***P. japonicum***

3 Plants twining, the stems sprawling or climbing; leaves hastate to triangular cordate . . . . ***P. convolvulus***

3 Plants not twining or climbing; leaves usually tapered to the petiole, never hastate-lobed 4

4 Inflorescence of 1 or few, many-flowered, dense, pedunculate spikes or spike-like racemes..5

4 Flowers borne singly or in few-flowered clusters; inflorescence not pedunculate 9

5 Plants perennial; inflorescence terminal............................6

5 Plants annual; inflorescence terminal and axillary 8

6 Flowers bright pink to rose-red; plants aquatic or semi aquatic (rarely terrestrial), stems prostrate to ascending, rooting at the nodes; basal leaves absent; stem leaves floating or emergent, petiolate, elliptic to lanceolate, mostly 1.5-6 cm wide .... *P. amphibium*

6 Flowers white or pale pinkish; plants terrestrial (though often growing in wet places, erect from a tuberous caudex; basal leaves present, petiolate; stem leaves subsessile, narrowly lanceolate to linear, strongly reduced upward, to 1 cm wide 7

7 Flowers crowded in a dense raceme mostly 1-2 cm wide, 1.5-4 times longer than wide; lower flowers not replaced by bulblets; plants mostly 15-35 cm tall .... *P. bistortoides*

7 Flowers in a spike-like raceme 0.4-0.8 cm wide, mostly over 4 times longer than wide; lower flowers replaced by bulblets; plants mostly 5-15 cm tall . ... *P. viviparum*

8 Lower leaf surface glabrous to short hairy, neither glandular-punctate not white-woolly; stipule-sheath prominently ciliate with bristles ca 1 mm long; perianth usually distinctly pink or rose colored .... *P. persicaria*

8 Lower leaf surface glandular punctate beneath or sometimes white-woolly; stipule-sheath without bristles (sometimes obscurely short-ciliate with soft hairs); perianth usually greenish or merely pink-tinged .. *P. lapathifolium*

9 Leaves and flowers densely crowded and forming a leafy-bracteate terminal spike; plants mainly less than 10 cm tall .... *P. kelloggii*

9 Flowers more remote, in small axillary clusters or solitary, or if in terminal spikes the bracts much reduced, not leafy; plant size various.......... 10

10 Stems round, with 8-16 conspicuous ribs; leaf blade venation pinnate with conspicuous secondary veins; anthers pale yellow 11

10 Stems 4-sided, with obscure ribs or ribs none; leaf blade venation parallel with obscure secondary veins; anthers pink to purple 14

11 Stamens 3; plants bluish green when fresh, sometimes dark brown or black when dry .... *P. ramosissimum*

11 Stamens 5-8; plants yellowish green or green fresh and dry 12

12 Inflorescences terminal and axillary; pedicles 1-2 mm long; hypanthium less than ¼ the length of the perianth; petioles 0-1.5 mm long .... *P. argyrocoleon*

12 Inflorescences axillary; pedicels 1.5-5 mm long; hypanthium ¼-½ the length of the perianth; petioles 0.3-9 mm long.............................13

13 Plants widespread and common; tepals usually with white or pinkish margins .... *P. aviculare*

13 Plants known from Carter Dugway, n. slope Uinta Mtns.; tepals mostly with greenish or yellowish margins .... *P. achoreum*

14 Leaves mostly at least ½ as wide as long; ocreae 1-4 mm long; perianth 1.8-
    2.5 mm long .... *P. minimum*
14 Leaves narrower than above; ocreae 4-12 mm long; perianth 2.5-4.5 mm
    long .... *P. douglasii*

Polygonum achoreum S. F. Blake — leathery knotweed — Abundant along the
dirt road leading into Carter Dugway, Daggett Co. at 8,700-8,800 ft.

Polygonum amphibium L. [*P. coccineum* Muhlenberg; *P. natans* (Michaux) A.
Eaton; *Persicaria amphibia* (L.) S. Gray; *P. coccinea* (Muhlenberg) Greene] —
water knotweed, water smartweed — Occasional across our area; in mud and
shallow water of reservoirs and lake margins, river bottoms, streams and
marshes, with bullrush, cattail, salt cedar, and other riparian plants; 7375-
9346 ft.

Polygonum argyrocoleon Steudel — silversheath knotweed — Specimens seen
are from the margins of Pelican Lake and Stewart Lake (Uintah Co.) and the
floodplain of the White River (Rio Blanco Co.), and plotted for Moffat Co.
(TPD); 4,765-5,200 ft. Stems decumbent to erect, 15-100 cm long.

Polygonum aviculare L. [*P. buxiforme* Small; *P. aviculare* var. *littorale* (Link) Koch;
P. monspeliensis Persoon] — dooryard-grass, prostrate knotweed, dishwater-
grass, devil's shoe-strings — Widespread and common; sidewalks, roadsides,
and otherwise disturbed and trampled areas, usually in at least moderately
mesic places; 4,600-8,700 ft; June-Oct. Stems prostrate to erect, 5-200 cm long.
This is a variable cosmopolitan species complex. Material long designated in
this country as *P. aviculare* might include *P. arenastrum* Bordeau (Styles 1962);
as the older name *P. arenastrum* would take precedence over *P. aviculare* if the
2 taxa are treated conspecifically. It seems better to retain the traditional
name *P. aviculare* until our material is examined by monographers. Flowers
and others (1959) reported as *P. erectum* L. for Williams Ranch (area now
inundated by Flaming Gorge Reservoir) belongs here.

Polygonum bistortoides Pursh [*P. bistortoides* var. *linearifolium* Small; *Bistorta
bistortoides* (Pursh) Small] — American bistort — Widespread and common in
the Uinta Mtns. and apparently uncommon in Strawberry Valley and on
Strawberry Ridge; wet and dry meadow, sagebrush-grass, and alpine communities,
and openings in lodgepole pine and Engelmann spruce forests; 7,050-11,500 ft;
June-Aug. In an alpine plant community classification in the Uinta Mtns., this
was the 5th most frequently encountered species (Brown 2006).

Polygonum convolvulus L. [*Fallopia convolvulus* (L.) A. Love] — black bindweed —
Introduced from Europe; The one specimen seen (Goodrich 3181 USUUB) is
from a shepherds corral in upper Blind Stream, were the species did not
persist, also reported by Graham (1937) from a garden at Lapoint (Graham
9987). To be expected in waste places along fences and in gardens.

Polygonum douglasii Greene — Douglas knotweed — With 2 well-marked vars.:
1 Flowers deflexed, the base with a persistent peg-like stipe 0.1-0.2 mm long;
   pedicel and stipe roseate .... **var. *douglasii***
1 Flowers erect, the base sessile, dehiscing without a peg-like stipe; pedicel
   prevailing green .... **var. *johnstonii***

Var. *douglasii* [*P. douglasii* var. *latifolium* (Engelmann) Greene] — Douglas
knotweed — Widespread in dry meadow, silver sage, bitterbrush-sagebrush,

forb-grass communities, parklands of aspen, lodgepole pine and spruce-fir, and in rocky openings in coniferous forests; thriving on pocket gopher disturbance; 7,500-10,300 ft; July-Sept.

Var. *johnstonii* Munz (*P. sawatchense* Small) — Sawatch knotweed — Widespread in pinyon-juniper, mt. brush, sagebrush, and forb-grass communities; 7,000-8,500 ft; June-Sept.

In addition to the above 2 taxa, var. *engelmannii* (Green) Kartesz and Gandhi is listed for Utah in TPD.

Polygonum japonicum (Houttuyn) S. L. Welsh (*P. cuspidatum* Siebold & Zuccharini, *Reynoutria japonica* Houttuyn) — fleece-flower, Japanese knotweed — The few specimens seen are from a roadside near the Duchesne River, near Rock Creek road junction, and along ditches and canals at Lapoint and Vernal, cultivated, adventive, and persisting; Aug.-Sept.

Polygonum kelloggii Greene — Kellogg knotweed — Occasional to locally common in the Uinta Mtns. including Cold Spring Mtn., and reported by Graham (1937) from the Tavaputs Plateau; mt. brush, silver sagebrush, subalpine meadow, aspen, lodgepole, and spruce-fir communities, usually in ephemerally wet places and in rock crevices; 7,400-10,600 ft; June-Sept. Treated as *P. polygaloides* Meisner in de Candolle in FNA where our plants might key to ssp. *confertiflorum* (Nuttall ex Piper) J. C. Hickman and ssp. *kelloggii* (Greene) J. C. Hickman (*P. kelloggii* var. *kelloggii*). Treated as *P. kelloggii* in IMF (2A: 230) where our plants might key to var. *kelloggii* and var. *watsonii* (Small) Reveal) based on shape and color of achenes.

Polygonum lapathifolium L. (*P. incanum* Schmidt; *Persicaria lapathifolia* (L.) S. Gray] — willow-weed, curly top knotweed — Widespread in riparian and lacustrine communities in water on drying mud and sandbars; 4,570-6,235 ft.

Polygonum minimum Watson — broadleaf knotweed — The 3 specimens seen are from Notch Mtn. & Provo Peak area, western Uinta Mtns. from rocky slopes at 10,160-11,100 ft.

Polygonum orientale L. — kiss-me-over-the-fence — A single specimen (Neese 15047) is from a garden in Vernal, where (according to the owner, Celestia Rasmussen) it has been self-seeding for many years.

Polygonum persicaria L. [*Persicaria maculata* (Rafinesque) S. F. Gray] — spotted ladysthumb — Specimens seen are from Daggett, Duchesne and Uintah Cos., to be expected elsewhere; occasional weed of disturbed places along floodplains and ditches and in gardens and pastures; July-Oct.

Polygonum ramosissimum Michaux — bushy knotweed — Apparently widespread; specimens seen are from wetlands and upland communities; 5190-8,050 ft; Aug.-Sept. Stems erect, usually much branched, 10-100 (200) cm long.

Polygonum viviparum L. [*Bistorta vivipara* (L.) S. Gray] — alpine bistort, viviparous bistort — Common in the Uinta Mtns. in wet or ephemerally moist meadow, spruce-fir and lodgepole pine parkland, and alpine communities; 8,100-11,900 ft; June-Sept.

Rumex L. — Dock; Sorrel

Annual or perennial glabrous or scabrous herbs; leaves alternate and basal, simple, entire or nearly so (or hastate in *R. acetosella*); tepals (4) 6 in 2 series with 3 in each series, subequal at flowering, the outer ones not enlarged, mostly inconspicuous, the inner 3 enlarged and wing-like on the mature fruit, and then referred to as valves; valves enveloping the achene, membranous, sometimes bearing a grain-like swelling (callus) on the midvein; stamens 6; styles 3, with numerous thread-like branches; achenes sharply 3-angled.

1 Plants mostly unisexual, montane, 15-50 (70) cm tall; valves small, without callosities; stems slender 2

1 Plants bisexual, montane or of low elevations, of various stature; valves usually conspicuously enlarged and/or bearing callosities; stems often stout 3

2 At least a few (often several) of the leaf blades strongly hastate, 1-4 cm long; pedicel jointed immediately below the flower .... *R. acetosella*

2 Leaf blades not hastate, the larger ones 4-7 cm long or longer; pedicels jointed below the middle .... *R. paucifolius*

3 Valves 1-3 cm wide at maturity, some usually over 1 cm long........... 4

3 Valves less than 1 cm wide at maturity and not over 1 cm long 5

4 Valves 8-18 mm long; basal tuft of leaves well developed in young plants; rhizomes bearing enlarged tubers .... *R. hymenosepalus*

4 Valves 15-30 mm long; basal tuft of leaves lacking even in young plants; rhizomes lacking tubers .... *R. venosus*

5 Valves conspicuously toothed or with bristly margins, the teeth or bristles from 1/2 as long to longer than the width of the body, with callosities... 6

5 Valves entire or minutely toothed with rounded teeth, with or without callosities 8

6 Plants annual, 5-50 (80) cm tall; valves bristly, many of the bristles longer than the width of the body, each valve with a narrow callus as high or higher than wide; pedicels jointed at the base; leaves linear-oblong, elliptic, to lanceolate, to 2 (3.5) mm wide .... *R. maritimus*

6 Plants perennial, height various but often over 50 cm tall; valves either sharply dentate and not bristly or else usually only 1 of the 3 valves with a callus; pedicels jointed below the middle but usually distinctly above the base 7

7 Leaf blades truncate to cordate basally, less than 3 times as long as wide; valves more or less bristly, usually only 1 of the 3 with a narrow callus; plants 60-120 cm tall .... *R. obtusifolius*

7 Leaf blades narrowed to the petiole, over 3 times as long as wide; valves sharply-dentate, each with a rounded callus; plants known from Pelican Lake and along the floodplains of the Green and Duchesne Rivers........ .... *R. stenophyllus*

8 Valves cordate at the base, the larger ones 5-8 mm wide, at least 1 of the 3 with a callus; plants (0.6) 1-2 m tall; leaves mostly basal, the larger blades 20-30 cm long, to 15 cm wide, truncate to cordate basally .... *R. patientia*

8 Valves not cordate, often less than 5 mm wide, or none of them with a callus; plants and leaves various.................................. 9

9 Most fruits with all 3 valves with reticulate callosities; veins on the lower surface of leaves conspicuously raised and scabrous, the intervals granular-scabrous; lower leaves larger than the mid-stem leaves, often with undulate

crisped margins; larger petioles often over 4 cm long; stems rarely branched below the inflorescence; plants (30) 50-150 cm tall .... **R. crispus**

9   Valves without callosities (or if with callosities as in *R. salicifolius,* then plants different from above in several other features) 10

10   Lower leaves mostly smaller than those of mid-stem, usually withering and deciduous before the upper leaves, the petioles rarely over 4 cm long, the blades linear, linear-elliptic, or rarely narrowly lanceolate, (4) 5-8 times as long as wide, mostly gradually tapered to the petiole; stems often branched below the inflorescence or at least some of the larger leaves with smaller leaves in the axils; valves with or without callosities; plants common, 20-60 cm tall .... **R. salicifolius**

10   Basal leaves larger than the progressively reduced stem leaves, the longer petioles well over 4 cm long, the blades oblong-ovate to oblong-lanceolate, mostly truncate or cordate at the base, the larger ones 10-30 cm; stems simple below the inflorescence, without smaller leaves in the axils of larger ones; valves without callosities; plants uncommon, 50-150 cm tall ....
.... **R. occidentalis**

Rumex acetosella L. [*Acetosella vulgaris* (Koch) Fourreau] — sheep sorrel — Introduced from Europe; occasional from Strawberry Valley and across the Uinta Mtns. to the Yellowstone drainage, and infrequent e. of there, also E. Tavaputs Plateau; sagebrush, forb-grass, aspen, snowbank, and open lodgepole pine and Engelmann spruce communities; 7,400-11,000 ft; July-early Sept.

Rumex crispus L. — curly dock — Introduced from Europe; widespread, somewhat weedy along roadsides, ditches, fence lines, margins of ponds, and in gardens, fields, floodplains, and riparian communities; 4,700-7,500 (10,100) ft; June-Aug.

Rumex hymenosepalus Torrey — canaigre — Specimens seen are from Daggett and Uintah Cos.; desert shrub, sagebrush, and juniper communities, often on sandy clay to heavy clay soils of the Duchesne River Formation and other formations that weather to badlands; 4,700-5,600 ft; May-June.

Rumex maritimus L. (*R. fueginus* Phil.; *R. persicarioides* L. misapplied) — golden dock — Widespread in drying mud and in shallow water at margins of ponds, lakes, floodplains of riparian communities, and along ditches and streams; 4,700-7,700 ft; July-early Oct. Our plants belong to var. *fueginus* (Philippi) Dusen.

Rumex obtusifolius L. — bitter dock — The one specimen seen (Goodrich 21203) is from a trail head in Uinta Canyon, probably introduced via hay.

Rumex occidentalis S. Watson [*R. aquaticus* var. *fenestratus* (Greene) Dorn] — western dock — The 5 specimens seen are from Rock Creek, Uinta and Whiterocks Canyons, Beaver Creek and Cart Creek, Uinta Mtns.; riparian communities; 7,600-8,500 ft; Aug. Our plants belong to var. *occidentalis.*

Rumex patientia L. — patience dock — Introduced from Europe; Graham 9874 from Post Canyon, E. Tavaputs Plateau, 7,800 ft, July is cited by Graham (1937). Not listed for our area in AUF (598) or in TPD.

Rumex paucifolius Nuttall [*Acetosella paucifolia* (Nuttall) Love] — alpine sorrel, alpine sheep sorrel — Infrequent to occasional; Wasatch Co. (Strawberry Valley to Current Creek and W. Tavaputs Plateau) and Daggett and Summit Cos.

(Uinta Mtns.); dry meadow, sagebrush, aspen, and streamside communities, and roadsides; 7,500-9,650 ft; June-Aug.

Rumex salicifolius Weinmann — willowleaf dock — Most common from Strawberry Valley and across the Uinta Mtns., infrequent on the E. Tavaputs Plateau and low elevations in the Uinta Basin; aspen, tall forb, sagebrush-aspen, spruce, and rarely greasewood-sagebrush and saltgrass-willow communities, margins of ponds and reservoirs, and along roads; (5,300) 7,500-11,000 ft; June-Aug. Plants without callosities on the valves have been referred to as var. *utahensis* (Rechinger f.) Reveal. Those with callosities at least on some of the valves belong to var. *lacustris* (Greene) J. C. Hickman. Names of taxa from outside our area that have been applied to our plants include: *R. denticulatus* Torrey (*R. salicifolius* var. *montigenitus* Jepson), *R. mexicanus* Meissner in de Candolle, *R. salicifolius* var. *mexicanus* (Meissner) C. L. Hitchcock, and *R. triangulivalvis* Danser.

Rumex stenophyllus Ledebour — slenderleaf dock — Introduced from Eurasia; locally common to abundant along the Green River from Jensen to s. of Ouray and up the Duchesne River to Myton, and at Pelican Lake, floodplains, lake margins, and disturbed marsh communities; 4,700-4,900 ft; Aug.-Sept.

Rumex venosus Pursh — veiny dock, winged dock, wild begonia — Apparently widespread, the 10 specimens seen are from desert shrub, greasewood, rabbitbrush-badlands wyethia, and sagebrush communities, locally abundant along roadsides; apparently most abundant in sandy soils, but also reported for clay soils; 5,100-6,400 ft; May-June.

POLYPODIACEAE   Common Fern Family

Stem consisting of a creeping to short and erect or branching caudex, bearing scales or hairs; (aerial stems lacking); leaves 1-3 times pinnate (ours), unfolding with fiddle-head shape as they mature with few to many pinnae (leaflets); reproduction by spores, the spores grouped in sporangia, the sporangia generally grouped into a sorus; sori marginal or on the lower surface of leaves, naked or more often covered by a membranous indusium. As treated here Polypodiaceae includes Aspleniaceae (*Asplenium*), Dennstaedtiaceae (*Pteridium*), Dryopteridaceae (*Athyrium, Cystopteris, Polystichum, Woodsia*), and Pteridaceae (*Cryptogramma, Cheilanthes, Pellaea*).

---

1   Leaves once pinnate or bipinnate with fewer than 40 ultimate segments . 2

1   Leaves 2-3 times pinnate, mostly with more than 30 ultimate segments. . 4

2   Base of plant with dense tufts of brown-woolly hair-like scales   ... ***Pellaea***

2   Base of plant not brown-woolly 3

3   Leaves 20-50 cm long; leaflets 25-50 per side of the rachis   ... ***Polystichum***

3   Leaves 3-15 cm long; leaflets fewer . . . . ***Asplenium***

4   Leaf blades (excluding petioles) (25) 30-100 cm long or longer; plants usually of moist woods or wet places. 5

4   Leaf blades (excluding petioles) 2-25 (35) cm long; plants mostly in rock crevices of cliffs and ledges or on talus slopes 6

5   Basal pair of primary leaflets (pinnae) often the largest, the other pairs progressively reduced toward the tip of the leaf . . . . ***Pteridium***

5   Middle pair of primary leaflets larger than the basal or apical pairs . . . . . .
    . . . . **Athyrium**

6   Plants with 2 types of strongly contrasting leaves, 1 type sterile with
    flattened, rounded ultimate segments with lobed or dentate or serrate
    margins, the other type fertile with linear, revolute, entire ultimate
    segments; the fertile leaves taller than the sterile ones   . . . **Cryptogramma**

6   Plants not as above 7

7   Leaves strongly villose-tomentose beneath with tawny or rust colored hairs
    . . . . **Cheilanthes**

7   Leaves glabrous or glandular or sparsely pubescent, but not villose-
    tomentose 8

8   Petioles wiry, persisting for some years, the old persistent bases 1-8 cm
    long, generally more numerous than the leaves of the current season; veins
    of the ultimate leaflets not prominent; indusium various . . . . **Woodsia**

8   Petioles delicate, not persisting; veins of the ultimate leaflets conspicuous,
    ending in the tips of the marginal teeth; sori cupped in a quickly deciduous
    indusium . . . . **Cystopteris**

Asplenium L. — Spleenwort

Leaves clustered on a thickened rootstock beset with slender dark scales;
petioles slender, wiry, green to brown; sori elongate, borne on veinlets;
indusium hyaline, flap-like, attached along the vein that bears the sorus, and
opening along the side toward the mid vein of the leaf segment.

---

1   Leaf blades irregularly forked, with 2-5 narrow segments 10-20 mm long,
    the segments entire or with a few slender teeth and with a single compound
    sorus running its whole length . . . . **A. septentrionale**

1   Leaf blades uniformly pinnate, with about 7-25 opposite or offset pairs of
    leaflets, the leaflets 3-8 mm long, about as broad as long, crenulate or
    denticulate and with few to several sori . . . . **A. viride**

Asplenium septentrionale (L.) Hoffman — grass-fern — The 5 specimens seen
are from Red Canyon Rim in Daggett Co. (T. F. Wieboldt 1460A UTC), eastern
Uinta Mtns. in Uintah Co.(Goodrich 21900, 23044 25271 BRY) and Haystack
Mtn. in Summit Co. (M. D. Windham 91-203 BRY), also plotted for Moffat Co.
(TPD); crevices of ledges and other rock outcrops in quartz-rich sandstones of
the Uinta Mtn. Group; 7,400-10,250 ft.

Asplenium viride Hudson. — green spleenwort — The 9 specimens seen are
from the s. slope of the Uinta Mtns. from North Fork Duchesne drainage to
Dyer Mt.; crevices of limestone outcrops and talus; 9,800-10,500 ft. Listed as *A.
trichomanes-ramosum* L. in TPD and FNA (2: 239).

Athyrium Roth — Lady-fern

Athyrium filix-femina (L.) Roth ex Mertens — common lady-fern — Leaves
closely clustered on a short, stout rootstock that is densely clothed with dark
marcescent petiole bases, (25) 30-100 cm long or rarely longer; petiole flattened
below, 3-10 mm wide, bearing brown or blackish scales toward the base, blade
2-3 times pinnate with about 20-35 pairs of opposite or offset primary leaflets;

the larger ones 4-5 cm long, 1-5 cm wide, each with 12-20 pairs of toothed to pinnatifid mostly offset pinna; indusium often deciduous; sori borne toward the center of the ultimate leaflets. The few specimens seen are from the Uinta Mtns.; sphagnum bogs and other wet places in woods; 7,320-8,000 ft.

Cheilanthes Sw. — Lip-fern

Cheilanthes feei Moore — Fee lip-fern — Leaves few to several and congested on a short thick caudex-like rhizome, the rhizome densely beset with long narrow hyaline scarious brown scales; petioles mostly 3-10 cm long, dark purplish-brown, the blade 3-13 cm long, 1.5-4 cm wide, 3-4 times pinnate, loosely and copiously villose-tomentose beneath usually with tawny or rusty hairs, primary leaflets mostly in 6-12 opposite or offset pairs, the ultimate segments 1-1.5 mm long, rounded, the margins involute; sori borne on ends of veins just within the margins of the ultimate leaflets, the point of attachment covered by the reflexed margins that form a continuous unmodified or partly scarious indusium. The few specimens seen are from scattered locations along canyons of the Green River and flanks of the Uinta Mtns., and Harrington 1459 is from Blue Mtn., Moffat Co. (Bradley 1950), also plotted for Rio Blanco Co. in TPD; crevices of rock outcrops and cliffs; 7,000-7,400 ft.

Cryptogramma R. Brown — Rockbrake

Cryptogramma crispa (L.) R. Brown ex Hooker (*C. acrostichoides* R. Brown) — American rockbrake, parsley-fern — Leaves congested on a stout rootstock, some sterile other fertile as described in the key, 5-25 cm long or the fertile ones to 35 cm long including the petiole; old persistent petioles and scales 4-6 mm long clothing the rootstock; ultimate segments of sterile leaves to 8 mm long and 4 mm wide, those of the fertile leaves 4-12 mm long, 1-2 mm wide, sori eventually covering the whole lower surface of the revolute pinnacle. Occasional, Uinta Mtns.; crevices of ledges and cliffs and among rocks on talus slopes; mostly above 9,500 ft to timberline. Our plants belong to var. *acrostichoides* (R. Br.) C. B. Clarke. The difference in fertile and sterile leaves sets this apart from other ferns of our area.

Cystopteris Bernhardi — Bladder-fern

Cystopteris fragilis (L.) Bernhardi — brittle bladder-fern — Leaves scattered on a scaly, thickened rootstock, delicate, 5-35 0 cm long, the blades 1.5-8 cm wide, (1) 2-3 times pinnate with 8-18 pairs of opposite or offset primary leaflets; numerous, with prominent veins on the lower side, with sori borne along the veins, ultimate leaflets cupped in a deciduous indusium. Widespread; occasional in crevices of rocks in ledges and cliffs, talus slopes, boulder fields, and often but not always on the shady side of rocks; 6,500 ft to above timberline.

Pellaea Link — Cliff-brake

Rhizomes branching, forming a caudex appearing brown-woolly with dense, brown scales; leaves evergreen, firm, 1-2 times pinnate, glabrous or sparsely hairy; petioles slender and wiry, reddish-brown or blackish-purple, the bases often persistent; sori borne on vein ends just within the reflexed margins of leaflets that form a continuous indusium, the mature sporangia often conspicuously exserted. The features listed in the key for separating *P. breweri* and *P. glabella* seem to be rather difficult to apply to some specimens.

---

1 Petioles with a series of cross-grooves, the persistent bases more numerous than the active leaves; middle and lower leaflets usually bifid, often with unequal segments and mitten-like .... **P. breweri**

1 Petioles mostly without cross-grooves, the persistent bases usually fewer than the active leaves; leaflets not bifid, the lower ones often with 1-2 pairs of secondary leaflets .... **P. glabella**

Pellaea breweri D. C. Eaton — Brewer cliff-break — The 5 specimens seen are form the Bluffs on the W. side of North Fork Duchesne drainage, Uinta Mtns. and Racetrack and Water Hollow drainages of the W. Tavaputs Plateau; listed for Round Top Lookout, Moffat Co. in Bradley (1950); rock crevices of cliffs and ledges of limestone and conglomerate at 9,600-10,300 ft. Leaves appearing thinner and not as firm as in *P. glabella*.

Pellaea glabella Mettenius ex Kuhn [*P. suksdorfiana* Butters; *P. atropurpurea* (L.) var. *suksdorfiana* (Butters) Morton] — dwarf cliff-break, smooth cliff-break — Specimens seen are from canyons of the Yampa and Green Rivers in and near Dinosaur National Monument where abundant on Weber Sandstone, Dry Fork drainage of the e. Uinta Mtns. and Smith and Morehouse drainage on the n. slope, w. Uinta Mtns.; cracks of cliffs and ledges of sandstone and less often in limestone. Both ssp. *simplex* (Butters) A. Love & D. Love and ssp. *occidentalis* (E. E. Nelson) Windham are included in our area in distribution maps in FNA (2: 183). Leaves leathery compared to those of *P. breweri*.

Polystichum Roth — Holly-fern

Polystichum lonchitis (L.) Roth — mtn. holly-fern — Evergreen fern with short petioles and elongate nearly linear blades 20-50 cm long; leaflets 25-50 per side of the leaf, spinulose-serrate, each leaflet with a large tooth or lobe at the base; sorus round, with a peltate indusium. Known from among rocks in Hades Canyon, 9,500 ft, Duchesne Co. (Flowers 3215 UT).

Pteridium Gleditsch Ex Scopoli — Bracken; Bracken fern

Pteridium aquilinum (L.) Kuhn in Decken — western bracken fern — Plants arising from deep-seated elongate, branching rhizomes, often forming colonies; leaves (30) 50-200 cm tall overall, the erect petioles usually shorter than the blade, the blade 2-3 times pinnate, glabrous or sparsely hairy above, more or less densely villose or villose-puberulent beneath; primary leaflets widely spreading, often at right angles to the rachis; sori borne on the veins of the ultimate leaflets, protected by the narrowly in-rolled indusium margin of the leaf segments and on the inner side by a delicate, hyaline indusium. Locally abundant in the Uinta Mtns.; rather moist to wet places in woods; 7,000-10,000 ft. Our plants belong to ssp. aquilinum var. pubescens Underwood. In large dosages, bracken fern is poisonous to horses, cattle, and to a lesser degree to sheep. Horses can be successfully treated with thiamine but not cattle (Kingsbury 1964). Apparently bracken poisoning has been rare in the Uinta Basin area. This is probably because of the very localized distribution of this plant in the area.

Woodsia R. Brown — Woodsia

Leaves usually tufted on a thickened short rhizome that is covered with yellowish to dark brown scales and beset with few to numerous persistent, slender, wiry petioles from previous years; sori laminar, provided with an inconspicuous inferior indusium consisting of septate hairs (*W. scopulina*) or a small sac that soon ruptures irregularly to form a small disk with spreading, unequal segments (*W. oregana*), the differences in the indusium not detected without high magnification.

---

1   Leaves glabrous or rarely glandular, 7-25 cm long overall, the petiole about 1 mm thick, dark reddish brown toward the base, the blade 4-15 cm long, 1-4.5 cm wide, with 7-17 pairs of primary leaflets . . . . **W. oregana**

1   Leaves glandular and with glandless, septate hairs, at least on 1 side, 8-35 cm long, the petiole mostly 1-2 mm thick, the blade up to 22 cm long, and up to 7 cm wide, with 9-25 pairs of primary leaflets . . . . **W. scopulina**

Woodsia oregana D. C. Eaton — Oregon woodsia — The few specimens seen are from the Lake Fork and Uinta drainages of the s. slope of the Uinta Mtns., Red Canyon, Browns Park and Dinosaur National Monument; cracks of rock outcrops and sandy soil at 6,400-8,500 ft. Small plants of *W. scopulina* are much like those of *W. oregana*. The 2 apparently hybridize (IMF 1: 219).

Woodsia scopulina D. C. Eaton — Rocky Mt. woodsia — The 14 specimens seen are from the Provo River drainage east to the Whiterocks drainage on the s. slope of the Uinta Mtns. and to the W. Fork Smiths Fork of the n. slope; usually in rock crevices, talus or other rocky places in various plant communities at 8,500-12,000 ft.

PORTULACACEAE Purslane Family

More or less succulent, usually glabrous annual or perennial herbs; leaves alternate, opposite, or basal, simple, entire; flowers bisexual, regular or nearly so; sepals 2 (ours); petals separate; stamens 7-12 (ours); style simple with 5-6 stigmas; ovary superior or partly inferior, 1-celled; fruit a circumscissile capsule; seeds few-many, often smooth and shining.

Portulaca L. — Purslane

Portulaca oleracea L. — purslane, mother of millions — Glabrous, succulent annual; stems prostrate, 2-30 cm long, freely branched and radially spreading; leaves 5-40 mm long; 3-20 mm wide; flowers sessile, solitary, axillary and in small clusters at the tips of branches; sepals 2, ovate, 3-4 mm long, united below, the tube fused to the lower part of the ovary; petals 5 (4-6), yellow, about equal to the sepals; stamens 5-12; ovary partly inferior, 1-chambered; the stigmas 4-6; capsule 4-9 mm long, circumscissile below the persistent sepals; seeds black. Common garden weed and in other places of cultivation, perhaps not persisting without cultivation; not expected over 7,000 ft; May-Oct. After weeding, the succulent stems are capable of taking root before they dry. Successful weeding requires removing all plant parts and placing them where they cannot take root again. Apparently the seeds remain viable in the soil for many years.

POTAMOGETONACEAE Pondweed Family

Potamogeton L. — Pondweed

Aquatic in shallow or sometimes deep ponds and lakes and slow-moving reaches of streams, perennial, rhizomatous herbs; leaves alternate or some opposite, simple, entire or minutely toothed; stipules prominent, but sometimes deciduous; flowers bisexual, sessile or subsessile in whorls, on axillary peduncles; perianth inconspicuous, of 4 oval, short-clawed segments; stamens 4, fused with the claws of the perianth segments; pistils 4, free, each 1-chambered; the short style persistent as a beak on the achenes or lacking. The stipules are often referred to in the keys. They are often best observed on younger leaves as they often quickly shred into thread-like fibers and fall away. The genus is difficult for several reasons: immature plants and those that lack flowers or fruit are almost impossible to key to species, and even with these parts both submersed and floating leaves are important for identification (IMF 6: 24).

---

1 Leaves 0.2-5 mm wide, linear to filiform, all submerged; peduncles slender or short; spikes usually interrupted with well-spaced whorls of flowers or less than 1.5 cm long; leaves or stipules sheathing 2

1 Some of the leaves over 5 mm wide, all submerged or more often some floating; peduncles comparatively stout; spikes rather densely flowered throughout and not much interrupted, 1-6 cm long, submerged or emerged; leaves not sheathing; stipules not sheathing or weakly so in *P. crispus* . . . 6

2 Leaves sheathing at the base, the blade diverging from the stem well above the point of attachment (node), the sheath 5-30 mm long; stipules fused with the sheathing part of the leaves and exserted beyond the sheath as a ligule; spikes 10-50 mm long, usually with interrupted or well-spaced whorls or flowers; peduncles 1-15 (25) cm long. 3

2 Leaves not sheathing; stipules free from the leaves and sheathing the stems, the sheath 3-15 mm long; spikes 1-15 mm long 4

3 Style lacking, the achenes beakless, the stigma persisting on the achene as a small flat, wart-like disk; leaves 0.2-5 mm wide, 1-7 nerved, obtuse to minutely apiculate; sheaths 0.5-2 cm long . . . . *P. filiformis*

3 Style persisting on the achenes as a beak to 0.5 mm long; leaves to 0.8 mm wide, 1-3 nerved, mostly long tapering; sheaths 1-3 cm long . *P. pectinatus*

4 Spikes 1-5 mm long; peduncles 3-15 (20) mm long; sheaths of stipules 5-15 mm long; achenes with an irregularly toothed or wavy dorsal ridge . . . . .
. . . . *P. foliosus*

4 Spikes 6-20 mm long; peduncles (4) 10-50 (80) mm long; sheaths of stipules 3-8 mm long or lacking; achenes with an obscure dorsal keel 5

5 Stems simple or sparingly branched . . . . *P. friesii*

5 Stems usually much branched at least distally . . . . *P. pusillus*

6 Leaves sessile or the blades gradually tapering to short petioles, the petioles not over 5 mm long, all submerged, or if some floating then the floating ones only slightly if at all different from the submerged ones 7

6 Floating and sometimes submerged leaves conspicuously divided into expanded blades and elongate petioles, the petioles over 5 mm long . . . 11

7   Leaves all sessile, auriculate-clasping at the base, or if not clasping then
    wavy and finely serrate, all submerged, linear or lanceolate; stems not
    reddish. 8

7   At least the upper leaves usually short-petiolate, sometimes all sessile, but
    not clasping, not wavy or serrate, some floating or all submerged, more or
    less narrowly elliptic, linear or narrowly lanceolate. 10

8   Leaves finely serrate, linear or nearly so, merely sessile or slightly clasping,
    3-12 mm wide, 2 or more pairs of opposite ones often present; stipules 0.3-
    0.8 cm long, more or less sheathing but soon shredding into filiform fibers;
    spikes 1-2 cm long, submerged . . . . *P. crispus*

8   Leaves entire, conspicuously auriculate-clasping, all alternate; spikes 1.5-4
    cm long, elevated just above the surface of the water 9

9   Stems usually zigzagged; leaves oblong-lanceolate, (5) 10-25 (35) cm long,
    (10) 20-30 mm wide; stipules 3-10 cm long, rigid, usually persistent;
    peduncles 10-30 (60) cm long; achenes 4-5 mm long . . . . *P. praelongus*

9   Stems not zigzagged; leaves lanceolate to ovate-lanceolate, 1.5-10 cm long,
    5-20 mm wide; stipules 1-2 cm long, soon shredding into fibers; peduncles
    1.5-25 cm long; achenes 2.5-3.5 mm long . . . . *P. richardsonii*

10 Leaves (5) 7-15 (20) mm wide; stipules (1) 1.5-2.5 (4) cm long; herbage often
    reddish or reddish brown especially when out of water . . . . *P. alpinus*

10 Leaves (10) 20-40 (50) mm wide; stipules (2.5) 3-8 cm long; herbage not
    particularly reddish; plants often with floating strongly petiolate leaves
    with expanded blades . . . . (see *P. gramineus* and *P. illinoensis* below)

11 Submerged leaves apparently reduced to bladeless petioles, 10-20 cm long,
    1-2 mm wide; floating leaves lance-ovate to ovate-elliptic, often cordate at
    the base, (3) 5-10 cm long, (1.5) 2.5-5 (6.5) cm wide, long-petiolate; stipules
    4-10 cm long, strongly fibrous; achenes 3-5 mm long . . . . *P. natans*

11 Submerged leaves expanded into distinct blades, the blades (1) 3-50 mm
    wide; broadly linear to ovate, sessile or petiolate 12

12 Submerged leaves sessile, (1) 3-10 (12) mm wide, with 3-9 veins; stipules 0.5-
    3 cm long; floating leaf blades (1.5) 2-5 (7) cm long, 1-2 (2.5) cm wide;
    achenes 2-2.8 mm long . . . . *P. gramineus*

12 Submerged leaves subsessile to long-petiolate, with 7-15 (19) veins; if
    subsessile or short-petiolate then (10) 20-40 (50) mm wide; stipules (2.5) 3-
    8 cm long; floating leaf blades (3) 5-12 cm long, (1.5) 2-6 cm wide; achenes
    3-4 mm long 13

13 Submerged leaf blades subsessile or with petioles to about 4 cm long, the
    blades (10) 20-40 (50) mm wide, elliptic to oblong-elliptic or oblanceolate;
    floating leaf blades (when present) 2-6 cm wide, with petioles 2-9 cm long;
    achenes not reddish . . . . *P. illinoensis*

13 Submerged leaf blades on petioles 2-10 (13) cm long, 9-20 (30) cm long, 1-2
    (3.5) mm wide, linear-lanceolate to lance-elliptic; floating leaf blades (1.5)
    2-4 cm wide, on petioles up to 20 cm long; achenes reddish in age *P. nodosus*

Potamogeton alpinus Balbis. — northern or, reddish pondweed — Specimens
seen are from the Uinta Mtns., confluence of the Green and Yampa Rivers, and
near Neola, to be expected elsewhere; ponds and slow moving streams; 5,060-
10,100 ft.

Potamogeton crispus L. — curly-leaved pondweed — Introduced from Europe, naturalized; the one specimen seen (Welsh and Moore 18694) is from Daggett Co.

Potamogeton filiformis Persoon [*Stuckenia filiformis* (Persoon) Borner] — fine-leaf pondweed — Widespread; occasional; aquatic; 5,400-8,200 ft. Most of our plants belong to var. *occidentalis* (J. W. Robbins) Morong.

Potamogeton foliosus Rafinesque — leafy pondweed — The few specimens seen are from widely scattered locations; 7,600-8,600 ft. Most or perhaps all plants of our area belong to var. *foliosus.*

Potamogeton friesii Ruprecht — Fries pondweed — Known from Summit Co. (IMF 6: 33 & AUF 939).

Potamogeton gramineus L. — variable-leaf pondweed — The 5 specimens seen are from the Uinta Mtns.; 8,200-10,300 ft. Extremely variable, forming hybrids with nearly any pondweed it comes in contact with and especially with *P. nodosus* and *P. illinoensis.* See IMF (6: 39) for further discussion.

Potamogeton illinoensis Morong. — Illinois pondweed — No specimens seen. The distribution map in FNA excludes our area.

Potamogeton natans L. — floating-leaf pondweed — One specimen (Goodrich 15270) is from 9,700 ft in Ashley Creek drainage in the Uinta Mtns, Uintah Co., listed for Duchesne Co. (AUF 940)

Potamogeton nodosus Poiret (*P. americanus* C. & S.; *P. fluitans* Roth.) — longleaf pondweed — Known from along the Green River at the confluence with the Yampa River (N. Holmgren-444 DINO), near Ouray (Folks 137 UTC), and Stewart Lake (Jensen & Dargan 153 UTC); 4,700-5,100 ft. See IMF (6: 37-39) for discussion of nomenclatural problems.

Potamogeton pectinatus L. [*Stuckenia pectinata* (L.) Borner] — fennel-leaf pondweed, sago-pondweed — The 4 specimens seen are from Uintah Co. from 4,800-7,800 ft. Listed for Summit in AUF (940), Harrington 1598 is reported for Irish Canyon, Moffat Co. by Bradley (1950). Expected to be widespread in our area; aquatic and stranded on mud of drawdown basins. Seeds of this species are highly selected by ducks (FNA 22: 71).

Potamogeton praelongus Wulfen — whitestem pondweed — Listed for Duchesne Co. (AUF 940). The one specimen seen (S. L. Welsh and Neese 19845 BRY) is from Lily Lake in the Yellowstone Drainage at 9350 ft. Plotted for Wasatch Co., Utah and Garfield Co., Colorado in TPD. The distribution map in FNA (22: 69) excludes our area.

Potamogeton pusillus L. (*P. berchtoldii* Fieber in Berchtold) — baby pondweed — The 2 specimens seen are from Strawberry Reservoir (Jensen & Dargen 89 UTC) and E. McKee Draw n. of Vernal (Goodrich 21888), also Harrington 3706 is reported for Blue Mt. (Bradley 1950); 7,500-8,160 ft. Our plants belong to ssp. *pusillus.*

Potamogeton richardsonii (Benn.) Rydberg [*P. perfoliatus* L. ssp. *richardsonii* (Benn.) Hulten] — Richardson pondweed — Specimens seen from are from East Greens Lake (S. Goodrich 24347), Sheep Creek Lake (S. Goodrich 22383), and Pelican Lake (E. Neese 14637) at 4,800-9,350 ft elevation; also listed for Summit, Co. in AUF (941). The distribution map in FNA (22: 69) excludes our area.

PRIMULACEAE Primrose Family

Herbs, often scapose, leaves simple, basal, alternate, or opposite; flowers bisexual, radially symmetrical, 5 (4-9) merous; stamens equal in number to and opposite with the petals; pistil 1; ovary superior or inferior in a few species, with 1 style; fruit a capsule with free central placentation.

1   Leaves cauline, numerous, overlapping, opposite below, alternate above; flowers axillary .... *Glaux*

1   Leaves all basal; flowers in a terminal umbel 2

2   Flowers 3-5 mm long; plants annual .... *Androsace*

2   Flowers at least 15-28 mm long; plants perennial 3

3   Flowers nodding at anthesis; corolla lobes strongly reflexed; anthers exserted .... *Dodecatheon*

3   Flowers erect; corolla lobes spreading but not reflexed; anthers included .... *Primula*

Androsace L. — Rock-jasmine

Low scapose annuals; herbage often puberulent with simple or branched, sometimes gland-tipped hairs; leaves in a basal rosette; flowers 2-numerous, in umbels subtended by an involucre of small bracts; calyx top-shaped to hemispheric, 5-lobed, the tube becoming chartaceous in fruit; corolla funnel-form to bell-shaped, with a short tube, with 5 spreading to reflexed lobes; capsule thin-membranous.

1   Leaves abruptly narrowed and petiolate; corolla less than 3 mm long, the lobes reflexed; calyx hemispheric, about 2 mm long, not keeled, the lobes 3-veined; plants glabrous or sparingly glandular puberulent above; seeds yellow, 0.2-0.3 mm long .... *A. filiformis*

1   Leaf blades tapering to the base, not distinctly petiolate; corolla often over 3 mm long, the lobes spreading to erect; calyx more or less turbinate, keeled, the lobes 1-veined; plants generally puberulent; seeds dark brown, 0.7-1 mm long. 2

2   Bracts subtending the umbel not more than 3 times as long as wide, abruptly acute at the apex; calyx lobes mostly erect, the tips incurved at maturity; corolla scarcely exceeding the calyx tube .... *A. occidentalis*

2   Bracts subtending the umbel often more than 3 times as long as wide, mostly long-tapered at the apex; calyx lobes erect to spreading, the tips not incurved; corolla somewhat longer than the calyx tube . *A. septentrionalis*

Androsace filiformis Retzius — slender rock-jasmine — Known from above Current Creek Dam site in Wasatch Co. (J. D. Brotherson 1782 BRY). Plotted for Rio Blanco Co. in TPD. Listed for subalpine wet places (Weber 1984).

Androsace occidentalis Pursh — western rock-jasmine — Bradley (1950) cited the following specimens: Harrington 3916 CS from Yampa Canyon near the mouth of Hells Canyon at 5,200 ft. Plotted for Moffat and Rio Blanco Cos. (TPD) and listed for open grassy slopes and ledges and canyonsides (Weber 1984).

Androsace septentrionalis L. — pygmy-flower rock-jasmine — Widespread and common in pinyon-juniper, sagebrush, ponderosa pine, Douglas-fir, aspen, lodgepole pine, aspen and conifer parklands, spruce-fir and alpine communities in moist and dry places from 7,000-12,500 ft; June-Aug. Quite variable in height, length of scapes, length of pedicels, and other features (AUF 604).

Dodecatheon L. — Shooting-star

Perennial scapose herbs from short rhizomes bearing fleshy fibrous roots; leaves basal, petiolate, entire to obscurely toothed; scapes 1 or 2; flowers (1) few or many, nodding on slender pedicels of an involucrate umbel; calyx lobes reflexed at flowering, erect in fruit; corolla lobes strongly reflexed; stamens opposite the corolla lobes and united around the style, the connectives between the anthers usually deep purple; style thread-like, slightly exceeding the stamens, the stigma globose or nearly so; capsule ovoid.

---

1   Filaments united and forming a column that is exserted for 1.8-3.6 mm beyond the corolla; flowers 5-merous; stigma about as wide as the style . . . . ***D. pulchellum***

1   Filaments free, not forming a column, not exserted beyond the corolla or exserted up to 1.2 mm; flowers 4-merous; stigma typically broader than the style . . . . ***D. alpinum***

Dodecatheon alpinum (A. Gray) Greene — alpine shooting-star — Occasional; Uinta Mtns.; bogs, meadows, seeps, springs, along streams, and around edges of lakes and ponds; (7,200) 9,400-11,500 ft; June-Aug.

Dodecatheon pulchellum (Rafinesque) Merrill (*D. pauciflorum* Greene; *D. radicatum* Greene). With 2 vars. as follows:

1 Leaves 4-15 cm long, 0.5-1.7 (2.5) cm wide; anthers 3-4.5 (5) mm long . . . . **var. *pulchellum***

1 Leaves 10-48 cm long, 1.5-8 cm wide; anthers 4.5-6.5 mm long . . . . **var. *zionense***

Var. *pulchellum* — beautiful shooting-star — Common across the Uinta Mtns., apparently rare on the Tavaputs Plateau; meadows and along streams, tolerant of but not restricted to alkaline or saline places; 5,000-10,000 ft; May-July.

Var. *zionense* (Eastwood) Welsh — Zion shooting-star — Seeps, hanging gardens and drip-lines in Desolation Canyon and sandstone canyons of the Eastern Uinta Mtns. Our plants are referable to subvar. *huberi* S. L. Welsh — Huber shooting-star — Vars. *pulchellum* and *zionense* overlap in all features listed in the key. However, the overall trend in size of features of the two seems to warrant recognition of 2 taxa.

Glaux L. — Saltwort; Sea Milkwort

Glaux maritima L. — common sea milkwort — Perennial herbs from rhizomes; stems 3-30 cm long, erect, simple or branched; herbage glabrous and glaucous; leaves entire, somewhat succulent, sessile or nearly so, mostly linear, occasionally oval, 4-25 mm long, 1-7 (10) mm wide; flowers solitary and sessile in the leaf-axils; calyx petal-like, bell-shaped, 3-5 mm long, pink or white, the 5 lobes ovate; petals lacking; stamens 5, arising at the base of the ovary and alternate with the calyx lobes; capsule ovoid, 2-3 mm long. Occasional in Daggett and Moffat Cos., apparently rare in Duchesne Co., no specimens seen from Uintah or Rio Blanco Cos.; wet or moist alkaline or saline meadows, seeps, along ditches and streams; up to about 7,000 ft; June-Sept.

Primula L. — Primrose

Perennial scapose glabrous herbs; leaves simple, entire or nearly so; flowers borne in umbels, 5-merous; calyx and corolla united, corolla salverform, the lobes emarginate to bilobed; stamens attached in the upper 1/3 of the corolla tube, included.

---

1   Leaves 6-30 cm long, not glaucous beneath; calyx 8-15 mm long, not glaucous; corolla tube 8-15 mm long, the lobes 5-13 mm long; plants stipitate-glandular throughout, malodorous, somewhat fleshy, widespread . . . . *P. parryi*

1   Leaves 1-4 cm long, glaucous beneath with a whitish crust-like bloom; calyx about 5-6 mm, glaucous like the lower surface of the leaves; corolla tube 6-8 mm long, yellow, the lobes about 5 mm long, blue or pale lavender; plants not or inconspicuously stipitate-glandular, rare . . . . *P. incana*

Primula incana M. E. Jones — silvery primrose — Specimens with well marked locations are from Sheep Creek Park, wet meadow, 8,600 ft. Garrett 6143 (UT) is rather ambiguous as to location but is likely from the same locality; June-Aug. Widespread in North America, rare in Utah.

Primula parryi A. Gray — Parry primrose — Uinta Mtns., occasional in rocky places, near springs or streams, but also in or at the edge of boulder fields away from water, common in late persisting snowbeds where snow depth exceeds 3 ft most of the winter; mostly near or above timberline, occasionally down to 8,000 ft; June-Sept.

PYROLACEAE Wintergreen Family

Suffrutescent or herbaceous perennials; leaves simple, alternate, opposite or appearing whorled, evergreen or much reduced and lacking chlorophyll; flowers bisexual, regular or irregular; sepals and petals 4 or 5, more or less distinct; stamens twice as many as petals; pistil 1; ovary superior, 4 or 5 celled; style 1; fruit a capsule.

---

1   Leaves evergreen, not strictly basal, whorled; the blades about 3 times as long as wide, serrate; flowers corymbose; styles short, inconspicuous . . . . *Chimaphila*

1   Leaves deciduous, mostly basal not whorled; the blades mostly less than 2 times as long as wide, entire or serrulate; flowers racemose or solitary; style usually conspicuous 2

2   Flowers solitary, terminal; petals 7-12 mm long . . . . *Moneses*

2   Flowers 2-several, racemose; petals 3-7 mm long . . . . *Pyrola*

Chimaphila Pursh — Pipsissewa

Chimaphila umbellata (L.) Bart. — prince's pine, common pipsissewa — Low shrubs from creeping rootstocks, decidedly woody at the base, 10-25 (30) cm tall; leaves whorled, thick, evergreen, 3-9 cm long, about 5-15 mm wide, dark green above, pale beneath, the margins serrate; petals separate, 5-6 mm long, white to pink; fruit a capsule, 5-7 mm in diameter. Occasional to locally abundant in coniferous forests and mixed coniferous forest-aspen stands in the Uinta Mtns. where especially common on the north slope from Eagle Creek and eastward in lodgepole pine stands where stand replacing fires have been common; 7,000-9,000 ft; June-Aug. Our plants belong to var. *occidentalis* (Rydberg) Blake.

Moneses Salisbury.

Moneses uniflora A. Gray (*Pyrola uniflora* L.) — single delight — Plants perennial, from slender creeping rhizomes; leaves basal or on the lower 1 of the stem; scapes or stems 3-15 cm tall, with (1) 2 bracts near the middle; flowers solitary and terminal; sepals about 1/4 as long as the petals, usually reflexed; petals about 7-12 mm long, separate, white; stamens 10; style 2-4 mm long, straight; fruit a capsule, nearly globose, 6-7 mm thick. Occasional across the west end of the Uinta Mtns. and east to Whiterocks River, Specimens also seen from Trout Creek and Water Hollow of Wasatch Co., and Slab Canyon of the West Tavaputs Plateau; 8,100-10,550 ft; spruce-fir communities, often along streams or in seeps and springs; July-Aug.

Pyrola L. — Shinleaf; Wintergreen

Perennial, glabrous herbs from slender rhizomes; leaves simple, entire or serrate, all basal or crowded toward the base of the plant; scape or stem with 1-3 bracts; inflorescence a raceme; calyx persistent, united, 5-lobed; petals 5, separate, deciduous; stamens 10, the anthers awned; ovary superior, 5-celled; style straight or strongly curved; fruit a globose capsule.

1 Styles not over 2 mm long, straight, without a collar or ring below the peltate-lobed stigma; petals 2-4 mm long; racemes not secund ... *P. minor*
1 Style 3-8 mm long, petals 4-7 mm long 2
2 Racemes secund; style 3-4 mm long, straight, without a collar or ring below the peltate-lobed stigma; petals white or greenish-white ...... *P. secunda*
2 Racemes not secund; style curved or bent to 1 side, with a collar below the lobed stigma 3
3 Petals pale yellowish; style 3-6 mm long; leaf blades 1-2.5 (3.5) cm long ..
.... *P. chlorantha*
3 Petals pinkish or purplish-red; style 5-8 mm long; leaf blades (1) 3-8 cm long
.... *P. asarifolia*

Pyrola asarifolia Michaux (*P. uliginosa* Torrey) — liver-leaf wintergreen — Occasional across the Uinta Mtns.; one specimen seen from the E. Tavaputs Plateau; wet places, mostly in woods; 7,000-9,000 ft; late June-Aug.

Pyrola chlorantha Swartz (*P. virens* Schreber in Schweigger) — green wintergreen — Occasional across the Uinta Mtns.; moist places, often in woods along streams; 7,300-9,600 ft; June-Aug.

Pyrola minor L. — lesser wintergreen — Occasional across the Uinta Mtns. along streams in coniferous forests; 7,300-10,000 ft; June-Aug.

Pyrola secunda L. [*Orthilia secunda* (L.) House] — one-sided wintergreen — Common in the Uinta Mtns.; one specimen seen from Blue Mt., and 1 from Bruin Point, W. Tavaputs Plateau; woods, often along streams; 7,300-11,000 ft; late June-Aug.

RANUNCULACEAE Buttercup Family

Annual or perennial herbs; leaves all basal or alternate, simple or compound, entire to variously toothed or lobed, without stipules, the petiole often dilated at the base; flowers unisexual or bisexual, regular or irregular; sepals 3-15, often deciduous; petals lacking or present, separate; stamens numerous, rarely few; ovaries superior, l-numerous; fruit of achenes, follicles, or berries.

1	Leaves simple, entire, or crenate 2
1	Leaves deeply dissected to compound 4
2	Plants annual; flowers not at all showy, petals and sepals 2-3 mm long; inflorescence slender, spike-like in fruit .... *Myosurus*
2	Plants perennial; petals 3 mm long or longer; inflorescence a head or head-like 3
3	Leaves 3-4.5 cm wide; flowers white or cream; fruit a head of follicles .... *Caltha*
3	Leaves less than 3 cm wide; flowers yellow; fruit of achenes .. *Ranunculus*
4	Petals or sepals yellow 5
4	Petals and sepals not yellow 7
5	Plants vines .... *Clematis orientalis*
5	Plants not vines 6
6	Petals with spurs 10-22 mm long .... *Aquilegia*
6	Petals without spurs .... *Ranunculus*
7	Plants aquatic .... *Ranunculus aquatilis*
7	Plants not aquatic 8
8	Leaves opposite or whorled 9
8	Leaves alternate or basal 11
9	Plants vines .... *Clematis*
9	Plants not vines 10
10	Leaves ternately or palmately divided or compound .... *Anemone*
10	Leaves pinnately compound .... *Clematis hirsutissima*
11	Stems with a solitary flower .... *Trollius*
11	Flowers not solitary... 12
12	Leaves palmately lobed or divided 13
12	Leaves ternately or pinnately compound; flowers regular, white, cream, or pale blue 15

13 Flowers irregular, mostly dark blue or purple; fruit of 2-5 follicles 14

13 Flowers with 5 similar sepals; petals lacking; stamens numerous, white; fruit a cluster of achenes . . . . **Trautvetteria**

14 Upper sepal spurred . . . . **Delphinium**

14 Upper sepal hooded . . . . **Aconitum**

15 Flowers showy, the petals and petaloid sepals mostly over 5 mm long, the petals with slender spurs; fruit of 5 follicles . . . . **Aquilegia**

15 Flowers hardy showy, petals and sepals (if present) 2-5 mm long, petals not spurred; fruit not of follicles 16

16 Ultimate segments of the leaves 2-10 cm long 1-8 cm wide; fruit a white or red berry . . . . **Actaea**

16 Ultimate segments of leaves smaller; fruits greenish achenes . **Thalictrum**

Aconitum L. — Monkshood

Aconitum columbianum Nuttall in Torrey & A. Gray — monkshood — Plants perennial, 60-150 cm tall; leaves palmately parted into 3-5 divisions, these cleft or toothed, the blade 5-15 cm wide; flowers in a simple or few-branched raceme, mostly blue, occasionally white, bisexual, irregular; sepals 5, petaloid and equal to or longer than the 2-5 petals, the upper sepal with a 1-2 cm long hood. Widespread; occasional along streams and around seeps and springs; 7,000-9,500 ft; late June-early Sept. Our plants belong to var. *columbianum*.

Actaea L. — Baneberry

Actaea rubra (Aiton) Willdenow (*A. arguta* Nuttall) — baneberry — Plants perennial, 20-80 cm tall, from thick rootstocks; leaves-ternately decompound; inflorescence a terminal raceme; sepals 3-5, quickly deciduous; petals 4-10, shorter than the stamens; fruit a red or white, shiny berry, 6-8 mm long. Strawberry Valley, Uinta Mtns. and Tavaputs Plateau at isolated locations were locally common along streams and other moist places, often in woods; 7,000-8,500 ft; late May-early July.

Anemone L. — Anemone; Windflower

Perennial herbs with erect scape-like stems; basal leaves long petiolate, palmately parted or divided; stem leaves in a pair or a whorl forming an involucre below the inflorescence; sepals petaloid, 4-20; petals none; stamens numerous; fruit a head of compressed, pubescent, often woolly achenes.

---

1 Sepals over 2 cm long; styles to 20-35 mm long, plumose . . . . **A. patens**

1 Sepals not over 2 cm long; styles not over 5 mm long, not plumose 2

2 Blades of basal leaves 1.5-3 cm wide with rounded lobes . . . . **A. parviflora**

2 Blades of basal leaves (2) 3-10 cm wide, dissected into linear or nearly linear lobes . . . . **A. multifida**

Anemone multifida Poiret (*A. cylindrica* A. Gray misapplied; *A. globosa* (Torrey & Gray) Nuttall ex Pritzel) — cutleaf anemone, globeflower — Common from Strawberry Valley and across the Uinta Mtns., rarely collected from the

490 RANUNCULACEAE • Buttercup Family

Tavaputs Plateau; in numerous plant communities including sagebrush-grass, aspen and coniferous forests, also alpine ridges and talus slopes, in wet or dry places; 7,300-11,200 ft; July-Aug. Apparently with 3 vars. in our area:

- 1 Fruiting style more or less straight, 1.5-1.8 mm long ...... var. *multifida*
- 1 Fruiting style recurved, hooked, or coiled 2
- 2 Fruiting style 1.5-2 mm long, mostly hooked; plants apparently widespread
  .... var. *tetonensis* (Porter ex Britton) C. L. Hitchcock
- 2 Fruiting style 4-6 mm long, hooked and often coiled at the tip; plants of Duchesne Co. (IMF 2A: 108) var. *stylosa* (A. Nelson) B. F. Dutton & Keener

Anemone parviflora Michaux — small-flowered anemone — The few specimens seen are from the S. Fork of Rock Creek on the south slope, and head of Mill Creek of the north slope of the Uinta Mtns.; moist meadows and openings in spruce forests; July. Our specimens are glabrous or nearly so and those of *A. multifida* are usually densely villose.

Anemone patens L. [*A. nuttalliana* de Candolle; *Pulsatilla ludoviciana* (Nuttall) A. Heller; *P. patens* (L.) Miller] — pasque-flower, wild crocus, lions-beard — Most of the specimens seen are from the Uinta Mtns. e. of the Whiterocks drainage of the s. slope and from Daggett Co.; aspen, sagebrush-grass and lodgepole pine communities; mostly about 8,500 ft; and 2 alpine specimens (Lambert & Woods 20 USUUB from Yellowstone Pass at 11,500 ft) and another from Henrys Fork, apparently rare on the E. Tavaputs Plateau; May-June. Our plants are more or less equal to those of Siberia and as such belong to var. *multifida* Pritzel. At the species level, our plants would be referable to *A. nuttalliana* de Candolle (Hitchcock et al. 1964).

Aquilegia L. — Columbine

Perennial herbs; leaves 2-3 times ternate; flowers bisexual, regular, showy; sepals 5, petaloid; petals 5, the blade little if at all larger than the sepals, with a hollow spur that extends back through the sepals; stamens numerous; fruit of follicles. Small plants of the Cathedral Bluffs have glabrous, glaucous leaves and small leaflets as in *A. barnebyi* and short, yellow or pinkish tinged spurs as in *A. flavescens*. These plants might belong to an undescribed taxon.

- 1 Flowers white or blue 2
- 1 Flowers pinkish, reddish, or yellowish at least in part 4
- 2 Stamens surpassing the petal blades by 13-24 mm plants common, widespread ....*A. coerulea*
- 2 Stamens surpassing the petal blades by 1-5 mm 3
- 3 Plants usually glandular, known from Dinosaur National Monument ....*A. micrantha*
- 3 Plants not glandular, known from the West Tavaputs Plateau ....*A. scopulorum*
- 4 Leaves not glandular... 5
- 4 Leaves glandular 6
- 5 Ultimate median leaflets (20) 28-45 mm long, (20) 28-55 mm wide; leaflets not glaucous; spurs 10-17 (20) mm long, usually yellowish; plants of the western Uinta Mtns. .... *A. flavescens*

5 Ultimate median leaflets 7-15 (22) mm long, 7-14 (20) mm wide; leaflets glaucous on both sides; spurs 17-22 (28) mm long, pinkish or reddish; plants of the Tavaputs Plateau . . . . *A. barnebyi*

6 Spurs 25-27 mm long, reddish pink; sepals reddish pink, 12-15 mm long; plants from Firewater Canyon tributary to Desolation Canyon *A. atwoodii*

6 Spurs 10-25 mm long, scarlet, magenta, or less commonly pinkish; sepals colored as the spurs 7

7 Sepals erect, 7-11 (15) mm long; petal blades 6-11 mm long; spurs 10-15 (20) mm long, sometimes with a greenish or yellowish tip; plants of the Tavaputs Plateau . . . . *A. elegantula*

7 Sepals spreading, 11-14 mm long; petal blades 4-7 (10) mm long; spurs 15-20 mm long; plants in stream canyons of the eastern Uinta Mtns. . . . . *A. micrantha*

Aquilegia atwoodii S. L. Welsh — Atwood columbine, Firewater columbine — The 7 specimens seen are all from Firewater Canyon tributary to Desolation Canyon at 4800 ft. Perhaps this is a glandular phase of *A. barnebyi.* Included in *A. fosteri* (S. L. Welsh) S. L. Welsh in IMF (2A: 68).

Aquilegia barnebyi Munz — oil shale columbine — Endemic, infrequent across the Tavaputs Plateau from the Avintaquin drainage to the Piceance Basin; pinyon-juniper, Douglas-fir, and bristlecone pine communities, often on dry, open, shale or marl limestone slopes; 5,500-7,000 ft; mid June-early July. Graham 9456, listed as *A. chrysantha* A. Gray by Graham (1937), belongs here.

Aquilegia coerulea James in Long — Colorado columbine — Common in Strawberry Valley and across the Uinta Mtns., and Tavaputs Plateau; aspen, riparian, wet meadow, coniferous forest, and alpine tundra communities and in boulder fields; 7,500-11,700 ft; mid June-mid Aug. Plants of this species hybridize with those of *A. elegantula* and *A. flavescens* (Arnow and others 1980, IMF 2A: 66). Most of the material from the aspen belt with white or pale blue sepals is assignable to var. *ochroleuca* Hooker. Some of the high elevation plants of the Uinta Mtns. with medium to deep blue sepals belong to var. *coerulea.*

Aquilegia elegantula Greene — elegant columbine — E. Tavaputs Plateau and Range Creek of the W. Tavaputs Plateau; occasional in aspen, coniferous forest, and wet meadow communities; 7,800-9,000 ft; June-July.

Aquilegia flavescens S. Watson — yellow columbine — The few specimens seen are from N. Fork Duchesne Canyon and Hades Canyon; 7,400-9,000 ft; Aug.

Aquilegia micrantha Eastwood (*A. pallens* Payson) With 3 vars. as follows:

1 Sepals and spurs white or cream colored, sometimes suffused with pale blue or pink; spurs 20-33 mm long . . . . **var. *micrantha***

1 Sepals and spurs red, magenta, or pink; spurs 17-25 mm long. 2

2 Sepals and spurs pink or red; petal blades cream white to pale yellow; stamens 12-17 mm long . . . . **var. *loriae***

2 Sepals and spurs magenta; petal blades bright yellow; stamens 16-22 mm long . . . . **var. *grahamii***

Var. *grahamii* (S. L. Welsh & Goodrich) N. H. Holmgren & P. K. Holmgren — Graham columbine — Endemic, rare or locally common in cracks of rock outcrops and drip-lines below cliffs of Weber Sandstone in Brush Creek and Ashley Creek drainages, Uinta Mountains. Graham 10009 CM! listed by Graham

(1937) as *Aquilegia formosa* Fischer in de Candolle from a crack of a cliff above the stream in Brush Creek Gorge at 6,500 ft) belongs here.

Var. *loriae* (S. L. Welsh & N. D. Atwood) N. H. Holmgren & P. K. Holmgren — Lori's columbine — Canyons of the Green and Yampa rivers in Dinosaur National Monument were disjunct from Kane Co., Utah.

Var. *micrantha* — alcove columbine — Specimens seen are from canyons associated with the Green and Yampa Rivers in and near Dinosaur National Monument. The report by Graham (1937) for Little Brush Creek Canyon is based on a specimen (Hermann 4911 MO!) that belongs to *A. coerulea*.

Aquilegia scopulorum Tidestrom — rock columbine — Locally common on semi-barrens of Green River Formation, West Tavaputs Plateau at 7400-9420 ft. Our plants belong to var. *goodrichii* S. L. Welsh.

Caltha L. — Marsh-marigold

Caltha leptosepala de Candolle [*C. rotundifolia* (Hutch) Greene; *Psychrophila leptosepala* (de Candolle) Weber] — marsh-marigold — Perennial, glabrous, fleshy, scapose herbs, 5-40 cm tall; leaves all basal or rarely a reduced one on the stem, the blades elliptic or orbicular, 3-10 cm long, 1.5-5 cm wide, crenate, undulate, or nearly entire, the petiole about as long to twice as long as the blade; sepals 5-15, 10-20 mm long, whitish or somewhat bluish, petaloid; petals lacking; stamens numerous; fruit a head of follicles 12-20 mm long each with several seeds. Strawberry Valley and Uinta Mountains and upper E. Tavaputs Plateau (Goodrich 24454 from Bogart Canyon); common to abundant in wet and moist meadows, along streams and around seeps and springs; (8,000) 9,000-12,000 ft; June-Aug. Our plants belong to var. *leptosepala*. The upright, robust plants with multiple scapes of Strawberry Valley are strikingly different from low-growing plants of the Uinta Mtns. with single or few scapes.

Clematis L. — Clematis, Virgins-bower

Perennial herbs or vines; leaves opposite, compound, flowers unisexual or bisexual; with 4-5, petaloid sepals; petals lacking; stamens numerous, the outer ones often with dilated, petaloid filaments; fruit of many congested achenes with elongate, plumose styles.

---

1   Plants herbaceous, not vines; sepals and stamens erect, the sepals united at least near the base .... *C. hirsutissima*

1   Vines, sepals and stamens spreading, the sepals not united 2

2   Sepals white, about 1 cm long, flowers often numerous in corymbose cymes
    .... *C. ligusticifolia*

2   Sepals yellow or blue, rarely white, over 1 cm long.................... 3

3   Flowers yellow .... *C. orientalis*

3   Flowers purple, lavender or blue, rarely white 4

4   Leaves mostly biternate (with 3 primary and 3 secondary divisions), with up to 9 leaflets .... *C. columbiana*

4   Leaves with 3 leaflets, once compound .... *C. occidentalis*

Clematis columbiana (Nuttall) Torrey & A. Gray [*C. pseudoalpina* (Kuntze) A. Nelson in Coulter & Rose; *Atragene columbiana* Nuttall] — Rocky Mtn. clematis — E. and W. Tavaputs Plateaus and w. of Rock Creek in the Uinta Mtns.; mostly in Douglas-fir but also in aspen, mt. brush, and sagebrush communities; 7,000-10,050 ft; May-Aug.

Clematis hirsutissima Pursh [*Coriflora hirsutissima* (Pursh) W. A. Weber] — sugarbowls, hairy clematis — Widespread but no specimens seen from the central Uinta Mtns.; occasional in a variety of plant communities including sagebrush-grass, aspen-tall forb, riparian, and grass-forb; 7,500-9,600 ft; late May-early July.

Clematis ligusticifolia Nuttall in Torrey & A. Gray — western white clematis, western virgin-bower — Widespread; usually in riparian communities where it forms great entanglements on trees, shrubs, fences, and other upright objects; up to 7,500 ft; June-Aug.

Clematis occidentalis (Hornemann) de Candolle [*C. columbiana* (Nuttall) Torrey & A. Gray misapplied; *Atragene occidentalis* Hornemann] — western blue virgins-bower — Uinta Mtns.; mostly along drainages or canyon bottoms in aspen and coniferous forests, and occasionally in mt. brush and sagebrush communities; 7,400-8,200 ft; late May-mid June.

Clematis orientalis L. [Possibly the correct name for this plant is *C. aurea* Nelson & Macbride; *Viticella orientalis* (L.) Weber] — oriental clematis or virgins bower — Apparently introduced, cultivated, escaping; the one specimen seen (A. Holmgren & Jensen 13955 UTC) is from Green River Bridge at Jensen, plotted for Moffat Co. in TPD.

Delphinium L. — Larkspur

Perennial herbs; leaves alternate, variously palmately lobed or divided; inflorescence a terminal raceme; flowers bisexual, irregular; sepals 5, petaloid, the upper one with a spur; petals in 2 pairs, the upper pair united into a spur that is enclosed in the sepal spur, the lower pair with expanded blades and narrow claws; stamens many; fruit of follicles, these usually 3.

---

1 Stems mostly less than 50 cm tall, easily detached from the bulb-like root or cluster of roots, not ashy puberulent . . . . *D. nuttallianum*

1 Stems mostly over 50 cm tall, or if shorter then ashy-puberulent, not easily detached from the fibrous woody roots. 2

2 Stems and leaves ashy-puberulent, not glandular; flowers bright blue; sinuses of the lower petals obsolete; plants mostly growing at the lower edge of and below the pinyon-juniper belt . . . . *D. geyeri*

2 Stems not ashy-puberulent, sometimes glandular; flowers dark blue or purple; sinuses of lower petals 1-2.5 mm deep; plants of aspen and coniferous forests belts. 3

3 Rachis thinly to densely glandular-hairy; sepals blue-purple in part; stems 1-several; plants common in mtns. . . . . *D. occidentale*

3 Rachis glabrous or pubescent but not glandular; sepals blue; stems usually 1; plants apparently rare and limited to the extreme e. part of our area . . . . *D. ramosum*

Delphinium geyeri Greene — Geyer larkspur — Daggett, Duchesne, Uintah, and eastern Summit Cos.; occasional to locally common in desert shrub, sagebrush, and juniper communities, up to about 7,000 ft; late May-early June.

Delphinium nuttallianum Pritzel ex Walpers — low larkspur, twolobe larkspur — Common across our area; desert shrub, sagebrush, and pinyon-juniper communities, occasionally in openings of aspen and coniferous woods; 4,900-10,400 ft; May-June. Our plants have been referred to *D. bicolor* Nuttall, *D. dumetorum* Greene, *D. menziesii* de Candolle, and *D. nelsonii* Greene.

Delphinium occidentale S. Watson [*D. cucullatum* A. Nelson; *D. reticulatum* (A. Nelson) Rydberg] — western larkspur — Common to locally abundant in Strawberry Valley and western Uinta Mtns., w. end of W. Tavaputs Plateau, and infrequent in the eastern Uinta Mtns. in mtn. big sagebrush, tall forb, aspen, coniferous forest, and streamside communities and on open rocky slopes; 8,000-10,500 (11,000) ft; late June-Aug.

Delphinium ramosum Rydberg — mtn. larkspur — Entering the e. edge of our area in Rio Blanco Co.

Myosurus L. — Mousetail

Annual low herbs (mostly 1-7 cm tall); leaves all basal, linear to narrowly spatulate, entire or few-toothed; scapes 1-several; flowers solitary, minute (about 1-3 mm long), terminal; sepals 5, spurred at the base, the spur about equal or shorter than the blade, deciduous; petals lacking or 5, as small as the sepals; fruit an achene, the achenes numerous, crowded and forming a dense, cylindrical aggregate fruit on the short to elongating receptacle.

---

1   Beak of achenes attached at the apex of the achene and projecting beyond the achene 0.7-2 mm, spreading-ascending and giving the 0.5-2 cm long aggregate fruit a slightly bristly appearance . . . . ***M. apetalus***
1   Beak of achenes attached to the back of the achene and projecting beyond the achene 0.2-0.3 mm, erect and of the same angle as the achene; the 0.5-3 (6) cm long aggregate fruit with smooth appearance . . . . ***M. minimus***

Myosurus apetalus A. Gray (*M. aristatus* Bentham ex Hooker; *M. minimus* ssp. *montanus* Campbell) — mousetail — Specimens seen are from the eastern Uinta Mtns., Blue Mt., Cold Spring Mtn., and Douglas Mt.; edge of ponds, in ephemeral ponds, dry rocky, ephemeral creek beds, and damp drainages at 7,000-9,075 ft; June-July. See Arnow and others (1980) for a discussion of the above synonymy, nomenclature, and other taxonomic aspects of this species. Our plants belong to var. *montanus* (G. R. (Campbell) Whittemore.

Myosurus minimus L. — tiny mousetail — The 4 specimens seen are from Daggett Co.; drying mud ephemeral pools; 5,700-8,040 ft; June-Aug. Without showy flowers, this and the preceding species are rather easily overlooked. They are likely more common and widespread than current collections indicate. Mature achenes provide the most reliable means of separation of the two.

Ranunculus L. — Buttercup; Crowfoot

Annual or perennial herbs; leaves all basal or alternate, simple, entire to variously lobed, dissected or compound; flowers solitary or few; sepals 5, usually quickly deciduous; petals usually 5, but sometimes less or more, yellow (white in an aquatic species); stamens 10 or more, rarely fewer; fruit of achenes, these borne in heads.

1   Some or all of the basal leaves entire, serrate or shallowly lobed, the lobes not extending over 1/4 the length of the blade 2

1   Basal leaves divided over 1/2 the way to the base; achenes glabrous. 7

2   Leaves all simple, entire or crenate-serrate, but not lobed; achenes mostly glabrous. 3

2   At least some of the stem leaves lobed to dissected or compound; achenes pubescent 5

3   Plants without stolons; basal leaf blades (1) 3-15 cm long; petals 7-11 mm long . . . . *R. alismaefolius*

3   Plants with well-developed strawberry-like stolons; leaf blades to 3.5 cm long; petals 2-5 mm long 4

4   Leaf blades all entire, some usually 2 or more times longer than wide; petals 2-4 mm long . . . . *R. flammula*

4   Leaf blades usually crenate, about as wide as long; petals 4-5 mm long . . .
. . . . *R. cymbalaria*

5   Blades of basal leaves entire or 3-lobed, not crenate, mostly 2-4 times as long as wide; petals (6) 8-15 mm long; plants glabrous except for the achenes
. . . . *R. glaberrimus*

5   Blades of basal leaves crenate to crenate-serrate, mostly about as wide as long; petals various; plants glabrous or pubescent 6

6   Cauline leaf segments 3-10 mm wide or wider; basal leaf blades cuneate to truncate but not cordate; petals mostly 5-7 mm long; plants common and widespread . . . . *R. inamoenus*

6   Cauline leaf segments 1-3 mm wide; basal leaf blades truncate to cordate; petals 8-10 mm long or lacking; plants rarely collected in our area . . . . . .
. . . . *R. cardiophyllus*

7   Plants aquatic or sometimes stranded on mud of drying pools. 8

7   Plants not aquatic 12

8   Petals white; submerged leaves dissected into filiform segments . . . .
. . . . *R. aquatilis*

8   Petals yellow; submerged leaves various. 9

9   Plants annual, often growing in mud; stems fistulose, often over 3 mm thick, not rooting at the nodes; leaves sometimes over 3 cm wide; achenes with short, blunt beaks, in a cylindrical cluster . . . . *R. sceleratus*

9   Plants perennial, aquatic; stems not fistulose, less than 3 mm thick, freely rooting at the nodes; leaves less than 3 cm wide; achenes with sharp-pointed beaks, in a ovoid or subglobose cluster 10

10   Lobes of leaves 3, rounded, not extending over 1/2 the way to the base of the blade, entire or with 1-2 shallow teeth; petals 3-4.5 mm long; beak of achene 0.15-0.25 mm long; receptacles glabrous . . . . *R. hyperboreus*

10   Lobes of leaves 3 or more, cut to near the base of the blade, often again lobed or divided; receptacles hairy. 11

11 Leaf blades 2-5 cm long or more; petals 7-12 mm long; achenes 1.8-2 mm long, with rugose sides and corky-thickened margins, beak 1-2 mm long. . . . . *R. flabellaris*

11 Leaf blades 1-2 cm long; petals 3-7 mm long; achenes 1-1.5 mm long, smooth-sided, without thickened margins, beak 0.6-0.8 mm long . . . . *R. gmelinii*

12 Plants scapose, more or less finely tomentose, 2-8 cm tall; sepals persistent after anthesis; achenes tomentose, persistent and forming a bur-like aggregate fruit . . . . *R. testiculatus*

12 Plants not as above 13

13 Basal leaves distinctly compound with 1 or more well developed rachis segments; receptacles hairy; achenes including beaks averaging over 3.5 mm long 14

13 Basal leaves simple, sometimes divided to the base with the divisions or lobes again variously lobed but without a rachis; receptacles glabrous or hairy; achenes smaller except in *R. acriformis* and *R. uncinatus* 16

14 Style more than 1 mm long even in bud; achene beak 2-4 mm long; basal leaves with 1 or 2 rachis segments, the lateral leaflets often sessile or nearly so . . . . *R. orthorhynchus*

14 Style less than 1 mm long at flowering; achene beak to about 1.6 mm long; basal leaves with only 1 rachis segment, some of the lateral leaflets usually with petiolules 15

15 Stems slightly if at all fistulose; petals 6-17 mm long, generally over 8 mm long, or if shorter then numerous; sepals erect to spreading; beak of achene recurved . . . . *R. repens*

15 Stems mostly fistulose; petals 4-8 mm long, mostly 5; sepals usually reflexed; beak of achene straight or nearly so . . . . *R. macounii*

16 Plants 20-40 (60) cm tall 17

16 Plants 2-20 cm tall; achenes 1-3 mm long 21

17 Petals 5.5-18 mm long. 18

17 Petals 0.5-5 mm long. 20

18 Herbage spreading hirsute; achenes including beaks over 3 mm long; plants known from below 9,000 ft . . . . *R. acriformis*

18 Herbage glabrous or sometimes sparsely pubescent; achenes including beaks less than 3 mm long; plants from above 9,000 ft 19

19 Leaves cut to the base, with more than 15 linear segments; beak of achene 1.2-2.5 mm long . . . . *R. adoneus*

19 Leaves not cut entirely to the base, the lobes fewer than 15 and not linear; beak of achene 0.8-1.5 mm long . . . . *R. eschscholtzii*

20 Achenes including beaks 3-5 mm long; aggregate fruit 4-6 mm long with (5) 10-25 achenes; receptacles glabrous . . . . *R. uncinatus*

20 Achenes including beaks 1-1.5 mm long; aggregate fruit 4.5-9.5 mm long, with 90-200 achenes; receptacles hairy . . . . *R. sceleratus*

21 Lobes of leaves 3, entire, mostly over 3 times longer than wide; achenes pubescent; plants with clavate tuberous roots . . . . *R. jovis*

21 Lobes of leaves 3 or more, if only 3 then less than 3 times as long as wide; achenes glabrous except in some *R. eschscholtzii*; roots fibrous 22

22 Petals 3-5 mm long 23

22 Petals 5-19 mm long 25

23 Plants annual, not alpine . . . . **R. sceleratus**

23 Plants perennial, alpine 24

24 Basal leaf blades 0.8-2 cm long, 1.4-3 cm wide; stamens about 50 . . **R. grayi**

24 Basal leaf blades 0.4-1 cm long, 0.6-1.7 mm wide; stamens 10-20 . . . .
. . . . **R. pygmaeus**

25 Beak of achene strongly bent; herbage usually pilose; basal and stems leaves dissimilar; petals 5-8 (10) mm long . . . . **R. pedatifidus**

25 Beak of achene more or less straight; herbage mostly glabrous; basal and stem leaves similarly dissected; petals 5-18 mm long 26

26 Leaves cut to the base, with more than 15 linear segments; beak of achene 1.2-2.5 mm long . . . . **R. adoneus**

26 Leaves not cut entirely to the base, the lobes fewer than 15 and not linear; beak of achene 0.8-1.5 mm long . . . . **R. eschscholtzii**

Ranunculus acriformis A. Gray — mtn. sharp buttercup — Strawberry Valley (Goodrich 27615) Middle Mt. near Three Corners area (Peterson (83-357), and Harrington 2090 listed for Cold Spring Mt. (Bradley 1950); meadows; 7,600-8,580 ft; June-Aug. Plants with appressed hairs belong to var. *acriformis* and plants with spreading hairs belong to var. *montanensis* (Rydberg) L. D. Benson. Plants of Strawberry Valley belong to var. *montanensis*. Perhaps those from the Three Corners area and Cold Spring Mt. belong to var. *acriformis*.

Ranunculus adoneus A. Gray — alpine buttercup, snow buttercup — Known from a few isolated locations in the Uinta Mtns. in meadows and open, rocky slopes, locally abundant in late-melting snowbeds; 9,500 ft to above timberline; June-Aug., flowering at the edge of snowmelt in late melting snowbeds.

Ranunculus alismaefolius Geyer ex. Bentham (*R. calthaeflorus* Greene) — plantain buttercup — Common from Strawberry Valley to the Whiterocks drainage in the Uinta Mtns., and uncommon on the E. Tavaputs Plateau, to be expected elsewhere; sagebrush, aspen, coniferous forest, and meadow communities; 8,000-9,500 ft; May-July. Our plants belong to var. *montanus* S. Watson.

Ranunculus aquatilis L. — white water-buttercup — Aquatic, widespread in ponds, ditches, and slow-moving streams. Two vars. are found in our area.

1 Floating leaves broadly lobed; submerged leaves capillary . **var. aquatilis**
1 Leaves all submerged and capillary . . . . **var. diffusus**

Var. aquatilis — hispidulous water-buttercup — The one specimen seen (Goodrich 27151) is from a pond in a lodgepole pine forest near Meeks Cabin Reservoir at 8740 ft elevation.

Var. diffusus Withering [*R. longirostris* Gordon; *R. trichophyllus* Chaix; *Batrachium circinatum* (Sibthorp) Fries ssp. *subrigidum* (Drew) Love & Love; *B. longirostre* (Godron) F. W. Schultz; *B. trichophyllum* (Chaix) F. W. Schultz] — threadleaf water-buttercup — Widespread and common in ponds, lakes and slow-moving reaches of streams; 4,670-9,000 ft; late June-Aug.

Ranunculus cardiophyllus Hooker — heartleaf buttercup — Uinta Mtns. and E. Tavaputs Plateau; infrequent in meadows and edge of meadows with sagebrush, aspen, and coniferous trees; 7,500-8,600 ft; June-July. Our plants belong to var. *cardiophyllus*.

Ranunculus cymbalaria Pursh [*R. c.* var. *saximontana* Fernald; *Halerpestes cymbalaria* (Pursh) Greene] — alkali buttercup — Widespread and common; edge of ponds, lakes, along water courses, wet meadows, around seeps and springs, often where the ground is ephemerally wet, tolerant of alkali; 5,000-8,500 ft; June-Aug.

Ranunculus eschscholtzii Schlechtendal. — subalpine buttercup — Across the Uinta Mtns.; occasional in meadows, snowbeds, along streams, and among boulders; 10,200-11,100 ft; June-July. Often flowering at the edge of snowmelt. Uinta Mtn. plants belong to var. *eschscholtzii*.

Ranunculus flabellaris Rafinesque — yellow water-buttercup — The 2 specimens seen (Brotherson 2427; Goodrich 15024) are from Rock Creek, Uinta Mtns.; aquatic; 7,600 ft; July-Aug.

Ranunculus flammula L. — creeping spearwort — Uinta Mtns.; common; edge of ponds, in ephemeral ponds, drawdown basins of reservoirs, and along streams in aspen and coniferous woods; 7,600-11,000 ft; (June) July-Aug.

Ranunculus glaberrimus Hooker — sagebrush buttercup — Common across much of the n. slope and eastern Uinta Mtns. (no specimens seen from Duchesne Co.) and apparently uncommon on the E. Tavaputs Plateau; mt. big sagebrush, aspen, lodgepole pine, ponderosa pine, and pinyon-juniper communities; (5,600) 7,000-9,000 ft; April-June. Often flowering at the edge of snowmelt. Our plants belong to var. *ellipticus* Greene.

Ranunculus gmelinii de Candolle — small yellow water-buttercup — The few specimens seen are from morainal ponds in the Uinta Mtns at 8,900-9,800 ft elevation.

Ranunculus grayi Britton [*R. gelidus* Karelin & Kirilow ssp. *grayi* (Britton) Hulten, listed under *R. karelinii* Czernjaew in TPD] — ice cold buttercup, arctic buttercup — The few specimens seen are from the alpine of the Uinta Mtns from fellfields and talus at 11,500-12,650 ft.

Ranunculus hyperboreus Rottboel (*R. natans* C. A. Meyer misapplied) — far northern buttercup — Most specimens seen are from the eastern Uinta Mountains from shallow open water of quaking bogs and shallow water and mud of beaver ponds. One specimen (Harrington 2153 CS) is from Cold Springs Mt., Moffat Co. 8,400-9,000 ft. July-Aug.

Ranunculus inamoenus Greene [*R. alpeophilus* A. Nelson; *R. inamoenus* var. *alpeophilus* (A. Nelson) L. D. Benson.] — pleasant buttercup — Strawberry Valley, Uinta Mtns., W. Tavaputs Plateau in Carbon Co. and E. Tavaputs Plateau; common in aspen, coniferous forest, oak, sagebrush, meadow and streamside communities; 7,500-11,200 ft; late May-early Aug. Our plants belong to var. *inamoenus*.

Ranunculus jovis A. Nelson — Utah buttercup — Strawberry Valley, w. Uinta Mtns., Blue Mtn. including Ruple Point Trail (T. Naumann 417 DINO), and MacLeod 866 (CS) is reported for Whisky Springs, Moffat Co.; sagebrush, meadow, aspen and fir communities; 7.800-9,000 ft; May-June. The disjunct Ruple Point population is associated with *Fritillaria pudica* which is also isolated here where they are also associated with *Claytonia lanceolata* and *Orogenia lineare*. The convergence of these plants with edible roots might indicate they might have been planted here by native Americans.

Ranunculus macounii Britton — Macoun buttercup — Strawberry Valley; flanks of the Uinta Mtns., and Tavaputs Plateau; occasional to common in riparian, and meadow-sagebrush-grass communities and around seeps and springs; 5,500-7,300 ft; May-Aug.

Ranunculus orthorhynchus Hooker — straight-beak buttercup — Strawberry Valley and Uinta Mtns. (no specimens seen from Uintah Co.), locally common in aspen, streamside, and meadow communities; 7,800-9,600 ft; June-Aug. Our plants belong to var. *platyphyllus* A. Gray.

Ranunculus pedatifidus J. E. Smith in Rees. — Northern buttercup — The one specimen seen (K. Ostler 640) is from alpine tundra at about 11,200 ft, Uinta Mtns. in Summit Co. Our plants belong to var. *affinis* (R. Br.) L. D. Benson.

Ranunculus pygmaeus Wahlenberg — pygmy buttercup — Known from late-persisting, alpine snowbeds near the crest of the Uinta Mountains at Bald Mtn. and head of Burnt Fork on the north slope and Gilbert Bench on the south slope; 11,550-12,300 ft. July-Aug. Plants commonly 2-5 cm tall in our area.

Ranunculus reptans L. — creeping buttercup — The few specimens seen are from scattered locations. Plants of the cultivated and escaping var. pleniflorus Fernald are found along ditches in valleys. Plants of this var. often have numerous petals that are shorter than those of var. *repens*.

Ranunculus sceleratus L. [*Hecatonia scelerata* (L.) Fourreau] — blister buttercup — Widespread; infrequent or occasional in mud, ephemeral ponds, slow-moving water, and along streambanks; 4,900-7,000 ft; June-Oct. Our plants belong to var. *multifidus* Nuttall in Torrey & A. Gray.

Ranunculus testiculatus Crantz [*Ceratocephala orthoceras* de Candolle, *C. testiculata* (Crantz) Roth] — bur buttercup — Introduced from Eurasia, adventive, weedy, and apparently spreading rapidly in our area; various plant communities, often on disturbed ground but also spreading into native plant communities; not expected much over 7,800 ft; April-early June. Included in subgenus *Ceratocephala* (Moench) L. D. Benson which is sometimes treated as a separate genus.

Ranunculus uncinatus D. Don — little buttercup — To be expected in the Colorado part of the Uinta Basin, not known from Utah.

Thalictrum L. — Meadowrue

Perennial herbs; leaves alternate, ternately decompound; flowers small; sepals 4 or 5, green or petal-like (not over 5 mm long); petals none; stamens many; fruit of achenes. The leaves are like those of *Aquilegia*.

---

1 Stems leafless or with a small leaf near the base, 5-30 cm tall .. **T. alpinum**

1 Stems with leaves, 25-80 cm tall 2

2 Ultimate leaflets 2-5 mm long, about as wide; aggregate fruit of 2-5 (6) achenes; plants of the Piceance Basin, of open sunny places **T. heliophilum**

2 Some of the ultimate leaflets over 5 mm long; aggregate fruit of 7-15 or more achenes; plants not restricted as above, often of shady places 3

3   Flowers bisexual; stigmas 5, usually shorter than the sepals at anthesis; filaments clavate .... *T. sparsiflorum*

3   Flowers unisexual; stigmas usually 4, usually longer than the sepals at anthesis; filaments filiform 4

4   Achenes strongly flattened, 2.5-3.6 mm wide; lower side of leaflets rather inconspicuously veined .... *T. fendleri*

4   Achenes terete or subterete, 1.4-2.2 mm wide; lower side of leaflets with conspicuous raised veins .... *T. venulosum*

Thalictrum alpinum L. — alpine meadowrue — The few specimens seen are from the n. slope of the Uinta Mtns. and from a single location in S. Fork Rock Creek on the s. slope; wet meadows on or near limestone; 8,600-9,200 ft; July-Aug.

Thalictrum fendleri Engelmann — fendler meadowrue — Widespread; common in aspen woods, occasional in riparian and coniferous forest communities; 7,500-10,000 ft; June-July.

Thalictrum heliophilum Wilken & DeMott — sun-loving meadowrue — A narrow endemic, Cathedral Bluffs and Roan Cliffs; Green River Shale barrens; 8,200-8,600 ft; July-Aug.

Thalictrum sparsiflorum Turczaninow ex Fischer & Meyer — few-flowered meadowrue — Uinta Mtns.; infrequent in riparian communities, lake margins, and open coniferous woods; 7,400-11,000 ft; June-July.

Thalictrum venulosum Trelease — veiny meadowrue — The few specimens seen are from widely scattered locations in the Uinta Mtns. in Duchesne, Summit, and Uintah Cos.; sagebrush, mtn. brush, spruce-fir, and talus communities at 8,000-10,800 ft. This is distinguished from *T. fendleri* by turgid (not flattened), thick-walled achenes with the base slightly if at all oblique.

Trautvetteria Fischer & C. A. Meyer — False Bugbane, Teasel-rue

Trautvetteria caroliniensis (Walter) Vail — false Bugbane, teasel-rue — Perennial herb 2.5-100 cm tall from widely spreading rhizomes; leaves basal and cauline, palmately lobed into 5-11 segments, irregularly serrate; inflorescence more or less dichotomously loosely branched; sepals 3-5 mm long, deciduous at anthesis; petals lacking; stamens numerous, white and showy; aggregate fruits of numerous achenes. The one specimen seen (B. Albee 6238) is from floodplain of the North Fork Duchesne River at Mill Flat at 7,600 ft.

Trollius L. — Globeflower

Trollius albiflorus (A. Gray) Rydberg (*T. laxus* Salisbury var. *albiflorus* A. Gray) — American globeflower — Perennial glabrous herb; stems 20-60 cm tall with 2-4 leaves; leaf blades nearly orbicular in outline, palmately parted or divided, the lobes cleft to incised; sepals 5-15, petaloid, white, 15-20 mm long, quickly deciduous; petals 5-8, 3-5 mm long, linear; stamens numerous; fruit of follicles 8-12 mm long, each many seeded. Occasional across the Uinta Mtns. along streams, in meadows and lodgepole pine and Engelmann spruce forests; 9,000-10,800 ft; June-July. The head of follicles is similar to that of *Caltha*.

RHAMNACEAE Buckthorn Family

Shrubs; leaves simple, alternate; flowers small, regular or nearly so, sepals united at the base; petals free; stamens the same number as the petals and opposite to the petals; fruit capsule-like or a berry-like drupe.

---

1 Leaves with 3 conspicuous nerves running the entire length of the blades; shrubs 0.2-1 (2) m tall, widespread .... *Ceanothus*

1 Leaves with l prominent nerve (midrib) running the entire length of the blades; shrubs (1) 2-3 m tall, known from the E. Tavaputs Plateau in Colorado .... *Rhamnus*

Ceanothus L. — Ceanothus; Buckbrush; Buckthorn; Wild-lilac

Leaves with 3 major nerves; flowers bisexual in a crowded inflorescence; sepals 5, calyx adnate by a disk to the lower part of ovary; petals hooded and clawed; stamens 5; ovary usually 3-lobed, 3-celled with a 3-cleft style; fruit capsule-like, separating at maturity into 3 nutlets.

---

1 Leaves serrulate for nearly the whole length, 2-8 (10) cm long *C. velutinus*

1 Leaves entire or serrulate only at the apex, 1-3 cm long 2

2 Leaves whitish below from dense, fine, matted pubescence (use strong magnification); branches spinescent; plants of the Uinta Mtns. . *C. fendleri*

2 Leaves not densely pubescent below except on veins; branches not spinescent; plants of the Tavaputs Plateau. .... *C. martinii*

Ceanothus fendleri A. Gray — Fendler ceanothus — Occasional or locally common on the s. slope of the Uinta Mtns. and in Daggett Co. on the n. slope, Blue Mtn.; sagebrush, ponderosa pine, and lodgepole pine communities; 7,000-9,000 ft; June-July. Fendler ceanothus is a vigorous sprouter, and a single plant can expand to a diameter of 4.5 m or more within a few years after fire.

Ceanothus martinii M. E. Jones — Martin ceanothus — W. Tavaputs Plateau and at least Piceance Basin of E. Tavaputs Plateau; occasional in sagebrush and mt. brush communities, and often dominant in leeward snowbeds; 7,500-9,000 ft; June-July.

Ceanothus velutinus Douglas — snowbush ceanothus — Locally abundant on the n. slope of the Uinta Mtns. from the Provo River and Bear River drainages and from Skull Creek east to the Green River, known from a single location on the s. slope of the Uinta Mtns., Blue Mt., extreme w. of W. Tavaputs Plateau, upper Bitter Creek of E. Tavaputs Plateau; mt. brush, ponderosa pine, aspen, Douglas-fir, and lodgepole pine communities; 7,000-9,000 ft; June-Aug.

Rhamnus L. — Buckthorn

Rhamnus smithii Greene — Smith buckthorn — Unarmed shrubs (1) 2-3 m tall; leaves 2-7 cm long, alternate, elliptic to oblong-lanceolate or ovate-oblong, serrate to crenulate; flowers, unisexual, 1-3 in the axils of leaves; sepals 4, free; petals 4, about 1 mm long; ovary free from the calyx; fruit drupe-like, about 8 mm long, about as wide, glabrous, blackish. Occasional; known from the Colorado portion of the E. Tavaputs Plateau.

ROSACEAE Rose Family

Shrubs, trees, or herbs, annual or perennial; leaves alternate, basal, or rarely opposite; stipules usually present; flowers bisexual or rarely unisexual, radially symmetrical, solitary or in clusters; floral tube flat, saucer shaped, or tubular, commonly lined with a glandular disk, the sepals, petals and stamens arising at or near the summit of the disk; sepals 5 (4-10), often alternating with sepal-like bracts in the herbaceous species and then appearing to be 10; petals 5 (4-10), lacking, or numerous, separate; stamens 1-numerous; pistils 1-many; ovary superior or partly to completely inferior, 1-5 chambered; styles 1-5, free or rarely fused; fruit an achene, drupe, drupelet, follicle, or pome.

Several cultivated fruit trees (apple, apricot, cherry, peach, pear, and plum) belong to this family. These are not included in this treatment as they seldom persist outside of cultivation. However, they sometimes do, and particularly apricot, plum, and some forms of apple.

---

1 Plants herbaceous, or if woody at the base then not over 60 cm tall 2
1 Plants shrubs, mostly over 60 cm tall. 10
2 Leaves simple, all basal. 3
2 Leaves compound 4
3 Leaf blades crenate; plants from above 10,500 ft . . . . ***Dryas***
3 Leaf blades entire; plants of lower elevations . . . . ***Petrophytum***
4 Leaves with 3 leaflets 5
4 Leaves with more than 3 leaflets. 7
5 Leaflets entire except for 3-5 teeth at the apex . . . . ***Sibbaldia***
5 Leaflets with more than 5 teeth 6
6 Petals white; plants stoloniferous; fruit (strawberry) of achenes on a fleshy (red when ripe) receptacle . . . . ***Fragaria***
6 Petals yellowish, red, or purplish; plants mostly not stoloniferous; fruit not as above . . . . ***Potentilla***
7 Leaflets on small petiolules; petals lacking; sepals 4, purplish; stamens strongly exserted, purplish . . . . ***Sanguisorba***
7 Leaflets mostly sessile; petals present; sepals 5, alternating with sepal-like bracteoles and appearing 10; stamens not especially exserted 8
8 Achenes pubescent; styles persistent, 3-50 mm long, sometimes plumose at least on the upper half . . . . ***Geum***
8 Achenes glabrous or papillose glandular; styles readily deciduous, 0.8-3 mm long, not plumose 9
9 Basal leaves with 20 or more leaflets . . . . ***Ivesia***
9 Basal leaves with fewer than 20 leaflets . . . . ***Potentilla***
10 Leaves compound 11
10 Leaves simple. 14
11 Stems armed with sharp bristles or prickles. 12
11 Stems unarmed 13
12 Leaflets 5-11, mostly less than 3.5 cm long; petiole and rachis mostly free of bristles; petals pink or red . . . . ***Rosa***

12 Leaflets 3-5, some usually over 3.5 cm long; petioles and rachis with bristles; petals white . . . . **Rubus**

13 Leaflets 9-13, over 3 cm long; plants mostly over 2 m tall . . . . **Sorbus**

13 Leaflets 3-7, less than 3 cm long; plants not over 2 m tall . . . . **Potentilla fruticosa**

14 Leaves lobed or parted 15

14 Leaves entire, crenate or serrate, not lobed or parted 18

15 Leaves over 8 cm wide; fruit raspberry-like . . . . **Rubus parviflorus**

15 Leaves less than 8 cm wide; fruit not as above 16

16 Leaves divided into linear entire, densely pubescent segments . . . . **Potentilla fruticosa**

16 Leaves not as above 17

17 Leaves truncate or cordate, the lobes with teeth . . . . **Physocarpus**

17 Leaves wedge-shaped at the base, the lobes entire . . . . **Purshia**

18 Leaves entire, linear, seldom over 1 cm wide 19

18 Leaves crenate, or serrate, if only inconspicuously serrate then leaves not linear and often over 1 cm wide 20

19 Flowers without petals; fruit an achene; style plumose, 1-8 cm long . . . . **Cercocarpus**

19 Flowers with conspicuous petals; fruit a pome; style not plumose, less than 1 cm long . . . . **Peraphyllum**

20 Twigs armed with thorns . . . . **Crataegus**

20 Twigs unarmed. 21

21 Leaves serrate for about their whole length, some usually over 4 cm long or over 2 cm wide . . . . **Prunus**

21 Leaves entire toward the base, seldom over 4 cm long or 2 cm wide 22

22 Leaves with rounded bases; petals over 5 mm long; fruit a berry-like pome . . . . **Amelanchier**

22 Leaves with a wedge-shaped base; petals lacking or less than 3 mm long; fruit not as above. 23

23 Flowers numerous in panicles; petals about 2 mm long; fruit a woody capsule; styles not elongate, not plumose; bark of older branches shredding in long strips . . . . **Holodiscus**

23 Flowers solitary or in small clusters; petals lacking; fruit an achene, the styles 3-10 cm-long, plumose; bark of older branches not shredding as above . . . . **Cercocarpus**

Amelanchier Medicus — Serviceberry

Unarmed shrubs; leaves simple, alternate, serrate above, rounded and usually entire at the base; flowers racemose; floral cup campanulate more or less adnate basally to the ovary with flared disk; sepals 5; petals 5, white; stamens 12-20; pistils 2-5; ovary 2-5 celled, inferior; fruit a pome, reddish to purplish, often glaucous.

1   Leaves (1.5) 2-5 cm long; petals (5) 9-15 (25) mm long; fruit glabrous or
     pubescent, purplish and more or less fleshy at maturity; styles and seeds 4-
     5 .... *A. alnifolia*
1   Leaves 0.5-2 (3) cm long; petals 5-10 mm long; fruit mostly pubescent,
     orange, yellowish, or purplish and dry at maturity; styles and seeds (2) 3-4
     (5) .... *A. utahensis*

Amelanchier alnifolia Nuttall [*A. florida* Lindley misapplied; *A. pallida* Greene; *A. pumila* (Nuttall) Roemer] — Saskatoon serviceberry — Widespread; common in sagebrush, mt. brush, ponderosa pine, and aspen communities and along water-courses; 6,000-8,500 (9,000) ft; May-July.

Amelanchier utahensis Koehne (*A. prunifolia* Greene) — Utah serviceberry — Widespread, usually on drier sites and at lower elevations than *A. alnifolia* and occasionally in rock outcrops in desert shrub communities. Hardly distinct from *A. alnifolia,* and separated by more of a trend in features than by any one decisive feature.

Cercocarpus H.B.K. — Mountain Mahogany

Shrubs or small trees; leaves simple, alternate; flowers 1-several in axillary clusters; floral tube turbinate, 3-8 mm long, persistent around but free of the ovary, bearing 15 or more stamens, 5-lobed; petals lacking; pistil 1; fruit a hardened achene, the style persistent, much elongate, plumose.

1   Leaves toothed on the upper 1/2, deciduous .... *C. montanus*
1   Leaves entire, persistent. 2
2   Leaves 3-12 (18) mm long, linear, the margins usually tightly rolled and
     concealing some of the lower surface; shrubs 0.5-2.5 m tall, intricately
     branched; styles 1-3 cm long .... *C. intricatus*
2   Leaves 10-42 mm long, elliptic to oblong, the margin revolute, but some of
     the lower leaf-surface usually visible; shrubs or small trees 2-5 m tall; styles
     4.5-8 cm long .... *C. ledifolius*

Cercocarpus intricatus S. Watson [*C. arizonicus* M. E. Jones; *C. ledifolius* var. *intricatus* (S. Watson) M. E. Jones] — littleleaf mtn. mahogany, dwarf mtn. mahogany — Rather abundant in the sandstone formations of the Reaves and Glades of Daggett Co., in isolated stands on limestone on Cold Spring Mtn., also on Nugget and other sandstone formations along the s. flank of the Uinta Mtns., Split Mt. and e. into Colorado, no specimens seen from Duchesne Co.; May-June.

Cercocarpus ledifolius Nuttall — curl-leaf mtn. mahogany — In the Uinta Mtns. most abundant on the n. slope and particularly abundant along the rim and slopes of Cart Creek to Jackson Creek, and upper elevations of Death Valley, and on Diamond Mtn. and Cold Spring Mtn., on the s. slope of the Uinta Mtns. occasional to abundant in canyons and especially on Weber sandstone, on the W. Tavaputs Plateau most abundant in the Timber Canyon drainage and decreasing common to rare eastward, a few isolated shrubs as far e. as Sowers Canyon; apparently rare on the E. Tavaputs Plateau; 7,500-9,000 ft; May-June.

In IMF (3A: 130) N. H. Holmgren assigns our plants to var. *intermontanus* N. H. Holmgren. Dorn (2001) lists them under var. *intercedens* Schneider. However, the type specimen for this taxon (from Utah Co.) might be a hybrid between *C. ledifolius* var. *intermontanus* and *C. intricatus* (IMF 3A: 130).

Cercocarpus montanus Rafinesque (*C. betuloides* Nuttall in Torrey & A. Gray misapplied) — alderleaf mtn. mahogany — Widespread and common on many geologic formations, highly adapted to calcareous areas including the Uinta and Green River formations, also common on Weber Sandstone; pinyon-juniper and mt. brush communities; 6,000-8,500 ft; May-June. Occasionally hybridizing with *C. ledifolius*, the hybrids often tree-like with persistent leaves as in *C. ledifolius* and with expanded-and serrate leaves as in *C. montanus*. The common name of true mtn. mahogany sometimes applied to this plant is unfortunate. It implies the other species of the genus are not mtn. mahoganies. Also "true" seems to imply some relation to *Swietenia mahogani* Jacquin (the tree from which mahogany lumber is made).

Crataegus L. — Hawthorn; Haws

Shrubs or small trees, usually armed with spines or thorns; leaves simple, alternate, serrate or lobed; flowers in cymose corymbs; floral cup saucer-shaped or campanulate, adnate to the ovary; sepals 5, reflexed after anthesis; petals 5, whitish, soon deciduous; stamens 5-25; ovary inferior, of 1-6 carpels, 1-5 celled; styles 1-5, distinct; fruit a pome.

---

1 Stamens about 20; nutlets not pitted or deeply concave on the ventral surface; fruit not bright red or scarlet when ripe; leaves usually not over 4 cm long, crenate-serrate, the teeth small, or entire near the base . . . .
. . . . *C. saligna*

1 Stamens about 10; nutlets pitted or deeply concave on the ventral surface; fruit usually scarlet or bright red when mature, but darker when old; leaves (2) 3-6 (9) cm long, irregularly and rather sharply serrate. 2

2 Leaf blades less than 2 times as long as wide, often distinctly lobed; spines usually numerous, often over 3 cm long; fruit about 7-8 mm in diameter .
. . . . *C. erythropoda*

2 Leaf blades about 2 times as long as wide, not lobed or scarcely so; spines usually few and mostly less than 3 cm long; fruit about 10 mm in diameter
. . . . *C. douglasii*

Crataegus douglasii Lindley (*C. rivularis* Nuttall) — Douglas hawthorne — Widespread; canyons and following water courses into valleys; 5,900-8,000 ft; May-June. Two vars. are found in our area as follows:

  1 Petals 3-3.8 mm long, 2-4.2 mm wide; leaves slender, at least 2-4 times longer than broad; fruit 6-8 mm thick when dried; plants of Duchesne, Uintah, and Rio Blanco Cos. . . . . **var. *duchesnensis*** S. L. Welsh

  1 Petals 4.5-8 mm long, 5-7.8 mm wide; leaves commonly less than twice longer than broad; fruit 8-12 mm thick when dried; plants of Daggett and possibly Uintah Cos. and eastward . . . . . . **var. *rivularis*** (Nuttall) Sargent

Crataegus erythropoda Ashe — Cerro hawthorne — Upper Piceance Creek (Goodrich & Huber 28590), listed by Graham (1937) for along the White River,

4 mi below the mouth of Piceance Creek; plotted in TPD for Moffat and Rio Blanco Counties; May-June.

Crataegus saligna Greene — willow hawthorne — Known from along the White River in Rio Blanco Co. and plotted in Uintah Co. in TPD.

Dryas L. — Dryad; Mountain-avens

Dryas octopetala L. — mtn. avens, white dryad — Low herbs or subshrubs, the woody stems at or below ground level; leaves basal, the blades 1-4 cm long, 0.3-1 cm wide, lanceolate to lance-oblong, crenate, dark green above, white to-mentose beneath, often strongly revolute; scapes 1-11 cm long; flowers solitary, regular, bisexual; petals 8-10, white or rarely yellowish, 9-15 mm long; fruit an achene with a persistent, plumose style, the styles forming a whiskered head. Locally common in alpine communities of the Uinta Mtns., usually in wind-swept areas; 11,400-13,000 ft; June-July. Our plants belong to var. *hookeriana* (Juzepezak.) Breitung.

Fragaria L. — Strawberry

Plants perennial, herbaceous, with stoloniferous stems; leaves basal, trifoliate, coarsely crenate-serrate; flowers showy, solitary or in bracteate cymes; floral cup saucer-like; sepals 5, ovate to lanceolate, alternate with sepal-like bracteoles; petals 5, white; stamens 20-25; pistils numerous; achenes born on an enlarged, fleshy and juicy, red receptacle.

---

1   Terminal tooth of leaflet usually longer than the 2 adjacent lateral teeth; petioles spreading hairy; flowering stems usually equaling or surpassing the leaves; leaves yellow-green, not glaucous above .... *F. vesca*
1   Terminal tooth of leaflets usually shorter than the 2 adjacent lateral teeth; petioles with appressed or appressed-ascending hairs; flowering stems usually shorter than the leaves; leaves rather bluish-green above, often somewhat glaucous above .... *F. virginiana*

Fragaria vesca L. (*F. bracteata* A. Heller) — woodland strawberry — Widespread; occasional to common in woods and openings and along streams in mountains 7,000-10,500 ft; June-July. Our plants belong to var. *bracteata* (A. Heller) Davis.

Fragaria virginiana Duchesne. [*F. americana* (T. C. Porter) Brill.; *F. glauca* (S. Watson) Rydberg; *F. virginiana* var. *glauca* S. Watson; *F. virginiana* var. *platypetala* (Rydberg) H. M. Hall] — Virginia strawberry — widespread; occasional to common in woods and meadows; 7,500 -10,800 ft; June-July. With 2 vars.:

1 Petioles and scapes appressed-pilose; petals 5-11 mm long ....
     .... **var. *glauca*** S. Watson
1 Petioles and scapes spreading-pilose; petals 8-12 mm long ....
     .... **var. *platypetala*** (Rydberg) H. M. Hall

Geum L. — Avens

Perennial herbs with irregularly pinnate or trifoliate leaves; flowers solitary or bracteate-cymose; floral cup saucer-shaped or campanulate; sepals 5, alter-nating with sepal-like bracteoles, and thus sepals appearing to be 10; petals 5, yellow, pink or purplish; stamens numerous; pistils numerous; fruit an achene, pubescent, with a straight, bent, or jointed, sometimes elongate and plumose style, the terminal segment often deciduous.

1   Sepals and petals reddish or the petals pale yellow, forming a bowl-shaped flower 10-15 mm long; flowers nodding on long pedicels; styles 2.5-5 cm long, plumose except at the tip, not strongly bent; stem with a single pair of opposite leaves . . . . *G. triflorum*

1   Petals bright yellow, sepals green or purple tinged, 4-10 mm long, not forming a bowl-shaped flower; flowers not nodding; styles about 0.3-0.5 cm long, glabrous, pubescent, or subplumose at the tip; stem leaves alternate
. . . . 2

2   Leaf segments rarely over 1 cm wide; plants growing near or above timberline, seldom over 30 cm tall; achenes not forming a bur-like head .
. . . . *G. rossii*

2   Larger leaf segments well over 1 cm wide; plants of wet places below timberline, usually over 30 cm tall; the dense achenes with their slender styles forming a bur-like head when mature 3

3   Lower (persistent) segment of the style glabrous or slightly hirsute near the base, not glandular, terminal segment of basal leaves somewhat larger than the main lateral lobes but with similar cuneate base . . . . *G. aleppicum*

3   Lower (persistent) segment of the style slightly glandular pubescent; terminal segment of basal leaves much larger than the main lateral lobes and usually rounded to nearly cordate at the base . . . . . . *G. macrophyllum*

Geum aleppicum Jacquin. (*G. decurrens* Rydberg; *G. strictum* Aiton) — Aleppo avens, yellow avens — The 4 specimens seen are from Browns park, Moffat Co. (Harrington 1572), Steer Ridge of E. Tavaputs Plateau (R. Foster & D. Foster 481), Chepeta-Taylor Canyons of the E. Tavaputs Plateau (Goodrich 24584), and Slab Canyon, W. Tavaputs Plateau (Goodrich 23160), plotted for Rio Blanco Co. in TPD; Douglas-fir, meadow, and riparian communities; 5,300-8,500 ft; July-Aug.

Geum macrophyllum Willdenow — large-leaf avens — Widespread; occasional in moist and often shady places; 6,000-9,400 ft; June-Aug.

Geum rossii (R. Brown) Sere (*G. turbinatum* Rydberg; *Acomastylis rossii* (R. Brown) Greene; *Sieversia turbinata* Greene) — alpine avens — Common to abundant across the alpine of the Uinta Mtns. and extending down into Krummholz and upper edge of Engelmann spruce forests; June-Aug. Our plants belong to var. *turbinatum* (Rydberg) C. L. Hitchcock. Often confused with *Potentilla ovina* var. *decurrens.* In an alpine plant community classification study in the Uinta Mtns., this was the 2nd most frequently encountered forb and the forb of highest percent cover (Brown 2006). It is found in many alpine plant communities, but it is most abundant in areas where snow cover is moderately deep for much of the winter. Brown (2006) found alpine avens with 73% consistency but only about 3% average canopy cover in curly sedge fellfield communities and 100% consistency with 21% average canopy cover in alpine avens communities This plant is mostly lacking in wetland communities.

Geum triflorum Pursh [*G. ciliatum* Pursh; *Erythrocoma triflora* (Pursh) Greene; *Sieversia ciliata* (Pursh) D. Don] — prairie smoke, purple avens, old man's beard, old man's whiskers — Strawberry Valley and Uinta Mtns. where locally common in sagebrush-grass, meadow, and mt. brush communities and open woods; 7,500-10,000 ft; May-June. The reddish flowers and long, plumose styles sets this apart from other members of the genus in our area.

Holodiscus Maxim. — Mountain spray

Holodiscus dumosus (Nuttall) A. Heller (*H. microphyllus* Rydberg) — mtn. spray — Shrubs 25-150 cm tall; bark of older stems shredding in gray strips; twigs of the season pubescent, light reddish-brown; leaves 5-25 mm long, usually with a few teeth on the upper 1/2, wedge-shaped and entire toward the base, the upper surface green, the lower whitened with pubescence; inflorescence a narrow raceme-like panicle, 3-10 cm long; floral tube saucer-shaped; sepals 5; petals 5, about 1-3 mm long, white or pinkish; fruit a villose achene, these usually 5 per flower. Widespread; most common on rocky canyon walls; 6,000-9,000 (10,600) ft; June-Aug.

Ivesia Torrey & A. Gray

Plants perennial from a woody caudex; subscapose; herbage with sessile and stalked glands; basal leaves pinnately compound with sessile, crowded leaflets divided into 2-5 segments that are palmately or pinnately lobed into linear-oblanceolate ultimate segments; stem leaves 1 or 2, greatly reduced; flowers in a capitate solitary head-like inflorescence, or few to several dense head-like cymose clusters; floral tube campanulate; sepals alternating with linear bracteoles; petals shorter than the sepals, yellow or white, deciduous; stamens 5; fruit of 1-6 achenes with deciduous, filiform styles.

1   Flowers yellow when fresh; stems erect or ascending . . . . *I. gordonii*
1   Petals white; stems radiate-decumbent 2
2   Stamens 5; plants alpine . . . . *I. utahensis*
2   Stamens 20; plants of saline or alkaline lowlands . . . . *I. kingii*

Ivesia gordonii (Hooker) Torrey & A. Gray (*Horkelia gordonii* Hooker) — Gordon ivesia — Common to abundant in rocky places, semi barrens of Red Pine Shale, limestone gravelly soils near and above timberline especially toward the western end of the Uinta Mtns., and rare or lacking in high elevations in these mountains east of the Yellowstone drainage on the south slope, but also found on sandstone at mid-elevations eastward in these mountains to Diamond and Blue Mtns., and semi-barrens of Bridger Formation near Lone Tree, Wyoming; 7,600-12,000 ft; June-Aug. Plants with open inflorescences are common on sandy ground near the head of Rhoads Canyon and Cold Spring area. Some specimens from Diamond Mt. are densely pubescent.

Ivesia kingii S. Watson — saline ivesia — Known from saline or alkaline ground sub-irrigated by artesian wells west of Roosevelt, Duchesne Co. Uinta Basin populations (disjunct from the Great Basin) were discovered by Jim Spencer in 2011.

Ivesia utahensis S. Watson — Utah ivesia — Known from a few alpine locations on the west end of the north slope of the Uinta Mountains one of which, in

West Fork Blacks Fork, is in a long persisting snowbed. This endemic plant of northern Utah is more common to the west in the Wasatch Mountains.

Peraphyllum Nuttall in Torrey & A. Gray — Wild crab apple

Peraphyllum ramosissimum Nuttall — wild crab apple, bitter-apple — Shrubs to 2 m tall; leaves 2-4 cm long, narrow oblanceolate, entire or serrulate; sepals triangular, reflexed; sepals 5, spreading to deflexed; petals 7-10 mm long, white to pink ovary inferior, of 2 carpels, but 4-celled; styles 2 or 3; fruit a fleshy, 4-seeded globose pome 10-15 mm in diameter, yellowish to reddish-brown. Locally common to abundant on the E. Tavaputs Plateau in mt. brush communities. The small crabapple-like fruit is attractive when ripe, but the taste is not agreeable (AUF 643). It is likely highly selected by wildlife.

Petrophytum (Nuttall) Rydberg — Rock Spiraea

Petrophytum caespitosum (Nuttall) Rydberg (*Spiraea caespitosa* Nuttall in Torrey & A. Gray; also spelled *Petrophyton*) — mat rock spiraea, Rocky Mt. rock spiraea — Mat-forming half shrubs, the woody stems mostly in contact with the ground or rocks; leaves basal, 3-17 mm long, 1.5-4.5 mm wide, gray-pubescent, entire; scapes 0.5-12 cm tall; flowers in dense, short head-like or open panicles; sepals 1-2 mm long; petals 1-3 mm long, white; stamens numerous; fruit a follicle. Known from isolated locations in the Uinta Mtns and abundant on Limestone Ridge of Cold Spring Mtn., rare or perhaps lacking on much of the Tavaputs Plateau; face of or in the cracks of limestone and sandstone outcrops; 7,000-9,000 ft; July-Sept.

Physocarpus Maxim. — Ninebark

Shrubs with exfoliating bark; leaves simple, palmately lobed, alternate, usually with stellate hairs; inflorescence of terminal corymbs; flowers perfect; floral cup turbinate to campanulate; sepals 5; petals 5, white; stamens 20-40, pistils superior, 1-5, more or less united; fruit a woody capsule with follicle-like parts, seeds 2-4.

---

1 Inflorescence with 1-6 flowers; leaf blades 0.3-2 cm long . . . . . *P. alternans*
1 Inflorescence with more than 6 flowers 2
2 Leaves mostly 0.7-2.5 cm long; mature carpels swollen, not flattened; plants rare . . . . *P. monogynus*
2 Leaves mostly 2-8 cm long; mature carpels flattened; plants common . . . . *P. malvaceus*

Physocarpus alternans (M. E. Jones) J. T. Howell [*Opulaster alternans* (M. E. Jones) A. Heller] — dwarf ninebark — Specimens seen are from Red Canyon, Sheep Creek, and Boars Tusk in Daggett Co., Ashley Gorge, Big Brush Creek Gorge, and Dry Fork Canyon of the s. slope Uinta Mtns., near Gates of Lodore, Moffat Co., vicinity of White River near the Utah-Colorado border, and from Indian Canyon of the W. Tavaputs Plateau; rocky canyons often in talus below cliffs, and shaly or marly, gravelly slopes; 5,610-8,560 ft; May-June.

Physocarpus malvaceus (Greene) Kuntze [*Opulaster malvaceus* (Greene) Kuntze] — mallow ninebark — Western Uinta Mtns e. to Rock Creek and Bear River drainages, and W. Tavaputs Plateau; locally common to abundant in aspen and Douglas-fir woods and occasionally with other trees; 7,000-10,000 ft; June-July.

Physocarpus monogynus (Greene) Kuntze — mtn. ninebark — The one specimen seen (S. L. Welsh & K. Taylor 15095) is from Range Creek, W. Tavaputs Plateau, aspen-Douglas-fir community at 7,000 ft; June-July. The location of this specimen is far removed from the main body of this species, and it might be a small-leaved specimen of *P. malvaceus*.

Potentilla L. — Cinquefoil

Plants shrubs or annual to perennial herbs; leaves alternate, pinnately or palmately compound; flowers bisexual, solitary or in terminal and axillary cymes; floral tube saucer-shaped or bowl-shaped; sepals 5, alternating with 5 sepal-like bracts; petals 5, early deciduous; stamens 10-40; pistils numerous, borne on a receptacle; ovary superior; fruit an achene, glabrous, style soon deciduous, minute, arising from the apex or side of the achene.

---

1 Plants shrubs . . . . *P. fruticosa*

1 Plants herbaceous 2

2 Petals red or purple; plants of bogs and ponds, rhizomatous, often with trailing and sometimes floating stems; leaflets (3) 5-7 pinnate **P. palustris**

2 Petals yellow, rarely white; plants of various habitats, with trailing stems only in *P. anserina* and then usually with more than 7 leaflets per leaf . . . 3

3 Flowers solitary on slender scapes; plants with creeping stolons; leaves pinnate with 5-25 main leaflets interspersed by smaller ones, bicolored, silver-white beneath . . . . *P. anserina*

3 Flowers not as above; plant not creeping; leaves various 4

4 Leaves pinnately compound with 5 or more leaflets 5

4 Leaves ternate, palmate or subpalmate, if subpalmate then usually with only 5 leaflets 12

5 Petals 3-4 mm long; mature achenes with a wedge-shaped appendage on the inner margin . . . . *P. paradoxa*

5 Petals mostly over 4 mm long; achenes not as above 6

6 Ovaries and achenes with styles basally attached, the style thickened at the base, about 1 mm long; leaflets not tomentose, often fan-shaped; herbage glandular; petals not over 2 mm longer than the sepals 7

6 Ovaries and achenes with style attached near the apex; leaflets tomentose beneath (except in *P. ovina* var. *decurrens*); petals various, but often over 2 mm longer than the sepals. 8

7 Branches of inflorescence erect or nearly so; stems usually with dense, brown villose hairs . . . . *P. arguta*

7 Branches of inflorescence spreading; stems usually not densely hairy . . . .
. . . . *P. glandulosa*

8 Styles thickened and warty-papillose; calyx glandular. 9

8 Styles filiform not papillose. 10

9   Leaflets cut about halfway way to the midrib, the lobes 2-5 mm long; stem
    pubescence more or less spreading; anthers 0.5-0.8 mm long; achenes 1.1-
    1.4 mm long .... *P. pensylvanica*

9   Leaflets cleft well over halfway to the midrib, the lobes 6-14 mm long; stem
    pubescence appressed; anthers 0.4-0.5 mm long; achenes 0.7-1.2 mm long
    .... *P. bipinnatifida*

10 Plants 20-60 cm tall, usually ascending with flowers elevated well above the
    basal leaves .... *P. hippiana*

10 Plants seldom over 20 cm tall, often decumbent or flowers not much
    elevated above the leaves... 11

11 Leaflets mostly 5 or 7, sub-pinnate or nearly palmate, mostly over 1 cm
    long .... *P. concinna* var. *proxima*

11 Leaflets mostly more than 7 or not over 1 cm long .... *P. ovina*

12 Basal leaves lacking or withering at or before anthesis; plants not
    tomentose, annual, biennial or short-lived perennial................. 13

12 Basal leaves mostly well developed and persistent at anthesis; stems
    branched only in the inflorescence, perennial....................... 16

13 Petals 5-12 mm long; leaflets 5-7 or more; achenes strongly reticulate ...
    .... *P. recta*

13 Petals 1.3-4 mm long; leaflets 3-5; achenes smooth or rugulose 14

14 Pubescence of the lower stem mostly of strongly spreading, straight or
    nearly so, unicellular, pustular based hairs, a few hairs sometimes
    glandular; stamens (15) 20; petals mostly ¾ as long as to equaling the sepals;
    leaflets 3-5; achenes usually undulate-corrugate longitudinally ....
    .... *P. norvegica*

14 Pubescence of the lower stem, softer, glandular, multicellular, or
    tomentose; stamens 10-15; petals usually less than 3/4 as long as the sepals;
    leaflets mostly 3; achenes smooth or nearly so 15

15 Pubescence of the lower stem tomentose or more or less lanate, not
    glandular, the hairs unicellular; calyx eglandular .... *P. rivalis*

15 Pubescence of the lower stem including some glandular and/or
    multicellular hairs; calyx mealy-glandular .... *P. biennis*

16 Styles thickened and somewhat warty-papillose at the base, usually
    thickened to just below the stigma, to 1 mm long; leaflets 3 or 5, 1-3 cm
    long, tomentose at least below; petioles strigose and tomentose; plants
    alpine on the Uinta Mtns., 3-10 (15) cm tall 17

16 Styles slender, 1.5-3 mm long; leaflets sometimes more than 5, if tomentose
    beneath then sometimes over 3 cm long; petioles various; plant on but not
    restricted to alpine elevations of the Uinta Mtns.................... 18

17 Some of the leaflets subpalmate with 5 leaflets .... *P. rubricaulis*

17 Leaves ternate .... *P. nivea*

18 Upper leaf-surfaces and calyx glandular; stems 1-10 cm tall; petioles to 5
    cm long .... *P. concinna*

18 Upper leaf-surfaces and calyx glabrous, sericeous, or tomentose but not
    glandular;-stems and petioles various but often longer than above..... 19

19 Leaflets entire on the lower 1/2, toothed distally, cuneate, not bicolored,
    green to gray-green, or glaucous blue-green; anthers mostly 0.5-0.7 mm
    long; stems decumbent to ascending . . . . *P. diversifolia*
19 Leaflets toothed on the lower 1/2 as well as distally; anthers 0.8-1.3 mm
    long; stems ascending to erect . . . . *P. gracilis*

Potentilla anserina L. [*Argentea anserina* (L.) Rydberg] — silverweed cinquefoil
— Widespread but no specimens seen from the E. Tavaputs Plateau; occasional
or common along floodplains, drawdown basins of reservoirs, and in lowland
meadows, tolerant of alkali; 4,725-6,100 ft; May-Sept.

Potentilla arguta Pursh [*Drymocallis arguta* (Pursh) Rydberg; *D. convallaria*
Pursh] — acute cinquefoil — The one specimen seen (E. Neese & J. L. Neese
9842) is from Hoyt Peak at the w. end of the Uinta Mtns., snowbed community
at 10,100 ft. Our plants belong to var. *convallaria* (Rydberg) F. T. Wolf which is
transitional to *P. glandulosa*. Listed for Daggett and Duchesne Cos. in AUF (646),
but we have seen no specimens from these counties.

Potentilla biennis Greene — biennial cinquefoil — The few specimens seen are
all from Strawberry Valley and Duchesne Co.; along roads, in canyons, and
stream sides often where the ground has been disturbed; June-July.

Potentilla bipinnatifida Douglas ex Hooker — tansy cinquefoil — Known from
along Wyoming Highway 410 at 5 miles south of Mtn. View where it was
common in a borrow pit (N. Holmgren & P. Holmgren 9954). In AUF (650-651)
this is treated as a questionable entity for Utah with two specimens as follows:
S. Goodrich 19414 from Water Hollow, Wasatch Co. and S. Goodrich 3539 from
Lake Fork Canyon, Duchesne Co.

Potentilla concinna Richards — elegant cinquefoil, pretty cinquefoil — With 3
vars.:

   Var. bicrenata (Rydberg) S. L. Welsh & B. C. Johnston (*P. bicrenata* Rydberg)
— bicrenate cinquefoil — The 3 specimens seen are from upper Sowers Canyon
and Antelope Ridge, W. Tavaputs Plateau. Leaflets toothed at apex only.

   Var. concinna [*P. modesta* Rydberg, *P. concinna* var. *modesta* (Rydberg) S. L.
Welsh & B. C. Johnston misapplied] — modest cinquefoil — Occasional on the
W. Tavaputs Plateau and E. Uinta Mtns. with mtn. big sagebrush and in open
woods at 7,850-10,200 ft elevation. Leaflets toothed to below the middle.

   Var. proxima (Rydberg) S. L. Welsh & B. C. Johnston (*P. proxima* Rydberg) —
near cinquefoil — Known in our area from the n. flank of the Uinta Mtns. in
Summit Co. and perhaps Daggett Co. Included as a synonym of *P. concinna* in
IMF (3A: 100). Stems often trailing.

Potentilla diversifolia Lehmann (*P. glaucophylla* Lehmann; *P. perdissecta* Rydberg)
— varileaf cinquefoil — Uinta Mtns.; occasional in open woods, wet and dry
meadows, and along streams, and frequent in alpine communities; 9,000-
12,000 ft; June-Aug. Plants with leaflets merely toothed or cut 70 percent to
the midrib belong to var. diversifolia. Plants with leaflets cut 70-90 percent to
the midrib have been found on the n. slope of the Uinta Mtns. These plants
belong to var. *perdissecta* (Rydberg) C. L. Hitchcock — mtn. meadow cinquefoil
— Grading into *P. gracilis* and possibly into *P. hippiana*. In an alpine plant
community classification study in the Uinta Mtns., this was the most frequently
encountered forb species (Brown 2006).

Potentilla fruticosa L. [*Dasiphora fruticosa* (L.) Rydberg; *Pentaphylloides fruticosa* (L.) O. Schwarz] — shrubby cinquefoil, brush cinquefoil — common from Strawberry Valley and across the Uinta Mtns. apparently rare on the Tavaputs Plateau; mtn. big sagebrush, streambank, and wet meadow, alpine and other communities; 7,000-12,000 ft; June-Aug.

Potentilla glandulosa Lehmann [*Drymocallis glandulosa* (Lindley) Rydberg] — gland cinquefoil, wedge-leaf cinquefoil — Strawberry Valley and across the Uinta Mtns., no specimens seen from the Tavaputs Plateau; occasional and locally abundant in various plant communities including mt. brush, aspen, tall forb, coniferous forest, and in rocky places; 7,000-10,000 ft; May-Aug. Our plants belong to var. *pseudorupestris* Rydberg) Breitung [*P. g. intermedia* (Rydberg) C. L. Hitchcock], which is intermediate to *P. arguta* Pursh.

Potentilla gracilis Douglas ex Hooker — slender cinquefoil — Intergrading into *P. diversifolia* and *P. hippiana*. Three intergrading vars.:

1 Leaflets divided 2/3 or more the way to the midrib into more or less linear lobes, these over 5 mm long . . . . **var. *elmeri***
1 Leaflets cut 2/3 or less the way to the midrib, the segments various but usually not linear or less than 5 mm long or both 2
2 Leaflets strongly bicolored, green above and densely white-pubescent beneath; plants common . . . . **var. *pulcherrima***
2 Leaflets about equally greenish on both sides . . . . **var. *fastigiata***

Var. *elmeri* (Rydberg) Jepson (*P. brunnescens* Rydberg; *P. gracilis* var. *brunnescens* (Rydberg) C. L. Hitchcock) — meadow cinquefoil — The several specimens seen are from Strawberry Valley, Goslin Mtn., W. Tavaputs Plateau (Argyle Canyon), from near Lapoint, and Blue Mt.; moist and wet meadows and along streams, tolerant of some alkali; 5,400-8,500 ft; June-July.

Var. *fastigiata* (Nuttall) S. Watson [*P. glabrata* (Lehmann) C. L. Hitchcock; *P. nuttallii* Lehmann; *P. viridescens* Rydberg] — glabrate cinquefoil — The few specimens seen are from the Strawberry valley and Uinta Mtns.; aspen-willow and boggy meadow communities; 7,380-8,000 ft; June-Aug.

Var. *pulcherrima* (Lehmann) Fernald (*P. pulcherrima* Lehmann; *P. filipes* Rydberg) — beautiful cinquefoil — Widespread; occasional to common in sagebrush-grass, aspen, pine-spruce, mt. brush, and meadow communities and rocky slopes; 7,400-11,000 ft; June-Aug. Intergrading somewhat into *P. hippiana.*

Potentilla hippiana Lehmann — woolly cinquefoil — With 2 vars.:

1 Calyx primarily tomentose, sometimes with a few coarse hairs intermixed; leaflets mostly about equally gray or white on both sides, mostly with fewer than 10 teeth . . . . **var. *effusa***
1 Calyx primarily sericeous, sometimes also tomentose; leaflets usually greenish above, often with more than 10 teeth . . . . **var. *hippiana***

Var. *effusa* (Douglas ex Lehmann) Dorn [*P. effusa* (Douglas) Lehmann] — branched cinquefoil — This densely white-pubescent plant known from British Columbia to Manitoba and south to New Mexico reaches Utah in Daggett Co. where it is occasional to common in mtn. big sagebrush, dry subalpine meadow, and lodgepole pine communities at 7,650-9,000 ft. This taxon is reported to intergrade into *P. hippiana* with which it is sympatric (Hitchcock et al. 1961). However, plants of the 2 taxa appear to be readily separable within our area.

Var. hippiana (*P. gracilis* var. *hippianoides* S. L. Welsh & N. D. Atwood) Occasional to common in the Uinta Mtns. and apparently rare on the E. Tavaputs Plateau in sagebrush-grass communities, subalpine dry meadows, and open rocky ground; 7,400-10,600 ft; June-Aug.

Potentilla nivea L (*P. hookeriana* Lehmann misapplied) — snow cinquefoil — Apparently rare, specimens seen are from the alpine of the n. slope and Provo River drainage of the Uinta Mtns. July-Aug.

Potentilla norvegica L. (*P. monspeliensis* L.) — Norwegian cinquefoil, rough cinquefoil — Strawberry Valley, Uinta Mtns., and E. Tavaputs Plateau; occasional in wet meadows, sagebrush-grass communities, pine and spruce woods, lake margins, and alluvial bottoms; 7,600-11,000 ft; July-Aug.

Potentilla ovina Macoun — sheep cinquefoil — With 2 vars. in our area:

1 Leaflets densely to uniformly sericeous, grayish-green or gray, often with a lower layer of sparse tomentum, 5-10 mm long, often with more than 6 teeth; leaf rachis 2-6 cm long .... **var. ovina**
1 Leaflets glabrous to sparsely sericeous-strigose, usually green, never tomentose, 10-20 mm long, often with 3-5 teeth; leaf rachis 3-12 cm long
.... **var. decurrens**

Var. decurrens (S. Watson) S. L. Welsh & Johnston [*P. decurrens* (S. Watson) Rydberg; *P. plattensis* Nuttall misapplied] — decurrent cinquefoil — Uinta Mtns.; common in dry and moist meadows, pine and spruce woods, krummholz, and alpine tundra, rarely in mtn. big sagebrush communities; 8,100-12,000 ft; June-Aug.

Var. ovina — sheep cinquefoil — Occasional at scattered locations in the Uinta Mtns. where usually associated with limestone or other basic substrates; 8,530-11,100 ft; June.

Potentilla palustris (L.) Scopoli (*Comarum palustre* L.) — purple cinquefoil, marsh cinquefoil — The 4 specimens seen are from morainal land forms of S. Fork Ashley Creek drainage, Uinta Mtns. in bogs and margins of ponds; 9,640-9,740 ft. July-Aug.

Potentilla paradoxa Nuttall — paradox or contrary cinquefoil — Specimens seen are from mud flats of the Flaming Gorge Reservoir and drying pools and mud flats at Myton and along the of the Green River at Browns Park and Leota Bottoms at 4,700-6040 ft; June-Aug.

Potentilla pensylvanica L. — Pennsylvania cinquefoil — Scattered locations in Strawberry Valley, Uinta Mtns., and W. Tavaputs Plateau in sagebrush-grass and pinyon-juniper communities and exposed windswept ridges; 6,000-10,000 ft; June-Aug. Our plants belong to var. *pensylvanica*.

Potentilla recta L. — sulphur or erect cinquefoil — Introduced from Eurasia; the one specimen seen (Huber 3359) is from the mouth of Whiterocks Canyon in a riparian community at 6,800 ft. Reported for Ouray National Wildlife Refuge, plotted for Uinta Co. Wyoming in TPD, and, reported for Cache, Weber, and Utah Cos., Utah (Corbin 2007). This weedy species can be expected to spread in our area.

Potentilla rivalis Nuttall [*P. leucocarpa* Rydberg; *P. pentandra* Engelmann; *P. rivalis* var. *millegrana* (Engelmann) S. Watson] — brook cinquefoil, river cinquefoil — Five specimens seen are from the floodplain of the Green and White rivers and Neese 14839 is from mud at the margin of Crouse Reservoir on Diamond Mt.; 4,655-7,225 ft; June-Sept.

Potentilla rubricaulis Lehmann [*P. saximontana* Rydberg; *P. pensylvanica* var. *paucijuga* (Rydberg) S. L. Welsh & B. C. Johnston] — Rocky Mtn. cinquefoil, redstem cinquefoil — Scattered locations in the Uinta Mtns. on slopes and ridges above timberline at 10,500-11,200 ft and one from Brownie Canyon at 7,900 ft; July-Aug. Reported for Zenobia Peak (Bradley 1950) but specimens we have seen from there belong to *P. ovina*. Reports of *P. quinquefolia* Rydberg might be based on specimens of *P. rubricaulis* or of *P. nivea*.

Prunus L. — Cherry; Plum

Trees or shrubs; leaves alternate, mostly serrate, usually with glands either along the margins of the leaf base or at the petiole apex; floral cup mostly campanulate; sepals 5; petals 5; stamens 20-30; pistil 1; ovary superior; fruit a drupe.

---

1   Flowers and fruit borne in elongated racemes; fruit cherry-like with a round stone; plants native .... *P. virginiana*

1   Flowers and fruit borne singly or 2-4 in umbellate clusters; fruit a plum with a flattened stone; plants cultivated and escaping, usually along ditch banks and fence lines .... *P. americana*

Prunus americana Marsh — American plum, wild plum — Cultivated, persisting, occasionally escaping along fencelines, ditches and natural drainages at lower elevations, not expected above 7,500 ft; May-June.

Prunus virginiana L. [*P. melanocarpa* (A. Nelson) Rydberg; *Padus virginiana* (L.) Miller] — chokecherry — Widespread and common in canyons and mtn. sides; 6,500-9,000 ft; May-June. Our plants belong to var. *melanocarpa* (A. Nelson) Sargent. Young racemes have been mistaken for the catkins of birch. The fruit has been used for making jelly and syrup.

Purshia de Candolle — Bitterbrush

Shrubs; twigs of the season pubescent, often glandular; leaves 3-5 lobed, green above, whitish-pubescent beneath; flowers spicy-fragrant; floral cup turbinate to narrowly campanulate; sepals 5; petals 5-10 mm long, yellow when fresh; pistils 1 to 12; fruit an achene.

---

1   Leaves commonly with 5 lobes; styles 4 or more, plumose and 2-4.5 cm long in fruit .... *P. stansburiana*

1   Leaves with 3 lobes; styles 1, not plumose, less than 1 cm long ....
.... *P. tridentata*

Purshia stansburiana (Torrey) Henrickson (*Cowania mexicana* D. Don, *C. stansburiana* Torrey). — Stansbury cliffrose — Known from Desolation Canyon and persisting in plantings on Gilsonite Ridge and elsewhere; May-June. If treated as part of the *C. mexicana* D. Don complex, our plants are assignable to *C. m.* var. *stansburiana* (Torrey) S. L. Welsh.

Purshia tridentata (Pursh) de Candolle — bitterbrush — Widespread; abundant in the Uinta Mtns., less common on parts of the Tavaputs Plateau; sagebrush, pinyon-juniper, and mtn. brush communities and rarely extending down into salt desert shrub communities on sandy soils; 5,600-9,000 ft; May-June. Mule deer show high preference for bitterbrush in the fall and early winter before the leaves fall. It is used later in the year, but likely at less intensity.

Rosa L. — Rose

Shrubs or woody vines; branches typically armed with prickles; leaves alternate, pinnately compound with 3-11 toothed leaflets; floral tube globose or ellipsoid, often constricted near the apex; sepals 5; petals 5 (in our native species); pistils few-numerous, the ovaries superior; fruit hardened achenes enclosed in a fleshy red, orange or purple floral tube (hip). A number of cultivated species are planted in our area. Some of these are persistent. None are included here.

1   Leaflets commonly double serrate with the teeth usually gland-tipped, the larger teeth 3 or more times longer than the small ones, some of the small ones essentially reduced to glands; flower small as in *R. woodsii* and usually solitary as in *R. nutkana* .... *R. manca*

1   Leaflets mostly singly serrate, the teeth often without glands 2

2   Flowers usually 2-5 in small cymes; sepals 2-3.5 mm wide at the base; petals to 2.5 cm long, floral tube 3-5 mm wide at flowering, 6-12 (20) mm long in fruit .... *R. woodsii*

2   Flowers usually solitary; sepals 3-6 mm wide at the base; petals 2.5-4 cm long; floral tube 5-9 mm wide at flowering and 15-22 mm long in fruit.... .... *R. nutkana*

Rosa manca Greene — Mancos rose — Two specimens seen as follows: W. Barmore sn from Douglas-fir above Hell's Half Mile in Split Mtn. Gorge and Goodrich 22716 from burned lodgepole pine forest on Alma Taylor Plateau.

Rosa nutkana Presl (*R. melina* Rydberg) — Nootka rose — Widespread and occasional in the Utah portion of our area in cottonwood, mtn. brush, aspen, Douglas-fir, and spruce-fir communities; 6,000-10,000 ft or higher; May-Aug.

Rosa woodsii Lindley [*R. chrysocarpa* Rydberg; *R. puberulenta* Rydberg; *R. ultramontana* (S. Watson) A. Heller] — woods rose — Widespread and common; in numerous plant communities, often-along waterways; 4,700-10,000 ft and perhaps higher; May-Aug. The closely related and possibly synonymous *R. acicularis* Lindley was listed for our area by Graham (1937), and IMF (3A: 134) indicates this circumboreal species might show up in the Uinta Mountains. However, 2 taxa are not apparent in specimens observed for this treatment.

Rubus L. — Raspberry; Thimbleberry

Shrubs; stems biennial from a perennial base, with or without prickles; leaves of first years shoots different from those of flowering stems of the 2-year-old shoots, alternate, simple or ternately to pinnately compound; flowers solitary or clustered; floral cup saucer-shaped or campanulate; sepals 5; petals 5, white; stamens 15 or more; ovaries superior; fruit a cluster of weakly coherent drupelets.

1   Stems armed with prickles; leaves compound .... **R. idaeus**
1   Stems unarmed; leaves simple .... **R. parviflorus**

Rubus idaeus L. (*R. melanolasius* Dieck; *R. sachalinensis* H. Lev; *R. strigosus* Michaux; *Oreobatus deliciosus* James) — red raspberry — Widespread; canyons and mountains, stream sides, often-rocky places; 7,000-9,000 ft or higher; May-July. This is the same species that is cultivated for raspberries. Native plants are referable to var. *strigosus* (Michaux) Maximowicz. The cultivated material is mainly referable to ssp. *idaeus*.

Rubus parviflorus Nuttall [*Rubacer parviflorum* (Nuttall) Rydberg — western thimbleberry — Strawberry Valley and across the Uinta Mtns. at least to Whiterocks Canyon; canyons usually in woods; 7,000-9,500 ft; June-July.

Sanguisorba L. — Burnet

Sanguisorba minor Scopoli — small burnet — Perennial herbs from a simple or branched caudex, monoecious or dioecious; stems 1-several, erect, 20-60 cm tall; leaves pinnately compound; leaflets mostly 9-21, 6-20 mm long, crenate to serrate; flowers mostly unisexual, the pistillate borne above the staminate, sessile in a compact terminal head-like cluster, the cluster 8-20 (30) mm long, to about 1 cm wide, globose ovoid or oblong; floral tube ovoid; sepals 4, 3-5 mm long, green or red; petals lacking; stamens 10 or more; pistils 2, the ovaries superior with a terminal style and branched stigma; fruit of 1-2 achenes enclosed in a 4 angled, woody, narrowly winged floral tube 5-8 mm long. Introduced from Eurasia, used in seeding rangelands in pinyon-juniper, sagebrush, and perhaps other communities, usually persisting in abundance for only a few years after seeding; May-July.

Sibbaldia L.

Sibbaldia procumbens L. — sibbaldia — Perennial herbs, densely tufted or even mat forming, from short rhizomes; stems 5-10 cm long; leaves ternate compound, the leaflets 3-20 (30) mm long, wedge-shaped, rounded and 3-5 toothed at the apex, mostly basal, those of the stem much reduced; flowers in dense axillary and terminal cymes; floral cup saucer-shaped; sepals 5, 2.5-4 mm long, spreading at flowering time, erect at fruiting, alternating with 5 sepal-like bracts; petals 5, about half as long as the sepals, yellow to nearly white; stamens 5; ovaries superior, the deciduous style about 1 mm long; fruit of 5-10 (20) ovoid achenes, these about 1.5 mm long, short stipitate on a slightly enlarged receptacle. Uinta Mtns.; occasional in fir, lodgepole pine, spruce, meadow, krummholz, and alpine communities; 9,000-12,000 ft; July-Aug. In a classification of alpine plant communities of the Uinta Mountains, sibbaldia was found in about ½ of the plant communities and with 80%-100% consistency in snowbed communities and mostly 50% or less consistency in other communities and totally lacking in fell field communities (Brown 2006).

Sorbus L. — Mountain-ash

Sorbus scopulina Greene — mtn.-ash — Shrubs about 3-5 m tall; young twigs with long soft hairs; leaves pinnately compound with 11-15 leaflets; leaflets 2-9 cm long, 0.7-3 cm wide, serrate; flowers small, congested into a flat-topped inflorescence; floral cup turbinate; sepals 5; petals 5, 4-6 mm long, white; ovary inferior; styles 2-5, the stigmas capitate; fruit a pome, red or orange, 6-10 mm in diameter. Infrequent in Strawberry valley and glacial and stream canyons of the Uinta Mtns. Reported (Graham 1937) for W. Fork Douglass Creek, E. Tavaputs Plateau. This is more common in the Wasatch Range than in our area.

RUBIACEAE Bedstraw Family

Galium L. — Bedstraw; Cleavers

Annual or perennial herbs, sometimes shrubby at the base; stems 4-angled or sometimes terete; leaves opposite or apparently whorled, entire, sessile or short-petiolate; stipules free or nearly so, often as large as the leaves, thus with the leaves appearing whorled; flowers unisexual or bisexual, small, 3-4 merous, in axillary or terminal cymes or cymose panicles, rarely solitary; calyx obsolete; corolla white or greenish-white or cream, usually rotate or cup-shaped, the tube shorter than the 3-4 flaring lobes; ovary inferior and 2-lobed; styles 2, free or fused basally, each with a minute globose stigma; fruit a schizocarp with 2 globose or ellipsoid, indehiscent, dry mericarps.

1 Stems usually many from a woody caudex, more or less woody below and often freely branched; flowers unisexual; fruits densely covered with long straight hairs; plants of dry places below 7,400 ft . . . . **G. multiflorum**

1 Stems 1-several from a taproot or rhizome, not woody at the base; flowers bisexual; fruits-glabrous or variously pubescent; habitat various 2

2 Flowers more or less numerous in a terminal compound, conspicuous inflorescence; stems erect; plants perennial from rhizomes; fruits glabrous or pubescent; stems various. 3

2 Flowers 1 or few together, axillary; inflorescence rather inconspicuous; stems often prostrate or ascending and scrambling over other vegetation; fruits often pubescent; stems often hispid on the angles 5

3 Ovaries and fruits with short, hooked hairs, or some of them nearly glabrous; margins of leaves and angles of stems with retrorse scabrous hairs; stems rather weakly ascending, tending to recline; plants about 10-30 cm tall, rather rare, known from the Piceance Basin . . . . **G. mexicanum**

3 Ovaries and fruits glabrous; leaves and stems glabrous or somewhat scabrous, but then the hairs not noticeably retrorse; stems rather stout, erect; plants 20-80 cm tall, not restricted as above 4

4 Flowers bright yellow; leaves 6-8 per whorl, 1-nerved . . . . **G. verum**

4 Flowers white; leaves 4 per whorl, 3-nerved . . . . **G. boreale**

5 Leaves 2-4 per whorl, often only 2 at some of the upper nodes; stems erect or ascending; flowers solitary in leaf-axils; fruit with hooked hairs; plants annual . . . . **G. bifolium**

5 Leaves 4-6 (8) per whorl; stems often trailing to ascending, often scrambling over other vegetation; flowers usually more than one from at least some of the leaf axils 6

6 Margins and sometimes midnerves of leaves as well as angles of stems retrorse-hispid . . . . **G. aparine**

6   Leaves not pubescent as above; stems retrorse hispid or glabrous....... 7
7   Fruit glabrous, nearly black at maturity; leaves 1-6 mm wide .. *G. trifidum*
7   Fruit with hooked bristles; leaves 5-15 mm wide .... *G. triflorum*

Galium aparine L. — catch bedstraw, cleavers, goose-grass — Indicated for Strawberry Valley (Albee et al. 1988) and found at Jones Hole and Little Hole. Most plants of the Intermountain area belong to var. *echinospermum* (Wallroth) Farwell.

Galium bifolium S. Watson — twinleaf bedstraw, low mt. bedstraw — Widespread; occasional in mt. brush, aspen, forb, and spruce-fir communities; 6,800-10,600 ft; June-Aug.

Galium boreale L. — northern bedstraw — Common in Strawberry Valley, and across the Uinta Mtns., apparently rare on the Tavaputs Plateau; mt. brush, aspen, riparian, and various coniferous forest communities and meadows; 7,000-10,500 ft; June-Aug.

Galium mexicanum H.B.K. — Mexican bedstraw — The one specimen seen (Neese et al. 11978) is from 6 mi w. of Rio Blanco, Piceance Basin, sandstone talus; pinyon-juniper belt; 7,200 ft; July. The specimen has leaves that are only 3-10 mm long, 1-3 mm wide and widest at or above the middle. It is referable to ssp. *asperrimum* (A. Gray) Dempster. Ssp. *asperulum* (A. Gray) Dempster with somewhat larger leaves is known from the Wasatch Range.

Galium multiflorum Kellogg (*G. coloradoense* Wight) — many-flowered bedstraw, shrubby bedstraw — Occasional or locally common in Uintah Co. e. of Whiterocks River, and in Rio Blanco and Moffat Cos.; desert shrub, pinyon-juniper, and dry mt. brush communities, often on rocky or sparsely vegetated slopes; 5,700-7,400 ft; May-June, with fruit through Aug. Plants with leaves 5-10 times longer than wide belong to var. *coloradoense* (Wight) Cronquist. Those with leaves 1.5-6 times longer than wide belong to var. *multiflorum*.

Galium trifidum L. (*G. brandegei* A. Gray) — small bedstraw, small cleavers — Strawberry Valley and across the Uinta Mtns.; willow-streamside, aspen, meadow, and riparian communities, and lake-margins; 6,000-11,000 ft; June-Aug. Our plants belong to var. *subbiflorum* Wiegand.

Galium triflorum Michaux — sweet-scented bedstraw, fragrant bedstraw — Widespread; occasional in riparian and meadow communities; 7,600-8,200 ft; July-Aug.

Galium verum L. — yellow bedstraw — Introduced from Europe, cultivated, sometimes escaping. The two specimens seen (Goodrich 26103 and 27883) were collected from a roadside and pastureland at the mouth of Birch Creek, Daggett Co.; 7,610-7,700 ft. July-Aug. This appears to be an invasive species that could impact agricultural production.

RUPPIACEAE Ditch-grass Family

Ruppia L. — Ditch-grass

Ruppia maritima L. — ditch-grass — Aquatic herbs with filiform branching stems, 60-100 cm long; leaves opposite or alternate, thread-like, 2-10 cm long, not over 0.5 mm wide, with a basal stipular sheath 6-10 mm long; flowers bisexual, minute, 1-3 in axillary, pedunculate spikes, the peduncles 2-30 mm long, straight or coiled; perianth lacking; stamens (1) 2; pistils 4 (8); fruit small, indehiscent, drupe-like, borne on a slender stipe that progressively elongates as the fruit matures and often becoming twisted, the stipes 1-2.5 cm long in fruit. The 2 specimens seen (Neese 14655, 15319) are from Pelican Lake. Aug.

SALICACEAE Willow Family

Dioecious dwarf shrubs to large trees; leaves alternate, simple, entire, serrate, crenate, rarely lobed, usually stipulate, but the stipules often readily deciduous; flowers borne in aments (catkins) without a perianth, each subtended by a small, scale-like bract (commonly referred to as a scale); staminate flowers of (1) 2-many stamens; pistillate flowers of a single pistil with 2-4 carpels and as many stigmas; placentation parietal or basal; fruit a sessile or stipitate capsule with 2-4 valves; seeds numerous, small, covered with long white hairs, dispersed easily by wind.

---

1 Trees with pendulous aments; leaf-buds covered by several, usually resinous scales, each flower subtended by a cup-shaped disk, without obvious glands; stamens 6-many; scale-like bracts subtending the flowers laciniate or fimbriate (except in P. alba) otherwise glabrous or ciliate .... *Populus*

1 Trees, shrubs, or dwarf shrubs with mostly ascending to erect aments; leaf buds covered by a single non-resinous scale; each flower subtended by 1 or 2 basal glands, but without a disc; stamens (1) 2-8, rarely more; scale-like bracts subtending the flowers entire or occasionally shallowly toothed, often densely pubescent .... *Salix*

Populus L. — Cottonwood; Poplar

Small to large trees; leaf buds covered by several overlapping scales, resinous in most taxa; aments pendulous, mostly appearing before the leaves, and often soon deciduous, the scale-like bracts quickly deciduous, deeply lobed to laciniate, often dilated (entire or nearly so and not dilated in *P. alba*); each flower subtended by a cup-like disk; stamens 6-60 or more, the filaments free; inserted on the disk; capsules pedicellate, with 2-4 valves, glabrous in our taxa.

---

1 At least some of the mature leaves deeply 3-5 lobed and aceriform, often densely tomentose beneath; bracts of flowers entire or shallowly toothed, long pilose-ciliate; twigs of the season and winter buds often white-woolly; stigma lobes slender; plants introduced, cultivated and escaping .. *P. alba*

1 Leaves not deeply lobed, not aceriform, merely toothed, glabrous or nearly so; scales of flowers deeply lobed to lacerate 2

2 Bark white and smooth except blackened and rough where scarred and with age, covered with a whitish powdery bloom; bracts of flowers more or less persistent, deeply lobed or cleft, ciliate with long white hairs; leaves orbicular to reniform-cordate; bud scales shiny but hardly resinous; stamens 6-14; capsules 4-6 mm long, with 2 carpels; stigma lobes slender; plants not confined to watercourses .... *P. tremuloides*

2   Bark turning gray or brown and roughly furrowed on older trunks; bracts of flowers laciniate-fringed, otherwise glabrous or inconspicuously short hairy; stamens 12-60 or more; capsules mostly over 6 mm long, with 2-4 carpels; stigmas broadly dilated; plants mostly cultivated or growing along water courses or edges of lakes and ponds 3

3   Leaves 0.6-1.3 times as long as wide, deltoid to rhombic or ovate; petioles compressed laterally, often 0.6-1.2 times as long as the blade. 4

3   Leaves (1) 1.2-7 (10) times longer than wide, ovate to lanceolate; petioles terete or dorsiventrally compressed, commonly 0.2-0.6 times as long as the blades. 6

4   Bud scales and twigs of the season pubescent; leaf blades commonly with 4-10 (15) fine to coarse teeth on each side; branches widely spreading and the crown often as broad or broader than the tree is tall; plants native, sometimes cultivated, mostly common along the Green River and its major tributaries . . . . ***P. deltoides***

4   Bud scales and twigs mostly glabrous; leaf blades commonly with 15-25 (30) fine teeth on each side; branches ascending to erect and the crown mostly longer than wide; plants introduced, cultivated, sometimes persisting. . . 5

5   Crowns narrow and columnar with branches comparatively small and strongly ascending to erect (in the trees planted in our area); leaf blades rhombic-ovate, cuneate at the base, seldom over 7 cm long; capsules 2-valved . . . . ***P. nigra***

5   Crown not columnar, the branches comparatively large and spreading-ascending; leaf blades more or less deltoid or broadly ovate, broadly cuneate at the base, some regularly over 7 cm long; capsules with 2 or more valves . . . . ***P. × canadensis***

6   Leaf blades (1.8) 2.5-6 (9.5) times longer than wide; petioles 0.2-0.3 (0.4) times as long as the blades, dorsiventrally compressed; carpels 2 . . . . ***P. angustifolia***

6   Leaf blades 1-2.4 times as long as wide; petioles 0.3-0.6 times as long as the blades, subterete or somewhat flattened; carpels 2 or 3; plants of apparent hybrid origin, intergrading into *P. angustifolia* on one hand and into *P. fremontii* and other broadleaf poplars on the other . . . . ***P. × acuminata***

Populus × acuminata Rydberg (*P. × intercurrens* Goodrich & S. L. Welsh) — lanceleaf cottonwood — Scattered locations along ephemeral and permanent streams, edges of ponds and lakes, often in mouths of canyons where the parental types (narrowleaf cottonwood and broadleaved species) come together and in isolated locations in absence of or at long distance from either of the supposed parents; 5,000-7,800 ft. A few trees in Sowers canyon are well isolated from either of the parental species. The name *P. × acuminata* is used here in a broad sense to apply to crosses of *P. angustifolia* with all broad-leaved species.

Populus alba L. — white poplar — Introduced from Eurasia, cultivated, escaping, more or less naturalized; fence lines, ditch banks, abandoned homesteads, and neglected fields; up to about 6,500 ft; May-June. The bark is white and smooth on young and mid-aged stems and black and fissured toward the base of older trees.

Populus angustifolia M. E. Jones — narrowleaf cottonwood — Widespread; common and locally dominant along water courses, often in canyons; 5,000-7,500 8,000) ft; May-June.

Populus × canadensis Moench — Carolina poplar, gray poplar — Probably originated in France as a cross between *P. deltoides* Marsh. and *P. nigra,* widely cultivated, persisting at old homes and occasionally escaped along ditches or other water courses; up to about 7,000 ft; May-June. Also distributed as Carolina popular are trees of a staminate clone (*P. × eugenei* Simon-Louis) that is intermediate between *P. deltoides* and Lombardy popular (IMF 2B: 119).

Populus deltoides Bartram ex Marshall [*Populus fremontii* S. Watson; *P. wislizeni* (S. Watson) Sargent] — cottonwood — Common to abundant along the floodplain of the Green River and up to about 6,000 ft or perhaps 7,000 ft along its major tributaries, and at scattered locations away from drainages, perhaps occasionally cultivated. Most of our trees are referable to *P. deltoides* var. *fremontii* (S. Watson) Cronquist — Fremont cottonwood — *P. deltoides* var. *wislizeni* (S. Watson) Dorn — Rio Grande cottonwood — is also reported for e. Utah, w. Colorado and s. Wyoming (IMF 2B: 122). These two taxa are separated by length of fruiting pedicels [2-5 (7)] mm in the former and [(5) 7-15] in the latter and width and shape of the floral disk (4-9 mm wide and cup-shaped in the former and 2-4 mm wide and saucer-shaped in the latter).

Populus nigra L. — black poplar — Introduced from Europe for shade and windbreaks. Trees of this species include those from a staminate clone with strongly ascending branches that produce a narrow, nearly cylindrical crown. Trees of this clone belong to var. *italica* Duroi (Lombardy poplar).

Populus tremuloides Michaux [*P. aurea* Tidestrom; *P. tremula* (L.) ssp. *tremuloides* (Michaux) Love & Love] — aspen, quaking aspen, quakey — widespread and common to dominant and forming large clones and aggregates of clones in canyons and on mtn. sides; 7,000-10,000 ft or cultivated at lower elevations; May-June.

Salix L. — Willow

Depressed, mat-forming dwarf shrubs to large trees; buds covered with 1 non-resinous scale; aments erect to spreading, rarely drooping, developing before (precocious), with (coetaneous), or after (serotinous) the leaves, the bracts mostly entire, occasionally with a slightly toothed apex; flowers with 1, occasionally 2 minute glands near the base; stamens (1) 2-8 (12), the filaments free or united toward the base, inserted on the base of the bracts; capsules sessile or stipitate, glabrous or pubescent.

Identification of the willows is compounded by unisexual plants, aments that are sometimes precocious and mostly early deciduous, and variations between the usually smaller leaves of the flowering branches that often lack or have inconspicuous stipules and the usually much larger leaves and stipules of vegetative branches and particularly of vigorous young shoots. Thus, herbarium specimens of each species present specimens of 3 or 4 phases (pistillate, staminate, flowering twigs with or without the deciduous aments, and vegetative twigs). Vigorous young shoots sometimes add a 5th dimension. At times whole plants in the field present only 1 or 2 of the various phases.

To facilitate identification of plants of the different phases, pistillate, staminate and vegetative features have been included in many of the leads in the key. Thus, some of the leads are rather long and features not applicable to a particular specimen will need to be skipped. An alternative approach to lengthy leads is separate keys for the different sexual and vegetative phases. Many such keys have been written, but these sometimes also contain a mixing of vegetative and sexual features. Reference: Dorn. R. D. 2005. *Salicaceae.* in Holmgren et al. *Intermountain Flora,* Vol. 2 Part 3: 126-160.

---

1   Plants shrubs or dwarf shrubs not over 1 (1.5) m tall, subalpine to alpine 2

1   Plants shrubs or trees, mostly over 1.5 m tall, of valleys to montane 3

2   Plants depressed dwarf shrubs 1-10 (20) cm tall mostly alpine, often forming mats, the stems creeping on or below the ground surface . . . . **KEY 1**

2   Plants (10) 20-100 cm tall or taller, subalpine or alpine, not forming mats on the ground, the stems ascending to erect . . . . **KEY 2**

3   Leaves (8) 10-20 (32) times longer than wide; plants often strongly colonial, spreading underground and forming patches and occasionally thickets. . 4

3   Leaves less than 8 times as long as wide 5

4   Floral bracts usually somewhat pointed at the tip, hairy; expanded leaves usually hairy, dull green to grayish, usually not veiny; our most common and widespread lowland willow . . . . ***S. exigua***

4   Floral bracts usually somewhat rounded at the tip, glabrous or sometimes hairy at the base or on the margins; expanded leaves mostly glabrous or glabrate, deep green and usually prominently veiny . . . . ***S. melanopsis***

5   Bracts persistent, dark brown to blackish, or if pale green or pale brown in age then silky pilose with the hairs exceeding the bracts by 1-2 mm and the capsules pubescent (rarely glabrous in unusual specimens); stamens 2 per flower, the filaments glabrous or pilose in a few species; plants shrubs or occasionally tree-like, native . . . . **KEY 3**

5   Bracts of at least the pistillate aments quickly deciduous, pale green or yellowish tan in age, short pubescent, the hairs hardly if at all exceeding the bract by more than 1 mm; capsules glabrous; stamens more than 2 per flower, or if only 2 then plants introduced trees, the filaments pilose; plants mostly trees or tree-like except in S. lucida, mostly of valleys and lower montane 6

6   Plants native, shrubs or small trees; stamens 3-9 per flower; stipes of capsules mostly 1-2 mm long, obviously longer than the gland . . . . **KEY 4**

6   Plants introduced trees; stamens 2, capsules sessile or the stipes mostly less than 1 mm long and hardly longer than the gland . . . . **KEY 5**

— KEY 1 —

1   Bracts of aments pale green or yellowish, glabrous dorsally; filaments 1.5-2 mm long; style obsolete or to 0.2 mm long, shorter than the stigmas; leaves elliptic to orbicular, 1.4-2.6 times longer than wide, glaucous and strongly reticulate-veined beneath, the tips mostly rounded or obtuse ***S. reticulata***

1   Bracts of aments blackish, pilose dorsally; filaments over 2 mm long; styles 0.5 mm long or-longer, longer than the stigmas; leaves elliptic or narrow elliptic, (1.25) 2.3-4.7 times longer than wide, glaucous or not, not strongly reticulate veined beneath, the tips mostly pointed 2

2   Leaves 2-5 (7) mm wide, 2-4.7 times longer than wide, sessile or the petiole to 3 mm long; plants seldom over 3 cm tall, aments 0.5-2.2 cm long . . . . . .
. . . . ***S. cascadensis***

2   Leaves 5-20 mm wide, mostly 2-3 times longer than wide, with petiole 3-13 mm long; plants mostly 5-10 (20) cm tall; aments (1) 2-4 cm long . . . .
. . . . ***S. arctica***

− KEY 2 −

1   Mature leaves glabrous, dark green and shiny above, strongly glaucous and glabrous or with a few hairs beneath; twigs of the season glabrous or scattered pubescent, dark chestnut to lustrous purplish-black; aments precocious or coetaneous, sessile or nearly so or rarely on a stalk to 0.5 (1) cm long, this not bearing nor subtended by bract-like leaves; filaments of stamens glabrous . . . . ***S. planifolia***

1   Mature leaves pubescent on both sides, but sometimes glabrate or glabrous in age; aments coetaneous or sub-serotinous, borne on stalks to 2 (4) cm long, these usually bearing and subtended by bract-like leaves; style and stigmas collectively up to 1.5 mm long 2

2   Capsules glabrous; filaments of stamens glabrous; lower surface of leaves not glaucous but often more densely pubescent and thus lighter than the upper surface; twigs of the season glabrous or thinly villous-puberulent .
. . . . ***S. wolfii***

2   Capsules mostly pubescent at least until mature; filaments of stamens sometimes pilose; leaves often glaucous beneath; twigs of the current season densely pubescent 3

3   Bracts of aments pale green when young, tan in age; capsules 3-5 mm long, pubescent even in age, crowded and nearly sessile so as to mostly conceal the rachis at the center of the aments, the stipes seldom over 0.5 mm long; pistillate aments 0.8-2 (2.5) cm long, 8-10 mm wide; staminate aments about 0.8-1 (1.2) cm long, 5-6 mm wide, the filaments densely pilose at the base and for 1/2 to 3/4 their length, the pilose portion often equaling or exceeding the scale, the anthers usually less than 0.5 mm long; petioles 1-4 mm long, seldom exceeding the bud even on vegetative twigs . . . .
. . . . ***S. brachycarpa***

3   Bracts of aments brown to blackish, sometimes light brown to whitish-tan but not green even when young; capsules (4) 5-7 (8) mm long, sometimes glabrate in age, dense but often not so crowded as to conceal the rachis at the center of the ament, the stipes 0.5-2 mm long; pistillate aments (1.8) 2.5-5 cm long, 11-15 mm wide; staminate aments 0.8-2 (4) cm long, sometimes over 6 mm wide, the filaments glabrous or pilose but usually not so conspicuously pilose as above, the anthers mostly over 0.5 mm long; petioles (1) 2-6 (10) mm long, equaling or often exceeding the bud, especially on vegetative twigs . . . . ***S. glauca***

− KEY 3 −

1   Capsules glabrous; leaves not both glaucous and pubescent on the lower surface when fully expanded 2

1 Capsules pubescent or plants keyed both ways; leaves glaucous and pubescent on the lower surface when fully expanded 6

2 Leaves glaucous beneath, not or scarcely pubescent when fully expanded 3

2 Leaves not glaucous beneath, although sometimes lighter colored from pubescence; pubescent at-least in part on both sides when fully expanded, but sometimes glabrate in age (vegetative specimens of S. lucida might be keyed here but this has glabrous as well as non-glaucous leaves and the petioles often have wart-like glands near the base of the blade, and the plants are commonly 3-6 m tall or taller) 5

3 Aments sessile or on a stalk, the stalk to 0.5 (1) cm long, not bearing not subtended by bract-like leaves; pubescence of aments straight or nearly so; leaves mostly entire, often slightly revolute; twigs dark chestnut to lustrous purplish-black, essentially glabrous; plants often less than 1.5 m tall and also keyed in KEY 2 .... *S. planifolia*

3 Aments usually stalked, the stalk usually subtended by or bearing 1-4 bract-like leaves; pubescence of aments crisped-villous; leaves serrate, serrulate, or entire, not at all revolute; twigs variously colored, glabrous or those of the current season more often pubescent; plants often over 1.5 m tall (small, vegetative specimens of S. amygdaloides might be keyed here)......... 4

4 Styles 0.7-1.5 (1.8) mm long; filaments free or connate or for less than ¼ the length; leaves of fertile and vegetative twigs often less than 3 times longer than wide, evidently crenulate-serrate or nearly entire; bark of older twigs not ashy gray or whitish; plants apparently uncommon, mostly montane
.... *S. monticola*

4 Styles 0.2-0.7 mm long; filaments often connate for ¼ or more of the length; leaves of vegetative twigs 2-5 times longer than wide, serrulate or entire; bark of older twigs usually ashy gray or white; plants widespread, mostly of valleys and lower montane .... *S. eriocephala*

5 Aments precocious or coetaneous (1.5) 2-5 cm long, with dense crisped-villous, tangled hairs; leaves glabrous in age or with inconspicuous hairs, entire or sometimes serrulate; plants (1.5) 2-4 m tall .... *S. boothii*

5 Aments coetaneous, 0.8-1.5 (3) cm long, with hairs straight or nearly so; leaves permanently pubescent throughout on both sides even in age, the hairs readily conspicuous at 10X, entire; plants 0.6-1.5 (2) m tall, also keyed in KEY 2 .... *S. wolfii*

6 Twigs strongly blue glaucous, the bloom sometimes deciduous, but then the twigs glabrous or sometimes puberulent; larger leaves mostly 3-5 times longer than wide, sericeous beneath; capsules densely pubescent 7

6 Twigs not glaucous or those of the current season often pubescent, or leaves not sericeous; the larger leaves various but sometimes wider than above; capsules pubescent or glabrous.................................... 8

7 Pistillate aments 2-5 cm long; capsules sessile or the stipes to 1 mm long, the style and stigmas together 0.8-1.3 mm long; staminate aments about 2 cm long, the filaments glabrous; aments sessile or nearly so, with or more often without subtending bract-like leaves, precocious or sub-coetaneous; bracts of the aments blackish; leaves permanently silvery, silky-sericeous to sub-tomentose beneath, dark green and glabrous above in age ....
.... *S. drummondiana*

7   Pistillate aments 1-2 cm long; capsules stipitate, the stipes 2-3 mm long, the
    style and stigmas together about 0.5 mm long; staminate aments 8-15 mm
    long, the filaments pilose on the lower 1/2; aments borne on 2-10 mm long,
    bracteate-leafy stalks; coetaneous or sub-precocious; bracts of the aments
    dark at the tip and pale below; leaves sericeous when unfolding, sparsely
    or moderately sericeous especially beneath when fully expanded, glabrate
    in age especially above .... *S. geyeriana*

8   Plants shrubs 0.6-3 m tall, mid-montane to above timberline, the stems less
    than 4 cm thick; leaves mostly less than 2 cm wide, occasionally wider on
    vegetative twigs, elliptic to narrowly lanceolate .... **KEY 2**

8   Plants shrubs or small trees, commonly 3-4 m tall or taller, but sometimes
    shorter when young, of valleys or montane, the stems of mature plants
    often 4-10 cm thick or thicker, leaves sometimes over 2 cm wide, oblong,
    obovate, oblanceolate, or elliptic 9

9   Bracts of aments pale green or tan to light brown in age, silky pilose, the
    hairs exceeding the bract by about 1 mm; aments coetaneous; capsules long-
    beaked, loosely arranged so as to expose much of the rachis; filaments of
    stamens 3-6 mm long; leaves mostly elliptic, occasionally lanceolate or
    obovate; twigs of the season with mostly appressed or ascending hairs or
    occasionally glabrous .... *S. bebbiana*

9   Bracts of aments black or purplish black, reddish or pale only at the base,
    pilose, the hairs exceeding the bract by about 2 mm; pistillate aments
    precocious, or sub-precocious, the capsules not long beaked, densely
    arranged and mostly concealing the rachis; staminate aments strictly
    precocious, the filaments about 10 mm long; leaves obovate or oblanceolate;
    twigs of the season with mostly widely spreading hairs .... *S. scouleriana*

— KEY 4 —

1   Leaf blades glaucous and pale beneath, commonly 2-7 cm long (longer on
    young plants); plants usually small trees or tree-like with a solitary or few
    trunk(s), not known from above 6,000 ft; petioles without glands; bracts of
    aments 1-2 mm long; staminate aments 2-10 cm long, often over 3.5 times
    longer than wide; bud scales with free overlapping margins, usually pointed
    at the tip .... *S. amygdaloides*

1   Leaf blades not glaucous beneath, about equally colored on both sides,
    commonly 5.5-11 cm long longer on young shoots); plants mostly multi-
    stemmed shrubs; commonly found from 5,000-8,000 ft; petioles of larger
    leaves often with wart-like glands near the base of the blade; bracts of
    aments 3-4 mm long; staminate aments 1-3.5 cm long, 2-3.5 times as long
    as wide; bud scales fused, without free overlapping margins .... *S. lucida*

— KEY 5 —

1   Pistillate aments mostly over 3 cm long, the capsules 3-6 mm long; trees
    with spreading or upright branches, not weeping, frequently escaping from
    cultivation. 2

1   Pistillate aments 1-2.5 (3) cm long, the capsules 1-2.5 mm long; staminate
    aments to 4 cm long; trees seldom if ever escaping from cultivation, not
    listed in the text. 3

2   Twigs orange-yellow (this feature especially noticeable in fall, winter, and spring when leaves are absent); our specimens apparently all staminate .
.... ***S. × rubens***

2   Twigs greenish, grayish, or brownish, but not orange-yellow; pistillate and staminate trees common .... ***S. fragilis***

3   Trees with spreading and ascending branches, mostly with uniform globose crown (umbrella willow, globe willow) .... ***S. matsudana*** Koidzumi

3   Trees weeping, with slender, greatly elongate, pendulous branches 4

4   Leaves mostly 3-15 mm wide, mostly deciduous in Oct.; twigs often bright yellow; capsules sessile (weeping willow) .... ***S. babylonica*** L.

4   Leaves 15-22 mm wide, often persisting into Nov.; twigs greenish or yellow-green; capsules with stipe exceeding the gland; plant hybrids of *S. babylonica* × *S. fragilis* (Niobe or Wisconsin weeping willow) ... ***S. × blanda*** Andersson

Salix amygdaloides Andersson — peach-leaf willow — Many specimens seen from Uintah Co., only 1 each from Daggett and Duchesne Cos., also in Moffat Co. No specimens seen from Rio Blanco Co. but to be expected there; occasional along water courses, around ponds, and other wet places; up to about 5,600 ft; May-June.

Salix arctica Pallas (*S. anglorum* Chamisso var. *antiplasta* Schneider; *S. petrophila* Rydberg) — arctic willow — The few specimens seen are from the w. end of the Uinta Mtns.; near or above timberline, commonly found on limestone but not restricted to limestone. Our plants belong to var. *petraea* Andersson

Salix bebbiana Sargent — Bebb willow — Widespread in scattered locations along water courses; mostly 6,000-8,800 ft; May-June.

Salix boothii Dorn [*S. myrtillifolia* Andersson misapplied; *S. pseudocordata* (Andersson) Rydberg misapplied; *S. pseudomyrsinites* Andersson misapplied] — Booth willow — Widespread; common to dominant along water courses and other wet places in mountains; 7,000-9,000 (perhaps 10,000) ft; May-June. This is the common willow of our area with bright yellow twigs.

Salix brachycarpa Nuttall — barrenground willow, shortfruit willow — Common in the Rock Creek drainage and to the west and rare to the e. of this drainage in the Uinta Mtns.; one specimen from the W. Tavaputs Plateau in Wasatch Co., and 1 from the E. Tavaputs Plateau in Grand Co.; limestone or other basic substrates, dry open ground in the spruce-fir belt, rocky or talus slopes and in moist or wet places; 6,800-10,600 ft; June-Aug.

Salix cascadensis Cockerell — Cascade willow — Uinta Mtns.; occasional at and above timberline, frequent association with sibbaldia, alpine pussytoes, and Drummond rush indicate this is a snowbed species, as found in other mtn. ranges (Brown 2006); June-Aug.

Salix drummondiana Barratt in Hooker (*S. subcoerulea* Piper) — Drummond willow — Uinta Mtns., Strawberry Valley, and W. Tavaputs Plateau; occasional and locally abundant along water courses; 7,000-10,300 ft (rarely higher); May-June. This is usually more common along steep gradient, rocky streams than is *S. geyeriana*.

Salix eriocephala Nuttall [*S. lutea* Nuttall; *S. rigida* Muhlenberg misapplied; *S. r.* var. *watsonii* (Bebb) Cronquist misapplied] — graybark willow — Widespread; one of our most common valley and lower montane willows; along water courses and other wet places; up to 7,400 (7,700) ft; May-June. Most plants of the our area can be expected to belong to var. *watsonii* (Bebb) Dorn with var. *ligulifolia* (C. R. Ball) Dorn perhaps sparingly present (IMF 2B: 154).

Salix exigua Nuttall (*S. interior* Rowlee misapplied; *S. melanopsis* Nuttall misapplied) — coyote willow, dusky w., narrow-leaf w. — Widespread; our most common lowland willow; water courses, around ponds, fields and pastures where water is allowed to flood and other moist places; up to about 7,500 ft (8,900 ft on basic substrates); May-July. Specimens with pubescent capsules belong to var. exigua. Those with glabrous capsules (most of ours) belong to var. *stenophylla* (Rydberg) Schneider.

Salix fragilis L. — crack willow — Introduced from Eurasia, cultivated and escaping along ditches and other water courses; up to about 7,000 ft; May-June.

Salix geyeriana Andersson — Geyer willow — Uinta Mtns., Strawberry Valley, and W. Tavaputs Plateau; common along water courses and other wet places; 7,200-9,500 ft; May-July. More common in low gradient reaches of streams in meadows than S. drummondiana.

Salix glauca L. (*S. pseudolapponum* Seemann in Engler) — grayleaf willow, glaucous willow — Uinta Mtns.; common along water courses, talus slopes, edge of boulder fields, snowbank areas, and dry alpine tundra but then often near krummholz trees; 9,100-12,000 ft; June-Aug. Depauperate, alpine specimens are sometimes mistaken for S. brachycarpa.

Salix lucida Muhlenberg [*S. caudata* (Nuttall) A. Heller; *S. lasiandra* Bentham] — whiplash willow, long-leaf willow — Widespread (no specimens seen from the E. Tavaputs Plateau); occasional or common along water courses and other wet places; 5,000-8,000 ft; May-June. Our plants belong to var. caudata (Nuttall) Cronquist — caudate willow — Dorn (IMF 2B: 138) suggests *Salix lucida* and *S. lasiandra* could be maintained as separate species.

Salix melanopsis Nuttall — dusky willow — Dorn (IMF 2B: 140) includes n. Utah within the range of this species. It might be expected in the Uinta Mountains. Some specimens from Sheep Creek on the north slope of the Uinta Mountains apparently belong here.

Salix monticola Bebb ex Coulter (*S. pseudomonticola* Ball misapplied) — mtn. willow — Rare or at least rarely collected; known from along water courses and other wet places, specimens seen are from the E. Tavaputs Plateau. Reports from the Uinta Mtns. appear to be based on specimens of S. boothii.

Salix planifolia Pursh [*S. phylicifolia* var. *monica* Bebb) C. K. Schneider; *S. phylicifolia* ssp. planifolia (Pursh) Hiitonen] — planeleaf willow — Uinta Mtns.; common to dominant along water courses, often along streams in subalpine meadows, wet meadows and other wet places; 9,500-12,000 ft or rarely down to 7,400 ft; May-July. Dorn (IMF 2B: 149) suggests var. *monica* (Bebb) C. K. Schneider and var. *planifolia* are better treated as ecotypes than as distinct taxa.

Salix reticulata L. [*S. nivalis* Hooker; *S. n.* var. *saximontana* (Rydberg) Schneider; *S. saximontana* Rydberg] — snow willow — Uinta Mtns.; the 26 specimens seen are from all drainages of the Uinta Mtns. from the peaks of the Provo River Drainage east to Leidy Peak; alpine communities, often below talus and boulder fields; 11,000-12,750 ft; June-Aug.

Salix × rubens Schrank — golden willow — Introduced from Europe, cultivated and infrequently used in wildland plantings up to about 7,200 ft. When planted above about 7,000 ft this tree commonly fails to attain full height. Usually readily identified by the bright (nearly florescent) orange-yellow twigs that are particularly conspicuous when leaves are absent.

Salix scouleriana Barratt in Hooker (*S. flavescens* Nuttall) — Scouler willow — Uinta Mtns. and Strawberry Valley, no specimens seen from the Tavaputs Plateau; occasional at scattered locations on well-drained slopes in timber and-along water courses and other moist places; 7,600-9,300 ft; May-June.

Salix wolfii Bebb in Rothrock — Wolf willow — Common to abundant on the n. slope of the Uinta Mtns. and less common on the s. slope, one specimen is from Willow Creek drainage, W. Tavaputs Plateau; along water courses and other wet places; (8,100) 9,000-10,100 ft; June-July.

SANTALACEAE Sandalwood Family
Comandra Nuttall — Bastard Toadflax

Comandra umbellata (L.) Nuttall (*C. pallida* de Candolle) — bastard toadflax — Perennial glabrous, glaucous herbs 10-40 cm tall, with horizontal rhizomes fastened at their ends to the underground parts of other perennial plants; leaves alternate, simple, sessile, entire, 15-35 mm long, linear, lanceolate to oblong; flowers regular, bisexual, in small terminal or crowded axillary clusters appearing terminal, the clusters forming a corymbose inflorescence, petals lacking; perianth 5-parted, rose, pink, green, or white, 3-4 mm long; stamens 5; ovary inferior; fruit drupaceous, 1-seeded, globose or ovoid. Widespread and occasional to frequent in many plant communities, but most often in sagebrush and mt. brush communities; 5,000-9,600 ft; May-July. Our plants belong to var. *pallida* (A. de Candolle) M. E. Jones.

SAXIFRAGACEAE Saxifrage Family

Perennial herbs; leaves simple (sometimes deeply lobed), alternate or basal; flowers usually bisexual, regular or nearly so; the 4-5 sepals, 4-5 petals, and stamens usually borne on a well developed floral tube or cup, this often with a disk; the sepals sometimes appearing as lobes of the floral tube; fruit a capsule or 2 follicles.

1 Petioles essentially lacking or tapered gradually from the blades and shorter than the blades . . . . *Saxifraga*

1 Petioles abruptly distinct from the blades, often as long or longer than the blade 2

2 Leaves entire; inflorescence of a single flower . . . . *Parnassia*

2 Leaves with teeth or lobes or both; inflorescence with more than 1 flower 3

3 Leaves all basal 4

3 Leaves not all basal 6

4 Leaf blades coarsely toothed but not lobed, glabrous *Saxifraga odontoloma*

4 Leaves lobed and toothed, usually with some hair. 5

5   Inflorescence a raceme, each node with only 1 flower .... *Mitella*
5   Inflorescence a panicle with at least lower nodes with more than 1 flower
     .... *Heuchera*
6   At least some leaves parted or divided to the base or nearly so 7
     .... *Lithophragma*
6   Leaves lobed to about ½ the way to the base 8
7   Leaf blades with 9-12 lobes, the lobes sharply dentate; plants 20-30 cm tall,
     of Cathedral Bluffs, Rio Blanco Co. .... *Sullivantia*
7   Leaf blades with 3-7 lobes and the lobes not dentate; plants 1-15 (20) cm
     tall, of the Uinta Mtns. .... *Saxifraga*

Heuchera L. — Alumroot

Heuchera parvifolia Nuttall [*H. parvifolia* var. *major* Rosendahl, Butters & Lakela; *H. parvifolia* var. *microcarpa* (Rosendahl) Butters & Lakela; *H. parvifolia* var. *utahensis* (Rydberg) Garrett] — common alumroot — Scapose perennial herb from a branched caudex; (12) 20-70 cm tall, usually with gland-tipped or simple hairs; basal leaves orbicular or reniform, (1) 2-6 cm wide, wider than long, shallow to deeply 5-7 lobed, the lobes more or less cleft and crenate; inflorescence a narrow, racemose or spike-like panicle, somewhat expanded at maturity; flower tube yellow-green; sepals less than 2 mm long, recurved; petals about 1 mm long, white; stamens 5; styles 2, divergent; fruit a capsule, (4) 6-10 mm long. Widespread; occasional in pinyon-juniper, mtn. big sagebrush, mtn. brush, ponderosa pine, aspen, and Douglas-fir communities, usually in rocky places; 7,000-8,400 ft (rarely to timberline); May-Aug.

Lithophragma (Nuttall) Torrey & A. Gray — Woodland Star

Perennial herbs from bulblet-bearing, fibrous roots or rhizomes; herbage often stipitate-glandular, often reddish or purplish; leaves opposite or alternate, the lower ones the largest, on slender petioles, the blades orbicular to reniform, palmately 3-5 lobed, parted, or divided, or compound with 3-5 wedge-shaped segments that are again lobed or parted; upper leaves reduced, sometimes sessile; flowers in simple or compound racemes; sepals 5; petals 5, ternately or palmately cleft or divided with 3-5 lobes; stamens (7-9) 10.

1   Plants with few to several purple bulblets in the inflorescence and usually
     in axils of the upper leaves; inflorescence with (1) 2-4 normal flowers,
     sometimes branched, with dark purple stipitate glands; lower pedicels 1.5-
     3 times longer than the floral cup; floral cup campanulate, 3-4 mm long at
     anthesis, 4-6 mm long in fruit including the lobes (sepals); leaves glabrous
     or sparsely pubescent; petals mostly 5-lobed .... *L. glabrum*
1   Plants without bulblets in the inflorescence or leaf-axils; inflorescence a
     simple raceme, with whitish or pale purplish stipitate glands, mostly with
     4-7 flowers; lower pedicels 0.5-1.5 times longer than the floral cup; floral
     cup turbinate, (3) 4-6 mm long at anthesis, 6-7 mm long in fruit including
     the lobes; leaves moderately to densely pubescent; petals mostly 3-lobed
     .... *L. parviflora*

Lithophragma glabrum Nuttall [*L. bulbiferum* (also spelled *L. bulbifera*) Rydberg] — fringe-cup woodland star — Strawberry Valley, Uinta Mtns., Blue Mtn, Cold Spring Mtn. no specimens seen from the Tavaputs Plateau; occasional in mtn. big sagebrush, mt. brush, and aspen communities; 6,600-9,000 ft; April-June.

Lithophragma parviflora (Hooker) Nuttall — small-flower woodland star — Specimens seen are from Strawberry Valley and 1 from near Whiterocks, Albee et al. (1988) show one location in each of Duchesne and Uintah Cos. n. of Neola and Whiterocks, but this appears to be rare or lacking in much of the Uinta Mtns.; reported for Dinosaur National Monument (Bradley 1948; Holmgren 1962); sagebrush and aspen communities; 7,500-8,000 ft; April-June.

Mitella L. — Miterwort

Perennial herbs from rhizomes, scapose or the stem with 1 or 2 reduced, bract-like leaves; plants often stipitate-glandular especially in the inflorescence; basal leaves more or less cordate-orbicular, more or less palmately lobed and crenate or serrate; inflorescence a spike-like raceme; sepals 5; petals 5, mostly lobed or dissected; stamens 5; ovary more or less inferior; fruit a capsule.

---

1   Inflorescence strongly secund, seldom with more than 1 flower per node, the pedicels 0.5-2 mm long; petals 2-3 lobed at the tip, rarely entire; stamens opposite the sepals; calyx lobes whitish, oblong, over 1 mm long; leaf margins with low rounded lobes, these mucronate or not, often stiff-ciliate; scapes often hirtellous or hirsute as well as glandular toward the base ... . *M. stauropetala*

1   Inflorescence seldom strongly secund, often with 2 flowers at some of the lower nodes, the pedicels 2-7 mm long; petals pinnately dissected into 4-10 filiform segments; stamens alternate with the sepals; calyx lobes greenish, triangular, less than 1 mm long; leaf margins with acute or rounded teeth, the teeth mucronate with a somewhat glandular short point, not stiff-ciliate; scapes glandular or glabrate but not hirtellous or hirsute toward the base . . . . *M. pentandra*

Mitella pentandra Hooker — five star miterwort — Specimens seen are from the Uinta Mtns. and Strawberry Valley, and 1 from the W. Tavaputs Plateau; moist woods and along streams; 7,800-10,800 ft; June-Aug. Separated by a series of small features from M. stauropetala, but still easily confused especially in specimens with a somewhat secund inflorescence. The petals are strikingly different. However, they are inconspicuous and often early deciduous.

Mitella stauropetala Piper (*M. stenopetala* Piper) — small-flower miterwort — The few specimens seen are from Strawberry Valley and Rock Creek and N. Fork Duchesne drainages on the s. slope of the Uinta Mtns. and from the Bear River drainage and west on the n. slope, apparently lacking in the central and eastern Uinta Mtns.; along streams and other wet places; 7,000-10,000 ft; May-Aug.

Parnassia L. — Grass-of-Parnassus

Perennial herbs, scapose the scape sometimes with a solitary, sessile bract; flowers solitary, terminal; sepals 5; petals 5; fertile stamens 5, with a cluster of more or less fused often fleshy staminodes opposite each of the petals; stigmas 4, but appearing simple and 4-lobed.

1   Petals fringed on the lower 1/2; petioles (1.5) 3-16 cm long; leaf blades truncate to cordate basally, 12-45 mm wide, wider than long; bract mostly borne above the middle of the scape .... *P. fimbriata*
1   Petals entire; petioles 0.7-4 cm long; leaf blades cuneate or obtuse basally (rarely cordate or truncate), 5-20 mm wide, mostly longer than wide; bract mostly borne below the middle of the scape .... *P. palustris*

Parnassia fimbriata Koenig — Rocky Mt. parnassia; fringed grass-of-Parnassus — Specimens seen are from the Uinta Mtns. in Duchesne and Summit Cos.; infrequent and locally common along streams, around springs, lake shores and other wet places; 7,600-9,300 ft; July-Aug.

Parnassia palustris L. (*P. parviflora* de Candolle) — marsh parnassia — Infrequent across the Uinta Mtns. and apparently rare on the E. Tavaputs Plateau; riparian communities and other wet places; about 7,000-9,300 ft; July-Aug.

Saxifraga L. — Saxifrage

Plants perennial herbs; herbage usually stipitate-glandular at least in part; leaves alternate or basal, entire or variously toothed or lobed; sepals and petals each usually 5, arising together with the stamens from the apex of a floral tube or receptacle; stamens 10, the filaments slender or expanded and sometimes petal-like; styles 2 (3-5); fruit a capsule.

1   Leaves all basal, the blades subentire to coarsely toothed but not lobed, commonly over 15 mm long, usually distinctly petiolate; flowers mostly more than 10; plants common and widespread 2
1   Leaves cauline, at least in part, or if all basal then less than 15 mm long, not distinctly petiolate, or if so then the blades lobed but not toothed; flowers 1-5, or rarely more; plants rather infrequent, of high elevations in the Uinta Mtns. 3
2   Inflorescence open; leaf blades coarsely and sharply dentate, orbicular or reniform; petioles 1.2-23 cm long, usually longer than the blades; plants 16-67 cm tall .... *S. odontoloma*
2   Inflorescence congested, occasionally interrupted; leaf blades entire or obscurely crenate, rhombic, obovate, or ovate, cuneate at the base; petioles 0.3-2.5 cm long, usually shorter than the blades; plants 3-20 (30) cm tall ..
    .... *S. rhomboidea*
3   Petioles evident, 0.5-4.5 cm long, usually longer than the blades; blades orbicular or reniform, palmately lobed or trilobata, cordate or truncate basally, some usually over 5 mm wide 4
3   Leaves sessile or essentially so; blades linear to oblanceolate, entire or 3- to 7-lobed, not cordate or truncate basally, 0.5-5 mm wide 5
4   Bulblets present in upper leaf-axils and inflorescence but not at petiole bases of lowest leaves; hypanthium about 1 mm long, not turbinate .....
    .... *S. cernua*

4   Bulblets not present in upper leaf-axils and inflorescence, often present at petiole bases of lower leaves; hypanthium 2-3 mm long, turbinate . . . . **S. rivularis**

5   Leaves toothed or lobed; petals white; sepals erect or somewhat spreading, about 1/2 as long as the floral tube; plants glandular, not stoloniferous. . 6

5   Leaves entire; petals yellow, but fading whitish; sepals spreading to reflexed, more than 1/2 as long as the floral tube, the floral tube sometimes obsolete. 7

6   Plants depressed caespitose; leaves lobed, those of the stem generally less deeply lobed than the basal ones; petals gradually narrowed to the base, clawless or shortly clawed; filaments longer than the sepals . **S. caespitosa**

6   Plants not depressed-caespitose; leaves merely toothed, those of the stem often more prominently toothed than the basal ones; petals abruptly narrowed to a short claw; filaments shorter than the sepals; known above timberline on the Uinta Mtns. . . . . **S. adscendens**

7   Plants stoloniferous; sepals ascending; herbage stipitate-glandular; leaves ciliate . . . . **S. flagellaris**

7   Plants not stoloniferous; sepals spreading to reflexed; herbage not stipitate-glandular (or only the stem sparingly so); leaves glabrous 8

8   Petals 4-5 mm long, short-clawed; sepals 2-3 mm long, glabrous; leaves 4-8 mm long, 0.5-1.5 mm wide; stems with minute stipitate glands, not pilose; plants 2-6 cm tall . . . . **S. serpyllifolia**

8   Petals 7-10 (15) mm long, not clawed; sepals 4-5 mm long, often pilose; leaves often over 8 mm long and some over 1.5 mm wide; stems often rusty pilose; plants 6-20 cm tall . . . . **S. hirculus**

Saxifraga adsendens L. [*Muscaria adscendens* (L.) Small] — wedge-leaf saxifrage — Uinta Mountains in alpine tundra and talus slopes, to 12,600 ft; July-Aug. Our plants belong to var. *oregonensis* (Rafinesque) Breitung. Otherwise known in Utah from the LaSal Mtns.

Saxifraga caespitosa L. (*Muscaria delicatula* Small; *M. micropetala* Small; *M. monticola* Small) — tufted saxifrage — The specimens seen are from the head of the Lake Fork, Yellowstone, Uinta, W. Fork Whiterocks, Burnt Fork, and Henrys Fork drainages, Uinta Mtns.; alpine tundra, rocks stripes, and fell fields; 11,000-13,000 ft; July-Aug.

Saxifraga cernua L. — nodding saxifrage — Infrequent; central Uinta Mtns. at 11,300-13,000 ft; alpine communities usually among rocks; 12,800 ft; July Aug. Otherwise known in Utah from the LaSal Mtns.

Saxifraga flagellaris — stoloniferous saxifrage — Uinta Mtns.; the 2 specimens seen are from Kings Peak (Neese & Neese 15339A) and 0.4 miles e. of Gilbert Peak (Huber 4112); alpine communities; 12,600-13,700 ft; July-Aug. Otherwise known in Utah from the LaSal Mountains.

Saxifraga hirculus L. — yellow marsh saxifrage — The few specimens seen are from Sheep Creek Park and Hickerson Park, Uinta Mtns.; boggy meadows; 8,600 ft; July-Aug. Unknown elsewhere in Utah, but circumboreal and common to the n. Disjunct from Montana and Colorado.

Saxifraga odontoloma Piper [*S. arguta* D. Don misapplied; *Micranthes arguta* (D. Don) Small misapplied; *M. odontoloma* (Piper) Weber] — brook saxifrage —

Common and locally abundant across the Uinta Mountains, apparently restricted to a few locations at higher elevations on the E. and W. Tavaputs Plateaus in wet or moist soil around springs along streams, often in woods; 7,500-11,000 ft.

Saxifraga rhomboidea Green [*Micranthes rhomboidea* (Piper) Weber] — Diamond-leaf saxifrage — Uinta Mtns.; occasional in wet and dry rocky meadows, coniferous forests, and alpine tundra; 7,500-12,000 ft; June-Aug.

Saxifraga rivularis L. (*S. debilis* Engelmann ex A. Gray; *S. hyberboria* R. Brown). — pygmy saxifrage — Uinta Mtns.; occasional in rocky places; 10,000-13,000 ft; July-Aug.

Saxifraga serpyllifolia Pursh [*S. chrysantha* A. Gray; *Hirculus serpyllifolius* (Pursh) W. A. Weber] — gold-bloom saxifrage — The 6 specimens seen are from the Uinta Mtns. from Tokewanna peak, Kings Peak, Bald Mtn. (Blacks Fork-Smiths Fork drainages), and the divide between the Uinta and Henrys Fork drainages; alpine communities; 11,200-12,650 ft; July-Aug. Our plants belong to var. *chrysantha* (A. Gray) Dorn.

Sullivantia Torrey & A. Gray

Sullivantia hapemanii (J. M. Coulter & Fisher) J. M. Coulter [*S. purpusi* (Brand.) Rosendahl — Purpus sullivantia — Perennial herbs, from small rootstocks, 15-30 cm tall; leaves mostly basal, but stems usually with at least 1 well-developed leaf and 1-few smaller leaves, the basal ones on long petioles that are 2-4 times longer than their blades; leaf blades, orbicular, palmately lobed, the lobes irregularly and sharply dentate; inflorescence open with spreading branches, glandular; calyx 2-4 mm long, the tube longer than the sepals; petals 2-3 mm long, whitish; stamens 10; ovary half-inferior; styles 2; fruit a 2-beaked capsule. The 2 specimens seen are from Cathedral Bluffs, Rio Blanco Co., from Green River Shale barrens; 8,400 ft. Reported for wet rocks from 7,000-7,500 ft (Harrington 1954). Plants of Colorado belong to var. *purpusi* (Brand) Soltis. This var. is endemic to Colorado.

SCROPHULARIACEAE Figwort Family

Perennial herbs, a few chlorophyll-containing hemiparasites or saprophytes (*Castilleja, Orthocarpus, Cordylanthus,* and *Pedicularis* included in Orobanchaceae in recent treatments); leaves basal, alternate or opposite, simple, entire or variously toothed to dissected, without stipules; flowers perfect, irregular (regular or nearly so in *Limosella, Veronica, Verbascum,* and *Synthyris*); calyx of 4-5 distinct or united sepals (2-lobed or bract-like in *Besseya, Cordylanthus,* and *Synthyris*); corolla 4-5 lobed or rarely lacking (*Besseya*), often strongly 2-lipped, beaked or galeate; stamens attached to the corolla tube, typically 4 and didynamous or 4 fertile and 1 sterile (*Penstemon* and *Scrophularia*), or 2 (*Synthyris* and *Veronica*) or 5 (*Verbascum*); ovary 1, superior, 2-loculed; fruit capsular, sometimes poricidal. A diverse family included in Orobanchaceae as listed above, Plantaginaceae (*Collinsia, Limosella, Linaria, Penstemon, Synthyris, Veronica*), Phrymaceae (*Mimulus*) with *Scrophularia* and *Verbascum* remaining in Scrophulariaceae in recent treatments.

1   Principal leaves all basal..2
1   Leaves not all basal...6
2   Leaves entire or with a few inconspicuous teeth; flowers solitary or 2; stamens 3
2   Leaves with numerous small teeth or pinnatifid; flowers few to several on usually bracteate scapes; stamens 2...............................5
3   Corolla yellow; flowers on capillary scapes or peduncles well elevated above the leaves .... *Mimulus primuloides*
3   Corolla bluish, white or pink, not much if at all exceeding the leaves 4
4   Flowers sessile among the leaves, showy; leaves not constricted to a petiole, not over 3 cm long; plants of dry places .... *Penstemon*
4   Flowers on scapes 0.8-2 (3) cm tall, not especially showy; leaf blades more or less constricted to a petiole, the petiole longer than the blade, the largest ones often over 3 cm long including the petiole; plants of wet places .... *Limosella*
5   Leaves with numerous small teeth, not pinnatifid; corolla lacking *Besseya*
5   Leaves pinnatifid; corolla blue .... *Synthyris*
6   Corolla tube strongly spurred at the base, yellow, 10-24 mm long; plants introduced, perennial, from spreading roots .... *Linaria*
6   Corolla tube not spurred, if pouched at the base then not yellow and smaller than above; plants various..7
7   Leaves opposite, bracts of the inflorescence sometimes alternate 8
7   Leaves alternate..13
8   Calyx united, the tube longer than the teeth, enveloping and exceeding the capsule; corolla yellow except in *Mimulus lewisii* and sometimes in *M. rubellus* .... *Mimulus*
8   Calyx divided or, if united then the tube equal-to or shorter than the lobes, rarely exceeding the mature capsule; corolla not yellow...............9
9   Stamens 2; capsules flattened .... *Veronica*
9   Stamens more than 2; capsules not flattened......................10
10 Leaf blades serrate...11
10 Leaf blades entire or obscurely crenate..........................12
11 Leaves sessile .... *Penstemon*
11 Leaves with petioles 1-3 (5) cm long .... *Scrophularia*
12 Corolla usually over 1 cm long .... *Penstemon*
12 Corolla less than 1 cm long .... *Collinsia*
13 Leaves and bracts all entire or shallowly toothed 14
13 Leaves lobed or divided 18
14 Basal leaves 6-50 cm long, 2.5-14 cm .... *Verbascum*
14 Leaves shorter or narrower than above or both shorter and narrower ..15
15 Leaves entire 16
15 Leaves serrate...17
16 Bracts greenish; corolla yellow .... *Orthocarpus*
16 Bracts red or yellow; corolla greenish .... *Castilleja*

17 Basal leaves much larger than the stem leaves, rounded or cordate at the base, with petioles about as long as the blades .... *Besseya*

17 Leaves not as above .... *Pedicularis*

18 Leaves pinnately or bipinnately lobed to divided with mostly 10 or more serrate or crenate lobes or segments .... *Pedicularis*

18 Leaves divided into linear or acute, entire segments, these rarely more than 5 per leaf 19

19 Bracts less showy than the corolla; plants annual .... *Cordylanthus*

19 Bracts mostly more showy than the corolla; plants perennial ... *Castilleja*

Besseya Rydberg — Alpine Kittentails

Besseya wyomingensis (A. Nelson) Rydberg [*B. cinerea* (Rafinesque) Pennell misapplied] — Wyoming kittentails — Perennial bracteate-scapose herb; stems mostly 10-25 cm tall; basal leaves mostly 5-15 cm long, the blades elliptic or lanceolate to ovate, crenulate to serrate, about equaling the petioles; cauline leaves bract-like, sessile, alternate, ovate to lanceolate, 1-2 cm long; inflorescence a densely flowered spike 2-6 cm long, 1-1.6 cm wide, densely pubescent; calyx 2-lobed, about 4 mm long; stamens 2; stamens and style exserted; capsule compressed, orbicular, about equaling the calyx. Known from alpine tundra along the summit of Big Ridge, 10 mi n. of Tabiona; about 11,000 ft; July-Aug. Otherwise in Utah in the Raft River Mtns. Our plants apparently are generally more diminutive than those of the foothills and high plains of Idaho and eastward to South Dakota and Nebraska.

Castilleja Mutis ex L. F. — Paintbrush

Plants perennial or annual, herbaceous; leaves alternate, sessile, entire, toothed, or parted; bracts subtending the flowers brightly colored, red, orange, yellow, or reddish-purple, entire or 3-lobed; flowers borne in terminal spikes or spike-like racemes; calyx 4-cleft, variously colored; corolla scarcely or not at all petaloid, usually greenish at least in part, often less showy than the bracts, 2-lipped, the upper lip straight-beaked, elongate, much exceeding the inconspicuous lower lip, which is reduced to 3 thickened teeth-like lobes. This genus is included in Orobanchaceae in recent works.

1 Plants annual; stems solitary; leaves and bracts all entire .... *C. exilis*

1 Plants perennial; stems usually not solitary 2

2 Bracts predominantly yellowish 3

2 Bracts predominantly red to purple, rarely yellowish or somewhat pink or orange 6

3 Upper leaves and bracts deeply lobed or divided, the lobes linear or narrowly lanceolate 4

3 Leaves and bracts entire, the bracts broad, obtuse; plants of the Uinta Mtns. .... *C. rhexifolia* var. *sulphurea*

4 Plants small, less than 15 cm tall, of tundra habitats in the Uinta Mtns.; inflorescence viscid villous, sometimes tinged with purple ... *C. pulchella*

4 Plants mainly well over 15 cm tall, of low to mid-elevation habitats well below tundra communities 5

5   Calyx decidedly more deeply cleft below than above; plants from the
    Tavaputs Plateau and Daggett Co .... *C. flava*

5   Calyx slightly more deeply cleft above than below; plants widespread,
    yellow phases of .... *C. chromosa*

6   Calyx more prominently colored than the relatively inconspicuous mostly
    green subtending bract, divided deeper below than above; plants of mid-
    altitudes .... *C. linariifolia*

6   Calyx less conspicuous than the brightly colored subtending bract and
    largely concealed by it, divided sub equally above and below; distribution
    various.. 7

7   Leaves entire or the upper ones shallowly lobed at the tip 8

7   At least the middle and upper cauline leaves mostly deeply cleft, with linear
    spreading lobes 9

8   Bracts broadly lanceolate to ovate, shallowly lobed, magenta or purple,
    rarely red; galea 8-12 mm long .... *C. rhexifolia*

8   Bracts lanceolate to narrowly lanceolate, mostly divided to below the
    middle, red or reddish-orange; galea 12-18 mm long .... *C. miniata*

9   Plants strongly glandular pubescent, mostly from above 9,000 ft ....
    .... *C. applegatei*

9   Plants not strongly glandular pubescent, from below 9,000 ft.......... 10

10 Stems with an elongate lower part devoid of chlorophyll and with scaly
    leaves that is sometimes buried in sand or rock crevices, erect or
    decumbent; corolla 2.3-4.3 cm long, the upper portion of the galea and the
    lower lip usually plainly exserted from the calyx; plants of the Tavaputs
    Plateau and from the vicinity of Dinosaur National Monument *C. scabrida*

10 Stems green nearly throughout, the lower scaly portion short and compact,
    mainly erect or ascending; corolla 2-3.6 cm long, the lower lip and most of
    the galea largely concealed by the calyx; plants widespread in many plant
    communities .... *C. chromosa*

Castilleja applegatei Fernald (*C. viscida* Rydberg) — sticky paintbrush —
Widespread in the Uinta Mtns.; occasional and locally common in rocky places
of spruce, fir, lodgepole pine, aspen, montane meadow, and alpine communities
including alpine rock stripes, talus, and fell fields; 9,200-11,700 ft; June-Aug.
Our plants belong to var. *viscida* (Rydberg) Ownbey.

Castilleja chromosa A. Nelson — desert paintbrush — Widespread and common
in greasewood, shadscale, mixed desert shrub, sagebrush, pinyon-juniper, and
mt. brush communities; 4,700-8,500 ft; late April-July. One of the most frequently
collected plants of our area. Desert paintbrush includes plants with yellow
bracts and plants with red bracts.

Castilleja exilis A. Nelson — marsh paintbrush — Specimens seen are from
Duchesne, Uintah, and Daggett Cos.; somewhat alkaline or saline places; seeps,
marshes, wet meadows, marshes, ditch banks, lake margins, and mudflats, in
saltgrass, sedge, rush, cattail, or other wet land communities; 4,700-7,000 ft;
July-Aug. Our plants have been referred to *C. minor* A. Gray. A close relationship
for these 2 taxa is indicated in IMF (4: 496), and *C. minor* could be the name of
priority for our plants.

Castilleja flava S. Watson — yellow paintbrush — Common on the n. slope of the Uinta Mtns and Tavaputs Plateau; uncommon and lacking on much of the s. slope of the Uinta Mtns.; occasional in sagebrush communities with Salina wildrye, little rabbitbrush, mt. brush, snowberry, and in openings of Douglas-fir communities; 7,200 -10:000-ft; June-Aug. Our plants belong to var. *flava.* Flowering abundantly after fire for at least a few years.

Castilleja linariifolia Bentham in de Candolle — Wyoming paintbrush — Wide-spread; occasional in pinyon-juniper, mt. brush, mtn. big sagebrush, aspen, ponderosa pine, and Douglas-fir communities, 6,000-9,000 ft; June-Sept.

Castilleja miniata Douglas — scarlet paintbrush — Uinta Mtns. and higher elevations of the Tavaputs Plateau; occasional in mesic or wet, sometimes rocky meadows and along streams in aspen, spruce-fir communities; 10,000-11,300 ft; July-early Sept.

Castilleja pulchella Douglas (*C. occidentalis* Torrey misapplied) — beautiful paintbrush — Fairly common across the crest of the Uinta Mtns.; alpine or subalpine fell fields, rock stripes, ridges, and meadows were mostly associated with curly sedge, kobresia, mountain avens, alpine avens, and Uinta clover (Brown 2006); 10,800-12,500 ft; July-Sept.

Castilleja rhexifolia Rydberg (*C. leonardii* Rydberg; *C. sulphurea* Rydberg) — split-leaf paintbrush — Strawberry Valley and Uinta Mtns.; mostly in openings in spruce-fir forests and in alpine communities at 8,800-11,500 ft; July-early Sept. Both var. *rhexifolia* (red inflorescences) and var. *sulphurea* (Rydberg) Atwood (*C. septentrionalis* Lindley misapplied) with mostly yellow inflorescences are present in the Uinta mountains where plants with red and plants with yellow inflorescences sometimes mix freely.

Castilleja scabrida Eastwood — rough paintbrush — Occasional on the E. & W. Tavaputs Plateaus, Blue Mt., and near Dinosaur National Monument, apparently absent from most of the Uinta Mtns.; characteristically in pinyon-juniper communities with sagebrush, occasionally in Salina wildrye-shadscale and mt. brush communities in sandy soil or sandstone outcrops, or in mixed sand and clay from Green River Formation; 5,000-8,200 ft; late April-July. Our plants belong to var. *scabrida*. This is likely to be confused with *C. chromosa* from which it is distinguished by the mostly overlapping features listed in the key. The elongate lower part of stems devoid of chlorophyll and with scaly leaves is likely the most striking difference.

Collinsia Nuttall — Blue-eyed Mary

Collinsia parviflora Lindley — blue-eyed Mary — Small tap-rooted annual, stems 1-15 cm tall, simple or few-branched from the base, with 2-5 usually remote pairs of leaves; leaves opposite, simple, entire, sessile or narrowed to a petiole-like base, narrowly oblong to elliptic, rounded at apex, 5-30 mm long, 1-10 mm wide; flowers axillary, solitary; calyx 3-6 mm long, the lobes acute, narrowly triangular, slightly accrescent and enclosing the capsule in fruit; corolla tubular, 2-lipped, 4-6 mm long, the central lower lobe shorter and narrower than the lateral 2. Widespread and in many communities; 6,600-9,000 ft; April-early July. One of our most common early spring flowers.

Cordylanthus Nuttall ex Bentham in de Candolle — Birdsbeak

Tap-rooted annuals, the root often yellow; leaves pinnately or palmately divided, the segments linear to filiform; flowers clustered in few-flowered heads near the ends of branchlets; front of calyx cleft to the base, bract-like, 2-lobed apically, subtended by an opposing bract, an additional leaf-like bract present below each flower or spike; corolla tubular below, expanding midway to a slightly expanded or pouch-like throat, 2-lipped, the lips about equal, the upper forming a hood-like beak. This genus is included in Orobanchaceae in recent works.

1 Corolla predominantly dingy yellow, abruptly dilated midway into a pouch-like throat, obscurely puberulent with hairs to 0.1 mm long; herbage puberulent, not glandular .... *C. ramosus*
1 Corolla predominately purple, gradually widened to a funnelform throat, evidently densely white-pilose on the lower throat with hairs mostly 0.5-1 mm long, herbage glandular-pubescent .... *C. kingii*

Cordylanthus kingii S. Watson [*Adenostegia kingii* (S. Watson) Greene] — King birdsbeak — Uinta Basin and E. and W. Tavaputs Plateaus in Utah, apparently lacking in Colorado; usually on dry clay soils in desert shrub, pinyon-juniper, sagebrush, and mt. brush communities; 5,000-7,300 ft; June-Oct. Our plants with more flowers and larger floral bracts than typical for the species belong to ssp. *densiflorus* T. I. Chuang & Heckard.

Cordylanthus ramosus Nuttall ex Bentham [*Adenostegia ramosa* (Nuttall) Greene] — much-branched birdsbeak — Specimens seen are from the vicinity of Fruitland, Farm Creek, Brush Creek, Diamond, and Blue Mtns., and near Manila; infrequent but locally common in pinyon-juniper, sagebrush-grass, and mt. brush communities; 6,600-8,200 ft; late June-Aug.

Limosella L. — Mudwort

Limosella aquatica L. — mudwort — Fibrous rooted perennial herb; leaves all basal, petiolate, exceeding the flowers, the petioles elongating when submerged, the blades floating, elliptic, 3-18 mm long, 1-7 mm wide; scapes slender, bearing a single flower, or sometimes stoloniferous and bearing a rooting tuft of leaves and flowers at the tip; corolla white, regular, 5-lobed, about equaling the calyx; capsule subglobose, about 3 mm long. Probably widespread but frequently overlooked; the few specimens seen are from along the Strawberry and Green rivers and scattered locations across the Uinta Mtns.; in mud or quiet shallow water, reservoir and pothole margins, and wet fresh-water meadows; 5,000-9,920 ft; late June-Nov.

Linaria Mill — Toadflax

Perennial herbs from creeping roots; leaves alternate or the lower ones opposite, sessile, entire; calyx of 5, essentially distinct segments; corolla yellow, the tube spurred at the base, strongly bilabiate, the throat nearly closed by the palate; stamens 4, didynamous.

1 Leaves linear, 2-4 mm wide, narrowed to a short petiole-like base, not clasping; flowers in a compact spike; calyx 2.5-3.2 mm long; corolla 10-14 (18) mm long; plants 20-60 (80) cm tall .... *L. vulgaris*

1   Leaves lance-ovate to ovate, 10-35 mm wide, broad and more or less clasping at the base; flowers in a more or less loose raceme; calyx 5-9 mm long; corolla 14-24 mm long; plants 40-70 (100) cm tall ...... *L. dalmatica*

Linaria dalmatica (L.) Miller — Dalmatian toadflax — Introduced from the Mediterranean Region, invasive (listed as a noxious weed) along roadsides and advancing into native plant communities at several, widely scattered locations from 5,300-7,620 ft; June-July.

Linaria vulgaris Hill — butter and eggs — Introduced from Eurasia; the one specimen seen (Goodrich 21084) is from the edge of Twin Potts Reservoir, Duchesne Co. Observed north of Altonah in 2011, and along the Mirror Lake Highway 6.5 miles s. of Evanston. This weedy plant is expected elsewhere. It dominates subalpine slopes in Colorado, and is reasonably expected to have the same capability in the Uinta Mountains. Early detection of new infestations and rapid control measures will be needed to prevent the spread of this invasive species.

Mimulus L. — Monkey-flower

Plants herbaceous, often glandular, annual or perennial; leaves opposite, entire or toothed; flowers axillary, solitary; calyx 5 angled or keeled, the midrib of each lobe prominent, lobes usually much shorter than the tube; corolla slightly to strongly 2-lipped, usually spotted in the throat; stamens 4, didynamous; capsule cylindrical, enclosed in the slightly to strongly accrescent calyx.

---

1   Corolla less than 0.8 (1.1) cm long; plants strictly annual. 2
1   Corolla 1.2 cm or longer; plants perennial (sometimes annual in M. *guttatus*) 4
2   Leaves mostly ovate, the blade abruptly narrowed to an evident slender petiole; leaves and calyx ribs green . . . . *M. floribundus*
2   Leaves elliptic or lanceolate to oblanceolate, often narrowly so, narrowing gradually to a sessile base; leaves and especially calyx ribs red-tinged . . . 3
3   Corolla 7-10 mm long, exceeding calyx by 2-3 mm; internodes mostly exceeding leaves; stem usually single except in larger individuals. . . . . *M. rubellus*
3   Corolla 4-7 mm long, exceeding calyx by 1-1.5 mm; internodes-mostly shorter than the leaves; stems usually branched from base even in depauperate individuals . . . . *M. suksdorfii*
4   Corolla rose-pink, 3-4.8 cm long, the throat yellow, rose-spotted *M. lewisii*
4   Corolla yellow, 1.2-3.5 cm long, usually red spotted 5
5   Leaves clustered toward the base of the stem, the inflorescence scapose, 1 (2) flowered, the flowers 1.2-1.8 mm long, borne well above the leaves . . . . *M. primuloides*
5   Stems evidently leafy with flowers in axils of leaves; inflorescence with more than 2 flowers 6
6   Calyx regular or nearly so, calyx lobes similar, narrowly acute to acuminate, about 1/3 as long as the tube; corolla inconspicuously 2-lipped, the throat

open; herbage viscid glutinous; ill smelling, soft, mat-forming perennial from slender rhizomes .... *M. moschatus*

6   Calyx bilaterally symmetrical, the upper lip longer and usually broader than the other 4, at least the lateral lobes broadly triangular, obtuse, 1/5 or less as long as the tube 7

7   Corolla-throat more or less open; calyx teeth blunt and relatively short, sometimes nearly obsolete, not folded inward in fruit; corolla 10-20 mm long .... *M. glabratus*

7   Corolla-throat closed by the well developed palate; lateral calyx teeth more or less acute and tending to fold inward in fruit 8

8   Plants 2 dm or less tall; stems 1-3 (5) flowered, relatively slender, usually 2.5 mm or less in diameter, usually decumbent, rooting at the nodes and forming mounds; leaves seldom exceeding 2.5 cm in length .... *M. tilingii*

8   Plants variable in size, but often taller than 2 dm, stems usually 5-many flowered, often some robust and 3 mm or more in diameter, mostly erect, not forming mounds; commonly some leaves more than 2.5 cm long ....
.... *M. guttatus*

Mimulus floribundus Douglas in Lindley — mealy monkey-flower, floriferous monkey-flower — Diminutive annual of vernally moist, often disturbed sites. The 4 specimens seen are from drying mud of an ephemeral pool, Rye Grass Draw, Taylor Flats, 5,700 ft, Daggett Co. (Neese 14813), below Tolivers Canyon in Browns Park at 6,800 ft (Thorne & Zupan 6301), burned pinyon-juniper community n. of Elkhorn Guard Station, s. slope of the Uinta Mtns. (Goodrich 27453A), and Dutch Oven Springs, Douglas Mt., 7,270 ft (M. MacLeod 15b DINO).

Mimulus glabratus H. B. K. — glabrous monkey-flower — Usually growing in perennial springs and streams; listed for Daggett, Duchesne, and Wasatch Cos. (AUF698). Our plants are likely referable to ssp. utahensis Pennell. Apparently much less common in our area than *M. guttatus*.

Mimulus guttatus Fischer ex de Candolle — common monkey-flower, yellow monkey-flower — Widespread; locally common along streambanks, in boggy meadows, shallow water, and rock crevices near springs and seeps. Specimens seen are from 6,000-10,000 ft; late May-June.

Mimulus lewisii Pursh — Lewis monkey-flower — Widespread but widely scattered in the Uinta Mtns., specimens seen are from the Bear River, Duchesne, Lake Fork, Rock Creek, Yellowstone, Uinta, Sheep Creek, and Whiterocks drainages; among rocks in seeps, along mt. streams and lake margins, and in wet meadows, mostly in Engelmann spruce and lodgepole pine, and alpine communities; 8,000-11,700 ft; July-Aug.

Mimulus moschatus Douglas in Lindley — musk monkey-flower — Specimens seen from Strawberry Valley, Hole In The Wall Canyon, Lake Fork, Provo River, and Rock Creek of the s. slope and western Uinta Mtns, Eagle Creek and Skull Creek in Daggett Co.; about seeps and springs, along streams and in wet meadows, often with willows and sedges; 7,380-9,000 ft; June-Aug.

Mimulus primuloides Bentham — primrose monkey-flower — Weight 125 RM! USUUB! is from Gilbert Basin, Uinta Mtns., Duchesne Co. Subsequent specimens (Goodrich 22749, 22748, Huber 4162) are from the same general location. Huber

5282 is from North Fork Park of the Uinta drainage; moist sedge-grass meadows and bare soil within meadows and along streams; 10,100-11,020 ft; July-Aug. The Weight collection was made in 1928. The later Gilbert Creek collections were made in 1988 and 1999. It appears this plant is persistent despite limited distribution in the Uinta Mtns. The North Fork Collection was made in 2013. Perhaps this plant is more widespread than the limited collections indicate.

Mimulus rubellus A. Gray — reddish monkey-flower — Uncommon or perhaps overlooked, the 6 specimens seen are from s. slope of the Uinta Mtns. (Pole Creek, Red Pine Canyon, Highway 191, and Jones Hole, Blue Mtn.), Highway 208 w. of Duchesne, and sw. corner of Dinosaur National Monument; desert shrub, pinyon-juniper, sagebrush, and mtn. brush communities; 5,100-8,085 ft; May-June. Diminutive annual; flowers axillary, or solitary in depauperate plans, the corolla yellow or purplish. Reported to have ciliate calyx lobes, but some specimens from our area seem to lack this feature.

Mimulus suksdorfii A. Gray — Suksdorf monkey-flower — The 6 specimens seen are from Strawberry Valley, Highway 191 north of Vernal, Diamond Mtn., Blue Mtn., and Kennedy Wash-Bonanza area; desert shrub, Utah Juniper, big sagebrush, silver sagebrush, and ponderosa pine communities, usually in sandy soil; 5,180-7,700 ft; May-July. Much branched, resembling the preceding species (but somewhat smaller in all its parts). Neese and Peterson 5348A (*M. suksdorfii*) and Neese and Peterson 5348 (*M. rubellus*) are from the same location (16 mi n. of Vernal on Highway 191).

Mimulus tilingii Regel — subalpine monkey-flower, Tiling monkey-flower — Specimens seen are from near Daniels Pass, drawdown basin of Strawberry Reservoir, Head of Lake Fork drainage, and near Chepeta Lake of the Whiterocks drainage; wet meadow and willow communities; 8,000-11,200 ft; July-Aug. Usually but not always separable from *M. guttatus* by the mound-like habit, more slender stems, and smaller leaves.

Orthocarpus Nuttall — Owl-clover

Taprooted annuals; leaves simple and entire below, gradually becoming bracteate and 3-lobed in the inflorescence; inflorescence a simple or branched spike or spike-like raceme, obscurely glandular pubescent; corolla yellow, 2-lipped, the upper lip beak-like, the lower somewhat pouched.

---

1 Upper lip of the corolla evidently exceeding the lower, hooked at the tip; stems usually freely corymbosely branched, the lateral branches about equaling or slightly exceeding the terminal one .... *O. tolmei*

1 Upper and lower lips about equal, the upper inconspicuously curved at the tip; stems strict, simple-or-less often sparingly branched, the terminal branch overtopping the lateral ones .... *O. luteus*

Orthocarpus luteus Nuttall — yellow owl-clover — Occasional and locally common along the flanks of the Uinta Mtns., to Blue Mt. and apparently rare on the E. Tavaputs Plateau, no specimens seen from the W. Tavaputs Plateau; moderately dry to wet places, rocky or gravelly slopes in sagebrush-grassland, lodgepole, aspen, and ponderosa communities and in grassy meadows with sedges, rushes, and willows; 7,400-9,600 ft; late June-Sept.

Orthocarpus tolmei Hooker & Arnot — Tolmei owl-clover — Known from Strawberry Valley, Current Creek drainage, and west end of the s. slope of the Uinta Mtns. as far east as Rock Creek; locally common in sagebrush-grass, mtn. brush, aspen, Douglas-fir, and spruce communities; 6,000-10,200 ft; late June-Sept.

Pedicularis L. — Lousewort

Perennial herbs; leaves crenate to (usually) pinnately lobed to dissected; inflorescence spicate, bracteate; calyx irregular, cleft into 5, 4, or 2 lobes; corolla 2-lipped, the lower lip of 3 into 5, 4, or 2 lobes; corolla 2-lipped, the lower lip of 3 spreading lobes, the upper hooded, enclosing the stamens, the tip truncate or variously elongated and curved; fruit an asymmetrical capsule.

1 Leaves crenate or serrate 2
1 Leaves pinnately lobed to bipinnately divided 3
2 Corolla white . . . . ***P. racemosa***
2 Corolla rose, red, or purplish . . . . ***P. crenulata***
3 Inflorescence dense, subcapitate, equaling or shorter than the leaves; corolla limb dull dark purple, the tube paler . . . . ***P. centranthera***
3 Inflorescence elongate, much surpassing the leaves; corolla white to yellow, or pinkish purple 4
4 Corolla pinkish purple, upper lip much elongated into an upward curving resembling an elephant's trunk . . . . ***P. groenlandica***
4 Corolla white to yellow. 5
5 Leaves mostly basal, stem leaves reduced upward; hood of upper lip narrowed to a beak 1-2 mm long . . . . ***P. parryi***
5 Leaves mostly cauline, the lower and middle stem leaves well developed; hood of upper lip truncate, blunt 6
6 Corolla 2.5-3.5 cm long, marked with reddish lines; hooded upper lip with 2 slender acute lateral teeth near the tip . . . . ***P. procera***
6 Corolla 1.8-2.5 cm long, without reddish lines; hooded upper lip without lateral teeth . . . . ***P. bracteosa***

Pedicularis bracteosa Bentham in Hooker (*P. paysoniana* Pennell) — bracteate lousewort — Across much of the Uinta Mtns. but no specimens seen from the Ashley Creek drainage; occasional in moderately moist to wet soil or rocky places of spruce, lodgepole and Douglas-fir communities and margins of meadow adjacent to forests; 8,200-11,100 ft. Also reported from the Tavaputs Plateau (IMF 4: 472); July-Aug. Our plants belong to var. *paysoniana* (Pennell) Cronquist.

Pedicularis centranthera A. Gray in Torrey — dwarf lousewort — The few specimens seen are from the Avintaquin drainage n. to Hwy. 40 and head of the Antelope drainage in Duchesne Co. and from the E. Tavaputs Plateau; pinyon-juniper communities; 6,900-7,760 ft; late April-June.

Pedicularis crenulata Bentham — meadow lousewort — Approaching our area on the outwash plain of the Uinta Mtns. in Uinta Co. Wyoming.

Pedicularis groenlandica Retz. — elephant-head — Common across the Uinta Mtns. where it is mostly restricted to boggy, cold, wet meadows and streambanks, at 7,800-12,500 ft, apparently absent from Strawberry Valley, the Tavaputs Plateau, and Diamond Mtn.; late June-Aug. Often growing with *Carex aquatilis* in meadows frequently covered with frost on summer mornings.

Pedicularis parryi A. Gray — Parry lousewort — Occasional in the central Uinta Mtns. at 10,500-12,500 ft, and common in Strawberry Valley down to 7,500 ft; riparian and meadow communities with sedges, grasses, and silver sagebrush, in openings of aspen, spruce, and lodgepole pine forests, and in alpine communities.

Pedicularis procera A. Gray (*P. grayi* A. Nelson) — Gray lousewort — Our one specimen (Huber and Goodrich 5307) is from a small aspen stand of northerly aspect at the head of the Piceance drainage. Reported for the E. Tavaputs Plateau in Utah (IMF 4: 473); to be expected in moderately moist aspen and coniferous forests and edges of mt. meadows; 7,600 ft and above.

Pedicularis racemosa Douglas in Hooker — leafy lousewort — Uinta Mtns.; occasional and locally abundant in spruce-fir and lodgepole pine communities, and infrequently with aspen; 9,000-10,600 ft; July-Aug. Our plants belong to var. *alba* (Pennell) Cronquist.

Penstemon Mitch. — Beardtongue; Penstemon

Glabrous to glandular-pubescent, perennial herbs or subshrubs; leaves basal and opposite, simple, entire or occasionally toothed, the lower petiolate, becoming sessile upward, reduced and more or less bract-like in the inflorescence; inflorescence of axillary cymes, often verticillate; calyx 5-cleft; corolla tubular to funnel form, more or less bilabiate; fertile stamens 4, didynamous, a fifth sterile filament (staminode) present; anthers glabrous or variously bearded; stigma globose; capsule 2-valved, many-seeded.

---

1   Corolla bright red . . . . *P. eatonii*

1   Corolla some shade of white, pink, purple, or blue 2

2   Cauline leaves, at least some, sharply and prominently toothed 3

2   Cauline leaves entire (rarely obscurely toothed in *P. grahamii* and *P. eriantherus*) 5

3   Anthers copiously woolly, plants indigenous on limestone, w. Uinta Mtns. . . . . *P. montanus*

3   Anthers glabrous; plants introduced 4

4   Anther sacs dehiscing across the apex only, horseshoe shaped; filaments hairy near the anthers; cauline leaves sessile but neither auriculate-clasping nor connate-perfoliate; corolla purple to violet, the tube gradually expanding into the throat . . . . *P. venustus*

4   Anther sacs dehiscing throughout, becoming opposite, not horseshoe shaped; filaments-glabrous; mid and upper cauline leaves auriculate-clasping to connate-perfoliate; corolla whitish, pale pink, or lavender-pink with prominent red-violet guide-lines on the lower lip, the tube abruptly expanding into a strongly inflated throat . . . . *P. palmeri*

5   Plants caespitose or mat forming, mostly less than 10 cm tall; flowers subsessile, borne among the leaves 6

5 Plants not caespitose, not mat-forming; flowers held above the leaves on erect stems 7

6 Plants without evident stems; flowers born among densely clustered tufts of leaves . . . . *P. acaulis*

6 Plants with short but conspicuous leafy flower-bearing, creeping stems . . . . *P. caespitosus*

7 Anther sacs pubescent, sometimes sparsely so . . . . **Key 1**

7 Anther sacs glabrous . . . . **Key 2**

— KEY 1 —
Anther sacs pubescent

1 Corolla puberulent or glandular externally 2

1 Corolla glabrous externally 5

2 Staminode sparingly villose with white tortuous hairs; plants of eastern Daggett Co. and adjacent Wyoming from below 7,000 ft . . . . . . *P. gibbensii*

2 Staminode with yellow or golden hair; mostly from above 7,000 ft 3

3 Staminode densely bearded for over ½ the length; plants 4-20 cm tall, from above 10,000 ft in the Uinta Mtns. . . . . *P. uintahensis*

3 Staminode sparsely bearded mostly for less than ½ the length; plants 15-100 cm tall, mostly from below 10,000 ft 4

4 Anther sacs with flexuous hair equal to or longer than the width of the sac, spreading at maturity at about a 90° angle; plants 15-50 cm tall . . . . *P. scariosus*

4 Anther sacs with straight hairs mostly shorter than the width of the sac, spreading at maturity to about 180°; plants 25-70 (100) cm tall . . . . *P. subglaber*

5 Leaves, stems, and inflorescence densely puberulent throughout; corolla deep blue-purple, mostly 18-22 mm long; plants widespread in pinyon-juniper, Wyoming big sagebrush and salt desert communities *P. fremontii*

5 Lower leaves and stem mostly glabrous, at least not densely puberulent; inflorescence glabrous or glandular puberulent 6

6 Hairs of the anther sac slender, flexuous, mostly longer than the length of the anther sac. 7

6 Hairs of the anther sac mostly shorter than the length of the sac 8

7 Corolla pale blue or pale lavender-blue; inflorescence relatively broad, the lower peduncles and pedicels often exceeding 2 cm, somewhat spreading, the inflorescence thus not markedly secund; anther sac usually densely woolly with long tangled hairs that somewhat obscure the surface; plants known from the Tavaputs Plateau . . . . *P. comarrhenus*

7 Corolla blue to dark blue; inflorescence narrow, the peduncles and pedicels mostly shorter than 1.5 cm, somewhat appressed, the inflorescence strongly secund; anther sac often somewhat less densely woolly, the surface not obscured; plants widespread . . . . *P. strictus*

8 Hairs of the anther sac somewhat flexuous, mostly as long or longer than the width of the sac . . . . *P. scariosus*

8 Hairs of the anther sac stiff, straight, shorter than the width of the sac . . 9

9   Leaves crisped margined; plants of the West Tavaputs Plateau in Carbon Co.
    .... *P. cyanocaulis*

9   Leaves not crisped margined 10

10 Plants mostly less than 20 cm tall, known from the margin of our area near
    Lonetree, Wyoming; corolla 15-21 mm long; leaves 2-8 (11) mm wide;
    anthers sparsely bearded with hairs less than 0.2 mm long; staminode
    bearded at apex only .... *P. paysoniorum*

10 Plants mostly taller than 20 cm; usually different than above in other
    respects 11

11 Middle and upper leaves broadly ovate, clasping; inflorescence a nearly
    cylindrical thyrse.................................... *P. cyananthus*

11 Middle and upper stem leaves lanceolate to narrowly lanceolate, scarcely
    clasping; inflorescence secund .... *P. subglaber*

— KEY 2 —
Anther sacs glabrous
(occasionally scabrous or short-ciliate along the line of dehiscence)

1   Anther sacs dehiscing at top only, across the apex and connective; anthers
    horseshoe-shaped 2

1   Anther sacs dehiscing throughout or distally......................... 3

2   Corolla 14-21 mm long, the lobes blue, the tube lavender; anthers sacs about
    1 mm long; middle stem leaves mostly less than 7 mm wide, many of them
    obtuse; plants mostly less than 3 dm tall .... *P. leonardii*

2   Corolla 20-27 mm long, lavender throughout; anther sacs ca 1.5 mm long;
    middle-stem leaves mostly greater than 8 mm wide, elliptic, sharply acute,
    noticeably firm and plane; plants mostly over 3 dm tall ... *P. platyphyllus*

3   Plants essentially glabrous throughout 4

3   Plants pubescent to glandular-pubescent, either on leaves and stems or in
    inflorescence 9

4   Inflorescence verticillate, the fascicles of flowers separated; leaves thin;
    plants of mtns.. 5

4   Inflorescence a more or less continuous thyrse, if somewhat verticillate the
    fascicles little separated; leaves thick, leathery; plants of lower elevations,
    mostly of pinyon-juniper and salt desert communities 6

5   Corollas 6-10 (12) mm long, the throat mostly less than 2 mm wide; tufts of
    basal leaves usually prominent; flowers tightly congested in the fascicle(s),
    sessile or nearly so, the individual pedicels not evident, usually declined;
    staminode bearded with yellow-brown hairs at apex (in distal 1.5 mm); calyx
    3-6 mm long, the lobes acuminate; anther sacs mostly 0.5-0.7 mm long ...
    .... *P. procerus*

5   Corollas 12-15 (17) mm long, the throat mostly wider than 2.5 mm; basal
    leaves absent or poorly developed; fascicles not so tightly congested, the
    individual pedicels usually evident; flowers usually horizontal or ascending;
    staminode yellow-bearded at apex in distal 3-5 mm; calyx 2.5-3.5 mm long,
    the lobes, triangular to ovate, abruptly narrowed to an acute or mucronate
    tip; anther sacs 0.9-1 mm long .... *P. watsonii*

6   Leaves narrowly lanceolate to linear, mostly much more than 4 times longer

than wide; plants of ne. Uintah and Moffat Cos., restricted to sandy soils . . . . **P. angustifolius**

6 Leaves ovate, broadly lanceolate, or broadly spatulate, usually much less than 4 times longer than wide. 7

7 Corolla pink; basal rosette lacking or poorly developed; staminode not much widened at apex, glabrous or merely short-barbellate; plants endemic near Roosevelt, Myton, and Randlett, Duchesne and Uintah Cos. on clay semi-barrens of the Uinta formation . . . . **P. flowersii**

7 Corolla blue to blue-purple or occasionally lavender pink; basal rosette well developed, the basal leaves petiolate, often broadly spatulate; staminode prominently dilated at apex, bearded with hairs 0.2-2 mm long. 8

8 Corolla glabrous externally, the throat and lobes often prominently veined within on all sides with wine red guidelines; staminode hairs mostly 0.5-2 mm long; plants widespread . . . . **P. pachyphyllus**

8 Corolla finely glandular puberulent externally, without prominent guidelines; staminode hairs mostly 0.2-0.5 mm long; plants known from Willow Creek, E. Tavaputs Plateau . . . . **P. carnosus**

9 Corolla glabrous externally 10

9 Corolla glandular-puberulent externally. 12

10 Corolla 6-10 (12) mm long, tufts of basal leaves usually prominent; flowers tightly congested in the fascicles(s), sessile or nearly so, the individual pedicels more-or-less concealed; flowers usually declined; staminode bearded with yellow-brown hairs in distal 1.5 mm . . . . **P. procerus**

10 Corolla (10) 12-22 mm long; tufts of basal leaves lacking or poorly developed; fascicles usually not so congested as to obscure individual pedicels; flowers horizontal to ascending; staminode bearded in the distal 3-5 mm. 11

11 Calyx 2.5-3.5 mm long, the lobes triangular to ovate, abruptly narrowed to an acute or mucronate tip, narrowly scarious-margined; middle stem leaves mostly abruptly narrowed to a cordate or rounded base, often somewhat clasping, broadest near the base . . . . **P. watsonii**

11 Calyx mostly 4-6 mm long, the lobes tapering to a caudate or long-acuminate tip, the margins broadly and prominently scarious; middle stem leaves mostly gradually narrowed at base, not clasping, widest near the middle . . . . **P. rydbergii**

12 Corolla flattened beneath, plicate with 3 longitudinal ridges, dark blue above but with a nearly white longitudinal strip beneath; leaves all cauline, finely puberulent; Daggett and Moffat Cos.. . . . . **P. radicosus**

12 Corolla rounded beneath, not markedly ridged 13

13 Staminode bearded on the distally 1 mm or less, sometimes with a few scattered hairs to about midway. 14

13 Staminode densely bearded for ½ or more of the length 15

14 Corolla 10-16 (18) mm long, blue-purple . . . . **P. humilis**

14 Corolla 20-27 (30) mm long, dingy white, maroon, or black-purple . . . . **P. whippleanus**

15 Corolla 10-15 mm long, the limb not strongly bilabiate . . . . . . **P. goodrichii**

15 Corolla (14) 17-37 mm long, mostly conspicuously bilabiate 16

16 Corolla throat constricted at the orifice, broadest near the middle; palate copiously and prominently bearded with long crinkled yellow hairs; staminode shaggy-villous at the tip and on upper surface with yellow hairs greater than 2 mm long; corolla lavender-blue with dark purple guidelines; Daggett Co. and adjacent Wyoming .... *P. eriantherus*

16 Corolla not constricted at the orifice, broadest at the throat; not as-above in both color and distribution 17

17 Corolla 26-34 mm long, lavender or lavender-pink with red-purple guidelines, the throat strongly inflated, 10-13 mm broad; staminode prominently exserted, curved at the tip, densely bearded on all sides nearly to the base with short golden hairs; endemic to shale barrens of the Green River Formation in e. Uintah and adjacent Rio Blanco Cos. ... *P. grahamii*

17 Corolla less than 20 mm long, blue to blue-purple, the throat less than 8 mm broad, little or not at all inflated; staminode neither prominently exserted nor curved, bearded only on the upper surface in the distal 2/3 or less; plants of central Duchesne Co. 18

18 Lower leaves and stem glabrous to puberulent; inflorescence densely glandular pubescent; stems mostly 10-20 cm tall; longer basal leaves 4-7 cm long; staminode slender throughout .... *P. moffatii*

18 Leaves, stems and inflorescence densely cinereus-puberulent, not glandular; stems 2.5-10 (12) cm tall; longer basal leaves 2-3.5 (4) cm long; staminode widened at tip .... *P. duchesnensis*

Penstemon acaulis Williams (*P. yampaensis* Pennell) — stemless penstemon — Daggett Co. and adjacent Uinta and Sweetwater Cos., Wyoming and in Brown's Park, in Moffat Co. in the Yampa and Vermillion Creek drainages, in pinyon-juniper, black sagebrush and Wyoming big sagebrush communities, 5,800-7,700 ft; May-early July. Plants of this species have the most condensed habit of any of the genus. Two varieties intergrade in the Browns Park area:

1 Leaves linear, 6-15 (20) mm long, 0.8-1.3 (1.5) mm wide; plants of Brown's Park and west of Flaming Gorge Reservoir to Phil Pico Mtn. . **var. acaulis**

1 Leaves, at least some, more broadly linear to oblanceolate, 15-25 (30) mm long, 1.5-2.6 (4.5) mm wide; plants of Browns Park east to near Greystone, Moffat Co. .... **var. yampaensis** (Pennell) Neese

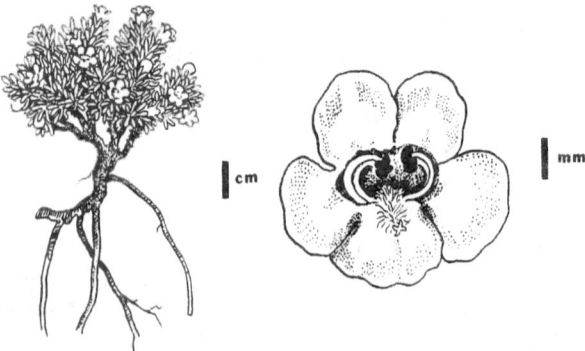

Penstemon acaulis Williams — stemless penstemon

Penstemon angustifolius Pursh — narrowleaf penstemon — Our plants belong to var. *vernalensis* N. Holmgren that is endemic to an area from Vernal, Maeser, and Steinaker Reservoir se. to the Utah-Colorado border and adjacent corners of Rio Blanco and Moffat Cos. and in Brown's Park, Daggett Co.; locally common on stabilized aeolian sand and other sandy sites in sagebrush and juniper communities; 5,000-5,800 ft; May-June. Specimens from Brown's Park (Neese & Peterson 5547, 5561), which are relatively small and have slightly smaller flowers, have been referred to *P. arenicola* A. Nelson, but calyx, corolla, and anther size, as well as habit, fall better within the range of variation of *P. angustifolius* var. *vernalensis*, to which they are referred in this treatment. Unfortunately, label notes that indicated them to be growing on white tuffaceous substrates are in error. They were collected on adjacent aeolian sand. *Penstemon arenicola*, a species of nearby Wyoming, is reported to occur in Daggett and Uintah Cos. (IMF 4: 414) and reported for Vermillion Creek Bridge in Moffat Co. (Bradley 1950). Plants of these taxa differ in the following features:

1 Anther sacs less than 1 mm long (mostly about 0.8 mm); corolla (10) 12-14 mm long; palate sparsely white-bearded .... *P. arenicola*
1 Anther sacs over 1 mm long (mostly 1.2-1.3 mm); corollas 15-20 mm long; palate glabrous .... *P. angustifolius*

Penstemon caespitosus Nuttall — mat penstemon — Widespread; locally common in sagebrush, pinyon-juniper, and mt. brush/grassland communities; 6,500-8,600 ft; June-July. With 2 well marked vars. separated as follows:

1 Leaves linear or linear-lanceolate, mostly 1-2 mm wide, more than 5 times longer than wide, green, scabrous-puberulent but only slightly cinereous; Uinta Mtns. in Duchesne, n. Uintah, and Daggett Cos., also in Strawberry Valley, Wasatch Co................................ **var. caespitosus**
1 Leaves lanceolate to spatulate, mostly 2-4 mm wide, about 3 times as long; Tavaputs Plateau in Duchesne Uintah, and Rio Blanco Cos. ....
.... **var. perbrevis** (Pennell) N. Holmgren

Penstemon carnosus Pennell — fleshy penstemon — East Tavaputs Plateau; desert shrub and juniper communities, usually on white or gray semi-barrens of Uinta or Green River Formations; 5,140-6,500 ft; May-June. Reported for Daggett Co. (AUF 708) based on a specimen of *P. pachyphyllus*.

Penstemon comarrhenus A. Gray — dusty penstemon — A species of generally more s. distribution, it enters our area at upper elevations of the E. and W. Tavaputs Plateaus; uncommon in aspen, Douglas-fir, mtn. brush, and mesic sagebrush-rabbitbrush communities. Our material approaches *P. strictus* in key characters; southward these species are more sharply differentiated.

Penstemon cyananthus Hooker — Wasatch penstemon — Entering the Uinta Basin only at its w. edge and on Bear Mt., Daggett Co. (T. Smith sn); montane to subalpine, in aspen, coniferous forest, sagebrush, and grassland communities. Common to the w. of our area in the Wasatch Mtns.; June-Aug.

Penstemon cyanocaulis Payson — bluestem penstemon — West Tavaputs Plateau in Carbon Co. (IMF 4: 442).

Penstemon duchesnensis (N. Holmgren) Neese (*P. dolius* M. E. Jones ex Pennell var. *duchesnensis* N. Holmgren) — Duchesne penstemon — Locally common in about a 10 mile radius of Duchesne on gravel slopes, road cuts, and desert shrub and pinyon-juniper communities; 5,400-6,600 ft; mid May-June. The low gray-puberulent plants with blue flowers form large colorful patches along roadsides in years of favorable precipitation.

Penstemon eatonii A. Gray — firecracker penstemon — Entering the Basin on the Tavaputs Plateau and extending as far e. as the Avintaquin and Currant Creek drainages, and to Blind Stream in the Uinta Mtns., a small population also occurs on a road-cut along Highway 40 near the Utah-Colorado line; locally common in sagebrush, mtn. brush, and pinyon-juniper communities and on rocky slopes and road cuts at (5,100) 7,300-7,500 ft; May-June. This penstemon is cultivated as an ornamental and used in rangeland seedings, and it can be expected to have an expanding distribution. Our plants belong to var. *eatonii*.

Penstemon eriantherus Pursh (*P. cleburnei* M. E. Jones) — Green River Basin penstemon — The several specimens seen are from the north flank of the Uinta Mtns. from Table Mtn.-Sage Creek on the west to Antelope Flat on the east with several specimens north of the Uinta Mtns. in Uinta and Sweetwater Cos., Wyoming; desert shrub, sagebrush, and Utah juniper communities and on badlands of Morrison, Hillard Shale, and Bridger Formations; 6,100-7,900 (9,100) ft; May-June. Our plants belong to var. *cleburnei* (M. E. Jones) Dorn.

Penstemon flowersii Neese & S. L. Welsh — Flowers penstemon — Endemic to the Uinta Basin; known from about 15 mi e. and w. of the Duchesne-Uintah Co. line near Myton, locally common; shadscale communities on semi barren, gravelly, clay slopes and benches of the Uinta Formation; 4,900-5,400 ft; May.

Penstemon fremontii Torrey & A. Gray — Fremont penstemon — Endemic to our area and adjacent Wyoming, common almost throughout the Uinta Basin, but infrequent in Daggett Co.; arid benches and slopes in shadscale, mixed desert shrub-grassland, Wyoming big sagebrush, and pinyon-juniper communities, on soils varying from clay to gravely sand; 5,000-8,000 ft; May-June.

Penstemon gibbensii Dorn — Gibben penstemon — White tuffaceous bluffs above the Green River at 5500-5600 ft. in eastern Daggett Co. Also known from Carbon and Sweetwater Cos., Wyoming.

Penstemon goodrichii N. Holmgren — Goodrich penstemon — Endemic, known only from between Roosevelt and Maeser and n. to Tridell in Duchesne and Uintah Cos.; uncommon to locally abundant in shadscale, sagebrush, and juniper communities in red and gray clays and sandy clays of the Duchesne River Formation; 5,600-6,200 ft; late May-June. This penstemon is highly variable in abundance from year to year depending on abundance and timing of precipitation.

Penstemon grahamii Keck — Graham penstemon, Green River Shale penstemon — Endemic, sporadically present on white to tan, semi-barren, shale slopes and ridges of the Parachute member of the Green River Formation in se. Uintah Co., on Raven Ridge near Mormon Gap in Rio Blanco Co., and in the vicinity of Sand Wash along the Green River near the Carbon-Uintah Co. line; growing with shadscale, greesebush, Salina wildrye, and scattered pinyon-juniper; 5,600-6,300 ft (about 5,000 ft at Sand Wash); May.

Penstemon humilis Nuttall — low penstemon — widespread, but apparently lacking from the E. Tavaputs Plateau; occasional and locally common in montane sagebrush-grassland, also in pinyon-juniper, mt. brush, aspen, and ponderosa pine communities; 6,000-9,000 ft; late May-June. Our plants belong to var. *humilis*.

Penstemon grahamii Keck
Graham penstemon, Green River Shale penstemon

Penstemon leonardii Rydberg — Leonard penstemon — Western Uinta Mtns. in Summit and Wasatch Cos. and Timber Canyon drainage of the W. Tavaputs Plateau in sagebrush communities and on steep, rocky, semi-barren slopes at 9,000-10,000 ft; late June-Aug.

Penstemon moffatii Eastwood — Moffat penstemon — Known in the Uinta Basin from near Starvation Reservoir; rare, but sometimes locally abundant in pinyon-juniper woodlands. Distribution of this seldom-collected species is mostly s. of our area in desert shrub communities, clay soil; May-June. Wolf & Dever 5101 DINO! from Round Top Mt., referred here by Bradley (1950) belongs to *P. humilis*.

Penstemon montanus Greene — cordroot penstemon — Known from limestone talus in S. Fork Rock Creek, Uinta Mtns. in Duchesne Co. at 9,500-10,000 ft; July-Aug.

Penstemon pachyphyllus Rydberg — thick leaf penstemon — Locally common in salt desert shrub, sagebrush, pinyon-juniper, and mt. brush communities, 5,300-8,800 ft. May-June. Specimens from the Basin referred to *P. osterhoutii* belong here. Two intergrading somewhat geographically separated varieties are present; most material can be separated as follows:

1 Staminode sparsely bearded with hairs 0.2-1 mm long, these not obscuring the broadly dilated tip; plants mostly of Daggett and northern Uintah Cos. and adjacent Colorado . . . . **var. mucronatus** (N. Holmgren) Neese
1 Staminode densely bearded with tangled hairs 1-2 mm long that obscure the apex; plants mostly of Wasatch, Duchesne and southern Uintah Cos. and likely east into Rio Blanco Co. . . . . **var. pachyphyllus**

Penstemon palmeri A. Gray — Palmer penstemon, balloon flower — Native to s. Utah, Arizona, California, and Nevada, used in roadside seedings and in burned areas and becoming relatively abundant for a few years. Persistence of this species in our area is yet to be determined.

Penstemon paysoniorum Keck — Payson penstemon — Known from the margin of our area near Lonetree, Wyoming on eroding badlands of Bridger Formation at 7,450 ft (Goodrich 26059).

Penstemon platyphyllus Rydberg — broadleaf penstemon — Our only record (Harrison 406H) is from Indian Canyon 28 mi sw. of Duchesne on bare eroding slopes, 8,200 ft. otherwise endemic to the Wasatch Mtns. The specimen might have been from a waif that did not persist.

Penstemon procerus Douglas — little-flower penstemon, meadow penstemon — Common in Strawberry Valley and across the Uinta Mtns. and extreme w of the W. Tavaputs Plateau, apparently rare on the E. Tavaputs Plateau in montane and alpine meadows, silver sagebrush-grassland, and coniferous forest communities; mostly above 8,500 (rarely as low as 7,600) ft; late June-Aug. Materials from the Uinta Basin which are transitional between *P. procerus* and *P. rydbergii* have been recognized at infraspecific level within both species. Our plants belong to var. *procerus.*

Penstemon radicosus A. Nelson — matroot penstemon — Uncommon; known from Daggett Co. from Phil Pico Mt. to Goslin Mt.) and on Cold Spring Mtn and Diamond Mt. in Moffat Co.; mt. brush, sagebrush, grass, serviceberry, and pinyon-juniper communities; 6,200-8,200 ft; June-July.

Penstemon rydbergii A. Nelson — Rydberg penstemon — Common in Strawberry Valley and across much of the Uinta Mtns. and W. Tavaputs Plateau to the Avintaquin drainage, and apparently rare on the E. Tavaputs Plateau; meadows, parklands, open rocky slopes, and leeward snowbeds; July-Aug. See *P. procerus.* MacLeod 13A DINO!, apparently the basis of the Holmgren (1962) report for Blue Mt., belongs to *P. procerus.* Our plants belong to var. *rydbergii.*

Penstemon scariosus Pennell — plateau penstemon — This is a variable taxon comprising a series of more or less habitat specific phases, these differing in plant size and habit, leaf size and shape, flower size and color, and degree of glandularity. Only a portion of the variation has been recognized at infraspecific rank. The following key serves to separate most of the variation, but not all material will "key" satisfactorily: Perhaps reports of *P. saxosorum* Pennell for our area are based on specimens of *P. scariosus*.

Small flowered plants of Yellow Creek and Piceance Basin appear to be part of the *P. scariosus* complex. These might be worthy of separate taxonomic recognition.

1 Corolla glandular externally, blue-purple, stems often decumbent, curved upward, the leaves unilaterally oriented; upper stems, pedicels, and calyx often strongly suffused with purple; anther sacs less than 1.9 mm long; leaves narrow, mostly much less than 1 cm broad; Plants of Blue Mtn. and the Yampa Plateau .... **var. *cyanomontanus***

1 Corolla glabrous externally, blue or blue-lavender; stems mostly erect, straight, the leaves not unilaterally oriented; inflorescence usually not suffused with purple; anther sacs 1.7-2.2 mm long; leaves narrow or not; distribution otherwise... 2

2 Leaves lanceolate, mostly over 7 mm wide; tuft of basal leaves prominent at flowering time; corolla blue, often longer than 23 mm; plants of the W. Tavaputs Plateau and Uinta Mtns. .... **var. *garrettii***

2 Leaves linear to linear-lanceolate, rarely over 7 mm wide; tuft of basal leaves withered or at least not prominent at flowering; corolla pale blue to lavender or pinkish, usually less than 22 mm long; plants of extreme e. Uintah Co. and closely adjacent Colorado on shale barrens of the Green River Formation .... **var. *albifluvis***

Var. *albifluvis* (England) N. Holmgren (*P. albifluvis* England) — White River penstemon — Occasional and locally common on the E. Tavaputs Plateau; shadscale-*Elymus salinus* communities, usually growing with *Cirsium barnebyi*, *Eriogonum ephedroides*, *Machaeranthera grindelioides* and scattered pinyon-juniper; 5,000-6,800 ft; May-July. Plants in the Big Pack Mt. area are transitional to var. *garrettii*.

Var. *cyanomontanus* Neese — Blue Mtn. penstemon — Common in crevices of exposed sandstone bedrock on the summit of Blue Mt. and adjacent Colorado; sagebrush-grassland, and ponderosa pine communities. Populations from sandstone and quartzite substrates on Diamond Mt. to Mosby Mt. are similar but usually somewhat less glandular-pubescent; 7,000-8,200 ft; June-July. Welsh 480, referred to *P. cyanocaulis* Payson (Welsh 1957) belongs here.

Var. *garrettii* (Pennell) N. Holmgren (*P. garrettii* Pennell) — Garrett penstemon — Occasional to locally common in the Uinta Mtns. and on the W. Tavaputs Plateau; sagebrush, snowberry, and Purshia-grasslands, and in openings of pinyon-juniper, mt. brush, ponderosa pine, Douglas-fir, and spruce communities, often on calcareous substrates, colonizing on roadcuts in the Uinta Mtns.; 6,700-10,000 ft; May-Aug. A small and dark colored phase from Green River shale near the head of Buffalo-Willow Creeks of the Strawberry River drainage may deserve taxonomic recognition. Populations from the e. end of the Uinta Mtns. are transitional to the previous variety.

Penstemon strictus Bentham — Rocky Mtn. penstemon — Widespread; occasional to locally abundant; apparently more common on the Tavaputs Plateau than in the Uinta Mtns.; sagebrush, pinyon-juniper, mt. brush, aspen, and coniferous forest communities; mostly 7,000-9,500 ft, rarely down to 6,000 ft (Daggett Co.) or to 10,000 ft (near Mt. Bartles); June-Aug.

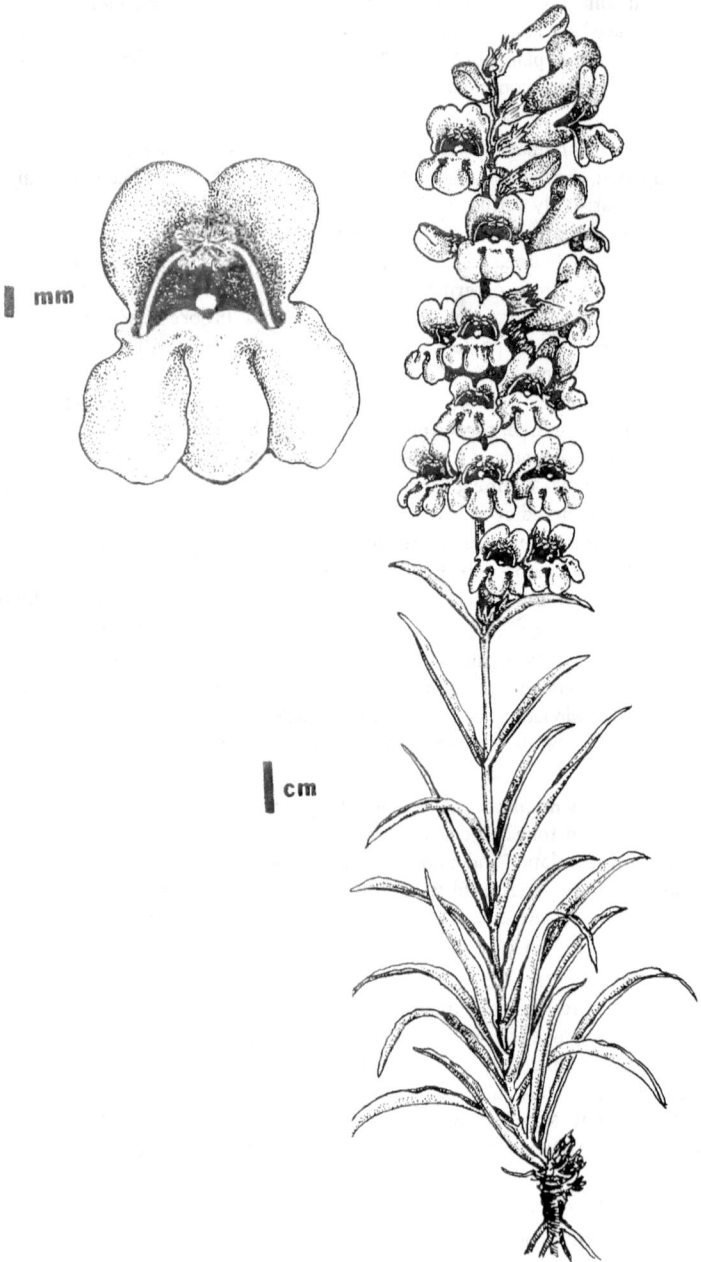

Penstemon scariosus var. *garrettii* (Pennell) N. Holmgren
Garrett penstemon

Penstemon subglaber Rydberg — glabrous penstemon — Occasional or locally abundant in the Uinta Mtns. and the far w. portion of the Tavaputs Plateau, apparently more common at the w. end of the Uinta Basin; sagebrush-grass, lodgepole pine, aspen, and coniferous forest communities and on road cuts; 7,000-10,000 ft; mid June-Aug. This is a tall and showy species that sometimes grows in great masses on road cuts. Occasionally the flowers are pink. Tall and with secund inflorescence as in *P. strictus,* but anthers with shorter hairs and calyx 4-8 mm long compared to 2.5-5 mm in *P. strictus.*

Penstemon uintahensis Pennell — Uinta Mtn. penstemon — Endemic to the Uinta Mtns. from the Yellowstone drainage and eastward on the s. slope and Smiths Fork and eastward on the n. slope; mostly on Precambrian quartzite, but the type specimen (from Dyer Peak) and a few others are from limestone; near and above timberline on morainal ridges, fell fields, talus slopes, and rock strips, in alpine meadow, spruce-sedge, and tundra communities; 10,500-12,200 ft; July-Aug.

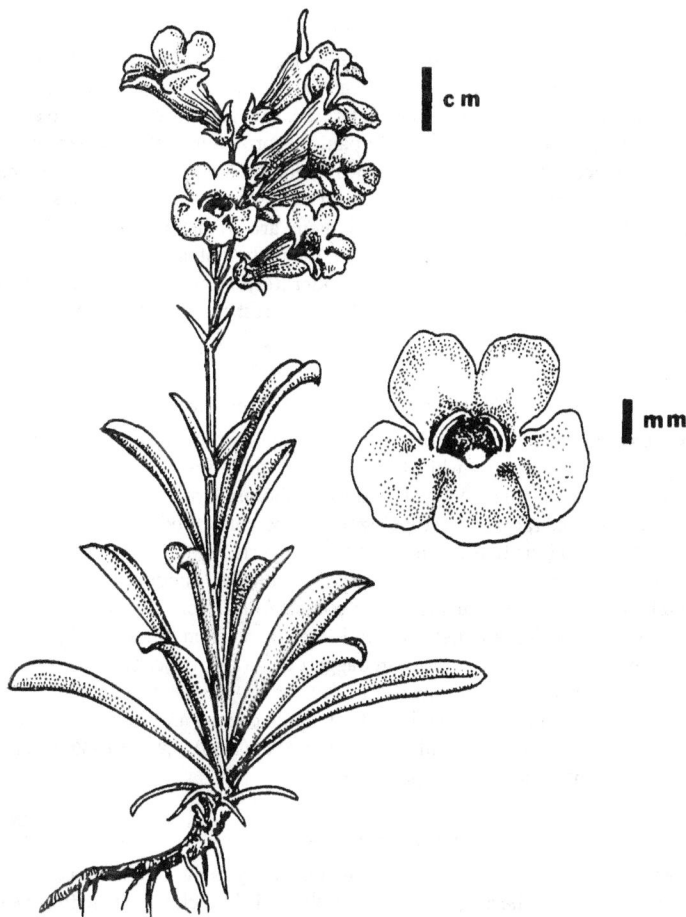

Penstemon uintahensis Pennell — Uinta Mtn. penstemon

Penstemon venustus Douglas — lovely penstemon — Known in Utah from a single specimen (Goodrich 19243), the single large plant growing on a rocky roadside about 2 mi w. of Mt. Home; a Pacific Northwest species apparently introduced in roadside seeding. Subsequent visits to this site indicate the plant did not persist.

Penstemon watsonii A. Gray — Watson penstemon — Widespread; occasional to frequent in the w. portion of the Uinta Mtns. and on the Tavaputs Plateau and Yampa Plateau; mtn. big sagebrush-grassland, mt. brush, and aspen communities at 7,000-9,200 ft; mid June-Aug. Becoming abundant following fire in big sagebrush communities.

Penstemon whippleanus A. Gray — Whipple penstemon — Occasional across the Uinta Mtns., rare along the rim of the Tavaputs Plateau; parkland, aspen, coniferous forest, and alpine communities at 8,600-11,000 ft; July-Aug. Plants with whitish flowers appear to be more common at high elevations in the Uinta Mountains.

Scrophularia L. — Figwort

Scrophularia lanceolata Pursh — lanceleaf figwort — Perennial herbs; stems several, erect, mostly unbranched except in the inflorescence, to about 1.5 m tall; leaves opposite, petiolate, lanceolate, sharply serrate, the blades to about 12 cm long; inflorescence a thyrsoid panicle; corollas urceolate, greenish and usually tinged with yellow, red, or brown, 2-lipped, 5-lobed, the lobes rounded, the upper 2 the longest, projecting forward, the lower middle 1 sharply reflexed; capsule ovoid. Our few specimens are from Strawberry Valley, s. slope of the Uinta Mtns., Blue Mt., W. Tavaputs Plateau, and the Piceance Basin; infrequent or occasional in moderately moist sites in sagebrush, snowberry, mt. brush, aspen, and coniferous forest communities; 7,000-10,400 ft; June-July.

Synthyris Bentham — Kittentails

Synthyris pinnatifida S. Watson — cutleaf kittentails — Low scapose perennial beginning to flower early while leaves are still immature; leaves all basal, pinnate-pinnatifid, to 18 cm long at maturity; inflorescence a scapose, bracteate, densely flowered raceme overtopping the leaves, condensed at first, elongating in fruit; corolla blue to blue-purple, 4 lobed, nearly regular, about 6 mm long; stamens dark blue, exserted; capsule suborbicular, moderately flattened, notched at apex. Infrequent; specimens seen are from Big Ridge, Wedge Hollow, Burnt Ridge, and head of Blind Stream and from limestone dip slopes of the n. slope of the Uinta Mtns. in Daggett Co., a single location is shown for the Bear River Drainage in Albee et al. (1988); Douglas-fir, Engelmann spruce and lodgepole pine forests, and heads of alpine cirques and leeward slopes in snowbeds, listed for wet meadows in Albee et al (1988); 8,900-10,880 ft; June-July. Our plants belong to var. *pinnatifida*.

Verbascum L. — Mullein

Verbascum thapsus L. — common mullein, woolly mullein, miners candle — Robust perennial; herbage densely woolly with branching hairs, first-year leaves forming a rosette, oblanceolate to spatulate, tapered to a short petiole, to 30 or more cm long, those of the flowering stem decurrent, reduced upward, becoming sessile, lanceolate; stem stout, typically simple but sometimes branching in the inflorescence, to 2 dm tall; inflorescence a densely flowered, bracteate spike-like raceme; corolla yellow, slightly irregular, rotate, Widespread introduced Eurasian weed; roadsides and other disturbed sites, up to about 8,000 ft; June-Sept.

Veronica L. — Speedwell; Brooklime

Perennial or annual herbs, often of wet places; leaves opposite, simple, sessile or short-petiolate; calyx deeply 4-lobed; corolla white to lavender or (usually) blue, sub rotate, 4 lobed, the upper lobe broader, the lower narrower than the lateral ones; stamens 2; capsule compressed, suborbicular or obcordate to 2-lobed, the style arising from the notch between the 2 valves.

---

1 Leaves sessile 2

1 Leaves at least short-petiolate. 5

2 Flowers in axillary racemes; capsules glabrous . . . . . *V. anagallis-aquatica*

2 Flowers in a terminal raceme; capsules pubescent 3

3 Style 0.1-0.4 mm long, shorter than the notch of the fruit . . . *V. peregrina*

3 Style 0.8-3 mm long, longer than the notch of the fruit 4

4 Style 0.8-1.3 mm long; inflorescence congested, head-like during early flowering, elongating in fruit . . . . *V. wormskjoldii*

4 Style 2.2-3 mm long; inflorescence elongate, scarcely head-like even when young . . . . *V. serpyllifolia*

5 Herbage glabrous; calyx to about 3 mm long in fruit . . . . *V. americana*

5 Herbage glandular puberulent or hirsute; calyx enlarging in fruit to as much as 7-10 mm 6

6 Leaves and bracts 3-5 lobed . . . . *V. hederaefolia*

6 Leaves and bracts toothed . . . . *V. biloba*

Veronica americana Schweinitz ex Bentham in de Candolle — American speedwell, American brooklime — Widespread along slow-moving reaches of streams, in seeps and springs; specimens seen are from 6,000-8,500 ft; June-Aug.

Veronica anagallis-aquatica L. — water speedwell — Introduced from Eurasia and Africa; widespread, naturalized along water courses and in mud or shallow water near springs, seeps, and canals; valley bottoms to 8,200 ft; June-Aug.

Veronica biloba L. [*Pocilla biloba* (L.) W. A. Weber] — twolobe speedwell — Introduced from Asia. Only 4 specimens were seen from our area in preparation for the 1986 edition of this flora. Since then this plant has been found in numerous locations. It is now widespread in sagebrush, pinyon-juniper, aspen, and Douglas fir communities where it is mostly associated with disturbance; 6,000-8,500 ft; May-July.

Veronica hederifolia L. — Ivyleaf speedwell — Introduced from Europe, generally in disturbed sites. Know from Little Hole Campground Daggett County under shade of trees at 5,545 ft elevation. Widespread in N. America and to be expected elsewhere in our area.

Veronica peregrina L. — purslane speedwell — Widespread across our area, especially common on Diamond Mt. also middle elevations in the Uinta Mtns., no specimens seen from the W. Tavaputs Plateau; wet to drying soil, sometimes in rocky places, margins of reservoirs and ephemeral ponds, swales, canal banks, and meadows, often in areas of disturbance; 5,800-9,000 ft; June-Aug. Our plants belong to var. *xalapensis* (H.B.K.) St. John & Warren.

Veronica serpyllifolia L. [*V. humifusa* Dickson; *Veronicastrum serpyllifolia* (L.) Fourreau — thyme-leaf speedwell — Strawberry Valley and Uinta Mtns.; lake margins, wet meadows, and along streams, often under willows at 7,500-11,000 ft; June-Aug. Our plants belong to var. *humifusa* (Dickson) Vahl.

Veronica wormskjoldii Roemer & Schultes — alpine speedwell, Wormskjold speedwell — Strawberry Valley and Uinta Mtns.; common in wet meadows and along streams and lake margins with sedges and willows, often in rocky places, in spruce, fir, lodgepole, and alpine communities; 8,100-11,200 ft; June-Aug.

SELAGINELLACEAE Selaginella Family

Selaginella Beauvios — Spikemoss

Evergreen, low growing, usually matted, perennial herbs; stems prostrate or some ascending to erect; stems covered with scale-like leaves (in a similar way to those of juniper); leaves not over 5 mm long; flowers lacking; reproduction by spores, the spores borne in the axils of leaf-like sporophylls, inconspicuous and usually hidden in the sporophylls.

---

1 Sporophylls and leaves awnless or with a mucro less than 0.3 mm long, notably ciliate at 10x .... ***S. mutica***

1 Sporophylls and leaves awn-tipped, not or only slightly ciliate 2

2 Awn-tip of leaves (0.3) 0.5-2 mm long; leaves of the lower or convex side of the stems somewhat longer than those of the corresponding level on the upper or concave side of the stems .... ***S. densa***

2 Awn-tip of leaves 0.2-0.4 mm long; leaves of a given level on the stem alike .... ***S. watsonii***

Selaginella densa Rydberg — lesser spikemoss — Uinta Mtns. and Cold Spring Mtn; rocky places and alpine tundra; (6,000) 8,000-12,000 ft, perhaps higher.

Selaginella mutica D. C. Eaton — bluntleaf spikemoss, awnless spikemoss — The few specimens seen are from rocky canyons along the Green River and Ashley Creek Gorge.

Selaginella watsonii Underwood — Watson spikemoss — Uinta Mtns.; occasional to common; rocky places and alpine tundra; (7,000) 8,000-13,000 ft.

SOLANACEAE Nightshade Family, Potato Family

Plants herbaceous or woody; leaves alternate; flowers bisexual, usually regular; sepals mostly 5, more or less united; corolla united, more or less 5-lobed; stamens 5, alternate with the corolla lobes, inserted on the corolla tube; ovary superior; fruit a berry or a capsule. The cultivated potato and tomato are members of the family. All of the taxa discussed below are considered to contain poisonous principles (Kingsbury 1964).

1   Plants woody, often over 1 m tall .... *Lycium*

1   Plants herbaceous, often annual, mostly less than 1 m tall 2

2   Corolla 5-10 cm long or longer, the limb 3 cm wide or wider; leaf blades 10-20 cm long, nearly as wide, on stout petioles .... *Datura*

2   Corolla shorter or limb narrower or both than above; leaf blades various but often smaller or sessile or nearly so 3

3   Leaves pinnately lobed 4

3   Leaves entire or toothed but not lobed 5

4   Flowers numerous in secund racemes; corolla greenish yellow with purple veins and throat; calyx 10-12 mm long and enlarging to 25-30 mm long in fruit .... *Hyoscyamus*

4   Flowers, corolla, and calyx not as above .... *Solanum*

5   Corolla less than 2 cm long .... *Solanum*

5   Corolla 2 cm long or longer 6

6   Corolla narrowly funnelform, the tube 4 or more times longer than wide; calyx 5-7 mm long in flower and to 10 mm in fruit .... *Nicotiana*

6   Corolla tube less than twice as long as wide; calyx 7-10 long in flower and 25-35 (40) mm long in fruit .... *Physalis*

Datura L. — Thorn-apple; Jimson Weed

Datura stramonium L. — Jimson weed — Coarse, weedy, ill-scented annual herbs, 30-100 cm tall, glabrous or nearly so; leaves coarsely toothed, to shallowly lobed; calyx 3-6 cm long, the lobes about 4-7 mm long, the persistent portion 4-6 mm long; corolla white to violet, 5-8 (11) cm long; fruit a large, usually spiny capsule that tardily divides into 4 valves. Reported by Graham (1937) from a garden in Lapoint and sparingly planted as an ornamental currently in Vernal. Introduced from tropical America, naturalized in much of the moister and warmer areas of the United States, probably not persistent in the climate of the Uinta Basin. *Datura wrightii* Regel (Indian apple) with corolla 8-20 (23) cm long; persistent portion of the calyx mostly (10) 15-20 mm long, and entire to coarsely sinuate dentate leaves has been planted as an ornamental. This is native to areas well s. of our area. It might not persist here.

Hyoscyamus L. — Henbane

Hyoscyamus niger L. — black henbane — Coarse weedy annual or biennial, viscid-villose herbs; stems 30-100 cm tall; leaves 6-20 cm long, ovate to lanceolate, irregularly lobed, cleft or pinnatifid, sessile, sometimes clasping or the lower ones short-petiolate; flowers numerous in terminal secund racemes as well as in axils of upper leaves; corolla 1.5-2 cm long, whitish with reddish or purplish veins; stamens exserted; fruiting calyx 2-2.5 cm long, fruit a capsule completely included in the calyx, about 1-1.4 cm long. Introduced from Europe, adventive but rather infrequent in much of Uinta Basin proper, common in the drawdown basin of Flaming Gorge Reservoir, to be expected to expand in range; roadsides, drawdown basins, and other disturbed sites; not expected over 8,000 ft; June-Aug. Once abundant along the road from Lonetree west to Henrys Fork but later nearly eradicated by persistent weed control.

Lycium L. — Desert-throne; Wolfberry

Lycium barbarum L. (*L. halimifolium* Miller) — matrimony vine, tea vine — Sprawling or more or less erect glabrous shrubs, 0.75-2 (6) m tall; leaf blades 2-6-cm long, oblong to elliptic or lanceolate; calyx about 4 mm long; corolla about 5-10 mm long, purple to reddish; stamens slightly exserted; fruit a reddish orange fleshy berry to about 1 cm in diameter. Introduced from China; cultivated mostly at old homesteads, persisting and escaping along fences and roadsides, occasionally forming thickets and becoming a pest on neglected pastureland; not expected above 6,000 ft; May-Oct.

Nicotiana L. — Tobacco

Nicotiana attenuata Torrey ex S. Watson — coyote tobacco, wild tobacco — Annual viscid-puberulent, malodorous herbs 30-60 cm tall; lower leaves usually with petioles at least ¼ as long as the blade, leaf blades 3-10 cm long, ovate to lanceolate, the upper ones linear; calyx 6-9 mm long; flowers in terminal paniculate racemes; corolla 2-4 cm long, white or greenish white; stamens included; fruit a capsule, about 1 cm long, not completely included in the calyx. Widespread; this is a seedbank species of desert shrub, pinyon-juniper, and Douglas-fir communities. The seedbank is activated following fire in pinyon-juniper and Douglas-fir communities. This plant grows in great abundance for a few years after fire then the seedbank often becomes dormant until the next disturbance. Apparently the seeds remain viable for 100 years or longer.

Physalis L. — Ground-cherry; Husk-tomato

Physalis longifolia Nuttall (*P. subglabrata* Mackenzie & Bush.; *P. virginiana* Miller misapplied) — common ground-cherry, popweed — Perennial herbs from rhizomes 20-80 cm tall, glabrous or with a few long hairs; leaf blades 4-10 cm long, entire or undulate; corolla yellowish with a brown center. Probably introduced from the eastern half of the United States, more or less weedy; gardens, ditch banks, roadsides, edge of fields, usually where the ground is frequently disturbed; up to about 7,000 ft; June-Sept. Our plants belong to var. *longifolia*.

Solanum L. — Nightshade

With features listed above for the family and: stamens with short filaments and large anthers which are more or less united around the style, usually slightly exserted; fruit a many-seeded berry, enveloped or subtended by the persistent calyx.

---

1 Plants with numerous long straight prickles especially on petioles, stems and calyces; leaves 1-2 times pinnatifid; corolla 18-28 mm long, yellow . . . . ***S. rostratum***

1 Plants not armed with prickles 2

2 Plants climbing, vine-like, perennial, slightly woody toward the base; leaves hastate or 2 lobed at the base; corolla lavender, about 1 cm long; berries glossy red . . . . ***S. dulcamara***

2 Plants annual herbs not climbing; leaves various; corolla white or bluish-white, sometimes purple spotted; berries greenish or yellowish 3

3   Leaves pinnately lobed to divided .... *S. triflorum*

3   Leaves entire or shallowly lobed.................................... 4

4   Herbage with spreading, glandular hairs; calyx conspicuously enlarged in fruit; berry yellow to dark green at maturity, but black in dried specimens .... *S. sarrachoides*

4   Herbage with appressed or slightly spreading, non-glandular hairs; calyx not much if at all enlarged in fruit; berry blackish at maturity .. *S. nigrum*

Solanum dulcamara L. — bittersweet nightshade, climbing nightshade — Introduced from Eurasia; specimens seen are from roadsides, ditch banks, and fence lines at Vernal, Whiterocks, and Neola, to be expected elsewhere in wet places at lower elevations; June-Sept.

Solanum nigrum L. — black nightshade — Listed for Uintah Co. (AUF 732) likely based on E. Neese & D. Nelson 15025 that belongs to *S. sarrachoides,* and reported (Flowers and others 1960) for Linwood (area now inundated by Flaming Gorge Reservoir), but no supporting specimen was found at UT. Based on the discussion of this species in IMF (4: 68) we expect all specimens from our area identified as *S. nigrum* belong to *S. sarrachoides*. However, this is cosmopolitan species that could show up in our area anytime.

Solanum rostratum Dunal — buffalo bur, Kansas thistle, Texas thistle — Native of the Great Plains; The 2 specimens seen are from a construction site in Tridell (J. Goodrich sn) and a flower garden in Vernal; July-Sept. The Vernal specimen might have come as a contaminant in commercial bird seed.

Solanum sarrachoides Sendtner in Martius — hairy nightshade, hoe nightshade — Introduced from S. America, more or less weedy; ditch banks, roadsides, and gardens; 5,600-6,000 ft; July-Sept.

Solanum triflorum Nuttall — cutleaf nightshade — widespread; along roads, in gardens, and other disturbed places, and in various plant communities; 6,000-8,200 ft; July-Sept.

SPARGANIACEAE Bur Reed Family

Sparganium L. — Bur Reed

Plants monoecious, aquatic or growing in wet mud of late drying pools, perennial, herbaceous, from rhizomes with fibrous roots; stems erect or floating, leaves alternate, 2 ranked, narrowly elongate, erect or floating, expanded and sheathing at the base, usually transverse-septate at the base; flowers unisexual, in sessile or pedunculate globose heads; staminate heads above the pistillate, 1-several, the flowers numerous, each consisting of 3-5 stamens, and 3-5 minute, scale-like bracts; pistillate heads 1-2, arising from the axils of leaf-like bracts, each flower consisting of a solitary pistil subtended by 3-6 free, oblanceolate to fan-shaped, membranous, inconspicuous perianth segments; ovary mostly 1 celled, narrowed at the base; style simple or forked; fruit a beaked achene.

---

1   Stigmas 2, rarely 1 in occasional flowers; leaves 7-15 mm wide; inflorescence branched; achenes rounded, to 5 (6) mm wide, abruptly narrowed to the beak; plants 50-200 cm tall, erect, to be expected at low elevations, no specimens seen .... *S. eurycarpum*

1   Stigmas 1, rarely 2 in occasional flowers; plants mostly less than 100 cm tall, erect or floating; leaves various; inflorescence mostly simple; achenes essentially fusiform, to about 3 mm wide, gradually tapered to the beak ..
    2

2   Staminate heads solitary (rarely 2); anthers at least half as wide as long, 0.3-0.8 mm long; achene beak 1-1.5 mm long including the stigma; leaves floating or erect and aerial, 2-6 (8) mm wide .... *S. natans*

2   Staminate heads mostly 2 or more; anthers less than half as wide as long, 0.8-1.5 mm long; achene beak 1.5-5 mm long 3

3   Achene beak 1.5-2 mm long including the stigma; leaves 1-8 mm wide, mostly submerged or floating; mature pistillate heads 1-2 cm wide; stigmas 0.8-1.2 mm long .... *S. angustifolium*

3   Achene beak 3-4 (5) mm long including the stigma; leaves (3) 5-12 (15) mm wide, mostly erect and aerial; mature pistillate heads 2-3 cm across; stigmas 1-2 mm long .... *S. emersum*

Sparganium angustifolium Michaux — narrow leaf bur-reed — Uinta Mtns.; occasional; floating or submersed in pools, shallow ponds, and margins of lakes, or stranded on mud especially in drawdown basins of reservoirs. The few specimens seen are from 8,700-10,500 ft.

Sparganium emersum Rehmann [*S. multipedunculatum* (Morong.) Rydberg] — European bur-reed — widespread; occasional in shallow water of ponds, marshy meadows, and lakes, and stranded on mud in dried ponds at 5,600-8,500 ft. Our plants belong to var. *multipedunculatum* (Morong) Reveal.

Sparganium natans L. [*S. minimum* (Hartman) Fries] — small bur-reed — Uinta Mountains; infrequent or at least seldom collected; floating or emersed in pools and shallow ponds.

TAMARICACEAE Tamarisk Family

Tamarix L. — Tamarisk

Tamarix chinensis Louriero (*T. ramosissima* Ledebour; *T. gallica* L. misapplied; *T. pentandra* Pallas misapplied) — tamarisk, five stamen tamarisk, salt-cedar — Shrubs or small trees 2-7 m tall; bark of branches reddish-brown; current years twigs green and same color as leaves; leaves scale-like, appressed to the branches and often appearing more as a part of the branches than definite leaves, 1-3 mm long, entire, alternate; inflorescence of numerous spikes, the spikes sometimes paniculate, slender, elongate; flowers 5-merous, minute but numerous and thus conspicuous; sepals erose-denticulate; petals 1-2 mm long, white to pink; fruit a capsule, 3-4 mm long; seeds tufted with hairs. Introduced from Eurasia, widespread, common to abundant on floodplains, drawdown basins of reservoirs, roadsides and other areas where the soil is wet for a part of the year, tolerant of alkali; not expected over 7,000 ft; May-Sept.

TYPHACEAE Cattail Family

Perennial herbs from creeping rhizomes; stems tall and erect; leaves alternate, long, linear, sheathing at the base; inflorescence of spikes or spike-like racemes, the staminate portion above the pistillate and pale, the pistillate portion brown to dark brown, cylindrical; flowers unisexual, densely crowded; perianth of hair-like bristles (these providing the wind-buoyant structure by which the seeds are dispersed); stamens 2-7, the filaments united; ovary superior, styles 1-2; fruit a nutlet-like achene.

Typha L. — Cattail

---

1   Pistillate part of the spike light cinnamon brown, 15-28 cm long, 1.5-2.5 cm thick, separated from the staminate part by a naked stalk 1-5 (8) cm long; leaves rounded on the back, 6-9 (12) mm wide; plants mostly 2.5-3 (4) m tall .... *T. domingensis*

1   Pistillate part of the spike dark brown, 10-19 cm long, 1.5-3.5 cm tick, contiguous with the staminate part or separated by an interval of up to 1.5 cm long; leaves flat 6-20 mm wide; plants 1-3 m tall .... *T. latifolia*

Typha domingensis Persoon — southern cattail — Sparingly collected from Uintah Co. and White River drainage in Rio Blanco Co. This plant forms large stands on the floodplain of the Green River near Ouray. Reported for Daggett Co. (Flowers and others 1960) but no specimens seen from that Co. Plants of this taxon have often been incorrectly referred to as *T. angustifolia* L., which is not known from our area. See *T. latifolia.*

Typha latifolia L. — broadleaf cattail — Widespread; forming clones around ponds and in wet lowlands and up to about 7,000 (rarely 9,000) ft. In addition to the features listed in the key, this taxon is distinguished from *T. domingensis* by a smooth appearance of the pistillate part of the spike. Minutely protruding stigmas and bracts give a roughened appearance to the spike of *T. domingensis.* Apparently *T. latifolia* is not as salt tolerant as *T. domingensis* (Arnow and others 1980). Hybrids of these 2 taxa are referred to as *T. glauca* Godron. Wolf & Dever 5054 DINO! reported as *T. angustifolia* (Welsh 1957) belongs here.

ULMACEAE Elm Family

---

1   Leaves entire or with a few teeth; plants native, shrubs or small trees .... *Celtis*

1   Leaves distinctly serrate; plants introduced, trees .... *Ulmus*

Celtis L. — Hackberry

Celtis reticulata Torrey (*C. douglasii* Planchon; *C. occidentalis* L. misapplied) — hackberry — Shrubs or small trees; bark becoming corky and wart-like on the trunk but often remaining smooth on the larger branches; leaves alternate, simple, 2-10 cm long, broadly ovate to ovate-lanceolate, rounded to cordate and unequal at the base, entire or rarely with a few teeth, smooth to scabrous above, glabrous to hairy beneath, somewhat palmately-veined; flowers bisexual or unisexual, axillary, solitary or in small clusters; calyx 4-6 parted or lobed; corolla none; stamens 4-5; ovary superior, 1-celled, style none; stigmas 2; fruit a drupe with thin flesh and hard-shelled seed. Occasional and locally common in canyons of the Green River in and near Dinosaur National Monument and

Desolation Canyon, with distribution like that of Fraxinus anomala; May-June. In view of many intergradations found in Colorado material, Harrington (1954) treated this taxon as a part of *C. occidentalis* L.

Ulmus L. — Elm

Ulmus pumila L. — Siberian elm — Trees to 30 m tall; bark gray, deeply fissured on older trees; leaves alternate; petioles 3-10 mm long; leaf blades narrowly elliptic to lanceolate, 2-7 cm long or to 12 cm long on sucker shoots, often slightly oblique at the base, singly or doubly serrate, often with tufts of white hairs in the axils of the veins of the lower surface; flowers appearing before the leaves, bisexual; calyx purplish, 4-9 lobed, about 2-3 mm long; corolla lacking; stamens exserted; fruit a flat, winged, suborbicular, glabrous samara to about 15 mm long, notched. Introduced from the Old World, often escaping and persisting in towns, along roads, and in riparian areas where in some places it becomes a dominant tree. This is a weedy and messy tree due to the abundance of seed and brittle branches, but it has been often planted in our area because it grows fast and provides shade in only a few years, and it is cold tolerant, likely planted during early years of settlement more than in recent times; probably not growing much above 7,000 ft elevation; April-May. The seeds germinate in cracks of sidewalks and other difficult sites as well as in favorable places.

URTICACEAE Nettle Family

Monoecious or dioecious herbs; leaves simple; flowers inconspicuous, the perianth of 4 greenish segments; staminate flowers of 4 stamens and sometimes with a rudiment of a pistil; pistillate flowers of a single pistil; fruit an achene enclosed by the perianth segments.

---

1   Leaves entire, alternate; stipules lacking; plants annual, with procumbent to ascending stems 5-40 cm long, pubescent with non stinging hairs . . . . ***Parietaria***

1   Leaves toothed, opposite, stipular; plants perennial, erect, 30-100 (200) cm tall, with hairs stinging . . . . ***Urtica***

Parietaria L. — Pellitory

Parietaria pensylvanica Muhlenberg ex. Willdenow — Pennsylvania pellitory, hammerwort — Annual herbs with procumbent, sprawling, or ascending, weak stems, pubescent with simple hairs, those of the inflorescence minute and often hooked; leaves with slender petioles, the blades 5-50 (70) mm long, 5-25 mm wide, rhombic-ovate to lanceolate; flowers sessile or nearly so, clustered in the axils of leaves, the clusters subtended by involucral bracts 2-5 mm long; staminate, pistillate, and sometimes bisexual flowers intermixed. The 3 specimens seen are from Big Brush Creek and Browns Park; rocky ledges along a creek and drying mud of an ephemeral pools; 5,600-5,700 ft; June-Aug.

Urtica L. — Nettle

Urtica dioica L. (*U. strigosissima* Rydberg) — stinging nettle — Plants perennial herbs, from rhizomes, armed with stinging hairs, and usually otherwise pubescent, mostly unisexual; leaves opposite, (5) 7-15 cm long, with (5) 10-15 mm long stipules, the blades narrowly to broadly lanceolate, serrate; flowers clustered in axillary panicles or branching spikes. The following 2 vars. belong to ssp. *gracilis* (Aiton) Selander (*U. gracilis* Aiton):

1 Lower surface of leaves glabrous or essentially so except for occasional stinging hairs; stems not obscured by pubescence; plants of the E. Tavaputs Plateau .... **var. *procera***
1 Lower surface of leaves more or less densely hirtellous; stems usually obscured by pubescence; plants widespread but apparently not of the E. Tavaputs Plateau .... **var. *occidentalis***

Var. *occidentalis* S. Watson [*U. holosericea* Nuttall; *U. dioica* var. *holosericea* (Nuttall) C. L. Hitchcock; *U. dioica* ssp. *holosericea* (Nuttall) R. F. Thorne] — western stinging nettle — Widespread in various plant communities, mostly on disturbed ground and along streams and ditches; 6,200-8,500 ft; July-Sept.

Var. *procera* (Muhlenberg) Weddell [*U. lyallii* S. Watson; *U. dioica* var. *lyallii* (S. Watson) C. L. Hitchcock; *U. d.* ssp. *gracilis* (Aiton) Selander] — stinging nettle — The few specimens seen are from the E. Tavaputs Plateau; sagebrush-rabbit-brush, mt. brush, and wet meadow communities; 7,000-8,200 ft; Aug.-Sept.

VALERIANACEAE Valerian Family

Perennial herbs, with faint or rather strong foul odor; leaves simple and entire to pinnatifid or pinnate compound, opposite; flowers mostly bisexual, in panicles or flat-topped clusters; calyx small, 9-17 lobed, obscure at flowering, the lobes narrow, elongate and plumose at fruiting; corolla united, funnelform to nearly salverform, the limb mostly 5 lobed, more or less 2-lipped, the tube sometimes pouched at the base; stamens 3; ovary inferior, 3-chambered but 2 of these obsolete and only 1 bearing seed; styles 1, with a 3-lobed stigma; fruit an achene.

Valeriana L. — Valerian

1 Basal leaves gradually tapering to an indistinct petiole, 7-40 cm long.....
  .... ***V. edulis***
1 Basal leaves sharply differentiated into a petiole and blade, the blades 2-8 cm long 2
2 Corolla 4-7 mm long usually puberulent externally near the base; stems usually puberulent or granular; stem leaves entire to pinnatifid; plants 15-45 cm tall .... ***V. acutiloba***
2 Corolla 1.5-4 mm long, glabrous; lower part of stems glabrous; stem leaves pinnatifid to pinnately compound; plants 30-90 cm tall .... ***V. occidentalis***

Valeriana acutiloba Rydberg [*V. capitata* Pallas ex Link ssp. *acutiloba* (Rydberg) F. G. Meyer] — sharp-leaf valerian — The 5 specimens seen are from Range Creek and Bruin Point of the W. Tavaputs Plateau, and western Uinta Mtns.; mt. brush communities and openings in spruce-fir woods and alpine snowbeds; 6,800-10,900 ft; June-Aug. Our plants belong to var. *pubicarpa* (Rydberg) Cronquist.

Valeriana edulis Nuttall (*V. furfurescens* A. Nelson) — edible valerian — Occasional in Strawberry Valley and Uinta Mtns., infrequent on the Tavaputs Plateau; many plant communities including: riparian, meadow, aspen, and open coniferous forest, and rarely sagebrush, also rarely in cirque basins; 7,500-10,700 ft; June-Aug.

Valeriana occidentalis A. Heller — western valerian — Infrequent from Strawberry Valley to the Avintaquin drainage on the W. Tavaputs Plateau and to Rock Creek in the Uinta Mtns. and disjunct on Cold Springs Mt.; riparian, aspen, and tall forb communities; 7,500-9,000 ft; May-July.

VERBENACEAE Vervain Family

Annual or perennial herbs; leaves opposite; flowers bisexual, in bracteate spikes; corolla united; stamens 4; ovary superior, 2- or appearing 4-celled; style 1, 1-2 lobed; fruit of 2-4, 1-seeded nutlets.

---

1   Plants annual; leaves irregularly toothed and cleft, often with 1 or 2 pairs of pinnatifid lower primary segments that are smaller than the terminal one; spikes terminal on stems and branches; calyx and corolla 5-lobed; nutlets 4 . . . . **Verbena**
1   Plants perennial; leaves with 1-4 pairs of teeth above the middle; spikes lateral; calyx and corolla 4-lobed; nutlets 2 . . . . **Phyla**

Phyla Loureiro

Phyla cuneifolia (Torrey) Greene [*Lippia cuneifolia* (Torrey) Steudel] — wedgeleaf — Stems 20-100 cm long, creeping often rooting at the nodes; leaves 1-3 cm long, thick and rigid, canescent-strigose; peduncles about 1-5 cm long; spikes globose and head-like in flower, elongating in fruit; corollas about 4 mm long, white, pink or purple-rose. Graham (1934) cited Doutt 895 for along the bank of the Green River at Ouray. N. Folks 183 (USUUB) is also from near Ouray, and Atwood & Fontane 32464 (BRY) is from Chandler Canyon.

Verbena L. — Verbena, Vervain

Herbs (ours) with opposite, mostly toothed, lobed or dissected leaves; flowers borne in dense, bracteate spikes, each flower in the axil of a narrow bract; calyx 5-angled and unequally 5-toothed; corolla united, salverform or funnelform with unequal lobes; only 1 lobe of the style stigmatic; fruit enclosed in the dry calyx; readily separating into 4 linear-oblong nutlets.

---

1   Stems prostrate or decumbent, mostly 10-50 cm long, radiating over the ground from the rootcrown or somewhat upright and few to solitary in small plants . . . . **V. bracteata**
1   Stems erect, 40-150 cm tall . . . . **V. hastata**

Verbena bracteata Lagasca & Rodregues (*V. bracteosa* Michaux) — prostrate vervain — Stems usually several and radiating from a taproot, mostly decumbent or prostrate, or solitary and somewhat upright in small plants; leaves irregularly toothed and cleft, passing abruptly into the lance linear bracts of the spikes, the bracts 2 or more times longer than the flowers; corollas almost hidden in the bracts, bluish, pinkish or rarely white, the tube about 4 mm long, the limb 2-3 mm wide. Widespread; mostly along roads or other disturbed ground; up to about 7,500 (8,000) ft; June-Oct.

Verbena hastata L. — blue vervain, swamp verbena — Stems solitary to several, erect, from fibrous roots; leaves serrate and infrequently lobed as well; bracts inconspicuous; corolla blue or violet, the tube 3-4 mm long, the limb 2-5 mm wide. Specimens seen are from wetlands created by seepage from irrigation west of Lapoint and south of Tridell, to be expected elsewhere; July-Sept.

VIOLACEAE Violet Family
Viola L. — Violet

Perennial herbs; leaves simple, alternate, sometimes opposite or basal; stipules present; flowers bisexual, mostly bilaterally symmetrical, 5-merous, solitary or clustered; sepals separate or slightly fused at the base, lanceolate or linear, persistent in fruit; petals separate, the lower ones usually the largest and pouched or spurred at the base, the upper ones usually strongly arched backwards; stamens 5, united around the pistil; ovary superior; style 1; fruit a rounded capsule that expels the seeds rather forcefully upon opening.

---

1   Petals predominately yellow on the face, often marked with purple on the back; flowering stems mostly with well-developed leaves at least on the lower 1/2; leaves seldom cordate at the base 2

1   Petals predominately bluish or white on the faces, sometimes with a yellow base, with or without purple on the back; flowering stems with or without leaves; leaves often with a cordate base 3

2   Leaf blades coarsely veined, generally not greater than 4 cm long, usually deltoid, truncate at the base, coarsely few-toothed or lobed, not regularly serrate or dentate, often glaucous and more or less purplish at least along the veins; upper petals deep purple on the back; capsules puberulent . . . .
. . . . **V. purpurea**

2   Leaf blades not coarsely veined, lanceolate or elliptic-lanceolate, tapered or abruptly narrowed to the petiole, entire to finely serrate or dentate, (2) 4-10 cm long, generally not glaucous, not purplish even along the veins; upper petals often yellow on the back except along the veins; capsules glabrous or puberulent . . . . **V. nuttallii**

3   Flowering stems without leaves, or leaves borne on the lower 1/4 of the stem; the scape often with a pair of opposite or alternate bracts about the middle; style head glabrous 4

3   Flowering stems with well-developed leaves, some born on the upper 3/4 of the stem; style head bearded. 5

4   Flowering stems and leaves arising separately from slender caudices or rhizomes that are mostly less than 1.5 mm thick; plants producing stolons; leaf blades weakly crenate to subentire; petals pale lavender, rarely violet or white, the lateral pair glabrous or sparsely bearded toward the base . .
. . . . **V. palustris**

4   Flowering stems and leaves arising together from a short, stout, mostly erect rootstock often over 1.5 mm thick; plants without stolons; leaf blades

prominently crenate; petals mostly light to deep violet, the lateral pair bearded with crinkled white hairs .... *V. nephrophylla*

5   Petals predominately bluish, often white toward the base; flowering stems with leaves on the lower 3/4; spur of the lowest petal 3-7 mm long, prolonged behind base of sepals .... *V. adunca*

5   Petals predominately white on the faces, sometimes yellow at the base; flowering stems with leaves on the upper 3/4; spur of the lowest petal less than 3 mm long .... *V. canadensis*

Viola adunca Smith — blue violet — Our most common blue or white flowered violet; widespread; riparian and wet meadow communities and other wet places; 7,000 ft to slightly above timberline; May-July.

Viola canadensis L. (*V. rugulosa* Greene) — Canada white violet — Apparently uncommon but widespread; the few specimens seen are from the E. & W. Tavaputs Plateaus riparian and Douglas-fir communities; April-Sept.

Viola nephrophylla Greene — bog violet — Widespread; occasional in moist or wet places in many plant communities; 5,000-10,000 ft; May-June.

Viola nuttallii Pursh (*V. praemorsa* Douglas; *V. vallicola* A. Nelson) — yellow prairie violet — Widespread (but no specimens seen from the E. Tavaputs-Plateau); sagebrush and mt. brush communities and in shade of woods and along streams; 5,000-10,000 ft; May-June.

Viola palustris L. [*V. macloskeyi* Lloyd var. *pallens* (Banks) C. L. Hitchcock misapplied; *V. pallens* (Banks) Brain. misapplied by Graham (1937)] — marsh violet — Western Uinta Mtns. with no specimens seen from Daggett or Uintah Cos.; meadow and streamside communities; 8,000-10,000 ft; June-July. Our plants belong to var. *brevipes* (Baker) R. J. Davis.

Viola purpurea Kellogg [*V. utahensis* Baker & Clausen; *V. venosa* (S. Watson) Rydberg] — goosefoot violet, pine violet — Strawberry Valley and western Uinta Mtns. in the Provo River and North Fork Duchesne drainages sagebrush, mt. brush, aspen, and spruce-fir communities, and in snowbed areas at higher elevations; May-July. Like *V. nuttallii*, but supposedly different by the overlapping criteria given in the key. In general, our plants are like those described for the central Wasatch Front (Arnow and others 1980) in that some plants are extreme and readily placed in one or other of the taxa, but other plants combine the characters traditionally used for separation.

VISCACEAE (LORANTHACEAE) Mistletoe Family

Monoecious or dioecious herbs or shrubs, parasitic on gymnosperms; stems brittle, with swollen jointed nodes, frequently much branched; leaves opposite, simple, entire, reduced to scales; flowers inconspicuous, unisexual, without petals, solitary or clustered at the nodes or in axillary spikes or cymes; sepals 2-4 (5), free, erect, persisting in fruit; stamens 2-4, free or fused to base of sepals; ovary inferior; style 1; stigma entire; fruit fleshy, berry-like, 1-seeded.

---

1   Plants parasitic on species of Juniperus; stems generally terete; pistillate flowers mostly with 3 tepals; fruit subglobose, sessile, white to pinkish at maturity .... *Phoradendron*

1   Plants parasitic on coniferous trees other than Juniperus; stems generally
    angled; pistillate flowers mostly with 2 sepals; fruit somewhat flattened,
    borne on recurved pedicels, greenish or grayish . . . . ***Arceuthobium***

Arceuthobium Bieberstein — Dwarf-mistletoe

With features of the family and as listed in the key. Reference: Hawksworth
and Wiens (1972).

---

1   Stems with at least some of the branching whorled, the branches more or
    less spreading in all directions from the nodes; staminate flowers whorled;
    flowering in Feb.-March; parasitic on *Pinus contorta* and *P. ponderosa* . . . . .
    . . . . ***A. americanum***
1   Stems opposite, or if more than 2 per node then the extra branches arising
    in the same plane as the primary ones and the several branches fan-like;
    staminate flowers not whorled; flowering at various times; not or rarely
    parasitic on *Pinus contorta* or *P. ponderosa* 2
2   Plants parasitic on *Pseudotsuga* (rarely on *Abies* and *Picea*), less than 4 cm
    long, scattered near the apex of the host branch; basal diameter of main
    shoots 0.1-1.5 mm; flowering in Feb.-March; host plant often forming a
    witches broom (a dense mass of distorted branches) . . . . ***A. douglasii***
2   Plants parasitic on *Pinus,* often over 4 cm long, not specific for the apex of
    the host branch; basal diameter of main shoots 1.5-4 mm; flowering in Aug.-
    Sept.; host plant not forming a witches broom. 3
3   Plants olive green to brownish, parasitic on *Pinus edulis* . . . ***A. divaricatum***
3   Plants yellow green or light gray, parasitic on *Pinus flexilis* and *P. longaeva*
    . . . . ***A. cyanocarpum***

Arceuthobium americanum Nuttall ex Engelmann in A. Gray — lodgepole pine
dwarf mistletoe — Infesting lodgepole pine and rarely infecting ponderosa
pine except where it is associated with lodgepole pine, rarely on limber pine.

Arceuthobium cyanocarpum Coulter & Nelson [*A. campylopodium* Engelmann
forma *cyanocarpum* (A. Nelson) Gill] — limber pine dwarf mistletoe — Infrequent
or occasional on limber pine and bristlecone pine.

Arceuthobium divaricatum Engelmann [*A. campylopodium* Engelmann f. *divaricatum*
(Engelmann) Gill] — pinyon dwarf mistletoe — Occasional on pinyon pine.

Arceuthobium douglasii Engelmann — Douglas-fir dwarf mistletoe — Occasional
to common on Douglas-fir, rarely on species of *Abies* and *Picea* where these
species are associated with Douglas-fir.

Phoradendron Nuttall — Mistletoe

Phoradendron juniperinum Engelmann in A. Gray — juniper mistletoe — With
features of the family and as listed in the key. Parasitic on *Juniperus*. No
specimens seen from our area. Mapped for s. of our area in TPD.

ZANNICHELLIACEAE

Zannichellia L. — Horned Pondweed

Zannichellia palustris L. — horned pondweed — Aquatic, submerged herbs; stems slender, branching; leaves opposite or whorled, linear, entire, not over 0.5 mm wide, stipules scarious, free from the leaf bases; flowers unisexual, both staminate and pistillate ones from the same axil, subtended or enclosed in a hyaline bract, inconspicuous, the staminate of a single stamen, the pistillate of 2-10 carpels, styles persistent; fruit a nutlet 2-4 mm long. The few specimens seen are from along the Green River and from Blue Mt.; shallow water; to be expected across our area.

ZYGOPHYLLACEAE Caltrop Family

Tribulus L. — Puncture vine

Tribulus terrestris L. — puncture vine — Annual, prostrate plants; stems radiating out over the ground from the taproot, 30-100 cm long or perhaps longer; leaves once pinnately compound, opposite, 2-5 cm long, with 4-8 pairs of opposite leaflets, these 3-15 mm long; flowers pedunculate, from axils of leaves; petals 5, separate, yellow, 3-5 mm long; stamens 10, style 1; stigma 5 lobed; fruits (bur-like or mace-like) with hard, sharp spines to 6 mm long. Introduced from the Old World, weedy along roads, sidewalks, parking lots, and on other disturbed ground at lower elevations. This pestiferous weed appears to be spreading in our area. The bur-like fruits flatten tires and cause painful injury to livestock and to people especially when bare-footed.

GLOSSARY

Abortive. Imperfectly developed or undeveloped.

Acaulescent. Apparently without a stem, that is, with the main basal leaves and slender, leafless flowering stems appearing above ground level.

Accrescent. Increasing in size after flowering; enlarging with age.

Aceriform. Similar to the leaf of a maple.

Achene. A dry, one-seeded fruit with a firm close-fitting wall that does not open by any regular dehiscence.

Acorn. The leathery fruit of an oak, containing a single large seed and enclosed basally in a cup formed from bracts.

Acuminate. With a long, tapering point set off rather abruptly from the main body (such as of a leaf).

Acute. With a pointed end forming an acute angle, that is, less than a right angle.

Adnate. Fused or attached to another structure from the beginning of development.

Adventive. Said of an introduced plant that is beginning to spread into a new locality or region.

Alpine. Above timberline.

Alternate. With a single structure of each kind occurring at each level of the axis.

Ament. A usually early deciduous spike or raceme of small, bracteate, apetalous, unisexual flowers.

Androgynous. With staminate flowers borne above the pistillate ones.

Annual. A plant completing its life cycle in a year or less.

Anther. The portion of the stamen that produces pollen.

Anthesis. The time when the flower expands and opens, or the process of expansion and opening.

Anthocyanin. Any of a common class of pigments having colors ranging from lavender to purple. These pigments are affected by the acidity or alkalinity of the cell sap, and they change color in approximately the same way as litmus paper, tending toward red in an acid medium and blue in a basic (alkaline) medium.

Anthrocyanous. With anthrocyanin pigments.

Antrorse. Directed upward or forward.

Apex. The uppermost part.

Apical. Of the apex.

Apiculate. Terminated by an abrupt, short, flexible point.

Apomictic. Reproducing by apomixis.

Apomixis. Reproduction by seed that has developed from an ovule that was not fertilized.

Appressed. Lying tightly against another (usually larger) organ.

Aquatic. Growing in water.

Arachnoid. With slender, tangled hairs resembling the threads of a spider web.

Areole. Diminutive of area; a small, clearly marked space. The term is used most frequently in reference to the small, special spine-bearing areas on the stem of a cactus.

Aristate. With an awn or stiff bristle.

Articulate. With conspicuous segments or joints.

Ascending. Arising at an oblique angle (or on a curve).

Attenuate. With a long, tapering point, this usually set off rather abruptly from the main body of the object, for example, a leaf blade.

Auricle. A small projecting lobe or appendage, generally at the base of an organ.

Auriculate. Having an auricle.

Awl-shaped. With a narrow flattened body tapering gradually upward into a point. See subulate.

Awn. A bristle-like part or appendage.

Awned. With an awn.

Axil. The adaxial angle between two organs, particularly the angle between the upper side of a leaf and the stem.

Axillary. In the axil.

Banner. The upper and usually largest petal in the papilionaceous corolla of a plant of the pea family (of the common type having markedly bilaterally symmetrical flowers). Known also as a *standard*.

Barbate. Bearded with stiff, fine hairs.

Barbellate. Diminutive of barbate

Barbed. With a rigid barb, like the barb of a fish hook.

Basal leaves. Leaves at the base of an herbaceous plant, arising from several nodes separated by exceedingly short internodes occurring at about ground level.

Basifixed. Attached at the base.

Beaked. Ending in a firm elongated slender structure.

Bearded. With a tuft of hairs.

Berry. A fleshy or pulpy fruit with more than one seed and formed from either a superior or an inferior ovary. The seeds are embedded in pulpy tissue.

Bidentate. With two teeth.

Biennial. Completing the life cycle in 2 years. Biennial plants usually produce only basal leaves above ground the first year and both basal leaves and flowering stems the second.

Bifid. Forked, that is, ending in two parts.

Bilabiate. Two-lipped. A bilabiate corolla has petals or lobes in two sets commonly with two in the upper and three in the lower.

Bilateral. Two-sided.

Bilaterally symmetrical. Capable of division into only two similar sections, which are mirror images of each other.

Bilocular. With 2 locules.

Bipinnate. Pinnate and the primary leaflets again pinnate.

Bipinnatifid. Pinnatifid with the primary divisions again pinnatifid.

Bisexual. With both sexes represented in the same individual or organ, such as with stamens and pistils in the same flower.

Bladdery. Inflated with thin walls.

Blade. The broad, usually flat, part of a leaf.

Bloom. A usually waxy, whitish or bluish powder covering the surface of a leaf, stem, fruit, or other organ.

Bract. A leaf subtending a reproductive structure, such as a flower or a cluster of flowers.

Ovuliferous scale. A scale or diminutive leaf subtending an ovule.

Bracteate. With bracts.

Bracteolate. With bractlets.

Bractlet. A minor bract.

Bristle. A stiff hair.

Bud. A growing structure at the tip of a stem or a leaves enclosed by scales or immature leaves (vegetative bud); a young flower that has not yet opened.

Bulb. An underground bud covered by fleshy scales.

Bulbiferous. With bulbs.

Bulblet. A small underground bulb or bulb-like structure produced on the stems, usually in the axils of leaves, or sometimes in the places where flowers ordinarily occur.

Bulbous. Bulb-like.

Bur. Any rough prickly covering of a seed or seeds.

Caducous. Falling early, as leaves in *Opuntia*. In the poppy family, the caducous sepals fall away when the flower opens.

Caespitose. Growing in tufts or mats.

Callosity. A thickened hard structure.

Callus. A hard or tough, usually swollen area.

Calyx. A cup; the sepals of a flower; the outermost series of flower parts.

Calyx tube. A tube formed from the lower portions of the sepals.

Campanulate. Bell-shaped, that is, the shape of an inverted church bell, rounded at the attachment and with a broad flaring rim.

Canescent. Grayish-white or hoary, densely covered with white or gray fine hairs, these usually short.

Capillary. Like an elongated delicate hair or thread.

Capitate. In a dense cluster or head.

Capsular. Pertaining to or formed like a capsule.

Capsule. A dry, many-seeded fruit made up of more than one carpel and splitting open (dehiscent) lengthwise at maturity.

Carinate. With a keel.

Carpel. A specialized leaf that forms either all or part of a pistil.

Carpophore. A receptacle or part of a receptacle that is prolonged between the carpels.

Cartilaginous. With the texture of cartilage, that is, tough and firm but somewhat flexible.

Caruncle. An appendage at the attachment point (hilum) of a seed.

Caryopsis. The fruit of a grass, that is, a one-seeded, indehiscent fruit with the pericarp adnate to the seed.

Catkin. See ament.

Caudate. With a tail-like structure.

Caudex, caudices (plural). A largely underground stem-base that persists from year to year and produces new leaves and flowering stems each year.

Caulescent. With a leafy stem that has visible internodes. Antonym: *acaulescent*.

Cauline. Of or on the stem.

Chaff. Dry, membranous scales or bracts.

Chaffy. Resembling chaff.

Chamber. A room. Applied to the cavities (locules or "cells") of an anther or an ovary.

Channeled. With lengthwise grooves, the grooves usually deep.

Chartaceous. Papery in texture.

Chlorophyll. The green pigment of most plants, and essential enzyme of photosynthesis.

Cilia. Hairs along the margin of a structure, placed like the eyelashes on a human eyelid.

Ciliate. With cilia along the margin. Diminutive: *ciliolate*.

Cinereous. The color of ashes.

Circumscissile. Opening by a horizontal circular line, the top coming off like a lid.

Clavate. Gradually enlarged upward after the manner of a baseball bat or the traditional giant's club, which is, either tapering gradually upward or with an enlarged knob at the summit.

Claw. The narrow stalk at the base of a petal, this resembling the petiole of a leaf.

Cleft. Indented about half way or a little more than half way to the base or to the midrib.

Climbing. Supported by clinging.

Clone. An aggregate of stems produced asexually from one sexually produced seedling as is common in aspen.

Colonial. In colonies. Used primarily in reference to plants occurring in clumps or stands connected by rhizomes.

Column. A slender aggregation of coalescent stamen filaments, as in some members of the mallow family.

Coma. A tuft of hairs on the end of a seed.

Commissure. The surface along which two or more locules are joined to each other.

Compound. Composed of two or more similar elements. For example, a panicle is a compound of spikes, racemes, or corymbs — that is, composed of two or more of anyone of these elements; a compound leaf is composed of two or more leaflets; a compound pistil is composed of two or more coalescent carpels.

Compressed. Flattened, particularly from side to side (laterally compressed).

Concave. Saucer-like; shallowly hollowed.

Cone. A reproductive structure composed of an axis (branch) bearing sporophylls or other seed-bearing or pollen-bearing structures.

Cone scale. Scale of a cone subtending a naked ovule.

Confluent. Running together; blending into one.

Coniferous. Cone-bearing.

Connective. The middle part of an anther connecting the two pollen sacs or two pairs of pollen sacs.

Connivent. Standing together (i.e. stamens with their tips ending against each other).

Convex. Rounded outward.

Cordate. Of a conventional heart-shape, the length greater than the width, the petiole attached in the basal sinus, the apex acute; applied also to the basal indentation of a leaf or other structure.

Coriaceous. Leathery.

Corm. A bulb-like structure formed by enlargement of the stem base. It is sometimes coated with one or more membranous layers.

Corniculate. With little horns or crests.

Corolla. The petals of a flower, the inner 1 or more series of the perianth.

Corolla lobe. A lobe of the limb of a united corolla.

Corolla tube. The hollow cylinder formed by coalescence of petals.

Corrugation. A ridge or grove of a corrugated (wrinkled or folded) surface.

Corymb. A flat-topped cluster of flowers; fundamentally like a raceme but with the pedicels of the lower flowers longer and the pedicels of the upper flowers gradually shorter. As a result of this arrangement, at a middle stage of development there may be fruits on the outside, buds in the center, and flowers in between. (The opposite of the arrangement in a cyme.)

Corymbose. Arranged in corymbs.

Cotyledon. One of the first leaves developed in the embryo in the seed.

Creeping. The stem growing along the ground and producing adventitious roots.

Crenate. With rounded teeth or scalloped. Diminutive: *crenulate*.

Crisped. Irregularly curled or crooked.

Culm. The stem of a grass. The term is also applied sometimes to sedges.

Cultivar. A variety originating and persistent under cultivation, usually derived and maintained by selective breeding or vegetative propagation (a cultivated variety).

Cuneate, cuneiform. Wedge-shaped; essentially a narrow isosceles triangle with the distal corners (those away from the petioles) rounded off. The petiole of a cuneate leaf is attached at the sharp angle.

Cupulate. With or subtended by a small cup.

Cupule. A little cup, the term being applied particularly to the cup of an acorn, which is an involucre formed from coalescent bractlets.

Cyme. A broad, more or less flat-topped cluster of flowers with terminal or central flowers developing before the outer ones.

Cymose. With cymes or with cyme-like clusters of flowers.

Deciduous. Falling off at the end of each growing season.

Decompound. More than once compound (or divided).

Decumbent. Reclining except at the apex.

Decurrent. Leaf bases that continue along the stem as wings or lines.

Deflexed. Abruptly bent or turned downward.

Dehisce. To split open along definite lines.

Dehiscence. The process of splitting open at maturity.

Deltoid. Of the shape of the Greek letter delta, that is, an equilateral triangle, the attachment being in the middle of one side.

Dentate. With angular teeth projecting at right angles to the edge of the structure (such as a leaf). Diminutive: *denticulate*.

Depauperate. Stunted, small. The term is applied particularly if the plant is undeveloped as compared with others of the species, but also if it is representative of a small species.

Depressed. Flattened from above as if pushed downward.

Diadelphous. 'In two brotherhoods,' the term being applied to stamens coalescent in two sets. The common usage is for members of the pea family in which nine of the 10 stamens are coalescent and the other stands alone.

Dichotomous. Forking, with two usually equal branches at each point of forking.

Dicotyledonous. With two cotyledons.

Didynamous. In two pairs, the pairs not being of the same length. The term is applied to stamens.

Diffuse. Spreading widely and diffusely in all directions.

Digitate. Resembling the fingers of a human hand, that is, with several similar structures arising at a common point. See palmate.

Dioecious. The flowers unisexual and the staminate on one individual and the pistillate another.

Disk. In the sunflower family or Asteraceae, the central portion of the compound receptacle bearing the disk flowers. Disk flower. In the sunflower family or Asteraceae, one of the flowers with tubular corollas. Discoid. Resembling a disk. In the sunflower family or Asteraceae a discoid head is one having only disk flowers and no ray flowers.

Dissected. Divided into narrow segments.

Divaricate. Spreading widely, divergent.

Divergent. Spreading away from each other.

Divided. Indented essentially to the base or midrib.

Dorsal. On the outer surface of an organ or the back.

Dorsiventral. A structure having a clear differentiation of a back and front or upper and lower side.

Double samara. A samara with two locules and two wings.

Double-serrate. Having small teeth set on larger teeth.

Drupe. A fruit with a fleshy exocarp and a hard, stony endocarp about each seed. Classical examples are the pitted fruits of plum, cherry, and peach.

Ellipsoid. Elliptical in outline with a three-dimensional body. See *elliptic*.

Elliptic, elliptical. In the form of an ellipse, that is, about one and one half times as long as broad, widest at the middle, and rounded at both ends.

Embryo. The new plant enclosed in the seed. It consists of an axis and the attached cotyledons and young secondary leaves.

Emersed. Above water, emergent. Growing in water but with the top of the plant extending above water.

Endemic. Restricted in occurrence to a particular geographical area.

Endosperm. A cell layer occurring in at least the immature seeds of flowering plants.

Entire. Without divisions or teeth of any kind.

Ephemeral. Lasting for a brief period, for example, one day.

Epicotyl. The portion of the embryo of a seed plant just above the cotyledon(s); the young stem.

Equitant. With the leaves folded around a stem after the manner of the legs of a rider around a horse or sometimes with the leaves folded around each other in rows, for example, in *Iris*.

Erose. With the margin appearing to have been gnawed.

Etiolated. White through failure to develop chlorophyll. Etiolated stems (for example, those developed in darkness or weak light) are elongated and spindly as well.

Even-pinnate. Without a terminal leaflet. See *odd-pinnate*.

Exfoliating. Separating into thin layers or strips.

Exserted. Projecting beyond the usual containing structure, as, for example, stamens projecting beyond the rim of a corolla or the midrib of a leaf that is projected beyond the margin of the blade.

Family. A group of related plants forming a category ranking above genus.

Fascicate. In a bundle or bundled together.

Fascicle. A bundle or cluster.

Fascicled. In bundles or clusters.

Fasciculate. In bundles or clusters.

Fastigiate. Erect and close together.

Fertile. Productive. Capable of producing fruit or spores.

Fibrous root system. A root system with several major roots about equal and arising from approximately the same point.

Filament. A thread; the stalk of a stamen.

Filiform. Thread-like, that is, long, slender, and cylindrical.

Fimbriate. Fringed, that is, resembling the fringe on the sleeve of an early American buckskin shirt.

Fistulose. Hollow and cylindrical.

Flaccid. Limp, floppy, wilted.

Flexuous. Curved in first one direction, then the opposite, or weakened, easily bent or curved.

Floccose. With tufts of woolly hair.

Floral cup. A cup bearing on its rim sepals, petals, and stamens.

Floral tube. An elongated, slender floral cup.

Floret. A small flower; the flower of a grass and the two immediately enclosing bracts, that is, the lemma and palea.

Floriferous. Bearing flowers.

Flower. A complex strobilus formed at the end of a branch, including the receptacle and bearing sepals, stamens, petals, and pistils, or some of these.

Follicle. A dry fruit formed from a single carpel, containing more than one seed and splitting open along the suture. The term is sometimes applied to similar fruits splitting only along the midrib.

Fornice. A small scale or appendage in the tube or throat of a corolla such as in *Cryptantha*.

Free. Separate from other organs.

Free-central placentation. An arrangement of ovules in which they are attached to a central stalk of a 1-chambered ovary.

Fruit. A mature ovary with its enclosed seeds and sometimes with attached external structures (as with an inferior ovary, the floral cup or tube).

Funnelform. In the shape of a funnel.

Fusiform. In the shape of a spindle, that is, widest at the middle and tapering gradually to each pointed end, the body being circular in cross-section.

Galea. A hood formed from a portion of the perianth derived, for example, from the two upper petals coalescent indistinguishably into one (in some members of the mint and snap-dragon families) or from the upper sepal (in the monkshood).

Galeate. With a galea.

Gamopetalous. With all the petals coalescent. See *sympetalous*.

Genus, genera (plural). A group of related species or sometimes a single species.

Gibbous. Swollen or distended on one side.

Glabrate, glabrescent. At first hairy but later becoming glabrous.

Glabrous. Not hairy, not glandular.

Gland. A secreting organ. Usually glands are recognized by their secretion, which accumulates into droplets or lumps. Often plant glands are on the tips of

Glandular. Bearing glands, small cellular organs secreting the secretion is visible.

Glaucous. Covered with a white or bluish powder or bloom, this often composed of finely divided particles of wax. Example: the bloom on a plum.

Globose, globular. Spherical.

Glochid. A sharp hair or bristle tipped with a barb.

Glochidiate. Barbed at the tip.

Glomerate. In compact clusters.

Glume. One of the two chaff-like bractlets at the base of a grass spikelet. The glumes do not enclose flowers.

Glutinous. Covered with sticky material.

Gynaecandrous. With pistillate flowers borne above the staminate ones.

Gynobase. An enlarged or elongated portion of the receptacle bearing the pistil.

Habit. The general appearance of a plant.

Habitat. The type of locality or the set of ecological conditions under which the plant grows.

Hair. A slender cellular projection. See *trichome*.

Hastate. More or less sagittate (arrow-head shaped) but with divergent basal lobes.

Haustoria. Organs through which a parasite extracts nourishment from its host (singular: haustorium).

Head. A cluster of sessile or essentially sessile flowers or fruits at the apex of a peduncle. Essentially a spike with a very short axis. The marginal flowers bloom first, the central last.

Helicoid cyme. A coiled inflorescence. The main stem terminates in a flower and a single bud just below grows out as a stem but terminates also in a flower, the process being repeated many times and always in the same direction.

Hemispheric. With the shape of a sphere but cut in half. Herb. A non-woody plant, or not woody above ground level.

Herbaceous. Not woody.

Herbarium. A collection of pressed plant specimens.

Hetero-. A prefix, meaning different (with two or more kinds).

Hip (of a rose). A floral cup that usually becomes enlarged and fleshy at fruiting time. The true fruits are the achenes inside.

Hirsute. With fairly coarse more or less stiff hairs.

Hirtellous. Minutely hirsute.

Hispid. With rigid or stiff bristles or bristly hairs.

Homomorphic. All of one kind. With similar morphology.

Homostylic. With styles of one kind.

Hood. A hood-like structure often formed from a petal or a sepal or more than one of either.

Hyaline. Thin and membranous, being transparent or translucent.

Hybrid. Produced by dissimilar parents.

Hypocotyl. The portion of the axis of an embryo of a seed plant just below the cotyledon(s).

Imbricate. Overlapping like the shingles on a roof.

Incised. Cut deeply and sharply into narrow, angular divisions.

Included. Not protruding beyond the surrounding structures.

Indigenous. Native in a particular region.

Indusium. An epidermal outgrowth or reflexed and modified leaf margin which covers the sori of many ferns.

Inferior. Below. An inferior ovary is attached below the other flower parts.

Inflorescence. The flowering area or segment of a plant.

Inserted. Attached upon another structure.

Internode. The portion of the stem between two nodes.

Introduced. Brought in from another area.

Involucel. The involucre of a secondary umbel (of a compound umbel).

Involucral. Pertaining to an involucre.

Involucrate. With an involucre.

Involucre. A series of bracts surrounding a flower cluster or sometimes a single flower.

Involute. Rolled inward.

Irregular. Used in botany to indicate a bilaterally symmetrical structure, for example, a bilaterally symmetrical flower.

Keel. A ridge along the outside of a fold, like the keel of a boat.

Key. An outline prepared for use in identifying plants by a process of elimination.

Laciniate. Cut into narrow lobes or segments.

Lanate. With long woolly hair.

Lanceolate. Lance-shaped; four to six times as long as broad, broadest toward the basal (attachment) end, sharply angled at both ends and especially the apical end, the sides being curved at least along the broad part.

Lateral. At the side or sides, or along the margins.

Leaflet. A leaf-like segment of a compound leaf.

Legume. A dry, several-seeded fruit of the pea family formed from a single carpel and dehiscent on both margins (the suture and the midrib).

Lemma. The lower member of the pair of bracts surrounding a grass flower, this bract enclosing not only the flower but the other bract (palea).

Lenticular. Lens-shaped, that is, biconvex.

Ligule. A membranous scale at the juncture of the sheath and the blade of a leaf.

Limb. The expanded and spreading parts of a sympetalous corolla or the broad portion of a petal; a branch.

Linear. Long and narrow, the sides being parallel and the length at least eight times the width.

Lip. A lobe, usually of a flower.

Lobe. A projecting segment of a leaf or other organ that is too large to be called a tooth.

Lobed, lobate. In the broad sense, from moderately to deeply indented toward the base or midrib; in the restricted sense, indented significantly but less than half way to the base or midrib. The broad sense includes lobed, cleft, parted, and divided.

Locule. The cavity of an anther or an ovule, that is, a pollen chamber or a seed chamber.

Loment. A legume divided by constriction into a series of segments, each containing one seed.

Longitudinal. Lengthwise.

Malpighian hair. A hair with two branches and almost no stalk, these appearing to be a single straight hair with the attachment point at the middle.

Marcescent. Persisting on the plant after withering.

Megagametophyte. A female gametophyte; the usually larger gametophyte developed from a megaspore, as opposed to the smaller from a microspore.

Megaspore. The larger type of spore developed by a plant with spores of two sizes or kinds.

Membranous. Thin, soft, pliable, and translucent.

Mericarp. A portion of a fruit that appears to be a whole fruit.

Merous. A suffix denoting parts or numbers, for example, 3-merous or trimerous, meaning with 3 parts in each series, such as 3 sepals or 3 petals.

Midrib. The middle vein of a leaf or other structure.

Monadelphous. Referring to stamens with their filaments coalescent into a single tube.

Monocotyledonous. With only one cotyledon.

Monoecious. With unisexual flowers, the staminate and the pistillate occurring on the same individual.

Morphology. The study of form and structure of plants.

Mucro. A short pointed structure terminal upon the organ (such as a leaf) that bears it and of about the same texture as the supporting organ (such as a leaf blade).

Mucronate. With a mucro.

Muricate. Roughened by the presence of short, hard points.

Naturalized. Established thoroughly after introduction from another region.

Node. A "joint" of a stem, including points of leaf and bud attachment.

Nut. A hard, relatively large, indehiscent, l-seeded fruit. Diminutive: *nutlet*.

Obcordate. A conventional heart-shape but with the attachment at the point instead of at the indentation.

Oblanceolate. Lanceolate but with the narrowest part toward the attachment.

Oblique. With the sides unequal or slanting.

Oblong. With the length roughly two to three times the width, the sides parallel, and with equal and more or less obtuse ends.

Obovate. Ovate but with the narrow part toward the point of attachment.

Obovoid. Ovoid, but with the attachment at the small end.

Obsolescent. Rudimentary or having nearly disappeared; becoming extinct.

Obsolete. Rudimentary or having practically or wholly disappeared.

Obtuse. Blunt, forming an obtuse angle, or somewhat rounded instead of pointed.

Ochroleucous. Yellowish-white.

Ocreae. Sheath around the stem (united stipules), subtending leaves in some of Polygonaceae.

Odd pinnate. With a terminal leaflet that makes the number of leaflets an odd rather than even number.

Opposite. Two organs (such as leaves) occurring at the same level or node and on the opposite sides of the supporting structure (such as a stem).

Orbicular. Circular or nearly so in outline.

Orifice. Opening.

Oval. Broadly elliptic.

Ovary. The lower part of the pistil, which contains the ovules or later the seeds.

Ovate. Egg-shaped; a two-dimensional object about one and a half times as long as broad, with rounded ends, and widest toward the base or attachment point.

Ovoid. Ovate, but a three-dimensional figure.

Ovulate. Producing ovules.

Ovule. The structure that after fertilization develops into a seed.

Palea. The upper bract of the pair that encloses the grass flower, this bract being subtended by the larger, lower bract or lemma.

Palmate. Descriptive of compound leaves in which leaflets arise at the same point at the apex of the petiole and radiate like fingers of a hand.

Palmately. In a palmate manner. A leaf may be palmately lobed, cleft, parted, or divided.

Palmatifid. Palmately divided and almost but not quite palmate.

Panicle. A cluster of associated spikes, racemes, or corymbs.

Paniculate. In a panicle; similar to a panicle.

Papilionaceous flower. A flower with a banner petal, 2 wing petals, and 2 partly connate keel petals.

Pappillose, papillate. Bearing minute rounded projections. (Papilla: an individual projection.)

Pappus. The specialized calyx of members of the sunflower family; composed of bristles or scales.

Parallel veined. With the principal veins parallel and usually close together.

Parasitic. Living upon food or water derived from another individual, ordinarily of another species. Noun: *parasite.*

Parted. Indented more than half way or nearly all the way to the midrib or base

Partly inferior (ovary). With only the basal portion of the ovary adnate to the floral cup.

Pectinate. Divided like the teeth of a comb.

Pedicel. The stalk or internode below a flower.

Pedicellate. Having a pedicel.

Peduncle. The stalk of a cluster of flowers or the next to the last internode below a single flower.

Pedunculate. Born on a peduncle.

Peltate. Shield-like; supported by a stalk attached near the center of the lower surface (thus resembling a coin balanced on the end of a pencil).

Pendulous. Hanging downward.

Pepo. An indehiscent, fleshy, 1-loculed or falsely 3-loculed, many seeded berry, usually with a hard rind, as in cucumbers and melons.

Perennial. Continuing to grow year after year.

Perfect. Descriptive of a flower having both functional pistils and functional stamens.

Perfoliate. Descriptive of a sessile leaf encircling the stem.

Perianth. A collective term for the calyx and the corolla.

Perianth tube. A tube formed by coalescence and adnation of the lower portions of the sepals and petals.

Pericarp. The wall of a matured ovary, that is, the wall of the fruit or the inner wall if the ovary is inferior.

Perigynium (singular, plural *perigynia*). The sac-like scale enclosing the achene in the genus *Carex*.

Persistent. Remaining attached longer than might be expected, for example, a calyx that remains on the receptacle or floral cup until fruiting time.

Petal. One member of the series of flower parts forming the corolla. See *corolla*.

Petaloid. Resembling a petal in color and texture.

Petiolate. Having a petiole.

Petiole. The stalk of a leaf supporting the expanded portion or blade.

Petiolate. With a petiole.

Petiolulate. With a petiolule.

Petiolule. The stalk of a leaflet.

Phloem. Conducting tissue that usually carries manufactured food downward to places of use or storage.

Phyllary. An involucral bract, the term often being used for the Asteraceae or sunflower family.

Pilose. Hairy, the hairs being elongated, slender, and soft.

Pinnate. With two rows of like parts (such as leaflets or veins) arranged on either side of a main axis (rachis or midrib). A pinnately compound leaf more or less has well-developed leaflets with flattened blades that are petioulate or at least jointed to the rachis.

Pinnately. In a pinnate manner. A leaf may be pinnately lobed, parted, cleft, or divided.

Pinnatifid. Deeply pinnately divided, the segments being not quite separate from each other. In general, a pinnatifid leaf as opposed to a pinnately compound leaf has segments that are not petioulate nor jointed to the rachis but rather are confluent with it.

Pistil. The organ of a flower that bears ovules and later seeds. It is composed of at least one ovary, stigma, and style, and is formed from one or more carpels.

Pistillate. Having pistils, that is, a flower that has pistils but no stamens.

Pith. The central soft tissue of a stem.

Plumose. Like a feather, the term being applied to hairs that have finer hairs attached along each side.

Pod. A dehiscent dry fruit.

Pome. A fleshy fruit with several seed chambers, this formed from an inferior ovary, the fleshy tissue being largely the floral cup, the seeds not embedded in pulp. Apples and pears are classical examples.

Poricidal. Dehiscing by pores.

Porrect. Directed outward and forward.

Precocious. Appearing early in the season, for example, flowers occurring before the leaves.

Prickle. A sharp, pointed outgrowth from the superficial tissues (epidermis and cortex) of a stem, as in a rose prickle.

Primary leaflet. A leaflet of the first degree in a bipinnate or more complex compound leaf.

Procumbent. Lying on the ground, but not rooting at the nodes.

Prostrate. Flat upon the ground.

Pseudoscape. False scape. Scape-like stem below a leaf or leaves.

Puberulent. Finely and minutely pubescent, the hairs short.

Pubescent. Hairy or downy, usually with fine soft hairs. Commonly the term is used to indicate hairiness of a generalized instead of a specialized type, and it is used loosely to cover any kind of hair. Noun: *pubescence*.

Pulvinate. Cushion-like.

Punctate. Dotted with depressed glands or colored spots.

Pustulate, pustular. Blistered.

Pyramidal. Pyramid-shaped.

Raceme. An inflorescence composed of pedicellate flowers arranged along an axis that elongates for an indefinite period. The lower flower blooms first and eventually the terminal bud forms the last flower.

Racemose. In racemes.

Rachilla. A secondary axis. See *rachis*.

Rachis. The primary axis of a pinnate leaf or of an inflorescence.

Radiate. Spreading from a common center; descriptive of a head that includes ray flowers (sunflower family).

Radially symmetrical. Capable of division into three or more similar sections.

Ray. A pedicel or peduncle within an umbel; a ray flower or its corolla.

Ray flower. In the sunflower family, one of the flowers with a ligulate (strap-shaped or flattened) corolla.

Receptacle. The apical area beyond a pedicel, that is, the portion that bears flower parts. The receptacle consists of several or many nodes and short internodes.

Recurved. Curved downward or backward.

Reduced. Small but probably derived from larger forerunners.

Reflexed. Bent or turned abruptly downward.

Regular. Uniform in shape or structure; radially symmetrical (especially as applied to a corolla).

Reniform. Kidney-shaped or bean-shaped, with the attachment in the indentation, the width greater than the length, the apex rounded.

Resinous. Producing resin, coated with a sticky substance.

Reticulate. In a network or a pattern that appears like a network.

Retrorse. Turned back or downward.

Retuse. With a shallow and rather narrow notch in a broad apex.

Revolute. With the margins or the apex rolled backward.

Rhizomatous. With a rhizome; rhizome-like.

Rhizome. A horizontal, underground stem.

Rhombic. More or less diamond-shaped and attached at one of the sharper angles.

Rib. A prominent raised nerve or vein.

Root. The underground portion of the main axis of a plant or the branches of the axis.

Rosette. A circular cluster.

Rostrate. With a beak.

Rotate. Spreading; wheel-shaped or saucer-shaped.

Rounded. Gently curved.

Rudiment. A vestigial organ.

Rugose. Wrinkled.

Rugulose. Diminutively rugose.

Sagittate. Arrow-head shaped.

Salverform. Descriptive of a sympetalous corolla with the slender basal tube abruptly expanded into a flat or saucer-shaped upper portion.

Samara. A dry, indehiscent fruit with a wing.

Saprophyte. A plant that lives upon dead organic matter such as the leaf mold on forest floors.

Saprophytic. Deriving nutrients from dead organic matter, adjective of saprophyte.

Scabrous. Rough to the touch, with minute rough projections.

Scale. A thin, membranous structure; a small more or less triangular leaf; a chaff-like bract; a flattened hair.

Scape. A flowering stem that bears no leaves, or only a small bract or a pair or whorl of bracts.

Scapose. With a scape or in the form of a scape.

Scarious. Dry, thin, membranous, non-green, and translucent.

Schizocarp. A fruit that splits into one-seeded sections (*mericarps*).

Scorpioid. Often used as descriptive of an inflorescence that is coiled in the bud stage, as in the forget-me-not or fiddle-neck. This is a helicoid cyme although it appears to be a spike or a raceme.

Scurfy. With scale-like particles on the surface (the scales resembling human dandruff).

Secondary leaf. A leaf produced above the cotyledon(s) or primary leaf or leaves.

Secund. Turned or directed to one side usually by twisting.

Seed. A matured ovule consisting of an integument (seed coat), an enclosed nucellus (sporangium), the remains of the megagametophyte, the endosperm (in flowering plants), and the embryo.

Sepal. One of the flower parts of the outer series, the sepals forming a calyx. See calyx.

Septate. Divided by partitions.

Sericeous. Silky with long, slender, soft, more or less appressed hairs.

Serotinous. Late; in some willows the catkins developing later than the leaves.

Serrate. With saw-like teeth, that is, angular and directed forward. Diminutive: *serrulate.*

Sessile. Without a stalk, that is, "sitting."

Seta. A bristle, a bristle-like hair.

Setose. Covered with bristles.

Sheath. A tubular cover. An example is the basal portion of a grass leaf, which surrounds the stem. *Sheathing.* Covering or enclosing.

Shrub. A woody plant not having a main trunk but several main branches. In general, shrubs are smaller than trees.

Silicle. A short silique, that is, one usually not more than two or three times as long as broad.

Silique. The elongate capsular fruit of the mustard family, which has two seed chambers separated by a false partition from the middle of one placenta to the middle of the other.

Silky. Covered densely with appressed, soft, straight hairs.

Sinuate. With a wavy margin, the margin winding strongly inward and outward.

Sinus. A cleft, recess, or embayment.

Sorus, sori (plural). A cluster of sporangia, as in ferns.

Spathe. A large bract enclosing an inflorescence at least when it is young. Spathes are either white or highly colored but usually not green.

Spatulate. Oblong or somewhat rounded with the basal end long and tapered; the shape of a spatula.

Species (singular and plural). A group of individual plants which are fundamentally alike.

Spicate. Spike-like.

Spike. An inflorescence in which the sessile flowers are arranged along an axis. The basal flower blooms first; the last one formed is at the apex.

Spikelet. The small bracteate spike of grasses and sedges. Also, diminutive of spike.

Spine. A sharp more or less woody or horny outgrowth from a leaf or a part of a leaf, sometimes representing the entire leaf.

Spinose. With spines; spine-like.

Sporadic. Of irregular occurrence here and there; not forming a continuous population.

Sporangium, sporagia (plural). A spore-case.

Spore. A simple, one-celled reproductive structure.

Sporocarp. A hard or leathery structure containing sporagia or a sporangium.

Spur. An elongated sac produced from a part of the flower, as in the larkspurs from a sepal or in the columbines from a petal.

Squarrose. With widely spreading or recurved tips.

Stamen. The pollen-producing structure of a flowering plant, consisting of an anther (which includes pollen sacs) and of a filament (stalk).

Staminate. With stamens but not pistils.

Staminode. A sterile stamen, that is, without an anther or at least not producing pollen.

Stellate. Star-shaped; descriptive of hairs branched so that the hair appears like a star.

Sterile. Not producing pollen or seeds.

Stigma. The apical portion of a pistil, that is, the portion receptive to pollen.

Stipe. A stalk. The term is applied to a stalk under the pistil of a flower.

Stipitate. With a stipe.

Stipulate. In reference to stipules.

Stipule. One of a pair of appendages at the base of the petiole or the leaf base at the point of attachment to the stem. These structures may be thin and scale-like, thickened and hard, green and leaf-like, or reduced to mere glands. Occasionally they are specialized as spines.

Stolon. A runner, that is, a branch that grows along the ground and produces adventitious roots.

Stoloniferous. With stolons.

Stramineous. Straw-colored.

Striate. With fine longitudinal lines or streaks.

Strigose. Covered with depressed, sharp, thin, straight hairs.

Strigulose. Pubescent with hairs between strigose and pilose.

Strobilus. A cone-like reproductive structure composed of a central axis or branch bearing sporophylls.

Style. The tubular upper or middle part of a pistil connecting the stigma and ovary.

Stylopodium. A swelling on the base of the style as in the members of the parsley family.

Submersed. **Submerged**. Growing in water and not extending above water.

Subspecies. A taxon rank between species and variety.

Subulate. Awl-shaped, that is, more or less flat, narrow, tapering gradually from the base to the sharp apex.

Succulent. Fleshy and juicy like a branch of a cactus, the structure (leaf or stem) much thicker than in most plants; applied as an adjective to fruits; applied as a noun to a plant with succulent parts, especially stems or leaves.

Suffrutescent. With the lower part of the stem just above ground level somewhat woody and living over from year to year.

Superior. Above. A superior ovary is free of and above and not adnate to a floral cup.

Symmetrical. Balanced, the parts similar to each other. The flower, for example, may be radially symmetrical, that is, capable of division into three or more similar parts, or bilaterally symmetrical, capable of division into only two similar parts.

Sympetalous. All the petals coalescent at least basally. See *gamopetalous*.

Synobasic. With the base united.

Synonym. A published name superseded by an earlier published name that designates the same taxon. The later name is disregarded and becomes a synonym of the earlier name.

Synonymous. With the same meaning.

Synonymy. See *synonym*. This refers to discarded names applied to a single taxon.

Taproot. The primary root extending downward as a continuation of the main axis of a plant with comparatively few or small secondary or lateral roots.

Taxon, taxa (plural). A category used in classification, for example, a variety, species, genus, or family. A living taxon is a reproducing natural population or system of populations of genetically related individuals. Ranks of plant taxa (divisions, classes, orders, families, genera, species, and varieties) depend upon the degree of their differentiation and isolation from each other.

Tendril. An elongated twining segment of a leaf or a branch, this usually supporting the stem.

Tepal. A collective term used for sepals and petals that are similar and not easily distinguished from each other; used in the buckwheat family for the sepals of two series, there being no corolla.

Terete. Slender and more or less cylindrical, approximately circular in any cross section, but of varying diameter.

Ternate. In three's.

Tessellate. Like a cobblestone pavement.

Thallus. A leaf-like plant body not differentiated into roots, stems, or leaves.

Thorn. A modified stem with a sharp point.

Throat. The opening of a sympetalous corolla or a symsepalous calyx, that is, the expanding part between the proper tube and the limb (spreading upper portion).

Tomentose. Woolly, that is, densely covered with matted hairs, which usually are not straight.

Toothed. With minor projections and indentations alternating along the margin.

Tortuous. Bent or twisted in different directions.

Torulose. Constricted between seeds.

Trailing. Prostrate but not rooting.

Transcorrugated. Corrugated in opposing directions.

Tree. A woody plant with a main trunk. Trees in general are larger than shrubs.

Trichome. Any hair-like outgrowth of the epidermis.

Trifid. Three-cleft to about the middle.

Trifoliate. With three leaflets.

Trigonous. Three-angled.

Truncate. "Chopped off" abruptly; ending abruptly.

Tuber. A thickened short underground branch of the stem serving as a storage organ containing reserve food. An example is a potato.

Tubercle. Diminutive of tuber, but not necessarily an underground structure; usually used in reference to processes or bumps on a surface.

Tuberculate. With tubercles, that is, processes or bumps.

Tuberous. With tubers.

Tubular. Forming an elongate hollow cylinder.

Turbinate. Top-shaped, that is, more or less in an inverted cone.

Turions. A small, bulb-like offset borne near or below the soil surface, as in some species of *Epilobium*.

Umbel. An inflorescence with the pedicels of the flowers arising from approximately the same point. A compound umbel includes ray-like branches (rays) that support smaller umbels of flowers.

Umbellate. Like an umbel, or in the form of an umbel.

Undulate. With the margin irregular and forming a wavy line, that is one that winds gently in and out.

Unilocular. With one locule.

Unisexual. Of only one sex. Descriptive of a flower having only stamens or only pistils, not both, or of a gymnosperm that produces only pollen or only ovules.

Urceolate. Urn-shaped.

Utricle. A small, 1-seeded, more or less indehiscent fruit that appears to be inflated or at least with a relatively thin pericarp more or less remote from the single seed.

Variety. The smallest taxon usually recognized. See *taxon*. A natural population or population system.

Vascular. Containing xylem and phloem (conducting tissues).

Vascular plant. A plant with vascular tissues of xylem and phloem.

Venation. The type of veining including parallel, palmate, and pinnate.

Ventral. The side toward the axis, for example, the upper side of a leaf. See dorsal.

Verticil. A whorl or a cycle.

Verticillate. Arranged in a whorl, cycle, or verticil.

Villous. With long, soft, more or less interlaced hairs.

Viscid. Sticky.

Wavy. See *undulate*.

Weed. An introduced, or less commonly a native plant, that grows where it is not wanted and tends to form dense patches and displace other plant species. An aggressive invader that causes economic losses to agriculture.

Weedy. Introduced and increasing with disturbance and tending to displace native vegetation.

Wing. A thin, membranous or leathery expansion on the surface of on organ; one of the two lateral petals of a papilionaceous corolla.

Woolly. Covered with long, matted hairs that are not straight. See *tomentose*.

Xylem. The principal cells forming the wood conducting elements, which usually carry water and dissolved salts and sometimes previously stored food upward from the roots to the leaves.

REFERENCES CITED

Albee, Beverly J.; Shultz, Leila M.; Goodrich, Sherel 1988. *Atlas of vascular plants of Utah.* Salt Lake City, UT: Utah Museum of Natural History, University of Utah. 670 p.

Allred, K W. *Poaceae.* In: Heil, K. D.; O'Kane, S. L.; Reeves, L. M.; Clifford, *A Flora of the Four Corners Region.* Missouri Botanical Garden. St. Louis, Missouri. 731-815.

Arnow, Lois 1981. Poa secunda *Presl vers.* P. sandbergii *Vasey (Poaceae).* Systematic Botany. 6(4): 412-421.

Arnow, Lois A. 1987. *Gramineae.* In: Welsh, S. L.; Atwood, N. D; Goodrich, S.; Higgins, L. C. eds. *A Utah Flora.* Great Basin Naturalist Memoirs. 894 p.

Arnow, Lois, A; Albee, Beverly J.; Wyckoff, Ann M. 1980. *Flora of the central Wasatch Front, Utah.* Salt Lake City, UT: University of Utah Printing Service. 663 p.

Baker, William L. 1983. *Some aspects of the presettlement vegetation of the Piceance Basin, Colorado.* Great Basin Naturalist. 43(4): 687-699.

Barneby, Rupert C. 1964. *Atlas of North American* Astragalus. Memoirs of the New York Botanical Garden. 13: 1-1188.

Barneby, Rupert, C. 1989. *Fabales. Intermountain Flora, vascular plants of the inter-mountain west, U.S.A, vol. 3 part B: Fabales.* The New York Botanical Garden. Bronx, NY. 279 p.

Beidleman, Richard G. 1957. *An annotated checklist of the flora and fauna of Dinosaur National Monument.* Boulder, CO: University of Colorado, Biology Department. 163 p.

Benson, Lyman. 1982. *The cacti of the United States and Canada.* Stanford, CA: Stanford University Press. 1044 p.

Bio/West Inc. 1984. [in conjunction with Endangered Plant Studies Inc. and Dr. Clayton White.] *A threatened and endangered species survey for proposed C02 and phosphate slurry pipeline Colorado, Utah, and Wyoming, PR-96-1.* Unpublished report on file at: Vernal District Office, Bureau of Land Management, Vernal, UT. 40 p. [Prepared for Mr. L. E. Johnson, Chevron Corporation, San Francisco, CA.].

Bradley, Ruth Ann Wolf. 1950. *The vascular flora of Moffat County, Colorado.* Boulder, CO: University of Colorado. 71 p. M.S. thesis.

Brown, G. D. 2006. *An alpine plant community classification for the Uinta Mountains, Utah.* Ogden, UT: U.S. Department of Agriculture, Forest Service, Intermountain Region. 140 p.

Bureau of Land Management. *Field report:* Sclerocactus glaucus *survey.* 1985. Unpublished report on file at: Vernal District Office, Bureau of Land Management, Vernal, UT. 6 p.

Corbin, B. L.2007. *Noteworthy discoveries: sulfur cinquefoil in Utah.* Sego Lily. 30: 8.

Correll, D. S.; Johnston, M. C. 1970. *Manual of the vascular plants of Texas.* Renner, TX: Texas Research Foundation. 1881 p.

Cronquist, Arthur. 1947. *Revision of the North American species of* Erigeron, *north of Mexico.* Brittonia. 6: 121-300.

Cronquist, Arthur. 1955. *Vascular plants of the Pacific Northwest, part 5: Compositae.* Publications in Biology, vol. 17. Seattle: University of Washington. 343 p.

Cronquist, Arthur. 1994. *Intermountain flora, vascular plants of the intermountain west, U.S.A, vol. 5, asterales.* Bronx, NY. The New York Botanical Garden. 496 p.

Cronquist, Arthur; Holmgren, Arthur H.; Holmgren, Noel H.; Reveal, James L. 1972. *Intermountain flora, vascular plants of the intermountain west, U.S.A, vol. 1: geology and botanical history of the region, its plant geography and a glossary, the vascular cryptogams and the gymnosperms.* New York: Hafner Publishing. 270 p.

Cronquist, Arthur; Holmgren, Arthur H.; Holmgren, Noel H.; Reveal, James L.; Holmgren, Patricia K. 1977. *Intermountain flora, vascular plants of the intermountain west, U.S.A, vol. 6, the monocotyledons.* New York, NY. Columbia University Press. 584 p.

Cronquist, Arthur; Holmgren, Noel H.; Holmgren, Patricia K. 1997. *Intermountain Flora, vascular plants of the intermountain west, U.S.A, vol. 3 part A. subclass Rosidae (except Fabales).* 446 p.

Cronquist, Arthur; Holmgren, Noel H.; Reveal, James L.1984. *Intermountain flora, vascular plants of the intermountain west, U.S.A, vol. 4: Subclass Asteridae (except Asteraceae).* New York Botanical Garden. 573 p.

Dorn, Robert D. 2001. *Vascular plants of Wyoming.* 3rd ed. Mountain West Publishing; Cheyenne, WY. 412 p.

Flora of North America Editorial Committee, eds. 1993+. *Flora of North America North of Mexico.* 16+ vols. New York and Oxford.

Flowers, Seville. *Vegetation of Flaming Gorge Reservoir Basin.* 1960. University of Utah Anthropological Papers. 48: 1-48. [Upper Colorado Series No.3].

Flowers, Seville; Hall, Heber H.; Groves, Gerald T. 1959. *Appendix: annotated list of plants found in Flaming Gorge Reservoir basin.* University of Utah Anthropological Papers. 48: 49-98; 1960. [Upper Colorado Series No.3].

Garcia, Sonia; McArthur, E. Durant; Pellicer, Jaume; et al. 2011. *A molecular phylogenetic approach to western North American endemic* Artemisia *and allies (Asteraceae): Untangling the sagebrushes.* American Journal of Botany. 98: 638-653.

Goodrich, Sherel; Neese, Elizabeth; Peterson, J. Scott; England, Larry. 1981. *Vascular plants of the Uinta Basin and Colorado.* Denver, CO: U.S. Department of the Interior, Bureau of Land Management. 58 p.

Goodrich, Sherel; MacArthur, E. Durant; Winward, Alma H. 1995. *A new combination and a new variety in* Artemisia tridentata. Great Basin Naturalist. 45: 99-104.

Gould, Frank W. 1947. *Nomenclatural changes in* Elymus *with a key to the California species.* Madrono. 9: 120-128.

Gould, Frank W. *Grass systematics.* 1968. New York: McGraw-Hill. 382 p.

Graham, Edward H. 1937. *Botanical studies in the Uinta Basin of Utah and Colorado.* Annals of the Carnegie Museum. 25: 1-432.

Grant, Verne. 1956. A synopsis of *Ipomopsis.* Aliso. 3: 351-362.

Harrington, H. D. 1954. *Manual of the plants of Colorado.* Denver, CO: Sage Books. 666 p.

Hawksworth, Frank G.; Wiens, Delbert. 1972. *Biology and classification of dwarf mistletoes* (Arceuthobium*).* Agricultural Handbook 401. Washington, DC: U.S. Department of Agriculture, Forest Service. 234 p.

Heil, Kenneth, D.; O'Kane, Steve L.; Reeves, Linda Mary; Clifford, Arnold. 2013. *Flora of the four corners region.* Missouri Botanical Garden Press. St. Louis, Missouri. 1098 p.

Higgins, Larry C. 1972. *The Boraginaceae of Utah.* Provo, UT: Brigham Young University Science Bulletin, Biological Series-Volume 16, No. 3. 83 p.

Higgins, Larry C. 1979. *Boraginaceae of the southwestern United States.* Great Basin Naturalist. 39: 293-350.

Hitchcock, A. S.; Chase, Agnes. 1950. *Manual of the grasses of the United States.* Miscellaneous Publication 200. Washington, DC: U.S. Department of Agriculture. 1051 p.

Hitchcock, C. Leo; Cronquist, Arthur. 1961. *Vascular plants of the Pacific Northwest, part 3: Saxifragaceae to Ericaceae.* Publications in Biology, Vol. 17. Seattle, WA: University of Washington. 614 p.

Hitchcock, C. Leo; Cronquist, Arthur. 1964. *Vascular plants of the Pacific Northwest, part 2: Salicaceae to Saxifragaceae.* Publications in Biology, Vol. 17. Seattle, WA: University of Washington. 597 p.

Hitchcock, C. Leo; Cronquist, Arthur. 1973. *Flora of the Pacific Northwest.* Seattle: University of Washington Press. 730 p.

Hitchcock, C. Leo; Cronquist, Arthur; Owenby Marion. 1969. *Vascular plants of the Pacific Northwest, part 1: vascular cryptograms, gymnosperms, and monocotyledons.* Publications in Biology, Vol. 17. Seattle, WA: University of Washington. 914 p.

Holmgren, Arthur H. 1962. *The vascular plants of the Dinosaur National Monument and the vascular plants of the Green River from the Flaming Gorge to Split Mountain Gorge.* Logan, UT: Utah State University and the National Park Service. 40 p.

Holmgren, Arthur H.; Reveal, James L. 1966. *Checklist of the vascular plants of the Intermountain Region.* Research Paper INT-32. Ogden, UT: U.S. Department of Agriculture, Forest Service, Intermountain Forest and Range Experiment Station. 160 p.

Holmgren, Noel H.; Holmgren, Patricia K.; Cronquist, Arthur. 2005. *Intermountain flora, vascular plants of the intermountain west, U.S.A. vol. 2 part B: subclass dilleniidae.* Bronx, NY: The New York Botanical Garden. 488 p.

Holmgren, Noel H.; Holmgren, Patricia K.; Reveal, James, L, and Collaborators. 2012. *Intermountain Flora, vascular plants of the intermountain west, U.S.A, vol. 2 part A: subclasses magnoliidae-carophyllidae.* The New York Botanical garden Press. 731 p.

Holmgren, Patricia K.; Holmgren, Noel H. 1998 [continuously updated]. *Index herbariorum: A global directory of public herbaria and associated staff.* New York Botanical Garden's Virtual Herbarium. http://sweetgum.nybg.org/ih/.

Huber, A. A. 1995. *A comparative floristic study of limestone and principally quartzite substrates of the Uinta Mountains, Utah.* Provo, UT: Brigham Young University. 2 vol. 359 p. M.S. thesis.

Isley, Duane. 1981. *Leguminosae of the United States. III, subfamily Papilionoidae: tribes Sophoreae, Podalyrieae, Loteae.* Memoirs of the New York Botanical Garden. 25: 1-264.

Jameson, David W. 2013. *Juncaceae.* In: Heil, Kenneth, D.; O'Kane, Steve L.; Reeves, Linda Mary; Clifford, Arnold. 2013. *Flora of the four corners region.* Missouri Botanical Garden Press. St. Louis, Missouri. 605-611.

Kearney, Thomas; 1969. Peebles, Robert H. *Arizona Flora.* Berkeley, CA: University of California Press. 1085 p.

Kellogg, Elizabeth Anne. 1985. *A biosystematic study of the* Poa secunda *complex.* Journal of the Arnold Arboretum. 66: 201-242.

Kingsbury, J. M. 1964. *Poisonous plants of the United States and Canada.* Enge1wood Cliffs, NJ: Prentice-Hall. 626 p.

Komarkova, V. 1979. *Alpine vegetation of the Indian Peaks area, Front Range, Colorado Rocky Mountains.* J. Cramer, Vaduz, Liechtenstein.591 p.

Lewis, Mont E. 1970. *Alpine rangelands of the Uinta Mountains: Ashley and Wasatch National Forests.* Ogden, UT: U.S. Department of Agriculture, Forest Service, Intermountain Region. 75 p.

Maguire, Bassett. *Studies in the Caryophyllaceae--IV. 1950. A synopsis of North American species of the subfamily Si1enoideae.* Rhodora. 52: 233-245.

Munz, Philip A. 1973. *A California flora with supplement.* Berkeley, CA: University of California Press. 1681 p.

Neese, Elizabeth; Smith, Frank. 1982. *Threatened and endangered plant inventory for the oil shale RMP, Brookc1iffs Resource Area, Utah Bureau of Land Management, Vernal District.* Unpublished report on file at: Vernal District Office, Bureau of Land Management, Vernal, UT. 5 vol. [Prepared by Bio/West Inc., Logan, UT.].

Palmquist, E. C. 2011. *Phylogeny and evolutionary history of* Anticlea vaginata *Rydb. (Melanthiaceae): A hanging garden endemic.* Final Report to the Colorado Plateau Cooperative Ecosystems Unit.80 p.

PammeL, L. H. 1913. *The grasses of the Uintah Mountains and adjacent regions.* Proceedings of Iowa Academy of Science. 20: 133-149.

Peterson, J. Scott; Baker, William L. 1982. *Inventory of the Piceance Basin, Colorado: threatened and endangered plants, plant associations, and the general flora.* Unpublished report on file at: Craig District Office, Bureau of Land Management, Craig, CO. 5 vol. [Prepared by Colorado Natural Heritage Inventory, Denver, CO.].

Plummer, A. Perry; Monsen, Stephen B.; Stevens, Richard. 1977. *Intermountain range plant names and symbols.* General Technical Report INT-38. Ogden, UT: U.S. Department of Agriculture, Forest Service, Intermountain Forest and Range Experiment Station. 82 p.

Potter, L. D.; Fischer, N. Timothy; Toll, Mollie S.; Cully, Anne C. 1983. *Vegetation along Green and Yampa Rivers and response to fluctuating water levels, Dinosaur National Monument.* Contract No. CX-1200-2 B024. Albuquerque, NM: University of New Mexico. 179 p.

Refsdal, C. H. 1996. *A general floristic inventory of southwest Wyoming and adjacent northeast Utah.* Laramie, WY: Rocky Mountain Herbarium, Department of Botany, University of Wyoming. 309 p.

Rollins, Reed C. 1983. *Interspecific hybridization and taxon iuniformity in* Arabis *(Cruciferae)*. American Journal Botany. 70(4): 625-634.

Rollins, Reed C. 1984. *Studies in the Cruciferae of Western North America, II*. Contributions from the Gray Herbarium of Harvard University. 214: 1-18.

Rollins, Reed C.; Shaw, Elizabeth A. 1973. *The genus* Lesquerella *(Cruciferae) in North America*. Cambridge, MA: Harvard University Press. 288 p.

Shultz, John S. 1982. *Report on the search for rare and endangered plant species in the Uinta Basin of Utah on Quintana Minerals Corporation Land Holdings*. Unpublished report on file at: Vernal District Office, Bureau of Land Management, Vernal, UT. 54 p. [Prepared for Quintana Minerals Corporation, Houston, TX, by Western Wildland Resources, Logan, UT.].

Shultz, Leila M. 1983. *Systematics and anatomical studies of* Artemisia *subgenus Tridentatae*. Claremont, CA: Claremont College. 169 p. Ph.D. dissertation.

Shultz, Leila M. 1986. *Taxonomic and geographic limits of* Artemisia *subgenus tridentata (Beetle) McArthur (Asteraceae: Anthemideae)*. In Proceedings, Symposium on the Biology of *Artemisia* and *Chrysothamnus;* 1984 July 9-13; Provo, UT. General Technical Report INT-200. Ogden, UT: U.S. Department of Agriculture, Forest Service, Intermountain Research Station: 20-28.

Shultz, Leila M. 2009. *Monograph of* Artemisia *subgenus* Tridentatae *(Asteraceae — Anthemideae)*. Systematic Botany Monographs 89: 1-131.

Shultz, Leila M.; Mutz, Kathryn M. 1979. *Threatened and endangered plants of the Willow Creek drainage*. Unpublished report on file at: Vernal District Office, Bureau of Land Management, Vernal, UT. 2 vol. [Prepared by Meiiji Resource Consultants, Bountiful, UT.].

Stern, William T. 1966. *Botanical Latin*. New York: Hafner Publishing. 566 p.

Styles, Brian T. 1962. *The taxonomy of* Polygonum aviculare *and its allies in Britain*. Watsonia. 5: 77-214.

Tidestrom, Ivar. 1925. *Flora of Utah and Nevada*. Contributions from the U.S. National Herbarium. 25: 1-665.

Turner, Billie L. A. 1956. *Cytotaxonomic study of the genus* Hymenopappus *(Compositae)*. Rhodora. 58: 163-308.

USDA. NRCS. 2014. *The PLANTS Database* (http://plants.usda.gov, 24 April 2014). National Plant Data Team, Greenboro.NC 27401-4901 USA.

Vories, Kimery C. 1974. *A vegetation inventory and analysis of the Piceance Basin and adjacent drainages*. Gunnison, CO: Western State Collage. 243 p. M.S. thesis.

Weber, W. A. 1987. *Colorado flora: western slope*. Colorado Associated University Press. 529 p.

Welsh, Stanley L. 1957. *An ecological survey of the vegetation of the Dinosaur National Monument, Utah*. Provo, UT: Brigham Young University. 65 p. M.S. thesis.

Welsh, Stanley L. 1974. *Anderson's flora of Alaska and adjacent parts of Canada*. Provo, UT: Brigham Young University Press. 724 p.

Welsh, Stanley L. 1979. *Illustrated Manual of Proposed Endangered and Threatened Plants of Utah*. Denver, UT: U.S. Fish and Wildlife Service. 342 p.

Welsh, Stanley L. 1981. *Threatened and endangered plant survey of transmission corridors: Bonanza plant site to Vernal substation, Bonanza plant to Rangely substation, and Bonanza plant to Upalco substation.* Unpublished report on file at: Vernal District Office, Bureau of Land Management, Vernal, UT. 8 p. [Prepared for Desert Generation & Transmission Co-operative, Sandy, UT.).

Welsh, Stanley L.; Atwood, N. Duane; Goodrich, S.; Higgins, Larry C. 2008. *A Utah flora.* Provo, UT: Print Services, Brigham Young University. 1019 p.

Welsh, Stanley L.; Neese, Elizabeth. 1979a. *Survey of proposed endangered and threatened plant species, Moon Lake Project, Rio Blanco Co., Colorado and Uintah Co., Utah.* Unpublished report on file at: Vernal District Office, Bureau of Land Management, Vernal, UT. 21 p. [Prepared for Burns and McDonnell, Kansas City, MO., by Endangered Plant Studies, Inc., Orem, UT.).

Welsh, Stanley L.; Neese, Elizabeth. 1978b. *Survey of proposed endangered and threatened plant species, Moon Lake Project, Rio Blanco Co., Colorado and Uintah Co., Utah.* Unpublished report on file at: Vernal District Office, Bureau of Land Management, Vernal, UT. 18 p. [Prepared for Western Fuels Association, Inc., Lakewood, CO. by Endangered Plant Studies, Inc., Orem, UT.).

Welsh, S. L.; Atwood, N. D.; Goodrich, S.; Neese, E.; Thorne, K. H.; Albee, Beverly. 1981. *Preliminary index of Utah vascular plant names.* Great Basin Naturalist. 41: 1-108.

Welsh, S. L.; Atwood, N. D. 2009. *Plant endemism and geoendemic areas of Utah.* S. L. Welsh, Orem, UT.98 p.

Wilken, Dieter H. [Letter to Sherel Goodrich). 1985. February 19. 2 leaves. Location: Personal files, Sherel Goodrich, Vernal, UT.

Wilson, F. Douglas. 1963. *Revision of* Sitanion *(Triticeae, Gramineae).* Brittonia. 15: 303-323.

Woodbury, Angus M.; Durant, Stephen D.; Flowers, Seville. 1960. *A survey of vegetation in the Flaming Gorge Reservoir Basin.* University of Utah Anthropological Papers. 45: 1-121. [Upper Colorado Series No.2.)

INDEX

The index includes all family names (scientific and common), genus names (scientific and common), and plant species names (scientific only), used in the *Flora*. Synonyms are listed in *italics*.